# PHARMACEUTICAL PROCESS CHEMISTRY FOR SYNTHESIS

# CONTENTS

# 1

# INTRODUCTION

## 1.1 INSPIRATION

This project was first conceptualized at a most unlikely place: at a visit to an *Inspiring Impressionism* exposition at the Denver Art Museum in 2008. The exhibition focused on the impressionists as students of earlier masters. They immersed themselves in these earlier masterpieces and then incorporated the insights they had gained and added their own techniques to convey the same subject matter in profound new ways. My 20 years as a process chemist at Syntex and Roche are much like the years the impressionists spent camped out in front of the works of the masters. The insights gained could be conveyed by presenting the theory and concepts of process research and development, but there are many well-worn reference books that collectively accomplish that objective. My experience has been that process chemistry is a roller-coaster ride, with tremendous highs and lows, where you learn theory and concepts, as needed, on the fly, from your colleagues and from those reference books (while meeting seemingly unattainable milestones and timelines). The aim of this book is to convey some of this experience by immersing the reader in the process chemistry of some of the most valuable pharmaceuticals we are fortunate to have available today. The masterpieces in this book are the top-selling drugs in the United States in 2007–2008. These are Lipitor®, Nexium®, Advair Diskus®, Prevacid®, Plavix®, Singulair®, Seroquel®, Effexor XR®, Lexapro®, and Actos®, all "blockbuster" drugs, generating more than $1 billion in revenue for their owners each year (Figure 1.1).[1]

I have no previous detailed knowledge of the process chemistry of most of these drugs. Why choose these as the subject matter? First, there is currently intense interest in the process chemistry of these drugs. Second, if I had detailed unpublished knowledge about these drugs, I would be bound by a secrecy agreement to discuss only information already in the public domain. Third, having no financial stake in any of these drugs or their process technology, I can be completely (and refreshingly) objective. I am not "selling" the value of any target or proprietary technology to a patent agency or a pharmaceutical manufacturer.

After a detailed review of the process chemistry for Plavix® and Nexium®, these will not be included. The process chemistry for Plavix® is omitted because I have published and patented process work and have detailed knowledge of the manufacturing process for Ticlid®. The antiplatelet drug Ticlid® is an adenosine diphosphate (ADP) receptor inhibitor with the same thienopyridine core as Plavix® (Figure 1.2).[2] The process chemistry for Nexium® is omitted because Prevacid® and Nexium® have the same core and there is considerable overlap in their process chemistry. Advair Diskus® has two active ingredients: salmeterol and fluticasone. The process chemistry of salmeterol is included. The process chemistry of fluticasone would be better presented "in context" with the process chemistry of other valuable steroids.

With this format, will this book touch on every important aspect of process chemistry in the pharmaceutical industry? If you carefully studied the techniques used to create 10 masterpieces at the art museum would you become an art

*Pharmaceutical Process Chemistry for Synthesis: Rethinking the Routes to Scale-Up*, By Peter J. Harrington

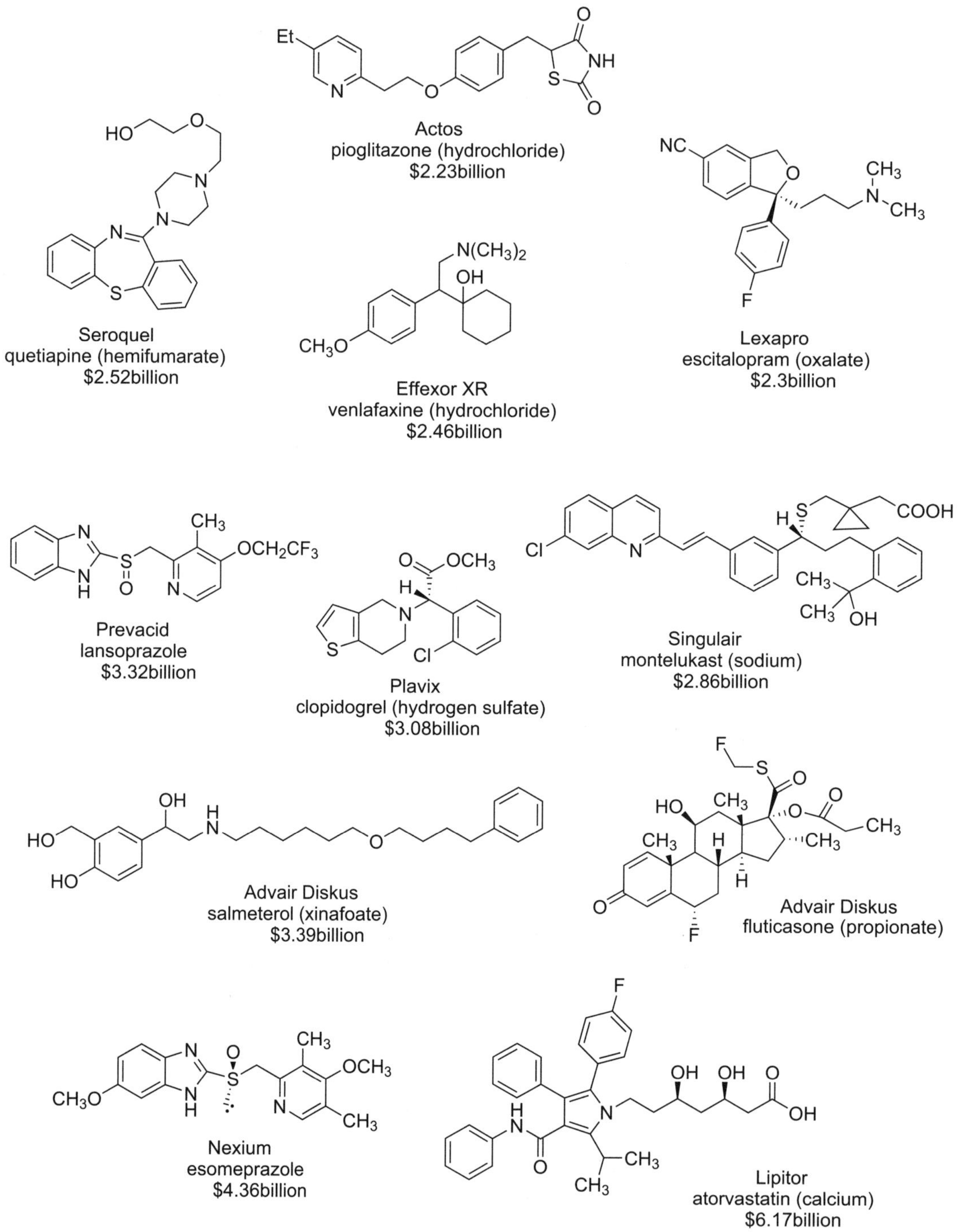


**FIGURE 1.1** The top-selling drugs in the United States in 2007.

expert? Most people would say no. Would you be better able to utilize the techniques in your own paintings? Most people would say yes. The scientific objective of this book is then twofold: to identify one "best" process for manufacturing these blockbuster drugs and to highlight the strategies and methodology that might be useful for expediting the process research and development of the blockbusters of the future.

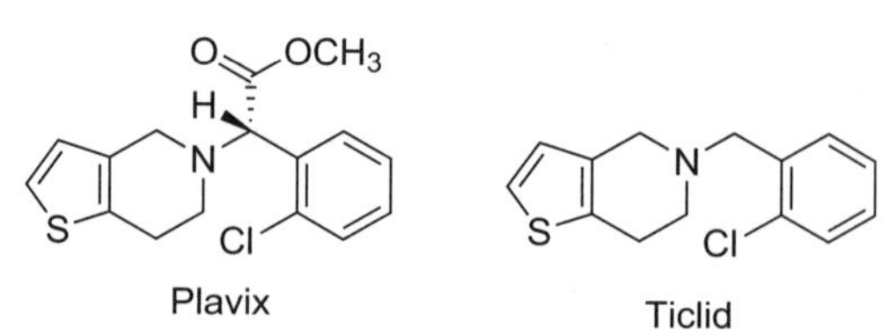


**FIGURE 1.2** The close structure similarity between the anti-platelet drugs Plavix® and Ticlid®.

## 1.2 INFORMATION SOURCES

This project must begin with meaningful and realistic objectives. A consistent strategy will be used to define, retrieve, and review the relevant literature. The process chemistry presented is based on published experimental data harvested from patents and journal publications. The majority of the information is taken from U.S., European (EP), and World (WO) patents. Other country-specific patents are included if they are cross-referenced several times, do not have a U.S./EP/WO equivalent, and are available in English, French, or German. Working with a finite production budget, information from Chinese (CN) and Japanese (JP) patents is taken from *Chemical Abstracts.* Journal articles are often published in tandem with patents and offer the same experimental procedures and data. Key journal articles offering information not found in the patent literature are included. The presentation is weighted to emphasize the process patents and publications and the marketplace information published in the past decade.

It is likely that at least a few details of the process chemistry of a valuable pharmaceutical may be carefully guarded as a trade secret. Speculation about unavailable data will be clearly marked as such. Legal questions such as who owns a particular patented process, how long they will own it, or how valid are their patent claims are important questions that should be directed to a legal expert. The answers to these questions are outside the scope of this book.

A quick SciFinder® search (January 1, 2009) for the Prevacid® structure, for example, revealed approximately 1700 references. A review using this number of references for each target cannot be accomplished in a realistic time frame. A solution to this is to structure search for the building blocks unique to each target. The building blocks selected for Prevacid® are shown in Figure 1.3. The building block structure searches provide the first generation of references. The cross-references from the first generation are then used and the process repeated until the cross-reference loop is completed. For Prevacid®, this structure search approach reduced 1700 references to a manageable 200 references. The structures searched are provided at the end of each chapter. No effort was made to update the chapters completed first.

Process chemistry is so multidimensional that there will inevitably be important points overlooked. I welcome your comments and suggestions for improving the content and format of future publications.

## 1.3 CONTENT AND FORMAT FOR PRESENTATION

The content of each chapter will vary according to the information harvested from the references. For example, one chapter emphasizes the manufacturing route selection while another focuses on conversion of the penultimate intermediate to the final target. This variable content accurately reflects the range of tasks assigned to process chemists. Your role in a process research and development team may be early route selection in one project. Your role may be late troubleshooting of a difficult crystallization to produce a target that filters well and meets crystal size and purity specifications in another. Your role might involve working closely with procurement specialists or engineers in the early route selection or with analytical and regulatory specialists on the difficult crystallization.

Just as the chemical transformations are central to the manufacturing process, the process chemist is the hub of manufacturing process research and development. The process chemist does not have to be an expert in the related specialties of marketing strategy, patent law, procurement, environmental health and safety, analytical chemistry, formulation, regulatory affairs, and engineering and facilities but he must be knowledgeable enough to identify questions best answered working in close collaboration with these experts. Answers will sometimes be offered to questions best answered by these experts with the understanding that the answer is meant to trigger a discussion with the expert.

Each chapter is written to stand alone. Chapters 2–9 can be read in any order. While the content for each chapter will vary, the same format will be used to present the available information. Each chapter begins with an *overview of current and past marketplace information* for the target. This discussion is included to emphasize that the process research and development team cannot work in a vacuum. The team should receive detailed updates at regular intervals on the market potential of the target, the timing of the delivery, and new clinical and post-launch data that may impact the market potential and timing of the delivery. This

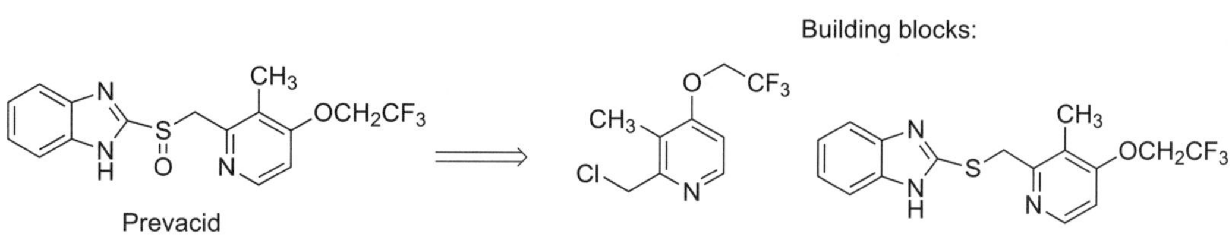


**FIGURE 1.3** Building blocks searched to provide references to process chemistry for Prevacid®.

information might come from a marketing or business development expert.

To minimize repetition, retrosynthetic analysis will not be used to stage the synthesis discussion. To emphasize the modularity of pharmaceutical manufacturing, the synthesis discussion in each chapter starts with identification of raw materials. These *raw materials* are usually commercially available or can be produced in a few steps from commercial materials.

Every process begins with *commercially available raw materials*. A price is provided for each raw material *that contributes at least one atom to the target* when that raw material first appears in the discussion. Since suppliers and prices for raw materials are in constant flux, all prices quoted are taken from the 2007–2008 Aldrich catalog. It is my intention that these prices will give a "snapshot" of a relative price and availability at this point in time. Quoting an Aldrich catalog price should suggest scheduling a preliminary communication with a procurement group. This communication would include estimates of the quantity and purity specifications, a preferred delivery date, and any special shipping and handling requirements. Other raw materials, for example, acids, bases, reagents used to create protecting groups or leaving groups, drying agents, filter aids, and decolorizing carbon are not priced since expensive materials might be replaced by less expensive alternatives.

The raw material prices are only intended for "back-of-the-envelope" calculations. Detailed cost calculations should include vendor-guaranteed raw material prices and labor and overhead (LOH) costs for the manufacturing site and are beyond the scope of this book.

*Aldrich catalog names* are used for all starting materials and *ChemDraw 11.0®* is used to generate names for all process intermediates. With the intention that each sentence can stand alone, full chemical names are used in the text in many cases. Process intermediates and products are each assigned a number to facilitate correlation of the names with the structures in schemes and figures. An example of a stand-alone sentence is taken from the Seroquel® discussion.

> The reaction of 11-chlorodibenzo[*b*,*f*][1,4]thiazepine **(25)** with 2-(2-(piperazin-1-yl)ethoxy)ethanol **(26)** (2.0 equivalents) in refluxing toluene is complete in 8 h.

Patent procedures often contain *data gaps*. These can be separated into two categories. A major data gap is missing information that would certainly have been generated but was not included in the process description. Examples of major data gaps are a missing quantity for one reagent of several or a missing volume for the reaction solvent. Major data gaps are clearly identified in the discussion, and where possible, an attempt is made to fill the gaps with information gleaned from another source. A minor data gap is information presented in a format that requires a translation. For example, reagent quantities might be quoted only in weights or volumes. This gap is filled by converting reagent quantities into *equivalents*. In process chemistry, an equivalent simply refers to the number of moles of reagent per mole of limiting reagent. Equivalents in this book are calculated to the nearest 0.1.

*Solvents* and *reaction temperatures* are critically important process characteristics. These are included in each reaction description. After selecting a best process, the process solvents used are revisited to emphasize the importance of minimizing the number of process solvents and to highlight the solvents commonly used in a pharmaceutical manufacturing plant. Temperatures in the range of 20–30°C, or "ambient," are standardized as 25°C in the reaction descriptions. Very low temperatures (<−70°C) require that expensive liquid nitrogen be available locally and that liquid nitrogen storage facilities be available on site. Expensive circulating fluid and energy are required to achieve and maintain very high reaction temperatures (>160°C). Examples of a reaction description and a process solvent review are taken from the Actos® discussion.

> The condensation of 4-(2-(5-ethylpyridin-2-yl)ethoxy)benzaldehyde **(19)** with thiazolidine-2,4-dione (1.2 equivalents) and pyrrolidine (1.0 equivalent) in methanol at 45°C is very efficient even after multiple precipitations and isolations for purity upgrade (95% yield). The process solvents are toluene, THF, ethanol, isopropanol, and water, all solvents commonly used in a pharmaceutical manufacturing plant.

It is assumed that all operations involving combustible organic materials are performed *under nitrogen* and that all chemical mixtures are *stirred*. This is not specifically stated in the procedures described.

When there are many similar procedures, they will be presented in a *parallel format* to facilitate comparison and highlight the differences. Material presented in parallel format is usually preceded by a summary of the trends and results. An example of parallel formatting is taken from the Effexor XR® discussion.

> A mixture of 1-(2-amino-1-(4-methoxyphenyl)ethyl)cyclohexanol **(34)**, 88% formic acid (5.0 equivalents), and 36% aqueous formaldehyde (3.1 equivalents) in water (96 L per kg **34**) is refluxed for 21 h.
>
> A mixture of 1-(2-amino-1-(4-methoxyphenyl)ethyl)cyclohexanol **(34)**, formic acid (6.3 equivalents), and paraformaldehyde (2.9 equivalents) in water (7.9 L per kg theoretical **34**) is refluxed for 24–48 h.

When the discussion leads to a choice between two very similar processes, the analysis may be taken to an even greater level of detail. An example of information on this next level is *volume throughput*. The discussion at this next level should be prefaced with the understanding that throughputs are rarely the focus of patent procedures, that some assumptions must be made, and that some questions (e.g., solubility and viscosity) can only be answered in the laboratory.

Nowhere is the phrase "time is money" more apt than in a manufacturing plant. Patent procedures typically quote reaction times in the range of 30 min to 24 h. I would suggest that a *reaction time* of 2 h is close to ideal, slow enough to allow for efficient heat transfer to or from the reaction vessel and to allow for sampling and an offline completion check. Any unusually long reaction times in key procedures will be identified and the potential for reducing these times may be addressed.

A great deal of process research and development effort is spent streamlining the transitions from one reaction to the next. For this reason, *workup procedures* are presented in detail to highlight potential scale-up problems. There may be product stability issues that will only become apparent during a scale-up or there may be a concentration at reduced pressure to a solid residue. When the workup description does not add to the discussion, it may be omitted or abbreviated to a "routine workup." In a routine workup, the reaction is quenched with water, dilute bicarbonate, or dilute brine and then extracted into an organic solvent (toluene, ethyl acetate, or dichloromethane). There may be several extractions. The combined organic layers are optionally dried ($MgSO_4$ or $Na_2SO_4$) and the solvent removed at reduced pressure to produce an oil or solid residue.

Drying agents such as sodium sulfate or magnesium sulfate are routinely used in the laboratory but rarely used at pilot plant scale. Drying agents used in the experimental procedure are omitted from the process descriptions in this book. The process chemist must use the water-wet solution or rely on (design in) an azeotropic distillation to remove water from the solution.

*Purity analysis* is critically important in process chemistry, yet often is not included in patent experimental procedures. The centrifuge may be filled to capacity with product but remember: *If the material does not meet specifications, the yield is zero.* To be consistent with this important tenet, yield and purity data are quoted when available. In the absence of purity data, the yield is quoted if the product is precipitated, chromatographed, crystallized, or distilled. Crude yields of early intermediates are included when other data suggest that the yield is an accurate reflection of efficiency of the reaction. HPLC area% data will be used for completion checks but not for purity analysis. Purity data for process intermediates are rounded to 0.1%. Purity data for the final drug substance, if available, are rounded to 0.01%.

*Physical data* such as boiling point or melting point are provided for process intermediates if those data are critical for determining the suitability of the process. For example, the crystallization and isolation of a solid with a low melting point (<50°C) may be more challenging. The distillation of an oil at high temperature and low pressure (>150°C at <1 mmHg) may not be a viable option.

Every effort will be made to identify undesirable reagents and intermediates. These include *carcinogens*, *lachrymators*, *sensitizers*, and *malodorous chemicals*. Information on these chemicals will be quoted from *material safety data sheets* (MSDS) to substantiate the objection to use of the chemical. The date accessed and online reference to the MSDS are not included in the references. *The most current version of the MSDS should be reviewed before working with any chemical.* An example of an MSDS review is taken from the Prevacid® discussion.

> Vanadium(V) oxide is considered to be a carcinogen.[82] All vanadium compounds should be considered toxic.[83] The toxicity depends on the valence state and the solubility of the compound. For example, vanadium(V) oxide ($V_2O_5$) is considered to be five times as toxic as vanadium(II) oxide ($V_2O_3$). The first concern in handling these vanadium catalysts is exposure to dust. For vanadium(V) oxide, the OSHA permissible exposure limit (PEL) for vanadium respirable dust is 0.5 mg/m$^3$ (ceiling) and for vanadium fume is 0.1 mg/m$^3$ (ceiling), and the ACGIH threshold limit value (TLV) is 0.05 mg/m$^3$.

The "no stone left unturned" level of detail is chosen to accurately reflect the day-to-day concerns and activities of a process chemist. It is also intended that the level of detail is sufficient to allow the reader to make an informed process decision without revisiting the original experimental description for additional details.

*Text boxes* are used to elaborate on the logic behind a process decision. They are largely the author's personal preferences honed by trial and error in the laboratory and pilot plant over 20 years. Text box topics include setting starting material specifications, solid addition to a reaction mixture, stability of intermediate mixtures produced during sequential reagent charges, compatibility of materials of construction with reaction conditions, concentration at reduced pressure, acceptable volume throughputs, estimating volume throughputs from gram-scale procedures and kilogram-scale procedures, identifying first/second-generation side products for workup design, distillation of high-boiling polar aprotic solvents, routine safety testing of lab distillation bottoms, self-accelerating decomposition temperature (SADT), alternatives to dichloromethane, "one-pot" procedures, the importance of hold points, mixtures of sulfonic acids and methanol, alignment of economic and environmental incentives, selecting reaction variables for design space studies, analysis of suspensions, why polymorphs are important, and deconvoluting polymorph literature. While the same text box topic could be inserted at many points in the book, each topic appears only once and where it is most relevant. An example of a text box is taken from the Singulair® discussion.

> Now that the challenges of producing 7-chloroquinaldine (**3**) are understood, a specification for 5-chloroquinaldine (**4**) in the starting material must be set and the fate of the side products from 5-chloroquinaldine (**4**) produced in the following step(s) must be determined. Our first inclination, as synthetic chemists, is to demand high-purity starting material. However, it would be prudent to invest some time up front to demonstrate efficient rejection of the side product from 5-chloroquinaldine (**4**). These data will empower us to use a lower grade of 7-chloroquinaldine (**3**) that will be available at a better price.

*Schemes* immediately follow the chemistry discussion. Since reagents and conditions are provided in the text and since many of the transformations can be performed using more than one combination of reagents and conditions, these are not included in the schemes. The highest yield or an appropriate yield for each transformation is provided under the reaction arrow. For example, see the scheme from the Lexapro® presentation (Scheme 1.1).

A section on *trade secrets, impurities, and analytical methods* is sometimes used to capture valuable process information that does not appear in the earlier chemistry review sections but might prompt valuable additional discussion.

Finally, the *best process available* offers criteria for selecting the process and uses the criteria to arrive at a single route as the standard for comparison. This best process is an amalgamation of the best available process steps and is intended to serve as a basis for further discussion rather than to end it.

For most of the targets, the method developed for generating the limited reference set intentionally minimizes the publications in other important areas, including *crystallization, polymorphism, particle size, storage stability,* and *formulation* of the final drug product. The Lexapro® presentation is expanded to include a detailed discussion on crystallization and polymorphism. The Lipitor® discussion includes a discussion of amorphous and crystalline polymorphs and the drying and storage stability of the final drug product.

A suitable formulation is most efficiently attained by the process chemist working in close collaboration with a formulation group. The involvement of the process chemist might end with developing crystallization, drying, and milling procedures to deliver the desired polymorph of the target to the formulation group with acceptable storage stability

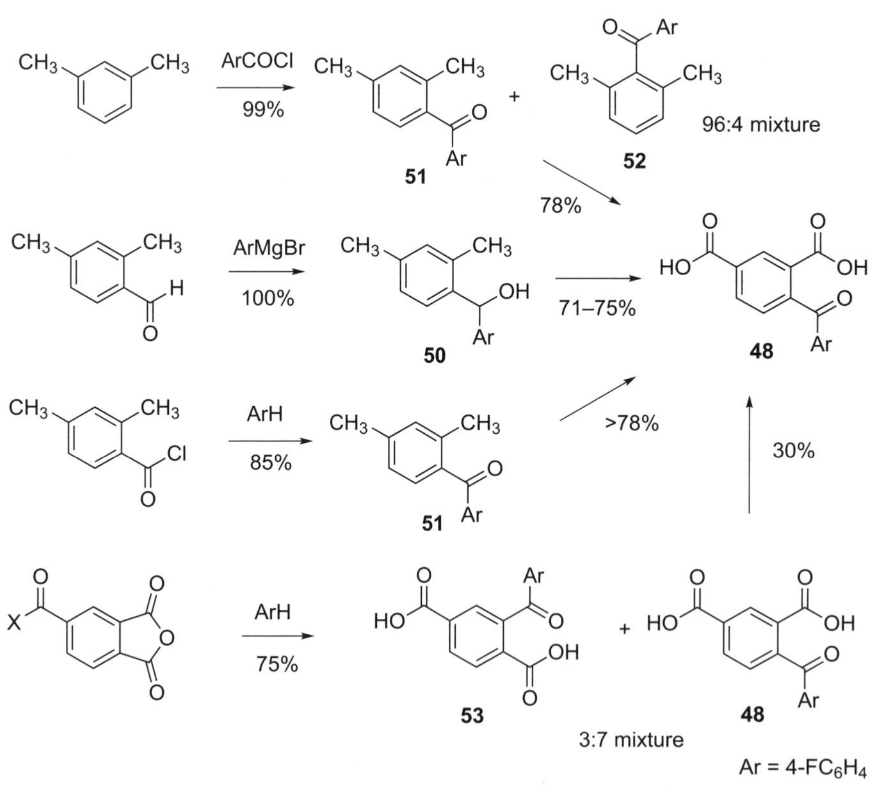


**SCHEME 1.1** A scheme from the Lexapro® presentation.

and a well-defined particle size range. Formulation is outside the scope of this book.

How reproducible are the patent experimental procedures at the heart of this project? Comparing similar procedures side by side certainly makes it easier to find inconsistencies. The inconsistencies are pointed out and corrections for typographical errors may be suggested. An example is taken from the Effexor XR® discussion.

> Palladium on carbon (10% w/w, 50% water-wet) (50 g Pd per kg **17**) is added to a mixture of 2-(1-hydroxycyclohexyl)-2-(4-methoxyphenyl)acetonitrile **(17)** and hydrochloric acid in methanol (8 L per kg **17**), presumably at 25°C. (*Note*: The amount of hydrochloric acid charged is quoted as "1–3 moles" or 10–29 equivalents. This is presumably a typographical error.)

If a quoted yield can't be reproduced is the best process still viable? *The underlying principle for selecting the process is still valid.* An optimistic process chemist would respond: if you can get 50%, you can get 80%. If you can get 80%, you can get 90%. All that is required is motivation and development time.

## 1.4 SPECIALIZATIONS: BIOTRANSFORMATIONS AND GREEN CHEMISTRY

Some readers will be disappointed that a particular specialization in process chemistry does not receive more attention. The presentation is weighted based solely on how many of the patents and publications deal with that specialization. For example, a chiral alcohol intermediate in the Singulair® discussion can be produced by a microbial reduction.

> There are five options for the asymmetric reduction: microbial reduction to (*R*)-alcohol **31** with the novel microorganism *Microbacterium* MB5614 (ATCC 55557) and a Mitsunobu inversion,[50,32] microbial reduction to (*S*)-alcohol **32** with *Mucor hiemalis* IFO 5834,[51] reduction to (*S*)-alcohol **32** with borane–THF catalyzed by an oxazaborolidine,[32] reduction to (*S*)-alcohol **32** with diisopinocampheylchloroborane,[43] and ruthenium-catalyzed transfer hydrogenation to produce (*S*)-alcohol **32**.[52] Since the microbial reduction patents provide only milligram-scale procedures and are more than 10 years old, we will focus on the chemical methods.

While the process chemist is not an expert in green chemistry, the process chemist plays a pivotal role in the *implementation* of green chemistry on a plant scale. The terms *green or greener* may be used to denote a process that is superior in its qualitative or quantitative adherence to one or more of the *Twelve Principles of Green Chemistry*.[3]

## 1.5 IMPACT ON PROCESS CHEMISTRY IN THE FUTURE

Rethinking the step-by-step *manufacturing process* is the overriding theme of this book. A secondary objective of this book is to increase awareness about the *process by which we transition from one supplier to multiple generic suppliers*. A long-standing interest in this transition dates back to the 1980's second-generation process research and development for (*S*)-naproxen, now sold as Aleve®.[4] After reading this book, it will be clear that there may be an incentive to regress to inferior process technology and that the regression is often accompanied by an increase in the environmental impact of manufacturing the drug. This regression is the inevitable consequence of the normal progression of patent protection for a new drug: the patents for the drug itself and the medicinal chemistry route(s) to the drug are followed, often over the course of many years, by a series of process patents from the manufacturing group. These process patents protect key steps in one or more finely honed manufacturing processes for many years beyond expiration of the drug patent. Unless groundbreaking new and directly applicable synthetic methodology is discovered in the 10 years after the drug manufacturing process was first put online, new manufacturing processes may offer little that is new and improved. Process regression is science in reverse, a step back for a society that celebrates and rewards innovation.

## 1.6 AUDIENCE

Synthetic chemists interested in manufacturing these top-selling drugs are the primary audience for this book. Another audience is graduate students with a specialization in organic synthesis. In many university interview trips in search of the next generation of process chemists, it became clear that most graduate students have no idea what a process chemist does. With instructor-added emphasis on synthetic strategy and control, this book could provide the core information for an interactive one-semester graduate course in process chemistry. Where is the academic value of learning process chemistry? Process research is mechanism based, it requires an in-depth analysis and understanding of reaction kinetics and thermodynamics, and it pushes the limits of established synthetic technology. Process research generates unexpected results, results considered improbable during the project planning phase, and results that are often the basis of valuable process patents.

Another intended audience for this book is process chemists always in search of *methods proven on scale-up*. Looking for a method for nitrile reduction to a primary amine? What better place to look than in the chapter on Effexor XR®. Methods are compared and contrasted for creating a chiral secondary alcohol from a ketone

(Singulair®), oxidation of a sulfide to sulfoxide (Prevacid®), and introducing an amino group using an ammonia surrogate (salmeterol of Advair Diskus®).

Discovery chemists seeking a strategy to protect their investment in a new drug might review the strategies generic manufacturers used to develop noninfringing processes. Generic drug manufacturers eager to design and implement new manufacturing processes can map out the company-specific patent strategies used to protect new drugs. The environmental chemist will find useful information on the environmental impact of drug manufacturing for these specific targets and for small-molecule drugs in general. Finally, the consumer activist will find useful information on the cost to produce these blockbuster drugs.

## ACKNOWLEDGMENTS

Thanks to Chemical Abstracts® for a grant of 115 tasks/1 year used for the structure searches. Journal articles were obtained through the interlibrary loan (ILL) program. Thanks to the ILL program coordinator, Sandra Richmond, at the Louisville Public Library for her time and support. Current and past marketplace information for each target was developed working in collaboration with Karen Ingish, reference librarian at the Louisville Public Library. A special thanks to Karen for her enthusiasm and her invaluable contribution. Thanks to Dr. Dave Johnston and Dr. Neal Anderson for their sage advice and support for this project. Finally, Rosemarie and Jack, my home team, there are no words of thanks I can offer to tell you how much I appreciate all that you did. This book is dedicated to you, Jack. No man could ask for a finer son.

At the beginning of this project, it was clear that this would be a journey of a thousand miles. You will be gratified with expectations met in some cases and surprised by unexpected selectivity in others. You will delight in the victory of efficiency of some manufacturing processes and be left dissatisfied with the state of affairs of others.

> A journey of a thousand miles begins with a single step.
>
> Lao-tsu (604–531 B.C.)

## REFERENCES

1. Accessed at www.drugs.com/top200.html.
2. (a) Harrington, P.J.; Sanchez, I.H. *Synth. Commun.* **1993**, *23*, 1307.(b) Harrington, P.J.; Sanchez, I.H.EP 522956(1/13/1993).
3. Accessed at www.epa.gov/greenchemistry.
4. Harrington, P.J.; Lodewijk, E.L. *Org. Process Res. Dev.* **1997**, *1*, 72.

# 2

# ACTOS® (PIOGLITAZONE HYDROCHLORIDE)

## 2.1 ACTOS® IN THE DIABETES MARKET

Pioglitazone hydrochloride (Actos®) and rosiglitazone maleate (Avandia®) are two thiazolidinedione (TZD) drugs used to treat patients with type II diabetes. Both are also marketed in combination with metformin or glimepiride, pioglitazone as Actoplus Met® and Duetact® and rosiglitazone as Avandamet® and Avandaryl®. Pioglitazone and rosiglitazone are agonists of peroxisome proliferation-activated receptors (PPAR), specifically PPAR-γ (Figure 2.1). These agonists improve glucose utilization and reduce glucose production in the liver by increasing insulin sensitivity in adipose and muscle tissue.

The statistics for the global diabetes epidemic are compelling. The global prevalence of diabetes for all age groups is estimated to rise from 2.8% (171 million people) in 2000 to 4.4% (366 million people) by the year 2030.[1] Another analysis estimated that 23.6 million people had diabetes in the United States in 2007.[2] The biggest increase in diabetes prevalence will be in the adult population and the vast majority (90–95%) of the adults diagnosed with diabetes are diagnosed as type II.

Global 2006 sales figures for pioglitazone were $2.8 billion. Pioglitazone was Takeda's best seller and accounted for 25% of their revenues. Global 2006 sales figures for rosiglitazone were $3.3 billion. Rosiglitazone was GSK's third best-selling drug that year. The U.S. figures for pioglitazone and rosiglitazone for 2006 were $1.9 billion and $1.7 billion, respectively, each up 20% from the 2005 figures. Both drugs had a promising future. But a lot has happened since then. Two meta-analyses were published back to back in the *Journal of the American Medical Association* on September 12, 2007. One reported that rosiglitazone increased the risk of heart attack by 42% while the other found that pioglitazone actually lowered the combined risks of heart attack, stroke, and death by 18%.[3,4] This was the first time a diabetes drug has been shown to reduce the risk of heart attacks. The U.S. figures for pioglitazone and rosiglitazone for 2007 showed a quick response to these meta-analyses: pioglitazone sales increased to $2.2 billion and rosiglitazone sales dropped to $1.1 billion.[5] Of course, this is just a snapshot in time and more studies are underway but, in 2008, pioglitazone was a very important target.

## 2.2 SYNTHESIS LEFT TO RIGHT

### 2.2.1 2-(5-Ethylpyridin-2-yl)ethanol (2)

Pioglitazone (**1**) has three distinct regions: the 2,5-dialkylpyridine, the *para*-substituted aryl ether, and the thiazolidine-2,4-dione (Figure 2.2). There is a chiral center at the 5-position of the thiazolidine-2,4-dione but this center is easily epimerized, so the synthetic challenge is to produce the racemate. Disconnection near the center, on either side of the ether oxygen, leads back to 2-(5-ethylpyridin-2-yl)ethanol (**2**). While many simple mono-, di-, and trimethylpyridines (picolines, lutidines, and collidines), and some ethylpyridines are obtained from coal tar, 2-(5-ethylpyridin-2-yl) ethanol (**2**) is not directly obtained from a natural source. It is a specialty chemical. A *Chemical Abstracts* structure search [5223-06-3] reveals less than 100 references, with

*Pharmaceutical Process Chemistry for Synthesis: Rethinking the Routes to Scale-Up*, By Peter J. Harrington

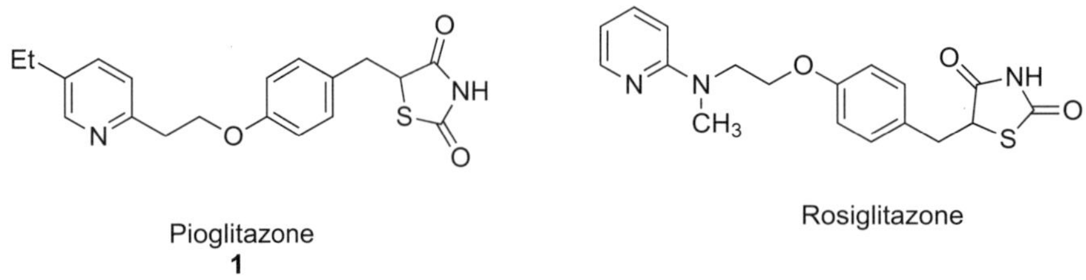


**FIGURE 2.1** Pioglitazone (**1**) and rosiglitazone.

the majority directly associated with pioglitazone process chemistry.

Since the alcohol **2** is a key component of pioglitazone, it is critically important to know who produces it, how and on what scale they produce it, and what is it produced from. The same questions should then be asked and answered for the material(s) used to produce **2**. At least two suppliers for **2** should be online. Taking this a step further, it would be preferable to have suppliers who have a long track record for reliability, perhaps suppliers in several continents. A search of *Chemical Abstracts* and a Google search "suppliers for 5223-06-3" provide lists of suppliers. The goal is not to identify the lowest price or the specific suppliers at this point but to make a case for the material as "readily available and inexpensive."

How is 2-(5-ethylpyridin-2-yl)ethanol (**2**) produced? Does the process involve operations that may raise safety concerns? Does it require special processing equipment? Conditions for the condensation of 5-ethyl-2-methylpyridine (**3**) with formaldehyde to produce 2-(5-ethylpyridin-2-yl) ethanol (**2**) were first described more than 60 years ago (**3**, trioxane, potassium persulfate, and *tert*-butylcatechol in ethanol at 220°C).[6] Perhaps it is produced today by an amine-catalyzed condensation of 5-ethyl-2-methylpyridine (**3**) with paraformaldehyde in water at 170°C.[7] The high temperatures and pressures and handling of aqueous formaldehyde waste (from paraformaldehyde or trioxane) are cost drivers.

High temperatures and pressures and aqueous formaldehyde waste are also associated with the manufacture of 5-ethyl-2-methylpyridine (**3**) (also known as "aldehyde-collidine") from paraformaldehyde, ammonium hydroxide, and ammonium acetate.[8] This ultimate starting material is readily available and amazingly inexpensive.[9] The similarities in materials and process conditions suggest that significant cost savings might be realized by producing both **2** and **3** at the same manufacturing site.

### 2.2.2 Construction of the Ether C–O Bond

There are two well-established approaches to construction of the ether C–O bond: $S_NAr$ displacement by alkoxide of a good leaving group on an aromatic activated by an electron-withdrawing group and Williamson ether synthesis using a primary alkyl toluenesulfonate or methanesulfonate and a phenoxide (Scheme 2.1).

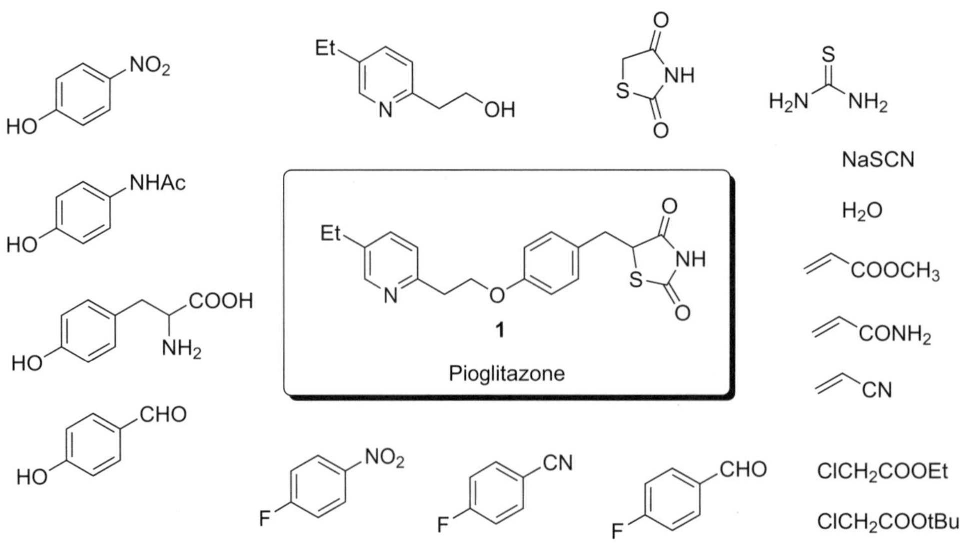


**FIGURE 2.2** Pioglitazone building blocks.

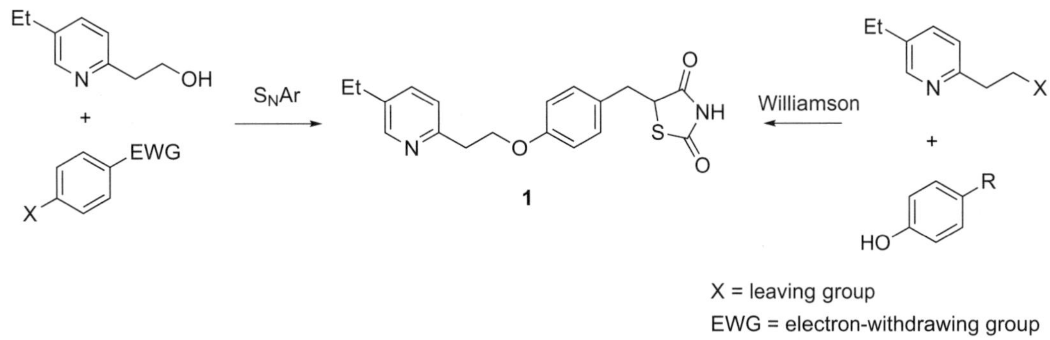


**SCHEME 2.1** Options for construction of an ether C–O bond in pioglitazone (**1**).

***2.2.2.1 $S_NAr$ Using 4-Fluoronitrobenzene*** The $S_NAr$ approach on a nitro-activated aromatic is well documented. Reaction of 2-(5-ethylpyridin-2-yl)ethanol (**2**) with 4-fluoronitrobenzene and sodium hydride in DMF at 25°C affords 5-ethyl-2-(2-(4-nitrophenoxy)ethyl)pyridine (**4**) (63%).[10,11] Other base and solvent combinations (powdered NaOH in DMSO, KOH in dichloromethane, NaOH in DMSO–water, and simply NaOH in water) eliminate the hazard associated with handling and quenching sodium hydride and avoid the formation of 4-dimethylaminonitrobenzene from DMF.[12–14] For example, the reaction of 2-(5-ethylpyridin-2-yl)ethanol (**2**) with 4-fluoronitrobenzene (1.06 equivalents) and aqueous sodium hydroxide (2.7 equivalents) in water at 30–35°C affords 5-ethyl-2-(2-(4-nitrophenoxy)ethyl)pyridine (**4**) (88%).[14]

This $S_NAr$ methodology is coupled with a Meerwein arylation via nitro group reduction and diazonium salt formation.[10,11] Reduction using 10% Pd on carbon (50% water wet) in methanol at 25°C and 1 atm hydrogen affords 4-(2-(5-ethylpyridin-2-yl)ethoxy)aniline (**5**), which requires no purification (93%). The nitro group is also reduced using Raney nickel and hydrogen or Raney nickel and hydrazine to eliminate the fire hazard associated with handling of the palladium catalyst after reduction. Aniline **5** is a low-melting solid that turns dark over time.[14]

The Meerwein arylation, first described in 1939, is the copper-catalyzed replacement of a diazonio group of an arenediazonium salt by an alkene or alkyne.[15–17] The Meerwein arylation is suggested to proceed via a free radical chain mechanism. The addition of the aryl radical to an alkene affords the more stable alkyl radical. This radical is then converted to an alkene by hydrogen atom abstraction or to a 1-aryl-2-haloalkane by halogen abstraction from copper (II) halide. Meerwein arylations of acrylic acids, acrylate esters, acrylonitriles, acrylamides, vinyl ketones, vinyl halides, and styrenes are all known with yields typically in the 40–70% range.

What can be described as typical Meerwein arylation conditions are used in one pioglitazone process. The arenediazonium salt is prepared by addition of sodium nitrite to the aniline in methanol–acetone–aqueous hydrobromic acid at <5°C. Methyl acrylate (5.9 equivalents) is added. Cuprous oxide is then added in small portions at 38°C. Aging the reaction at 38°C until complete, neutralization, concentration at reduced pressure, and an extractive workup affords methyl 2-bromo-3-(4-(2-(5-ethylpyridin-2-yl)ethoxy)phenyl)propanoate (**6**) as an oil (86% crude).[10,11] A similar procedure suggests that the yield after correction for purity could be much lower (47–48%) (Scheme 2.2).[14]

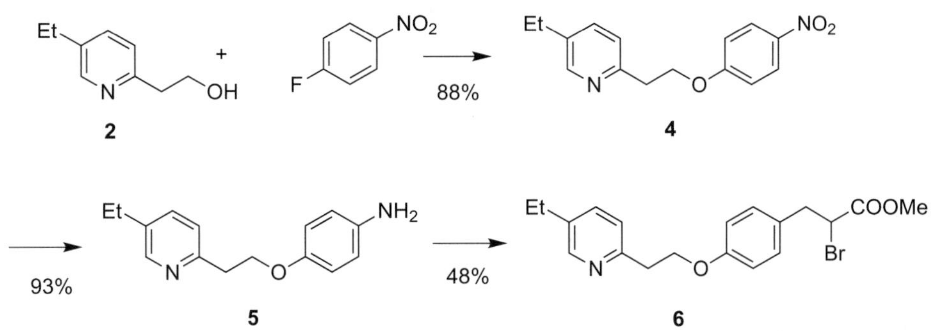


**SCHEME 2.2** Pioglitazone intermediate methyl 2-bromo-3-(4-(2-(5-ethylpyridin-2-yl)ethoxy)phenyl)propanoate (**6**) from 1-fluoro-4-nitrobenzene via $S_NAr$ and Meerwein arylation.

This procedure raises many scale-up concerns. The diazonium salt solution must be prepared and held at 5°C. If cooling is lost during or after the diazonium salt preparation due to an unplanned power outage, the diazonium salt will decompose. Solid copper(I) oxide is charged in small portions, potentially exposing the operators to the corrosive reaction vapors during the vigorous nitrogen gas evolution that accompanies each charge. Careful planning could mitigate these concerns: have a backup power supply ready and charge the copper catalyst as an aqueous suspension. There are still other concerns. Malodorous methyl acrylate is both toxic and irritating. There will be copper in the aqueous waste stream. The low purity of the crude oil **6** suggests that chromatography or at least a carbon treatment will be necessary before moving on to construction of the thiazolidinedione.

> There is one phrase in the above procedure that at first appears innocuous: *concentrate at reduced pressure*. We concentrate at reduced pressure on a rotary evaporator in the lab every day. We have the option of using a water-cooled coil condenser or cold finger condenser. We have a trap in the vacuum line to collect any volatiles not condensed in the rotary evaporator. We have a pump that can be set to deliver vacuum down as low as 5 mmHg, allowing us to evaporate at or near ambient temperature. All the wetted surfaces are glass or Teflon. The condensate is transferred to a waste can for appropriate disposal. The entire process from setup to waste disposal can be completed in less than an hour. The situation changes dramatically when you consider scale-up of a concentration at reduced pressure. First, there will likely be no option for chilling the condenser below −10 to −20°C. The condensate will be sent to a second reactor cooled to −10°C. The trap in the vacuum line, if there is one, will likely be a third reactor also cooled to −10°C. The vacuum pump will maintain, at best, 50–100 mmHg. We should certainly know the composition of the condensate and have a plan in place to recycle the expensive components. The scaled-up concentration at reduced pressure may take anywhere from 2 to12 h.

The composition of the condensate in the Meerwein arylation workup is a witch's brew of hydrobromic acid, methanol, acetone, water, methyl acrylate, and the by-product bromoacetone. Bromoacetone is a potent lachrymator and is a "show-stopper" for this process.[18] While an alternative acetone-free procedure starting with the isolated arenediazonium tetrafluoroborate[19] could eliminate this "show-stopper," the combined weight of all these concerns would certainly motivate a process research group to find an alternative route.

#### *2.2.2.2 Williamson Ether Synthesis*

*Preparation of a Sulfonate Ester* A Williamson ether synthesis will provide many more options for introducing and elaborating a *para*-substituent. The Williamson ether synthesis invariably begins with the conversion of the hydroxyl group of 2-(5-ethylpyridin-2-yl)ethanol (**2**) to a halide, methanesulfonate, or toluenesulfonate leaving group. These intermediates will not be "campaignable." They possess both a leaving group and a basic pyridine ring nitrogen and will be prone to elimination. The chloride, bromide, or iodide could be prepared by many well-established methods using, for example, thionyl chloride, phosphorus oxychloride, phosphorus tribromide, or triphenylphosphine–iodine. The methanesulfonate or toluenesulfonate esters are the preferred intermediates, since they can be prepared at or below ambient temperature. For example, the methanesulfonate ester **7** is prepared by slow addition of methanesulfonyl chloride to solution of 2-(5-ethylpyridin-2-yl)ethanol (**2**) and triethylamine in dichloromethane or toluene at 0–10°C. The reaction is complete in 1–4 h at 25°C. A water wash to separate triethylamine hydrochloride follows. At this point, the lab-scale and large-scale procedures diverge. In the lab, the solution is dried over sodium sulfate and concentrated at ambient temperature and reduced pressure. On large scale, it would be desirable to use the water-wet solution in toluene in the Williamson ether synthesis. A nearly quantitative yield of the methanesulfonate **7** is expected both in the lab and on large scale.

How dry does the toluene solution of the methanesulfonate **7** have to be after the water wash? Assume that the toluene solution would at least be saturated with water (0.05 wt%) after a perfect layer separation. And a near-perfect layer separation is far easier to achieve when draining a separatory funnel in the lab than when draining a large reactor using a sight glass. What is the best answer we can hope for? The answer is that we do not need to do anything to remove the water, as low levels of water are acceptable in the next step. Is this the case? Isolate some methanesulfonate **7** by a standard lab workup procedure: dry the toluene solution over sodium sulfate and concentrate it at 25°C and reduced pressure. Prepare a toluene stock solution of methanesulfonate **7** (store cold) and use this solution to screen the Williamson ether synthesis with different concentrations of water present. Even after the screen confirms that water is tolerated in the next step, it would be wise to be present during the phase split in the pilot plant to see just how easy it is to detect the interface as it enters the sight glass. The phase split will be easier to see (more precise and reproducible) if there is little or no interface emulsion, if the layers are of different colors, and/or if there is a trace of interface "rag."

If water cannot be tolerated in the next step, the options are quite limited. Removing the water by azeotropic distillation at atmospheric pressure will likely cause decomposition of thermally labile methanesulfonate **7**. Removing the water by azeotropic distillation at a reduced temperature and pressure is far less efficient because the vapor phase contains less water.[20]

There are several examples using dichloromethane as solvent in lab preparation of the methanesulfonate **7** or toluenesulfonate ester **8**. Since the lab yields are comparable in either solvent, preparation on large scale in dichloromethane offers no advantages over preparation in toluene.[21] Dichloromethane retains more water after a water wash (0.2 wt%) than toluene (0.05 wt%) and an azeotropic removal of water is less efficient (1.5 wt% $H_2O$ removed by azeotrope versus 13.5 wt% for toluene). Of course, both these points are moot if the wet dichloromethane solution of **7** is carried into the Williamson ether synthesis. The decision point then comes after the Williamson ether synthesis. Whether distilling at atmospheric or at reduced pressure, the recovery of dichloromethane (bp 40°C) will be less efficient than the recovery of toluene (bp 111°C). Keep in mind the high odor threshold (205–307 ppm) and the low permissible exposure limit (25 ppm time-weighted average (TWA) with 12.5 ppm 8-hour TWA action level) for dichloromethane when considering the amount of dichloromethane that will not be recovered.

*4-Nitrophenol* The first of many ethers accessed from **2** by Williamson ether synthesis is 5-ethyl-2-(2-(4-nitrophenoxy) ethyl)pyridine (**4**).[22] The sodium salt of 4-nitrophenol is first prepared using sodium hydroxide in methanol–toluene. After removal of the solvent by distillation and drying at 100–110°C under vacuum, the isolated sodium salt is reacted with isolated toluenesulfonate **8** in DMSO at 40°C. The procedure is not detailed enough to say the conditions are anhydrous. The yield is 70–75% from 2-(5-ethylpyridin-2-yl)ethanol (**2**). The same mixture of the salt of 4-nitrophenol and the toluenesulfonate **8** might also be produced under Williamson ether synthesis conditions by a toluenesulfonate exchange. The reaction of 4-nitrophenyl toluenesulfonate (**9**) with 2-(5-ethylpyridin-2-yl)ethanol (**2**) also affords 5-ethyl-2-(2-(4-nitrophenoxy)ethyl)pyridine (**4**).[23]

Reduction of the nitro group is accomplished with sodium sulfide in methanol–water (83%). Again Meerwein arylation conditions are used, this time with acrylamide in place of methyl acrylate. The arenediazonium salt is prepared by addition of sodium nitrite to the aniline **5** in methanol–acetone–aqueous hydrobromic acid at <5°C. Acrylamide (5.6 equivalents) is added. Freshly prepared cuprous bromide catalyst is then added at 30–35°C. Aging the reaction at 30–35°C until complete, concentration, aqueous bicarbonate–hexanes digestion, isolation, and purity upgrade by water and hexanes resuspension affords 2-bromo-3-(4-(2-(5-ethylpyridin-2-yl)ethoxy)phenyl)propanamide (**10**) as an oil in 64% crude yield (Scheme 2.3).[22]

The acrylamide process highlights a logic trap:

> Acrylamide is *not as bad as* ethyl acrylate.
>
> Therefore, the acrylamide process is better than the ethyl acrylate process.

But, why is acrylamide *not as bad as* ethyl acrylate? The response would be: because ethyl acrylate is a volatile liquid (bp 99.4°C at 760 mmHg) with a sharp, acrid odor and

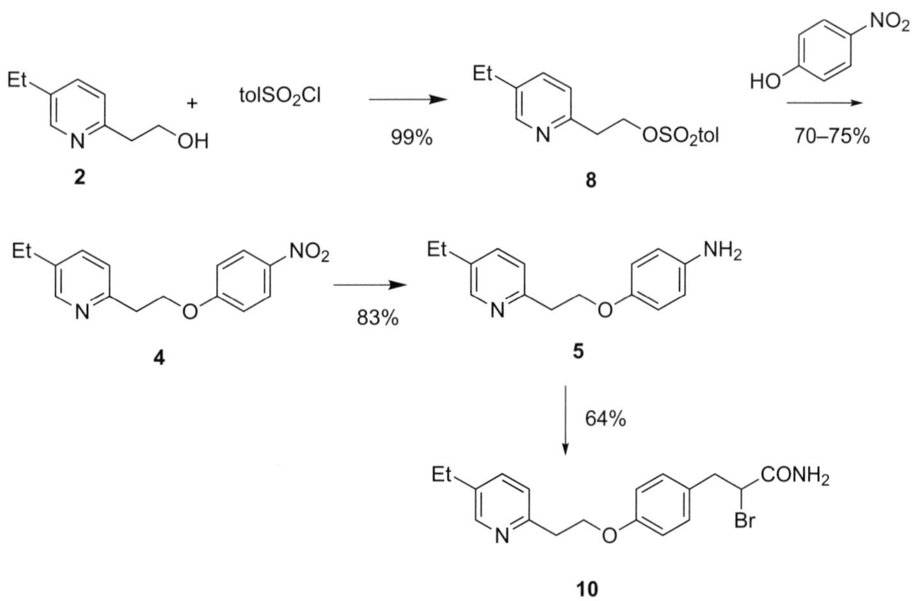


**SCHEME 2.3** Pioglitazone intermediate 2-bromo-3-(4-(2-(5-ethylpyridin-2-yl)ethoxy)phenyl)propanamide (**10**) from 4-nitrophenol via Williamson ether synthesis and Meerwein arylation.

acrylamide is a solid. This response focuses only on how the reagents' physical properties complicate the concentration after the Meerwein arylation. From the perspective of charging the vessel before the Meerwein arylation, there is less potential for operator exposure while charging the liquid ethyl acrylate than when charging the solid acrylamide. From an environmental health and safety perspective, the permissible exposure limit (PEL) for ethyl acrylate is 100 $mg/m^3$ as an 8-hour TWA while the PEL for acrylamide is just 0.3 $mg/m^3$. The process operations and the safety data sheets could be reviewed line by line and the debate continued but the bottom line is this: *ethyl acrylate and acrylamide are both unacceptable.*

*Acetaminophen* 4-(2-(5-Ethylpyridin-2-yl)ethoxy)aniline (**5**) can be produced from acetaminophen. Williamson ether synthesis of the isolated methanesulfonate **7** with the potassium salt of acetaminophen[24] in ethanol at 60°C affords the crystalline *N*-(4-(2-(5-ethylpyridin-2-yl)ethoxy) phenyl)acetamide (**11**) (51%).[25] Comparable yields are achieved with acetaminophen, benzyltributylammonium chloride, and potassium carbonate in dichloromethane–water at 25°C. The acetanilide **11** can be converted to the aniline **5** by acid hydrolysis (HCl in ethanol at reflux) (88%) or basic hydrolysis (KOH in ethanol at reflux) (80%). A toluenesulfonate exchange approach is possible in this context as well. 4-Aminophenol is N-protected and converted to the toluenesulfonate. The reaction of this toluenesulfonate with 2-(5-ethylpyridin-2-yl)ethanol (**2**) and sodium hydride in DMF followed by N-deprotection (KOH in THF) also affords 4-(2-(5-ethylpyridin-2-yl) ethoxy)aniline (**5**).[26]

It is preferable to hydrolyze the acetamide with hydrobromic acid and carry the aniline hydrobromide salt solution directly into the next step. Again, typical Meerwein arylation conditions are used, this time with acrylonitrile. The arenediazonium salt is prepared by addition of sodium nitrite to the aniline **5** in methanol–acetone–aqueous hydrobromic acid at <5°C. Acrylonitrile (4.8 equivalents) (recall the *not as bad as* logic trap) is added. The cuprous oxide catalyst is then added in small portions at 38°C. Aging the reaction at 38°C, concentration at reduced pressure, neutralization, and an extractive workup affords 2-bromo-3-(4-(2-(5-ethylpyridin-2-yl)ethoxy)phenyl)propanenitrile (**12**) as an oil in a remarkable 99% crude yield (Scheme 2.4). Since this oil has seen neither chromatography nor a carbon treatment, purity is difficult to assess.

The Williamson ether syntheses with acetaminophen (50–55%) and 4-nitrophenol (70–75%) were not run under identical conditions, nor were they run under conditions suitable for scale-up. While we cannot make an apples-to-apples comparison, these two approaches are, at best, comparable. Both the amide hydrolysis and the nitro group reduction are high yielding and do not require an isolation of 4-(2-(5-ethylpyridin-2-yl)ethoxy)aniline (**5**).

Taking a still broader perspective, there are five routes to 4-(2-(5-ethylpyridin-2-yl)ethoxy)aniline (**5**): $S_NAr$ with 1-fluoro-4-nitrobenzene and Williamson ether synthesis with 4-nitrophenol or acetaminophen and their toluenesulfonate exchange cousins (Scheme 2.5). How do you choose which is

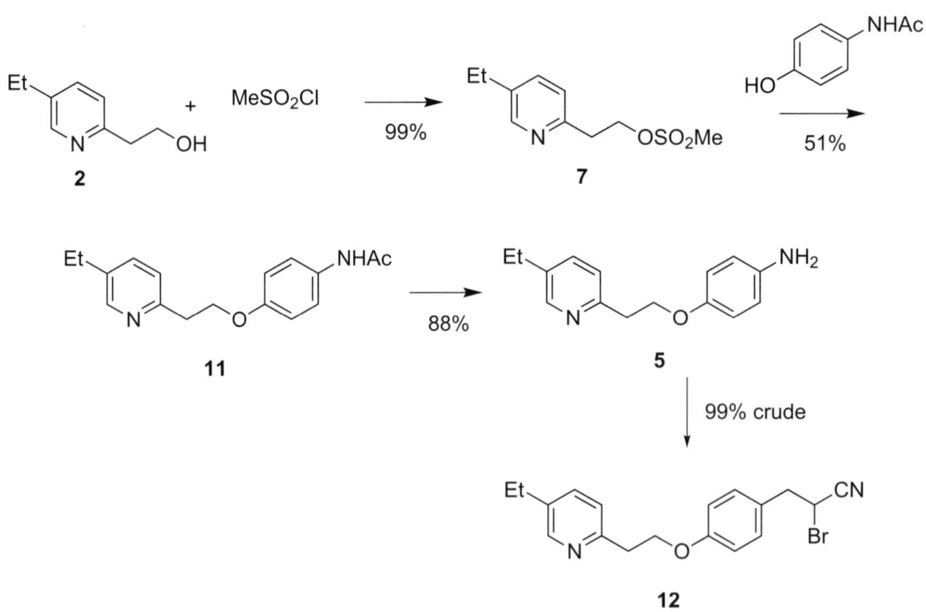


**SCHEME 2.4** Pioglitazone intermediate 2-bromo-3-(4-(2-(5-ethylpyridin-2-yl)ethoxy)phenyl)propanenitrile (**12**) from acetaminophen via Williamson ether synthesis and Meerwein arylation.

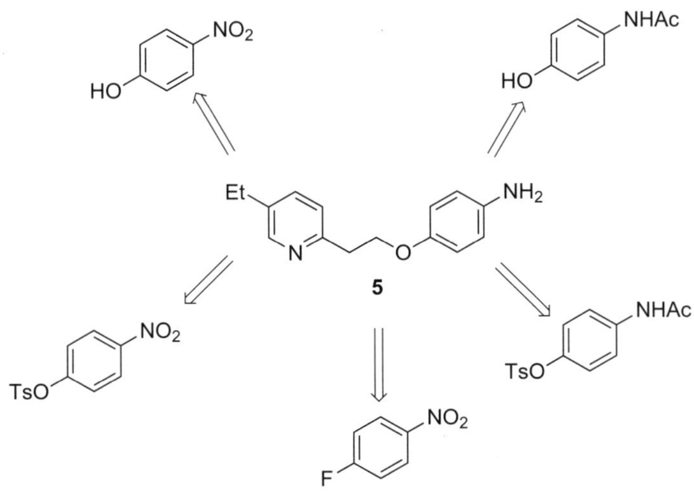


**SCHEME 2.5** Routes to pioglitazone intermediate 4-(2-(5-ethylpyridin-2-yl)ethoxy)aniline (**5**).

best? The question is not relevant to the final objective. They all lead to the Meerwein arylation and its associated negatives: excess methyl acrylate (acrylamide, acrylonitrile) in the distillate and aqueous waste, bromoacetone in the distillate and aqueous waste, a necessary carbon treatment or chromatography of the product, and a low overall yield.

*Tyrosine* An efficient Williamson ether synthesis with the hydroxyl group of tyrosine[27] starts with protection of the amino acid carboxyl and amino groups as the methyl ester and the benzaldehyde imine, respectively, to produce **13**. Ether formation is then accomplished using a dry toluene solution of the methanesulfonate **7**, potassium carbonate, and tetrabutylammonium bromide in toluene at 70°C. After deprotection of the imine and ester, 2-amino-3-(4-(2-(5-ethylpyridin-2-yl)ethoxy)phenyl)propanoic acid (**15**) is isolated in 63% yield based on tyrosine and 61% yield based on 2-(5-ethylpyridin-2-yl)ethanol (**2**). The yields are lower using other amino protecting groups (*t*-BuOC(O), 24%; BnOC(O), 21%; $CH_3CO$, 12%).

The Williamson ether synthesis can also be accomplished using *N*-acetyltyrosine isopropyl ester (**16**). The ether formation is accomplished with a dry toluene solution of the methanesulfonate **7** and *N,N*-diisopropylethylamine (Hunig's base) in isopropanol at reflux. After hydrolyzing the ester and amide, 2-amino-3-(4-(2-(5-ethylpyridin-2-yl)ethoxy)phenyl)propanoic acid (**15**) is isolated in 49% yield based on 2-(5-ethylpyridin-2-yl)ethanol (**2**). As is typically the case, the phenol **16** is used in excess (1.22 equivalents).[28]

The amino group is next converted to the bromide **17** via the diazonium salt. No yield is reported for this conversion. The yield from 2-amino-3-(4-(2-(5-ethylpyridin-2-yl)ethoxy)phenyl)propanoic acid (**15**) to pioglitazone (**1**) is 41%. The yield from ethyl 2-bromo-3-(4-(2-(5-ethylpyridin-2-yl)ethoxy)phenyl)propanoic acid (**18**) to pioglitazone (**1**) is 50%.[14] Assuming a 50% yield from 2-bromo-3-(4-(2-(5-ethylpyridin-2-yl)ethoxy)phenyl)propanoic acid (**17**) to pioglitazone (**1**), the yield for the amine-to-bromide transformation is 82% (Scheme 2.6). Thus, the yield and stability issues associated with the diazonium salt step in this process are comparable to the yield and stability issues associated with the earlier Meerwein arylation. Disadvantages of this route are the high cost of tyrosine,[29] the long linear sequence including two protection and deprotection steps, the challenges of scaling up the diazonium salt chemistry, and the low overall yield.

*4-Hydroxybenzaldehyde* The Williamson ether synthesis with 4-hydroxybenzaldehyde has been extensively studied

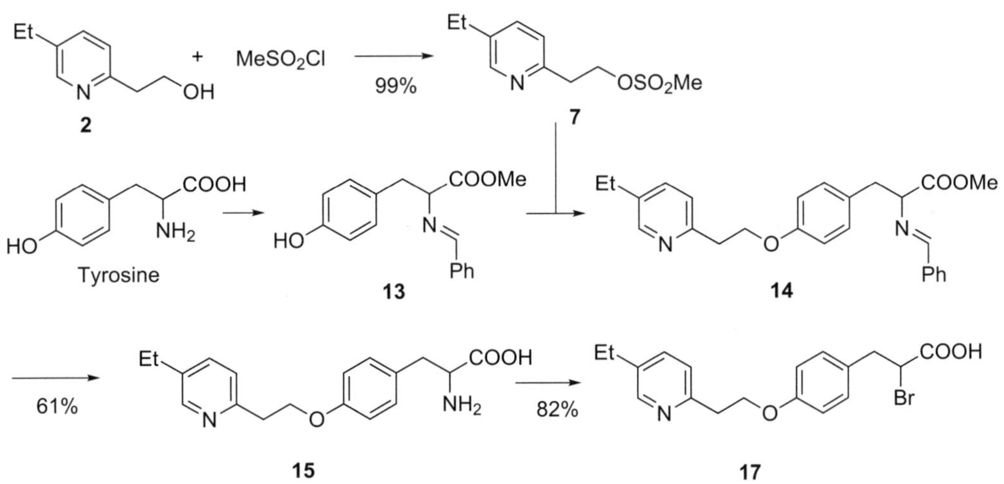


**SCHEME 2.6** Pioglitazone intermediate 2-bromo-3-(4-(2-(5-ethylpyridin-2-yl)ethoxy)phenyl)propanoic acid (**17**) from tyrosine via Williamson ether synthesis.

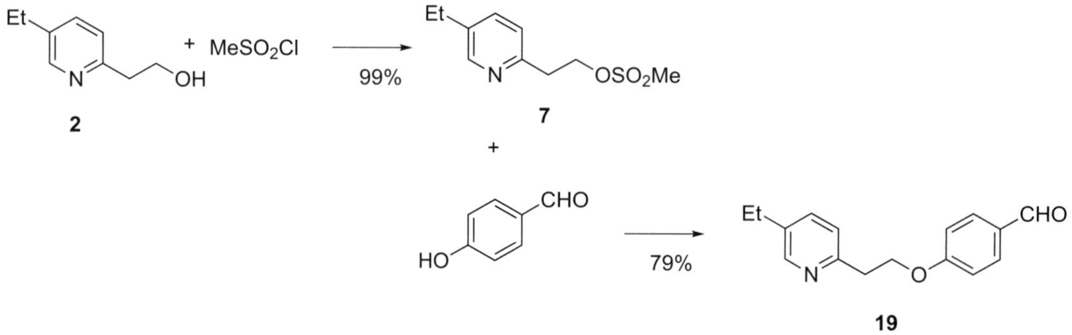


**SCHEME 2.7** Pioglitazone intermediate 4-(2-(5-ethylpyridin-2-yl)ethoxy)benzaldehyde (**19**) via Williamson ether synthesis.

and conditions ranging from phase transfer catalysis in water–organic solvent (1:1) to carefully anhydrous all afford the desired 4-(2-(5-ethylpyridin-2-yl)ethoxy) benzaldehyde (**19**). The reaction of the toluenesulfonate **8** with 4-hydroxybenzaldehyde can be run under phase transfer conditions using sodium hydroxide and benzyltributylammonium chloride in dichloromethane–water at 25°C.[30] The toluenesulfonate **8** is preferred in this case because it can be produced under the same phase transfer conditions. Thus, both the preparation of toluenesufonate **8** and the Williamson ether synthesis can be run in one pot. Unfortunately, the lab results for this two-phase, one-pot process are difficult to reproduce on scale. The reaction of the methanesulfonate **7** with 4-hydroxybenzaldehyde in toluene under phase transfer conditions using potassium carbonate and PEG 200 is very efficient at 80°C.[31]

When the Williamson ether synthesis is not run under phase transfer conditions, hydrophilic solvents such as ethanol and isopropanol are preferred. The phenoxide is generated using potassium hydroxide,[32] potassium *tert*-butoxide,[33] or potassium carbonate. In what are perhaps the best procedures for scale-up, reaction of the methanesulfonate **7** with 4-hydroxybenzaldehyde and potassium carbonate in ethanol–toluene or isopropanol–toluene–water at 77–80°C affords 4-(2-(5-ethylpyridin-2-yl)ethoxy)benzaldehyde (**19**).[34,35] Lower yields are reported using other solvents (toluene, 1,2-dichloroethane, THF, and acetonitrile). The crude aldehyde **19** produced using any of these protocols is typically of unacceptable quality, due to the competitive elimination of the sulfonate ester to 5-ethyl-2-vinylpyridine (**20**) and polymerization of the vinylpyridine. The crude 4-(2-(5-ethylpyridin-2-yl)ethoxy)benzaldehyde (**19**) generated from the toluenesulfonate **8** under phase transfer conditions is upgraded by silica gel chromatography (62% yield from **2**).[30] The crude 4-(2-(5-ethylpyridin-2-yl) ethoxy)benzaldehyde (**19**) generated from the methanesulfonate **7** is upgraded by carbon treatment[34] or by salt formation with hydrochloric, trifluoroacetic, maleic, or oxalic acid.[35] The yield of the free base from **2** after carbon treatment is 79% (79% purity by HPLC assay) and the yield of oxalate salt from **2** is 74% (99.7% purity by HPLC assay) (Scheme 2.7).

An exchange of the methanesulfonate ester to the phenol to give 4-formylphenyl methanesulfonate may compete with the desired methanesulfonate ester displacement. Addition of sodium or potassium iodide minimizes this transfer by converting the methanesulfonate to the iodide, which is then displaced. The reaction of isolated methanesulfonate **7** with 4-hydroxybenzaldehyde, potassium hydroxide, and 6.3 mol % potassium iodide in isopropanol at reflux affords 4-(2-(5-ethylpyridin-2-yl)ethoxy)benzaldehyde (**19**) (74%). Thus, short-circuiting the transfer does not increase the yield.[36]

#### *2.2.2.3 $S_NAr$ Using 4-Fluorobenzonitrile and 4-Fluorobenzaldehyde*

Does a nitrile or aldehyde activate a halogen at the 4-position of an aromatic toward $S_NAr$? If so, 4-(2-(5-ethylpyridin-2-yl)ethoxy)benzaldehyde (**19**) could be produced in a single step from 2-(5-ethylpyridin-2-yl)ethanol (**2**) (Scheme 2.8). This would lead to a more robust process by avoiding the methanesulfonate intermediate **7** and its propensity to eliminate.

A nitrile does activate the fluorine of 4-fluorobenzonitrile toward $S_NAr$ by alkoxide. The reaction of 4-fluorobenzonitrile with 2-(6-methylpyridin-2-yl)ethanol (**21**) and sodium hydride in THF at 25°C affords 4-(2-(6-methylpyridin-2-yl) ethoxy)benzonitrile (**22**) (50%). Nitrile **22** is reduced using Raney nickel in refluxing 75% formic acid to give 4-(2-(6-methylpyridin-2-yl)ethoxy)benzaldehyde (**23**) (64%).[30,37] The reaction of 4-fluorobenzonitrile with 2-(5-methyl-2-phenyl-4-oxazolyl)ethanol (**24**) and sodium hydride in THF at 25°C is even more efficient (91%).[38] The same conditions are used for nitrile reduction (65%). So, the fluoride is activated and the nitrile can be reduced. But, is 4-fluorobenzonitrile readily available and inexpensive?[39]

An aldehyde also activates fluorine at the 4-position of an aromatic ring toward $S_NAr$ by alkoxide. The reaction conditions are usually more vigorous than those for the reaction with the nitrile. The reaction of 4-fluorobenzaldehyde with 2-(methyl(pyridine-2-yl)amino)ethanol (**25**) and sodium

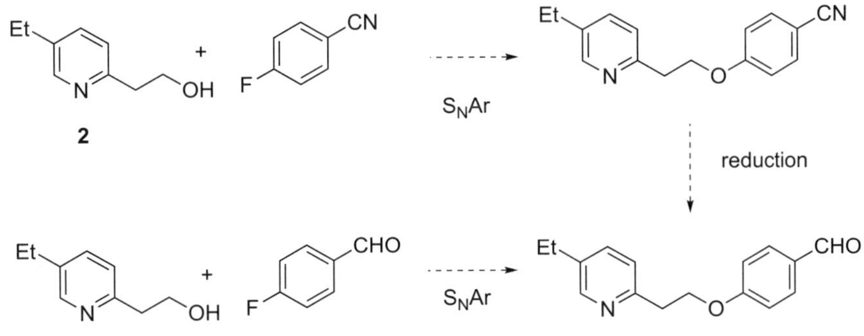


**SCHEME 2.8** Proposed routes to pioglitazone intermediate 4-(2-(5-ethylpyridin-2-yl)ethoxy)benzaldehyde (**19**) from 4-fluorobenzonitrile and 4-fluorobenzaldehyde.

hydride in DMF at 80°C affords 4-(2-(methyl(pyridin-2-yl) amino)ethoxy)benzaldehyde (**26**). The yield is not provided. The $S_NAr$ with this and many other *N*-methyl-*N*-heteroaryl aminoethanols can also be carried out using potassium carbonate as the base in DMSO at 100–120°C.[40] But is 4-fluorobenzaldehyde readily available and inexpensive?[41] 4-Fluorobenzaldehyde is certainly less expensive than 4-fluorobenzonitrile.

### 2.2.3 4-(2-(5-Ethylpyridin-2-yl)ethoxy)benzaldehyde (19) to Pioglitazone (1)

#### *2.2.3.1 Darzens Condensation and Tcherniac's Synthesis*

There are several options for elaborating the thiazolidinedione-containing side chain from the aldehyde **19**. A Darzens condensation with ethyl chloroacetate and sodium ethoxide in ethanol at 25°C yields the *cis*- and *trans*-glycidic esters (**27**).[42] Hydrogenolysis of the mixture using 10% Pd on C in methanol at 25°C and 1 atm hydrogen and methanesulfonylation of the resulting alcohol **28** affords the α-methanesulfonyloxy ester **29** (Scheme 2.9).

There are two methods for converting the α-bromo esters **6** and **18** (acid **17**, amide **10**, or nitrile **12**) and α-methanesulfonyloxy ester **29** to pioglitazone (**1**) (Scheme 2.10). In the first process, bromide displacement with thiocyanate (Tcherniac's synthesis) followed by hydrolysis of the thiocyanate **30** and cyclization affords pioglitazone (**1**) in low yield. The 2-thiocyanatopropanoic acid **30** can also be produced directly from the 2-aminopropanoic acid **15** and lithium thiocyanate by diazotization with isopentylnitrite in THF–acetic acid at 25°C (50%).[28] No yield is provided for the thiocyanate hydrolysis/cyclization to pioglitazone (**1**) but we can anticipate a yield of 70% based on earlier work preparing other 5-(4-oxobenzyl)thiazolidine-2,4-diones.[43] In the second and preferred process, reaction of **6** or **29** with thiourea and sodium acetate (Hantzsch's synthesis) generates a 2-imino-4-thiazolidinone **31** in ethanol or isopropanol at reflux, which is then hydrolyzed with dilute hydrochloric acid. The yield for the 2-imino-4-thiazolidinone formation from the crude α-bromo esters **6** and **18** (acid **17**, amide **10**, or nitrile **12**) is 50–56% and yield for the imine hydrolysis is high (90%).[14,22]

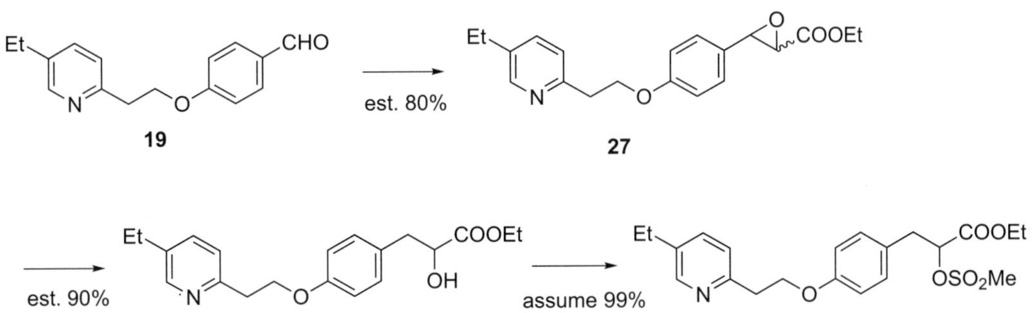


**SCHEME 2.9** Pioglitazone intermediate ethyl 3-(4-(2-(5-ethylpyridin-2-yl)ethoxy)phenyl)-2-methanesulfonyloxy)propanoate (**29**) via Darzens condensation.

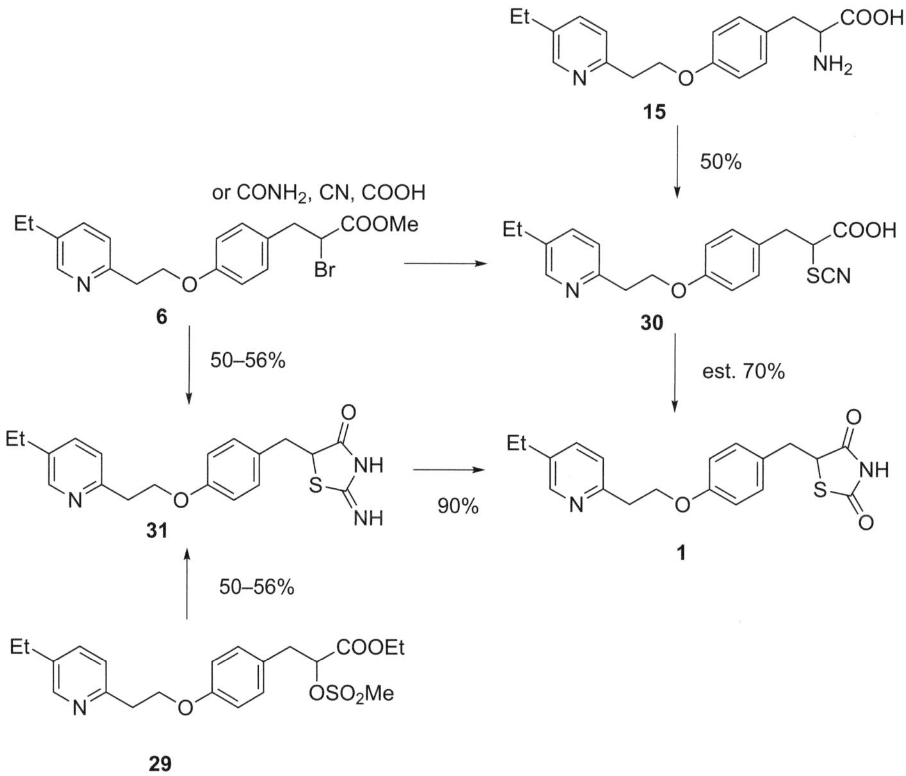


**SCHEME 2.10** Pioglitazone (**1**) via Hantzsch and Tcherniac methods of thiazolidine-2,4-dione synthesis.

The yield for the 2-imino-4-thiazolidinone formation from the α-methanesulfonyloxy ester **29** in the linear sequence of five steps converting 4-(2-(5-ethylpyridin-2-yl) ethoxy)benzaldehyde (**19**) to pioglitazone (**1**) is not available, since the purities of the crude oils obtained in each step are not reported. The overall yield for the five-step sequence is 33%. Assuming an 80% yield for the Darzens condensation and a 90% yield for the hydrogenolysis, the yield for the imine hydrolysis is 90%.[14,42] Assuming a 99% yield for the methanesulfonylation leaves a 51% yield for the 2-imino-4-thiazolidinone formation. This nicely fits the yield data for the α-bromo esters **6** and **18** (acid **17**, amide **10**, or nitrile **12**). While the Darzens route does circumvent the Meerwein arylation and its issues, this route has a long linear sequence and a low overall yield.

#### *2.2.3.2 Knoevenagel Condensation and Catalytic Reduction*

The same conversion of 4-(2-(5-ethylpyridin-2-yl)ethoxy) benzaldehyde (**19**) to pioglitazone (**1**) can also be accomplished in just two steps: Knoevenagel condensation with thiazolidine-2,4-dione and reduction of the double bond. The Knoevenagel condensation partner, thiazolidine-2,4-dione, is commercially available but expensive.[44] The condensation product, 5-(4-(2-(5-ethylpyridin-2-yl)ethoxy)benzylidene) thiazolidine-2,4-dione (**32**), conveniently precipitates as the (*Z*)-isomer when the condensation solvent is methanol or ethanol. The configuration is determined by irradiation of the single isomer to produce an equilibrium mixture, separation of the other isomer by silica gel chromatography, and assignment of the methine proton of each isomer by $^1H$ NMR. The (*Z*)-isomer methine hydrogen is further downfield (δ 7.74 for *Z*-isomer and δ 7.30 for *E*-isomer), suggesting that it is on the same side as the carbonyl group at the 4-position of the thiazolidinedione (Figure 2.3).[37] ChemNMR $^1H$ estimates these methine protons at δ 7.95 for the *Z*-isomer and δ 7.76 for the *E*-isomer.

The condensation of 4-(2-(5-ethylpyridin-2-yl)ethoxy) benzaldehyde (**19**) with thiazolidine-2,4-dione (1.2 equivalents) and pyrrolidine (1.0 equivalent) in methanol at 45°C is very efficient even after multiple precipitations and isolations for purity upgrade (95% yield) (Scheme 2.11).[34] Just one isolation is required to obtain the condensation product using thiazolidine-2,4-dione (2.7 equivalents) and piperidine (0.78 equivalents) in ethanol at reflux (73%, 99.5% pure).[32] 5-(4-(2-(5-Ethylpyridin-2-yl)ethoxy)benzylidene)thiazolidine-2,4-dione (**32**) can also be isolated from ethanol as the hydrochloride salt (81%, 99.6% pure by HPLC).[35] One key feature of almost every robust pharmaceutical manufacturing process is a highly reproducible purification of the penultimate intermediate. With these three detailed procedures, this process feature is certainly demonstrated for the Knoevenagel process.

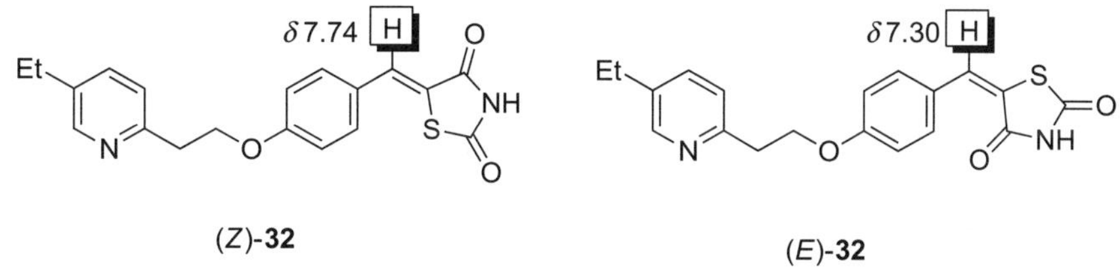


**FIGURE 2.3** Assignment of the configuration of 5-(4-(2-(5-ethylpyridin-2-yl)ethoxy)benzylidene)thiazolidine-2,4-dione (**32**) by $^1H$ NMR.

> Piperidine and morpholine are often used almost interchangeably in Knoevenagel condensations. When the condensation yields are comparable, exposure limits suggest that morpholine is a better choice. The OSHA permissible exposure limit for morpholine is 20 ppm or 70 mg/m$^3$ TWA with skin absorption designation. Although OSHA does not have a permissible exposure limit for piperidine, in the 1980s the American Industrial Hygiene Association recommended a level of 1 ppm TWA with skin absorption designation.[45]

While the low solubility of 5-(4-(2-(5-ethylpyridin-2-yl)ethoxy)benzylidene)thiazolidine-2,4-dione (**32**) in methanol and ethanol greatly simplifies the workup of the Knoevenagel condensation, it certainly makes the final reduction more challenging. Reductions run in DMF, tetrahydrofuran, or dioxane (a carcinogen[46]) require a high catalyst loading, high temperatures and pressures, and a hot filtration of the catalyst. Reduction of 30 g in dioxane (580 mL) at 110°C and 711 psi required 0.75 g of palladium (30 g of 5% palladium on carbon, 50% water wet).[34] With a reduction yield of 70%, the cost of palladium[47] metal to produce 1 kg of pioglitazone (**1**) crude by this process will be roughly $340. The cost for palladium on support will be still higher. Of course, the spent catalyst can be returned to the catalyst manufacturer for recovery of the metal and some cost savings. While the first inclination might be to focus on reducing the cost of the precious metal, another difficult hurdle for this reduction process will be to reduce palladium to an acceptable level in the final product. The oral PDE (permitted daily exposure) for palladium has been set by the EMEA at 100 μg/day (2 μg/kg/day in a 50 kg person).[48]

The 5-(4-(2-(5-ethylpyridin-2-yl)ethoxy)benzylidene)thiazolidine-2,4-dione (**32**) hydrogen chloride salt has higher solubility than the free base in methanol–water. Reductions of the hydrogen chloride salt in methanol–water give better yields (80–85%) but still require a high catalyst loading and high temperatures and pressures. Reduction of 22.5 g of the free base in methanol and 36% hydrochloric acid at 100°C and 121 psi hydrogen requires 1.125 g of palladium (11.25 g of 20% palladium on carbon, 50% water wet).[34] Reduction of 9.77 g of the isolated hydrochloride salt can be accomplished in methanol–water in 15 h at 60°C at 73–87 psi hydrogen using just 0.15 g of palladium (1.5 g of 10% palladium on carbon).[35] Either the first example did not require such vigorous conditions or the second example illustrates the value of isolating and upgrading the purity of the hydrogenation substrate!

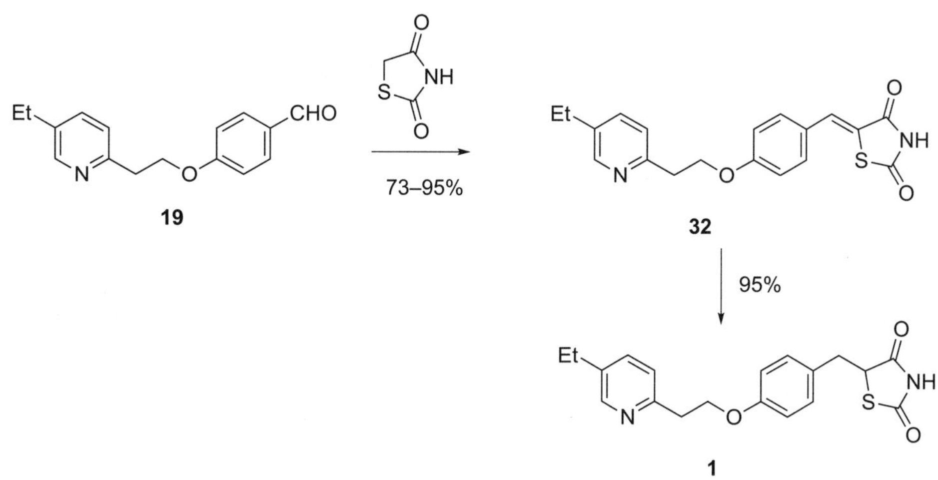


**SCHEME 2.11** Pioglitazone (**1**) from the Knoevenagel condensation of 4-(2-(5-ethylpyridin-2-yl)ethoxy)benzaldehyde (**19**) and thiazolidine-2,4-dione.

5-(4-(2-(5-Ethylpyridin-2-yl)ethoxy)benzylidene)thiazolidine-2,4-dione (**32**) has good solubility in formic acid. Reduction in formic acid can have a higher throughput in the hydrogenation reactor and yields are as high as 84%. Hydrogen pressures are 30–90 psi but reaction temperatures and catalyst loadings are still high.[49] Reduction of 50 kg of **32** in formic acid at 80°C and 29 psi hydrogen required 2 kg of palladium (40 kg of 10% palladium on carbon, 50% water wet). Under these conditions, formic acid is both solvent and reducing agent.[50] The by-product of reduction by formic acid is carbon dioxide. The reactor is vented every 30 min to purge the carbon dioxide and then refilled with hydrogen. The reduction described above required 20 h, *and 40 purges*, to go to completion! This is not a "green" option. Even the higher throughput in the hydrogenation reactor is questionable since the frequency of purges and the percentage of the volume in the hydrogenation reactor that must be used for headspace will be linked.

If formic acid is an effective reducing agent, is hydrogen gas required? A positive pressure of hydrogen is necessary to maintain palladium catalyst activity. Incomplete conversion (66%) is observed in the reduction of 2.5 g in 99% formic acid at reflux under nitrogen using 0.064 g of palladium (as 1.37 g of 10% palladium on carbon, 53% water wet).[49] The positive pressure of hydrogen is not necessary using a platinum catalyst. Reduction of 20 g of **32** in 99% formic acid at 84°C under nitrogen required 0.5 g of platinum (as 0.6 g of platinum oxide).[50] Unfortunately, even with a reduction yield of 97%, the cost of platinum[51] metal to produce 1 kg of pioglitazone (**1**) crude by this process will be roughly $1060. Other heterogeneous and homogeneous iridium, rhodium, ruthenium, and nickel catalysts apparently give inferior results.

> Hydrogenations at elevated pressures are typically run in 316 stainless steel reactors. The materials of construction were not specified for the lab autoclaves used in the hydrogenations of 5-(4-(2-(5-ethylpyridin-2-yl)ethoxy) benzylidene)thiazolidine-2,4-dione (**32**) as the hydrogen chloride salt in methanol–water or 5-(4-(2-(5-ethylpyridin-2-yl)ethoxy)benzylidene)thiazolidine-2,4-dione (**32**) in formic acid. Material compatibility tables[20] indicate that a 316 stainless steel reactor should not be used. A Hastelloy C reactor is a better choice (good for ammonium chloride, excellent for dilute hydrochloric acid, and excellent for formic acid). When considering scale-up of any process always identify special equipment needs and verify that the equipment is available at the manufacturing site.

A cobalt catalyst generated by reaction of cobalt(II) chloride, cobalt(II) acetate, or cobalt(III) chloride with sodium borohydride and dimethylglyoxime mediates the reduction of 5-(4-(2-(5-ethylpyridin-2-yl)ethoxy)benzylidene)thiazolidine-2,4-dione (**32**) by sodium borohydride.[52] Reduction of 5.0 g in tetrahydrofuran–water using a solution of sodium borohydride (1.25 equivalents) in water containing dilute HCl at 15°C required 42 mg of cobalt(II) chloride hydrate. With a reduction yield of 95%, the catalyst cost to produce 1 kg of pioglitazone (**1**) crude by this process is just $1.[53]

It is known that cobalt(II) chloride and sodium borohydride react rapidly to produce cobalt boride ($Co_2B$). It is also known that freshly prepared cobalt boride is an effective catalyst for the heterogeneous catalytic hydrogenation of alkenes. These facts together suggest that the reduction to pioglitazone (**1**) is a heterogeneous hydrogenation catalyzed by cobalt boride generated *in situ*. The required hydrogen is generated by decomposition of sodium borohydride in the aqueous reduction medium.[54] Why does this cobalt boride-catalyzed heterogeneous hydrogenation proceed at 15°C and ambient pressure when palladium-catalyzed variants require high catalyst loadings and high temperatures and pressures? Perhaps cobalt boride is less prone to catalyst poisoning by the divalent sulfur found in both the substrate and product.

Two issues associated with the use of a cobalt boride catalyst are the potential for operator exposure during handling of solid cobalt(II) chloride and the residual cobalt in the pioglitazone (**1**) isolated from the catalytic reduction. Cobalt (II) chloride is an animal carcinogen. The OSHA permissible exposure limit is 0.1 $mg/m^3$ (TWA) for cobalt metal dust and fume as cobalt. The ACGIH threshold limit value for inorganic cobalt compounds is 0.02 $mg/m^3$ (TWA) as cobalt. The cobalt solids charging procedure must be designed to meet these limits and air monitoring and analysis must confirm that the containment procedure is adequate. What would be an acceptable level of residual cobalt in a final drug substance has not been specifically addressed by the regulatory agencies. Cobalt is a naturally occurring element. As the core metal in vitamin $B_{12}$, it is an essential element in humans. The average daily intake of cobalt from food is 5–40 μg/day.[55]

A chemical reduction approach would eliminate the need for high-pressure equipment and avoid the safety issues associated with using hydrogen gas or hydride. For example, 5-(4-(2-(5-ethylpyridin-2-yl)ethoxy)benzylidene)thiazolidine-2,4-dione (**32**) is reduced by sodium hydrosulfite and potassium carbonate in water–DMF at 80°C (55%).[31] The isolated pioglitazone (**1**) (HPLC purity 99%) contains just 0.05 area% by HPLC of the starting material **32**. This is important since the starting material is difficult to separate from pioglitazone (**1**) downstream. Higher levels of starting material **32** are left using methanol (0.50–0.58%), ethanol (0.28–0.30%), tetrahydrofuran (0.38%), or dioxane (1.09%) as the cosolvent. The low yield and poor solubility of both 5-(4-(2-(5-ethylpyridin-2-yl)ethoxy)benzylidene)

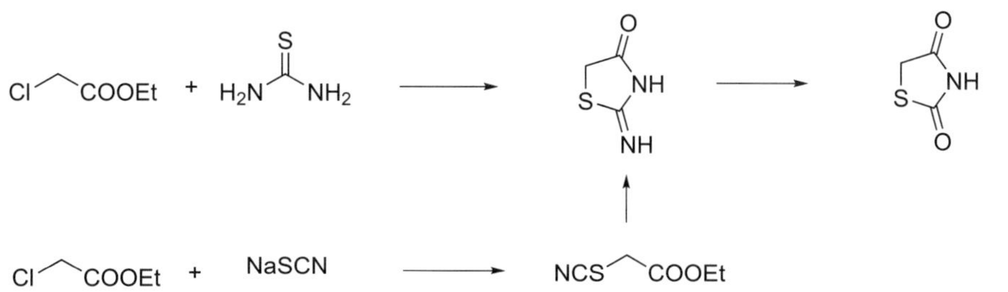


**SCHEME 2.12** Hantzsch and Tcherniac methods for synthesis of thiazolidine-2,4-dione.

thiazolidine-2,4-dione (**32**) and pioglitazone (**1**) result in an unacceptable volume throughput (just 9.7 g/L) for the hydrosulfite reduction.

> Always anticipating the need for maximum efficiency in a production train, a process chemist strives to achieve an acceptable volume throughput (here defined as g product/L reaction mixture) for each process step. A volume throughput of 100–250 g/L is typical in an efficient batch process.

5-(4-(2-(5-Ethylpyridin-2-yl)ethoxy)benzylidene)thiazolidine-2,4-dione (**32**) is reduced by free whole yeast cell culture or immobilized whole yeast cells from *Rhodotorula glutinis* CBS 4406 or *Rhodotorula rubra* CBS 6469. Yeast reductions show promise for producing enantiomerically enhanced glitazones, including pioglitazone (**1**).[56]

5-(4-(2-(5-Ethylpyridin-2-yl)ethoxy)benzylidene)thiazolidine-2,4-dione (**32**) is certainly highly valued as the penultimate intermediate to pioglitazone (**1**). There are at least six methods for the final double bond reduction: hydrogenation with palladium on carbon, transfer hydrogenation with formic acid and palladium, transfer hydrogenation with formic acid and platinum, reduction with sodium borohydride and cobalt boride, reduction with sodium hydrosulfite, and reduction with yeast.

### 2.2.4 Preparation of Thiazolidin-2,4-dione

With a demonstrated two-step sequence from 4-(2-(5-ethylpyridin-2-yl)ethoxy)benzaldehyde (**19**) to pioglitazone (**1**), the next step is to identify the bulk suppliers of thiazolidine-2,4-dione and gain an understanding of how and from what starting materials it is manufactured. The methods for ring construction parallel the methods used to convert methyl 2-bromo-3-(4-(2-(5-ethylpyridin-2-yl)ethoxy)phenyl)propanoate (**6**) to pioglitazone (**1**). A Hantzsch approach converts two inexpensive starting materials, ethyl chloroacetate and thiourea, to 2-imino-4-thiazolidinone (pseudothiohydantoin), which is then hydrolyzed to thiazolidine-2,4-dione. A Tcherniac approach also converts inexpensive starting materials, ethyl chloroacetate and sodium thiocyanate, to pseudothiohydantoin (Scheme 2.12).

## 2.3 SYNTHESIS LEFT TO RIGHT FROM 5-ETHYL-2-VINYLPYRIDINE (20)

The one constant in all the routes described so far is the starting material, 2-(5-ethylpyridin-2-yl)ethanol (**2**). It is used as the nucleophile in $S_NAr$ displacements and is converted to the electrophile, the methanesulfonate **7**, for the Williamson ether syntheses. We have seen that the Williamson ether syntheses afford less than optimal yields and require chromatography or carbon treatment due to the competitive elimination of methanesulfonate **7**. The elimination product, 5-ethyl-2-vinylpyridine (**20**), is a novel starting material for a parallel approach.[57,58] The parallel approach uses methodology already discussed to produce 5-(4-(2-(5-ethylpyridin-2-yl)-2-hydroxyethoxy)benzyl)thiazolidine-2,4-dione (**33**). Pioglitazone (**1**) is then produced by conversion of the hydroxyl group to a chloride and reductive dechlorination (Schemes 2.13 and 2.14).

The reaction of 5-ethyl-2-vinylpyridine (**20**) with *N*-bromosuccinimide in 25% aqueous *tert*-butanol at 25–30°C affords a mixture of bromohydrins **34** (95%). The same conditions (25% aqueous *tert*-butanol at 25–30°C) are also suitable for converting the bromohydrins **34** to the oxirane **35** by reaction with potassium carbonate (98%). Both the bromohydrins **34** and the oxirane **35** should be reactive with 4-hydroxybenzaldehyde in a Williamson ether synthesis. Recall that Williamson ether syntheses with 4-hydroxybenzaldehyde and the methanesulfonate **7** are best run in alcohol solvents, especially in ethanol, ethanol–toluene, or isopropanol–toluene–water. The Williamson ether synthesis with the bromohydrins **34**, and/or with the oxirane **35** generated *in situ*, can be carried out using potassium carbonate in 25% aqueous *tert*-butanol. Thus, in a remarkably efficient and convenient one-pot process, sequential bromohydrin formation, conversion to the oxirane **35**, and ether formation with 4-hydroxybenzaldehyde affords 4-(2-(5-ethylpyridin-2-yl)-2-hydroxyethoxy)benzaldehyde **36** (79% from vinylpyridine

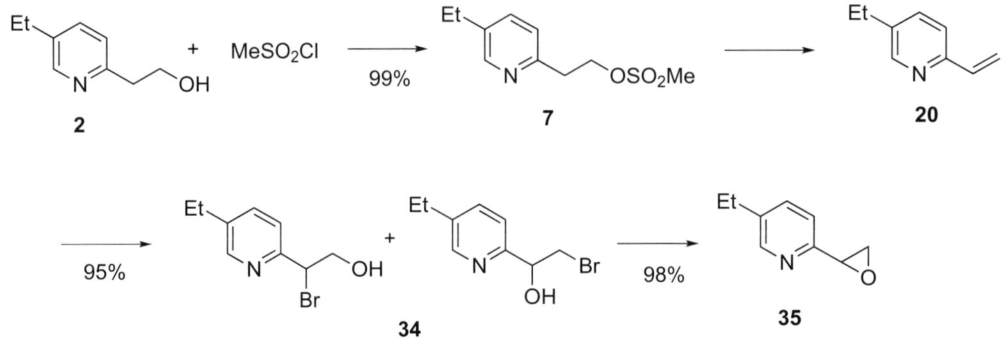

**SCHEME 2.13** 5-Ethyl-2-(oxiran-2-yl)pyridine (**35**) from 2-(5-ethylpyridin-2-yl)ethanol (**2**).

**20**). Aldehyde **36** requires a purity upgrade but the procedure is not specified.[58]

The process description to this point is encouraging but leaves us with many questions. What is the stability of the vinylpyridine **20** and of the oxirane **35**? What is the regioselectivity of the Williamson ether synthesis using the oxirane **35**? How stable is oxirane **35** under Williamson ether synthesis conditions? Reeder's *Organic Process Research & Development* publication[57] addresses these questions. 5-Ethyl-2-vinylpyridine (**20**), as a neat oil, is prone to polymerization even when stored cold. Oxirane **35** is more suitable for storage. It can be stored at ambient temperature in methyl *tert*-butyl ether solution for months without polymerization or decomposition. However, the onset temperature for decomposition of neat oxirane **35** is just 56°C.

The reaction of the potassium salt of 4-hydroxybenzaldehyde with oxirane **35** produces a mixture of regioisomers in DMF at 60–65°C (65%). Separation by chromatography on silica gel affords 43% of the desired 4-(2-(5-ethylpyridin-2-yl)-2-hydroxyethoxy)benzaldehyde (**36**) and 5% of the

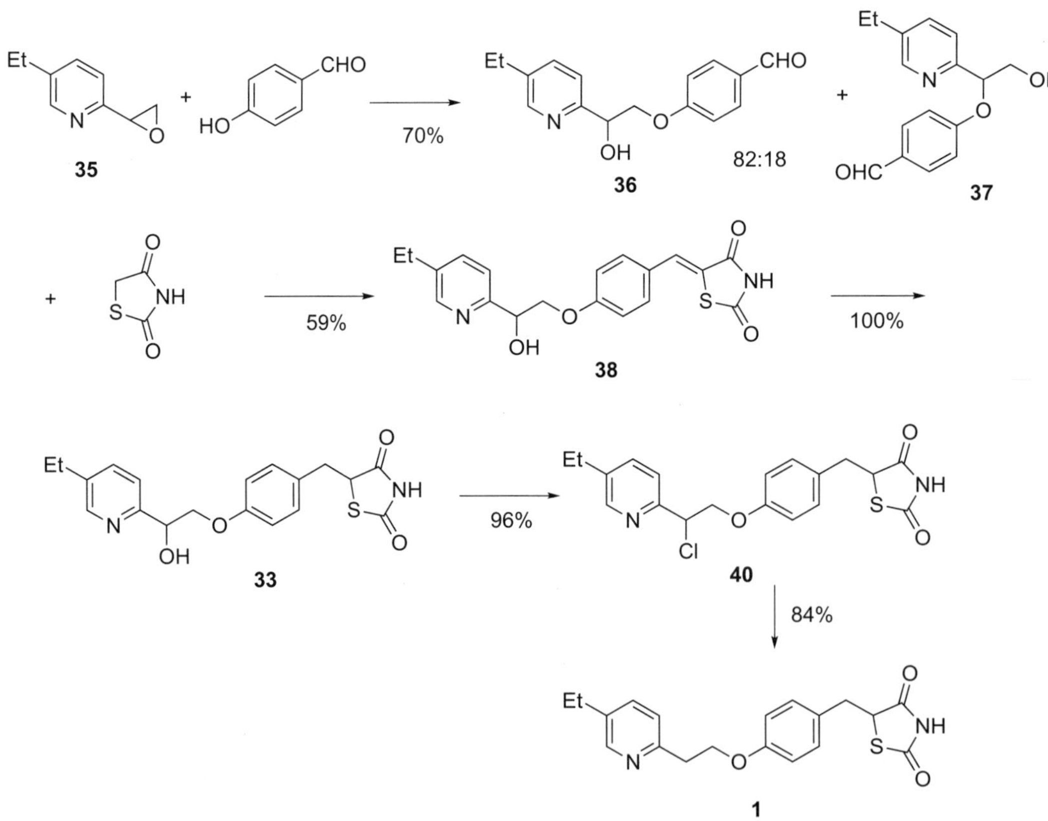

**SCHEME 2.14** Pioglitazone (**1**) from 5-ethyl-2-(oxiran-2-yl)pyridine (**35**).

regioisomer **37**. The regioselectivity can be improved by modifying the reaction conditions. For example, the Williamson ether synthesis with a magnesium(II)-assisted oxirane ring opening is highly regioselective in DMF at 70°C (62%, 80:1 mixture). However, both the nonmetal-assisted and metal-assisted conditions produce some tar that must be removed before proceeding. Phase transfer conditions using PEG 4000 and 1 N sodium hydroxide in toluene/methyl *tert*-butyl ether at 78°C do not produce tar but also do not offer high regioselectivity (57%, 82:18 mixture).

Knoevenagel condensation of the aldehyde **36** with thiazolidine-2,4-dione (1.1 equivalents) using pyrrolidine (1.1 equivalents) in methanol at 55°C affords (*Z*)-5-(4-(2-(5-ethylpyridin-2-yl)-2-hydroxyethoxy)benzylidene)thiazolidine-2,4-dione (**38**) (91%).[58] The same condensation with the 82:18 mixture of aldehydes **36** and **37** in methanol at 38–42°C gave a much lower yield (59%). Taking into account the unspecified losses incurred in upgrading aldehyde **36**, these results are in close agreement. At this point, we are just two reductions away from pioglitazone (**1**). Recall the facile hydrogenolysis of oxirane **27** from the Darzens condensation.[42] The analogous hydrogenolysis of the hydroxyl group of **38** is more difficult and could not be accomplished under conditions that reduced the double bond. The double bond is first reduced to afford **33** using the cobalt(II) chloride–sodium borohydride method in DMF–water (95%). It is striking that the reduction to **33** required higher loadings of cobalt(II) chloride (10.5 mol%) and sodium borohydride (3.24 equivalents) and a higher temperature (65–70°C) than the reduction of (*Z*)-5-(4-(2-(5-ethylpyridin-2-yl)ethoxy)benzylidene)thiazolidine-2,4-dione (**32**) to pioglitazone (**1**).[52] The higher catalyst and borohydride charges probably are in response to the reaction stalling prior to completion. At more typical loadings (0.06 mol% cobalt(II) chloride, 2.9 mol% dimethylglyoxime, 1.44 equivalents of sodium borohydride) in THF–water at 15–20°C, this reduction will go to completion if the pH is maintained $<11$ by addition of acetic acid (quantitative yield, $>95\%$ pure).[57] This same hydroxyated pioglitazone **33** is also directly available from pioglitazone (**1**) by conversion to pioglitazone *N*-oxide (**39**) and rearrangement using a modification of the Boekelheide method (71%).[59]

Removal of the hydroxyl group of **33** is accomplished by conversion to a chloride, methanesulfonate, or toluenesulfonate and reduction with zinc. Reaction with thionyl chloride in dichloromethane at reflux affords the chloride **40** as the hydrochloride salt (96%). Reduction of **40** with zinc (2.0 equivalents) in methanol–acetic acid at 25°C affords crude pioglitazone (**1**), which is converted to its hydrochloride salt (84%). Some cleavage of the ether is observed: 5-ethyl-2-vinylpyridine (**20**) and 5-(4-hydroxybenzyl)thiazolidine-2,4-dione (**41**) are side products.

Since 5-ethyl-2-vinylpyridine (**20**) is produced from 2-(5-ethylpyridin-2-yl)ethanol (**2**), whether it is produced "in house" or not, 2-(5-ethylpyridin-2-yl)ethanol (**2**) is *someone's* starting material. This process has eight linear steps from 2-(5-ethylpyridin-2-yl)ethanol (**2**) to pioglitazone (**1**). 5-Ethyl-2-vinylpyridine (**20**) is prone to polymerization "upon standing even when stored cold with inhibitors."[57] The elimination problem with the methanesulfonate **7** is simply replaced by a regioselectivity problem with the oxirane **35**. Finally, it is very difficult to justify using this or any route that produces two significant new impurities and a zinc waste stream in the final step.

## 2.4 SYNTHESIS RIGHT TO LEFT

There are many "left-to-right" routes. Would there be any competitive advantage in going from right to left? There will be no advantage with the current methodologies for three reasons: (1) moving the Williamson ether synthesis back will likely necessitate protection and deprotection of the phenolic hydroxyl group during elaboration of the thiazolidinedione side chain, (2) the thiazolidinedione nitrogen may be reactive during the Williamson ether synthesis, and (3) this approach will deliver a relatively expensive phenol into the Williamson ether synthesis and the phenol is typically used in excess.

One right-to-left approach begins with 4-benzyloxybenzaldehyde (Scheme 2.15).[60] Darzens condensation of the aldehyde with *tert*-butyl chloroacetate[61] and potassium *tert*-butoxide in *tert*-butanol at 25°C yields the *cis*- and *trans*-glycidic esters **42** (88%). Hydrogenolysis with wet 10% palladium on carbon in ethyl acetate at 25°C now cleaves both the benzyl protecting group and the epoxide (92%). Williamson ether synthesis with isolated methanesulfonate **7** and potassium carbonate in acetonitrile at reflux stopped at 55% conversion. Difficulties encountered in this Williamson ether synthesis no doubt led use of the more expensive *tert*-butyl ester for the Darzens condensation. The alcohol is converted to the α-methanesulfonyloxy *tert*-butyl ester **45**, an analog of the α-methanesulfonyloxy ethyl ester prepared by the left-to-right Darzens route. The α-methanesulfonyloxy *tert*-butyl ester **45** is converted to pioglitazone (**1**) by condensation with thiourea and imine hydrolysis.

Other right-to-left approaches highlight the undesired involvement of the thiazolidinedione during the Williamson ether synthesis using 5-(4-hydroxybenzyl)thiazolidine-2,4-dione (**41**).[25] 5-(4-Hydroxybenzyl)thiazolidine-2,4-dione (**41**) can be prepared from tyrosine by the thiourea/imine hydrolysis method[25] or from 4-hydroxybenzaldehyde by the Knoevenagel/reduction method (Scheme 2.16).[62] The amino group of tyrosine is converted to the bromide **46** via the diazonium salt. Bromide **46** is then reacted with thiourea and sodium acetate in ethanol at reflux to afford 5-(4-hydroxybenzyl)-2-iminothiazolidin-4-one (**47**) (70% from tyrosine). The imine of **47** is hydrolyzed and 5-(4-hydroxybenzyl)

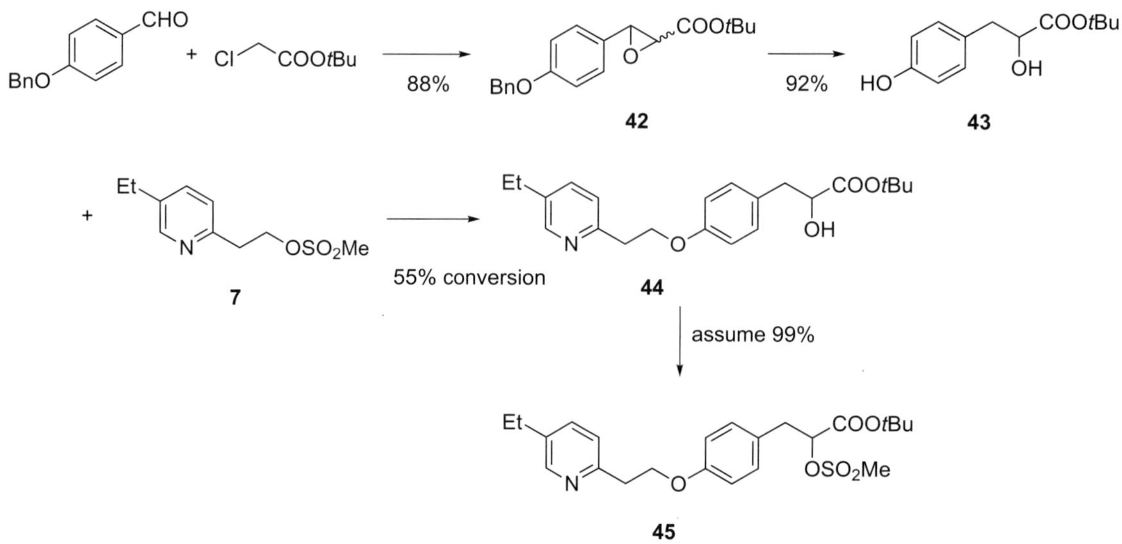


**SCHEME 2.15** Pioglitazone intermediate *tert*-butyl 3-(4-(2-(5-ethylpyridin-2-yl)ethoxy)phenyl)-2-(methanesulfonyloxy)propanoate (**45**) from 4-benzyloxybenzaldehyde via right-to-left Darzens condensation approach.

thiazolidine-2,4-dione (**41**) recrystallized (59% from **47**). The Knoevenagel condensation of 4-hydroxybenzaldehyde and thiazolidine-2,4-dione is accomplished using a mixture of lithium hydroxide and basic alumina in DMF at 25°C (64%). The double bond in the condensation product **48** is reduced using the cobalt(II) chloride–sodium borohydride method (75%) or with magnesium in methanol (60%).

The Williamson ether synthesis using 5-(4-hydroxybenzyl)thiazolidine-2,4-dione (**41**) with isolated toluenesulfonate **8** and potassium hydroxide in ethanol at reflux affords pioglitazone (**1**) in low yield (22%) (Scheme 2.17). The ratio of the phenol to the toluenesulfonate used in the Williamson ether synthesis is not specified. Earlier work constructing other 5-(4-oxybenzyl)thiazolidine-2,4-diones suggests that the low yield is likely due to competing N-alkylation to produce 3-(2-(5-ethylpyridin-2-yl)ethyl)-5-(4-hydroxybenzyl)thiazolidine-2,4-dione (**49**).[43]

## 2.5 PIOGLITAZONE (1) TO PIOGLITAZONE HYDROCHLORIDE

Pioglitazone (**1**) free base (colorless needles, mp 183–184°C) is most often recrystallized from DMF–water.[10,11] It can also be recrystallized from dioxane or tetrahydrofuran,[34] dioxane–methanol,[14] acetic acid–water, or acetic acid–ethanol.[30] Why is this important? Acetic acid and DMF are at the "end of the line" of organic solvents used for recrystallization. The poor solubility of pioglitazone (**1**) free base suggests that there will be no opportunity to reject

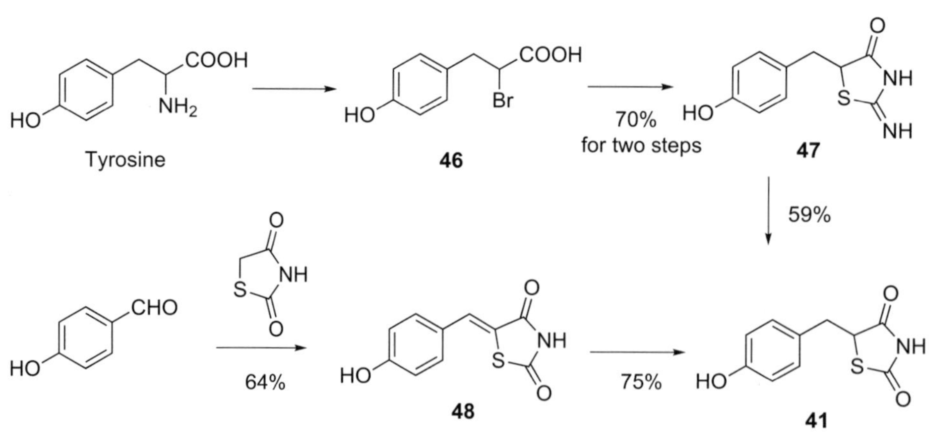


**SCHEME 2.16** Pioglitazone intermediate 5-(4-hydroxybenzyl)thiazolidine-2,4-dione (**41**) from tyrosine or 4-hydroxybenzaldehyde.

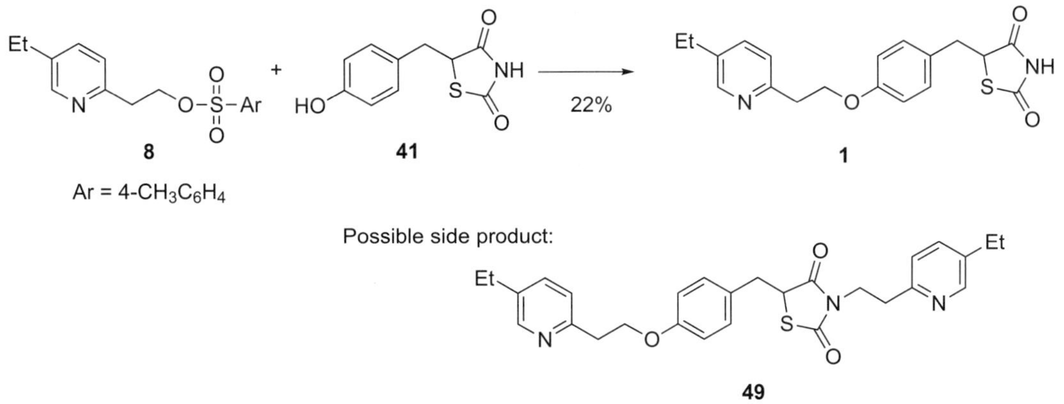


**SCHEME 2.17** Pioglitazone (**1**) from 5-(4-hydroxybenzyl)thiazolidine-2,4-dione (**41**) via right-to-left Williamson ether synthesis.

impurities, for example, (*Z*)-5-(4-(2-(5-ethylpyridin-2-yl)ethoxy)benzylidene)thiazolidine-2,4-dione (**32**), at or beyond the free base stage.

Pioglitazone (**1**) hydrochloride (colorless prisms, mp 192–193°C) is most often produced from the free base in ethanol by adding hydrogen chloride in ethanol or dilute aqueous hydrochloric acid.[31,34] It can also be produced in 1 N hydrochloric acid[34] and in methanol with concentrated hydrochloric acid.[52] Pioglitazone (**1**) hydrochloride is recrystallized from ethanol.[34]

## 2.6 USEFUL LAB PROCESSES NOT SUITABLE FOR LARGE-SCALE MANUFACTURING

Some reactions are often used in the lab, yet never scaled up. An example of this in the pioglitazone arena would be the Mitsunobu reaction for constructing 4-(2-(5-ethylpyridin-2-yl)ethoxy)benzaldehyde (**19**). The Mitsunobu reaction has been used to prepare small quantities of potential pioglitazone metabolites oxygenated at the carbons bound to the pyridine ring.[59] For example, 4-(2-(5-(1-(methoxymethoxy)ethyl)pyridin-2-yl)ethoxy)benzaldehyde (**50**) is prepared from 4-hydroxybenzaldehyde and 2-(5-(1-(methoxymethoxy)ethyl)pyridine-2-yl)ethanol (**51**) using triphenylphosphine and diethyl azodicarboxylate in THF at 25°C (62%) (Scheme 2.18). One needs to only fill in the equation on the right-hand side to understand why Mitsunobu reactions are not scaled up. The by-products of this process are triphenylphosphine oxide (1.6 g/g aldehyde **50**) and dicarbethoxyhydrazine (0.99 g/g aldehyde **50**) and these are removed by silica gel chromatography!

## 2.7 VIABLE ROUTES NOT TAKEN

At this point, pioglitazone processes not yet published or patented could be considered. Which ones are worth a look? Categorize the proposed routes. Some routes are *fill-the-gaps routes*, routes that use a known mechanism to add a new member to a family of routes. The Meerwein arylation with acrylamide and acrylonitrile are examples of *fill-the-gaps routes*. Other routes are *detour routes*, routes where novel transformations create a new pathway to an early intermediate. The $S_NAr$ route from 4-fluorobenzaldehyde to 4-(2-(5-ethylpyridin-2-yl)ethoxy)benzaldehyde (**19**) is a *detour route* from the Williamson ether route using 4-hydroxybenzaldehyde. Still others are *penultimate routes*, routes that focus on a novel penultimate intermediate and demonstrate conversion of this penultimate intermediate to the final product. 5-(4-Hydroxybenzyl)thiazolidine-2,4-dione (**41**) is the penultimate intermediate of one right-to-left approach (Scheme 2.19). We have seen many routes that fill the gaps and many that provide detours. There are typically just a few

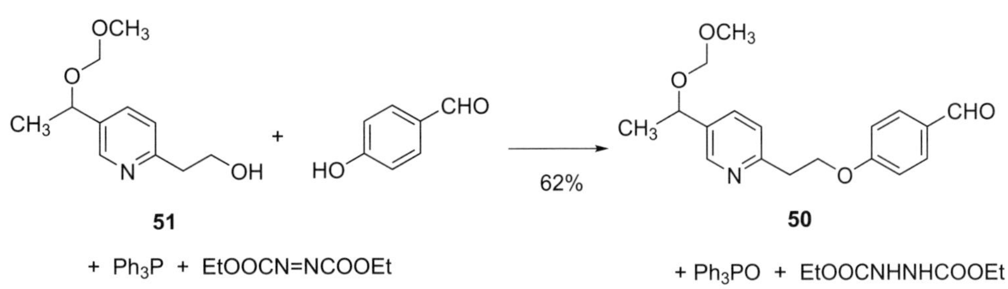


**SCHEME 2.18** Constructing potential pioglitazone metabolites using the Mitsunobu reaction.

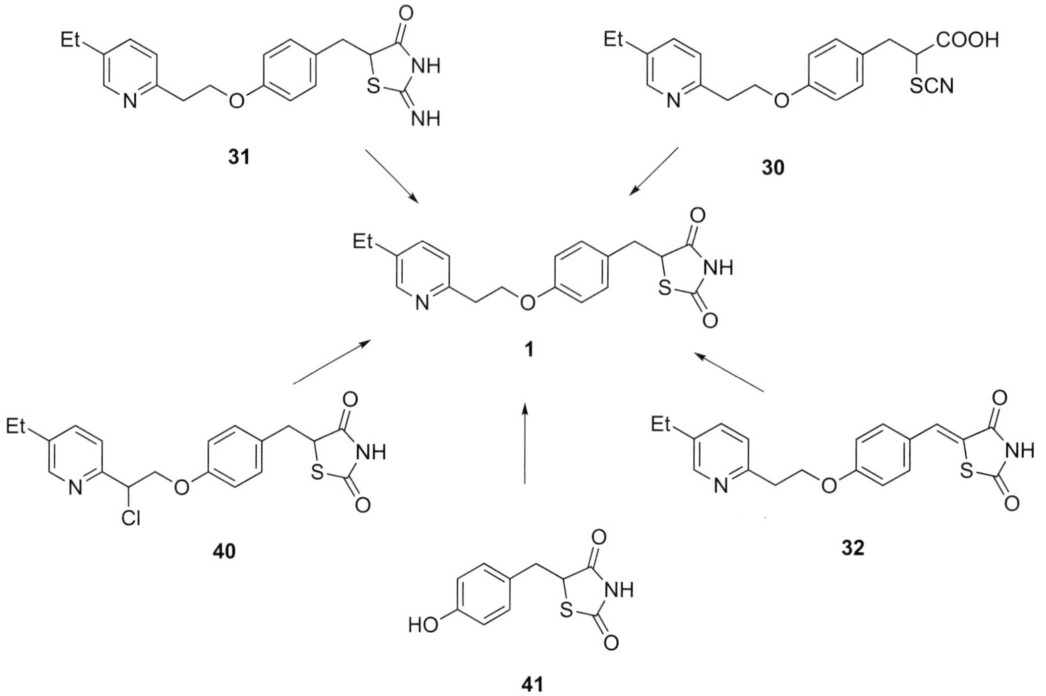


**SCHEME 2.19** Known penultimate intermediates of pioglitazone (**1**).

candidates for penultimate intermediate, so new penultimate routes are worth a close look.

With the routes categorized, the next step is to verify the justification to proceed based on the hoped-for improvement. For example, a new *detour route* from 2-(5-ethylpyridin-2-yl)ethanol (**2**) to 4-(2-(5-ethylpyridin-2-yl)ethoxy)benzaldehyde (**19**) uses 4-bromophenol (Scheme 2.20).[63] 1-Bromo-4-(2-(5-ethylpyridin-2-yl)ethoxy)benzene (**52**) is produced by Williamson ether synthesis and then converted to aldehyde **19** by a transition metal-catalyzed hydroformylation. The justification is reduced cost for the target aldehyde **19**. But this new route uses a more expensive starting material, 4-bromophenol, and requires an additional step to reach the target! An alternative route from 4-fluoronitrobenzene via Sandmeyer reaction on 4-(2-(5-ethylpyridin-2-yl)ethoxy)aniline (**5**) would be one step longer still. Good science does not always make business sense.

## 2.8 TRADE SECRETS AND THE CONTINUING REFINEMENT OF ANALYTICAL METHODS

Some processes or process improvements may not be patented. One way to check up on the competition is to identify

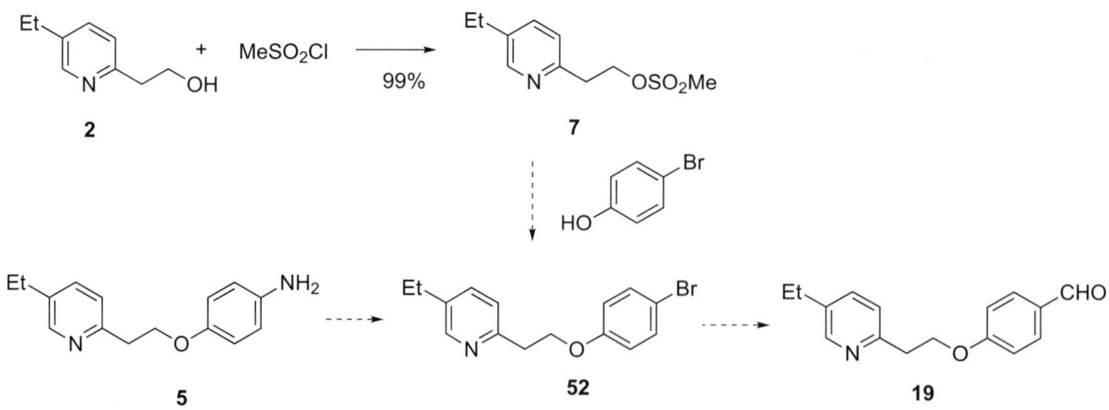


**SCHEME 2.20** Proposed detour routes to 4-(2-(5-ethylpyridin-2-yl)ethoxy)benzaldehyde (**19**).

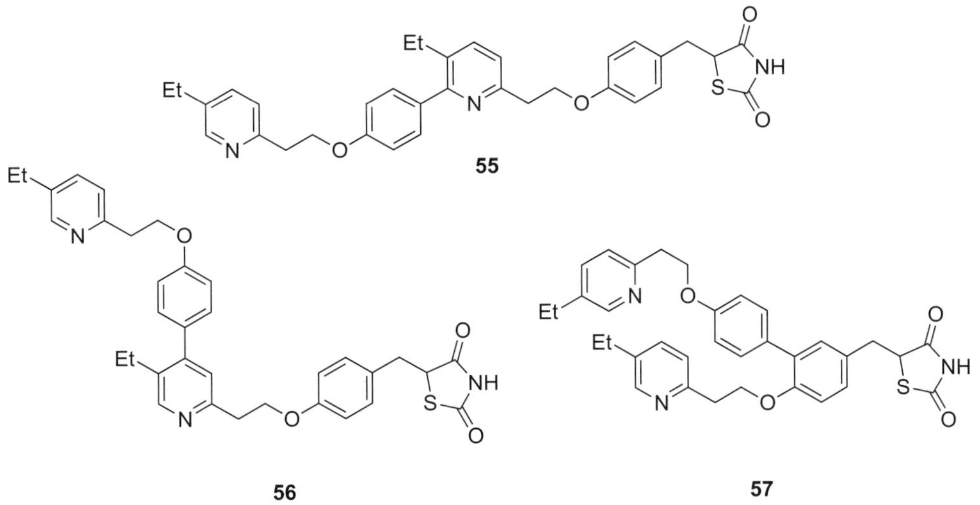


**FIGURE 2.4** Biaryl impurities isolated from bulk pioglitazone (**1**) produced by Meerwein arylation.

the impurities in their drug. Three impurities in bulk pioglitazone (**1**) free base were separated by HPLC and identified by $^{1}H$ and $^{13}C$ NMR, DEPT, IR, and mass spectrometry. They are 5-(4-hydroxybenzyl)thiazolidine-2,4-dione (**41**), 5-(4-fluorobenzyl)thiazolidine-2,4-dione (**53**), and 2-(2-(4-bromophenoxy)ethyl)-5-ethylpyridine (**54**).[64] With the difficult work of isolating and characterizing trace impurities complete, all that remains is to fit the pieces in the puzzle. 5-(4-Hydroxybenzyl)thiazolidine-2,4-dione (**41**) may come from cleavage of the ether linkage during acidic hydrolysis of the 2-iminothiazolidin-4-one. 5-(4-Fluorobenzyl)thiazolidine-2,4-dione (**53**) is produced from unreacted 4-fluoronitrobenzene left in the $S_NAr$ reaction producing 5-ethyl-2-(2-(4-nitrophenoxy)ethyl)pyridine (**4**). 2-(2-(4-Bromophenoxy)ethyl)-5-ethylpyridine (**54**) is likely produced by a competing Sandmeyer reaction during the Meerwein arylation. Three additional impurities found in bulk pioglitazone (**1**) free base were separated by HPLC and characterized by NMR. Impurities **55–57** are biaryls likely produced during Meerwein arylation (Figure 2.4). The major biaryl impurity **55** results from arylation at the pyridine 6-position.[65]

A convenient and fully validated HPLC method is available to separate all the process intermediates, especially the penultimate intermediate, from pioglitazone (**1**) produced by the Knoevenagel/reduction process.[66]

Chemists at Novo Nordisk recently developed an HPLC method for separation of the pioglitazone (**1**) enantiomers. With the pure enantiomers and a method for separating them in hand, we can determine just how easily pioglitazone (**1**) racemizes. At pH 2.5 and 37°C, the single enantiomer is converted to a 2:1 mixture in 8 days and is completely racemized in 30 days. The speed of racemization increases with pH. At pH 7.4 and 37°C, the single enantiomer is converted to the 2:1 mixture in just 10 h and is completely racemized in 48 h. With a half-life of just 4 h at physiological pH, the safety and efficacy of one enantiomer must be far superior to warrant developing a single-enantiomer version of pioglitazone (**1**).[67]

## 2.9 THE BEST PROCESS AVAILABLE TODAY

The best process converts three key raw materials, all available from multiple suppliers, to pioglitazone (**1**) hydrochloride in just five steps using well-established reactions for bond construction: Williamson ether synthesis, Knoevenagel condensation, and catalytic reduction of an alkene. 2-(5-Ethylpyridin-2-yl)ethanol (**2**) is converted to the methanesulfonate (**7**) in toluene solution (quantitative). The methanesulfonate is displaced by 4-hydroxybenzaldehyde in ethanol–toluene, isopropanol–toluene–water, or toluene with PEG 200 using potassium carbonate as base (79%). Knoevenagel condensation of the aldehyde **19** with thiazolidine-2,4-dione is accomplished in ethanol using a secondary amine as catalyst (95%). The double bond is reduced with sodium borohydride–cobalt boride in THF–water to produce pioglitazone (**1**) free base (95%), which is then converted to the hydrochloride salt in ethanol (95%) (Scheme 2.21). The overall yield from 2-(5-ethylpyridin-2-yl)ethanol (**2**) is 68%.

The penultimate step is both robust and efficient at rejecting impurities. It is likely that two of the intermediates, 2-(5-ethylpyridin-2-yl)ethyl methanesulfonate (**7**) and 4-(2-(5-ethylpyridin-2-yl)ethoxy)benzaldehyde (**19**), can be carried on with no purification. The process solvents are toluene, THF, ethanol, isopropanol, and water, all solvents commonly used in a pharmaceutical manufacturing plant.

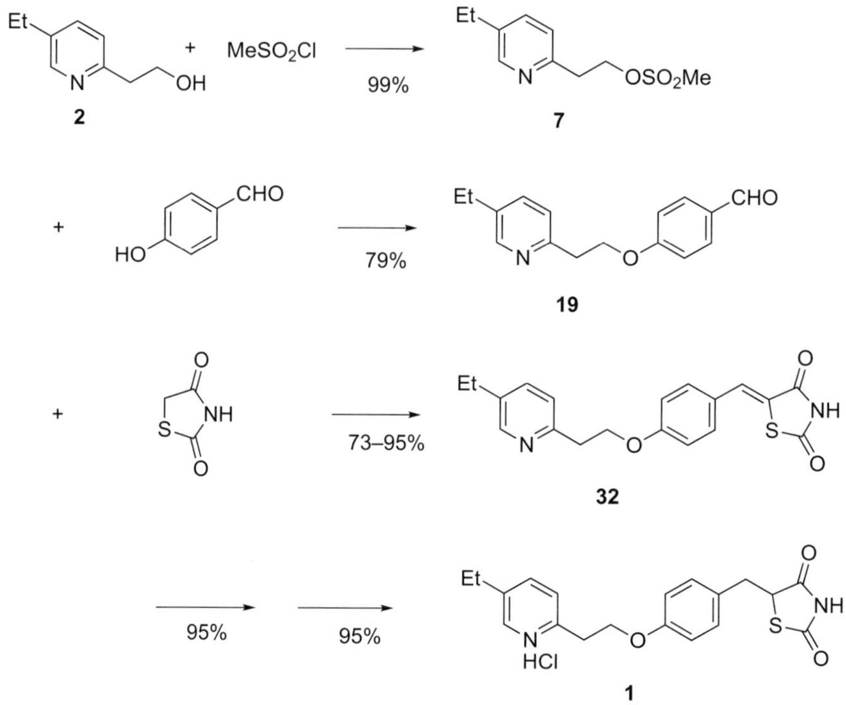


**SCHEME 2.21** The best manufacturing process for pioglitazone (**1**) hydrochloride (Actos®) in 2008.

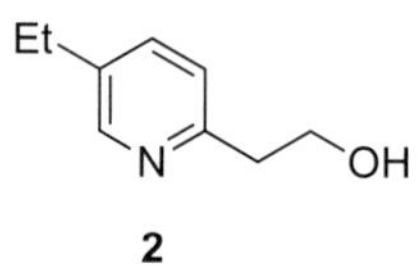


**FIGURE 2.5** Structures searched for pioglitazone (**1**) hydrochloride presentation.

## 2.10 STRUCTURES SEARCHED

A single structure search was used to generate all the information presented in this chapter (Figure 2.5).

## REFERENCES

1. Wild, S.; Roglic, G.; Green, A.; Sicree, R.; King, H. *Diabetes Care* **2004**, *27*, 1047.
2. Information taken from National Diabetes Information Clearinghouse (NDIC) at www.diabetes.niddk.nih.gov/dm/pubs/statistics.
3. Singh, S.; Loke, Y.K.; Furberg, C.D. *J. Am. Med. Assoc.* **2007**, *298*, 1189.
4. Lincoff, A.M.; Wolski, K.; Nicholls, S.J.; Nissen, S.E. *J. Am. Med. Assoc.* **2007**, *298*, 1180.
5. Information taken from www.drugs.com/top200.html.
6. Frank, R.L.; Blegen, J.R.; Dearborn, R.J.; Myers, R.L.; Woodward, F.E. *J. Am. Chem. Soc.* **1946**, *68*, 1368.
7. Yasuda, S.; Asahi, S.; Koito, S.Jpn. Kokai Tokkyo Koho JP 53147076 (1978)
8. Frank, R.L.; Pilgrim, F.J.; Riener, E.F. *Org. Synth., Coll.* Vol. *4* **1963**, 451.
9. The Aldrich 2007–2008 catalog price for 500 g is $24.90.
10. Meguro, K.; Fujita, T.U.S. Patent 4,687,777 (8/18/1987).
11. Meguro, K.; Fujita, T. EP 193256 (9/3/1986).
12. Balwant, D.P.; Mohansinh, R.D.; Onkarsingh, T.S. IN 2005MU01255 (8/17/2007).
13. Reddy, M.S.; Eswaraiah, S.; Ragupathi, A.; Reddy, E.S.; Rajendra, S. IN 2002MA00647 (8/3/2007).
14. Suri, S.; Sarin, G.S. WO 2006035459 (4/6/2006).
15. Rondestvedt, C.S., Jr. *Org. React.* **1960**, *11*, 189.
16. Rondestvedt, C.S., Jr. *Org. React.* **1976**, *24*, 225.
17. Zollinger, H. *Diazo Chemistry I*, VCH, Weinheim, 1994, pp. 243–253.
18. Cleland, G.H. *Org. Synth., Coll.* Vol. *6*, **1988**, 21.
19. Brunner, H.; Bluchel, C.; Doyle, M.P. *J. Organomet. Chem.* **1997**, *541*, 89.
20. McConville, F.X. *The Pilot Plant Real Book*, 2nd edition, FXM Engineering and Design, Worcester, MA, 2007. See page 6–26 for a discussion of the effect of pressure on azeotrope composition. See pages 10-10–10-27 for material compatibility tables.
21. Anderson, N.G. *Practical Process Research and Development*, Academic Press, San Diego, CA, 2000. See pages 81–111 for a discussion of process solvent selection.

22. Chhabada, V.C.; Pandya, R.R.; Rehani, R.B.; Thennati, R. WO 2004108721 (12/16/2004).
23. Oh, C.R.; Kim, Y.D.; Kim, M.S. KR 791399 (1/7/2008).
24. The Aldrich 2007–2008 catalog price for 1 kg is $74.70.
25. Halama, A.; Hejtmankova, L.; Lustig, P.; Richter, J.; Srsnova, L.; Jirman, J. WO 2002088120 (11/7/2002).
26. Kim, Y.D.; Oh, C.R.; Jang, J.C. KR 791400 (1/7/2008).
27. Duran Lopez, E. WO 2004099147 (11/18/2004).
28. Pospisilik, K.; Zhu, J.; Picha, F. WO 2004101560 (11/25/2004).
29. The Aldrich 2007–2008 catalog price for 1 kg is $424.00.
30. (a) Meguro, K.; Fujita, T.; Hatanaka, C.; Ooi, S. EP 257781 (7/20/1987).(b) Meguro, K.; Fujita, T.; Hatanaka, C.; Ooi, S. U.S. Patent 4,812,570 (3/14/1989).
31. Nambiar, S.; Pise, A.C. WO 2005049610 (6/2/2005).
32. (a) Arita, M.; Mizuno, Y. EP 506273 (3/16/1992).(b) Mizuno, Y.; Arita, M.U.S. Patent 5,554,758 (9/10/1996).
33. Hu, Z.; Zheng, Z. CN 1459450 (12/3/2003).
34. (a) Saito, Y.; Hikari, Y.; Mizufune, H.; Yamashita, M. EP 816340 (1/7/1998).(b) Saito, Y.; Mizufune, H.; Yamashita, M.U.S. Patent 5,952,509 (9/14/1999).(c) Saito, Y.; Mizufune, H.; Yamashita, M.U.S. Patent 6,100,403 (8/8/2000).
35. Fischer, J.; Fodor, T.; Levai, S.; Petenyi, E.; Perenyi, E. WO 2005058827 (6/30/2005).
36. Adiyaman, M.; Guner, D.; Yurdakul, A.; Ridvanoglu, N. WO 2004000810 (12/31/2003).
37. Momose, Y.; Meguro, K.; Ikeda, H.; Hatanaka, C.; Oi, S.; Sohda, T. *Chem. Pharm. Bull.* **1991**, *39*, 1440.
38. Sohda, T.; Mizuno, K.; Momose, Y.; Ikeda, H.; Fujita, T.; Meguro, K. *J. Med. Chem.* **1992**, *35*, 2617.
39. The Aldrich 2007–2008 catalog price for 25 g is $102.50.
40. Hindley, R.M. EP 306228 (11/17/1999).
41. The Aldrich 2007–2008 catalog price for 250 g is $161.50.
42. Thijs, L.; Zhu, J. WO 2005080387 (9/1/2005).
43. Sohda, T.; Mizuno, K.; Imamiya, E.; Sugiyama, Y.; Fujita, T.; Kawamatsu, Y. *Chem. Pharm. Bull.* **1982**, *30*, 3580.
44. The Aldrich 2007–2008 catalog price for 100 g is $128.00.
45. Chemical Sampling Information, www.osha.gov.
46. (a) National Toxicology Program. *Report on Carcinogens*, 11th edition, Public Health Service, U.S. Department of Health and Human Services, Atlanta, GA. (b) Chemicals Known to the State to Cause Cancer or Reproductive Toxicity, State of California, March 21, 2008.
47. Palladium is roughly $260/oz or $9/g today (9/08/2008).
48. EMEA Guideline on the Specification Limits for Residues of Metal Catalysts, www.emea.europa.eu/pdfs/human/swp/444600.
49. Dolitzky, B.-Z. WO 2003053367 (7/3/2003).
50. Bulliard, M.; Derrien, Y.; Pintus, T. WO 2002064577 (8/22/2002).
51. Platinum is roughly $1170/oz or $41/g today (9/10/2008).
52. Huber, J.E. WO 19931305 (7/8/1993).
53. The Aldrich 2007–2008 catalog price for 1 kg cobalt(II) chloride hexahydrate is $115.50.
54. Osby, J.O.; Heinzman, S.W.; Ganem, B. *J. Am. Chem. Soc.* **1986**, *108*, 67.
55. Agency for Toxic Substances and Disease Registry (ASTDR). *Toxicological Profile for Cobalt*, Public Health Service, U.S. Department of Health and Human Services, Atlanta, GA, 1992 (revised 2000).
56. (a) Hindley, R.M.; Woroniecki, S.R.U.S. Patent 5,726,055 (3/10/1998).(b) Hindley, R.M.; Woroniecki, S.R. EP 615549 (6/9/1999).
57. Carpenter, D.E.; Imbordino, R.J.; Maloney, M.T.; Moeslein, J.A.; Reeder, M.R.; Scott, A. *Org. Process Res. Dev.* **2002**, *6*, 721.
58. Pandey, B.; Lohray, V.B.; Lohray, B.B. WO 2004007490 (1/22/2004).
59. Tanis, S.P.; Parker, T.T.; Colca, J.R.; Fisher, R.M.; Kletzein, R. F. *J. Med. Chem.* **1996**, *39*, 5053.
60. The Aldrich 2007–2008 catalog price for 100 g is $90.30.
61. The Aldrich 2007–2008 catalog price for 100 g is $50.60.
62. Ujire, S.; Sridharan, M.; Ganesh, S. WO 2003029251 (4/10/2003).
63. The Aldrich 2007–2008 catalog price for 500 g is $164.00.
64. Kumar, Y.R.; Reddy, A.R.; Eswaraiah, S.; Mukkanti, K.; Reddy, M.S.N.; Suryanarayana, M.V. *Pharmazie* **2004**, *59*, 836.
65. Richter, J.; Jirman, J.; Havlicek, J.; Hrdina, R. *Pharmazie* **2007**, *62*, 580.
66. Radhakrishna, T.; Rao, D.S.; Reddy, G.O. *J. Pharm. Biomed. Anal.* **2002**, *29*, 593.
67. Jamali, J.; Bjomsdottir, I.; Nordfang, O.; Hansen, S.H. *J. Pharm. Biomed. Anal.* **2008**, *46*, 82.

# 3

# LEXAPRO® (ESCITALOPRAM OXALATE)

## 3.1 LEXAPRO® AND THE ANTIDEPRESSANT MARKET

Lexapro® (escitalopram oxalate) is one member of a class of antidepressant drugs called selective serotonin reuptake inhibitors (SSRIs). The SSRIs were the most commonly prescribed medications in the United States in 2008 and escitalopram oxalate is the best-selling SSRI in terms of U.S. sales and number of prescriptions. A meta-analysis of 12 new-generation antidepressants concluded that escitalopram oxalate and sertraline (Zoloft®) provide the best combination of efficacy and tolerability in the acute-phase treatment of adults with major depression.[1]

The active ingredient in Lexapro® (escitalopram) (Figure 3.1) is the (*S*)-enantiomer of the racemate in Celexa® (citalopram). The antidepressant activity of citalopram is almost exclusively associated with the (*S*)-enantiomer.[2] Anticipating the citalopram patent expiration in 2005, Lundbeck and Forest Laboratories began marketing Lexapro® and stopped marketing Celexa® in 2002. The selling point for Lexapro® in 2002 was a 5–10% lower price than the approximately $2 per pill of Celexa®. Combined sales figures for Lexapro® and Cipralex® (escitalopram oxalate outside the United States) from Lundbeck showed a steady growth rate from $806 million in 2004 to $1.44 billion (1 DKK = 0.1979 USD on September 18, 2009) in 2008. Forest Laboratories sales of Lexapro® were $2.3 billion in 2009.[3–5]

Depression affects about 121 million people worldwide. Major depression affects approximately 15 million Americans (about 8% of the 18-year-old and older population) each year. Approximately 12% of women will experience clinical depression during their lifetime. One in 33 children and 1 in 8 adolescents experience depression. Most people (80%) experiencing depression are not currently receiving treatment despite the fact that depression is one of the most treatable illnesses. Most people (80%) treated for depression show some relief within 4–6 weeks of beginning medication, psychotherapy, participation in a support group, or some combination of these treatment options. In 2002, depression accounted for 4.5% of the worldwide total burden of disease in terms of disability-adjusted life years. Depression is expected to be the second most common health problem in the world by the year 2020 for all ages and both sexes.[6–9]

There are many antidepressants on the market and patients often go through several antidepressants before finding one that provides relief and has tolerable side effects. Patients are advised to continue with an effective antidepressant that is working rather than consider changing to a cheaper (perhaps generic) alternative. The impressive increase in sales of Lexapro® suggests that it is and continues to be an effective antidepressant for many people.

Antidepressants have some serious side effects. They may increase the potential for suicidal thinking in children. This is more likely when starting treatment and when the dose is increased or decreased. In 2003, the British Medicines and Healthcare Products Regulatory Agency, the equivalent of the FDA in Britain, ordered physicians to stop prescribing many of the SSRIs, including Lexapro®, to children. In 2007, a review of 295 antidepressant trials with >77,000 patients also suggested a slight increase in suicidal thinking in young adults, aged 18–24. The FDA ordered that all antidepressants

*Pharmaceutical Process Chemistry for Synthesis: Rethinking the Routes to Scale-Up*, By Peter J. Harrington

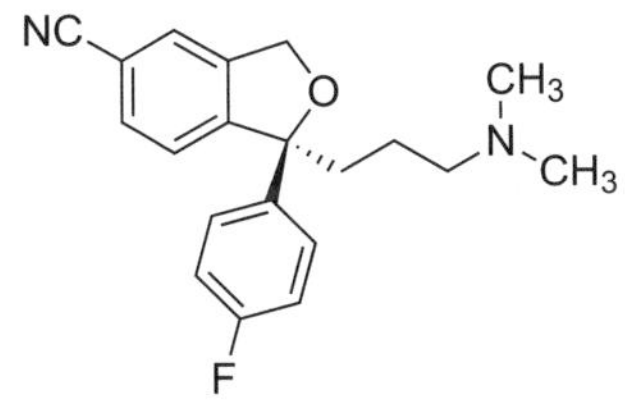


**FIGURE 3.1** Escitalopram (**1**).

carry a black box warning to alert physicians and patients to these risks.

By far the biggest concern for Lexapro® sales from an antidepressant marketing perspective is the impact of generic competition. The market share of generic antidepressants increased (by volume) from 20% in 2000 to 42% in 2004 and this market share was expected to increase to 75% by 2009. There is some activity on the generic escitalopram front. Cipla (Cipla Pharmaceuticals, Ltd.) and Ivax (a U.S. subsidiary of Teva Pharmaceuticals, Ltd.) won FDA approval for a generic version of escitalopram oxalate in May 2006. In July 2006, the Lundbeck patents on escitalopram oxalate were upheld and Teva was blocked from launching generic escitalopram oxalate in the United States until the patent expiration in March 2012. In July 2009, Lundbeck and Forest entered into a settlement agreement with Caraco (Caraco Pharmaceuticals Laboratories, Ltd.) and Sun Pharmaceuticals (Sun Pharmaceuticals Industries, Ltd.) in a pending patent infringement on the escitalopram patent in the United States. In the agreement, Caraco will be able to enter the U.S. market as of the date that any other third-party generic (other than the first filer or a generic authorized by Lundbeck and Forest) enters the market. Lundbeck gained a license to Sun Pharmaceuticals patents and patent applications relating to a manufacturing process for citalopram and escitalopram.[3–5],10

## 3.2 CONSTRUCTION OF THE 1,2,4-TRISUBSTITUTED AROMATIC COMPONENT

The generation of the quaternary carbon and only chiral center is the focal point for the synthesis of escitalopram (**1**). Expanding out from this focal point, we will discuss construction of the 1,2,4-trisubstituted aromatic, timing of and methods for introduction of the nitrile, 4-fluorophenyl, and *N*,*N*-dimethylaminopropyl substituents, and strategies for creating the required chirality. Despite the apparent simplicity of the target molecule, the many options weave a complex web of process chemistry that provides access to citalopram and escitalopram from nearly every conceivable penultimate intermediate (Figure 3.2).

There are at least twelve 5-substituted phthalides used as key value-added starting materials for the synthesis of escitalopram (Figure 3.3). The 5-substituent can be a cyano or a group that can be converted to the cyano group.

The phthalides are derived from phthalimide, phthalic acid, trimellitic anhydride, terephthalic acid, diethyl terephthalate, or *meta*-xylene. All except trimellitic anhydride require adding a third substituent and this is invariably introduced by electrophilic aromatic substitution. The substitution options include nitration or bromination of phthalic acid or phthalimide, bromination or benzoylation of *meta*-xylene or a derivative, and hydroxymethylation of terephthalic acid or diethyl terephthalate.

### 3.2.1 5-Aminoisobenzofuran-1(*3H*)-one (2)

Starting with phthalimide,[11] nitration (52%) followed by nitro group reduction with hydrogen, palladium on aluminum oxide, and ammonium vanadate in methanol at 25°C and 1 atm (100%) affords 4-aminophthalimide (**4**).[12,13] 4-Aminophthalimide (**4**) is reduced with zinc powder using a copper sulfate catalyst in aqueous sodium hydroxide at 70–80°C to produce 5-aminoisobenzofuran-1(3*H*)-one (**2**).[14]

### 3.2.2 5-Bromo-, 5-Chloro-, and 5-Hydroxyisobenzofuran-1(*3H*)-ones (5–7)

Of the 5-halo- and 5-hydroxyisobenzofuran-1(3*H*)-ones, 5-bromoisobenzofuran-1(3*H*)-one (**5**) is commercially available but is too expensive to be a viable starting material.[15] These value-added starting materials can all be produced from 5-aminoisobenzofuran-1(3*H*)-one (**2**) by diazotization and Sandmeyer reaction or hydrolysis of the diazonium salt (X = Br, 86%; X = Cl, 91%) (Scheme 3.1).[14]

The undesirable diazonium salt intermediate is avoided in an alternative sequence starting with inexpensive phthalic anhydride.[16] Bromination of disodium phthalate in water, quench of the excess bromine, acidification, extraction into an organic solvent, and dehydration during distillation affords 4-bromophthalic anhydride (**8**).[17] 4-Bromophthalic anhydride (**8**) is then reduced to a nearly 1:1 mixture (56:44) of 5- and 6-bromophthalides (**5** and **9**) with sodium borohydride in THF at 25°C. 5-Bromophthalide (**5**) is separated from the mixture by crystallization.

Slow addition of a solution of 4-bromophthalic anhydride (**8**) in THF to a suspension of sodium borohydride (0.56 equivalents) in THF at 5–15°C is followed by aging at 25°C for 1 h. Excess hydride is quenched with water and dilute hydrochloric acid. After layer separation and brine wash, some THF is distilled and the crude bromophthalide **5** is crystallized by adding water. Filtration, wash with THF, and recrystallization from THF–water affords 5-bromophthalide (**5**) (33%, >98% pure by HPLC). A very similar process in

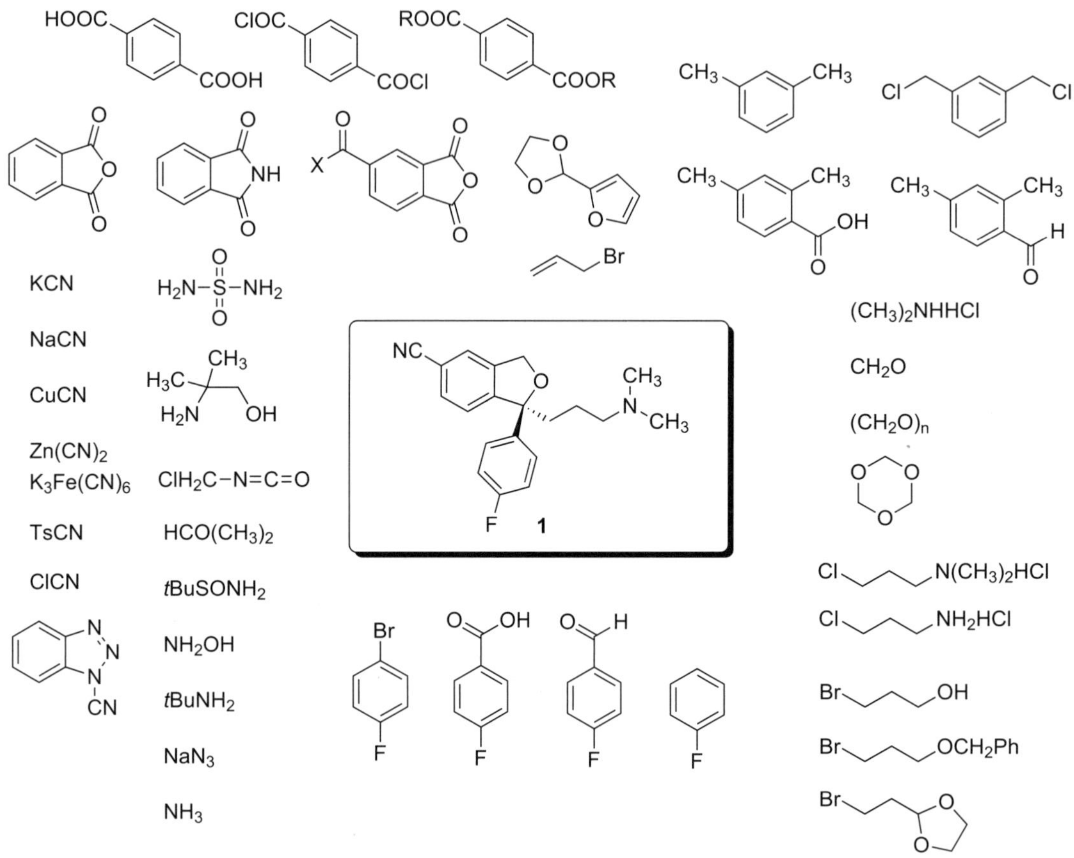


**FIGURE 3.2** Escitalopram building blocks.

1,2-dimethoxyethane affords 5-bromophthalide (**5**) (38%, >99% pure by HPLC) (Scheme 3.2).[18]

### 3.2.3 1-Oxo-1,3-dihydroisobenzofuran-5-carboxylic Acid (10)

1-Oxo-1,3-dihydroisobenzofuran-5-carboxylic acid (**10**) can be produced by the electrochemical reduction of trimellitic anhydride. Trimellitic anhydride[19] is taken up in a solution of ammonium carbonate in 25% aqueous ammonia and water is added to produce a solution. The solution is pumped through an electrolytic cell with lead cathode, lead oxide on lead anode, and a cation exchange membrane containing fluorine. Complete conversion is observed after electrolysis at 24°C for 24 h. The solution is concentrated at reduced pressure. The residue is dissolved in a small amount of water and acidified with 50% sulfuric acid at 80°C. After aging at 80°C for 1 h, the suspension is cooled to 5°C and the solid is filtered, washed with 0–5°C water, and dried to afford 1-oxo-1,3-dihydroisobenzofuran-5-carboxylic acid (**10**) (90% selectivity) (Scheme 3.3).[20]

1-Oxo-1,3-dihydroisobenzofuran-5-carboxylic acid (**10**) is also produced by the reaction of terephthalic acid with 1,3,5-trioxane or paraformaldehyde in fuming sulfuric acid (oleum). The reaction requires an oleum charge from 3.1 L (20% $SO_3$) to 4.6 L (27% $SO_3$) per kg terephthalic acid, 1.3–2.6 equivalents of formaldehyde, and a temperature of 125–150°C. The product is capable of reacting further with formaldehyde to produce isobenzofuro[5,6-*c*]furan-1,5(3*H*,7*H*)-dione (**11**). The crude product first isolated from the reaction mixture typically contains 85–90% of product **10** and 10–15% of dione **11**. Product **10** and dione **11** are easily separated by suspending the solid mixture in pH 7–8 water at 25°C (product soluble/dione insoluble) or perhaps in isopropanol at reflux (product insoluble/dione soluble).

A solution of terephthalic acid[21] in 27% oleum (3.1 L per kg terephthalic acid) is prepared at 25°C. 1,3,5-Trioxane[22] (2.6 equivalents of formaldehyde) is added to this solution at 10–25°C. The suspension is then heated to 130–135°C, becoming a solution at 90°C. After aging at 130–135°C for 4 h, the mixture is cooled and quenched with ice water at 25–35°C. Aqueous sodium hydroxide solution is added at 35–40°C (to pH 8). The dione **11** is filtered (volume throughput is 15 g product/L) using filter aid and the cake is washed with water. Hydrochloric acid is added to the combined liquors (to pH 1) and the suspension is heated to 35°C. The solid is filtered, washed with water at 40°C, and suspended in water at 45°C. The solid is filtered, washed

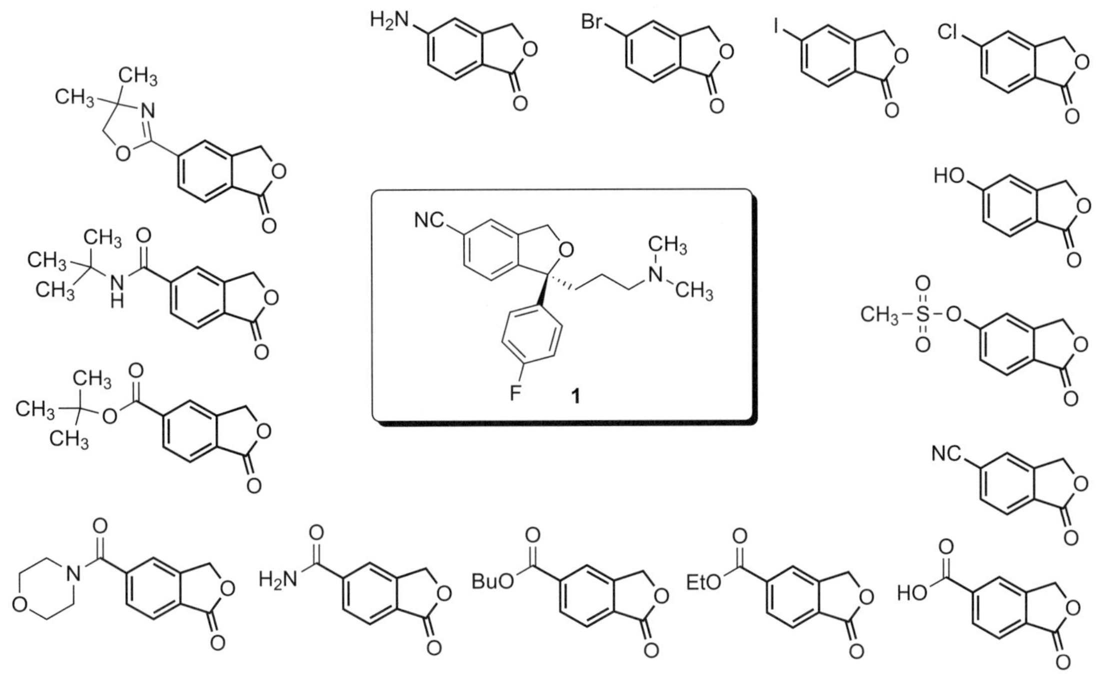

**FIGURE 3.3** 5-Substituted phthalide starting materials.

with water, and dried at 50°C and reduced pressure to afford 1-oxo-1,3-dihydroisobenzofuran-5-carboxylic acid (**10**) (77–81% yield, >95% pure by HPLC).[23]

The neutralization of the sulfuric acid results in a very low volume throughput. In an alternative process, a solution of terephthalic acid in 25–27% oleum (4.6 L per kg terephthalic acid) is prepared at 23°C. 1,3,5-Trioxane (2.6 equivalents of formaldehyde) is added to this solution at 15°C. The suspension is then heated to 130–133°C for several hours. The reaction mixture is cooled and glacial acetic acid is added at <25°C. Cold water is then added at <45°C. After aging the suspension for 1 h at 25°C, the solid is filtered (volume throughput is 33 g/L), washed with water, and suspended in water. Aqueous sodium bicarbonate is added (to pH 7.6–7.8). The dione **11** is filtered using filter aid. The filter cake is washed with water. Hydrochloric acid is added to the combined liquors (to pH 1). The resulting suspension is filtered and the solid is washed with water and dried at 50°C

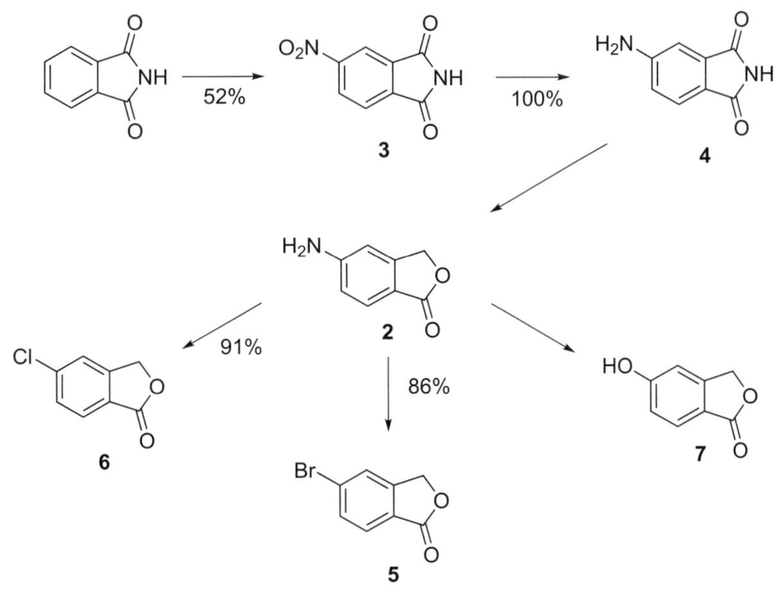

**SCHEME 3.1** 5-Bromo-, 5-chloro-, and 5-hydroxyisobenzofuran-1(3*H*)-ones (**5–7**) from 5-aminoisobenzofuran-1(3*H*)-one (**2**).

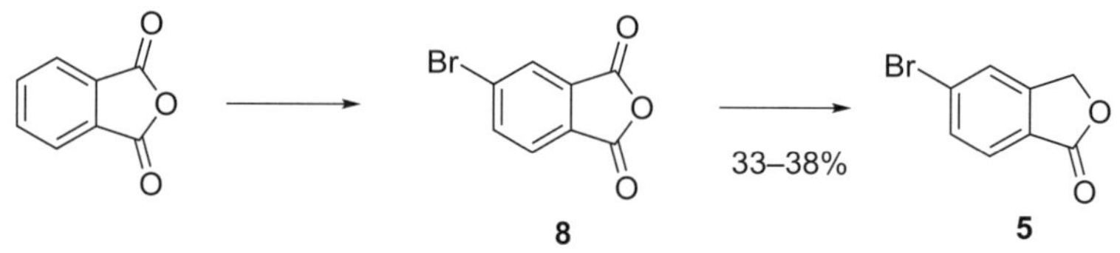


**SCHEME 3.2** 5-Bromoisobenzofuran-1(3*H*)-one (**5**) from phthalic anhydride.

and reduced pressure to afford 1-oxo-1,3-dihydroisobenzofuran-5-carboxylic acid (**10**) (76% yield, >94% pure by HPLC).[23]

In a 10 kg scale demonstration run, a reactor is charged with terephthalic acid, 20% oleum (18–24% sulfur trioxide in sulfuric acid) (3.1 L per kg terephthalic acid), and paraformaldehyde[24] (1.3 equivalents of formaldehyde). The mixture is aged at 125°C for 17 h. Water (13 L per kg terephthalic acid) and filter aid are added. The suspension temperature is set to 70°C and the solids are filtered and washed with water (volume throughput is 55 g/L). The wet solid is suspended in water, aqueous sodium hydroxide is added (to pH 7), and carbon is added. The dione **11**, carbon, and filter aid are filtered and washed with water. The filtrate is cooled to 65°C and 50% aqueous sulfuric acid is added (to pH 2). The resulting suspension is filtered, washed with water, and dried to afford 1-oxo-1,3-dihydroisobenzofuran-5-carboxylic acid (**10**) (83%). In a 13 kg demonstration run, reaction with 20–25% oleum (1.7 L per kg terephthalic acid) at 138–145°C for 4.5 h is quenched with water (6.7 L per kg terephthalic acid) and the suspension is filtered at 100°C rather than 70°C. The volume throughput is 104 g/L (82%).[25]

A process using 25–30% oleum and trioxane requires 140–150°C for 4–8 h. After cooling and quench with water at <80°C, the suspension is further cooled to 40°C. The solid is filtered, washed with water and isopropanol, and suspended in isopropanol at reflux. The suspension is cooled to 20–25°C and filtered and the solid is washed with isopropanol and dried at 60–70°C and reduced pressure to afford 1-oxo-1,3-dihydroisobenzofuran-5-carboxylic acid (**10**). No yield is provided.[26]

Better control over the preparation of the initial mixture can be achieved by preparing a solution of terephthalic acid in 20–25% oleum (2.3 L per kg terephthalic acid) at 150°C and adding a solution of paraformaldehyde (1.6 equivalents of formaldehyde) in 20–25% oleum (1.3 L per kg terephthalic acid) at 150°C over 1–2 h. The mixture is aged at 150°C for 4 h. Quench with water (6.3 L per kg terephthalic acid) at a rate keeping the temperature <100°C and the same workup procedure as the 10 kg scale run affords 1-oxo-1,3-dihydroisobenzofuran-5-carboxylic acid (**10**) (76%).[27]

There are several scale-up issues associated with the use of paraformaldehyde or 1,3,5-trioxane in the terephthalic acid process. A review of material safety data for 1,3,5-trioxane and paraformaldehyde might suggest that handling of 1,3,5-trioxane is preferred. Paraformaldehyde dust poses a significant explosion hazard. Paraformaldehyde is both a sensitizer and a cancer suspect agent and emits formaldehyde monomer that is also a sensitizer and a cancer suspect agent.[28] The OSHA airborne exposure limits for formaldehyde are 0.5 ppm (action level), 0.75 ppm (TWA), and 2 ppm (STEL). The ACGIH threshold limit value (TLV) is 0.3 ppm

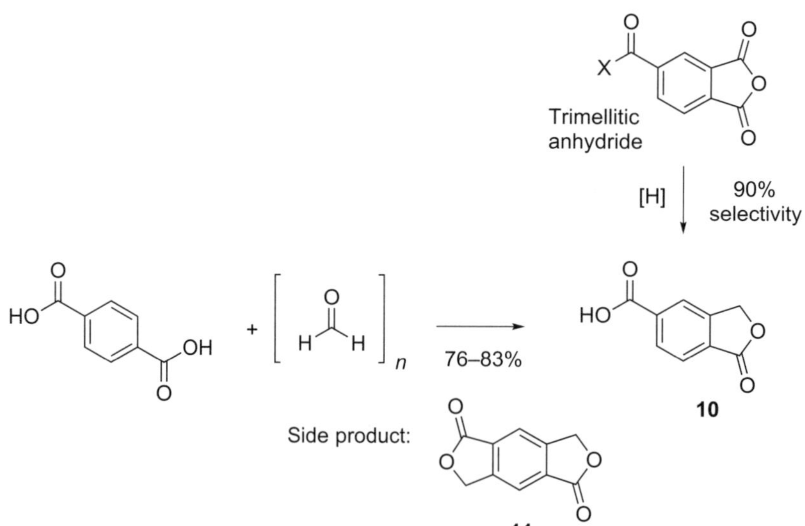


**SCHEME 3.3** 1-Oxo-1,3-dihydroisobenzofuran-5-carboxylic acid (**10**) from trimellitic anhydride and from terephthalic acid.

(STEL/ceiling). However, there is also concern that solid 1,3,5-trioxane (mp 61–62.5°C and bp 114.5–115.5°C) may collect on cold surfaces in the reactor headspace, condenser, and connecting lines.

Above and beyond the reagent storage and handling issue, formaldehyde containment is the primary concern. All the procedures require a solid charging of 1,3,5-trioxane or paraformaldehyde to oleum. How is this charge best done to avoid operator exposure to formaldehyde? What is the heat of mixing and how can it be controlled in a (presumably) viscous mixture? How will sulfur trioxide and formaldehyde emanating from the reaction mixture be contained and scrubbed? What is the safest way to store and dispose of the aqueous liquors containing formaldehyde? All these questions must be adequately answered before moving forward with production of 1-oxo-1,3-dihydroisobenzofuran-5-carboxylic acid (**10**) from terephthalic acid.

### 3.2.4 Alkyl 1-Oxo-1,3-dihydroisobenzofuran-5-carboxylates

***3.2.4.1 From 1-Oxo-1,3-dihydroisobenzofuran-5-carbonyl Chloride (12)*** 1-Oxo-1,3-dihydroisobenzofuran-5-carboxylic acid (**10**) can be converted to an ester via the acid chloride. The acid chloride formation is typically performed with thionyl chloride neat or in toluene solution using DMF or *N,N*-dimethylacetamide as a catalyst. Handling of the solid acid chloride can be avoided by a solvent exchange and transfer of the solution of the acid chloride into the next step.

The neat reaction of the carboxylic acid **10** with thionyl chloride (6–12 equivalents) and catalytic DMF or *N,N*-dimethylacetamide catalyst requires an aging time of several hours at 60°C or at least 1 h at reflux. The excess thionyl chloride is distilled at reduced pressure. Residual thionyl chloride is chased by dissolution of the residue in toluene and distillation of the toluene at reduced pressure. The dissolution and distillation are repeated two times to afford a solid (91%).[29–32]

Less thionyl chloride (1.2–2.0 equivalents) is required when the reaction is run in toluene solution. The reaction of the carboxylic acid **10** with thionyl chloride (2.0 equivalents) and catalytic DMF in toluene is complete in 3 h at reflux. The mixture is cooled to 25°C and an equal volume of heptane is added. The precipitated acid chloride **12** is filtered and dried at an unspecified temperature and pressure (88%).[26,33]

The reaction of the carboxylic acid **10** with thionyl chloride (1.5 equivalents) and catalytic DMF in toluene is complete in 3 h at reflux. Half the toluene is distilled and the resulting solution is cooled to 25°C and aged overnight. The precipitated acid chloride **12** is filtered and dried at an unspecified temperature and pressure (82%).[34]

The acid chloride **12** is refluxed in ethanol or isopropanol for 15–30 min to produce the ester. The ethanol solution is cooled to 25°C and the resulting precipitate is filtered, washed with ethanol, and dried to afford ethyl 1-oxo-1,3-dihydroisobenzofuran-5-carboxylate (**13**) (88%) (Scheme 3.4). The isopropanol solution is cooled to 0°C and the resulting precipitate is filtered, washed with cold isopropanol, and dried to afford isopropyl 1-oxo-1,3-dihydroisobenzofuran-5-carboxylate (**14**) (87%).[32,33]

***3.2.4.2 From 1-Oxo-1,3-dihydroisobenzofuran-5-carboxylic Acid (10) by Fischer–Speier Esterification*** Conversion of the carboxylic acid **10** to the ethyl ester **13** by Fischer–Speier esterification avoids the use of thionyl

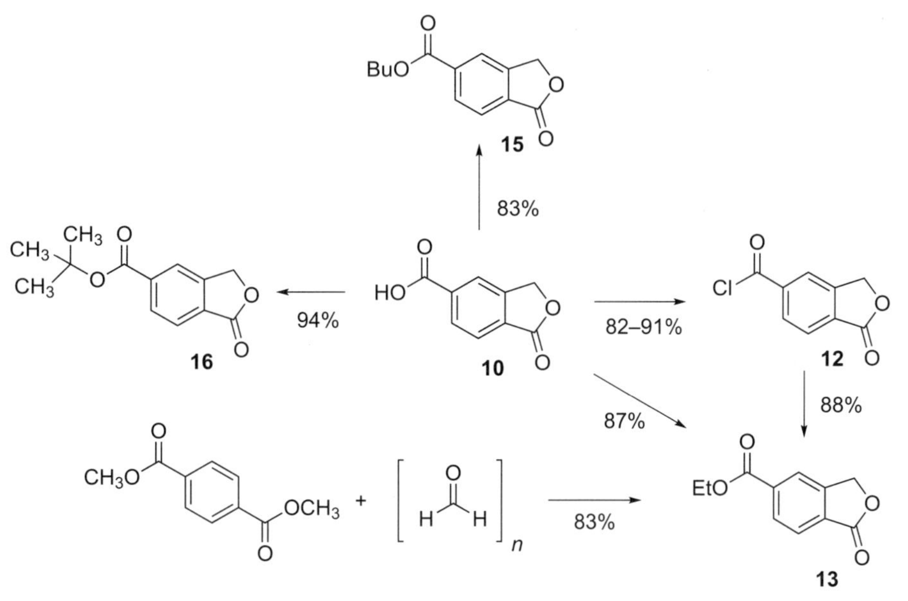


**SCHEME 3.4** Routes to 1-oxo-1,3-dihydroisobenzofuran-5-carboxylic acid esters.

chloride. Phosphorus oxychloride (0.35 equivalents) is added to a suspension of the carboxylic acid **10** in ethanol. The mixture is refluxed for 5 h and then cooled to 25°C. The resulting suspension is filtered and the solid is washed with ethanol and dried to afford ethyl 1-oxo-1,3-dihydroisobenzofuran-5-carboxylate (**13**) (87%).[33]

An alternative process converts 1-oxo-1,3-dihydroisobenzofuran-5-carboxylic acid (**10**) to the butyl ester **15** by Fischer–Speier esterification. A mixture of the water-wet carboxylic acid **10**, 1-butanol (2.1 equivalents), and 96% sulfuric acid (0.27 equivalents) in toluene is refluxed using a Dean–Stark trap for water removal until the pot temperature reaches 110°C. The mixture is distilled to recover a portion of the toluene, and then cooled to 80°C and diluted with heptanes. Triethylamine (0.16 equivalents) is added followed by activated carbon and dicalcite. The suspension is heated to 80°C and filtered. The liquors are washed with water, presumably at 80°C, and allowed to cool to 25°C. The resulting suspension is filtered and the solid is washed with heptanes and dried to afford butyl 1-oxo-1,3-dihydroisobenzofuran-5-carboxylate (**15**) (83%).[27]

***3.2.4.3 From an Active Ester*** 1-Oxo-1,3-dihydroisobenzofuran-5-carboxylic acid (**10**) can be converted to an ester via an active ester. *p*-Toluenesulfonyl chloride (2.0 equivalents) is added to a solution of the carboxylic acid **10** in pyridine, presumably at 25°C. The solution is aged at 25°C for 30 min. *tert*-Butanol (1.3 equivalents) is added and the solution is aged at 25°C for 3 days. Quench with ice water produces a suspension, which is filtered. The solid is presumably washed with water and recrystallized from isopropanol to afford *tert*-butyl 1-oxo-1,3-dihydroisobenzofuran-5-carboxylate (**16**) (94%).

***3.2.4.4 From the Terephthalic Acid–Formaldehyde Condensation Mixture*** An ester manufacturing process can be streamlined by going directly from the terephthalic acid–formaldehyde condensation into the esterification. Dimethyl terephthalate[35] is added to 20–25% oleum (2.1 L per kg terephthalic acid). Paraformaldehyde (1.3 equivalents of formaldehyde) is added and the mixture is aged at 125°C for 5 h. The mixture is cooled to 70°C and ethanol (8.0 L per kg terephthalic acid) is added, presumably while still controlling the temperature at 70°C. After aging for 90 min at 85–93°C, the mixture is cooled to 80°C and ice water (3.1 L per kg terephthalic acid) is added. The suspension is aged overnight, presumably at 25°C, and then cooled to 15°C. The resulting suspension is filtered (volume throughput is 94 g/L) and the solid is washed with water and taken up in water. Aqueous 28% sodium hydroxide is added (to pH 4). The suspension is filtered and the solid is washed with water and dried to afford ethyl 1-oxo-1,3-dihydroisobenzofuran-5-carboxylate (**13**) (83%). Terephthaloyl chloride[36] can be used in place of terephthalic acid.[27]

A similar process begins with the addition of sulfur trioxide (0.9 L per kg terephthalic acid) to terephthalic acid. 1,3,5-Trioxane (3.0 equivalents of formaldehyde) is added with cooling and the mixture is heated at 130°C for 15 min. The dark red solution is cooled to 100°C and more 1,3,5-trioxane (1.0 equivalent of formaldehyde) is added. After aging at 130°C for 105 min, the mixture is cooled to 100°C and quenched with ethanol (10 L per kg terephthalic acid) (volume throughput is 98 g product/L). Some of the ethanol is removed by distillation and ice water is added. The suspension is cooled, presumably to 25°C, and filtered. The wet solid is suspended in water and aqueous sodium hydroxide is added (to pH 4). The suspension is filtered and the solid is washed with water and dried to afford ethyl 1-oxo-1,3-dihydroisobenzofuran-5-carboxylate (**13**) (86%, 98.4% pure by GC) contaminated with 1.6% diethyl terephthalate.[37]

### 3.2.5 1-Oxo-1,3-dihydroisobenzofuran-5-carboxamides

Ethyl 1-oxo-1,3-dihydroisobenzofuran-5-carboxylate (**13**) is converted to amide **17** by reaction with ammonia in methanol (10 M) at 80°C for 20 h in a pressure reactor. The reactor is cooled to 25°C and vented and the mixture is quenched with ice. Hydrochloric acid is added (to pH 1). After aging for 2 h, presumably at 25°C, the suspension is filtered and the solid is washed with water and dried to afford 1-oxo-1,3-dihydroisobenzofuran-5-carboxamide (**17**) (93%) (Scheme 3.5).

Slow addition of a solution of the acid chloride **12** in THF to a dilute solution of ammonium hydroxide at 0–5°C also produces amide **17**. The resulting suspension is filtered and the solid is washed with water and dried (97%).[33]

Amides are also produced by reaction of the acid chloride **12** with *N-tert*-butyl amide, morpholine, and dimethylamine. A solution of the acid chloride **12** in THF is added to a solution of *tert*-butylamine (1.3 equivalents) and triethylamine (1.3 equivalents) in THF at 25°C. After aging for 1 h, the mixture is quenched with ice water. The suspension is filtered and the solid is washed with water and dried to afford *N-tert*-butyl-1-oxo-1,3-dihydroisobenzofuran-5-carboxamide (**18**) (91%). A similar result is observed using the crude acid chloride and *tert*-butylamine (2.1 equivalents) in place of the *tert*-butylamine–triethylamine combination (87%).[32,33]

A solution of the acid chloride **12** in THF is added to a solution of morpholine (1.3 equivalents) and triethylamine (1.3 equivalents) in THF at 0°C. The mixture is allowed to warm to 25°C over 1 h and then quenched with ice water. The THF is distilled at reduced pressure and the pH is adjusted to 2. The solution is cooled to 5°C and the resulting suspension is filtered. The solid is washed with water and dried to afford 5-(morpholine-4-carbonyl)isobenzofuran-1(3*H*)-one (**19**) (78%).[38]

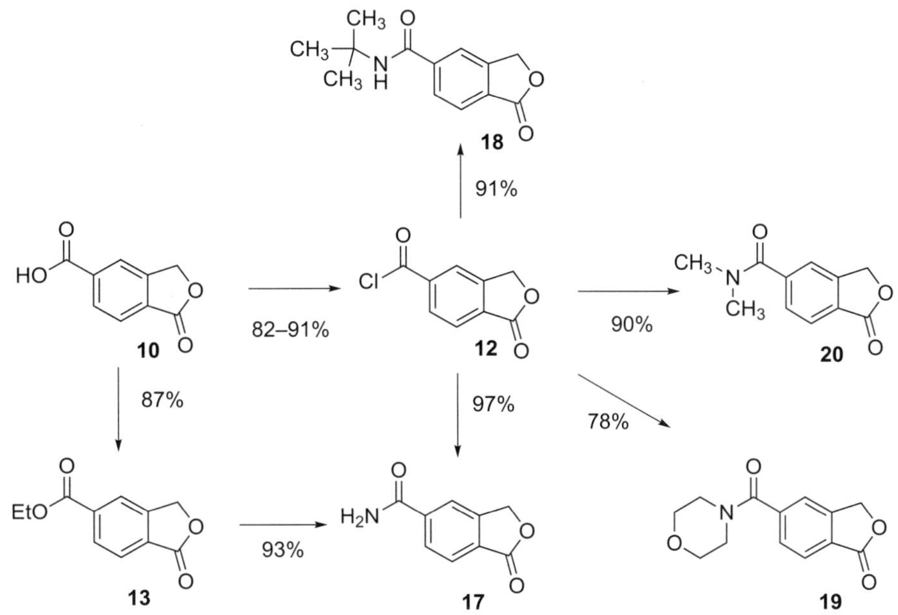


**SCHEME 3.5** Amides from 1-oxo-1,3-dihydroisobenzofuran-5-carboxylic acid (**10**) and ethyl ester **13**.

A solution of the acid chloride **12** in THF is added to a cold mixture of 40% aqueous dimethylamine and ice. The mixture is allowed to warm to 25°C over 1 h. The THF is distilled at reduced pressure. The resulting suspension is cooled to 5°C and filtered. The solid is washed with water and dried to afford *N,N*-dimethyl-1-oxo-1,3-dihydroisobenzofuran-5-carboxamide (**20**) (90%).[38]

### 3.2.6 5-(4,4-Dimethyl-4,5-dihydrooxazol-2-yl) isobenzofuran-1(*3H*)-one (21)

1-Oxo-1,3-dihydroisobenzofuran-5-carboxylic acid (**10**) is converted to an oxazoline in three steps: formation of the acid chloride **12** or active ester **22**, reaction with a β-hydroxylamine to produce amide **23**, and cyclization/dehydration of the amide. The amide can be prepared from the acid chloride **12** in THF or toluene–DMF, or from the active ester **22** in acetone. The amine is typically used in excess (2.2–3.4 equivalents) but anhydrous potassium carbonate can be added to reduce the amine charge. The oxazoline **21** is prepared from the amide **23** using thionyl chloride neat or in dichloromethane, THF, or acetone. While excess thionyl chloride is required for the neat reaction, as little as one equivalent is required when a solvent is used. The final isolation of the oxazoline **21** is often from water after adding base to pH 7–9. The yield for the three-step sequence is 61% when amide **23** and oxazoline **21** are isolated. The yield can be as high as 67% when the amide isolation is eliminated in a one-pot approach.

Thionyl chloride-free crude acid chloride **12** is dissolved in THF and the solution is added to a THF solution of 2-amino-2-methyl-1-propanol[39] (3.0 equivalents) at 5–10°C. After aging at 20°C for 2–20 h, the solvent is removed at 50°C and reduced pressure and the residue taken up in water. The water can also be added before the concentration. The suspension is filtered and the solid is washed with water and dried at 50°C and reduced pressure to afford the amide **23** (76–77% from acid **10**).[30,31]

In one oxazoline process, the oxazoline salt is isolated from toluene and the oxazoline **21** is isolated from water at pH 7–9. Amide **23** is added in portions to thionyl chloride (4.9 equivalents) at 0–10°C. The resulting mixture is aged at 28–30°C for 5 h. Excess thionyl chloride is distilled at 60°C and reduced pressure. Residual thionyl chloride is chased by suspension of the residue in toluene and distillation of the toluene at 60°C and reduced pressure. The residual solid is filtered, washed with toluene, and dried at an unspecified temperature and reduced pressure. The solid is suspended in water, 28% aqueous ammonia is added (to pH >7), and the suspension is filtered. The solid is washed with water and dried at 50°C and reduced pressure to afford 5-(4,4-dimethyl-4,5-dihydrooxazol-2-yl)isobenzofuran-1(3*H*)-one (**21**) (78%) (Scheme 3.6).[30]

In a similar oxazoline process, the oxazoline salt is isolated from THF and the oxazoline **21** is isolated from water at pH 7–9. The amide **23** is added in portions to thionyl chloride (5.2 equivalents) at −10 to −5°C. The resulting mixture is aged at −10 to −5°C for 1.5 h and at 25°C overnight, and then cooled to 0°C and diluted with THF at <8°C. The resulting suspension is aged at 5°C for 2 h and filtered. The solid is washed with THF and dissolved in water. Aqueous ammonia (25%) is added to pH 9. The resulting suspension is filtered and the solid is washed with water and dried at 50°C and reduced pressure to afford

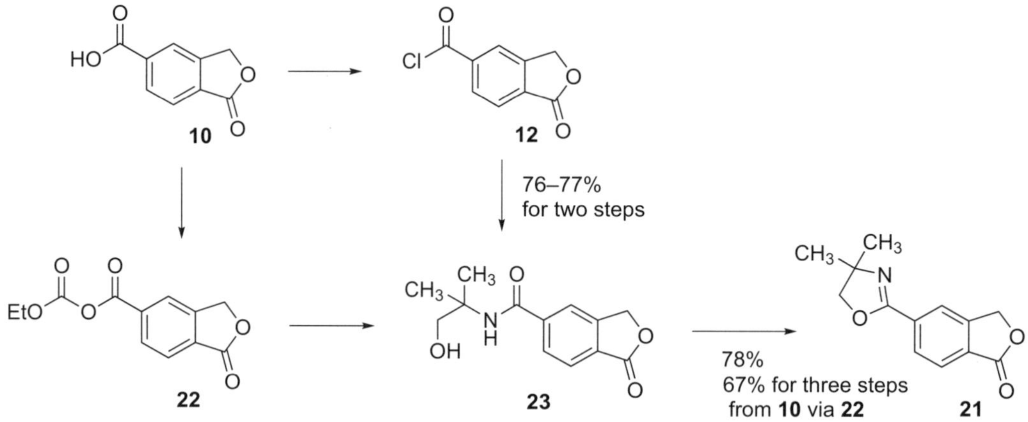


**SCHEME 3.6** 5-(4,4-Dimethyl-4,5-dihydrooxazol-2-yl)isobenzofuran-1(3*H*)-one (**21**) from 1-oxo-1,3-dihydroisobenzofuran-5-carboxylic acid (**10**).

5-(4,4-dimethyl-4,5-dihydrooxazol-2-yl)isobenzofuran-1(3*H*)-one (**21**) (80%).[31]

In a third process, the amide **23** is prepared in DMF–toluene and the oxazoline **21** is isolated from toluene–heptane. The crude acid chloride **12** in toluene is diluted with DMF at 75°C. The resulting solution is cooled to 20°C and added to a solution of 2-amino-2-methyl-1-propanol (2.2 equivalents) in toluene at 15–20°C. The mixture is aged at 15–20°C for 30 min. Hydrochloric acid is added (to pH 2–3) and the suspension aged at 15–20°C for 1 h. The solid is filtered, washed with water, and dried at 60°C and reduced pressure to afford the amide **23**. No yield is available. Amide **23** is then suspended in dichloromethane and thionyl chloride (1.1 equivalents) is added at 20–25°C. After aging for 1 h at 20–25°C, excess thionyl chloride is quenched with methanol. Triethylamine is added to neutralize the hydrogen chloride (to pH 4–5) and water is added. The layers are separated and the dichloromethane is distilled. Toluene is added to the residue and the suspension is heated to 60–70°C and filtered. The liquors are diluted with heptane, refluxed to produce a solution, and cooled to 15–20°C. The suspension is filtered and the solid is washed with heptane and dried at 50°C and reduced pressure to afford 5-(4,4-dimethyl-4,5-dihydrooxazol-2-yl)isobenzofuran-1(3*H*)-one (**21**). No yield is available.[26]

In a one-pot process, the amide **23** and oxazoline **21** are both prepared in THF, the amide isolation is eliminated, and the oxazoline **21** is isolated from water at pH 5. A solution of the crude acid chloride **12** in THF is added to a suspension of 2-amino-2-methyl-1-propanol (1.1 equivalents) and micronized anhydrous potassium carbonate (1.3 equivalents) in THF at 5–10°C. The amide formation is complete in 30 min. The suspension is cooled to 0–3°C and thionyl chloride (0.96 equivalents) is added. The reaction is quenched by adding water at 5–10°C. The THF is distilled at reduced pressure and 25% aqueous ammonia is added (to pH 5). The suspension is aged at 50°C for 1 h, at 20°C for 2 h, and at 10–15°C for 1 h and filtered. The solid is washed with water and dried at 40°C and reduced pressure to afford 5-(4,4-dimethyl-4,5-dihydrooxazol-2-yl)isobenzofuran-1(3*H*)-one (**21**) (60%).[30]

In a second one-pot process, the amide **23** is prepared from active ester **22**, amide **23** and oxazoline **21** are both prepared in acetone, and the oxazoline **21** is isolated from water at pH 7. Ethyl chloroformate (1.2 equivalents) is added to the carboxylic acid **10** in acetone at −10°C. Triethylamine (1.1 equivalents) in acetone is added at <−10°C. The mixture is allowed to warm to 10–13°C and aged for 30 min at 10–13°C. The mixture is cooled to −10°C and 2-amino-2-methyl-1-propanol (3.4 equivalents) is added. The mixture is allowed to warm to 15–20°C. The mixture is cooled to −5°C and thionyl chloride (3.4 equivalents) is added. The mixture is allowed to warm to 20°C and then concentrated at reduced pressure. Residual acetone is removed by suspending the residue in water and again concentrating the suspension at reduced pressure. More water is added, aqueous 25% ammonia is added (to pH >7), and the suspension is cooled to 5°C. The precipitate is filtered, washed with water, and dried at an unspecified temperature and reduced pressure to afford 5-(4,4-dimethyl-4,5-dihydrooxazol-2-yl)isobenzofuran-1(3*H*)-one (**21**) (67%).[30]

### 3.2.7 1-Oxo-1,3-dihydroisobenzofuran-5-carbonitrile (24)

All the efforts to produce a 1,2,4-trisubstituted aromatic component can coalesce on 1-oxo-1,3-dihydroisobenzofuran-5-carbonitrile (**24**). The amino group of 5-aminoisobenzofuran-1(3*H*)-one (**2**) might be converted to the nitrile by diazotization and Sandmeyer reaction with copper cyanide. The alternative cyanide exchange reaction of 5-bromoisobenzofuran-1(3*H*)-one (**5**) with copper cyanide or zinc cyanide might avoid the undesirable diazonium salt intermediate. Conversion of 1-oxo-1,3-dihydroisobenzofuran-5-carboxylic acid (**10**) to the nitrile can be accomplished in a single step, in two steps via the acid chloride **12**, in three steps via the acid chloride **12** and amide **17** or the ester **13** and amide **17**, or in four steps via the acid chloride **12**, amide **23**, and oxazoline **21**. 1-Oxo-1,3-dihydroisobenzofuran-5-carbonitrile (**24**) can be recrystallized from toluene or acetic acid.[40,41]

#### *3.2.7.1 From 5-Bromoisobenzofuran-1(3H)-one (5)*

1-Oxo-1,3-dihydroisobenzofuran-5-carbonitrile (**24**) is produced from 5-bromoisobenzofuran-1(3*H*)-one (**5**) and copper cyanide at high temperature (140–150°C). The conversion is more efficient using zinc cyanide and a palladium catalyst at 75°C. Nontoxic potassium hexacyanoferrate(III) can replace zinc cyanide in a palladium-catalyzed process.

A mixture of 5-bromoisobenzofuran-1(3*H*)-one (**5**) and copper(I) cyanide[42] (1.0 equivalent) (*Note*: The procedure reads $Cu(CN)_2$.) in NMP is heated at 140°C for 3 h. The NMP is distilled at reduced pressure. Water is added to the residue and the suspension is refluxed for a few minutes and cooled to 25°C. The suspension is filtered and the solid is presumably washed with water and dried to afford 1-oxo-1,3-dihydroisobenzofuran-5-carbonitrile (**24**) (66%, 97% pure by HPLC).[43]

A mixture of 5-bromoisobenzofuran-1(3*H*)-one (**5**) and sodium cyanide (1.0–2.0 equivalents), copper(I) iodide (5–30 mol%), potassium iodide (1.5–3.0 equivalents), and *N,N*-dimethylethylenediamine (1.0–1.5 equivalents) in an alkylbenzene is heated at 100–150°C for 20–48 h. The solvent is distilled at reduced pressure. Water is added to the residue and the suspension is filtered. The solid is washed with water and recrystallized from ethanol to afford 1-oxo-1,3-dihydroisobenzofuran-5-carbonitrile (**24**).[44]

Tetrakis(triphenylphosphine)palladium (4.8 mol%) is added to a degassed mixture of 5-bromoisobenzofuran-1 (3*H*)-one (**5**) and zinc cyanide[45] (1.0 equivalent) in DMF. After heating at 75°C for 3 h, DMF is distilled at reduced pressure and the residue is poured into water. The suspension is filtered and the solid is presumably washed with water and dried to afford 1-oxo-1,3-dihydroisobenzofuran-5-carbonitrile (**24**) (88%, 95% pure by HPLC). Comparable results are observed using sodium cyanide (1.0 equivalent) and a catalytic amount (13 mol%) of zinc cyanide (85%, 94% pure by HPLC) (Scheme 3.7).[43]

Nontoxic potassium hexacyanoferrate(III) is an attractive metal cyanide for the palladium-catalyzed cyanide exchange. A mixture of 5-bromoisobenzofuran-1(3*H*)-one (**5**), potassium hexacyanoferrate(III)[46] (0.15–0.30 equivalents), a palladium catalyst (0.1–5 mol%), and an alkali metal carbonate (0.5–2.0 equivalents) in *N,N*-dimethylacetamide is heated at 100–150°C for 1–6 h. *N,N*-Dimethylacetamide is distilled at reduced pressure and the residue is suspended in water. The suspension is filtered and the solid is washed with water and recrystallized from ethanol to afford 1-oxo-1,3-dihydroisobenzofuran-5-carbonitrile (**24**).[47]

#### *3.2.7.2 From 5-Iodoisobenzofuran-1(3H)-one (25)*

Tetrakis(triphenylphosphine)palladium (4.8 mol%) is added to a degassed mixture of 5-iodoisobenzofuran-1(3*H*)-one (**25**) and zinc cyanide (1.0 equivalent) in DMF. After heating at 75°C for 3 h, DMF is distilled at reduced pressure and the residue is poured into water. The suspension is filtered and the solid is presumably washed with water and dried to afford crude 1-oxo-1,3-dihydroisobenzofuran-5-carbonitrile (**24**) (75%, 93% pure by HPLC).[43]

> Most polar aprotic solvents used in pharmaceutical manufacturing are high boiling (DMF, 153°C; DMA, 164.5–166°C; NMP, 202°C; DMPU, 146°C at 44 mmHg). When designing a process in a polar aprotic solvent, use the minimum amount of solvent and work up the reaction by diluting with water and extracting with a low-polarity organic solvent (toluene, *n*-heptane). Recovery of the polar aprotic solvent by distillation at reduced pressure prior to the water–organic separation will require high vacuum and/or high batch temperature for a prolonged time in a manufacturing process. The high vacuum may be difficult to attain or maintain and the high temperature may cause some degradation of the product(s).

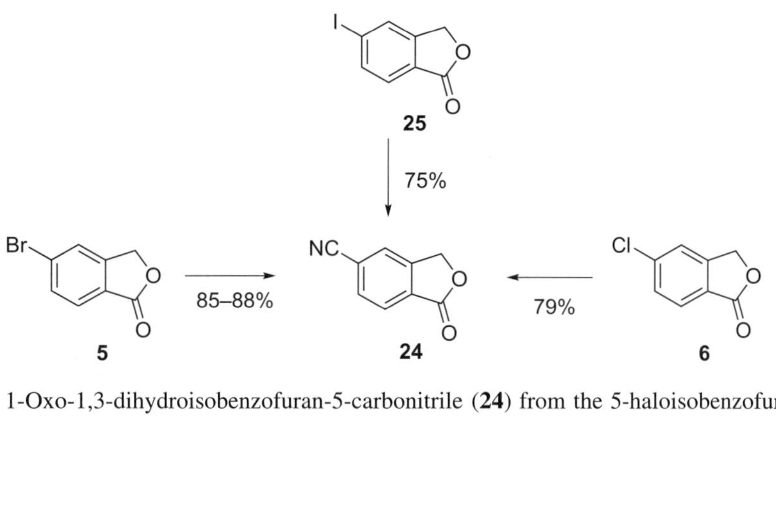


**SCHEME 3.7** 1-Oxo-1,3-dihydroisobenzofuran-5-carbonitrile (**24**) from the 5-haloisobenzofuran-1(3*H*)-ones.

***3.2.7.3 From 5-Chloroisobenzofuran-1(3H)-one (6)*** The chloride exchange requires a nickel-phosphine catalyst that is conveniently generated *in situ*. A mixture of nickel(II) chloride (7.5 mol%), triphenylphosphine (31 mol%), and acetonitrile is refluxed for 45 min and cooled to 25°C. Zinc powder (30 mol%) is added and the suspension is aged for 15 min. A solution of 5-chloroisobenzofuran-1(3*H*)-one (**6**) in THF is added and the suspension is aged for 10 min. Sodium cyanide (1.1 equivalents) is added and the suspension is heated at 70°C for 3 h. After cooling, presumably to 25°C, the suspension is diluted with acetonitrile and filtered. The liquors are concentrated at reduced pressure. Water is added to the residue and the suspension is refluxed and cooled to 25°C. The suspension is filtered and the solid is washed with water and dried to afford 1-oxo-1,3-dihydroisobenzofuran-5-carbonitrile (**24**) (79%).[43]

***3.2.7.4 From 1-Oxo-1,3-dihydroisobenzofuran-5-carbonyl Chloride (12)*** A mixture of 1-oxo-1,3-dihydroisobenzofuran-5-carbonyl chloride (**12**) and expensive sulfamide[48] (1.2 equivalents) in sulfolane is heated at 135°C for 3 h. The solution is allowed to cool, presumably to 85–90°C, and quenched with water. The solution is aged at 85–90°C for 5 min and then cooled to 60°C. The resulting suspension is filtered and the solid is washed with water and with acetic acid and dried to afford 1-oxo-1,3-dihydroisobenzofuran-5-carbonitrile (**24**) (96%) (Scheme 3.8).[49]

A solution of 1-oxo-1,3-dihydroisobenzofuran-5-carbonyl chloride (**12**) in (presumably) THF is added to a mixture of hydroxylamine hydrochloride[50] (2.3 equivalents) and triethylamine (2.3 equivalents) in THF at 10°C. After aging for 1 h, presumably at 10°C, the volatiles are removed by distillation at reduced pressure to afford the *N*-hydroxycarboxamide **26** (92%, 99.2% pure by HPLC). A mixture of *N*-hydroxycarboxamide **26** and thionyl chloride (21 equivalents) is refluxed for 6 h. Toluene is added and excess thionyl chloride and toluene are distilled at reduced pressure. The residue is suspended in toluene at reflux. The suspension is presumably cooled to 25°C and filtered. The solid is presumably washed with toluene and dried to afford 1-oxo-1,3-dihydroisobenzofuran-5-carbonitrile (**24**) (91% or 84% for the two steps from acid chloride **12**).[29]

The *N*-hydroxycarboxamide **26** can also be produced by adding the solution of acid chloride **12** (presumably in THF) to aqueous hydroxylamine[51] at 10°C over 1 h. After aging overnight, presumably at 25°C, the suspension is filtered and the solid is washed with water and dried to afford the *N*-hydroxycarboxamide **26** (92%, 99% pure by HPLC).[29]

***3.2.7.5 From 1-Oxo-1,3-dihydroisobenzofuran-5-carboxylic Acid (10)*** The conversion of the carboxylic acid to the acid chloride **12** and reaction of acid chloride **12** with sulfamide can be carried out in a one-pot process. Thionyl chloride (1.2 equivalents) is added to 1-oxo-1,3-dihydroisobenzofuran-5-carboxylic acid (**10**) and sulfamide (1.1 equivalents) in sulfolane, presumably at 25°C. The mixture is heated at 135–140°C for 2 h. The solution is allowed to cool, presumably to 85–90°C, and quenched with water. The solution is aged at 85–90°C for 15 min and then cooled to 35°C. The resulting suspension is filtered and the solid is washed with water, recrystallized from acetic acid, and dried to afford 1-oxo-1,3-dihydroisobenzofuran-5-carbonitrile (**24**) (77%).[49]

The nitrile **24** can also be produced using sulfamide starting with water-wet carboxylic acid **10**. The toluene–water azeotrope and toluene are distilled from a mixture of water-wet carboxylic acid in sulfolane and toluene. The resulting mixture is presumably cooled to 25°C and sulfamide (1.2 equivalents) and thionyl chloride (1.4 equivalents) are added. The mixture is heated at 135–140°C for 5 h. The solution is allowed to cool, presumably to 85–90°C, and quenched with water. The solution is aged at 85–90°C for 15 min and then cooled to 35°C. The resulting suspension is filtered and the solid is washed with water, recrystallized

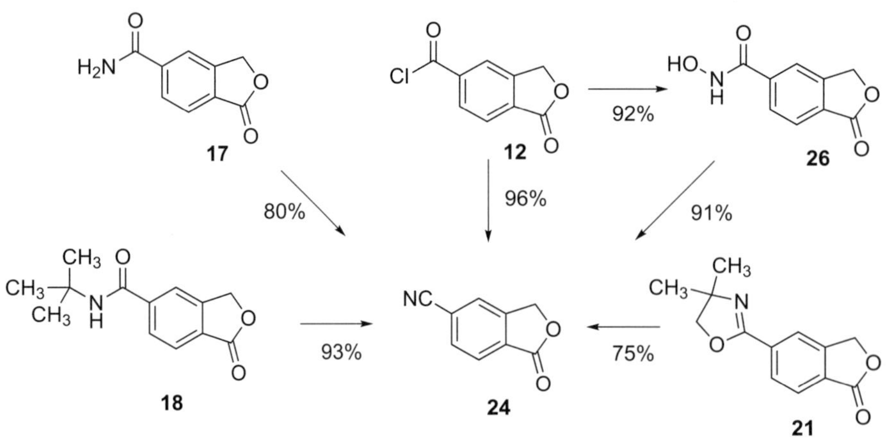


**SCHEME 3.8** 1-Oxo-1,3-dihydroisobenzofuran-5-carbonitrile (**24**) from amides, oxazolidine **21**, and acid chloride **12**.

from acetic acid, and dried to afford 1-oxo-1,3-dihydroisobenzofuran-5-carbonitrile (**24**) (68%).[49]

The reaction of 1-oxo-1,3-dihydroisobenzofuran-5-carboxylic acid (**10**) with ammonium carbonate and a dehydrating agent is a one-pot approach to nitrile **24** via carboxamide **17**. Ammonium carbonate (2.9–3.3 equivalents) is added over 3 h to a mixture of 1-oxo-1,3-dihydroisobenzofuran-5-carboxylic acid (**10**) and ethyl polyphosphate (5–10 kg per kg carboxylic acid) in acetonitrile at 0–3°C. The mixture is allowed to warm to 25°C and then aged at 25°C for 3 h. Acetonitrile is distilled until the pot temperature reaches 95°C, and then the mixture is aged at 95°C for 12 h. Water is added at an unspecified temperature and the suspension is cooled to an unspecified temperature and filtered. The solid is washed with water and dried to afford 1-oxo-1,3-dihydroisobenzofuran-5-carbonitrile (**24**) (93–95%, 98.2–99.2% pure by GC). The dehydrating agent, ethyl polyphosphate, is best prepared by reaction of phosphorus pentoxide with dry ethyl ether in refluxing alcohol-free chloroform over 4 days. The clear solution is decanted and concentrated on a rotary evaporator at 40°C over 36 h.[52,53]

When more accessible drying agents such as thionyl chloride, phosphorus oxychloride, phosphorus trichloride, or phosphorus pentachloride are used, the conversion of ammonium 1-oxo-1,3-dihydroisobenzofuran-5-carboxylate (**27**) to the nitrile **24** in a suitable solvent requires 0.5–4 h at 90–150°C.[54]

***3.2.7.6 From 1-Oxo-1,3-dihydroisobenzofuran-5-carboxamide (17)*** Thionyl chloride (1.5 equivalents) is added to a suspension of 1-oxo-1,3-dihydroisobenzofuran-5-carboxamide (**17**) in toluene. A DMF catalyst (0.13 equivalents) is added and the mixture is heated at 75°C for 6 h. The bulk of the toluene is distilled and the remaining solution is cooled to 25°C. The precipitate is filtered, washed with toluene and water, and recrystallized from toluene (80%).[33]

***3.2.7.7 From N-tert-Butyl-1-oxo-1,3-dihydroisobenzofuran-5-carboxamide (18)*** A suspension of *N-tert*-butyl-1-oxo-1,3-dihydroisobenzofuran-5-carboxamide (**18**) in thionyl chloride (13.7 equivalents) is refluxed for 30 min. Toluene is added and excess thionyl chloride and toluene are distilled. The residue is recrystallized from toluene (93%).[33]

***3.2.7.8 From 5-(4,4-Dimethyl-4,5-dihydrooxazol-2-yl)isobenzofuran-1(3H)-one (21)*** A DMF catalyst (13 mol %) is added to a suspension of 5-(4,4-dimethyl-4,5-dihydrooxazol-2-yl)isobenzofuran-1(3*H*)-one (**21**) and thionyl chloride (4.9 equivalents) and the solution is refluxed for 1 h. After cooling to 25°C, toluene is added and the suspension is filtered. The solid is washed with toluene and suspended in water. Aqueous ammonia (25%) is added to pH 8.0. The suspension is filtered and the solid is washed with water and dried at 60°C and reduced pressure to afford 1-oxo-1,3-dihydroisobenzofuran-5-carbonitrile (**24**) (75%). Analytical data suggest that this crude nitrile should be recrystallized from acetic acid or toluene to meet purity specifications.[31]

***3.2.7.9 Other Routes to 1-Oxo-1,3-dihydroisobenzofuran-5-carbonitrile (24)*** Almost every process for 1-oxo-1,3-dihydroisobenzofuran-5-carbonitrile (**24**) presented thus far uses either a metal cyanide or thionyl chloride. The use of a metal cyanide reagent results in an undesirable aqueous cyanide waste stream that also contains a heavy metal. Thionyl chloride is highly corrosive. Vapors cause severe irritation to the skin, eyes, and respiratory tract. The airborne exposure limit (ACGIH threshold limit value) for thionyl chloride is 1 ppm (ceiling).

One nitrile route that requires neither thionyl chloride nor a metal cyanide begins with bromination of 2,4-dimethylbenzoic acid.[55] 2,4-Bis(bromomethyl)benzoic acid (**27**) is cyclized to 5-(bromomethyl)isobenzofuran-1(3*H*)-one (**28**). Sommelet reaction with hexamethylenetetramine in acetic acid–water at reflux for 2 h affords 1-oxo-1,3-dihydroisobenzofuran-5-carbaldehyde (**29**) (81%). Aldehyde **29** is converted to oxime **30** by reaction with hydroxylamine hydrochloride and triethylamine in toluene at 65°C over 1 h. Dehydration of oxime **30** with acetic anhydride at 120–125°C over 3 h affords 1-oxo-1,3-dihydroisobenzofuran-5-carbonitrile (**24**) (69%) (Scheme 3.9).[56]

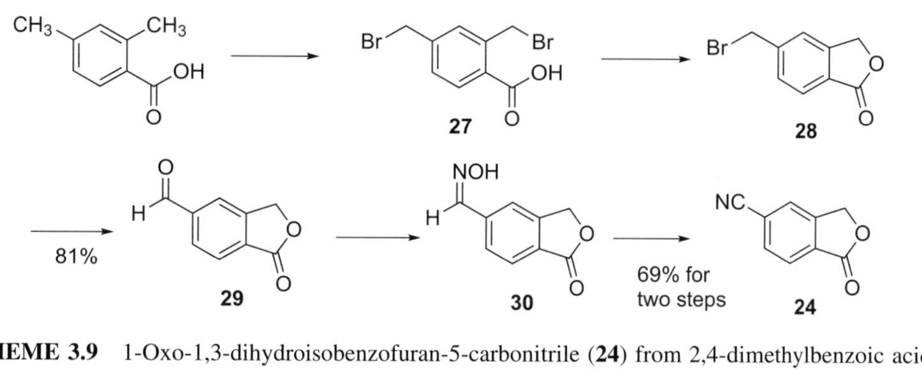


**SCHEME 3.9** 1-Oxo-1,3-dihydroisobenzofuran-5-carbonitrile (**24**) from 2,4-dimethylbenzoic acid.

## 3.3 CONSTRUCTION OF CITALOPRAM BY ALKYLATION OF 1-(4-FLUOROPHENYL)-1,3-DIHYDROISOBENZOFURAN-5-CARBONITRILE (31)

Strategies for converting a 5-substituted 1-oxo-1,3-dihydroisobenzofuran to citalopram introduce the 4-fluorophenyl substituent by addition of 4-fluorophenylmagnesium bromide to the oxo group at the 1-position (Scheme 3.10). *The Grignard reagent addition to produce a ketone is successful because the initial adduct is slow to ring open and reveal the ketone*. In one strategy, the ketone is reduced and the diol cyclized. The 3-dimethylaminopropyl substituent is then introduced by generation of an anion at the 1-position and alkylation with 3-chloro-*N*,*N*-dimethylpropan-1-amine.

4-Fluorophenylmagnesium bromide is typically used as a 15–20 wt% solution in THF. A suspension of magnesium turnings (1.05 equivalents) in THF is prepared and the magnesium is activated with iodine (<0.1 mol%). The suspension is heated to reflux and a solution of 1-bromo-4-fluorobenzene[57] in THF is added at reflux over 1 h. The mixture is aged for 30 min, and then cooled to 25°C and stored in the dark under dry nitrogen.[58]

### 3.3.1 1-(4-Fluorophenyl)-1,3-dihydroisobenzofuran-5-carbonitrile (31)

A large body of process research has targeted 1-(4-fluorophenyl)-1,3-dihydroisobenzofuran-5-carbonitrile (**31**) as a key intermediate. It can be prepared by the Grignard reagent addition to 5-bromoisobenzofuran-1(3*H*)-one (**5**), reduction of the ketone **33**, cyclization of the diol **34**, and a bromide–cyanide exchange. It can also be prepared in a parallel sequence by Grignard reagent addition to 1-oxo-1,3-dihydroisobenzofuran-5-carbonitrile (**24**), reduction of the ketone **37**, and cyclization of the diol **38**. A competing second Grignard reagent addition to ketone **33** or **37** limits the efficiency of these processes.

Processes that do not use the Grignard reagent addition to an isobenzofuran-1(3*H*)-one proceed via (1-(4-fluorophenyl)-1,3-dihydroisobenzofuran-5-yl)methanol (**40**). This is converted to 1-(4-fluorophenyl)-1,3-dihydroisobenzofuran-5-carbonitrile (**31**) by oxidation of the alcohol to the aldehyde **41**, oxime **42** formation, and dehydration. The processes either start with a commercially available and inexpensive aromatic core, such as *m*-xylene or trimellitic anhydride, and utilize Friedel–Crafts acylation or start with a value-added aromatic core, such as 1,3-bis(1-ethoxyethoxy)methyl)benzene (**43**), and utilize halogenation, halogen–metal exchange, and aldehyde condensation chemistry.

#### *3.3.1.1 From 5-Bromoisobenzofuran-1(3H)-one (5)*

A 33 wt% solution of 4-fluorophenylmagnesium bromide (1.4 equivalents) in THF is added to a suspension of 5-bromoisobenzofuran-1(3*H*)-one (**5**) in dichloromethane at −6 to −2°C. After an unspecified aging time, presumably at −6 to −2°C, the reaction is quenched with 20% aqueous ammonium chloride. The layers are separated. The organic layer is diluted with methanol and sodium borohydride (0.69 equivalents) is added at <25°C over 1 h. After aging at 25°C for 1 h, excess hydride is quenched by adding water and the layers are separated. The organic layer is washed with 10% hydrochloric acid and with water and then dried and concentrated at reduced pressure to afford crude diol **34**. Cyclization of the diol with *p*-toluenesulfonic acid (8.2 mol%) in toluene at reflux is driven by collection of water from the distillate using a Dean–Stark apparatus. After collecting the expected amount of water, the mixture is cooled to 25°C, washed with 10% aqueous sodium hydroxide and with water, and then dried and concentrated at reduced pressure to afford 5-bromo-1-(4-fluorophenyl)-1,3-dihydroisobenzofuran (**35**)

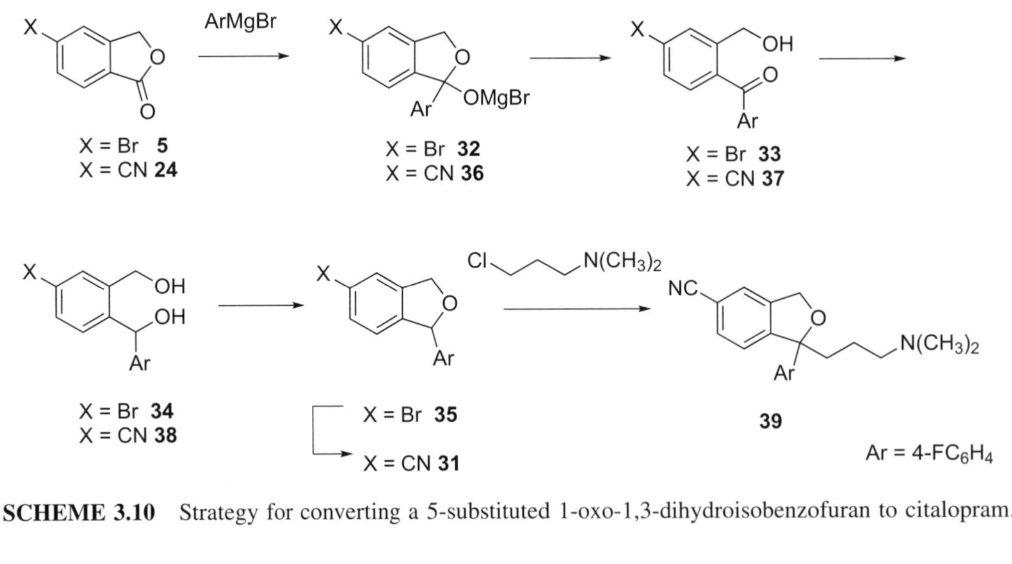


**SCHEME 3.10** Strategy for converting a 5-substituted 1-oxo-1,3-dihydroisobenzofuran to citalopram.

(69–73%, 90–92% pure by HPLC). Similar results (65–73%, 92% pure by HPLC) are observed using dichloroethane or chloroform as the cosolvent in place of dichloromethane. Lower yields and purities (56–58%, 78–83% pure by HPLC) are observed using toluene or chlorobenzene as the cosolvent or using no cosolvent.[59]

A boron complex **44** of the diol can be isolated. A 33 wt% solution of 4-fluorophenylmagnesium bromide (1.4 equivalents) in THF is added to a suspension of 5-bromoisobenzofuran-1(3*H*)-one (**5**) in dichloromethane at −6 to −2°C. After an unspecified aging time, presumably at −6 to −2°C, the reaction is quenched with 20% aqueous ammonium chloride. The layers are separated. The organic layer is diluted with methanol and sodium borohydride (0.69 equivalents) is added at <25°C over 1 h. After aging at <25°C for 1 h, the solvents are distilled at reduced pressure. The residue is suspended in hexane. The suspension is filtered and the solid is presumably dried to afford boron complex **44** (51–54%, 98% pure by HPLC). Cyclization to the isobenzofuran with dilute hydrochloric acid at 65–70°C is complete in 4–5 h. The mixture is cooled and separated between water and toluene. The water layer is extracted with toluene. The toluene extracts are concentrated at reduced pressure to afford 5-bromo-1-(4-fluorophenyl)-1,3-dihydroisobenzofuran (**35**) (92–98%, 98% pure by HPLC).[59]

Quench of the Grignard reagent addition with aqueous ammonium chloride before proceeding to the borohydride reduction is not necessary. A 20 wt% solution of 4-fluorophenylmagnesium bromide (1.1 equivalents) in THF is added to a suspension of 5-bromoisobenzofuran-1(3*H*)-one (**5**) in THF at −10 to 0°C. The mixture is aged at −10 to 0°C for 3 h and then a suspension of sodium borohydride (1.4 equivalents) in ethanol is added at <10°C. The mixture is aged at 10°C for 1 h and quenched with dilute hydrochloric acid. The suspension is aged at 25°C for 30 min and the layers are separated. The aqueous layer is extracted with toluene and the combined organic layers are washed with brine, dried, and concentrated at <60°C and reduced pressure to afford crude diol **34**. Cyclization of the diol with *p*-toluenesulfonic acid (8.2 mol%) in toluene at reflux is driven by collection of water from the distillate using a Dean–Stark apparatus. After collecting the expected amount of water, the mixture is cooled to 25°C and washed with water. The water wash is back-extracted with toluene. The combined organic layers are washed with 5% aqueous sodium carbonate and with brine and then concentrated at <60°C and reduced pressure to afford crude 5-bromo-1-(4-fluorophenyl)-1,3-dihydroisobenzofuran (**35**). Purity upgrade by distillation (bp 170–175°C at 1 mmHg) is an option in the lab.

Reaction of the crude bromide **35** with copper(I) cyanide (1.8 equivalents) in DMF is complete in 4–5 h at reflux. The mixture is cooled to 40–50°C and aqueous ammonia is added. The suspension is aged for 30 min, presumably at 25°C, and filtered. The liquor layers are separated. The organic layer is washed with 10% aqueous ammonia and the aqueous layer is back-extracted with toluene. The combined organic layers are concentrated at 50–60°C and reduced pressure to afford the crude nitrile. The crude nitrile is recrystallized from isopropanol and presumably dried to afford 1-(4-fluorophenyl)-1,3-dihydroisobenzofuran-5-carbonitrile (**31**) (75–89% from **5**, 98% pure by HPLC) (Scheme 3.11). Comparable or lower yields are reported using methanol or ethanol as the solvent for the sodium borohydride slurry transfer, benzenesulfonic acid or sulfuric acid as the cyclization catalyst, *N*,*N*-dimethylacetamide or pyridine as the solvent for the cyanide exchange, and methanol or isopropanol–DMF for the final crystallization.[60–62]

### *3.3.1.2 From 1-Oxo-1,3-dihydroisobenzofuran-5-carbonitrile (24)*

A 40 wt% solution of 4-fluorophenylmagnesium bromide (1.4 equivalents) in THF is added to a suspension of 1-oxo-1,3-dihydroisobenzofuran-5-carbonitrile (**24**) in dichloromethane at −6 to −2°C. After an unspecified aging time, presumably at −6 to −2°C, the reaction is quenched with 20% aqueous ammonium chloride. The

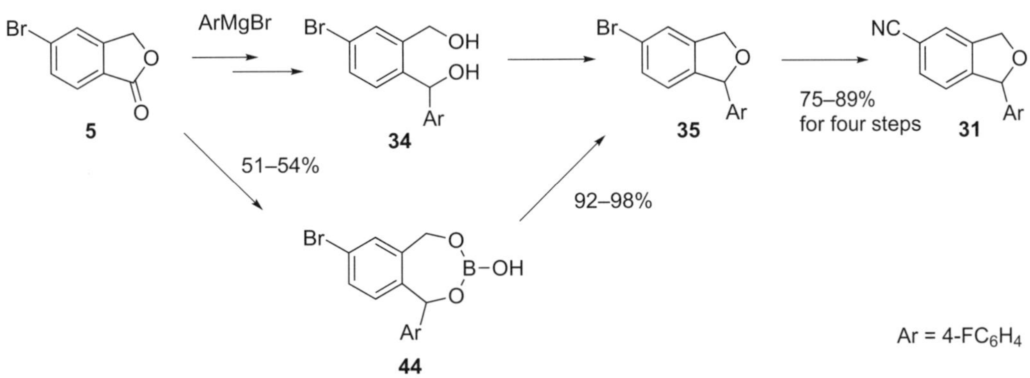


**SCHEME 3.11** 1-(4-Fluorophenyl)-1,3-dihydroisobenzofuran-5-carbonitrile (**31**) from 5-bromoisobenzofuran-1(3*H*)-one (**5**).

layers are separated. The organic layer is diluted with methanol and sodium borohydride (0.52 equivalents) is added at <25°C over 1 h. After aging at 25°C for 1 h, excess hydride is quenched by adding water and the layers are separated. The organic layer is washed with 10% hydrochloric acid and with water and then dried and concentrated at reduced pressure to afford crude diol **36**. Cyclization of the diol with *p*-toluenesulfonic acid (8.2 mol%) in toluene at reflux is driven by collection of water from the distillate using a Dean–Stark apparatus. After collecting the expected amount of water, the mixture is cooled to 25°C, washed with 10% aqueous sodium hydroxide and with water, dried, and concentrated at reduced pressure. The residue is suspended in isopropanol. The suspension is filtered and the solid is presumably washed with isopropanol and dried to afford 1-(4-fluorophenyl)-1,3-dihydroisobenzofuran-5-carbonitrile (**31**) (87–93%, 99.3% pure by HPLC) (Scheme 3.12). Similar results (85–86%, 99.1–99.4% pure by HPLC) are observed using dichloroethane or chloroform in place of dichloromethane as the cosolvent. Lower yields and purities (70–78%, 94–98% pure by HPLC) are observed using toluene or chlorobenzene as the cosolvent.[59]

A boron complex **45** of the diol can be isolated. A 40 wt% solution of 4-fluorophenylmagnesium bromide (1.4 equivalents) in THF is added to a suspension of 1-oxo-1,3-dihydroisobenzofuran-5-carbonitrile (**24**) in dichloromethane at −6 to −2°C. After an unspecified aging time, presumably at −6 to −2°C, the reaction is quenched with 20% aqueous ammonium chloride. The layers are separated. The organic layer is diluted with methanol and sodium borohydride (0.52 equivalents) is added at <25°C over 1 h. After aging at <25°C for 4–6 h and at 5–10°C for 2 h, the suspension is filtered. The solid is washed with cold dichloromethane and dried at <40°C and reduced pressure to afford boron complex **45** (65–68%, 98–99% pure by HPLC). Cyclization to the isobenzofuran with dilute hydrochloric acid at 65–70°C is complete in 4–5 h. The mixture is cooled and separated between water and toluene. The water layer is extracted with toluene. The combined toluene layers are washed with water and concentrated at reduced pressure to afford 1-(4-fluorophenyl)-1,3-dihydroisobenzofuran-5-carbonitrile (**31**) (98–99% pure by HPLC). The yield range provided (107–112%) exceeds the theoretical yield. In another patent from the same authors, the solid isolated from the borohydride reduction mixture, washed with water and dried at <40°C, is identified as the diol **36**.[59]

Quench of the Grignard reagent addition with aqueous ammonium chloride before proceeding to the borohydride reduction is not necessary. A 21 wt% solution of 4-fluorophenylmagnesium bromide (1.1 equivalents) in THF is added to a suspension of 1-oxo-1,3-dihydroisobenzofuran-5-carbonitrile (**24**) in THF at <5°C. The mixture is allowed to warm to 25°C and aged at 25°C overnight. The mixture is diluted with ethanol. Sodium borohydride (2.0 equivalents) is added in portions at 25°C and the mixture is aged at 25°C overnight. The mixture volume is reduced by 60–70% by distillation of THF and ethanol at reduced pressure. Water is added and the resulting solution extracted with ethyl acetate. The extracts are concentrated at reduced pressure to afford the crude diol. The diol is dissolved in 60% phosphoric acid and the solution is heated at 80°C for 3 h. The mixture is presumably cooled to 25°C and extracted with toluene. The extracts are concentrated at reduced pressure. The residue is recrystallized from 99% ethanol to afford 1-(4-fluorophenyl)-1,3-dihydroisobenzofuran-5-carbonitrile (**31**) (29%). Results from similar processes suggest that this low yield may be due to poor recovery in the crystallization from 99% ethanol.[63]

It is likely that the initial Grignard adduct **36** opens to the ketone **37** and that some double addition occurs. How are these processes designed to minimize the amount of the double addition side product **46**? How much side product **46** is formed? How is the side product **46** efficiently separated from the ketone **37**? These are questions addressed during process development.

Process development begins with the preparation of an authentic sample of the double addition product and the

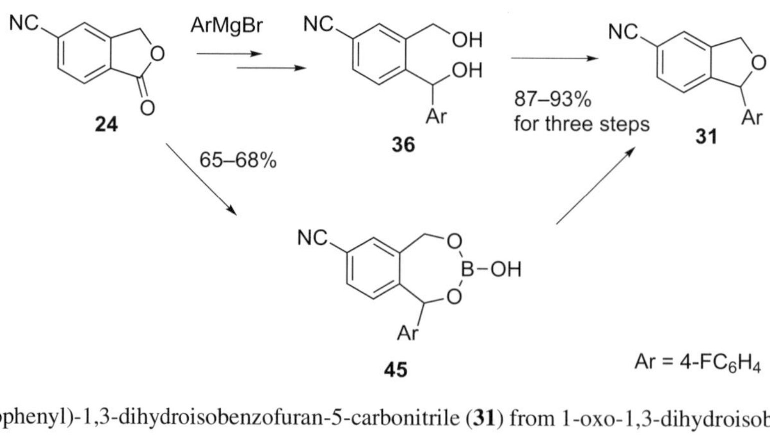


**SCHEME 3.12** 1-(4-Fluorophenyl)-1,3-dihydroisobenzofuran-5-carbonitrile (**31**) from 1-oxo-1,3-dihydroisobenzofuran-5-carbonitrile (**24**).

development of an analytical method for separating the desired product and the double addition product. 4-Fluorophenylmagnesium bromide in THF (20 wt%) (3.4 equivalents) is added to a suspension of 1-oxo-1,3-dihydroisobenzofuran-5-carbonitrile (**24**) in THF at 25–35°C. The aging time after the addition is complete is not specified. Quench with aqueous ammonium chloride, layer separation, and concentration of the organic layer at reduced pressure affords the crude double addition product **46**. The purity is upgraded by chromatography (68% from **24**, 97.4% pure by HPLC). With an authentic sample and the analytical method in hand, the next objectives for process development are to identify the lowest temperature and the minimum amount of Grignard reagent required to achieve complete conversion in <4–6 h.[58]

The temperature identified is −5 to −10°C. 4-Fluorophenylmagnesium bromide in THF (20 wt%) is added to a suspension of 1-oxo-1,3-dihydroisobenzofuran-5-carbonitrile (**24**) in THF at −5 to −10°C. The reaction is monitored during the Grignard reagent solution addition by withdrawing, quenching, and analyzing samples of the reaction mixture. Complete conversion is observed after 1.4 equivalents of Grignard reagent is added. The charges of Grignard reagent solution between samples are progressively smaller (0.68, 0.40, 0.20, and 0.09 equivalents) as the reaction approaches completion. The aging times after each charge and during the sample preparation and analysis are not specified. The reaction is quenched by adding 15% aqueous ammonium chloride. The phases are separated and the aqueous phase is extracted with THF. *The combined organic layers contain 14–16% of the double addition side product* **46**. Water, sodium borohydride (1.1 equivalents), and aqueous sodium hydroxide (1.7 mol%) are added at <15°C. After the ketone reduction is complete, the mixture is warmed to 25°C and the layers are separated. The organic layer is distilled at 50°C and reduced pressure. Ethyl acetate is added and the mixture is again concentrated at reduced pressure. The residual liquid is extracted with ethyl acetate and extracts containing diol **38** and double addition side product **46** are used in the next step.

Phosphoric acid (60%) is added and the biphasic mixture is refluxed for 2 h. The mixture is presumably cooled to 25°C, water is added, and the phases are separated. The aqueous layer is extracted with ethyl acetate. The organic layers are washed with brine, treated with carbon, and concentrated at 50°C and reduced pressure. The residue is suspended in methyl *tert*-butyl ether and the suspension is aged (25°C for 30 min and 0–5°C for 15 h) and filtered. The solid is washed with methyl *tert*-butyl ether and dried at 50°C and reduced pressure to afford 1-(4-fluorophenyl)-1,3-dihydroisobenzofuran-5-carbonitrile (**31**) (71%, 98.2–98.5% pure by HPLC). The yield is slightly lower (66%) using isopropanol in place of methyl *tert*-butyl ether for the final isolation.[58]

1-Oxo-1,3-dihydroisobenzofuran-5-carbonitrile (**24**) can be converted to 3-(chloromethyl)-4-(4-fluorobenzoyl)benzonitrile (**47**). Ketone reduction with sodium borohydride and *in situ* cyclization also affords 1-(4-fluorophenyl)-1,3-dihydroisobenzofuran-5-carbonitrile (**31**) (21%).[64]

***3.3.1.3 From (1-(4-Fluorophenyl)-1,3-dihydroisobenzofuran-5-yl)methanol (40)*** The poor selectivity in the Grignard reagent addition to the 1-oxo-1,3-dihydroisobenzofuran no doubt prompted the search for alternative routes of 1-(4-fluorophenyl)-1,3-dihydroisobenzofuran-5-carbonitrile (**31**). Friedel–Crafts acylation and oxidation methods are used to convert 2,4-dimethylbenzaldehyde, 2,4-dimethylbenzoyl chloride, *m*-xylene, or trimellitic anhydride to 4-(4-fluorobenzoyl)isophthalic acid (**48**). This key intermediate is reduced with borohydride and the resulting triol **49** is cyclized to afford (1-(4-fluorophenyl)-1,3-dihydroisobenzofuran-5-yl)methanol (**40**). (1-(4-Fluorophenyl)-1,3-dihydroisobenzofuran-5-yl)methanol (**40**) is also accessible from a diether derived from α,α′-dichloro-*m*-xylene by ring lithiation, addition of the aryllithium to 4-fluorobenzaldehyde, and deprotection/triol cyclization under acidic conditions. The conversion of the hydroxymethyl group of (1-(4-fluorophenyl)-1,3-dihydroisobenzofuran-5-yl)methanol (**40**) to the nitrile follows a now-familiar path: oxidation to the aldehyde **41**, conversion to oxime **42**, and dehydration of the oxime with acetic anhydride (Schemes 3.13 to 3.16).

*Routes to 4-(4-Fluorobenzoyl)isophthalic Acid (48)* Four routes to (1-(4-fluorophenyl)-1,3-dihydroisobenzofuran-5-yl)methanol (**40**) converge on 4-(4-fluorobenzoyl) isophthalic acid (**48**) as a key intermediate. In the first route, 4-fluorophenylmagnesium bromide in THF (1.1 equivalents) is added to 2,4-dimethylbenzaldehyde[65] in THF at 0–20°C. The mixture is aged at 0–20°C for 2 h and then quenched with saturated aqueous ammonium chloride. The layers are separated and the aqueous layer is extracted with toluene. The combined organic layers are washed with brine and concentrated at reduced pressure to afford (2,4-dimethylphenyl)(4-fluorophenyl)methanol (**50**) (100%). Oxidation with potassium permanganate (6.6–7.0 equivalents or 4.5–4.8 kg per kg **50**) affords 4-(4-fluorobenzoyl) isophthalic acid (**48**) (71–75%).[66]

In the second route, 4-fluorobenzoyl chloride[67] is added to a suspension of aluminum chloride (1.1 equivalents) in *m*-xylene[68] at 0–10°C. The mixture is aged at 0–10°C for 3 h and then quenched with 6 N hydrochloric acid. The layers are separated. The organic layer is washed with water, with 10% aqueous sodium hydroxide solution, and again with water, and concentrated at reduced pressure to afford (2,4-dimethylphenyl)(4-fluorophenyl)methanone (**51**) and (2,6-dimethylphenyl)(4-fluorophenyl)methanone (**52**) (99%, 96:4 mixture). Oxidation of the mixture with potassium

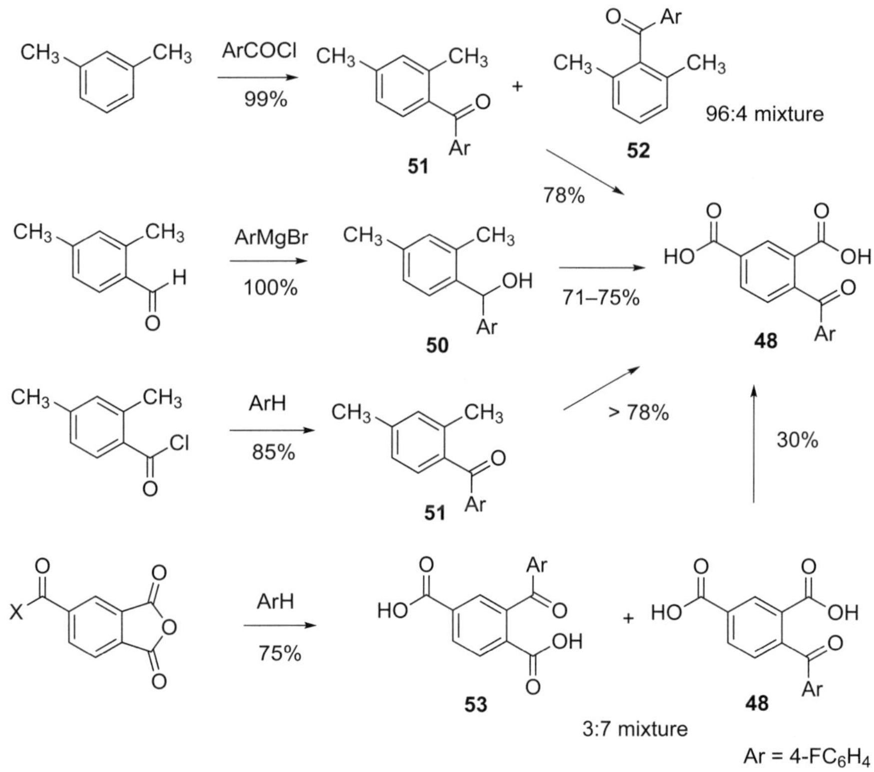


**SCHEME 3.13** Four routes to 4-(4-fluorobenzoyl)isophthalic acid (**48**).

permanganate (6.5 equivalents or 4.5 kg per kg substrate) affords 4-(4-fluorobenzoyl)isophthalic acid (**48**) (78%).[66]

In the third route, 2,4-dimethylbenzoyl chloride[69] is added to a suspension of fluorobenzene[70] (1.3 equivalents) and aluminum chloride (1.2 equivalents) in 1,2-dichlorobenzene at 0–20°C. The suspension is aged at 10–30°C for 1 h and at 80°C for 1 h. The suspension is cooled to 25°C and quenched with 6 N hydrochloric acid. Toluene is added and the layers are separated. The organic layer is washed with 5% aqueous sodium hydroxide solution and with water and concentrated at reduced pressure. Chromatography of the residue affords (2,4-dimethylphenyl)(4-fluorophenyl) methanone (**51**) (85%). Oxidation with potassium permanganate (6.5 equivalents or 4.5 kg per kg substrate) would presumably afford 4-(4-fluorobenzoyl)isophthalic acid (**48**) (>78% yield).[66]

In the fourth route, aluminum chloride (3.0 equivalents) is added to a mixture of trimellitic anhydride and fluorobenzene (1.8 equivalents) in 1,2-dichlorobenzene. The suspension is aged at 70–90°C for 8 h. The suspension is cooled to 25°C, quenched with 4 N hydrochloric acid, and extracted with methyl isobutyl ketone. The organic extracts are extracted with 5% sodium hydroxide solution. Hydrochloric acid (6 N) is added to the aqueous extract (to pH 7). The resulting suspension is filtered and the solid is washed with water and dried to afford 4-(4-fluorobenzoyl)isophthalic acid (**48**) and 2-(4-fluorobenzoyl)terephthalic acid (**53**) (75% yield, 7:3 mixture). 4-(4-Fluorobenzoyl)isophthalic acid (**48**) is isolated from the mixture by crystallization from 8:5 methanol–water (23% from trimellitic anhydride).[66]

Which route would be preferred? Oxidation with potassium permanganate is unattractive primarily because of the potential for operator exposure while charging potassium permanganate and handling the manganese dioxide byproduct (2–3 kg per kg product), which must be filtered, washed, transferred to a storage/shipping container, and

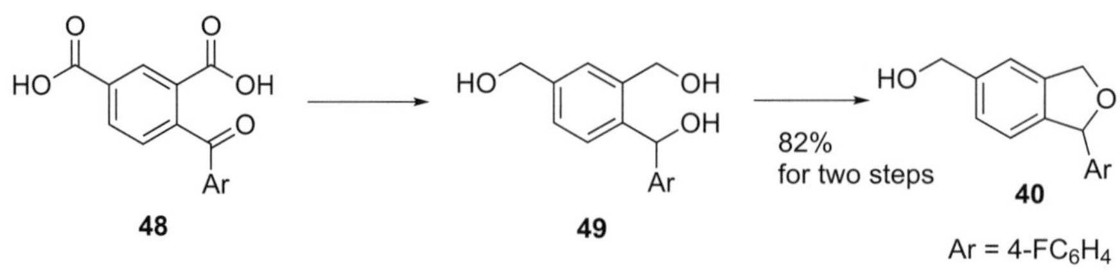


**SCHEME 3.14** (1-(4-Fluorophenyl)-1,3-dihydroisobenzofuran-5-yl)methanol (**40**) from 4-(4-fluorobenzoyl)isophthalic acid (**48**).

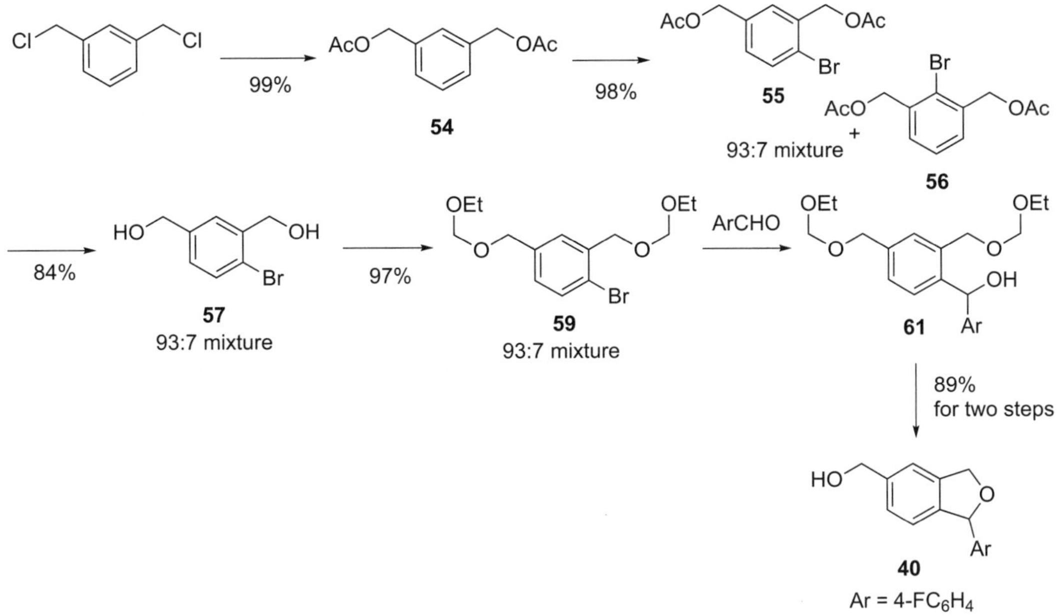


**SCHEME 3.15** (1-(4-Fluorophenyl)-1,3-dihydroisobenzofuran-5-yl)methanol (**40**) from $\alpha,\alpha'$-dichloro-*m*-xylene.

shipped for reprocessing. The OSHA permissible exposure limit (PEL) is 5 mg/m$^3$ ceiling for manganese compounds as manganese and the ACGIH threshold limit value is 0.2 mg/m$^3$ for elemental and inorganic compounds as manganese. Despite generation of a mixture in the Friedel–Crafts acylation and low overall yield, the last route is preferred since it has the two least expensive starting materials and requires no oxidation. Having selected this route, a more efficient separation of the 7:3 mixture of product **48** and side product **53** becomes an important process development priority.

*(1-(4-Fluorophenyl)-1,3-dihydroisobenzofuran-5-yl) methanol (40) from 4-(4-Fluorobenzoyl)isophthalic Acid (48)* The reduction of 4-(4-fluorobenzoyl)isophthalic acid (**48**) with sodium borohydride and boron trifluoride etherate is followed by cyclization with 30% sulfuric acid to produce (1-(4-fluorophenyl)-1,3-dihydroisobenzofuran-5-yl)methanol (**40**).

In a large-scale demonstration run, sodium borohydride (3.3 equivalents) is suspended in THF. Trimethyl borate (0.25 equivalents) is added at 20–30°C. (*Note*: The addition of trimethyl borate prevents gel formation.) A solution of 4-(4-fluorobenzoyl)isophthalic acid (**48**) in THF is added at 20–30°C. Boron trifluoride–THF complex (3.8 equivalents) is added at 35–42°C. The mixture is aged at 38–42°C for 3 h and at 48–50°C for 4 h. The mixture is cooled to 0–5°C and quenched by adding water at 0–25°C. The mixture is heated to 50–55°C and more water is added. The volume is reduced by distillation of THF at 50–85°C and atmospheric pressure. The mixture is cooled to 57°C and 30% sulfuric acid is

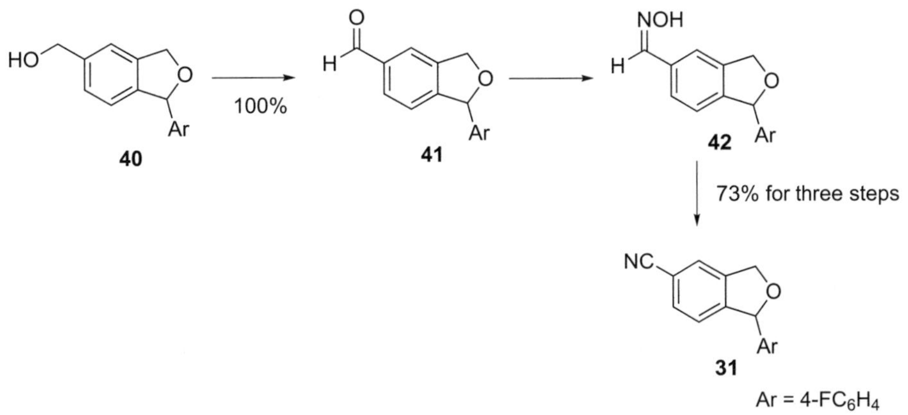


**SCHEME 3.16** 1-(4-Fluorophenyl)-1,3-dihydroisobenzofuran-5-carbonitrile (**31**) from (1-(4-fluorophenyl)-1,3-dihydroisobenzofuran-5-yl)methanol (**40**).

added. The mixture is aged at 60–65°C for 4 h. The mixture is cooled to 25°C and 25% aqueous sodium hydroxide is added. Toluene is added and the biphasic mixture is heated to 75–80°C. The layers are separated and the organic layer is washed with water at 70–80°C. Warm water is added to the organic layer, the biphasic mixture is cooled to 25–30°C, and heptane is added. The biphasic mixture is warmed to 40°C, slowly cooled to 5°C, and aged at 5°C for 1 h. The resulting suspension is filtered and the solid is washed with cold toluene–heptane and dried at 45°C and later at 60–70°C and reduced pressure to afford (1-(4-fluorophenyl)-1,3-dihydroisobenzofuran-5-yl)methanol (**40**) (82%). This process delivers 65 kg of **40** with a volume throughput of 50–55 g/L. The maximum volume is just prior to the THF distillation.[66]

> Gram-scale procedures may not accurately reflect the maximum throughput attainable. Kilogram-scale procedures are often developed to maximize the output in fixed equipment and do provide an estimate of the maximum throughput attainable. A rough volume throughput can be calculated for kilogram-scale processes by dividing the weight of product isolated by the total volume of solvents at the largest volume point in the procedure. Process development time should be invested to increase the volume throughput to at least 100 g/L.

*(1-(4-Fluorophenyl)-1,3-dihydroisobenzofuran-5-yl) methanol (40) from α,α′-Dichloro-m-xylene* An alternative route to (1-(4-fluorophenyl)-1,3-dihydroisobenzofuran-5-yl) methanol (**40**) starts with α,α′-dichloro-*m*-xylene. Both substituents on this starting material are in the correct oxidation state for the eventual target, eliminating the need for a permanganate oxidation or a hydride reduction.[71]

A mixture of α,α′-dichloro-*m*-xylene,[72] potassium acetate (2.4 equivalents), and benzyltriethylammonium chloride (15 mol%) in acetone is refluxed for 2.5 h. The suspension is cooled, presumably to 25°C, and filtered. The liquors are concentrated at reduced pressure. The residue is separated between toluene and water. The organic layer is washed with brine and concentrated at reduced pressure to afford 1,3-phenylenebis(methylene) diacetate (**54**) (99%).

Bromine (5.0 equivalents) is added at 15–20°C over 30 min to 1,3-phenylenebis(methylene) diacetate (**54**) and sodium acetate (5.0 equivalents) in glacial acetic acid. The mixture is aged at 20–30°C for 13 h. The mixture is quenched with 10% aqueous sodium sulfite with ice water cooling and extracted with ethyl acetate. The extracts are washed three times with 10% aqueous sodium bicarbonate and concentrated at reduced pressure to afford (4-bromo-1,3-phenylene)bis(methylene) diacetate (**55**) and (2-bromo-1,3-phenylene)bis(methylene) diacetate (**56**) (98%, 93:7 mixture).

The acetate esters of **55** and **56** are hydrolyzed with sodium hydroxide (2.7 equivalents) in methanol–water at 25°C over 1 h. The methanol is distilled at reduced pressure and dilute hydrochloric acid (to pH 7) and toluene are added. The biphasic mixture is aged at 80–85°C for 1 h and then cooled to presumably 25°C. The resulting suspension is filtered and the solid is dried to afford (4-bromo-1,3-phenylene)dimethanol (**57**) and (2-bromo-1,3-phenylene)dimethanol (**58**) (84%, 93:7 mixture).

The diols **57** and **58** are then protected by reaction with ethyl vinyl ether (2.5 equivalents) and a *p*-toluenesulfonic acid monohydrate catalyst in toluene at 25°C over 2 h. The mixture is transferred into 5% aqueous sodium carbonate. The layers are separated. The organic layer is washed with 5% aqueous sodium carbonate, dried over potassium carbonate, decanted, and concentrated at reduced pressure to afford the bis(1-ethoxyethoxy)methyl ethers **59** and **60** (97%, 93:7 mixture).

Butyllithium in hexane (1.57 M, 1.1 equivalents) is added to the 93:7 mixture of **59** and **60** in THF at −40 to −30°C. The mixture is warmed to −20°C, 4-fluorobenzaldehyde[73] (1.1 equivalents) is added, and the resulting mixture is allowed to warm to 15°C over 1 h. The mixture is quenched with 20% aqueous ammonium chloride. The layers are separated and the aqueous layer is extracted with toluene. The combined organic layers are washed with brine and concentrated at reduced pressure. Phosphoric acid (60%) is added to the residue and the solution is aged at 80–85°C and reduced pressure (70–100 mmHg) for 2 h. The solution is cooled to 10°C and the resulting suspension is filtered. The solid is washed thoroughly with ethanol and dried to afford (1-(4-fluorophenyl)-1,3-dihydroisobenzofuran-5-yl)methanol (**40**) (89%).[74]

*(1-(4-Fluorophenyl)-1,3-dihydroisobenzofuran-5-yl) methanol (40) from Ethyl 1-Oxo-1,3-dihydroisobenzofuran-5-carboxylate (13)* A 15 wt% solution of 4-fluorophenylmagnesium bromide (1.2 equivalents) in THF is added to a suspension of ethyl 1-oxo-1,3-dihydroisobenzofuran-5-carboxylate (**13**) in THF at <5°C. The mixture is allowed to warm to 25°C and aged at 25°C overnight. The mixture is diluted with ethanol. Sodium borohydride (2.0 equivalents) is added in portions at 25°C and the mixture is aged at 25°C for 4 h. The solvents are distilled at reduced pressure. The residue is suspended in saturated aqueous ammonium chloride, 4 N hydrochloric acid is added (to pH 7.2), and the mixture is extracted with ethyl acetate. The extracts are concentrated at reduced pressure to afford the diol **62**.

Diol **62** is dissolved in 60% phosphoric acid and the solution is heated at 80°C for 1.5 h. The mixture is presumably cooled to 25°C, diluted with water, and extracted with ethyl acetate. The extracts are concentrated at reduced pressure to afford crude ethyl 1-(4-fluorophenyl)-1,3-

dihydroisobenzofuran-5-carboxylate (**63**). The ester is hydrolyzed with sodium hydroxide in aqueous ethanol at reflux. The volume is reduced by 50% by distillation at reduced pressure and the resulting mixture is washed with ethyl acetate. Hydrochloric acid is added (to pH 1) and the suspension is cooled to 5°C and filtered. The solid is dried at an unspecified temperature and reduced pressure to afford 1-(4-fluorophenyl)-1,3-dihydroisobenzofuran-5-carboxylic acid (**64**) (66% for three steps from **13**). Hydride reduction of the ester **63** or acid **64** could be another source of (1-(4-fluorophenyl)-1,3-dihydroisobenzofuran-5-yl)methanol (**40**).[63]

*1-(4-Fluorophenyl)-1,3-dihydroisobenzofuran-5-carbonitrile (31) from (1-(4-Fluorophenyl)-1,3-dihydroisobenzofuran-5-yl)methanol (40)* Manganese dioxide (31 equivalents or 9.0 kg per kg of alcohol **40**) is added in three portions to the alcohol **40** in toluene at 10–30°C. Complete conversion to the aldehyde **41** is observed after 1 h at 25°C. Filter aid and anhydrous magnesium sulfate are added, the suspension is filtered, and the solids are washed with toluene. The combined liquors contain 1-(4-fluorophenyl)-1,3-dihydroisobenzofuran-5-carbaldehyde (**41**) (100%).[66]

The oxidation can also be accomplished with sodium hypochlorite using 4-hydroxy-TEMPO as a catalyst. Sodium bicarbonate (0.41 equivalents), tetrabutylammonium bromide (5.9 mol%), and 4-hydroxy-TEMPO (0.90 mol%) are added to a solution of (1-(4-fluorophenyl)-1,3-dihydroisobenzofuran-5-yl)methanol (**40**) in ethyl acetate. The mixture is cooled to 5°C, 13% aqueous sodium hypochlorite (1.1 equivalents) is added at 5–10°C, and the mixture is aged, presumably at 5–10°C, for 1 h. Water is added and the mixture is extracted with ethyl acetate. The extracts are washed with 5% aqueous sodium bicarbonate and with brine. Some silica gel is added and the suspension is filtered. The liquors are concentrated at reduced pressure to afford 1-(4-fluorophenyl)-1,3-dihydroisobenzofuran-5-carbaldehyde (**41**) (84%).[74]

The conversion of the hydroxymethyl group to the nitrile can be accomplished without isolation of the intermediate aldehyde or oxime. The suspension after oxidation with manganese dioxide (14 equivalents or 4.0 kg per kg of alcohol **40**) in xylene is filtered and the solid is washed with xylene. Hydroxylamine hydrochloride (1.0 equivalent) and triethylamine (1.0 equivalent) are added and the mixture is aged at 70–75°C for 1 h. Acetic anhydride (3.6 equivalents) is added and the mixture is aged at 130–140°C for 6 h. The mixture is presumably cooled to 25°C and water and 10% aqueous sodium hydroxide are added. The layers are separated and the organic layer is concentrated at reduced pressure. The residue is dissolved in xylene–heptane at 60°C and the solution is cooled to 25°C. The resulting suspension is filtered and the solid is dried to afford 1-(4-fluorophenyl)-1,3-dihydroisobenzofuran-5-carbonitrile (**31**) (73%).[66]

Alternative routes from (1-(4-fluorophenyl)-1,3-dihydroisobenzofuran-5-yl)methanol (**40**) to 1-(4-fluorophenyl)-1,3-dihydroisobenzofuran-5-carbonitrile (**31**) are also possible. For example, the alcohol could be converted to a bromide or chloride, the bromide or chloride could be replaced by an amino group, and the aminomethyl group could be oxidized to the nitrile (Scheme 3.17).[75]

### 3.3.2 Citalopram (39) by Alkylation of 1-(4-Fluorophenyl)-1,3-dihydroisobenzofuran-5-carbonitrile (31)

The alkylation of 1-(4-fluorophenyl)-1,3-dihydroisobenzofuran-5-carbonitrile (**31**) with 3-chloro-*N*,*N*-dimethylpropan-1-amine affords citalopram (**39**). 3-Chloro-*N*,*N*-dimethylpropan-1-amine is released from the commercially available hydrochloride salt.[76–83]

A 66 wt% aqueous solution of the hydrochloride salt[84] is added to a 13 wt% aqueous sodium hydroxide (1.3

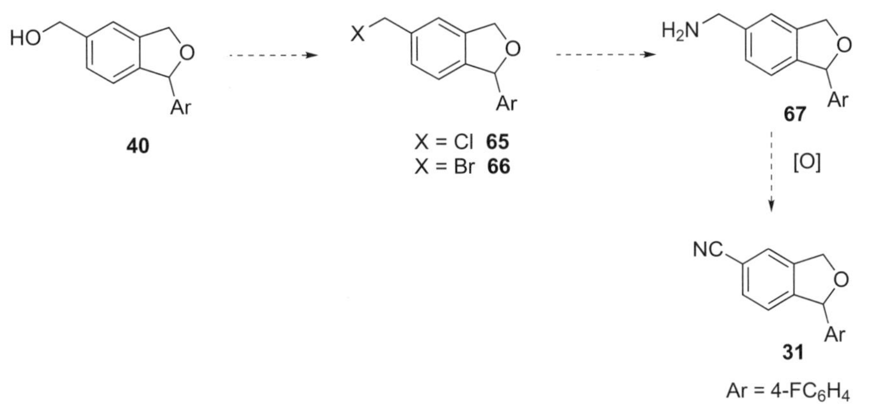


**SCHEME 3.17** Proposed route to 1-(4-fluorophenyl)-1,3-dihydroisobenzofuran-5-carbonitrile (**31**) from (1-(4-fluorophenyl)-1,3-dihydroisobenzofuran-5-yl)methanol (**40**).

equivalents) solution at 20–25°C. The mixture is extracted with toluene. The extracts are dried over potassium carbonate and 4 Å molecular sieves (28 g per kg HCl salt) and filtered to afford a solution of the free base. When methyl *tert*-butyl ether is used in place of toluene in a similar process, the organic extracts can be dried by distillation of the methyl *tert*-butyl ether–water azeotrope (bp 52.6°C). The remaining methyl *tert*-butyl ether (bp 55–56°C) is distilled at atmospheric pressure and the free base is distilled at a bath temperature of 80–90°C and unspecified reduced pressure (81%). It would be advisable to store 3-chloro-*N*,*N*-dimethylpropan-1-amine in solution and to time the preparation of the solution to minimize polymerization during storage.[66,85]

The alkylation reaction requires a strong base and a polar aprotic solvent. Sodium hydride, potassium *tert*-butoxide, and lithium diisopropylamide (LDA) are suitable bases. DMSO, DMF, 1,2-dimethylimidazolidinone (DMEU), and 1,2-dimethoxyethane are suitable solvents. The disposal or recovery of the polar aprotic solvent is an issue, especially when using expensive DMEU. A phase transfer catalyst or *N*,*N*,*N*′,*N*′-tetramethylethylenediamine (TMEDA) may provide an incremental increase in yield. The workup often includes an acid–base extraction to separate citalopram from nonbasic impurities. Citalopram (**39**) yields of 69–87% are typical and an 87% yield has been demonstrated on a 50 kg scale. Citalopram (**39**) free base can be precipitated as a solid (mp 89–91°C) but is more often isolated as a crude oil and converted to the hydrobromide salt.

A solution of LDA is prepared by addition of butyllithium (2.0 equivalents) in hexanes to diisopropylamine (1.9 equivalents) in 1,2-dimethoxyethane (DME) at −50°C. A solution of 1-(4-fluorophenyl)-1,3-dihydroisobenzofuran-5-carbonitrile (**31**) in DME is added at <−40°C over 45 min. After aging at −40 to −50°C for 20 min, 3-chloro-*N*,*N*-dimethylpropan-1-amine (3.2 equivalents) is added at −50°C. The mixture is allowed to warm to 25°C over 1 h and then aged at 50°C for 2 h. The mixture is presumably cooled to 25°C, quenched with water at 0–5°C, and extracted with toluene. Citalopram hydrochloride is extracted from the toluene extracts with 4 N hydrochloric acid. Toluene and aqueous 10 M sodium hydroxide are added to the aqueous acid extract (to pH 10). The layers are separated and the organic layer containing the free base is washed with water, dried, treated with carbon, and concentrated at reduced pressure to afford citalopram (**39**) (76–84%) (Scheme 3.18). A similar process is described using THF as the reaction solvent at −20°C (no yield available).[63,64]

A mixture of potassium *tert*-butoxide (1.6 equivalents) and DMSO is aged at 60–70°C for an unspecified time. The resulting solution is cooled to 25°C and a solution of 1-(4-fluorophenyl)-1,3-dihydroisobenzofuran-5-carbonitrile (**31**) in DMSO is added at 25–30°C. A solution of 3-chloro-*N*,*N*-dimethylpropan-1-amine (1.1 equivalents) in DMSO is added and the mixture is aged at 40°C for 50 min. The mixture is quenched with water at 0–5°C and extracted with toluene. Citalopram acetate is extracted from the toluene extracts with 20% aqueous acetic acid. The aqueous acid extract is diluted with isopropanol and aqueous ammonia is added at 5–10°C (to pH 8.5–9.0). The resulting suspension is filtered. The solid is washed with cold isopropanol–water (1:4) and with hexane and then dried at 40°C and 500–600 mmHg to afford citalopram (**39**) (73–75%, 99.5% pure by HPLC). Similar results (72–74%, 98.8–99.3% pure by HPLC) are observed using ethanol, methanol, DMF, DMSO, or acetone in place of isopropanol. Similar results (69–72%, 98.3–99.3% pure by HPLC) are observed using sodium hydride in place of potassium *tert*-butoxide and the same five solvent options.[86]

Similar results (74%, 99% pure by HPLC) are observed using 1.3 equivalents of potassium *tert*-butoxide and an alternative workup procedure. Aqueous ammonia is added to the acid extracts (to pH 7–7.5) and the free base is extracted into isopropyl ether. The isopropyl ether extracts are treated with carbon and the suspension is filtered. The liquors are concentrated at 45°C and reduced pressure to afford citalopram (**39**).[61]

Sodium hydride (1.1 equivalents of 50 wt%) is suspended in DMSO. The mixture is heated at 60–65°C for 30 min and then cooled to 25°C. A solution of 1-(4-fluorophenyl)-1,3-dihydroisobenzofuran-5-carbonitrile (**31**) in DMSO is added at 20–25°C. The mixture is aged at 20–25°C for 20 min and a solution of 3-chloro-*N*,*N*-dimethylpropan-1-amine (1.1 equivalents) in toluene is added at 25–30°C. The mixture is aged at 25°C for 30 min. The mixture is quenched by adding methanol, transferred into water, and extracted with toluene. Citalopram acetate is extracted from the toluene extracts with 20% aqueous acetic acid. Aqueous ammonia is added to the acid extracts (to pH 8–8.5) and citalopram free base is extracted with toluene. The toluene extracts are

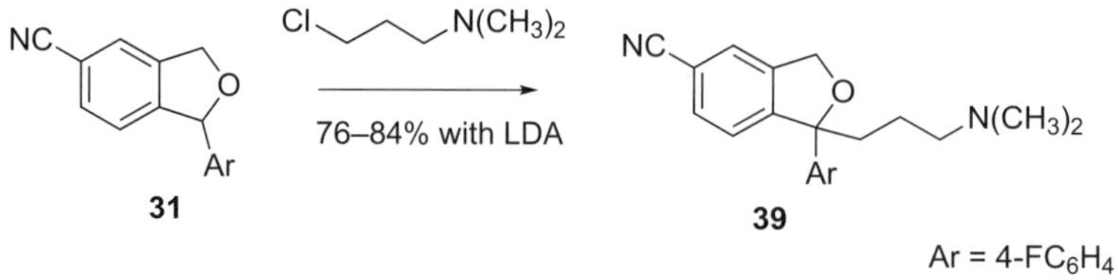


**SCHEME 3.18** Citalopram (**39**) from 1-(4-fluorophenyl)-1,3-dihydroisobenzofuran-5-carbonitrile (**31**).

treated with carbon and the suspension is filtered. Dilute aqueous hydrobromic acid is added to the toluene extracts at 25–30°C, the biphasic mixture is aged for 1 h at 25–30°C, and the layers are separated. The acidic aqueous layer is aged at 25°C for 10 h and at 10°C for 2 h. The resulting suspension is filtered and the solid is crystallized from isopropanol–water to afford citalopram (**39**) as the hydrobromide salt (80%).[87]

Citalopram (**39**) free base is isolated using an alternative workup procedure. Aqueous ammonia is added to the acid extracts (to pH 7–7.5) and the free base is extracted into isopropyl ether. The isopropyl ether extracts are treated with carbon and the suspension is filtered. The liquors are concentrated at 45°C and reduced pressure to afford citalopram (**39**) (87–89%, 99% pure by HPLC). The yield is lower (74%) when the concentration is stopped at 0.5 g citalopram (**39**)/mL isopropyl ether and the suspension is filtered. A purity upgrade for citalopram free base (**39**) by distillation (bp 175–181°C at 0.03 mmHg) is an option in the lab.[60–62]

> Several process intermediates and even citalopram (**39**) free base can be distilled. While the high temperature (170–180°C) and high vacuum (<0.1 mmHg) required make distillation unattractive for the production process, a bulb-to-bulb or Kugelrohr distillation in the lab can provide a great deal of valuable information in a very short time, typically <1 h. The distillation might provide the first pure sample of the target amine free base for salt screening. A second distillation of the distillate provides a first look at thermal stability. Even the distillation residue is valuable. The distillation residue ("bottoms") is a concentrated sample of a mixture of "heavies" (perhaps dimers), which can be used to expedite their separation, characterization, and eventual quantification.

A solution of 1-(4-fluorophenyl)-1,3-dihydroisobenzofuran-5-carbonitrile (**31**) in THF is added to a suspension of 60% sodium hydride (1.1 equivalents) in THF at 40–50°C. Tetrabutylammonium bromide (0.30 mol%) and a solution of 3-chloro-*N*,*N*-dimethylpropan-1-amine (1.3 equivalents) in methyl *tert*-butyl ether are added, presumably at 40–50°C, and the suspension is aged at 40–50°C for 10 min. 1,3-Dimethylimidazolidinone (DMEU) (5.3 kg per kg nitrile **31**) is added, presumably at 40–50°C, and the mixture is aged at 61–64°C for 6 h. The mixture is quenched with ice water and extracted with toluene. Citalopram acetate is extracted from the combined organic extracts using 20% aqueous acetic acid. The aqueous acidic extracts are neutralized with hydroxide solution and citalopram free base is extracted with toluene. The extracts are washed with water and concentrated at reduced pressure to afford citalopram (**39**) (79%). No reaction is observed in the control experiment in THF (no DMEU) with a phase transfer catalyst. Similar procedures in THF–DMF, toluene–DMF, or toluene–DMSO with a phase transfer catalyst or in THF–DMSO with no phase transfer catalyst gave lower yields (52–69%).[66,74]

In a large-scale demonstration run, a solution of 3-chloro-*N*,*N*-dimethylpropan-1-amine (2.1 equivalents) in toluene is added to a solution of 1-(4-fluorophenyl)-1,3-dihydroisobenzofuran-5-carbonitrile (**31**) in toluene. 1,3-Dimethylimidazolidinone (DMEU) is added. Sodium hydride (1.3 equivalents of 65 wt%) is then added at 25–30°C. More DMEU (total DMEU is 4.1 kg per kg nitrile **31**) is added as the mixture is warmed to 60°C over 4 h. The mixture is then aged at 60–63°C for 6 h. The mixture is quenched with water at 5°C and extracted with toluene. Citalopram hydrochloride is extracted from the combined organic extracts using 5% hydrochloric acid. The aqueous acidic extracts are neutralized with 25% aqueous sodium hydroxide and citalopram free base extracted with toluene. The extracts are washed with water and dried over potassium carbonate. Silica gel (150 g per kg of nitrile **31**) is added, the suspension is filtered, and the solid is washed with toluene. The liquors are concentrated at reduced pressure to afford citalopram (**39**) (87%). This process delivers 53 kg of product with a volume throughput of 35 g/L. The maximum volume is attained in the first toluene extraction after the water quench.[66]

What is the best option for scale-up of the alkylation reaction? The first consideration should be process safety. There is the potential for explosive decomposition during the generation of the dimsyl anion by reaction of DMSO with sodium hydride or potassium *tert*-butoxide at elevated temperatures. There is also the potential for explosion during a premature or poorly controlled water quench of a sodium hydride reaction. Lithium diisopropylamide might not be the least expensive option, but it is the safest.[88]

## 3.4 CONSTRUCTION OF CITALOPRAM (39) BY CYANIDE EXCHANGE IN THE FINAL STEP

The substrates for a final cyanide exchange are prepared from the 5-substituted isobenzofuran-1(3*H*)-ones by addition of 4-fluorophenylmagnesium bromide to the lactone to produce the ketone, addition of 3-dimethylaminopropylmagnesium chloride to the ketone to produce the alcohol, and cyclization of the diol. We have already discussed the limitations of the first Grignard reagent addition to an isobenzofuran-1(3*H*)-one. The yield of ketone substrate produced *in situ* will be roughly 80–85%. Development efforts often focus on identifying an expeditious way to get high-purity diol out of this less than perfect process without the use of chromatography.[85]

3-Dimethylaminopropylmagnesium chloride is prepared as a 30 wt% solution in THF–toluene. Aqueous sodium hydroxide (30%) is added to a suspension of 3-chloro-*N*,*N*-dimethylpropan-1-amine hydrochloride in toluene at <20°C. The phases are separated and the toluene layer is diluted with THF and dried over molecular sieves. Some (5%) of this solution is added to a suspension of magnesium (1.0 equivalent) in THF. Ethyl bromide (0.5 mol%) is added as an initiator. The suspension is heated to reflux and the remaining amine solution is added to the refluxing suspension over 1.5 h. After the addition is complete, the mixture is refluxed for 1 h, cooled to 25°C, and stored under dry nitrogen.[89]

### 3.4.1 3-(5-Bromo-1-(4-fluorophenyl)-1,3-dihydroisobenzofuran-1-yl)-*N*,*N*-dimethylpropan-1-amine (68)

The second Grignard reagent addition introduces an amine handle. Acid–base extraction in the workup can serve as a first-pass separation of the diol product **69** from nonbasic side products. A 17 wt% solution of 4-fluorophenylmagnesium bromide (1.6 equivalents) in THF is added to a suspension of 5-bromoisobenzofuran-1(3*H*)-one (**5**) and magnesium bromide in THF at <20°C over 2 h. The mixture is aged, presumably at <20°C, for an unspecified time. A solution of 3-dimethylaminopropylmagnesium chloride (solvent and equivalents not specified) is added at <20°C over >1 h. The mixture is quenched with saturated aqueous ammonium chloride. The layers are separated and the organic layer is washed, presumably with water, and concentrated at reduced pressure. The residue is taken up in toluene and extracted with 20% aqueous acetic acid. Aqueous sodium hydroxide is added to the aqueous acid extracts (to pH >7) and the mixture is extracted with toluene. The toluene extracts are concentrated at reduced pressure to afford crude diol **69** (57%). Aqueous phosphoric acid (50%) is added and the mixture is heated at an unspecified temperature, probably 80–90°C. The mixture is washed with toluene. Ammonium hydroxide is added (to pH >7) at <30°C and the mixture is extracted with toluene. The extracts are concentrated at reduced pressure to afford 3-(5-bromo-1-(4-fluorophenyl)-1,3-dihydroisobenzofuran-1-yl)-*N*,*N*-dimethylpropan-1-amine (**68**) (49% for three steps from 5-bromoisobenzofuran-1(3*H*)-one).[90]

The diol purity is higher when it is isolated as a hydrochloride salt. A 22 wt% solution of 4-fluorophenylmagnesium bromide (1.3 equivalents) in THF is added to a suspension of 5-bromoisobenzofuran-1(3*H*)-one (**5**) in toluene at <20°C over 1 h. The mixture is aged at <20°C for 30 min. A 26 wt% solution of 3-dimethylaminopropylmagnesium chloride (1.2 equivalents) in THF–toluene is added at <30°C over 2 h. The mixture is aged at 25°C for 16 h and then quenched with 6% aqueous hydrochloric acid. The biphasic suspension is aged at 25°C for 1 h and then filtered. The solid is washed with water and toluene and then dried at 40°C and reduced pressure to afford the crude hydrochloride of diol **69** (80%). The crude salt is dissolved in 2-butanol at 70°C and the hazy solution is filtered. The liquors are slowly cooled to 5°C. The suspension is filtered and the solid is washed with 2-butanol and dried at 40°C and reduced pressure to afford diol **69** as the hydrochloride salt (59%, 99.9% pure by HPLC). Aqueous phosphoric acid (60%) is added to the hydrochloride salt and the mixture is heated at 90°C for 1 h. The mixture is cooled to <10°C and toluene and water are added. Ammonium hydroxide (28%) is added, presumably to pH >7. The layers are separated and the aqueous layer is extracted with toluene. The toluene extracts are washed with water and concentrated at reduced pressure to afford 3-(5-bromo-1-(4-fluorophenyl)-1,3-dihydroisobenzofuran-1-yl)-*N*,*N*-dimethylpropan-1-amine (**68**) (95%, 56% for three steps from **5**, 99.0% pure by HPLC) (Scheme 3.19).[89]

The dimagnesium salt produced by the addition of the second Grignard reagent to the ketone can be directly precipitated by dilution of the reaction mixture with toluene or cyclohexane. A 19 wt% solution of 4-fluorophenylmagnesium bromide (1.1 equivalents) in THF is added to a suspension of 5-bromoisobenzofuran-1(3*H*)-one (**5**) in THF at −15 to −10°C. The mixture is aged at −15 to −10°C for 3 h. A 12 wt% solution of 3-dimethylaminopropylmagnesium chloride (1.2 equivalents) in 1:1 THF–toluene is added at −15 to 0°C over 2–3 h. The resulting suspension is aged at −15 to 0°C for 1 h, warmed to 10°C, and filtered. The solid is washed with toluene and air dried to afford the dimagnesium salt **70** (80%). The salt is suspended in water. Diol **69** is released by adding ammonium chloride and extracting with toluene. The extracts are treated with carbon and the suspension is filtered. The liquors are concentrated at reduced pressure and isopropyl ether is added to the residue. The suspension is filtered and the solid is presumably washed with isopropyl ether and dried at reduced pressure to afford the diol (70% from 5-bromoisobenzofuran-1(3*H*)-one, 98.5% pure by HPLC). The yields are lower (63% to **70** and 40% to **69**) in a very similar process using cyclohexane in place of toluene.[91]

Diol **69** can also be cyclized by formation and displacement of a methanesulfonate or *p*-toluenesulfonate. Methanesulfonyl chloride (1.3 equivalents) is added to diol **69** and triethylamine (2.9 equivalents) in toluene at −5 to 0°C over 3 h. After aging at 0°C for 1 h, the mixture is quenched with water. The layers are separated and the aqueous layer is extracted with toluene. The organic extracts are washed with water and concentrated at <60°C and reduced pressure to afford 3-(5-bromo-1-(4-fluorophenyl)-1,3-dihydroisobenzofuran-1-yl)-*N*,*N*-dimethylpropan-1-amine (**68**) (91%, 64% for three steps from **5**). The cyclization yield is comparable (96%) using *p*-toluenesulfonyl chloride and triethylamine in dichloromethane.[91]

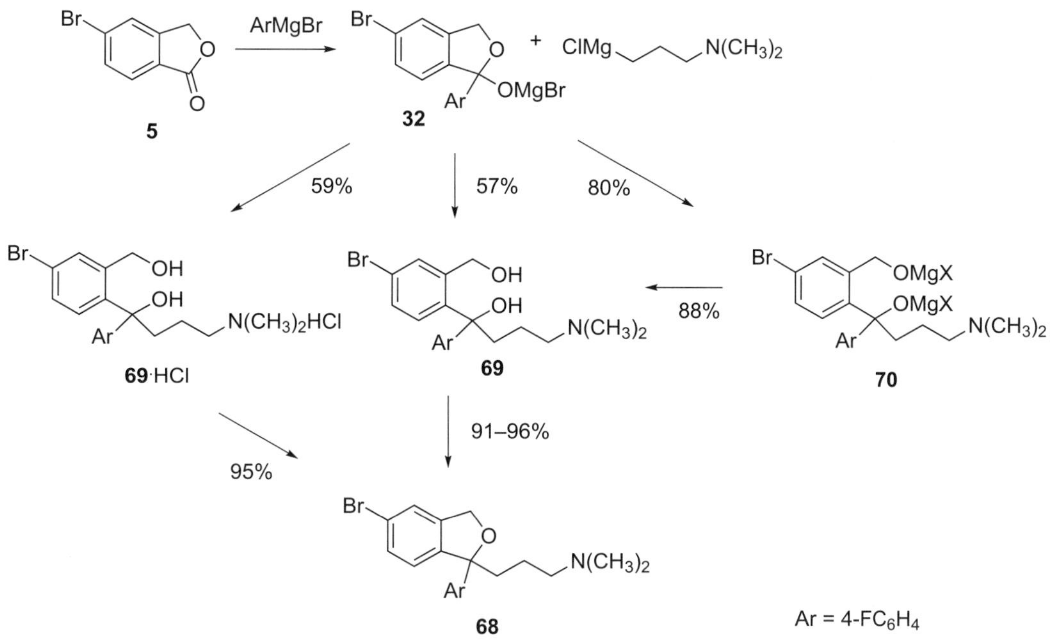


**SCHEME 3.19** 3-(5-Bromo-1-(4-fluorophenyl)-1,3-dihydroisobenzofuran-1-yl)-*N*,*N*-dimethylpropan-1-amine (**68**) from 5-bromoisobenzofuran-1(3*H*)-one (**5**).

### 3.4.2 3-(1-(4-Fluorophenyl)-5-iodo-1,3-dihydroisobenzofuran-1-yl)-*N*,*N*-dimethylpropan-1-amine (71)

Acid–base extraction in the workup can also serve as a first-pass separation of the 5-iodo diol **72** from nonbasic side products. A 20 wt% solution of 4-fluorophenylmagnesium bromide (1.1 equivalents) in THF is added to a suspension of 5-iodoisobenzofuran-1(3*H*)-one (**25**) in THF at <0°C. The mixture is aged at <0°C for 3 h. A 16 wt% solution of 3-dimethylaminopropylmagnesium chloride (1.2 equivalents) in THF is added at <0°C. The mixture is aged at 25°C for 2 h. The mixture is quenched with saturated aqueous ammonium chloride and THF is distilled at reduced pressure. Toluene is added and the organic layer is separated. Diol **72** hydrochloride is extracted from the organic layer with 1 M hydrochloric acid. Ammonium hydroxide (25%) is added to the aqueous acidic extract (to pH 9). Toluene is added and the organic layer is separated and used in the cyclization. Sulfuric acid (70%) is added and the mixture is aged at 25°C for 2 h. Ammonium hydroxide (25%) is added, presumably to pH 8–9, and the organic phase is separated, filtered, and concentrated at reduced pressure to afford 3-(1-(4-fluorophenyl)-5-iodo-1,3-dihydroisobenzofuran-1-yl)-*N*,*N*-dimethylpropan-1-amine (**71**). No yield is available.[92]

### 3.4.3 3-(5-Chloro-1-(4-fluorophenyl)-1,3-dihydroisobenzofuran-1-yl)-*N*,*N*-dimethylpropan-1-amine (73)

Another 1,3-dihydroisobenzofuran cyclization strategy is highlighted in a preparation of the 5-chloro-1,3-dihydroisobenzofuran. 5-Chloroisobenzofuran-1(3*H*)-one (**6**) is reacted with thionyl chloride (1.5 equivalents), boron trifluoride etherate (5 mol%), and benzyltriethylammonium chloride (8 mol%) in xylene at reflux for 20 h. The xylene is then distilled at reduced pressure and the residue distilled under high vacuum to afford 4-chloro-2-(chloromethyl)benzoyl chloride (**74**) (39%).[64]

An 18 wt% solution of 4-fluorophenylmagnesium bromide (1.1 equivalents) in THF is added to a solution of 4-chloro-2-(chloromethyl)benzoyl chloride (**74**) at <0°C. The mixture is aged at <0°C for 2 h. A 12 wt% solution of 3-dimethylaminopropylmagnesium chloride, presumably in THF, is added at 0°C. The mixture is aged, presumably at 0°C, for 2 h. Water and acetic acid are added (to pH 4–4.5) and the layers are separated. Ammonium hydroxide (25%) is added to the aqueous layer (to pH 8–8.5) and the mixture is extracted with toluene. The extracts are concentrated at reduced pressure to afford 3-(5-chloro-1-(4-fluorophenyl)-1,3-dihydroisobenzofuran-1-yl)-*N*,*N*-dimethylpropan-1-amine (**73**) (51%) (Scheme 3.20).[64]

### 3.4.4 1-(3-(Dimethylamino)propyl)-1-(4-fluorophenyl)-1,3-dihydroisobenzofuran-5-yl trifluoromethanesulfonate (75)

The trifluoromethanesulfonate is produced by the double Grignard reagent addition process from 5-hydroxyisobenzofuran-1(3*H*)-one (**7**). A 28 wt% solution of 4-fluorophenylmagnesium bromide (2.0 equivalents) in THF is added to a suspension of 5-hydroxyisobenzofuran-1(3*H*)-one (**7**) in THF at <8°C and the mixture is aged at 25°C overnight.

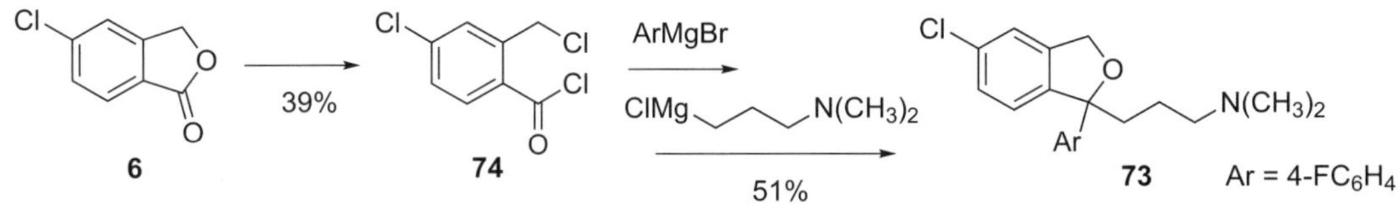


**SCHEME 3.20** 3-(5-Chloro-1-(4-fluorophenyl)-1,3-dihydroisobenzofuran-1-yl)-*N*,*N*-dimethylpropan-1-amine (**73**) from 5-chloroisobenzofuran-1(3*H*)-one (**6**).

A 22 wt% solution of 3-dimethylaminopropylmagnesium chloride (1.0 equivalent) in THF is added at <10°C and the mixture is aged at 25°C overnight. The mixture is quenched with water and aqueous ammonium chloride is added (to pH 7). The layers are separated. The aqueous phase is washed with ethyl acetate. Ammonium hydroxide (25%) is added (to pH 8–9) and the aqueous phase is extracted with toluene–ethyl acetate (3:2). The toluene–ethyl acetate extract is dried and treated with carbon. The suspension is filtered and the liquors are concentrated at reduced pressure to afford 1-(3-(dimethylamino)propyl)-1-(4-fluorophenyl)-1,3-dihydroisobenzofuran-5-ol (**76**) (48%). The diol cyclization, facilitated by the electron releasing group at the 5-position, likely occurs at 25°C in the pH 7 aqueous solution during the workup procedure. Trifluoromethanesulfonyl chloride (1.2 equivalents) is added to phenol **76** in dichloromethane at <5°C. After aging at 25°C overnight, triethylamine and water are added and the phases are separated. The aqueous layer is extracted with dichloromethane and the combined organic layers are dried and concentrated at reduced pressure to afford the crude trifluoromethanesulfonate (**75**) (93%) (Scheme 3.21).[92]

### 3.4.5 Citalopram (39) by Exchange of Bromide, Iodide, Chloride, or Trifluoromethanesulfonate for Cyanide in the Final Step

Citalopram (**39**) is produced by exchange of bromide, iodide, chloride, or trifluoromethanesulfonate for cyanide in the final step. Activation of the leaving group requires either temperatures that are difficult to attain and hold in a manufacturing plant (135–160°C) or catalysis by palladium or nickel. The yield of citalopram (**39**) can be as high as 92%. The manufacturing cost must include the cost for treatment of aqueous waste streams containing cyanide and copper or zinc salts.

***3.4.5.1 From 3-(5-Bromo-1-(4-fluorophenyl)-1,3-dihydroisobenzofuran-1-yl)-N,N-dimethylpropan-1-amine (68)*** Exchange of bromide for cyanide can be accomplished using copper(I) cyanide at 135–160°C neat or in sulfolane, pyridine, PEG 400, DMF, or mixtures of these solvents. Potassium iodide and/or dextrose are sometimes added but data on their beneficial effect are not readily available. Exchange of bromide for cyanide is also possible at lower temperatures (65–75°C) using zinc cyanide and a palladium catalyst in THF or DMF.

*Cyanide Exchange at 135–160°C* 3-(5-Bromo-1-(4-fluorophenyl)-1,3-dihydroisobenzofuran-1-yl)-*N*,*N*-dimethylpropan-1-amine (**68**) is reacted with copper(I) cyanide (1.8 equivalents) at 140–150°C for an unspecified time. The mixture is diluted with DMF and toluene and cooled to 80°C. Aqueous ethylenediamine (50% w/v) is added and the layers are separated. The organic layer is washed with aqueous EDTA (2% w/v) and with water. Toluene is distilled at reduced pressure and citalopram (**39**) is isolated from the remaining oil by distillation under high vacuum (64%).[93]

3-(5-Bromo-1-(4-fluorophenyl)-1,3-dihydroisobenzofuran-1-yl)-*N*,*N*-dimethylpropan-1-amine (**68**) is reacted with

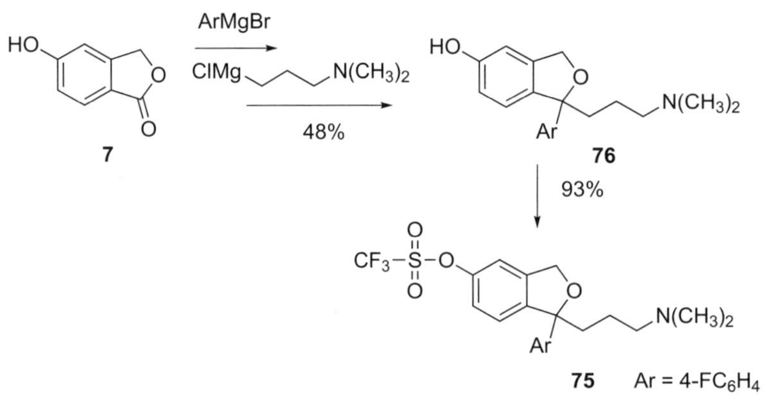


**SCHEME 3.21** 1-(3-(Dimethylamino)propyl)-1-(4-fluorophenyl)-1,3-dihydroisobenzofuran-5-yl trifluoromethanesulfonate (**75**) from 5-hydroxyisobenzofuran-1(3*H*)-one (**7**).

copper(I) cyanide (1.2 equivalents) in sulfolane (bp 285°C, 104°C at 0.2 mmHg) at 150°C for 5 h. The mixture is diluted with sulfolane and cooled to 80°C. Aqueous ethylenediamine (50% w/v) and toluene are added and the layers are separated. The organic layer is washed with aqueous EDTA (5% w/v) and with water. Citalopram (**39**) is isolated from the sulfolane solution by wiped film distillation (wiper temperature 245°C, pressure 0.7 mmHg). Yield data are not available.[94,95]

3-(5-Bromo-1-(4-fluorophenyl)-1,3-dihydroisobenzofuran-1-yl)-*N,N*-dimethylpropan-1-amine (**68**) is reacted with copper(I) cyanide (1.3 equivalents) in pyridine at 145–150°C for 8 h. The mixture is cooled to 25°C and poured into water. After aging at 25°C for 1 h, ethylenediamine is added. The mixture is aged at 25°C for 3 h and toluene is added. The suspension is filtered and the cake is washed with toluene. The combined liquor layers are separated and the aqueous layer is extracted with toluene. Citalopram acetate is extracted from the organic layer with 10% aqueous acetic acid. The aqueous acid extracts are treated with carbon and the suspension is filtered. The free base is recovered from the acid extract by adding ammonium hydroxide (to pH 8.5–9.0) and extracting with isopropyl ether. The isopropyl ether extract is concentrated at reduced pressure to afford citalopram (**39**) (70%, >96% pure by HPLC).[91]

3-(5-Bromo-1-(4-fluorophenyl)-1,3-dihydroisobenzofuran-1-yl)-*N,N*-dimethylpropan-1-amine (**68**) is reacted with copper(I) cyanide (4.2 equivalents), potassium iodide, and dextrose in DMF at 140–160°C for an unspecified amount of time. Some of the DMF (65%) is distilled at reduced pressure. Toluene and dilute ammonium hydroxide are added at >50°C. The suspension is cooled to 20°C and filtered. The solid is washed with toluene and dilute ammonium hydroxide. The layers in the liquors are separated. The toluene layer is washed with dilute ammonium hydroxide, then with water, and finally with 0.05% hydrobromic acid. The toluene layer is extracted with 20% acetic acid (pH 4.5–4.8). The aqueous acidic extract is washed with toluene. Sodium hydroxide (10%) is added to the aqueous acid solution (to pH >7) and the mixture is extracted with toluene. The extracts are treated with silica gel and, after dilution with *n*-hexane, with silica gel and carbon. The suspension is filtered and the liquors are concentrated at reduced pressure. The residue is dissolved in ethyl acetate and citalopram (**39**) is isolated as the oxalate. The oxalate salt is converted to the free base. The free base is dissolved in isopropanol and converted to the hydrobromide salt (50%).[90]

3-(5-Bromo-1-(4-fluorophenyl)-1,3-dihydroisobenzofuran-1-yl)-*N,N*-dimethylpropan-1-amine (**68**) is reacted with copper(I) cyanide (2.1 equivalents) and sodium cyanide (0.75 equivalents) in DMF at 154–159°C for an unspecified amount of time. The mixture is cooled to 80°C and transferred into 10% aqueous sodium cyanide. Aqueous ethylenediamine (41%) and toluene are added. The suspension is filtered and the liquor layers are separated. The aqueous layer is extracted with toluene. Citalopram acetate is extracted from the combined organic layers with 20% aqueous acetic acid. The free base is recovered from the acid extract by adding aqueous sodium hydroxide (to pH 9–10) and extracting with toluene. The toluene extract is treated with carbon. The suspension is filtered and the liquors are concentrated at reduced pressure to afford crude citalopram (**39**) (84%). A similar procedure on smaller (5 g) scale with sodium cyanide (1.3 equivalents) afforded crude citalopram (**39**) (58%, >97% pure by HPLC).[89,91]

3-(5-Bromo-1-(4-fluorophenyl)-1,3-dihydroisobenzofuran-1-yl)-*N,N*-dimethylpropan-1-amine (**68**) is reacted with copper(I) cyanide (2.0 equivalents) and potassium iodide (2.3 equivalents) in pyridine and PEG 400 (1:1) at 135–145°C for 27 h. The mixture is cooled to 100°C and transferred into aqueous ammonium hydroxide and toluene. The biphasic mixture is aged for 2 h, presumably at 25°C, and the layers are separated. Citalopram acetate is extracted from the organic layer with 20% aqueous acetic acid. The free base is recovered from the acid extract by adding ammonium hydroxide (to pH >7) and extracting with toluene. The toluene extract is washed twice with water, dried, and concentrated at reduced pressure to afford crude citalopram (**39**) (92%, 92–93% pure by HPLC). Lower yields (77–86%) are observed using pyridine or PEG 400 alone, DMF, or 2,6-lutidine–DMF (1:1) as the solvent(s).[96]

*Palladium-Catalyzed Cyanide Exchange* Tetrakis (triphenylphosphine)palladium (4.3 mol%) is added to 3-(5-bromo-1-(4-fluorophenyl)-1,3-dihydroisobenzofuran-1-yl)-*N,N*-dimethylpropan-1-amine (**68**) and zinc cyanide (0.63 equivalents) in DMF at 25°C. After aging at 75°C for 3 h, the mixture is presumably cooled to 25°C and poured into water and ethyl ether. The layers are separated and the aqueous layer is extracted with ethyl ether. The ether extracts are dried and concentrated at reduced pressure. The free base is converted to the oxalate salt in acetone (92%) (Scheme 3.22). A lower yield (82%) is observed in a similar process using sodium cyanide (2.0 equivalents), zinc cyanide (1.7 mol%), and the palladium catalyst (3.7 mol%) in THF at reflux.[92]

***3.4.5.2 From 3-(5-Chloro-1-(4-fluorophenyl)-1,3-dihydroisobenzofuran-1-yl)-N,N-dimethylpropan-1-amine (73)*** Exchange of chloride for cyanide requires catalysis by nickel. The nickel catalyst is prepared by reduction of nickel (II) chloride by zinc in the presence of a phosphine. A mixture of triphenylphosphine (16 mol%) and nickel(II) chloride (4.0 mol%) in acetonitrile is refluxed for 45 min. The mixture is cooled to 25°C and zinc powder (40 mol%) is added. The suspension is aged at 25°C for 15 min. A solution of 3-(5-chloro-1-(4-fluorophenyl)-1,3-dihydroisobenzofuran-1-yl)-*N,N*-dimethylpropan-1-amine (**73**) in acetonitrile

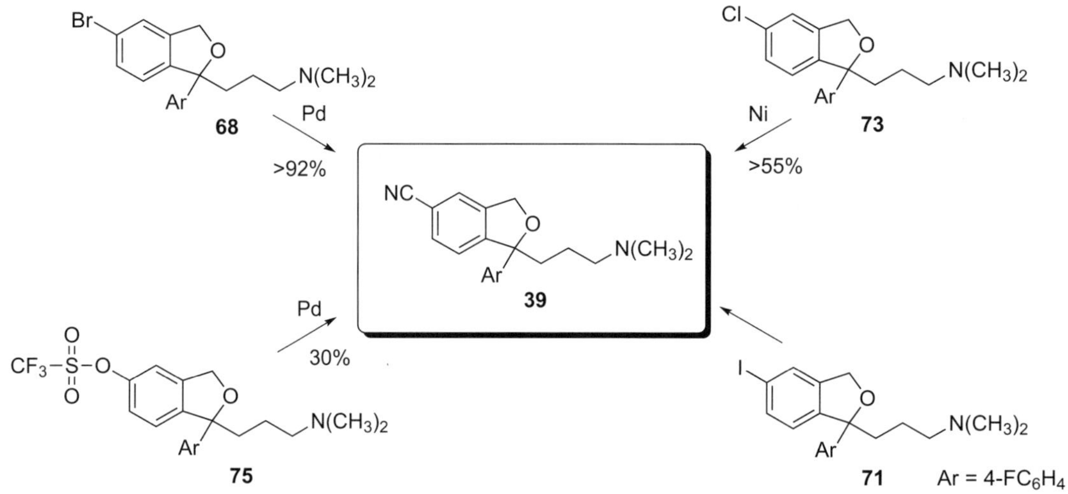


**SCHEME 3.22** Citalopram (**39**) by exchange of bromide, iodide, chloride, or trifluoromethanesulfonate for cyanide in the final step.

is added. The mixture is aged at 25°C for 10 min. Sodium cyanide (0.43 equivalents) (*Note*: The quantity is insufficient.) is added and the mixture is refluxed overnight. The suspension is cooled to 25°C, diluted with ethyl ether, and filtered. The liquors are washed with brine, dried, and concentrated at reduced pressure to afford crude citalopram (**39**). The free base is converted to the oxalate salt in acetone and the salt is upgraded by crystallization from ethanol (55%).[64,97]

***3.4.5.3 From 1-(3-(Dimethylamino)propyl)-1-(4-fluorophenyl)-1,3-dihydroisobenzofuran-5-yl Trifluoromethanesulfonate (75)*** Exchange of trifluoromethanesulfonate for cyanide requires catalysis by palladium and copper. A mixture of the trifluoromethanesulfonate **75**, sodium cyanide (2.0 equivalents), copper iodide (13 mol%), and tetrakis (triphenylphosphine)palladium (4.3 mol%) in acetonitrile is refluxed for 5 h. The mixture is cooled to 25°C overnight, diluted with ethyl acetate, and filtered. The liquors are washed with brine, dried, and concentrated at reduced pressure. Citalopram (**39**) is isolated from the residue by chromatography (30%).[92]

***3.4.5.4 Upgrading the Purity of Citalopram (39) Produced in the Cyanide Exchange*** Upgrading the purity of citalopram (**39**) produced by cyanide exchange is most often discussed in the context of the bromide-for-cyanide exchange. Important side products of the exchange are citalopram amide (**77**), desmethyl citalopram (**78**), and descyano citalopram (**79**) (Figure 3.4). To minimize formation of desmethyl citalopram (**78**), the exchange is stopped at ~95% conversion, leaving 5% of the starting material **68** to remove during the workup procedure. An acid–base extraction does not remove side products **77–79** or the starting bromide **68** since all possess an amine side chain. Crystallization of citalopram (**39**) free base to eliminate the impurities is inefficient. Crystallization of a citalopram salt to eliminate the impurities is also inefficient.[90]

A prevalent purity upgrade strategy is to remove the impurities by chemical modification. Unconverted bromide **68** is reduced to descyano citalopram (**79**). Citalopram amide (**77**) is dehydrated to citalopram (**39**). Desmethyl citalopram (**78**) is converted to a nonbasic derivative that is separated by acid–base extraction, or to a nonbasic derivative bound to a polymer support, or to citalopram (**39**).

The bromide starting material **68** can be converted to descyano citalopram (**79**) by catalytic transfer hydrogenation but even descyano citalopram (**79**) is difficult to eliminate by salt formation in a single pass. A mixture of citalopram (**39**) containing 0.59% of the starting bromide **68** and 1.95% descyano citalopram (**79**), 50% water-wet palladium on carbon (5%), and ammonium formate in ethyl

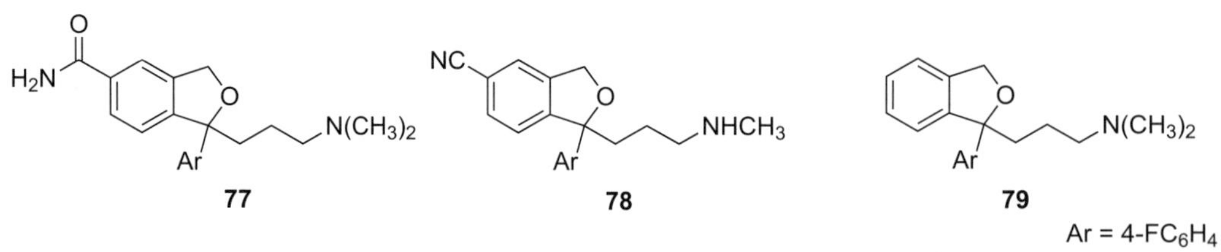


**FIGURE 3.4** Side products from the cyanide exchange reaction of 3-(5-bromo-1-(4-fluorophenyl)-1,3-dihydroisobenzofuran-1-yl)-*N*,*N*-dimethylpropan-1-amine (**68**).

acetate is refluxed for 7 h. Analysis reveals no remaining bromide **68** and 2.62% descyano citalopram (**79**). The suspension is cooled to 25°C and filtered. The liquors are washed with water and concentrated at reduced pressure. The isolated solid (a mixture of 95.4% of **39** and 2.62% of **79**) is suspended in acetone and oxalic acid is added. The resulting suspension is filtered and the solid is washed with cold acetone and dried. Analysis reveals 98.2% citalopram (**39**) oxalate and 1.18% descyano citalopram (**79**) oxalate.[98]

Desmethyl citalopram (**78**) is also difficult to eliminate by salt formation in a single pass. Oxalic acid is added to a solution of citalopram (**39**) crude containing up to 5% desmethyl citalopram (**78**) in acetone at 40°C. The mixture is heated to 50–55°C and then cooled to 25°C. The suspension is filtered and the solid is dried at 60°C and atmospheric pressure to afford citalopram (**39**) oxalate. The oxalate salt is converted to the free base and the free base is again converted to the oxalate salt to reduce the level of desmethyl citalopram (**78**) to <0.1%. The yield of citalopram (**39**) oxalate from citalopram (**39**) crude is 85%.[99]

Desmethyl citalopram (**78**) is efficiently removed from crude citalopram (**39**) by reaction with acetic anhydride to produce an acetamide **80**. Citalopram (**39**) and the acetamide **80** are separated by an acid–base extraction procedure. Acetic anhydride (4.0 equivalents) is added to a toluene solution of citalopram (**39**) containing 2.5% (*Note*: The procedure reads 25%.) of desmethyl citalopram (**78**). The mixture is heated at 60°C for 30 min. Water and 12 M hydrochloric acid are added (to pH 1). The layers are separated. The organic layer containing desmethyl citalopram acetamide (**80**) is discarded. Ammonium hydroxide is added to the aqueous acid solution (to pH 9) and citalopram free base is extracted with toluene. The extract is concentrated at reduced pressure to afford citalopram (**39**) containing <0.1% of desmethyl citalopram (**78**). The dehydration of citalopram amide (**77**) to citalopram (**39**) during the acetylation of desmethyl citalopram (**78**) is possible but not addressed.[100]

A solution of citalopram crude (**39**) (containing an unspecified amount of **78**) in toluene is stirred with methylisocyanate polystyrene resin at 25°C for 2 h. The resin is filtered and the liquors are concentrated at reduced pressure. Analysis of the residue reveals a 70% reduction in the amount of desmethyl citalopram (**78**).[89]

Aqueous formaldehyde (35%) (0.10 equivalents) is added to a mixture of citalopram crude (**39**) (containing 7% of **78**) and 98% formic acid. The mixture is aged at 85–95°C for 30 min. The mixture is cooled to 30°C and diluted with ethanol. Citalopram (**39**) is then isolated as the oxalate (79%, 99.7% pure by HPLC). No desmethyl citalopram (**78**) is detected in the citalopram oxalate.[93]

Citalopram amide (**77**) is efficiently converted to citalopram (**39**) by dehydration with phosphorus oxychloride. At the same time, desmethyl citalopram (**78**) can react with phosphorus oxychloride to produce an aqueous base-soluble phosphoramidate. Phosphorus oxychloride (0.67 equivalents) is added to a toluene solution of citalopram (**39**) crude containing 4.7% citalopram amide (**77**) and 0.72% desmethyl citalopram (**78**). The mixture is aged at 70°C for 1 h and then quenched with water. Hydrochloric acid is added (to pH 2.0–2.5) and the layers are separated. Ammonium hydroxide is added to the aqueous acid solution (to pH 9.0–9.5) and citalopram free base is extracted with toluene. The extracts are dried and concentrated at reduced pressure to afford citalopram (**39**) containing 0.05% citalopram amide (**77**) and 0.23% desmethyl citalopram (**78**). Starting with citalopram crude (**39**) containing 5.85% citalopram amide (**77**) and 7.43% desmethyl citalopram (**78**), the same process delivered citalopram (**39**) with the levels of **77** and **78** reduced to 0.36% and 0.45%, respectively. The levels of citalopram amide (**77**) and desmethyl citalopram (**78**) can be further reduced by crystallization from cyclohexane–isopropanol or cyclohexane–*n*-propanol but reducing the levels of these impurities to <0.1% remains a challenge.[96]

### 3.4.6 Citalopram (39) by Cyanation of an Arylmagnesium Bromide

3-(5-Bromo-1-(4-fluorophenyl)-1,3-dihydroisobenzofuran-1-yl)-*N*,*N*-dimethylpropan-1-amine (**68**) can be converted to a Grignard reagent. Reaction of the Grignard reagent with a cyanogen source (*p*-toluenesulfonyl cyanide, 1-cyanobenzotriazole, or cyanogen chloride) affords citalopram (**39**).

The Grignard reagent is prepared by the standard procedure. Bromoethane (6.7 mol%) is added to a suspension of magnesium in THF at 30–35°C. The activated suspension is heated to 55°C and a solution of 3-(5-bromo-1-(4-fluorophenyl)-1,3-dihydroisobenzofuran-1-yl)-*N*,*N*-dimethylpropan-1-amine (**68**) in THF is added over 7 h. The resulting solution is cooled to 20°C and stored under dry nitrogen.[101]

A solution of the Grignard reagent in THF is added to a solution of cyanogen chloride[102] (2.0 equivalents) in THF at −10°C. The mixture is warmed to 25°C and aged at 25°C overnight. The mixture is quenched with cold ammonium hydroxide and then warmed to 25°C to decompose the excess cyanogen chloride. Hydrochloric acid (1 M) is added (to pH 5) and the mixture is extracted with toluene. The extracts are washed with brine and concentrated at reduced pressure. Citalopram (**39**) is isolated from the residue by crystallization (presumably from isopropanol) (63%, >98% pure by HPLC). Chromatography is required to isolate citalopram (**39**) from the residue in similar procedures using *p*-toluenesulfonyl cyanide[103] (52%) and 1-cyanobenzotriazole (71%) (Scheme 3.23).[101]

Cyanogen chloride is broken down in the body to release cyanide. The exposure limit for cyanogen chloride is 0.3 ppm (ACGIH ceiling, NIOSH recommended ceiling). While

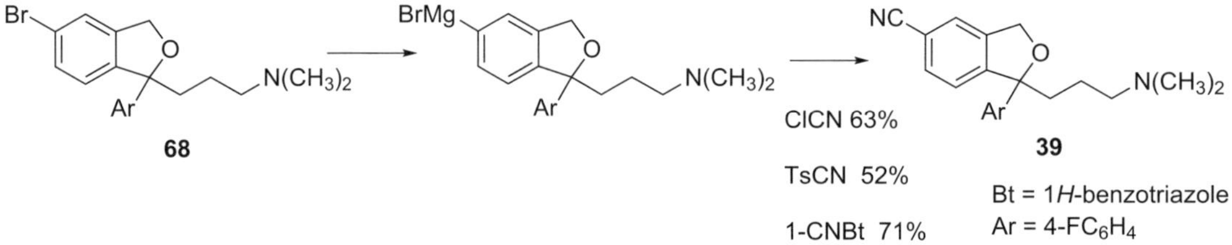


**SCHEME 3.23** Citalopram (**39**) by cyanation of a Grignard reagent from bromide **68** in the final step.

*p*-toluenesulfonyl cyanide and 1-cyanobenzotriazole would be preferred from a potential exposure perspective, they are both expensive specialty chemicals.

## 3.5 CONSTRUCTION OF CITALOPRAM (39) BY FUNCTIONAL GROUP TRANSFORMATION TO NITRILE IN THE FINAL STEP

Producing high-purity citalopram (**39**) from a cyanide exchange requires removing copper or zinc salts and cyanide to obtain citalopram crude and then reducing the levels of starting bromide **68**, desmethyl citalopram (**78**), citalopram amide (**77**), and descyano citalopram (**79**) to acceptable levels. These challenges prompted the identification of other nitrile surrogates, other functional groups that can be carried through a double Grignard reagent addition sequence to construct the isobenzofuran and can be transformed into the nitrile in the final step(s).

### 3.5.1 Citalopram (39) from 1-(3-(Dimethylamino) propyl)-1-(4-fluorophenyl)-1,3-dihydroisobenzofuran-5-amine (81)

An amine can be transformed into a nitrile by diazotization and Sandmeyer reaction with copper(I) cyanide and sodium cyanide. The required amine **81** is produced by the double Grignard reagent addition strategy.

A 23 wt% solution of 4-fluorophenylmagnesium bromide (3.0 equivalents) in THF is added to a suspension of 5-aminoisobenzofuran-1(3*H*)-one (**2**) in THF at <5°C. The mixture is aged at 25°C for 30 min. The mixture is cooled to 0°C and an 18 wt% solution of 3-dimethylaminopropylmagnesium chloride (1.0 equivalent) in THF is added at <5°C. The mixture is aged at <5°C for 30 min and at 25°C overnight. The mixture is quenched with cold aqueous acetic acid. The THF is distilled at reduced pressure and the remaining aqueous acidic solution is washed with ethyl acetate. Ammonium hydroxide is added (to pH 9) and the free base is extracted with ethyl acetate. The extracts are washed with water and concentrated at reduced pressure to afford diol **82** (58%). Diol **82** is cyclized in 60% aqueous phosphoric acid at 80°C over 2 h. The mixture is transferred into ice water. Ammonium hydroxide is added (to pH 9) and the free base is extracted with ethyl acetate. The extracts are washed with water, dried, and concentrated at reduced pressure to afford 1-(3-(dimethylamino)propyl)-1-(4-fluorophenyl)-1,3-dihydroisobenzofuran-5-amine (**81**). No yield is provided.

Diazotization is followed by Sandmeyer reaction with copper(I) cyanide (1.1 equivalents) and sodium cyanide (3.3 equivalents) in toluene–water at 50–60°C for 30 min. The layers are separated and the organic layer is washed with 10% aqueous sodium cyanide and concentrated at reduced pressure. Citalopram (**39**) is isolated from the residue by chromatography (32%) (Scheme 3.24).[104]

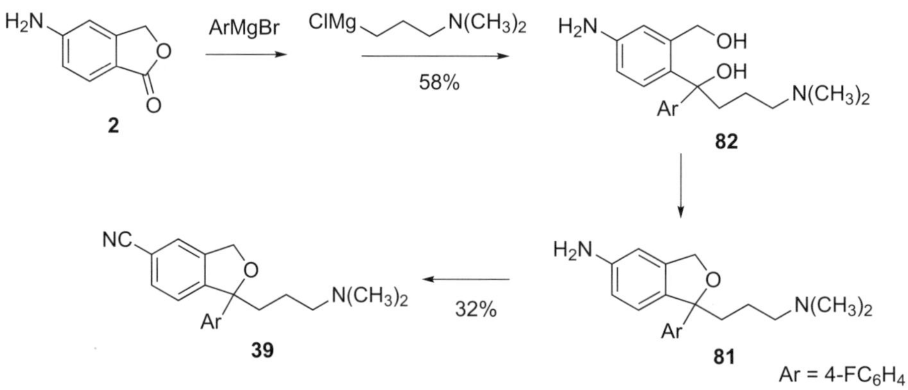


**SCHEME 3.24** Citalopram (**39**) from 5-aminoisobenzofuran-1(3*H*)-one (**2**).

### 3.5.2 Citalopram (39) from 1-(3-(Dimethylamino)propyl)-1-(4-fluorophenyl)-1,3-dihydroisobenzofuran-5-carboxylic Acid (83)

We have already seen conversion of a carboxylic acid to an acid chloride followed by reaction with sulfamide to produce a nitrile in one approach to 1-oxo-1,3-dihydroisobenzofuran-5-carbonitrile (**24**). This same two-step transformation can be used to convert 1-(3-(dimethylamino)propyl)-1-(4-fluorophenyl)-1,3-dihydroisobenzofuran-5-carboxylic acid (**83**) to citalopram (**39**). The required carboxylic acid (**83**) can be produced by the double Grignard reagent addition strategy starting with *tert*-butyl 1-oxo-1,3-dihydroisobenzofuran-5-carboxylate (**16**). It can also be produced from 3-(5-bromo-1-(4-fluorophenyl)-1,3-dihydroisobenzofuran-1-yl)-*N*,*N*-dimethylpropan-1-amine (**68**) by a halogen–metal exchange and capture of the arylmetal intermediate with carbon dioxide (Scheme 3.25).

A 21 wt% solution of 4-fluorophenylmagnesium bromide (1.2 equivalents) in THF is added to a suspension of *tert*-butyl 1-oxo-1,3-dihydroisobenzofuran-5-carboxylate (**16**) in THF at <5°C. The mixture is aged at 25°C for 3 h. The mixture is cooled to 5°C and a 16 wt% solution of 3-dimethylaminopropylmagnesium chloride (1.2 equivalents) in THF is added at <10°C. The mixture is aged at 25°C overnight. The mixture is quenched with cold aqueous ammonium chloride and the THF is distilled at reduced pressure. The remaining aqueous solution is extracted with ethyl acetate and the extract is washed with water and brine. Diol **86** is extracted from the organic layer into 2 M hydrochloric acid. Aqueous sodium hydroxide is added to the acidic extract (to pH >9) and diol **86** free base is extracted into ethyl acetate. The extract is washed with water and dried using a solid drying agent. The suspension is filtered and triethylamine (3.0 equivalents) is added to the liquors. The solution is cooled to 5°C and methanesulfonyl chloride (1.1 equivalents) in ethyl acetate is added. The mixture is aged for 1 h, presumably at 5°C. The mixture is then washed with 0.1 M sodium hydroxide, dried, and concentrated at reduced pressure. The residual *tert*-butyl 1-(3-(dimethylamino)propyl)-1-(4-fluorophenyl)-1,3-dihydroisobenzofuran-5-carboxylate (**85**) is dissolved in acetone and converted to the oxalate salt (43%). The *tert*-butyl ester is converted to the carboxylic acid by reaction with hydrogen bromide in acetic acid at 25°C.[32]

In the halogen–metal exchange process, butyllithium in hexanes (1.6 M, 2.7 equivalents) is added to a solution of 3-(5-bromo-1-(4-fluorophenyl)-1,3-dihydroisobenzofuran-1-yl)-*N*,*N*-dimethylpropan-1-amine (**68**) in THF at −78 to −65°C. The solution is allowed to warm to −30°C over 2 h and added to dry solid carbon dioxide. The mixture is aged at 25°C for 16 h and concentrated at reduced pressure. Water and 4 N hydrochloric acid are added to the residue (to pH 5.5) and the resulting mixture is extracted with toluene. The extracts are concentrated at reduced pressure to afford crude 1-(3-(dimethylamino)propyl)-1-(4-fluorophenyl)-1,3-dihydroisobenzofuran-5-carboxylic acid (**83**) (91%). A lower yield (73%) is observed in the analogous process via the arylmagnesium bromide. No yield is provided in the analogous process via a trialkylmagnesate.

Thionyl chloride (1.3 equivalents) is added to a solution of the carboxylic acid (**83**) and sulfamide (1.1 equivalents) in sulfolane and the solution is aged at 130°C for 2 h. The

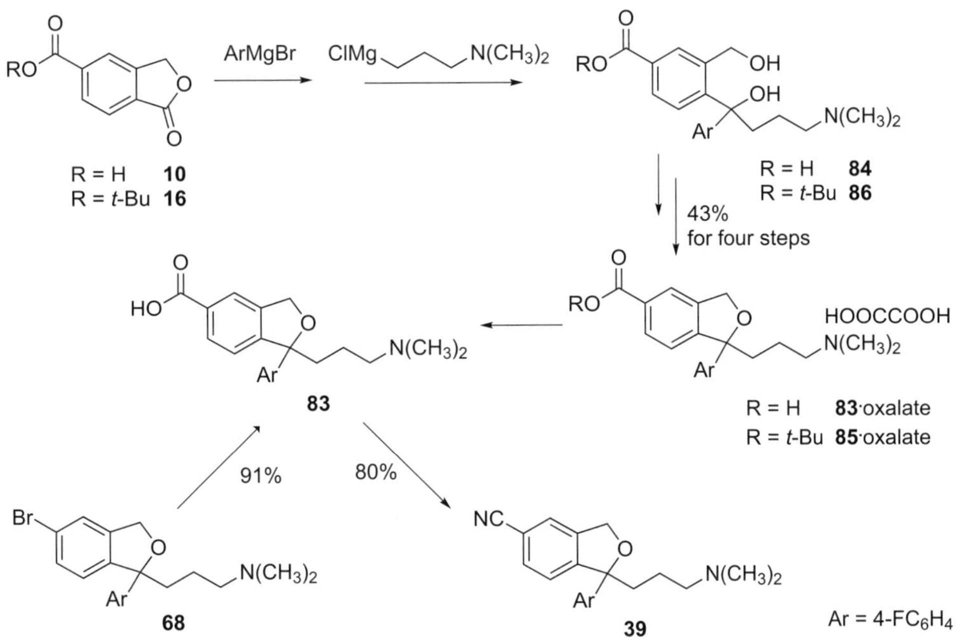


**SCHEME 3.25** Citalopram (**39**) from 1-(3-(dimethylamino)propyl)-1-(4-fluorophenyl)-1,3-dihydroisobenzofuran-5-carboxylic acid (**83**).

resulting mixture is cooled to 75°C and water is added. After aging at 75°C for 25 min, the mixture is cooled to 25°C and ammonium hydroxide (to pH 9) and *n*-heptane are added. The biphasic mixture is heated to 70°C and the layers are separated. The *n*-heptane layer is cooled to an unspecified temperature to afford citalopram (**39**) (80%).[105,106]

### 3.5.3 Citalopram (39) from 1-(3-(Dimethylamino)propyl)-1-(4-fluorophenyl)-1,3-dihydroisobenzofuran-5-carboxamide (88)

We have already seen conversion of a carboxamide to a nitrile with thionyl chloride and DMF in one approach to 1-oxo-1,3-dihydroisobenzofuran-5-carbonitrile (**24**). This same transformation can be moved later in the sequence to convert 1-(3-(dimethylamino)propyl)-1-(4-fluorophenyl)-1,3-dihydroisobenzofuran-5-carboxamide (**88**) to citalopram (**39**). The required amide is produced from the acid via the acid chloride.

Hydrolysis of the *tert*-butyl ester (**85**) oxalate salt with hydrogen bromide in acetic acid is complete at 25°C in 10 min. Acetic acid is distilled at reduced pressure. Residual acetic acid is removed by distillation with toluene at reduced pressure. The residue is dissolved in toluene and thionyl chloride (23 equivalents) and DMF (27 mol%) are added. After refluxing for 1 h, toluene is distilled at reduced pressure. The residue is dissolved in ethyl acetate and aqueous ammonium hydroxide and ice are added. The biphasic mixture is stirred at an unspecified temperature for 30 min. The organic layer is separated, washed with water and with brine, dried, and concentrated at reduced pressure to afford the crude carboxamide **88**. This is dissolved in thionyl chloride (11 equivalents) and the solution is refluxed for 2 h. Toluene is added and excess thionyl chloride and toluene are distilled at reduced pressure. The residue is dissolved in toluene. The solution is washed with 2 N sodium hydroxide and with water and concentrated at reduced pressure. The residual crude citalopram (**39**) is dissolved in acetone and converted to the oxalate salt (43% for five steps from *tert*-butyl ester **85**) (Scheme 3.26).[32]

Dehydration with phosphorus oxychloride also converts carboxamide **88** to the nitrile **39**. Carboxamide **88** is dissolved in acetonitrile at 60°C. Phosphorus oxychloride (1.0 equivalent) is added at a rate that maintains the mixture at reflux. When the addition is complete, the solution is cooled to 0°C and quenched by adding 20% aqueous sodium carbonate. The resulting suspension is filtered and the acetonitrile in the liquors is distilled at reduced pressure. Citalopram (**39**) is extracted from the residual aqueous mixture with ethyl acetate. The extracts are treated with carbon, the suspension is filtered, and the liquors are concentrated at reduced pressure to afford citalopram (**39**) (92%).[26]

### 3.5.4 Citalopram from *N-tert*-Butyl-1-(3-(dimethylamino)propyl)-1-(4-fluorophenyl)-1,3-dihydroisobenzofuran-5-carboxamide (89)

We have already seen conversion of an *N-tert*-butyl carboxamide to a nitrile with thionyl chloride in one approach to 1-oxo-1,3-dihydroisobenzofuran-5-carbonitrile (**24**). This same transformation can be moved later in the sequence to convert *N-tert*-butyl-1-(3-(dimethylamino)propyl)-1-(4-fluorophenyl)-1,3-dihydroisobenzofuran-5-carboxamide (**89**) to citalopram (**39**). The required *N-tert*-butyl carboxamide **89** is produced by the double Grignard reagent addition strategy starting with *N-tert*-butyl-1-oxo-1,3-dihydroisobenzofuran-5-carboxamide (**18**) (Scheme 3.27).

A 31 wt% solution of 4-fluorophenylmagnesium bromide (2.4 equivalents) in THF is added to a suspension of *N-tert*-butyl-1-oxo-1,3-dihydroisobenzofuran-5-carboxamide (**18**) in THF at <5°C. The mixture is aged at 25°C for 3 h. The mixture is cooled to 5°C and a 16 wt% solution of 3-dimethylaminopropylmagnesium chloride (1.2 equivalents) in THF is added at <10°C. The mixture is aged at 25°C overnight. The mixture is quenched with cold aqueous ammonium chloride and the THF is distilled at reduced pressure. The remaining aqueous solution is extracted with ethyl acetate and the extract is washed with water and brine. The diol **90** is extracted from the organic layer into 2 M hydrochloric acid. Aqueous sodium hydroxide is added to

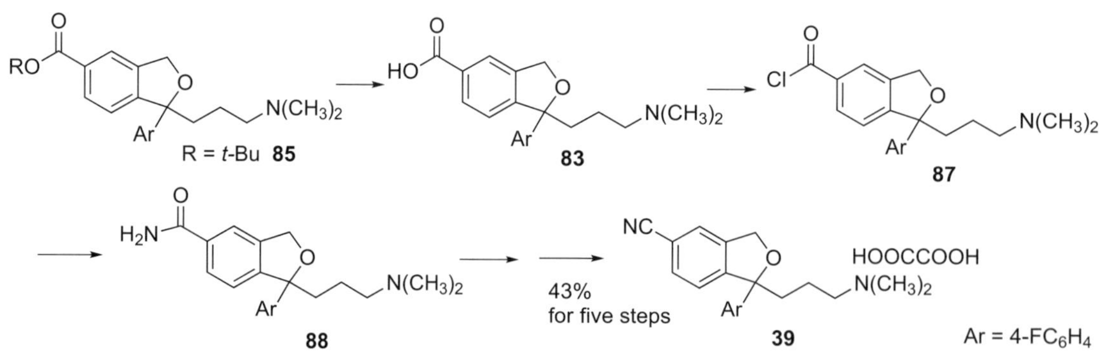


**SCHEME 3.26** Citalopram (**39**) from 1-(3-(dimethylamino)propyl)-1-(4-fluorophenyl)-1,3-dihydroisobenzofuran-5-carboxamide (**88**).

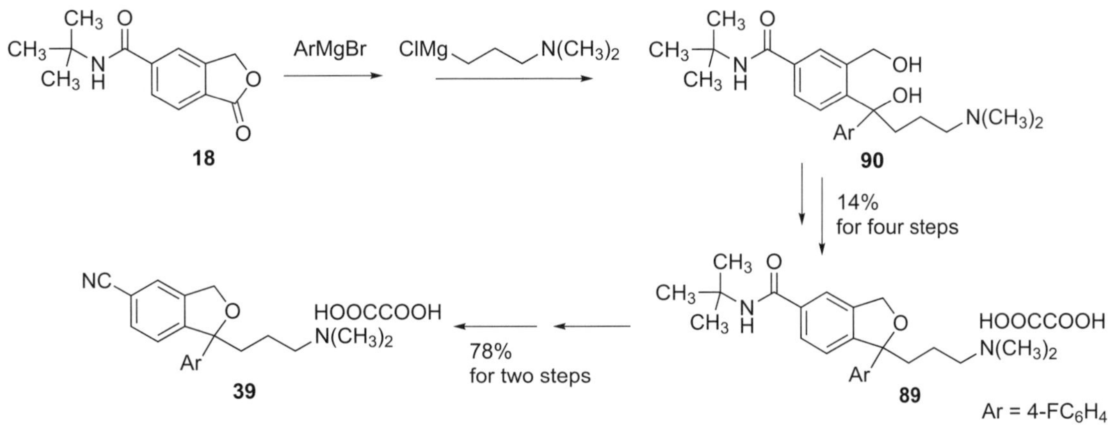


**SCHEME 3.27** Citalopram (**39**) from *N-tert*-butyl-1-oxo-1,3-dihydroisobenzofuran-5-carboxamide (**18**).

the acid extract (to pH >9) and the diol free base is extracted into ethyl acetate. The extract is washed with water and dried over a solid drying agent. The suspension is filtered and triethylamine (4.5 equivalents) is added to the liquors. The solution is cooled to 5°C and methanesulfonyl chloride (1.7 equivalents) in ethyl acetate is added. The mixture is aged, presumably at 5°C, for 1 h. The mixture is then washed with 0.1 M sodium hydroxide, dried, and concentrated at reduced pressure. The residue is dissolved in acetone and *N-tert*-butyl-1-(3-(dimethylamino)propyl)-1-(4-fluorophenyl)-1,3-dihydroisobenzofuran-5-carboxamide (**89**) is isolated as the oxalate salt (14%).

A solution of the oxalate salt in thionyl chloride (69 equivalents) (on a 1 g scale) is refluxed for 2 h. Toluene is added and excess thionyl chloride and toluene are distilled at reduced pressure. The residue is separated between ethyl acetate and cold aqueous ammonium hydroxide (12%). The organic layer is washed with water, dried, and concentrated at reduced pressure. The residue is dissolved in acetone and citalopram (**39**) is isolated as the oxalate salt (78%).[32]

### 3.5.5 Citalopram (39) from 1-(3-Dimethylamino)propyl)-1-(4-fluorophenyl)-1,3-dihydroisobenzofuran-5-carbaldehyde (91)

We have already seen the aldehyde-to-nitrile transformation, via oxime formation and dehydration, in approaches to 1-oxo-1,3-dihydroisobenzofuran-5-carbonitrile (**24**) and 1-(4-fluorophenyl)-1,3-dihydroisobenzofuran-5-carbonitrile (**31**). The aldehyde can be accessed by DIBAH reduction of an amide or by capture of the arylmagnesium bromide with DMF. Amide **92** is produced by the double Grignard reagent addition strategy starting with 5-(morpholine-4-carbonyl)isobenzofuran-1(3*H*)-one (**93**). After conversion of aldehyde **91** to the oxime **94** (96–97%), the oxime **94** is dehydrated with acetic anhydride in toluene or pyridine at >100°C. Citalopram (**39**) is isolated as the oxalate salt (66%) or free base (Scheme 3.28).

***3.5.5.1 1-(3-Dimethylamino)propyl)-1-(4-fluorophenyl)-1,3-dihydroisobenzofuran-5-carbaldehyde (91) by DIBAL Reduction of an Amide*** A 28 wt% solution of 4-fluorophenylmagnesium bromide (1.1 equivalents) in THF is added to a suspension of 5-(morpholine-4-carbonyl)isobenzofuran-1(3*H*)-one (**93**) in THF at <5°C. The mixture is aged at 25°C for 1.5 h. The mixture is cooled to 5°C and a 16 wt% solution of 3-dimethylaminopropylmagnesium chloride (1.1 equivalents) in THF is added at <10°C. The mixture is aged at 25°C overnight. The mixture is quenched with cold aqueous ammonium chloride and the THF is distilled at reduced pressure. The remaining aqueous solution is extracted with dichloromethane and the extract is washed with water and brine. Diol **94** is extracted from the organic layer into 2 M hydrochloric acid. Aqueous sodium hydroxide is added to the acid extract (to pH >9) and diol **94** is extracted into dichloromethane. The extract is washed with water and brine and dried over a solid drying agent. The suspension is filtered and triethylamine (1.3 equivalents) is added to the liquors. The solution is cooled to 5°C and methanesulfonyl chloride (0.73 equivalents) (presumably calculated to match the amount of **94** in solution) in dichloromethane is added. The mixture is aged, presumably at 5°C, for 1 h. The mixture is then washed with 0.1 M sodium hydroxide, dried, and concentrated at reduced pressure. The residual (1-(3-dimethylamino)propyl)-1-(4-fluorophenyl)-1,3-dihydroisobenzofuran-5-yl)(morpholino)methanone (**92**) is dissolved in acetone and converted to the oxalate salt (26%). The process using triethylamine (2.5 equivalents) and methanesulfonyl chloride (2.0 equivalents) for diol cyclization is used to produce 1-(3(-dimethylamino)propyl)-1-(4-fluorophenyl)-*N*,*N*-dimethyl-1,3-dihydroisobenzofuran-5-carboxamide (**96**) oxalate salt in 69% yield.

The morpholine amide (**92**) oxalate salt is presumably converted to a solution of the free base in toluene. The solution is cooled to 0°C and diisobutylaluminum hydride (DIBAH) (1.2 equivalents) is added at 0°C. The mixture is aged at 25°C for 2 h. Cold (0–5°C) water is carefully added to quench excess

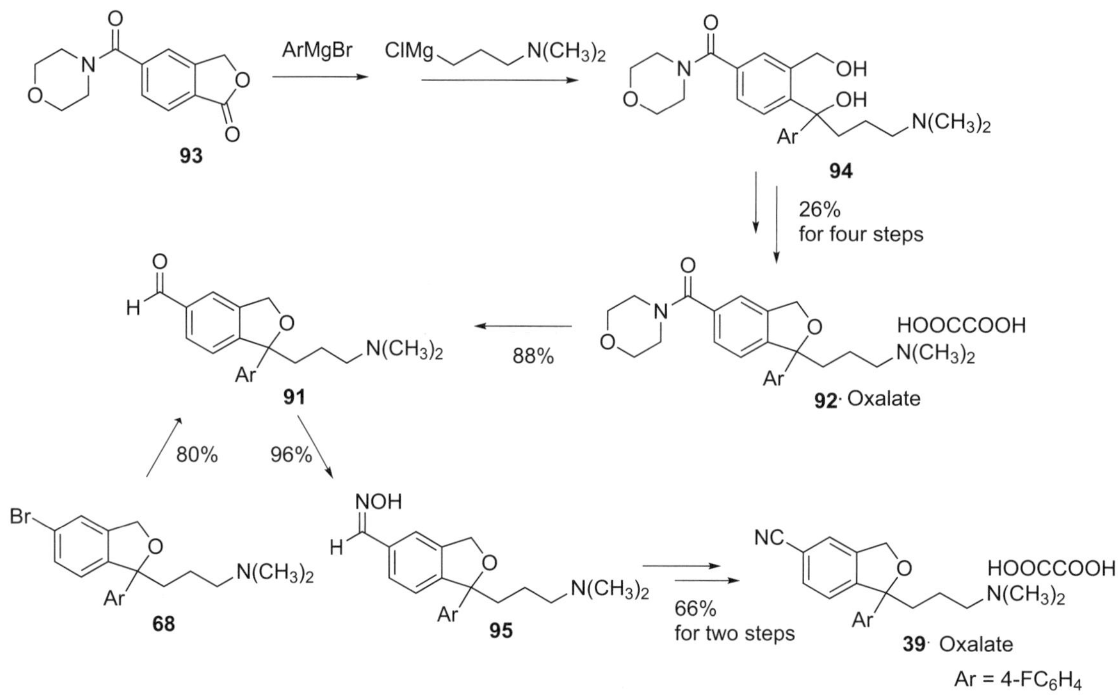


**SCHEME 3.28** Routes to 1-(3-dimethylamino)propyl)-1-(4-fluorophenyl)-1,3-dihydroisobenzofuran-5-carbaldehyde (**91**) and the transformation to citalopram (**39**).

hydride and the mixture is aged, presumably at 25°C, for 30 min. Solid potassium carbonate is added and the mixture is aged, presumably at 25°C, for 10 min. The suspension is filtered and the layers are separated. The organic layer is washed with water and concentrated at reduced pressure to afford aldehyde **91** (88%).[38]

***3.5.5.2 1-(3-Dimethylamino)propyl)-1-(4-fluorophenyl)-1,3-dihydroisobenzofuran-5-carbaldehyde (91) from a Grignard Reagent and DMF*** A 23 wt% solution of the Grignard reagent from 3-(5-bromo-1-(4-fluorophenyl)-1,3-dihydroisobenzofuran-1-yl)-*N*,*N*-dimethylpropan-1-amine (**68**) in THF is added to DMF (2.0 equivalents) at 20–25°C. The mixture is aged at 25°C for 3 h. The mixture is quenched with water and toluene. Acetic acid is added and the mixture is heated to 55–60°C. The organic layer is separated and washed with water. Analysis of the toluene solution reveals aldehyde **91** (83% contained yield). Aldehyde **91** is purified via the bisulfite adduct (96% recovery). Capture of the Grignard reagent with DMF is more efficient than capture of the aryllithium with DMF.[101]

***3.5.5.3 Citalopram (39) from the Aldehyde 91*** A solution of hydroxylamine hydrochloride (2.0 equivalents) in water is added to a solution of aldehyde **91** in ethanol. Aqueous sodium hydroxide (28%) is added (to pH 10). The mixture is aged, presumably at 25°C, for 14 h. Ethanol is distilled at reduced pressure. Ethyl acetate and water are added and the layers are separated. The organic layer is concentrated at reduced pressure to afford the oxime **95** (96%). Oxime **95** is dissolved in acetic anhydride and pyridine and the solution is refluxed for 2 h. The volatiles are distilled at reduced pressure. Toluene is added and the residual volatiles and toluene are distilled at reduced pressure. The residual crude citalopram (**39**) is dissolved in acetone and converted to the oxalate salt (66%). A comparable yield (83% yield of citalopram free base) is observed in dehydration of the oxalate salt of the oxime **95** using thionyl chloride in toluene.[38]

Oxime **95** is also produced by reaction with hydroxylamine sulfate (0.61 equivalents) (*Note*: This quantity is insufficient.) in toluene at reflux. Acetic acid is added during the reflux to maintain the pH at 4–5. The mixture is cooled and the layers are separated. Toluene and acetic anhydride (2.0 equivalents) are added to the aqueous layer containing oxime **95** (97% contained yield) and the mixture is refluxed for 3 h. The mixture is cooled to 85°C and quenched by slow addition of water. The biphasic mixture is cooled to 25°C and 30% sodium hydroxide is added (to pH 9–10). The layers are separated and the organic layer is washed twice with water at 60°C and concentrated at reduced pressure. Citalopram (**39**) is isolated from the residue by crystallization from isopropanol. No yield is provided.[101]

***3.5.5.4 Citalopram (39) from 5-(Dimethoxymethyl)isobenzofuran-1(3H)-one (97)*** A late-stage aldehyde-to-nitrile transformation is also prominently featured in

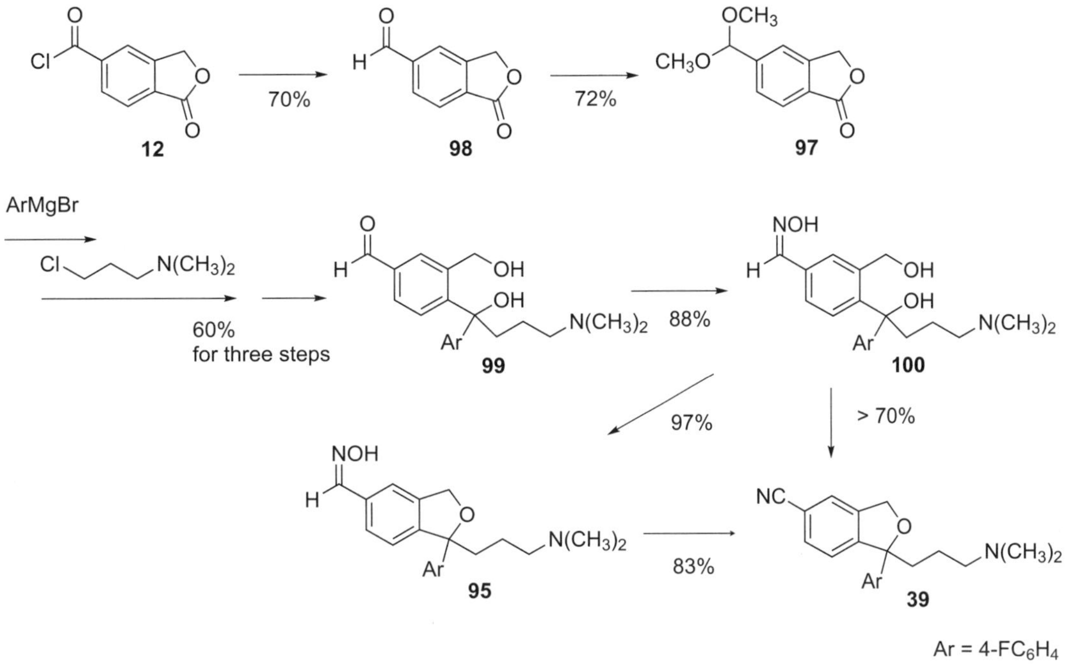


**SCHEME 3.29** Citalopram (**39**) from 5-(dimethoxymethyl)isobenzofuran-1(3*H*)-one (**97**).

another approach to citalopram (**39**) starting with 5-(dimethoxymethyl)isobenzofuran-1(3*H*)-one (**97**). The acetal is produced by reduction of the acid chloride and protection of the aldehyde. Workup of a double Grignard reagent addition affords the diol **99**. The aldehyde is converted to the oxime. Citalopram (**39**) is produced by diol cyclization and oxime dehydration (Scheme 3.29).

1-Oxo-1,3-dihydroisobenzofuran-5-carbonyl chloride (**12**) is reduced using 10% palladium on carbon poisoned with quinoline-S in toluene at 70°C and 56 psi hydrogen for 4 h. The suspension is cooled, transferred, heated to 100°C, and filtered. The solid is washed with hot toluene. The liquors are concentrated to a smaller volume, cooled to 10°C, and aged for 2 h. The resulting suspension is filtered and the solid is washed with cold toluene and dried at an unspecified temperature and reduced pressure to afford 1-oxo-1,3-dihydroisobenzofuran-5-carbaldehyde (**98**) (70%, 99% pure by HPLC).

The aldehyde is protected with methanol using *p*-toluenesulfonic acid as catalyst (12 mol%) at 35°C over 3 h. The suspension is filtered and the liquors are concentrated at reduced pressure. The residue is suspended in water, saturated aqueous bicarbonate is added (to pH 8), and the mixture is extracted with ethyl acetate. The extract is washed with water and with brine and then dried and concentrated at reduced pressure. Isopropanol is added to the residue and the suspension is aged for 3 h at 15°C. The suspension is filtered and the solid is dried to afford 5-(dimethoxymethyl)isobenzofuran-1(3*H*)-one (**97**) (72%).

A 13 wt% solution of 4-fluorophenylmagnesium bromide (1.0 equivalent) in THF is added to a suspension of 5-(dimethoxymethyl)isobenzofuran-1(3*H*)-one (**97**) in THF at <0°C. The mixture is aged at 0–5°C for 1 h. A 26 wt% solution of 3-dimethylaminopropylmagnesium chloride (1.6 equivalents) in THF is added at <10°C. The mixture is aged at 25°C overnight. The mixture is quenched with saturated aqueous ammonium chloride. The layers are separated and the aqueous layer is extracted with ethyl acetate. Methanol, water, and 6 M hydrochloric acid are added to the combined organic layers. The mixture is aged at 25–30°C for 30 min and then concentrated at reduced pressure to reduce the volume by half. Water is added and the mixture is washed with ethyl acetate. Ammonium hydroxide is added to the aqueous layer (to pH 10–10.5) and the mixture is extracted with ethyl acetate. The extracts are dried and concentrated at reduced pressure to afford diol **99** (60%, 98% pure by HPLC).

A solution of hydroxylamine hydrochloride (1.8 equivalents) in water is added to a solution of diol aldehyde **99** in 95% ethanol. The mixture is aged at 30°C for 30 min and then diluted with water. Ammonium hydroxide (25%) is added (to pH 9.5) and the mixture is extracted with ethyl acetate. The extract is dried and concentrated at reduced pressure to afford the diol oxime **100** (88%, 98% pure by HPLC).

Triethylamine (4.0 equivalents) and methanesulfonyl chloride (2.2 equivalents) are added to a solution of diol oxime **100** in dichloromethane at 0°C. The mixture is quenched with aqueous sodium hydroxide and the layers are separated. The organic layer is washed three times with water and then concentrated at reduced pressure. The residue is dissolved in acetone and citalopram (**39**) is isolated as the hydrobromide salt (70%).

The yield is comparable when the diol cyclization and oxime dehydration are accomplished separately. Reaction of the diol oxime **100** with 69% sulfuric acid in toluene at 80°C is complete in 1 h. The mixture is cooled to 25°C and transferred into ice water. The phases are separated. Sodium hydroxide (1 M) is added to the aqueous layer (to pH >7) and the mixture is extracted with toluene. The extract is concentrated at reduced pressure (97% yield, 95% pure by HPLC). Oxime **95** is then dehydrated with acetic anhydride at reflux. Acetic acid and excess acetic anhydride are distilled at reduced pressure. The residue is suspended in water. Ammonium hydroxide (25%) is added (to pH 9) and the mixture is extracted with ethyl acetate. The extract is washed with water, dried, and concentrated at reduced pressure. Citalopram (**39**) is isolated from the residue by chromatography (83%).[34]

### 3.5.6 Citalopram (39) from 3-(5-(4,4-Dimethyl-4,5-dihydrooxazol-2-yl)-1-(4-fluorophenyl)-1,3-dihydroisobenzofuran-1-yl)-*N,N*-dimethylpropan-1-amine (101)

We have already seen conversion of a 4,4-dimethyloxazoline to a nitrile with thionyl chloride in one approach to 1-oxo-1,3-dihydroisobenzofuran-5-carbonitrile (**24**). This same transformation can be used later in the sequence to convert 3-(5-(4,4-dimethyl-4,5-dihydrooxazol-2-yl)-1-(4-fluorophenyl)-1,3-dihydroisobenzofuran-1-yl)-*N,N*-dimethylpropan-1-amine (**101**) to citalopram (**39**). The required 4,4-dimethyloxazoline is produced by double Grignard reagent addition starting with 5-(4,4-dimethyl-4,5-dihydrooxazol-2-yl)isobenzofuran-1(3*H*)-one (**21**) (Scheme 3.30).

A 20 wt% solution of 4-fluorophenylmagnesium bromide (2.0 equivalents) in THF is added to a suspension of 5-(4,4-dimethyl-4,5-dihydrooxazol-2-yl)isobenzofuran-1(3*H*)-one (**21**) in THF at −15 to −10°C. The mixture is aged at 5–10°C for 1 h. The mixture is cooled to −5°C and a 30 wt% solution of 3-dimethylaminopropylmagnesium chloride (1.5 equivalents) in THF is added at −5 to −2°C. The mixture is aged at 5–10°C for 1 h. The mixture is quenched by addition of 15% aqueous ammonium chloride. After the acid–base extraction procedure, the final toluene extracts are washed with water and aged at 25°C for 3 h and at 5°C for 15 h. The resulting suspension is filtered and the solid is washed with toluene to afford crude diol **102** (71%). Triethylamine (4.2 equivalents) is added to a solution of diol **102** in dichloromethane. The solution is cooled to 5°C and methanesulfonyl chloride (1.5 equivalents) is added at 5–7°C. The mixture is aged at 25°C for 2 h and then cooled to an unspecified temperature. Aqueous sodium hydroxide (0.1 M) is added and the layers are separated. The organic layer is washed with water and concentrated at reduced pressure to afford crude 3-(5-(4,4-dimethyl-4,5-dihydrooxazol-2-yl)-1-(4-fluorophenyl)-1,3-dihydroisobenzofuran-1-yl)-*N,N*-dimethylpropan-1-amine (**101**) (96%).

Phosphorus oxychloride (2.0 equivalents) is added to a solution of oxazoline **101** in pyridine at 5–10°C. The mixture is then refluxed (115–116°C) for 3–4 h. The mixture is cooled to 10°C and quenched with water. Ammonium hydroxide (28%) is added (to pH 9) and citalopram free base is extracted with toluene. The extracts are treated with carbon. The suspension is filtered and the liquors are concentrated at reduced pressure. The residual crude citalopram (**39**) is dissolved in acetone and converted to the hydrobromide salt (65–67%).[30]

Citalopram (**39**) is also produced by reaction of the oxazoline **101** with thionyl chloride (47 equivalents) (on a 2–3 g scale) at reflux for 3 h. Excess thionyl chloride is distilled at reduced pressure. The residue is suspended in

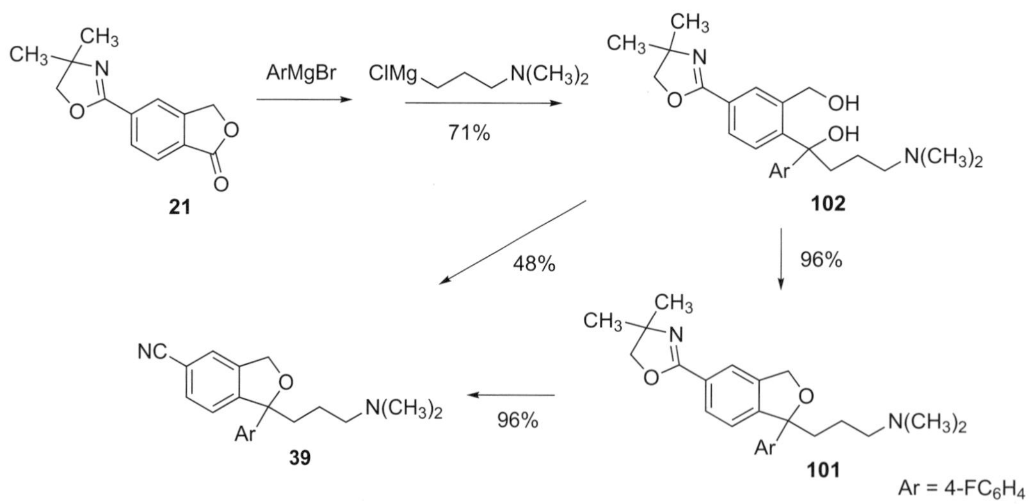


**SCHEME 3.30** Citalopram (**39**) from 5-(4,4-dimethyl-4,5-dihydrooxazol-2-yl)isobenzofuran-1(3*H*)-one (**21**).

water and toluene and 2 M sodium hydroxide are added (to pH 9). The layers are separated and the organic layer is washed with water and concentrated at reduced pressure to afford crude citalopram (**39**) (96%). No purity data are available.[30]

The diol cyclization and oxazoline-to-nitrile transformation can be accomplished under the same conditions. Reaction of diol **102** with phosphorus oxychloride (3 equivalents) in pyridine at 20–25°C for 1 h and then at 80–90°C for 4 h affords citalopram (**39**) (38%). A higher yield (48%) is observed in the reaction with thionyl chloride (1.5 equivalents) in pyridine at 20–25°C for 1 h followed by reaction with phosphorus oxychloride (2 equivalents) for 4 h. Details of the workup procedures and purities for citalopram (**39**) produced are not available.[26]

### 3.5.7 Citalopram (39) from 3-(5-(Aminomethyl)-1-(4-fluorophenyl)-1,3-dihydroisobenzofuran-1-yl)-*N*,*N*-dimethylpropan-1-amine (103)

In a process described earlier, the intermediate (1-(4-fluorophenyl)-1,3-dihydroisobenzofuran-5-yl)methanol (**40**) was produced by hydride reduction of 4-(4-fluorobenzoyl)isophthalic acid (**48**) followed by cyclization of the resulting triol **49**. Zinc in acetic acid selectively reduces the ketone 4-(4-fluorobenzoyl)isophthalic acid (**48**) to the alcohol. Cyclization then affords another perhaps useful lactone intermediate **104**. A link is suggested between intermediate **104** and citalopram (**39**) where the final transformation in the sequence is an oxidation of an aminomethyl group to the nitrile (Scheme 3.31). The experimental data for the sequence are incomplete. No data are provided on the introduction of the dimethylaminopropyl side chain by alkylation.[107]

## 3.6 CONSTRUCTION OF CITALOPRAM (39) BY ELABORATION OF AN ALCOHOL, ALDEHYDE, OR AMINE TO COMPLETE THE DIMETHYLAMINOPROPYL SIDE CHAIN IN THE FINAL STEP

It is suggested that 3-chloro-*N*,*N*-dimethylpropan-1-amine is prone to polymerization and that alkylation of 1-(4-fluorophenyl)-1,3-dihydroisobenzofuran-5-carbonitrile (**31**) with other electrophiles may be more efficient with respect to utilization of the electrophile. Introduction of an electrophile followed by transformation into the 3-dimethylaminopropyl chain constitutes a novel approach to citalopram (**39**). Despite the many patents on the subject, this general approach is described in detail in just a few specific cases. It is suggested that LDA is a suitable base for the alkylation and that the long list of suitable electrophiles includes ethyl acrylate, *N*,*N*-dimethyl acrylamide, allyl bromide, epichlorohydrin, 3-(dimethylamino)propanal, *N*,*N*-dimethyl-1-(oxiran-2-yl)methanamine, 4-bromo-1-butene, 4-bromo-1-butyramide, and protected forms of 3-halo-1-propanol and 3-halopropanal. Of course, 1-(4-fluorophenyl)-1,3-dihydroisobenzofuran-5-carbonitrile (**31**) could be replaced by other 5-substituted 1-(4-fluorophenyl)-1,3-dihydroisobenzofurans, creating a still more complex array of potential routes to citalopram. The discussion will be limited to the

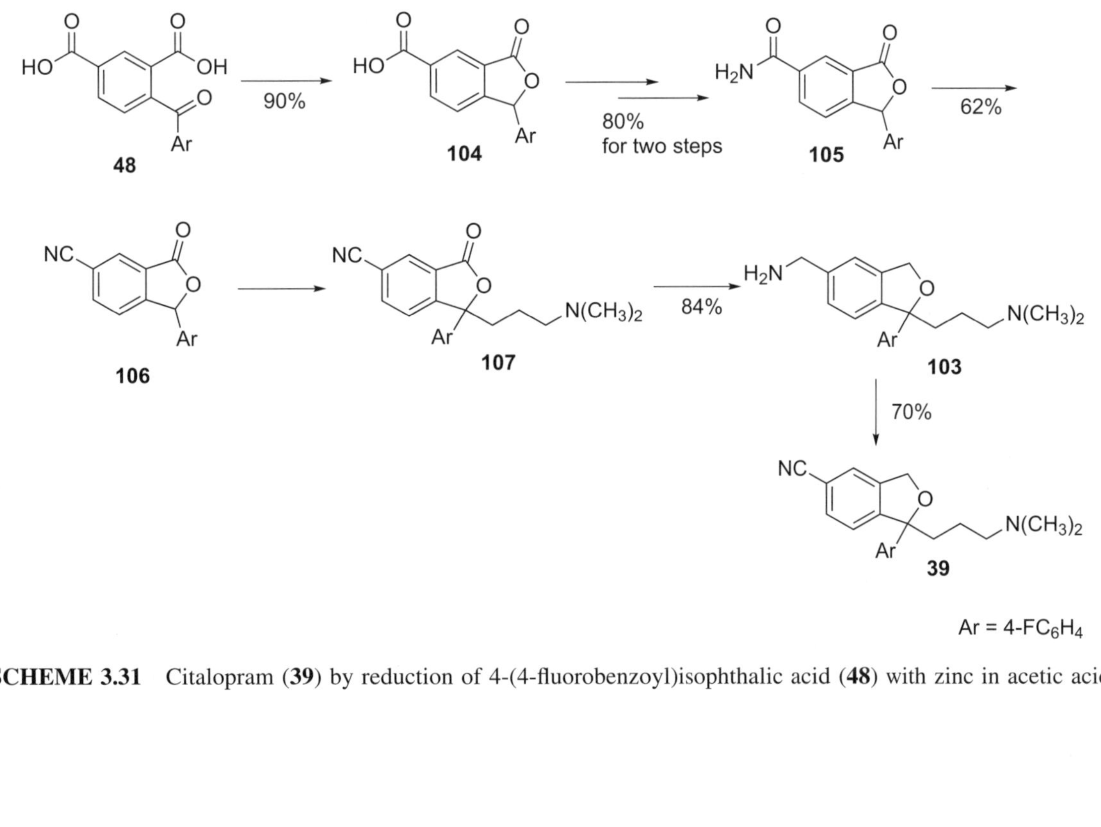


**SCHEME 3.31** Citalopram (**39**) by reduction of 4-(4-fluorobenzoyl)isophthalic acid (**48**) with zinc in acetic acid.

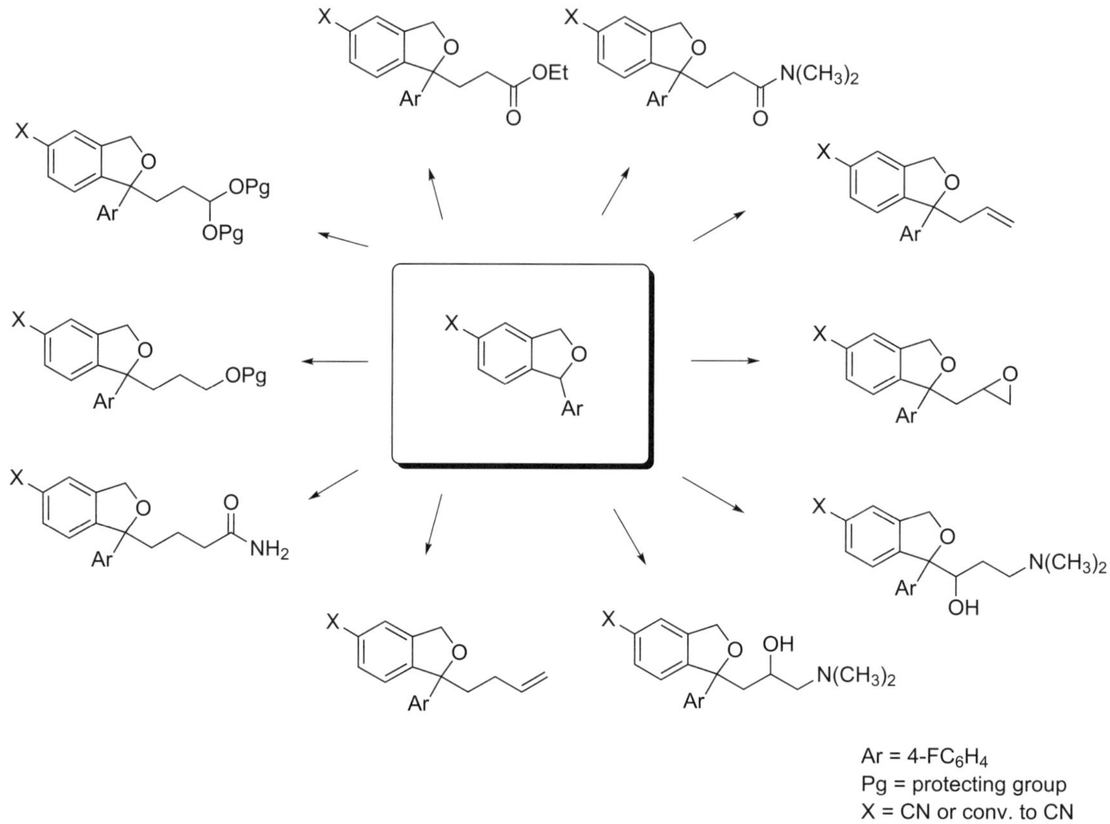


**SCHEME 3.32** Citalopram intermediates by alkylation of 1-(4-fluorophenyl)-1,3-dihydroisobenzofuran-5-carbonitrile (**31**) with other electrophiles.

alkylation of 1-(4-fluorophenyl)-1,3-dihydroisobenzofuran-5-carbonitrile (**31**) with a protected 3-bromo-1-propanol, a protected 3-bromopropanal, and 3-chloropropylamine and conversion of the alkylation products to citalopram (**39**) (Scheme 3.32).[108–114]

### 3.6.1 Citalopram (39) from 1-(4-Fluorophenyl)-1-(3-hydroxypropyl)-1,3-dihydroisobenzofuran-5-carbonitrile (108)

The protected bromopropanols are all specialty chemicals. Benzyl 3-bromopropyl ether is commercially available but expensive.[115] The tetrahydropyranyl (THP) ether and *tert*-butyl dimethylsilyl (TBDMS) ether are likely prepared from expensive 3-bromo-1-propanol.[116]

A solution of 1-(4-fluorophenyl)-1,3-dihydroisobenzofuran-5-carbonitrile (**31**) in THF is added to a solution of LDA (1.3 equivalents) in THF at −78°C. The mixture is aged at −78°C for 30 min and a solution of benzyl 3-bromopropyl ether (1.3 equivalents) in THF is added. The mixture is allowed to warm to 25°C and aged at 25°C for 2 h. The mixture is transferred into ice water and extracted with ethyl ether. The extracts are washed with water and with brine, dried, and concentrated at reduced pressure. 1-(3-Benzyloxy)propyl)-1-(4-fluorophenyl)-1,3-dihydroisobenzofuran-5-carbonitrile (**109**) is isolated from the residue by chromatography (60%). The benzyl ether is hydrogenolyzed by reaction with 1,4-cyclohexadiene (19 equivalents) and 5% palladium on carbon (61 mol%) in ethanol at reflux over 2 days. The suspension is cooled to 25°C and filtered. The liquors are concentrated at reduced pressure and **108** is isolated by chromatography (48% from **31**) (Scheme 3.33).

The same alkylation procedure is used to prepare THP-protected and TBDMS-protected alcohols **110** and **111**. The THP ether of **110** is hydrolyzed by reaction with *p*-toluenesulfonic acid monohydrate (7.7 mol%) in methanol at 25°C over 1 h. The TBDMS ether of **111** is deprotected with 1 M hydrochloric acid (3.3 equivalents) in methanol at 25°C over 1 h. In both cases, the mixture is concentrated at reduced pressure and 1-(4-fluorophenyl)-1-(3-hydroxypropyl)-1,3-dihydroisobenzofuran-5-carbonitrile (**108**) is isolated from the residue by chromatography (55% via **110**, 81% via **111**).[117]

Triethylamine (4.3 equivalents) and methanesulfonyl chloride (3.0 equivalents) are added to a solution of 1-(4-fluorophenyl)-1-(3-hydroxypropyl)-1,3-dihydroisobenzofuran-5-carbonitrile (**31**) in THF at 0–5°C and the mixture is aged at 25°C overnight. The mixture is diluted with toluene, washed with water and with saturated aqueous sodium

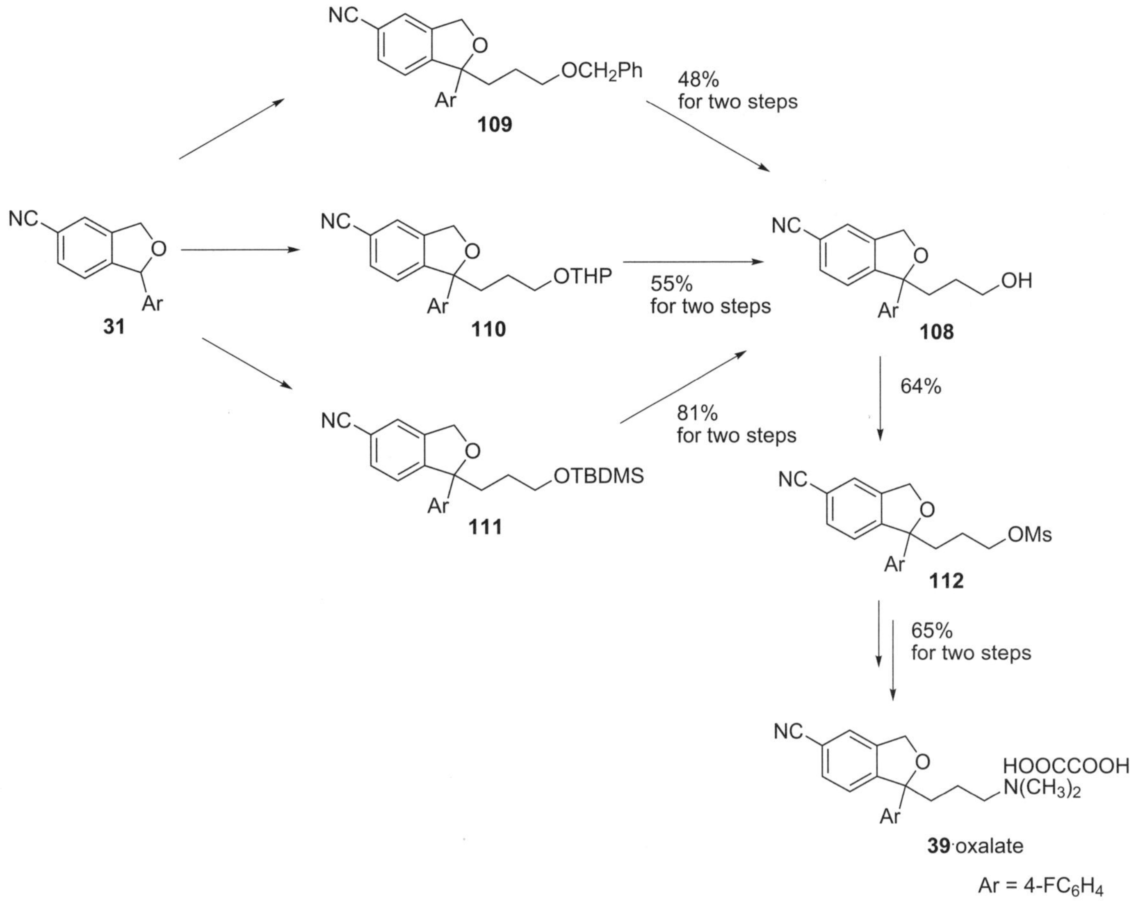


**SCHEME 3.33** Citalopram (**39**) from 1-(4-fluorophenyl)-1-(3-hydroxypropyl)-1,3-dihydroisobenzofuran-5-carbonitrile (**108**).

bicarbonate, and concentrated at reduced pressure. The methanesulfonate **112** is isolated from the residue by chromatography (64%). A solution of dimethylamine (37 equivalents) in ethanol is added to methanesulfonate **112** in ethanol–THF. The mixture is aged at 25°C for 1 h and at 60°C for 3 h. Excess dimethylamine, ethanol, and THF are distilled at reduced pressure. Aqueous 1 M sodium hydroxide is added to the residue and the mixture is extracted with ethyl ether. The extracts are washed with brine, dried, and concentrated at reduced pressure. Citalopram (**39**) is isolated from the residue by chromatography and then converted to the oxalate salt in acetone (65%).[117,118]

Triethylamine (2.1 equivalents) and *p*-toluenesulfonyl chloride (1.6 equivalents) are added to a solution of 1-(4-fluorophenyl)-1-(3-hydroxypropyl)-1,3-dihydroisobenzofuran-5-carbonitrile (**108**) in THF at 0–5°C and the mixture is aged at 25°C for 3 days. The mixture is diluted with toluene, washed with water and with saturated aqueous sodium bicarbonate, and concentrated at reduced pressure. The *p*-toluenesulfonate **113** is isolated from the residue by chromatography (42%).

Triethylamine (1.8 equivalents) and dimethylamine (1.3 equivalents) are added to a solution of *p*-toluenesulfonate **113** in DMF and the mixture is aged at 70°C overnight. The mixture is cooled to 25°C, transferred into ice water, and extracted with ethyl ether. The extracts are washed with water and with brine, dried, and concentrated at reduced pressure. Citalopram (**39**) is isolated from the residue by chromatography and then converted to the oxalate salt in acetone (70%).[117,118]

### 3.6.2 Citalopram (39) from 1-(4-Fluorophenyl)-1-(3-oxopropyl)-1,3-dihydroisobenzofuran-5-carbonitrile (114)

The side chain aldehyde **114** is available by alkylation of 1-(4-fluorophenyl)-1,3-dihydroisobenzofuran-5-carbonitrile (**31**) or by double Grignard reagent addition and diol cyclization from 1-oxo-1,3-dihydroisobenzofuran-5-carbonitrile (**24**). Commercially available 2-(2-bromoethyl)-1,3-dioxolane is the side chain precursor in both routes.

***3.6.2.1 1-(4-Fluorophenyl)-1-(3-oxopropyl)-1,3-dihydroisobenzofuran-5-carbonitrile (114) from 1-(4-Fluorophenyl)-1,3-dihydroisobenzofuran-5-carbonitrile (31)*** A solution of 1-(4-fluorophenyl)-1,3-dihydroisobenzofuran-

5-carbonitrile (**31**) in THF is added to a solution of LDA (1.3 equivalents) in THF at −78°C. The mixture is aged at −78°C for 30 min and a solution of 2-(2-bromoethyl)-1,3-dioxolane[119] (1.3 equivalents) in THF is added. The mixture is allowed to warm to 25°C and aged at 25°C for 2 h. The mixture is transferred into ice water and extracted with ethyl ether. The extracts are washed with water and with brine, dried, and concentrated at reduced pressure. 1-(2-(1,3-Dioxolan-2-yl)ethyl)-1-(4-fluorophenyl)-1,3-dihydroisobenzofuran-5-carbonitrile (**115**) is isolated from the residue by chromatography (86%).

The acetal is deprotected by reaction with 30% aqueous acetic acid at reflux over 5 h. The mixture is cooled and extracted with dichloromethane. The extracts are concentrated at reduced pressure to afford the crude aldehyde **114** (50%). Sodium cyanoborohydride (1.8 equivalents) is added to a mixture of crude aldehyde **114** and dimethylammonium chloride (1.8 equivalents) in methanol at 0–5°C. The mixture is warmed to 25°C and aged at 25°C overnight. The mixture is diluted with toluene and ethyl acetate and washed with water. The water wash is back-extracted with ethyl ether. The combined organic layers are concentrated at reduced pressure. Citalopram (**39**) is isolated from the residue by chromatography and then converted to the oxalate salt in acetone (82%).[117]

*3.6.2.2 1-(4-Fluorophenyl)-1-(3-oxopropyl)-1,3-dihydroisobenzofuran-5-carbonitrile (114) from 1-Oxo-1,3-dihydroisobenzofuran-5-carbonitrile (24)* A solution of 4-fluorophenylmagnesium bromide (1.1 equivalents) in ethyl ether is added to a suspension of 1-oxo-1,3-dihydroisobenzofuran-5-carbonitrile (**24**) in THF at 5–10°C. The mixture is warmed to 25°C and aged at 25°C for 5 h. A solution of the Grignard reagent from 2-(2-bromoethyl)-1,3-dioxolane in THF (1.5 equivalents) is added at 25°C and the mixture is aged at 25°C for 14 h. The mixture is quenched with aqueous ammonium chloride at 0°C. The organic layer is separated, washed with water, and concentrated at reduced pressure. The diol acetal **116** is isolated from the residue by chromatography (50%).

Methanesulfonyl chloride (1.7 equivalents) is added to a solution of diol acetal **116** and triethylamine (4.3 equivalents) in dichloromethane at 5–10°C. The mixture is warmed to 25°C and aged for 10 min. Aqueous sodium hydroxide (2%) is added at 10–20°C. The organic layer is separated, washed with water, and concentrated at reduced pressure to afford 1-(2-(1,3-dioxolan-2-yl)ethyl)-1-(4-fluorophenyl)-1,3-dihydroisobenzofuran-5-carbonitrile (**115**). The yield exceeds the theoretical yield. The acetal is hydrolyzed by reaction with 5 M hydrochloric acid in acetone. The mixture is aged at 25°C for 60 h. Acetone is distilled at reduced pressure. The residual aqueous mixture is extracted with ethyl acetate. The extracts are concentrated at reduced pressure. The deprotection with 5 M hydrochloric acid is repeated on the residue to again afford an ethyl acetate extract. The extract is washed with water and concentrated at reduced pressure to afford 1-(4-fluorophenyl)-1-(3-oxopropyl)-1,3-dihydroisobenzofuran-5-carbonitrile (**114**) (Scheme 3.34). The yield again exceeds the theoretical yield.[120]

### 3.6.3 Citalopram (39) from Didesmethyl Citalopram (117) and Desmethyl Citalopram (78)

*3.6.3.1 Didesmethyl Citalopram (117)* There are three approaches to didesmethyl citalopram (**117**). 1-(4-Fluorophenyl)-1-(3-hydroxypropyl)-1,3-dihydroisobenzofuran-5-carbonitrile (**108**) is converted to the methanesulfonate **112**. Methanesulfonate displacement by azide and reduction of the azide affords didesmethyl citalopram (**117**). Alternatively, 1-(4-fluorophenyl)-1-(3-oxopropyl)-1,3-dihydroisobenzofuran-5-carbonitrile (**114**) is condensed with *tert*-butylsulfinamide. Borohydride reduction and acid

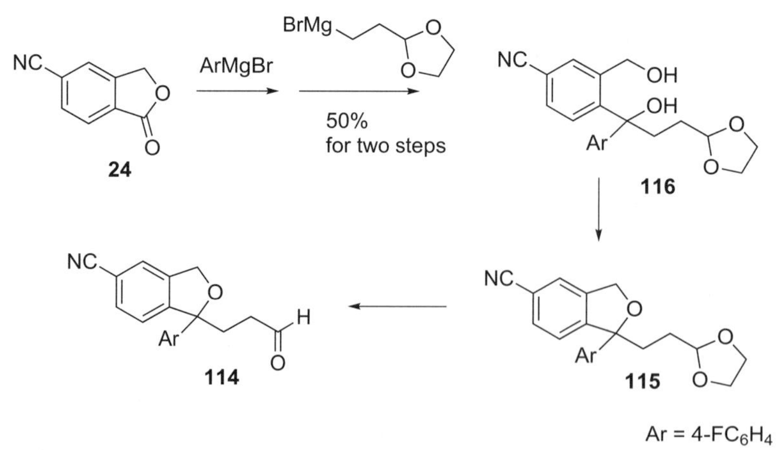


**SCHEME 3.34** 1-(4-Fluorophenyl)-1-(3-oxopropyl)-1,3-dihydroisobenzofuran-5-carbonitrile (**114**) from 1-oxo-1,3-dihydroisobenzofuran-5-carbonitrile (**24**).

cleavage affords didesmethyl citalopram (**117**). Finally, alkylation of 1-(4-fluorophenyl)-1,3-dihydroisobenzofuran-5-carbonitrile (**31**) with 3-chloropropylamine affords didesmethyl citalopram (**117**). Didesmethyl citalopram (**117**) is converted to citalopram (**39**) with formaldehyde and sodium cyanoborohydride or with formaldehyde and formic acid (Eschweiler–Clarke amine methylation) (Scheme 3.35).

A mixture of the methanesulfonate **112** and sodium azide (7.6 equivalents) in DMF is heated at 40°C for 3 h and then refluxed for 2 h. The mixture is cooled, transferred into water, and extracted with ethyl ether. The extracts are washed with water and with brine and then dried and concentrated at reduced pressure to afford crude azide **118** (45%). Azide reduction with hydrogen and palladium on carbon (6.4 mol%) in ethanol, presumably at 25°C and 1 atm, is stopped after 2 h. The suspension is filtered and the liquors are concentrated at reduced pressure to afford crude didesmethyl citalopram (**117**) (66%).[117,118]

(−)-*tert*-Butylsulfinamide[121] (1.2 equivalents) and titanium(IV) ethoxide (1.9 equivalents) in ethanol are added to a solution of 1-(4-fluorophenyl)-1-(3-oxopropyl)-1,3-dihydroisobenzofuran-5-carbonitrile (**114**) in THF. The mixture is aged at 25°C for 10 min and at 55°C for 1 h. The mixture is cooled to 5–10°C and quenched by adding brine and ethyl acetate. The resulting suspension is aged for 10 min and filtered. The liquor layers are separated and the organic layer is washed with brine and concentrated at reduced pressure. The residue is dissolved in THF, the solution is cooled to 5–10°C, and sodium borohydride (4.1 equivalents) in methanol is added. The mixture is aged for 14 h, presumably at 5–10°C. Water is added at 5–10°C and the mixture is extracted with ethyl acetate. The extract is washed with brine and concentrated at reduced pressure. The *tert*-butylsulfinamide **120** (a mixture of diastereoisomers) is isolated from the residue by chromatography (86%). Hydrochloric acid (10%) is added to a solution of the *tert*-butylsulfinamide **120** in methanol and the mixture is aged at 25°C for 16 h. Methanol is distilled at reduced pressure. Water, methyl *tert*-butyl ether, and aqueous potassium carbonate (to pH >7) are added and the layers are separated. The organic layer is washed with water and with brine and then dried and concentrated at reduced pressure to afford didesmethyl citalopram (**117**) (88%).[120]

The alkylation of 1-(4-fluorophenyl)-1,3-dihydroisobenzofuran-5-carbonitrile (**31**) with 3-chloropropylamine could reduce the sequence from 1-(4-fluorophenyl)-1,3-dihydroisobenzofuran-5-carbonitrile (**31**) to didesmethyl citalopram (**117**) by several steps. However, a great deal of development effort and our discussion has been expended to find a superior side chain precursor *because 3-chloro-N,N-dimethylpropan-1-amine is likely to polymerize during*

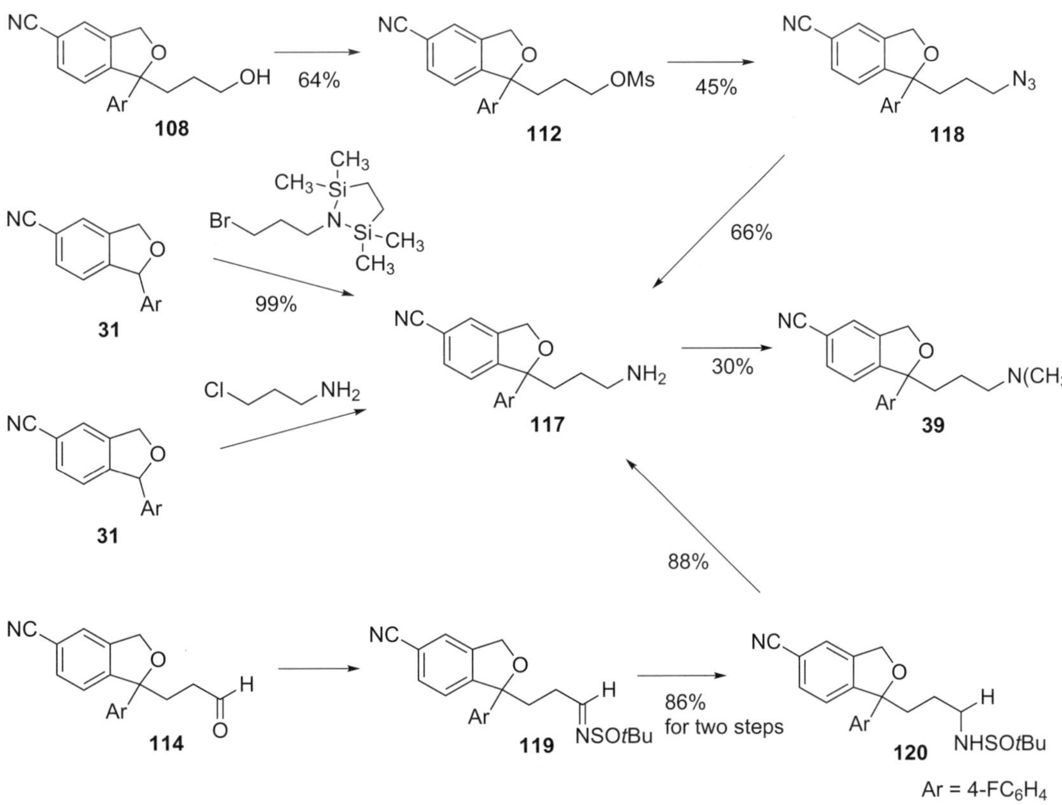


**SCHEME 3.35** Citalopram (**39**) from didesmethyl citalopram (**117**).

*storage*. Certainly 3-chloropropylamine will be even more prone to polymerization.

3-Chloropropylamine is released from the hydrochloride salt[122] using 15% aqueous hydroxide and toluene at 25–35°C for 35–40 min. The layers are separated and the aqueous layer is extracted twice with toluene. The combined organic layers are used directly in the alkylation step.

Potassium *tert*-butoxide (1.5 equivalents) is added to DMSO. The mixture is aged at 60–65°C for an unspecified time and then cooled to 25°C. A solution of 1-(4-fluorophenyl)-1,3-dihydroisobenzofuran-5-carbonitrile (**31**) in DMSO is added at 25–35°C and the mixture is aged for 15–20 min. The toluene solution of 3-chloropropylamine (1.7 equivalents) is added rapidly and the mixture is heated to 40–45°C and aged at 40–45°C for 60–70 min. The mixture is presumably cooled to 25°C and transferred into 0–5°C water. The layers are separated and the aqueous layer is extracted with toluene. The combined toluene layers are extracted with dilute hydrochloric acid (at pH 2). The aqueous acidic extract is washed with toluene. Aqueous sodium hydroxide is added to the aqueous layer (to pH 10–12) and the mixture is extracted with toluene. The toluene extracts are washed with water at 60–65°C and concentrated at reduced pressure to afford didesmethyl citalopram (**117**). No yield is provided. Alkylation in DMSO (no toluene) or acetone is also described.[123]

The storage stability of 3-chloropropylamine has been the subject of some debate. The available data suggest that 3-chloropropylamine polymerizes during storage at 0–5°C. The data also suggest that there is some increase in storage stability of 3-chloropropylamine in toluene or dichloromethane solution. These solutions may be useful in a laboratory process. However, the requirements for carefully choreographed release of the free base (<72 h before use), minimal storage time of the solution, use of the water-wet solution, and rapid addition of the solution to the reaction mixture all suggest that the alkylation of 1-(4-fluorophenyl)-1,3-dihydroisobenzofuran-5-carbonitrile (**31**) with 3-chloropropylamine is not viable as a manufacturing process.[124–126]

The potential for polymerization can be circumvented using a protection–deprotection strategy. Sodium hydride (1.1 equivalents) is added to DMSO. The mixture is aged at 65°C for 20 min and then cooled to 25°C. A solution of 1-(4-fluorophenyl)-1,3-dihydroisobenzofuran-5-carbonitrile (**31**) in DMSO is added and the mixture is aged for 20 min. 1(3-Bromopropyl)-2,2,5,5-tetramethyl-1-aza-2,5-disilacyclopentane[127] is added and the mixture is aged at 25°C for 1 h. The mixture is transferred into ice water and extracted with ethyl acetate. The extracts are dried and concentrated at reduced pressure. The residue is dissolved in 2 M hydrochloric acid and the solution is washed with ethyl ether. Ammonium hydroxide is added to the aqueous acid solution (to pH >7) and the mixture is extracted with ethyl acetate. The extracts are dried and concentrated at reduced pressure to afford didesmethyl citalopram (**117**) (99%). The yield is impressive but the protected 3-bromopropylamine is expensive.[128]

***3.6.3.2 Methylation of Didesmethyl Citalopram (117)*** Sodium cyanoborohydride (2.0 equivalents) is added to a mixture of didesmethyl citalopram (**117**) and 37% aqueous formaldehyde (2.0 equivalents) in methanol and the mixture is aged at 25°C for 3 h. More formaldehyde (1.0 equivalent) and sodium cyanoborohydride (1.0 equivalent) are added and the mixture is aged at 25°C for 1 h. Water is added and the mixture is extracted with ethyl ether. The extracts are dried and concentrated at reduced pressure. Citalopram (**39**) is isolated from the residue by chromatography (30%). A more efficient reductive methylation with formaldehyde and formic acid will be discussed in the context of escitalopram synthesis.[117,118]

***3.6.3.3 Desmethyl Citalopram (78)*** Desmethyl citalopram (**78**) might be prepared from 1-(4-fluorophenyl)-1-(3-hydroxypropyl)-1,3-dihydroisobenzofuran-5-carbonitrile (**108**) by conversion of the alcohol to a leaving group and displacement of that leaving group with methylamine. Desmethyl citalopram (**78**) might also be prepared from 1-(4-fluorophenyl)-1-(3-oxopropyl)-1,3-dihydroisobenzofuran-5-carbonitrile (**114**) by reductive amination of the aldehyde with methylamine. With citalopram (**39**) already in hand, desmethyl citalopram (**78**) is conveniently prepared by demethylation (Scheme 3.36).

1-Chloroethyl chloroformate (1.6 equivalents) is added to citalopram (**39**) in toluene–dichloromethane at <20°C. Diisopropylethylamine (0.47 equivalents) is added at <20°C. The mixture is aged at 20°C for 30 min and at 60–65°C for 2–3 h. Methanol is added at 60–65°C and the mixture is aged at 60–65°C for 2–4 h. Dichloromethane and methanol are distilled at reduced pressure. The mixture is cooled to 20–25°C and water is added. The layers are separated and the aqueous layer is washed with toluene. Ammonium hydroxide is added to the aqueous layer (to pH 9–10) and the mixture is extracted with toluene. The toluene extracts are washed with water and with brine and then dried and treated with carbon. The suspension is filtered and the liquors are concentrated at reduced pressure to afford desmethyl citalopram (**78**) (92%). This has been demonstrated on a 140 kg scale.[129]

The reductive methylation will be discussed in the context of escitalopram synthesis.

## 3.7 CITALOPRAM BY DIOL CYCLIZATION IN THE FINAL STEP

Double Grignard reagent addition processes to this point deliver diols with a range of substituents in the 5-position (X

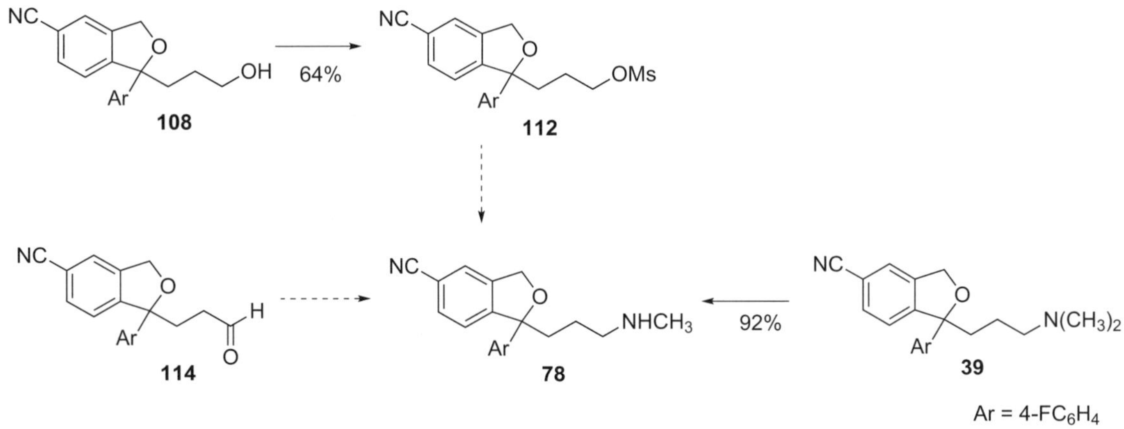


**SCHEME 3.36** Routes to desmethyl citalopram (**78**).

= Br, OH, NHCOOR, COOH, COOR, $CONH_2$, CONHR, $CONR_2$) despite the fact that many of these substituents are prone to attack by Grignard reagents. The obvious substituent missing from the list is X = CN. The addition of Grignard reagents to nitriles is a well-established method for producing ketones. Does 4-fluorophenylmagnesium bromide react with 1-oxo-1,3-dihydroisobenzofuran-5-carbonitrile (**24**) at the nitrile or the lactone carbonyl? The question has already been answered. 4-Fluorophenylmagnesium bromide reacts at the lactone carbonyl to produce ketone **37**. The subsequent addition of 3-dimethylaminopropylmagnesium chloride to the ketone **37** affords the diol **121**, which can be cyclized to produce citalopram (**39**) in what has to be the shortest sequence.[130–139]

### 3.7.1 4-(4-(Dimethylamino)-1-(4-fluorophenyl)-1-hydroxybutyl)-3-(hydroxymethyl)benzonitrile (121)

In a procedure dating back to 1986, a 29 wt% solution of 4-fluorophenylmagnesium bromide (1.2 equivalents) in THF is added to a suspension of 1-oxo-1,3-dihydroisobenzofuran-5-carbonitrile (**24**) in THF at 0–3°C over 3–h. The mixture is allowed to warm to 25°C and aged at 25°C overnight. The mixture is cooled to 10°C and a 29 wt% solution of 3-dimethylaminopropylmagnesium chloride (1.0 equivalent) in THF is added at 10–12°C over 6 h. The mixture is allowed to warm to 25°C and aged at 25°C overnight. The mixture is quenched with ice water and acetic acid is added (to pH 6.5–7.0). The THF is distilled at <50°C and 60 mmHg. Toluene is added followed by ammonium hydroxide (25%) (to pH 9). The biphasic mixture is heated to 45–50°C. The layers are separated and the aqueous layer is extracted with toluene. The combined toluene extracts are washed with water at 50°C, presumably cooled to 25°C, and then extracted with 20 wt% acetic acid. Ammonium hydroxide (25 wt%) is added to the aqueous acidic extract (to pH >9) and the mixture is extracted with toluene. The toluene extracts are washed with 50°C water four times to produce a solution of diol **121** ready for the next step. The contained yield is at least 40–45%. Pure diol **121** hydrobromide can be isolated from this solution (44% crude, 35% after purity upgrade). Diol **121** in this solution can also be cyclized to citalopram (**39**), which is isolated as the hydrobromide (41–42%).[140]

In a procedure from 2006, a solution of 4-fluorophenylmagnesium bromide (equivalents not specified) in THF is added to a suspension of 1-oxo-1,3-dihydroisobenzofuran-5-carbonitrile (**24**) in THF at −10°C over 2 h. A solution of 3-dimethylaminopropylmagnesium chloride (equivalents not specified) in THF is then added, presumably at −10°C. The mixture is aged at an unspecified temperature until complete. The mixture is quenched with 15 wt% aqueous ammonium chloride. The layers are separated and the organic layer is concentrated at reduced pressure. The residual oil is separated between toluene and water. The toluene layer is concentrated at reduced pressure to afford an oil. The contained yield of diol **121** in the oil is 80%. Some unexpected selectivity is suggested in the comparable contained yield of the diol **121** (84%) observed when a THF solution containing both 4-fluorophenylmagnesium bromide (1.8 equivalents) and 3-dimethylaminopropylmagnesium chloride (1.1 equivalents) is added at −10 to 0°C.[141,142]

Some of the relatively expensive THF can be replaced by toluene or dichloromethane. A 29 wt% solution of 4-fluorophenylmagnesium bromide (1.4 equivalents) in THF is added to a suspension of 1-oxo-1,3-dihydroisobenzofuran-5-carbonitrile (**24**) in dichloromethane at −4 to −2°C. The mixture is aged at an unspecified temperature until the starting material is consumed. A 24 wt% solution of 3-dimethylaminopropylmagnesium chloride (2.3 equivalents) in toluene–THF is added at 0–5°C over 6 h. The mixture is aged at −5 to 0°C for 4 h. The mixture is quenched with 20% aqueous ammonium chloride. The layers are separated and the organic layer is washed with water. Dichloromethane and

THF are distilled at reduced pressure and the resulting toluene solution is used in the diol cyclization. Citalopram (**39**) is isolated as the hydrobromide (65–67%).[143,144]

Based on the results from the earlier discussion of the reaction of 1-oxo-1,3-dihydroisobenzofuran-5-carbonitrile (**24**) with 4-fluorophenylmagnesium bromide, we should expect approximately 15% yield of side product **46** resulting from the double addition of the Grignard reagent. Since side product **46** has no amine side chain, it can be separated by an acid–base extraction procedure during the workup. Crude diol **121** isolated after acid–base extraction is likely to be >95% pure.

Crude diol **121** from an acid–base extraction procedure can be crystallized from isopropyl ether–*n*-heptane or ethanol–water. Crude diol **121** is dissolved in hot isopropyl ether. *n*-Heptane is added and the solution is allowed to cool to 5°C. The resulting suspension is filtered and the solid is dried to afford diol **121** (95% recovery, 99.9% pure by HPLC). Alternatively, crude diol **121** is dissolved in hot ethanol. Water is added and the solution is cooled, presumably to 25°C, and seeded. The resulting suspension is cooled to 5°C and filtered. The solid is dried to afford diol **121**, perhaps as a hydrate or solvate (94% recovery, 99.9% pure by HPLC).[145]

The side product **46** might also be separated after the workup by salt formation with hydrobromic acid in ethanol.[140,146]

### 3.7.2 Citalopram (39) from 4-(4-(Dimethylamino)-1-(4-fluorophenyl)-1-hydroxybutyl)-3-(hydroxymethyl) benzonitrile (121)

The diol cyclization can be accomplished by activation and $S_N1$ displacement of either the primary benzylic or the tertiary benzylic alcohol (phosphoric acid, sulfuric acid, hydrochloric acid, or *p*-toluenesulfonic acid at 70–85°C). Cyclization can also be accomplished by conversion of the primary benzylic alcohol to a leaving group (using methanesulfonyl chloride, *p*-toluenesulfonyl chloride, 2,5-dichloronitrobenzene, or diethylazodicarboxylate and triphenylphosphine) and $S_N2$ displacement of the leaving group at low temperature. Some nitrile hydrolysis is likely using aqueous acid at elevated temperatures. The yield of crude citalopram (**39**) by either cyclization method can be nearly quantitative. Finally, the primary benzylic alcohol can also be converted to a leaving group earlier in the sequence. The cyclization by $S_N2$ displacement immediately follows addition of the second Grignard reagent.

#### *3.7.2.1 Diol Cyclization by Activation with Strong Acid*

A mixture of the diol **121** in toluene and 70% sulfuric acid is heated at 80°C for 3 h. The mixture is cooled to 30°C and cold water and ammonium hydroxide (25%) are added (to pH 10). The mixture is aged at 50–60°C for 15 min and the layers are separated. The toluene layer is washed with warm water and then dried. The toluene solution is treated with silica gel and the suspension is filtered. The liquors are concentrated at up to 50°C and 30 mmHg. The residue is dissolved in acetone and citalopram (**39**) is isolated as the hydrobromide (53–57% from 1-oxo-1,3-dihydroisobenzofuran-5-carbonitrile (**24**)). Three crystallizations afford pure citalopram (**39**) hydrobromide (41–42% from 1-oxo-1,3-dihydroisobenzofuran-5-carbonitrile (**24**)).[140]

A mixture of diol **121** in toluene and 76% sulfuric acid is heated at 85–90°C for 4–5 h. The mixture is presumably cooled and cold water is added. Ammonium hydroxide (25%) is added at 10–15°C (to pH >7). The layers are separated. The acid–base extraction often used to upgrade the purity of diol **121** is instead performed at this point in the process. The toluene layer is washed with water and then extracted with 20% acetic acid. Ammonium hydroxide is added to the acidic extract at 5–10°C (to pH 8.5–9.0) and the mixture is extracted with toluene. The toluene extracts are washed with water, dried, and treated with carbon. The suspension is filtered and the citalopram (**39**) in the toluene solution is converted to the hydrobromide salt (65–67% from 1-oxo-1,3-dihydroisobenzofuran-5-carbonitrile (**24**)).[143]

A mixture of diol **121** in toluene and 69% sulfuric acid (0.36–0.37 g sulfuric acid/g diol **121**) is heated at 78–85°C for 1–2 h. The mixture is cooled and water is added. Ammonium hydroxide (25%) is added (to pH >9.5–10.5) and the mixture is heated to 55°C. The layers are separated. The toluene layer is washed with water and concentrated at <60°C and reduced pressure to afford citalopram (**39**) as an oil (97–99%).[147]

A mixture of diol **121** in acetonitrile and 69% sulfuric acid (0.36 g sulfuric acid/g diol **121**) is heated at 78–85°C for 3 h. The mixture is cooled and water and toluene are added. Ammonium hydroxide (25%) is added (to pH >9.5–10.5) and the mixture is heated to 55°C. The layers are separated. The aqueous layer is extracted with toluene at 55°C. The combined toluene layers are washed with water and concentrated at <60°C and reduced pressure to afford citalopram (**39**) as an oil (100%) (Scheme 3.37). The yield is lower (89–90%) when the sulfuric acid charge is increased (1.2–1.5 g sulfuric acid/g diol **121**). Excellent yields (95–100%) are also possible using hydrochloric acid or *p*-toluenesulfonic acid in place of sulfuric acid.[147]

A mixture of diol **121** in toluene and phosphoric acid is heated at 78–85°C for 4 h. (*Note*: If the phosphoric acid used is 85%, the phosphoric acid charge is 3.4 g/g diol **121**.) The mixture is cooled and water is added. Ammonium hydroxide (25%) is added (to pH >9.5–10.5) and the mixture is heated to 55°C. The layers are separated. The toluene layer is washed with water and concentrated at <60°C and reduced pressure to afford citalopram (**39**) as an oil (100%).[147]

A mixture of diol **121** in toluene and phosphoric acid is heated at 80°C for 2.5 h. (*Note*: If the phosphoric acid is 85%, the phosphoric acid charge is 3.4 g/g diol **121**.) The mixture

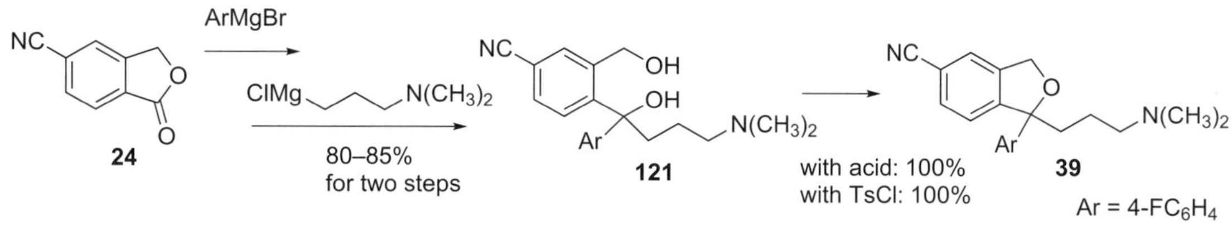


**SCHEME 3.37** Citalopram (**39**) from 1-oxo-1,3-dihydroisobenzofuran-5-carbonitrile (**24**) by the double Grignard reagent addition process.

is cooled to 50°C and 20% sodium hydroxide is added (to pH 10). The layers are separated and the toluene solution is washed with water and then dried. The toluene solution is treated with silica gel and the suspension is filtered. The liquors are concentrated at reduced pressure to afford citalopram (**39**) as an oil (>86%, 99.7% pure by HPLC).[145]

The double Grignard reagent addition reaction mixture can be carried directly into the cyclization reaction. The double Grignard reagent addition reaction mixture is added to 85% phosphoric acid at <10°C. The THF is distilled from the resulting mixture at atmospheric pressure. The remaining acidic solution is aged at 70–80°C for 2 h. The mixture is presumably cooled and toluene and water are added. The layers are separated and the aqueous acidic layer is washed with toluene. The aqueous acidic layer is treated with carbon at 25°C and the suspension is filtered. Toluene is added to the liquors. Ammonium hydroxide (30%) is added to the biphasic mixture at <15°C (to pH 9.5). Any salts that precipitate are filtered and the phases are separated. The toluene layer is concentrated at reduced pressure. The residue is dissolved in acetone and citalopram (**39**) is isolated as the hydrobromide salt (37%). One crystallization affords pure citalopram hydrobromide (27% from 1-oxo-1,3-dihydroisobenzofuran-5-carbonitrile (**24**)).[142]

Some hydrolysis of the nitrile is likely during diol cyclization with aqueous acid at elevated temperature. A mixture of diol **121** and 21% phosphoric acid is heated at 105°C for 14 h. (*Note*: The percent phosphoric acid presumes the phosphoric acid charged is 85%.) The mixture is cooled to 40°C and diluted with water. Ammonium hydroxide is added (to pH 8–10) and the mixture is extracted with ethyl acetate. The extract contains citalopram (**39**), citalopram amide (**77**) (3%), and diol **121** (0.8%).[148]

***3.7.2.2 Diol Cyclization by Displacement of a Sulfonate Ester*** A solution of methanesulfonyl chloride (1.2 equivalents) in dichloromethane is added at 0–5°C over 30 min to a solution of diol **121** and triethylamine (2.8 equivalents) in dichloromethane. The mixture is washed twice with 0.1 M sodium hydroxide. The organic layer is dried and concentrated at reduced pressure to afford citalopram (**39**) (81%). The citalopram can perhaps be upgraded by a reslurry procedure using *n*-heptane.[146,149]

A solution of methanesulfonyl chloride (1.5 equivalents) in dichloromethane is added at <10°C to a solution of diol **121** and triethylamine (3.0 equivalents) in dichloromethane. The mixture is aged at <10°C for 1 h. Water is added and the layers are separated. The organic layer is washed with water and concentrated at reduced pressure. The residue is dissolved in toluene and the toluene solution is washed with water and concentrated at reduced pressure. The residue is dissolved in acetone and citalopram (**39**) (in this case, escitalopram (**1**)) is isolated from the solution as the oxalate salt (49%).[150]

A solution of *p*-toluenesulfonyl chloride (1.1 equivalents) in toluene is added to a solution of the diol and triethylamine (2.1 equivalents) in toluene–acetonitrile (8:1) at <5°C. (*Note*: The procedure reads <50°C.) The mixture is aged at 10°C for 20 min. Water and ammonium hydroxide (25%) are added (to pH 9.5). The layers are separated and the aqueous layer is extracted with toluene at 45°C. The combined toluene layers are washed with water and concentrated at <50°C and reduced pressure to afford citalopram (**39**) (100%).[146]

2,5-Dichloronitrobenzene (1.2 equivalents) is added to a suspension of diol **121** and potassium carbonate (3.0 equivalents) in DMSO at 25°C. The mixture is aged at 100–105°C for 15 h. The mixture is cooled to 30°C and water is added. The resulting mixture is extracted with toluene. The toluene extract is washed with water and with 5% sodium hydroxide and then subjected to an acid–base extraction procedure. After concentration of the toluene solution at reduced pressure, the residual oil is dissolved in isopropanol–water and citalopram (**39**) (in this case, escitalopram (**1**)) is isolated as the oxalate salt (69%).[150]

Ethyl azodicarboxylate (3.9 equivalents) is added to a solution of diol **121** and triphenylphosphine (3.0 equivalents) in THF at 0°C. Sodium *tert*-butoxide (2.4 equivalents) is added, presumably in portions, at 0°C. The mixture is aged overnight, presumably at 25°C. The reaction is quenched by adding 1 M hydrochloric acid and the THF is distilled at reduced pressure. Toluene and water are added and the layers are separated. Ammonium hydroxide (30%) is added to the aqueous layer (to pH 9.4) and the resulting mixture is extracted with toluene. The toluene extract is concentrated at reduced pressure. The residue is dissolved in acetone and citalopram (**39**) is isolated from the solution as the hydrobromide salt. The purity of the salt is upgraded by crystallization from water (42% yield, 99.9% pure by HPLC).[151]

*3.7.2.3 Cyclization by Displacement of Chloride or Pivalate* Rather than introducing the leaving group last, the sequence can be redesigned to introduce the leaving group before the Grignard reagents are added, or before the second Grignard reagent is added. The cyclization immediately follows (cascades from) the addition of the second Grignard reagent.

A chloride leaving group can be introduced before adding the Grignard reagents. A mixture of 1-oxo-1,3-dihydroisobenzofuran-5-carbonitrile (**24**), thionyl chloride (8.0 equivalents), boron trifluoride etherate (4.1 mol%), and benzyltriethylammonium chloride (2.0 mol%) is refluxed for 17 h. Excess thionyl chloride is removed by distillation until the pot temperature reaches 95°C and the mixture is aged at 95°C for 24 h. 2-(Chloromethyl)-4-cyanobenzoyl chloride (**122**) is then distilled at an unspecified temperature and high vacuum (mp 44–44.5°C) (92%). A lower yield (73%) is observed using less thionyl chloride (1.5 equivalents) in xylene as the reaction solvent.

A 14 wt% solution of 4-fluorophenylmagnesium bromide (1.1 equivalents) in THF is added to a solution of 2-(chloromethyl)-4-cyanobenzoyl chloride (**122**) in toluene at <0°C. The mixture is aged at 0°C for 2 h. Hydrochloric acid (concentration not specified) is added. The mixture is aged, presumably at 25°C, for 30 min and the layers are separated. The aqueous layer is extracted with toluene. The combined organic layers are washed with saturated aqueous sodium bicarbonate and with water and concentrated at reduced pressure to afford crude 3-(chloromethyl)-4-(4-fluorobenzoyl)benzonitrile (**123**) (90%).

A 17 wt% solution of 3-dimethylaminopropylmagnesium chloride (1.1 equivalents) in THF is added to a solution of the crude 3-(chloromethyl)-4-(4-fluorobenzoyl)benzonitrile (**123**) in toluene–THF at <−5°C. The mixture is aged at 0°C for 2 h. Saturated aqueous ammonium chloride is added (to pH 9). The layers are separated and the aqueous layer is extracted with toluene. The combined organic layers are washed with water and concentrated at reduced pressure to afford crude citalopram (**39**) (95%). No purity data or purity upgrade is provided. The crude citalopram yield is lower (56% from **122**) when 3-(chloromethyl)-4-(4-fluorobenzoyl) benzonitrile (**123**) is not isolated (Scheme 3.38).[152]

A pivalate leaving group can be introduced after the first Grignard reagent has been added by capturing the released alkoxide with pivaloyl chloride. A 20 wt% solution of 4-fluorophenylmagnesium bromide (1.1 equivalents) in THF is added to a suspension of 1-oxo-1,3-dihydroisobenzofuran-5-carbonitrile (**24**) in THF at <5°C. The mixture is aged at 25°C overnight. Pivaloyl chloride (1.1 equivalents) is added, presumably at 25°C, and the mixture is aged at 60°C for 2 h. The mixture is transferred into cold aqueous ammonium chloride. Ethyl ether is added and the layers are separated. The organic layer is washed with 0.1 M sodium hydroxide and with water and then dried and concentrated at reduced pressure to afford 5-cyano-2-(4-fluorobenzoyl)benzyl pivalate (**124**) (88%). Crystallization from *n*-heptane–ethyl acetate (9:1) can provide a purity upgrade.

A solution of 3-dimethylaminopropylmagnesium chloride (2.2 equivalents) is added to a solution of 5-cyano-2-(4-fluorobenzoyl)benzyl pivalate (**124**) in THF at 0°C. The mixture is aged at 0°C for 1 h. Excess Grignard reagent is quenched with saturated aqueous ammonium chloride and the mixture is extracted with ethyl acetate. The ethyl acetate extract is dried and concentrated at reduced pressure to afford crude citalopram (**39**) (103%, 87% pure by HPLC) (Scheme 3.39).[153]

## 3.8 CLASSICAL RESOLUTION OF CITALOPRAM (39), DESMETHYL CITALOPRAM (78), AND DIDESMETHYL CITALOPRAM (117)

Citalopram (**39**) or any intermediate with the chiral center established and possessing a basic site capable of forming a salt with a chiral acid (or an acidic site capable of forming a salt with a chiral base) can potentially be resolved. All the known examples of resolution to produce escitalopram (**1**) form a salt between an amino group on the side chain and *O*,*O*-di-*p*-toluoyltartaric acid (DPTTA) (Scheme 3.40).

### 3.8.1 Escitalopram (1) by Resolution of Citalopram (39)

A suspension of citalopram (**39**) and (−)-(DPTTA) (1.0 equivalent) in acetonitrile is aged at 25°C for 15 min and then heated to 70–75°C. Methanol is added to produce a clear solution (acetonitrile–methanol volume ratio is 12.5:1). The solution is aged at 70–75°C for 30–45 min and then cooled to 30–35°C over 2–3 h. The resulting suspension is filtered and the *liquors* are concentrated at reduced pressure.

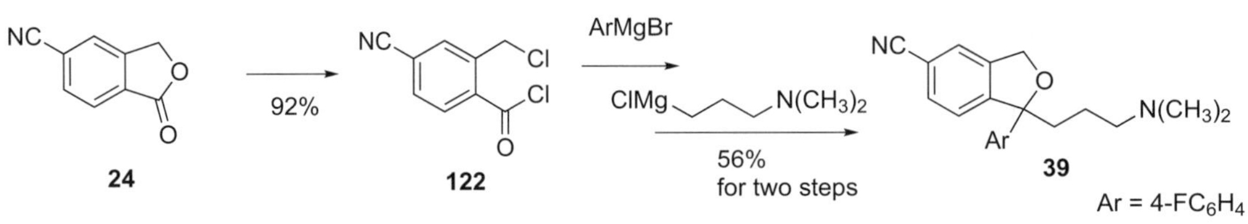


**SCHEME 3.38** Citalopram (**39**) from 2-(chloromethyl)-4-cyanobenzoyl chloride (**122**).

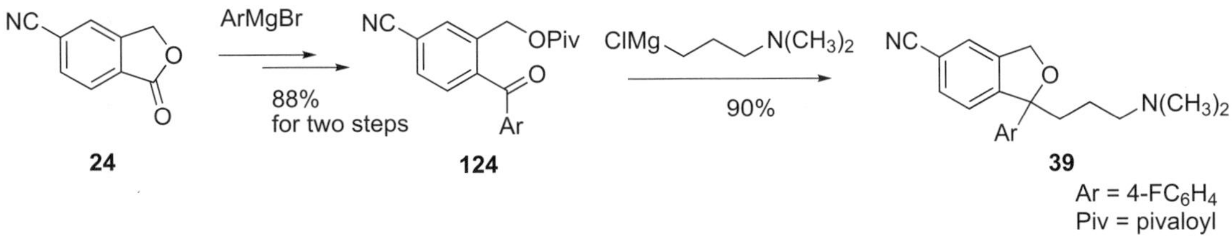


**SCHEME 3.39** Citalopram (**39**) from 5-cyano-2-(4-fluorobenzoyl)benzyl pivalate (**124**).

The residual solid is suspended in 1.6% aqueous sodium hydroxide (0.26 equivalents) and extracted with toluene. (*Note*: The amount of hydroxide suggests there is little salt recovered from the liquors.) The extracts are washed with water and concentrated at reduced pressure. After one repeat of the salt formation process, escitalopram (**1**) is isolated with an optical purity of 98.8%. No yield is provided.[123]

This process has been studied in greater detail with the focus on the *filtered salt*. A suspension of citalopram (**39**) and (−)-DPTTA (1.0 equivalent) in acetonitrile is heated to 70–75°C. Methanol is added to produce a clear solution (acetonitrile–methanol volume ratio is 12.5:1). The solution is aged at 70–75°C and then cooled to 25°C. The resulting suspension is aged at 25°C for 1 h and at 0–5°C for 70 min and then filtered. Salt **125** is washed with acetonitrile and dried. Analysis of isolated salt **125** by HPLC reveals a mixture with *S*:*R* = 55.4:44.6. Salt **125** is crystallized from acetonitrile–methanol (volume ratio 12.5:1). Analysis of isolated salt **125** by HPLC reveals a mixture with *S*:*R* = 55.7:44.3. Salt **125** is crystallized from acetonitrile–methanol (volume ratio 12.5:1) once more. Analysis of isolated salt **125** by HPLC reveals a mixture with *S*:*R* = 56.2:43.8. The salt **125** yield after three isolations is 60% based on citalopram (**39**). These results are consistent with the solubilities of the citalopram–DPTTA salts **125** in acetonitrile-–methanol: (*S*)-**125** 101 mg/mL, (*R*)-**125** 103 mg/mL, and (*RS*)-**125** 18 mg/mL. These results are also consistent with seeding experiments that led to the conclusion that the (*RS*)-**125** salt is the kinetically favored product.[124]

It is suggested that the results of the two processes are reconciled by correcting the sign on the resolving agent in the first process from (−) to (+). The salt isolated is then slightly enriched in the (*R*)-enantiomer (*R*:*S* = 55:45), with the (*S*)-enantiomer increasing in the liquors in each pass (*S*:*R* = 59.3:40.7 first pass, 83.3:16.7 second pass, 96.4:3.6 third pass). The yield of escitalopram (**1**) (*S*:*R* 96.4:3.6) after three passes is 5.5% based on citalopram (**39**). (*Note*: Some of the mass balance and *S*:*R* ratio data for the three passes are inconsistent.) An attempt to reproduce these results confirmed that (*S*)-enantiomer increases in the liquors in the first two passes (*S*:*R* = 64.4:35.6 first pass, 88.3:11.7 second pass). No precipitate formed in the third pass. While we could continue to compare and contrast data from the two groups, in the end the groups ultimately agree: the resolution of citalopram (**39**) under any of the conditions described is not suitable for manufacturing escitalopram (**1**).[124–126]

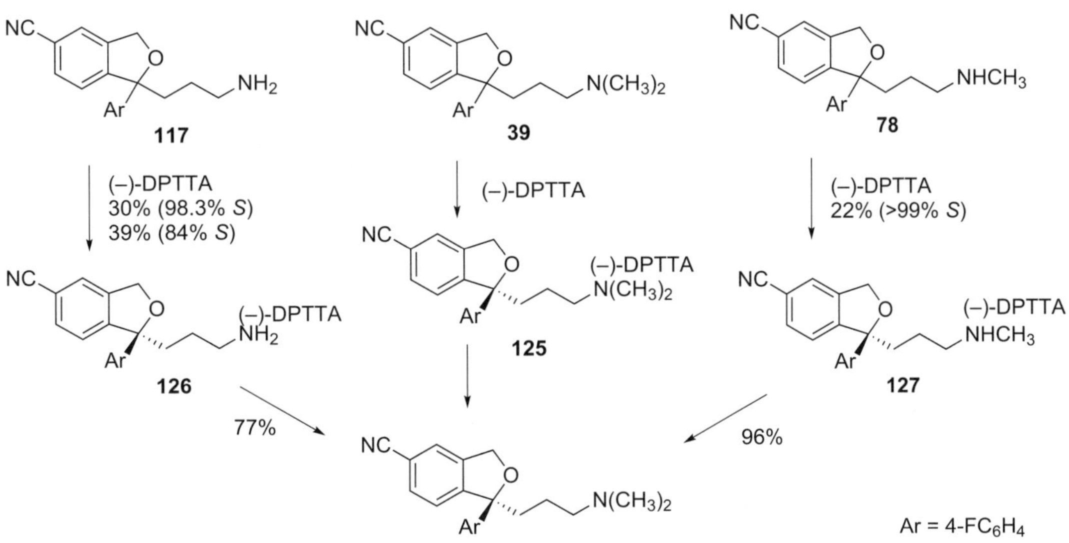


**SCHEME 3.40** Escitalopram (**1**) by resolution of citalopram (**39**), didesmethyl citalopram (**117**), and desmethyl citalopram (**78**).

### 3.8.2 Escitalopram (1) by Resolution of Didesmethyl Citalopram (117)

Resolution of didesmethyl citalopram (**117**) with (−)-DPTTA does not produce (*S*)-didesmethyl citalopram-(−)-DPTTA salt **126**, which is >97% (*S*)-enantiomer, in a single pass. A solution of (−)-DPTTA (1.0 equivalent) in acetonitrile is added to a solution of didesmethyl citalopram (**117**) in acetonitrile at 25°C. The resulting suspension is aged at 60–65°C for 45–60 min and then cooled to 0–5°C and aged for 1 h. The suspension is filtered and salt **126** is suspended in acetonitrile at reflux. Water is added to produce a clear solution (acetonitrile–water volume ratio is 15:1). The solution is cooled and aged at 0–5°C for 3 h. The resulting suspension is filtered and salt **126** is suspended in water. Base (presumably aqueous sodium hydroxide) is added (to pH 12) and the mixture is extracted with toluene. The toluene extracts are washed with water, dried, and concentrated at reduced pressure to afford (*S*)-didesmethyl citalopram (**117**) (30% yield, 98.3% *S*).[123]

A solution of didesmethyl citalopram (**117**) in acetonitrile is added to a mixture of (−)-DPTTA (1.0 equivalent) in acetonitrile at 27°C. The mixture is warmed to 55–60°C and water is added to produce a clear solution (acetonitrile–water volume ratio is 3.3:1). The solution is aged at 55–60°C for 45 min, cooled to 0–5°C and aged for 1 h, and warmed to 25°C and aged for 45 min. The warming/cooling cycle is repeated two times. The resulting suspension is filtered and the solid is washed with cold acetonitrile and dried at an unspecified temperature and reduced pressure to afford (*S*)-didesmethyl citalopram-(−)-DPTTA salt (**126**) (39%, 84% *S*). These results suggest that it will require *multiple crystallizations* to deliver (*S*)-didesmethyl citalopram (**117**) with acceptable optical purity.[126]

A solution of (*S*)-didesmethyl citalopram (**117**) and formaldehyde (4.9 equivalents) (37% aqueous solution) in 98% formic acid (4.9 equivalents) is refluxed (95–100°C) for 12 h. The solution is cooled, hydrochloric acid (4 M) is added (to pH <7), and the mixture is concentrated at reduced pressure. The residue is suspended in 1 M sodium hydroxide and the mixture is extracted with ethyl ether. The organic extracts are washed with water, dried, and concentrated at reduced pressure to afford escitalopram (**1**) (77%, 97.1% *S*).

Heating (*S*)-didesmethyl citalopram (**117**) with formic acid is likely to produce the formamide (**127**) as a side product. The amount of formamide **127** produced under these conditions is not provided. Formamide **127** should be easily separated from escitalopram (**1**) since it is not extracted into aqueous acid and does not form an oxalate salt.[123]

### 3.8.3 Escitalopram (1) by Resolution of Desmethyl Citalopram (78)

Resolution of desmethyl citalopram (**78**) affords (*S*)-desmethyl citalopram-(−)-DPTTA salt (**128**), which is >99% (*S*)-enantiomer, after two isolations. (−)-Di-*p*-toluoyl-L-tartaric acid (DPTTA) (0.96 equivalents) is added to a solution of desmethyl citalopram (**78**) in isopropanol at 45°C. The solution is aged at 45°C for 2–3 h. Water is added at 45°C (isopropanol–water = 5:1) and the mixture is aged at 45°C for 1 h. The salt **128** precipitates as the solution is cooled to 25–30°C. The suspension is filtered and the solid is crystallized from isopropanol–water (5:1) and dried to afford (*S*)-desmethyl citalopram-(−)-DPTTA salt (**128**) (22%, >99% (+)-enantiomer).[129]

> The patent process quotes a yield of 44%. Resolution yields will be quoted relative to the racemic starting material (22%) to highlight the necessity of a racemization of the unwanted enantiomer to achieve maximum process efficiency.

(*S*)-Desmethyl citalopram-(−)-DPTTA salt (**128**) is dissolved in water and dichloromethane. Aqueous sodium hydroxide (10%) is added (to pH 9–10) and the layers are separated. The aqueous layer is extracted with dichloromethane. The combined organic layers are washed with water, dried, and concentrated at reduced pressure to afford (*S*)-desmethyl citalopram (**78**). The (*S*)-desmethyl citalopram (**78**) is suspended in water. Formaldehyde (9.0 equivalents) (presumably as 37% aqueous solution) and formic acid (3.4 equivalents) are added and the mixture is aged at 80–90°C for an unspecified time. Hydrochloric acid (12 M) is added (to pH <7) and the mixture is washed with toluene. Aqueous sodium hydroxide (10%) is added (to pH 9–10) at <20°C and the mixture is extracted with dichloromethane. The dichloromethane extracts are washed with water, dried, and concentrated at reduced pressure. The residue is dissolved in acetone and escitalopram (**1**) is isolated from the solution as the oxalate salt. The salt is dried at 50–55°C and reduced pressure (96% yield from (*S*)-**128**). The reductive methylation yield is lower (77%) using paraformaldehyde and sodium borohydride in methanol.[129]

### 3.8.4 Resolution of 4-(4-(Dimethylamino)-1-(4-fluorophenyl)-1-hydroxybutyl)-3-(hydroxymethyl) benzonitrile (121)

A resolution can be moved to an earlier process intermediate provided the intermediate has an amino group and the resolved intermediate can be efficiently converted to escitalopram (**1**). 4-(4-(Dimethylamino)-1-(4-fluorophenyl)-1-hydroxybutyl)-3-(hydroxymethyl)benzonitrile (**121**) can be resolved with DPTTA in ethanol, isopropanol, 1-propanol, or acetonitrile and mixtures such as methanol–isopropanol or methanol–ethyl acetate.[154,155]

(+)-DPTTA (0.49 equivalents) is added to a solution of 4-(4-(dimethylamino)-1-(4-fluorophenyl)-1-hydroxybutyl)-3-(hydroxymethyl)benzonitrile (**121**) in isopropanol at 40°C. The mixture is presumably cooled to 25°C and aged at 25°C for 3 h. The resulting suspension is filtered and the solid is dried at an unspecified temperature and reduced pressure to afford the (*S*)-diol-(+)-DPTTA salt (**129**) (20% based on (*RS*)-diol **121**). Optical purity is provided only in terms of rotation ($[\alpha]_D = +12.33°$ ($c = 1$, $CH_3OH$)).[149]

It is suggested that the suspension of (*S*)-diol-(+)-DPTTA (**129**) generated in isopropanol may be poorly suited for filtration. The suspension is instead centrifuged and the solid dissolved in hot ethyl acetate. The suspension in ethyl acetate produced on cooling is filtered and the solid is washed with ethyl acetate and dried to afford the salt **129** (30% from (*RS*)-diol **121**). Optical purity is quoted only in terms of rotation ($[\alpha]_D = +10.02°$ ($c = 1$, $CH_3OH$)).[156]

Optical purity data are offered as percent of desired enantiomer, percent enantiomeric excess (% ee), percent chiral purity, or optical rotation. The definitive data on optical purity after a resolution are percent of the desired enantiomer (in this case, the (*S*)-enantiomer) as determined by chiral HPLC. Percent chiral purity is interpreted to be the same as percent desired enantiomer. Percent of desired enantiomer can be back-calculated from percent enantiomeric excess. Optical rotation data only *suggest* optical purity.

A solution of (+)-DPTTA (0.51 equivalents) in ethanol is added to a solution of 4-(4-(dimethylamino)-1-(4-fluorophenyl)-1-hydroxybutyl)-3-(hydroxymethyl)benzonitrile (**121**) in ethanol at 40°C over 1 h. The solution is seeded with (*S*)-diol-(+)-DPTTA salt (**129**), cooled to 25°C, and aged at 25°C overnight. The resulting suspension is cooled to 0°C and filtered. The solid is presumably washed with cold ethanol and dried to afford (*S*)-diol-(+)-DPTTA salt (**129**) (30% based on (*RS*)-diol **121**, 98.2% *S*).[157]

The variables that may have an effect on efficiency of the resolution include the solvent, equivalents of resolving agent, addition of a second achiral acid, the cosolvent (coming in with the diol), water, and aging time after the salt precipitates and before the suspension is filtered. The resolving agent is typically expensive and recoverable. The equivalents of resolving agent are first selected to maximize optical purity, then to maximize yield, and finally to minimize the amount of resolving agent that must be recovered. Addition of an achiral acid, such as acetic acid, may increase optical purity of the isolated salt. However, the achiral acid in the mixture may complicate recovery of the resolving agent. The final step in producing the (*RS*)-diol **121** may be a water wash of a diol solution (in toluene, dichloromethane, or ethyl acetate) followed by concentration to remove the solvent. Diol **121** will contain some residual solvent that will be carried into the resolution step. It will be important to determine the amount of residual solvent that prevents crystallization of the (*RS*)-diol **121** during storage at 25°C and that gives the highest optical purity and yield of the isolated salt **129**.

A solution of (+)-DPTTA (0.40 equivalents) and acetic acid (0.20 equivalents) in 1-propanol is heated to 40°C and transferred into a 40°C solution of 4-(4-(dimethylamino)-1-(4-fluorophenyl)-1-hydroxybutyl)-3-(hydroxymethyl)benzonitrile (**121**) in 1-propanol containing some toluene. (A total of 4.5 volumes of 1-propanol and 0.1 volumes of toluene are used.) The solution is seeded with (*S*)-diol-(+)-DPTTA salt (**129**) and aged at 40°C for 2 h. The suspension is cooled to 25°C in 2 h and filtered. The solid is washed with 1-propanol and dried to afford (*S*)-diol-(+)-DPTTA salt (**129**) (34% based on (*RS*)-diol **121**, 99.0% *S*). The yields and optical purities are comparable (33–35%, 98.7–99.1% *S*) using 0.01–0.05 volumes of water and no acetic acid. The yields are lower (29–30%) using no toluene cosolvent.

A solution of (+)-DPTTA (0.39 equivalents) in 1-propanol is heated to 40°C and transferred into a 40°C solution of 4-(4-(dimethylamino)-1-(4-fluorophenyl)-1-hydroxybutyl)-3-(hydroxymethyl)benzonitrile (**121**) in 1-propanol containing some toluene. (A total of 4.3 volumes of 1-propanol and 0.13 volumes of toluene are used.) The solution is seeded with (*S*)-diol-(+)-DPTTA salt (**129**) and aged at 40°C for 2 h. The suspension is cooled to 25°C in 2 h, aged at 25°C overnight, and filtered. The solid is washed with 1-propanol and dried to afford (*S*)-diol-(+)-DPTTA salt (**129**) (35% based on (*RS*)-diol **121**, >98% *S*). The yield is comparable and the optical purity is lower (36%, 97.2% *S*) using 0.50 equivalents of (+)-DPTTA and 4.5 volumes of 1-propanol. The yield is lower (30%, 99.0% *S*) using 0.75 equivalents of (+)-DPTTA, 4.5 volumes of 1-propanol, and overnight aging of the suspension at 25°C. The yield does increase but optical purity decreases (32%, 92.8% *S*) when this same mixture is aged at 25°C for 4 days.

A solution of (+)-DPTTA (0.50 equivalents) in acetonitrile is heated to 40°C and transferred into a 40°C solution of 4-(4-(dimethylamino)-1-(4-fluorophenyl)-1-hydroxybutyl)-3-(hydroxymethyl)benzonitrile (**121**) in acetonitrile containing some toluene. (A total of 3.0 volumes of acetonitrile and 0.13 volumes of toluene are used.) The solution is seeded with (*S*)-diol-(+)-DPTTA salt (**129**) and aged at 40°C for 2 h. The suspension is cooled to 25°C in 2 h, aged at 25°C overnight, and filtered. The solid is washed with acetonitrile and dried to afford (*S*)-diol-(+)-DPTTA salt (**129**) (25% based on (*RS*)-diol **121**, 99.2% *S*).

A solution of (+)-DPTTA (0.50 equivalents) in 1-propanol–dichloromethane (85:15) is heated to 40°C and transferred into a 40°C solution of 4-(4-(dimethylamino)-1-(4-fluorophenyl)-1-hydroxybutyl)-3-(hydroxymethyl)

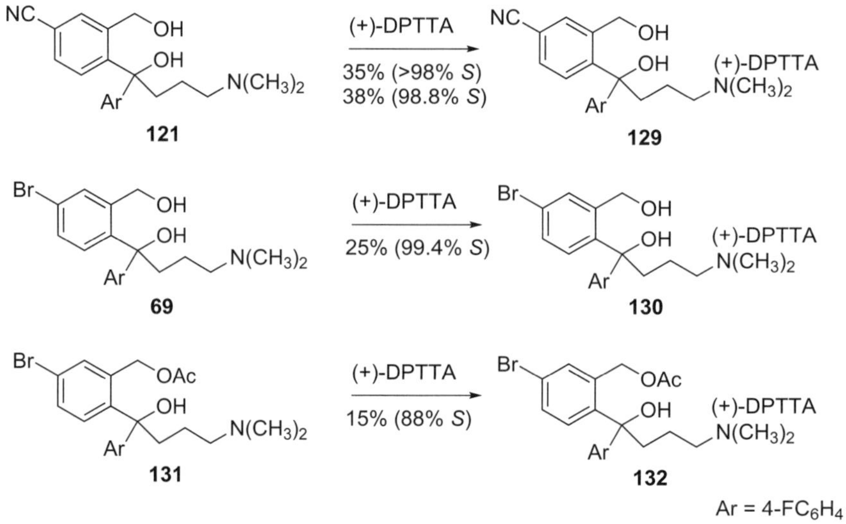


**SCHEME 3.41** Resolution of diol intermediates with (+)-DPTTA.

benzonitrile (**121**) in 1-propanol–dichloromethane (85:15). (A total of 4.5 volumes of 1-propanol–dichloromethane (85:15) are used.) The solution is seeded with (*S*)-diol-(+)-DPTTA salt (**129**) and aged at 40°C for 2 h. The suspension is cooled to 25°C in 2 h, aged at 25°C overnight, and filtered. The solid is washed with 1-propanol and dried at an unspecified temperature and reduced pressure to afford (*S*)-diol-(+)-DPTTA salt (**129**) (38% based on (*RS*)-diol **121**, 98.8% *S*) (Scheme 3.41). The yields and optical purities are comparable using 72:25, 90:10, and 95:5 1-propanol–dichloromethane as the solvent (34–39%, 97.7–99.1% *S*). (*Note*: There are some inconsistencies in the data. Molar yield is calculated as mol (*S*)-product **129**/mol (*RS*)-diol **121** starting material. But examples 5, 11, and 21 use only 0.25 equivalents of (+)-DPTTA and have yields of 29.4%, 31.8%, and 30.0%, respectively.)[157]

### 3.8.5 Resolution of 1-(4-Bromo-2-(hydroxymethyl)phenyl)-4-(dimethylamino)-1-(4-fluorophenyl)butan-1-ol (69)

(+)-DPTTA monohydrate (0.49 equivalents) is added to a solution of 1-(4-bromo-2-(hydroxymethyl)phenyl)-4-(dimethylamino)-1-(4-fluorophenyl)butan-1-ol (**69**) in isopropanol at 40–45°C. The mixture is presumably cooled to 25°C and aged at 25°C for 3 h. The resulting suspension is filtered and the solid is washed with isopropanol and dried at an unspecified temperature and reduced pressure to afford the (*S*)-diol-(+)-DPTTA salt (**130**) (28% based on (*RS*)-diol **69**). (*Note*: The procedure reports 74% yield.) A suspension of the salt in isopropanol is aged at 70°C for 1 h. The suspension is cooled to 25°C and filtered and the solid is dried to afford the (*S*)-diol-(+)-DPTTA salt (**130**) (25% based on (*RS*)-diol **69**, 99.4% *S*). (*Note*: The procedure reads 99.4% chiral purity.)

The (*S*)-diol-(+)-DPTTA salt (**130**) is cleaved with sodium hydroxide in water and the free base is extracted with toluene. The extracts are concentrated at reduced pressure to afford the (*S*)-diol (**69**) (95%). The (*S*)-diol cyclization with methanesulfonyl chloride and triethylamine has already been discussed in the context of citalopram synthesis (91% yield, 99.2% *S*). (*Note*: The procedure reads 99.2% chiral purity.)[91]

A monoester derivative of 1-(4-bromo-2-(hydroxymethyl)phenyl)-4-(dimethylamino)-1-(4-fluorophenyl)butan-1-ol (**69**) can also be resolved with (+)-DPTTA. The diol monoacetate **131** and (+)-DPTTA (1.0 equivalent) are dissolved in ethyl acetate at reflux. The solution is allowed to cool to 25°C and some hexanes are added to induce crystallization. The suspension is aged at 25°C over 65 h and then filtered and the solid is dried to afford (*S*)-diol monoacetate–(+)-DPTTA salt (**132**) (15% from the (*RS*)-diol monoacetate **131**, 88% *S*). The salt is upgraded using acetone–hexane (65% recovery, 97% *S*). Salt cleavage and ester hydrolysis affords the (*S*)-diol (**69**) (9.8% from the (*RS*)-diol **69**).[85]

The enantiomers of diols **121** and **69** and the dihydroisobenzofuran **68** can be separated by chiral chromatography. Reaction of (*S*) (**68**) (99.7% *S*) from the chromatography with zinc cyanide (2.0 equivalents) catalyzed by tetrakis(triphenylphosphine)palladium (5.0 mol%) affords escitalopram (**1**) (80%, 99.8% *S*).[158]

### 3.8.6 Recovery of (+)-DPTTA

The (*R*)-diol-(+)-DPTTA salt (**129**) is suspended in methyl *tert*-butyl ether and water. Ammonium hydroxide (25%) (2.5 equivalents) is added and the mixture is aged at 25°C for 10 min. The layers are separated and the aqueous layer is washed with methyl *tert*-butyl ether. Hydrochloric acid

(1 M) is added to the aqueous layer (to pH 2.4). The mixture is aged for 5 h, presumably at 25°C. The resulting suspension is filtered and the solid is washed with water and dried at 25°C and 20 mmHg to afford the (+)-DPTTA monoammonium salt (93.7%, 93.4% pure by HPLC). (+)-DPTTA is released from the monoammonium salt using Amberlite IR 120 in isopropanol (99%).[159]

## 3.9 CHEMOENZYMATIC RESOLUTION OF 4-(4-(DIMETHYLAMINO)-1-(4-FLUOROPHENYL)-1-HYDROXYBUTYL)-3-(HYDROXYMETHYL) BENZONITRILE (121)

The two strategies for chemoenzymatic resolution of the (*RS*)-diol **121** are based on the selective reactivity of an (*R*)-diol monoester or (*R*)-diol with a lipase. The (*R*)- and (*S*)-diols **121** are converted to monoacetates **133** and the (*R*)-diol monoacetate is selectively hydrolyzed using a lipase. Alternatively, the (*R*)-diol in the mixture of (*R*)- and (*S*)-diols **121** is selectively esterified with vinyl butyrate using an esterase. In both cases, the mixture of diol **121** and diol monoacetate **133** or monobutyrate **134** can be separated by crystallization.

### 3.9.1 Ester Hydrolysis with Lipase Extract from *Aspergillus niger*

A monoester derivative of 4-(4-(dimethylamino)-1-(4-fluorophenyl)-1-hydroxybutyl)-3-(hydroxymethyl)benzonitrile can be resolved using an esterase enzyme from crude lipase extract from *Aspergillus niger*. Acetyl chloride (17 equivalents) is added to 4-(4-(dimethylamino)-1-(4-fluorophenyl)-1-hydroxybutyl)-3-(hydroxymethyl)benzonitrile (**121**) at 30–35°C. The mixture is aged at 35°C for 5 min and excess acetyl chloride is distilled at reduced pressure. The residue is presumably the (*RS*)-diol monoacetate **133** hydrochloride salt. Less acetyl chloride (4.0 equivalents) is required when dichloromethane is used as a reaction solvent.[141,160]

The crude (*RS*)-diol monoacetate **133** hydrochloride salt is dissolved in methanol and the solution is diluted with pH 8 monobasic potassium phosphate buffer. Hydrochloric acid (2 M) is added (to pH 6), an identical volume of methanol is added, and the solution temperature is controlled at 20–25°C. Esterase enzyme derivative obtained from crude lipase extract of *Aspergillus niger* and immobilized on epoxy resin is added and the mixture is aged at 20–25°C for 70–80 h (55% ester hydrolysis). The pH is maintained at 6 during the aging time. The suspension is filtered and the immobilized enzyme is washed with methanol–water. Methanol is distilled from the liquors at reduced pressure. Aqueous sodium hydroxide is added to the remaining aqueous mixture (to pH 8.5). The mixture is diluted with brine and extracted with ethyl acetate. The extracts are concentrated at reduced pressure. The residue, a mixture of (*R*)-diol **121** and (*S*)-diol monoacetate **133**, is dissolved in ethyl ether containing a trace of ethyl acetate at reflux. The solution is cooled to 25°C. The resulting suspension is filtered and the solid is crystallized again from ethyl ether–ethyl acetate at 0–4°C to afford the (*S*)-diol monoacetate **133** (98% pure, 99.9% *S*). Hydrolysis with ammonium hydroxide in methanol affords the (*S*)-diol **121** (26% yield from (*RS*)-diol **121**) (Scheme 3.42). The volume throughput is <18 g/L during the 70–80 h hydrolysis.[160]

The yield is higher when the EDTA complex with (*RS*)-diol monoacetate **133** is used as the substrate for the enzyme-mediated hydrolysis. The crude (*RS*)-diol monoacetate **133** hydrochloride salt is dissolved in ethanol. The solution is diluted with water and disodium EDTA (1.6 equivalents) is added. The solution is partially concentrated at reduced pressure to precipitate the EDTA complex. The suspension is filtered and the EDTA complex is dissolved in ethanol and monobasic sodium phosphate buffer. Hydrochloric acid (1 M) is added (to pH 6). Esterase enzyme derivative obtained from crude lipase extract of *Aspergillus niger* and immobilized on epoxy resin is added and the mixture is aged

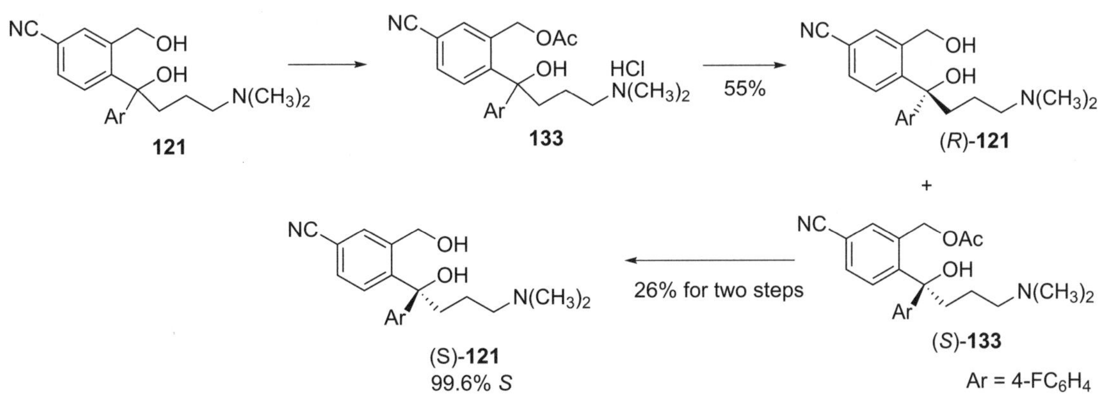


**SCHEME 3.42** Resolution of 4-(4-(dimethylamino)-1-(4-fluorophenyl)-1-hydroxybutyl)-3-(hydroxymethyl)benzonitrile (**121**) by ester hydrolysis.

at 20–25°C for an unspecified time. The pH is maintained at 6 during the aging time. The suspension is filtered and the immobilized enzyme is presumably washed with ethanol–water. Ethanol is distilled from the liquors at reduced pressure. Aqueous sodium hydroxide is added to the remaining aqueous mixture (to pH 8.5) and the mixture is extracted with ethyl acetate. The extracts are concentrated at reduced pressure. The residue is suspended in ethanol and the suspension is refluxed for 30 min and then cooled to 25°C. The suspension is filtered and the solid is dried to afford (*S*)-diol monoacetate **133** (37%, 95.9% *S*). (*Note*: The isolated product is identified as the hydrochloride salt in the patent procedure. The purity quoted is 95.9%. This is presumed to refer to the % (*S*)-enantiomer.) The volume throughput is <16 g/L during the hydrolysis.[141]

### 3.9.2 Ester Hydrolysis with *Candida antarctica*

Immobilized *Candida antarctica* lipase B is added to a solution of the (*RS*)-diol monoacetate **133** and water (4–5 equivalents) in acetonitrile. The suspension is aged at 30°C for 100 h. The suspension is filtered and the liquors are concentrated at reduced pressure. The (*S*)-diol monoacetate **133** is isolated from the residue by chromatography (40%, >90% ee). Ester hydrolysis with sodium hydroxide in methanol–water affords the (*S*)-diol **121** (>95%).[161,162]

### 3.9.3 Esterification with Lipoprotein Lipase Pseudomonas

Lipoprotein lipase pseudomonas is added to a solution of the (*RS*)-diol **121** and vinyl butyrate (2.0 equivalents) in 1,4-dioxane. The mixture is aged at 50°C for 192 h (41% conversion). More lipoprotein lipase pseudomonas is added and the aging time is extended to 504 h (63% conversion). The suspension is filtered and the solid is washed with 1,4-dioxane. The liquors are concentrated at reduced pressure. The residue is dissolved in THF and succinic anhydride (1.1 equivalents with respect to (*S*)-diol **121** in the mixture) is added, presumably at 25°C. The resulting suspension is filtered and the solid is washed with THF and dried to afford the succinate ester of the (*S*)-diol **121** (24%, 98% *S*). (*Note*: The purity quoted is 98% chiral purity.) The volume throughput is 17 g (*S*)-diol **121**/L during the lipase-mediated esterification. The (*S*)-diol **121** succinate ester is directly converted to escitalopram (**1**) with sodium hydride in DMF. Escitalopram (**1**) is isolated from acetone solution as the oxalate salt (69%, 97–98% *S*).[163,164]

### 3.9.4 Esterification with *Candida antarctica*

Immobilized *Candida antarctica* lipase B is added to a solution of the (*RS*)-diol **121** and vinyl acetate (1.5 equivalents) in acetonitrile. The suspension is aged at 30°C for 20 h. The suspension is filtered and the liquors are concentrated at reduced pressure. The (*S*)-diol **121** is isolated from the residue by chromatography (47%, >98% *S*).[161,165]

Immobilized *Candida antarctica* lipase B (Novozyme 435) (0.2 g/g (*RS*)-diol **121**) is added to a solution of (*RS*)-diol **121**, pyridine (1.0 equivalent), vinyl butyrate (2.0 equivalents), and benzoic acid (1.0 equivalent) in toluene at 40°C. The suspension is aged at 40°C for 20 h. The suspension is filtered and the solid is washed with toluene. The toluene liquors containing (*S*)-diol **121** benzoate salt (44%, 99.5% *S*) are extracted with water at 60°C. The combined water extracts contain (*S*)-diol **121** benzoate salt (42% from (*RS*)-diol **121**) (Scheme 3.43). This near-optimal process is one example taken from an extensive development program. Variables studied include the enzyme, additives, stoichiometry of additives, acylating agent, stoichiometry of acylating agent, solvent, temperature, and aging time.

Immobilized *Candida antarctica* lipase B (Novozyme 435) is added to a solution of (*RS*)-diol **121**, vinyl butyrate (1.0 equivalent), and pivalic acid (1.5 equivalents) in toluene at 35–45°C. The suspension is aged at 35–45°C for 21 h. The suspension is filtered and the recovered immobilized enzyme is washed with toluene. The recovered Novozyme 435 can be recycled with little change in activity in five cycles at 35°C (enzyme/substrate ratio = 0.30) but a significant loss of activity is observed in five cycles at 45°C (enzyme/substrate ratio = 0.20).[163]

Promising preliminary results indicate that the esterification with Novozyme 435 and vinyl butyrate can also produce (*S*)-diol **121** from the bromo, iodo, and chloro

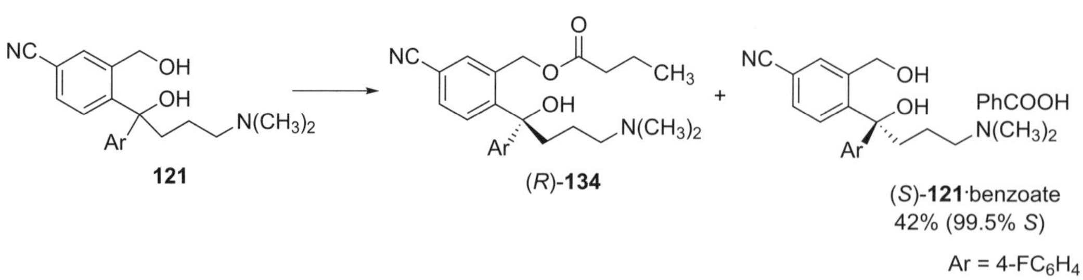


**SCHEME 3.43** Resolution of 4-(4-(dimethylamino)-1-(4-fluorophenyl)-1-hydroxybutyl)-3-(hydroxymethyl)benzonitrile (**121**) by esterification with vinyl butyrate.

(*RS*)-diols (**69**, **72**, and **135**, respectively) (98–99.7% *S* at 51–54% hydrolysis).[163]

## 3.10 ASYMMETRIC ADDITION OF 3-DIMETHYLAMINOPROPYLMAGNESIUM CHLORIDE TO 4-(4-FLUOROBENZOYL)-3-(HYDROXYMETHYL)BENZONITRILE (37)

An asymmetric addition of 3-dimethylaminopropylmagnesium chloride to 4-(4-fluorobenzoyl)-3-(hydroxymethyl)benzonitrile (**37**) would produce the (*S*)-diol **121** directly, eliminating the need for the resolution. The concept for inducing asymmetry is to use the neighboring hydroxymethyl group to anchor a boron complex containing a chiral amino alcohol. The close proximity of this complex to the ketone induces asymmetry during the addition of 3-dimethylaminopropylmagnesium chloride. Screening of conditions for the asymmetric addition is best done starting with the pure ketone **37**.

A 10 wt% solution of 4-fluorophenylmagnesium bromide (1.4 equivalents) in THF is added to 1-oxo-1,3-dihydroisobenzofuran-5-carbonitrile (**24**) in 1,2-dimethoxyethane at −10°C over 3 h. The mixture is aged at −10°C for 30 min and then quenched with aqueous ammonium chloride at 20°C. After a routine workup, the residue is suspended in diisopropyl ether. The suspension is cooled and filtered and the solid is washed and dried at 50°C and reduced pressure to afford 4-(4-fluorobenzoyl)-3-(hydroxymethyl)benzonitrile (**37**) (86%, 99.8% pure by HPLC). Ketone **37** can also be produced by deprotection of the pivalate ester **124** with sodium methoxide in methanol at 25°C.[153,165]

Of the many boron complexes and chiral amino alcohols tried, the best asymmetric induction is observed using the combination of diisopropoxymethylborane and 1*S*,2*S*-*N*-methylpseudoephedrine. Diisopropoxymethylborane (1.1 equivalents) is added to a solution of 4-(4-fluorobenzoyl)-3-(hydroxymethyl)benzonitrile (**37**) and 1*S*,2*S*-*N*-methylpseudoephedrine (1.0 equivalent) in toluene at 25°C. The mixture is aged at 70°C for 30 min and then cooled to 45°C and concentrated at 45 mmHg. Toluene is added to the residue and the mixture is cooled to −80°C. A 31 wt% solution of 3-dimethylaminopropylmagnesium chloride (2.0 equivalents) in THF is added at −80°C over 5 min and the mixture is aged at −80°C for 10 min. Analysis by HPLC indicates >98% conversion and 95% (*S*)-diol **121**. The yield after workup, chromatography, and conversion to the (+)-DPTTA salt **129** is 71% (99.5% *S*) (Scheme 3.44).[166]

## 3.11 ESCITALOPRAM (1) FROM (S)-4-(4-(DIMETHYLAMINO)-1-(4-FLUOROPHENYL)-1-HYDROXYBUTYL)-3-(HYDROXYMETHYL) BENZONITRILE (121)

With high-purity (*S*)-diol **121** available, it remains to demonstrate that the diol cyclization can be accomplished in high yield with no loss of optical purity. This is best accomplished by generation and displacement of a sulfonate ester.

Inversion with significant racemization is observed in cyclization using sulfuric acid at elevated temperature. Sulfuric acid (71%) is added to a solution of the (*S*)-diol **121** (*S*:*R* ratio = 99:1) in toluene and the mixture is aged at 78–85°C for 2 h. The mixture is cooled and diluted with water. Ammonium hydroxide (25%) is added (to pH 10.5–11.0), the mixture is aged at 57°C for 10 min, and the layers are separated. The organic layer is washed with water

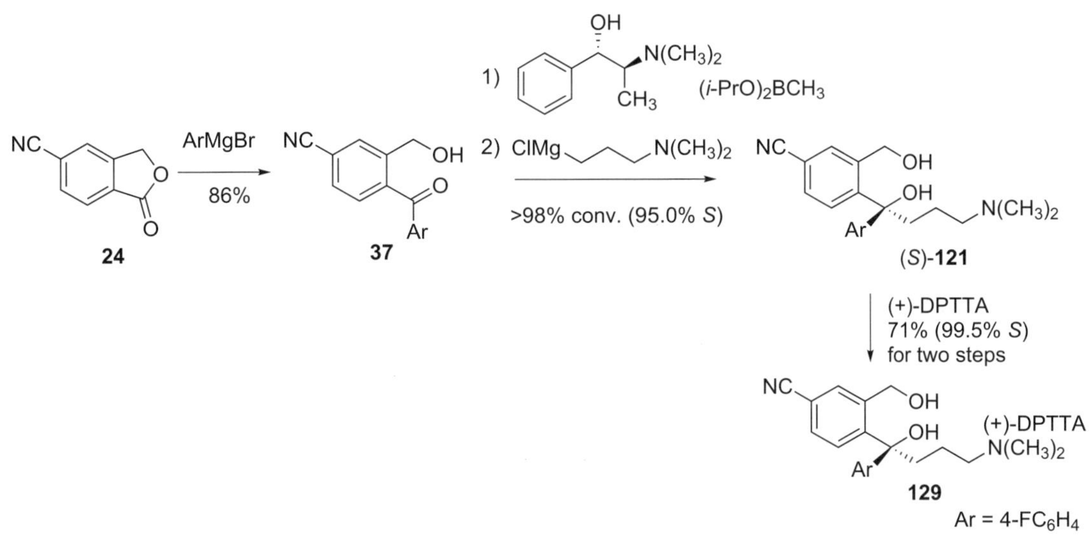


**SCHEME 3.44** Asymmetric addition of 3-dimethylaminopropylmagnesium chloride to 4-(4-fluorobenzoyl)-3-(hydroxymethyl) benzonitrile (**37**).

and concentrated at <60°C and reduced pressure to afford citalopram (**39**) (100%, 94.9% pure by HPLC, *S*:*R* ratio = 26:74).[147]

There are many examples of the diol cyclization with methanesulfonyl or *p*-toluenesulfonyl chloride and triethylamine. None provide the three key data points: percent (*S*)-enantiomer for the (*S*)-diol and percent (*S*)-enantiomer and yield for the crude escitalopram (**1**). Data compiled from five process patents suggest that the (*S*)-diol **121** can be converted to escitalopram (**1**) with little or no loss of optical purity and in high (90–95%) yield.

The following process is representative. Triethylamine (2.0 equivalents) is added to a solution of (*S*)-diol **121** (99% pure, 99.3% *S*) in toluene. A solution of *p*-toluenesulfonyl chloride (1.1 equivalents) in toluene is added at 25°C. The solution is aged at 25°C for 20 min. Water and ammonium hydroxide are added (presumably allowing the temperature to rise) and the mixture is aged for 2 min at 45°C. The layers are separated and the organic layer is washed with water, dried, and concentrated at reduced pressure to afford crude escitalopram (**1**) (95%). Crude escitalopram (**1**) is dissolved in ethanol and isolated as the oxalate salt. The salt is upgraded by crystallization from ethanol (64% from (*S*)-diol **121**, 99.3% *S*).[163]

The second process provides yield and percent (*S*)-enantiomer of escitalopram (**1**) crude but does not provide percent (*S*)-enantiomer of the (*S*)-diol **121**. A solution of methanesulfonyl chloride (1.2 equivalents) in dichloromethane is added to a solution of (*S*)-diol **121** in dichloromethane at 0°C. The mixture is aged at 15°C for 1 h. The mixture is washed with 1 M sodium hydroxide, dried, filtered, and concentrated at reduced pressure to afford crude escitalopram (**1**) (90%, >98% *S*).[161] Assuming that there is no optical purity upgrade in the diol cyclization, the optical purity of the (*S*)-diol starting material is also >98% *S*. The simplicity and similarity of the processes and the comparable yields of crude escitalopram (**1**) suggest that (1) the crude escitalopram (**1**) in the first process is 99–99.5% *S* and (2) there is little or no optical purity upgrade when escitalopram (**1**) is isolated as the oxalate salt.

The third process provides percent (*S*)-enantiomer of the (*S*)-diol **121** and of the isolated escitalopram (**1**) but no yield. A solution of the (*S*)-diol **121** in dichloromethane is produced from the (*S*)-diol-(+)-DPTTA salt (**129**) (>99.8% *S*). (*Note*: The purity quoted is >99.8% chiral purity.) Triethylamine (4.8 equivalents) and a solution of methanesulfonyl chloride (2.3 equivalents) in dichloromethane are added at 20–25°C. The mixture is aged at 20–25°C for 1 h. The mixture is washed with dilute ammonium hydroxide and with water. Escitalopram acetate is extracted into 10% acetic acid. The aqueous acidic extract is diluted with isopropanol and ammonium hydroxide is added (to pH >7). The mixture is aged at 20–25°C for 4–6 h and then cooled to 0–5°C. The resulting suspension is filtered and the solid is dried to afford escitalopram (**1**) (>99.8% *S*). (*Note*: The purity quoted is >99.8% chiral purity.) Methanol or ethanol can be used in place of isopropanol.[144]

The fourth process demonstrates conversion of the (*S*)-diol-(+)-DPTTA salt (**129**) to escitalopram (**1**) oxalate in high yield but does not provide the percent (*S*)-enantiomer of salt **129**. Salt **129** is suspended in water and dichloromethane. Ammonium hydroxide is added (to pH 9). The dichloromethane layer is separated, dried, and cooled to <5°C. Triethylamine (1.4 equivalents) is added at <5°C. *p*-Toluenesulfonyl chloride (1.1 equivalents) is added and the mixture is aged at <5°C for 1 h. The mixture is washed with water at pH 6 and again with water at pH 12 and then concentrated at reduced pressure. The residue is dissolved in acetone and oxalic acid (0.94 equivalents) is added. The resulting suspension is aged at an unspecified temperature for an unspecified time and then filtered. The solid is washed with cold acetone and dried to afford escitalopram (**1**) (93%, 99.4% *S*).[166]

The fifth process provides the percent (*S*)-enantiomer for the (*S*)-diol **121** and escitalopram (**1**) oxalate but has a lower yield of escitalopram oxalate. A solution of methanesulfonyl chloride (1.1 equivalents) in toluene is added to a solution of (*S*)-diol **121** (99.5% *S*) and triethylamine (2.8 equivalents) in toluene at 20–25°C. The mixture is aged at 20–25°C for an unspecified time. The mixture is washed with 0.1 M sodium hydroxide and then dried and concentrated at reduced pressure. The residue is dissolved in hot acetone and oxalic acid (1.2 equivalents) is added. After treating with carbon, the suspension is filtered hot and the liquors are allowed to cool to an unspecified temperature. The resulting suspension is filtered and the solid is crystallized again from acetone to afford escitalopram (**1**) oxalate (82%, 99.9% pure, 99.5% *S*).[144]

Finally, (*S*)-diol cyclizations using 2,5-dichloronitrobenzene or Mitsunobu conditions are described. Only the percent (*S*)-enantiomer of the escitalopram (**1**) oxalate is provided for the 2,5-dichloronitrobenzene process. Only the yield of escitalopram (**1**) is provided for the Mitsunobu process.[150,151]

## 3.12 OTHER ROUTES TO CITALOPRAM (39): DIELS–ALDER APPROACHES TO THE 1,2,4-TRISUBSTITUTED AROMATIC

All the pieces of a process to produce escitalopram (**1**) are on the table. Before selecting the process, there is one last general approach to citalopram (**39**) that has not been considered. The 1,2,4-trisubstituted aromatic can be constructed near the end of the process by [4 + 2]-cycloaddition of a 2-substituted butadiene or 3-substituted furan with a 2,5-dihydrofuran (Scheme 3.45). The sequence are long and there are no experimental data for any of the steps.[167]

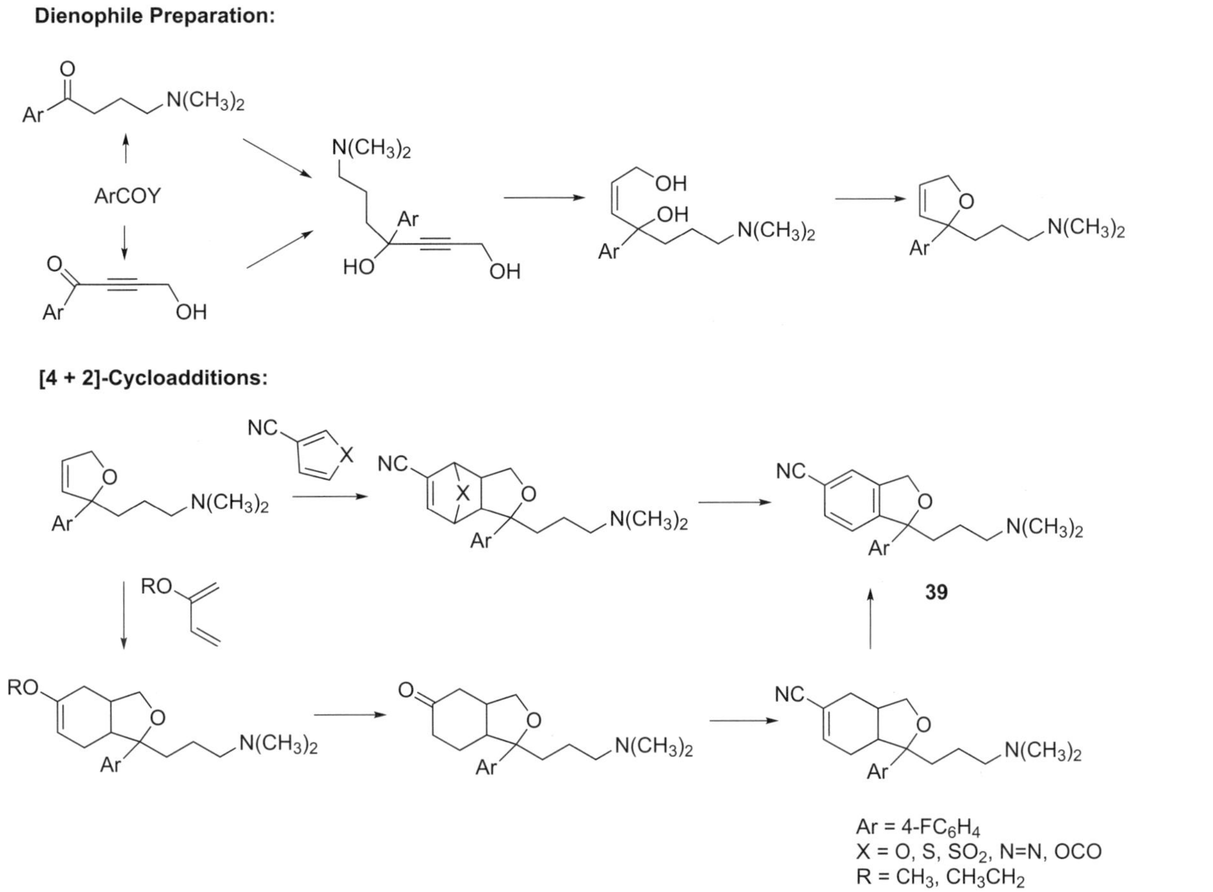


**SCHEME 3.45** Proposed route to citalopram (**39**) by [4 + 2]-cycloaddition of a 2-substituted butadiene and a 2,5-dihydrofuran.

The 1,2,4-trisubstituted aromatic can also be constructed by intramolecular [4 + 2]-cycloaddition of an alkene with a furan (Scheme 3.46). The linear sequence is six steps from specialty chemical 2-(furan-2-yl)-[1,3]-dioxolane to the established intermediate 1-(3-dimethylamino)propyl)-1-(4-fluorophenyl)-1,3-dihydroisobenzofuran-5-carbaldehyde (**91**). Perhaps 2-furaldehyde diethyl acetal[168] could be used in place of 2-(furan-2-yl)-[1,3]-dioxolane. Ring metallation and capture of the aryllithium with 4-fluorobenzaldehyde affords secondary alcohol **136**. Construction of the chiral carbon follows a now-familiar path: oxidation of secondary alcohol **136** to the ketone **137** and condensation of ketone **137** with 3-dimethylaminopropylmagnesium chloride. The (*R*)- and (*S*)-alcohols **138** are separated by chromatography. (*S*)-Alcohol **138** is converted to the allyl ether **139**. Heating allyl ether **139** induces the intramolecular [4 + 2]-cycloaddition. Hydrolysis of the cycloadduct under acidic conditions affords 1-(3-dimethylamino)propyl)-1-(4-fluorophenyl)-1,3-dihydroisobenzofuran-5-carbaldehyde (**91**).

A solution of 2-furan-2-yl-[1,3]-dioxolane in THF is added at −78°C to a solution of lithium diisopropylamide (1.0 equivalent) in THF–hexane and the mixture is aged at −78°C for 30 min. A solution of 4-fluorobenzaldehyde (1.0 equivalent) in THF is added at −78°C and the mixture is aged at −78°C for 1 h. The temperature is allowed to rise to 25°C and the mixture is aged at 25°C for 16 h. Hexane and THF are distilled at reduced pressure. The residue is dissolved in ethyl ether, washed with water and with brine, and then dried and concentrated at reduced pressure. (5-(1,3-Dioxolan-2-yl)furan-2-yl)(4-fluorophenyl)methanol (**136**) is isolated from the residue by crystallization from ethyl ether–*n*-hexane (97%).

The alcohol is oxidized with manganese dioxide (2.0 equivalents) in dichloromethane at reflux overnight. The suspension is cooled and filtered. The liquors are concentrated at reduced pressure. Ketone **137** is isolated from the residue by crystallization from methanol (99%).

A solution of ketone **137** in THF is added to a 35 wt% solution of 3-dimethylaminopropylmagnesium chloride (1.1 equivalents) in THF at 0°C over 2 h. The mixture is allowed to warm to 25°C and is aged at 25°C for 16 h. Saturated aqueous ammonium chloride and ethyl ether are added and the layers are separated. The aqueous layer is extracted with ethyl ether. The combined organic layers are washed with water and with brine and then dried and concentrated at reduced pressure. The tertiary alcohol is isolated from the

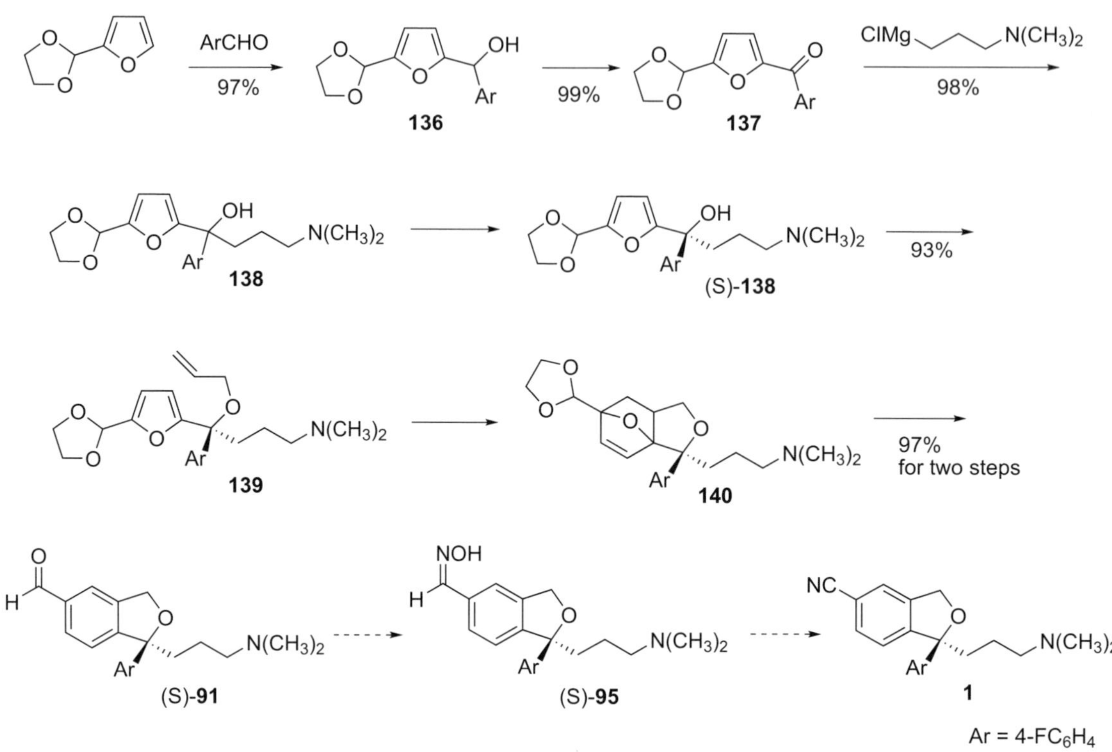


**SCHEME 3.46** Escitalopram (**1**) by intramolecular [4 + 2]-cycloaddition of an alkene with a furan.

residue by crystallization from *n*-heptane (98%). The (*R*)- and (*S*)-alcohols **138** can be separated by chiral chromatography.

A solution of the (*S*)-alcohol **138** in THF is added to a suspension of potassium hydride (3.0 equivalents) in THF and the mixture is refluxed for 2 h. The suspension is cooled to 25°C and the alkoxide solution is decanted from the excess potassium hydride. 18-Crown-6 (1.0 equivalent) is added. The mixture is refluxed for 20 min and then cooled to 25°C. Allyl bromide (1.2 equivalents) is added over 40 min and the mixture is aged for an unspecified time until the reaction is complete. Ethyl ether is added and the mixture is washed with water and with brine and then dried and concentrated at reduced pressure to afford the allyl ether **139** as an oil (93%). A solution of allyl ether **139** in toluene is aged at 85–95°C overnight. Acetic acid and 48% aqueous hydrobromic acid are added to the resulting solution of the *exo*-cycloadducts **140** in toluene. The biphasic mixture is aged at 25°C overnight. The mixture is transferred into aqueous sodium hydroxide–ice and the resulting mixture (aqueous is pH >7) is extracted with ethyl acetate. The extracts are washed with water and brine and then dried and concentrated at reduced pressure to afford (*S*)-1-(3-dimethylamino)propyl)-1-(4-fluorophenyl)-1,3-dihydroisobenzofuran-5-carbaldehyde (**91**) (97%).[169]

Recall that the (*RS*)-aldehyde **91** is converted to the oxime **95** (96–97%) and oxime **95** is dehydrated with acetic anhydride in toluene or pyridine at >100°C to produce citalopram (**39**) (66% oxalate salt).[38]

## 3.13 THE BEST PROCESS AVAILABLE TODAY

With this intricate web of process chemistry options, it might appear that the selection of a best process for escitalopram (**1**) oxalate will be equally complex. This is not the case. At the center of the web (and the center of escitalopram (**1**)) is the generation of the chiral center. The best enantioselectivity for an asymmetric Grignard reagent addition is 95%. An additional purity upgrade step will be required to produce escitalopram (**1**) by this approach. There are no asymmetric Grignard reagent addition processes leading to (*S*)-didesmethyl citalopram (**117**) or (*S*)-desmethyl citalopram (**78**).

Citalopram (**39**) has not been resolved. Assuming that more efficient resolutions of didesmethyl citalopram (**117**) or desmethyl citalopram (**78**) could be developed, the established approaches to both involve a more expensive side chain precursor and at least three additional steps for manipulation of functionality on that side chain (deprotection to release the alcohol, conversion to a leaving group, and displacement of the leaving group). Consider a "best case" scenario. Assume that these steps are all efficient (90% yield) and could be run without isolation of an intermediate. Assume the theoretical yield in the resolution (50%).

Assume that the yield for alkylation after resolution is 95%. With these assumptions, the yield of escitalopram (**1**) from 1-oxo-1,3-dihydroisobenzofuran-5-carbonitrile (**24**) is just 32%. Even with these "stretch goals," the overall yield is unacceptable for a manufacturing process. For every resolution, there must be a racemization.

> A manufacturing process that utilizes a resolution to produce a single enantiomer can deliver no more than 50% yield of product. This process should be replaced by an asymmetric process designed to deliver the desired enantiomer or a resolution process that includes a racemization of the unwanted enantiomer. If the asymmetric process cannot deliver the desired enantiomer without an additional purity upgrade, the resolution process will likely be the better option.

Racemization of (*R*)-didesmethyl citalopram (**117**) or (*R*)-desmethyl citalopram (**78**) requires breaking a bond to the chiral center. Breaking the more labile carbon–oxygen bond is the best option. Breaking this bond would likely produce a diol, but regenerating this bond (cyclizing the diol) with a primary or secondary amine on the side chain is not known.

The remaining strategy is to move back in the sequence and resolve a diol intermediate with a tertiary amine on the side chain. Any of the 10 (*RS*)-diols with a nitrile or substituent capable of being converted to a nitrile ($NH_2$, Br, I, COO*t*Bu, COOH, CONH*t*Bu, $CONR_2$, CH=NOH, oxazoline) are candidates for resolution but only the (*RS*)-diols with nitrile and bromide substituents (**121** and **69**) have been resolved. Both resolutions use DPTTA as the resolving agent. Assume that both resolutions (and the now-required racemizations) are equally efficient. Which is preferred? A considerable labor and overhead expense is incurred in a resolution. The sequence from the value-added (*S*)-diol intermediate to escitalopram (**1**) must be short and efficient. (*S*)-4-(4-(Dimethylamino)-1-(4-fluorophenyl)-1-hydroxybutyl)-3-(hydroxymethyl)benzonitrile (**121**) is converted to escitalopram (**1**) in a single high-yielding step. (*S*)-1-(4-Bromo-2-(hydroxymethyl)phenyl)-4-(dimethylamino)-1-(4-fluorophenyl)butan-1-ol (**69**) is converted to escitalopram (**1**) in two steps: a high-yielding cyclization and a cyanide exchange. On the basis of this argument alone, the process via (*S*)-4-(4-(dimethylamino)-1-(4-fluorophenyl)-1-hydroxybutyl)-3-(hydroxymethyl)benzonitrile (**121**) is preferred.

Revisiting the details of the cyanide exchange provides additional information in support of this decision. There are at least two good reasons to avoid the cyanide exchange. First, the use of cyanide and heavy metals would put an enormous burden on the waste treatment facility at the manufacturing plant. Perhaps replacing sodium cyanide and copper cyanide with nontoxic potassium hexacyanoferrate (III) could address this issue. Second, the cyanide exchange is not clear. Bromide **69**, desmethyl, descyano, and amide impurities must be removed in one or more additional operations to ensure that the escitalopram (**1**) will meet specifications.

The starting material for the manufacturing process is 1-oxo-1,3-dihydroisobenzofuran-5-carbonitrile (**24**). A process for 1-oxo-1,3-dihydroisobenzofuran-5-carbonitrile (**24**) is selected using four criteria. The ideal process has no expensive raw materials, the shortest sequence from raw materials to product, the highest yield for the sequence, and the lowest charges for reagents paraformaldehyde and/or metal cyanide. These reagents are inexpensive, but use of a large excess of either would increase the cost of waste treatment.

1-Oxo-1,3-dihydroisobenzofuran-5-carbonitrile (**24**) is likely produced in three steps. The reaction of terephthalic acid or diethyl terephthalate, paraformaldehyde (1.3 equivalents), and 20–25% oleum is quenched with ethanol to produce ethyl 1-oxo-1,3-dihydroisobenzofuran-5-carboxylate (**13**) (80–85%). An acceptable throughput (90–100 g/L) is possible when the sulfuric acid is neutralized after filtering the suspension. The ethyl ester **13** is converted to the amide **17** with ammonia in methanol at elevated temperature (93%). The amide **17** is then dehydrated with thionyl chloride (1.5 equivalents) and a DMF catalyst in toluene (80%) (Scheme 3.47). There are three solid isolations and three drying operations in this process. Impurities produced from diethyl terephthalate in the ethyl 1-oxo-1,3-dihydroisobenzofuran-5-carboxylate (**13**) are likely to be removed in the isolations of the amide **17** and nitrile **24**. Nitrile **24** is filtered from dry toluene, so there is no concern about carrying some residual solvent into the double Grignard reagent addition step to follow. The yield for the three-step sequence is 60–63%. The weaknesses of this process are the three solid isolations and drying operations (for producing a starting material!) and the potential for operator exposure during the paraformaldehyde solid transfer.

The shorter (two-step) sequences from 1-oxo-1,3-dihydroisobenzofuran-5-carboxylic acid (**10**) to 1-oxo-1,3-dihydroisobenzofuran-5-carbonitrile (**24**) use expensive sulfamide or specialty chemical ethyl polyphosphate. The three-step process (bromination, borohydride reduction, and cyanide exchange) from phthalic anhydride to 1-oxo-1,3-dihydroisobenzofuran-5-carbonitrile (**24**) has poor selectivity and a low yield in the borohydride reduction. The yield for the sequence is just 28–33%. The four-step sequence via 1-oxo-1,3-dihydroisobenzofuran-5-carboxylic acid (**10**), ester **13**, amide **17**, and nitrile **24** and the five-step sequence from 2,4-dimethylbenzoic acid offer no advantages to compensate for the additional steps.

The double Grignard reagent addition process converts 1-oxo-1,3-dihydroisobenzofuran-5-carbonitrile (**24**) to the

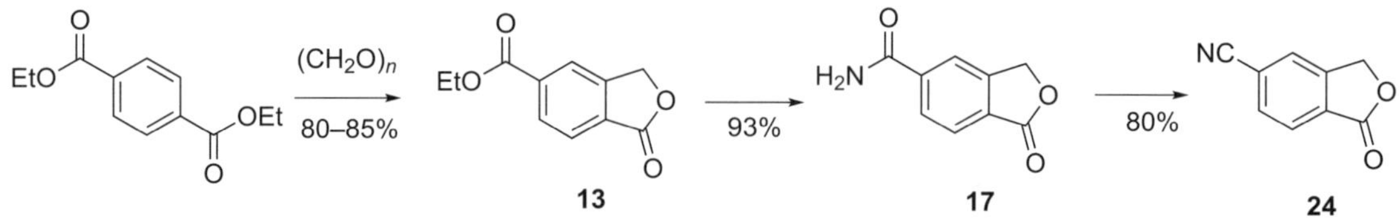


**SCHEME 3.47** The best available process for 1-oxo-1,3-dihydroisobenzofuran-5-carbonitrile (**24**).

(*RS*)-diol **121**. All available evidence suggests that the reaction with 25–30 wt% 4-fluorophenylmagnesium bromide in THF at −10 to 0°C for 2–10 h will produce a mixture of 4-(4-fluorobenzoyl)-3-(hydroxymethyl)benzonitrile (**37**) and the tertiary alcohol side product **46** (85:15). It is important to use enough of the Grignard reagent (1.2 equivalents) to achieve complete conversion of the starting 1-oxo-1,3-dihydroisobenzofuran. A 25–30 wt% solution of 3-dimethylaminopropylmagnesium chloride in THF is added, probably at −10 to 0°C, and the mixture is aged at 0–25°C until ketone **37** is consumed. An excess of the second Grignard reagent (1.4 equivalents) is used to ensure complete conversion of ketone **37** in 2–3 h. The mixture is quenched with cold water and acetic acid is added (to pH 6.5–7.0). The workup then follows the procedure published back in 1986. The THF is distilled, the (*RS*)-diol **121** and the side product **46** are separated by acid–base extraction using 20% acetic acid, and the (*RS*)-diol **121** is obtained as a concentrated solution in toluene (80–85% contained yield, >95% pure) (Scheme 3.48).

3-Chloro-*N*,*N*-dimethylpropan-1-amine is likely introduced into the Grignard reagent preparation as a concentrated solution in a dry solvent of low polarity. Methyl *tert*-butyl ether appears to be the best option since it can be dried by azeotropic distillation at 52–53°C. Methyl *tert*-butyl ether might also be used in the recovery of (+)-DPTTA.

There are three options for resolution of (*RS*)-diol **121**, chemical resolution with (+)-DPTTA, enzyme-mediated resolution, and chiral chromatography. The raw material cost driver for the chemical resolution is associated with the recovery of (+)-DPTTA after cleaving the (*S*)-diol-(+)-DPTTA salt (**129**) from the chemical resolution. (+)-DPTTA can be released from the salt **129**, isolated as the monoammonium salt, and converted to the acid by ion exchange (87%). This (+)-DPTTA recovery is likely to improve with continued process development. The primary raw material cost driver for the enzyme-mediated resolution is the cost of the immobilized enzyme. It is assumed that the immobilized enzyme will be recycled one or more times. It should be assumed that the price for the immobilized enzyme

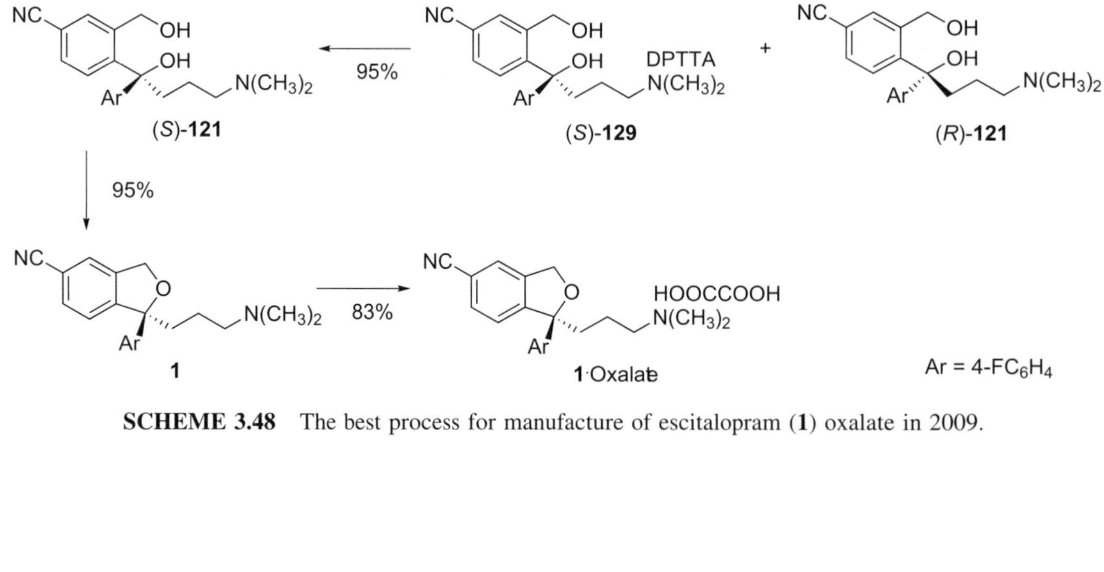


**SCHEME 3.48** The best process for manufacture of escitalopram (**1**) oxalate in 2009.

would decrease with increase in the amount required. Novozyme 435[170] can be recycled at 25°C, or perhaps 35°C. With insufficient information on the enzyme-mediated resolution and the prices for enzyme and column packing material at various tonnages, the chemical resolution is selected to complete the discussion. We should be prepared to revisit this decision when these competitive processes deliver (*S*)-diol **121** at a lower price.

The concentrated solution of (*RS*)-diol **121** in toluene is carried into the chemical resolution. The 1-propanol solvent for a chemical resolution is added and the 1-propanol–toluene azeotrope is distilled to remove most of the toluene. The alternative distillation of toluene followed by 1-propanol addition might result in crystallization of diol **121**(mp 98°C) and a bound or broken agitator. A 40°C solution of (+)-DPTTA (0.39 equivalents) in 1-propanol is added to the solution of 4-(4-(dimethylamino)-1-(4-fluorophenyl)-1-hydroxybutyl)-3-(hydroxymethyl)benzonitrile (**121**) in 1-propanol containing some toluene at 40°C. Cooling and seeding with (*S*)-diol-(+)-DPTTA salt (**129**) produces a suspension. The solid is filtered, washed with 1-propanol, and dried to afford (*S*)-diol-(+)-DPTTA salt (**129**) (35% based on (*RS*)-diol **121**, >98% *S*).

The mixture (*R*)-diol (major) and (*S*)-diol (minor) in the liquors must be recovered, converted back to the (*RS*)-diol **121**, and returned to the resolution. We know that reaction of the (*RS*)-diol **121** with acid (sulfuric acid, phosphoric acid, hydrochloric acid) affords citalopram (**39**) in high yield (>90%). We also know that reaction with concentrated aqueous acid at elevated temperature causes racemization of the chiral center. How can the conditions be modified to induce the racemization but avoid the cyclization?

Hydrochloric acid (1 M) is added to a solution of (*S*)- and (*R*)-diols (98.4:1.6) (**121**) in toluene. Sodium chloride solid is added to make the aqueous phase 1 M in sodium chloride and the mixture is aged at 25°C overnight. The resulting suspension is filtered. Analysis reveals that the solid is (*RS*)-diol **121** hydrochloride salt (*S*:*R* = 54:46) and the toluene layer in the liquors contains (*S*)- and (*R*)-diols (97.9:2.1) (**121**). The yield of (*RS*)-diol **121** is not provided. Perhaps increasing the temperature to 40–50°C during the aging time will increase the rate of racemization and provide the (*RS*)-diol **121** in high yield. The racemization with 1 M hydrochloric acid is offered with no experimental details in another patent.[141,171]

At this point we have a suspension of (*RS*)-diol hydrochloride salt in dilute hydrochloric acid–toluene. Consider two options. First, add hydroxide, separate the layers, and return the toluene solution of the (*RS*)-diol free base to the resolution flywheel. Alternatively, filter the (*RS*)-diol hydrochloride salt, wash the salt with toluene, dissolve the salt in water and toluene, and *then* add hydroxide, separate the layers, and return the toluene solution to the resolution flywheel. The first option has fewer operations but returns not only (*RS*)-diol **121** but also low-level impurities. As the flywheel cycles, these impurities would increase to unacceptable levels. The second option is selected. The (*RS*)-diol hydrochloride salt is filtered.

> In a perfect world, a resolution flywheel would have just two streams: starting materials going in and product coming out. In the real world of pharmaceutical manufacturing, a resolution flywheel must also have a waste stream to control the levels of side products.

The (*S*)-diol-(+)-DPTTA salt (**129**) is cleaved with sodium hydroxide in water and the free base is extracted with toluene. The extracts are dried by distillation of the toluene–water azeotrope and the resulting solution is cooled to 25°C. The contained yield of (*S*)-diol **121** is 95%. Triethylamine (2.0 equivalents) is added. A solution of *p*-toluenesulfonyl chloride (1.1 equivalents) in toluene is added at 25°C and the solution is aged at 25°C for <30 min. Water and ammonium hydroxide are added and the mixture is warmed to 45°C. The layers are separated and the organic layer is washed with water, dried, and concentrated at reduced pressure to afford a concentrated solution of crude escitalopram (**1**) (95% contained yield). The solution is diluted with ethanol and toluene is removed as the ethanol–toluene azeotrope. The solution is cooled, oxalic acid (1.0 equivalent) is added, and ethanol is distilled to produce a solution of escitalopram (**1**) oxalate in ethanol (4–5 L ethanol/kg escitalopram oxalate). The solution is cooled to 15°C at a rate of 0.2–0.5°C/min in the range from 80 to 40°C and seeded at intervals (75, 65, and 60°C). The resulting 15°C suspension is aged for 10 h and filtered. The solid is washed with 1.1–1.2 L ethanol/kg escitalopram oxalate and dried at 50–60°C and 90–100 mmHg for 12 h to afford escitalopram (**1**) oxalate (83% from crude escitalopram (**1**), 79% from (*S*)-diol **121**).[172]

Assuming a 90% recovery of (*RS*)-diol **121** from the resolution mother liquors, the process from 1-oxo-1,3-dihydroisobenzofuran-5-carbonitrile (**24**) to escitalopram (**1**) oxalate has five chemical steps and proceeds in 51–54% overall yield. An 80% recovery of (*RS*)-diol **121** would deliver a 44–47% overall yield. There are three solid isolations (*S*)-diol-(+)-DPTTA salt (**129**), escitalopram (**1**) oxalate, and (*RS*)-diol **121** hydrochloride salt). The wet (*RS*)-diol **121** hydrochloride salt need not be transferred as a solid or dried. The key raw materials are terephthalic acid, 1-bromo-4-fluorobenzene, and 3-chloro-*N*,*N*-dimethylpropan-1-amine hydrochloride. The process solvents are THF, toluene, methyl *tert*-butyl ether, 1-propanol, and ethanol, all solvents found in a pharmaceutical manufacturing plant.

With process selection completed, consider what opportunities might be available for purity upgrade in the event of a process deviation. Recrystallization of escitalopram (**1**)

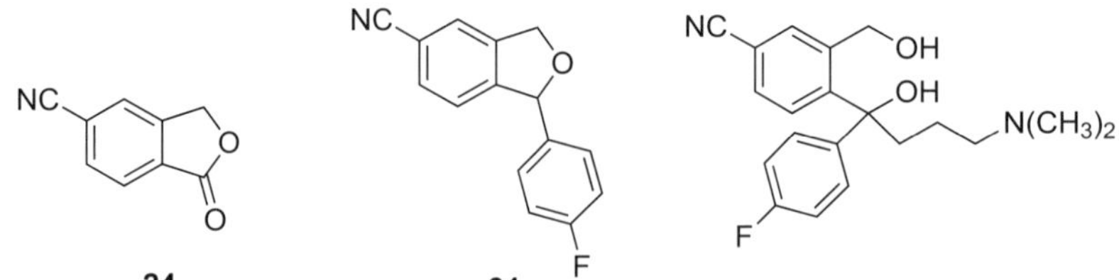


24 31 121

**FIGURE 3.5** Structures searched for escitalopram (**1**) presentation.

oxalate or (*S*)-diol-(+)-DPTTA salt (**129**) are viable options. (*RS*)-Diol **121** as free base (mp 98°C) and escitalopram (**1**) free base (mp 45.8°C) could be precipitated or crystallized by diluting a concentrated toluene solution with *n*-heptane. Filtration and drying of the low-melting escitalopram (**1**) free base is likely to be problematic.[173]

## 3.14 STRUCTURES SEARCHED

Three structure searches were used to generate all the information presented in this chapter (Figure 3.5).

## REFERENCES

1. Cipriani, A.; Furukawa, T.A.; Salanti, G.; Geddes, J.R.; Higgins, J.P.T.; Churchill, R.; Watanabe, N.; Nakagawa, A.; Omori, I.M.; McGuire, H.; Tansella, M.; Barbui, C. *Lancet* **2009**, *373*, 746.
2. Burke, W.J.; Kratochvil, C.J. *Prim. Care Companion J. Clin. Psychiatry* **2002**, *4*, 20.
3. Accessed at www.lundbeck.com/investor/Financials/Investor_kit/Investor_kit.pdf.
4. Lundbeck Second Quarter 2009 Report, Corporate Release No. 382, August 13, 2009.
5. Forest Laboratories, Inc. 2009 Annual Report.
6. Accessed at www.pbs.org/wgbh/takeonestep/depression/pdf/dep_stats.pdf and www.pbs.org/wgbh/takeonestep/depression/pdf/dep_facts.pdf.
7. Accessed at www.who.int/mental_health/management/depression/definition.
8. Accessed at www.who.int/whosis/whostat2007.
9. Accessed at www.dbsalliance.org/site/PageServer?pagename=about_statistics_depression.
10. Accessed at www.researchandmarkets.comreports/314820/impact_of_generics_on_the_antidepressant.htm.
11. The Aldrich 2007–2008 catalog price for 500 g (98%) is $19.70.
12. Huntress, E.H.; Shriner, R.L. *Org. Synth., Coll. Vol. 2*, **1943**, 459.
13. Lamatsch, B. U.S. Patent 6,087,480 (7/11/2000).
14. Tirouflet, J. *Bull. Soc. Sci. Bretagne Spec.* **1951**, (26), 35.
15. The Aldrich 2007–2008 catalog price for 25 g is $69.80.
16. The Aldrich 2007–2008 catalog price for 16 kg is $158.00.
17. Oren, J. IL 115814 (3/12/1999).
18. Oren, J. WO 2004089924 (10/21/2004).
19. The TCI 2006–2007 catalog price for 500 g is $18.95.
20. GB 1578989 (11/12/1980).
21. The Aldrich 2007–2008 catalog price for 500 g is $18.40.
22. The Aldrich 2007–2008 catalog price for 2 kg is $57.60.
23. (a) Dall'Asta, L.; Casazza, U.; Cotticelli, G. U.S. Patent 6,703,516 (3/9/2004). (b) Dall'Asta, L.; Casazza, U.; Cotticelli, G. U.S. Patent 6,458,973 (10/1/2002).
24. The Aldrich 2007–2008 catalog price for 3 kg (95%, prilled) is $40.10.
25. (a) Petersen, H.; Dahlberg Nielsen, P. U.S. Patent 6,888,009 (5/3/2005). (b) Petersen, H.; Dahlberg Nielsen, P. U.S. Patent 6,403,813 (6/11/2002).
26. (a) Greenwood, A.K.; McHattie, D.; Rechka, J.A.; Hedger, P. C.M.; Gamble, M.P. U.S. Patent 7,002,025 (2/21/2006). (b) Greenwood, A.K.; McHattie, D.; Rechka, J.A.; Hedger, P.C. M.; Gamble, M.P. WO 2001066536 (9/13/2001).
27. Pittelkow, T.; Castellin, A.; Sbrogio, F.; Dahlberg Nielsen, P.; Zanon, J.; Sogaard, S.; Humble, R.E. WO 2004083197 (9/30/2004).
28. Accessed at www.ehs.berkeley.edu/pubs/factsheets/49formaldehyde.html.
29. (a) Cotticelli, G.; Salvetti, R.; Zappa, M. U.S. Patent 7,482,476 (1/27/2009). (b) Cotticelli, G.; Salvetti, R.; Zappa, M. U.S. Patent 20070117991 (5/24/2007). (c) Cotticelli, G.; Salvetti, R.; Zappa, M. EP 1777221 (4/25/2007).
30. (a) Dall'Asta, L.; Casazza, U.; Petersen, H. U.S. Patent 6,365,747 (4/2/2002). (b) Dall'Asta, L.; Casazza, U.; Petersen, H. WO 2000023431 (4/27/2000).
31. (a) Petersen, H. U.S. Patent 6,911,548 (6/28/2005). (b) Petersen, H. WO 2001051477 (7/19/2001).
32. (a) Petersen, H. U.S. Patent 6,229,026 (5/8/2001). (b) Petersen, H. WO 1998019513 (5/14/1998).
33. (a) Petersen, H.; Dahlberg Nielsen, P. EP 1140886 (4/2/2003). (b) Petersen, H.; Dahlberg Nielsen, P. U.S. Patent 6,392,060 (5/21/2002). (c) Petersen, H.; Dahlberg Nielsen, P. WO 2000039112 (7/6/2000).
34. Guazzi, G. WO 2002048133 (6/20/2002).
35. The Aldrich 2007–2008 catalog price for 3 kg is $70.30.
36. The Aldrich 2007–2008 catalog price for 1 kg is $74.90.
37. Forney, L.S. U.S. Patent 3,607,884 (9/21/1971).

38. (a) Petersen, H. U.S. Patent 7,030,252 (4/18/2006). (b) Petersen, H. U.S. Patent 6,849,749 (2/1/2005). (c) Petersen, H. WO 1999030548 (6/24/1999).
39. The Aldrich 2007–2008 catalog price for 1 L ($d = 0.934$ g/mL) is $35.40.
40. Liu, Y.; Fanhao, F.X. *Huaxue Yu Shengwu Gongcheng* **2006**, *23*, 17.
41. Mahavir, A.S. IN 2003MU01074 (9/9/2005).
42. The Aldrich 2007–2008 catalog price for 500 g is $65.90.
43. (a) Petersen, H.; Rock, M.H. U.S. Patent 7,271,273 (9/18/2007). (b) Petersen, H.; Rock, M.H. WO 2001049672 (7/12/2001).
44. Xu, F.; Lin, X. CN 1962651 (5/16/2007).
45. The Aldrich 2007–2008 catalog price for 1 kg is $77.80.
46. The Aldrich 2007–2008 catalog price for 500 g (~99%) is $47.50.
47. Xu, F.; Lin, X. CN 1962652 (5/16/2007).
48. The Aldrich 2007–2008 catalog price for 25 g is $246.00.
49. (a) Weber, B. U.S. Patent 6,441,201 (8/27/2002). (b) Weber, B. WO 2000044738 (8/3/2000).
50. The Aldrich 2007–2008 catalog price for 3 kg (99%) is $130.00.
51. The Aldrich 2007–2008 catalog price for 250 mL of 50 wt% in water is $48.10.
52. Veera Reddy, A.; Rajendiran, C.; Reddy, I.; Jasti, V. WO 2005111010 (11/24/2005).
53. Cava, M.P.; Lakshmikantham, M.V.; Mitchell, M.J. *J. Org. Chem.* **1969**, *34*, 2665.
54. Wang, S.; Liu, X. CN 1379025 (11/13/2002).
55. The Aldrich 2007–2008 catalog price for 50 g is $122.50.
56. Ikemoto, T.; Kobori, K.; Iki, S. JP 2001106681 (4/17/2001).
57. The Aldrich 2007–2008 catalog price for 500 g is $129.00.
58. (a) Cotticelli, G.; Dall'Asta, L.; Di Lernia, G. U.S. Patent 7,411,077 (8/12/2008). (b) Cotticelli, G.; Dall'Asta, L.; Di Lernia, G. WO 2004080988 (9/23/2004).
59. (a) Babu, A.N.; Goud, V.S.; Gaonkar, S.L.; Manjunatha, S.G.; Kulkarni, A.K. U.S. Patent 20060116522 (6/1/2006). (b) Narahari Babu, A.; Srinivas Goud, V.; Gaonkar, S.L.; Manjunatha, S.G.; Kulkarni, A.K. WO 2005066185 (7/21/2005). (c) Narahari Babu, A.; Srinivas Goud, V.; Gaonkar, S.L.; Manjunatha, S.G.; Kulkarni, A.K. WO 2004020425 (3/11/2004).
60. Muddasani, P.R.; Nannapaneni, W.C. WO 2004016602 (2/26/2004).
61. Pulla, R.M.; Venkaiah, C.N. WO 2002066453 (8/29/2002).
62. Bogeso, K.P.; Toft, A.S. U.S. Patent 4,136,193 (1/23/1979).
63. (a) Petersen, H.; Bogeso, K.P.; Bech Sommer, M. EP 1032566 (2/12/2003). (b) Petersen, H.; Bogeso, K.P.; Bech Sommer, M. U.S. Patent 6,291,689 (9/18/2001). (c) Petersen, H.; Bogeso, K.P.; Bech Sommer, M. WO 1998019511 (5/14/1998).
64. Hilden, L.; Rummakko, P.; Grumann, A.; Pietikainen, P. WO 2004011450 (2/5/2004).
65. The Aldrich 2007–2008 catalog price for 25 g is $131.50.
66. (a) Ikemoto, T.; Gao, W.-G.; Igi, M. U.S. Patent 7,119,215 (10/10/2006). (b) Ikemoto, T.; Igi, M. U.S. Patent 6,946,564 (9/20/2005). (c) Ikemoto, T.; Gao, W.-G.; Arai, N.; Igi, M. EP 1428813 (6/16/2004). (d) Ikemoto, T.; Gao, W.-G.; Arai, N.; Igi, M. U.S. Patent 6,458,975 (10/1/2002). (e) Ikemoto, T.; Gao, W.-G.; Arai, N.; Igi, M. U.S. Patent 6,433,196 (8/13/2002).
67. The Aldrich 2007–2008 catalog price for 100 g is $101.00.
68. The Aldrich 2007–2008 catalog price for 8 L ($d = 0.868$ g/mL) is $205.00.
69. The Aldrich 2007–2008 catalog price (2,4-dimethylbenzoic acid) for 50 g is $122.50.
70. The Aldrich 2007–2008 catalog price for 500 g is $54.00.
71. Arai, N.; Katsura, T.; Mizuno, T. WO 2006001529 (1/5/2006).
72. The Aldrich 2007–2008 catalog price for 25 g is $58.20.
73. The Aldrich 2007–2008 catalog price for 250 g is $161.50.
74. (a) Ikemoto, T.; Igi, M. U.S. Patent 6,930,211 (8/16/2005). (b) Ikemoto, T.; Igi, M. EP 1095926 (10/6/2004).
75. Petersen, H. WO 2001068632 (9/20/2001).
76. Arai, N.; Ikemoto, T. JP 2007045807 (2/22/2007).
77. Arai, N.; Katsura, T.; Mizuno, T. WO 2006057261 (6/1/2006).
78. Arai, N.; Ikemoto, T.; Iki, M. JP 2003206284 (7/22/2003).
79. Arai, N.; Ikemoto, T.; Iki, M. JP 2003040882 (2/13/2003).
80. Muddasani, P.; Donthineni, L.; Podile, K.; Nannapaneni, V.C. IN 2003MA00331 (9/1/2006).
81. Muddasani, P.R.; Vattikuti, U.R.; Nannapaneni, V.C. IN 2003MA00049 (9/1/2006).
82. Stivanello, M. IT 2002MI1238 (12/9/2003).
83. Petersen, H.; Sommer, M.B.; Bogeso, K.P. IN 2001MA00237 (9/19/2008).
84. The Aldrich 2007–2008 catalog price for 500 g is $69.70.
85. Tse, H.L.A. WO 2003087081 (10/23/2003).
86. (a) Goud, V.S.; Gaonkar, S.L.; Thomas, S.; Manjunatha, S.G.; Kulkarni, A.K.; Babu, A.N. U.S. Patent 7,255,741 (8/14/2007). (b) Babu, A.N.; Goud, V.S.; Gaonkar, S.L.; Thomas, S.; Manjunatha, S.G.; Kulkarni, A.K. WO 2003080590 (10/2/2003). (c) Babu, A.N.; Goud, V.S.; Gaonkar, S.L.; Thomas, S.; Manjunatha, S.G.; Kulkarni, A.K. EP 1346989 (9/24/2003).
87. Venkaiah Chowdary, N.; Khadgapathi, P. WO 2004094399 (11/4/2004).
88. *Bretherick's Handbook of Reactive Chemical Hazards*, 5th edition, Butterworth-Heinemann, Ltd., Oxford, UK, 1995, pp. 347, 349.
89. (a) Malik, A.A.; Palandoken, H.; Stringer, J.A.; Huang, D.; Romero, A.; Dapremont, O. U.S. Patent 6,967,259 (11/22/2005). (b) Malik, A.A.; Palandoken, H.; Stringer, J.A.; Huang, D.; Romero, A.; Dapremont, O. WO 2003029236 (4/10/2003).
90. (a) Chodankar, N.; Bhobhe, A.; Oak, G.M.; Eappen, P. U.S. Patent 6,812,355 (11/2/2004). (b) EP 1288211 (3/5/2003).
91. Nannapaneni, V.C.; Muddasani, P.R. WO 2004065375 (8/5/2004).

92. (a) Petersen, H.; Rock, M.H.; Svane, H. U.S. Patent 20020077353 (6/20/2002). (b) Petersen, H.; Rock, M.H.; Svane, H. WO 2000013648 (3/16/2000).
93. Kaushik, V.K.; Srinivasa Rao, D.V.N.; Handa, V.K.; Sivakumaran, M. U.S. Patent 6,781,003 (8/24/2004).
94. (a) Villa, M.; Sbrogio, F.; Dancer, R. EP 1181713 (9/29/2004). (b) Villa, M.; Sbrogio, F.; Dancer, R. WO 2001045483 (6/28/2001).
95. (a) Castellin, A.; Volpe, G.; Sbrogio, F. U.S. Patent 6,855,834 (2/15/2005). (b) Castellin, A.; Volpe, G.; Sbrogio, F. WO 2001047877 (7/5/2001).
96. Rajamannar, T.; Srinivasu, K.; Patel, N.S.; Rajendran, C. U.S. Patent 7,148,364 (12/12/2006).
97. (a) Petersen, H.; Rock, M.H. U.S. Patent 6,750,358 (6/15/2004). (b) Petersen, H.; Rock, M.H.WO 2000011926 (3/9/2000).
98. Borase, A.P.; Patel, N.S.; Kilaru, S.; Thennati, R. U.S. Patent 7,019,153 (3/28/2006).
99. (a) Abbineni, J.B.; Bodepudi, H.B.; Chava, S.; Chunchu, V.R. R. U.S. Patent 7,365,216 (4/29/2008). (b) Chunchu, V.R.R.; Chava, S.; Bodepudi, H.B.; Abbineni, J.B. WO 2003072565 (9/4/2003).
100. Villa, M.; Sbrogio, F.; Dancer, R. U.S. Patent 6,455,710 (9/24/2002).
101. Bolzonella, E.; Castellin, A.; Nicole, A. WO 2001002383 (1/11/2001).
102. The Aldrich 2007–2008 catalog price cyanogen bromide for 500 g is $431.00.
103. The Aldrich 2007–2008 catalog price for 5 g is $117.50.
104. (a) Petersen, H.; Bregnedal, P.; Bogeso, K.P. U.S. Patent 6,258,842 (7/10/2001). (b) Petersen, H.; Bregnedal, P.; Bogeso, K.P. WO 1998019512 (5/14/1998).
105. Petersen, H.; Ahmadian, H.; Dancer, R. U.S. Patent 6,509,483 (1/21/2003).
106. Petersen, H.; Ahmadian, H. U.S. Patent 6,426,422 (7/30/2002).
107. (a) Petersen, H.; Dancer, R. EP 1296970 (3/19/2008). (b) Petersen, H.; Dancer, R. U.S. Patent 6,660,873 (12/9/2003). (c) Petersen, H.; Dancer, R. WO 2001085712 (11/15/2001).
108. (a) Petersen, H. U.S. Patent 6,806,376 (10/19/2004). (b) Petersen, H. EP 1265882 (1/14/2004). (c) Petersen, H. WO 2001068630 (9/20/2001).
109. (a) Petersen, H. EP 1265880 (11/10/2004). (b) Petersen, H. U.S. Patent 6,717,000 (4/6/2004). (c) Petersen, H. WO 2001068628 (9/20/2001).
110. (a) Petersen, H. U.S. Patent 6,992,198 (1/31/2006). (b) Petersen, H. WO 2001068631 (9/20/2001).
111. Petersen, H. U.S. Patent 6,762,308 (7/13/2004).
112. (a) Petersen, H.; Ahmadian, H. U.S. Patent 6,864,379 (3/8/2005). (b) Petersen, H.; Ahmadian, H. WO 2001068629 (9/20/2001).
113. Ikemoto, T.; Watanabe, Y. WO 2005113530 (12/1/2005).
114. Ikemoto, T.; Watanabe, Y. JP 2005263741 (9/29/2005).
115. The Aldrich 2007–2008 catalog price for 10 g is $154.50.
116. The Aldrich 2007–2008 catalog price for 100 g is $206.00.
117. (a) Petersen, H.; Rock, M.H.; Ahmadian, H. U.S. Patent 6,420,574 (7/16/2002). (b) Petersen, H.; Rock, M.H.; Ahmadian, H. WO 2001043525 (6/21/2001).
118. (a) Rock, M.H.; Ahmadian, H. U.S. Patent 6,768,011 (7/27/2004). (b) Rock, M.H.; Ahmadian, H. WO 2001051478 (7/19/2001).
119. The Aldrich 2007–2008 catalog price for 50 g is $101.50.
120. (a) Barberich, T. WO 2005079787 (9/1/2005). (b) Bush, L.R.; Currie, M.G.; Senanayake, C.H.; Fang, K.Q. WO 2003040121 (5/15/2003).
121. The Aldrich 2007–2008 catalog price for 25 g (racemic) is $393.00.
122. The Aldrich 2007–2008 catalog price for 250 g is $93.60.
123. (a) Sundaram, V.; Mathad, V.T.; Venkavala, P.J.; Elati, C.R.; Kolla, N.; Govindan, S.; Chalamala, S.R.; Gangula, S. WO 2005047274 (5/26/2005).(b) Elati, C.R.; Kolla, N.; Vankawala, P.J.; Gangula, S.; Chalamala, S.; Bhattacharya, A.; Vurimidi, H.; Mathad, V.T. *Org. Process Res. Dev.* **2007**, *11*, 289.
124. Dancer, R.J.; Lopez de Diego, H. *Org. Process Res. Dev.* **2009**, *13*, 23.
125. Elati, C.R.; Kolla, N.; Mathad, V.T. *Org. Process Res. Dev.* **2009**, *13*, 34.
126. Dancer, R.J.; Lopez de Diego, H. *Org. Process Res. Dev.* **2009**, *13*, 38.
127. The Aldrich 2007–2008 catalog price for 5 g is $58.80.
128. Jin, C.; Boldt, K.G.; Rehder, K.S.; Brine, G.A. *Synth. Commun.* **2007**, *37*, 901.
129. Rao, D.R.; Kankan, R.N.; Bidari, D.R.; Singh, M.P. WO 2008142379 (11/27/2008).
130. Kumar, N.; Nayyar, S.; Gurusami, K. IN 2006DE00279 (8/17/2007).
131. Gao, R. CN 1830952 (9/13/2006).
132. Liu, Z.; Huang, W.; Yuan, A.; Xiao, K.; Li, Y.; Zhuang, J. CN 1740167 (3/1/2006).
133. Gu, J.M.; Wang, Y.W.; Hu, X.R. *Acta Crystallogr. Sect. E: Struct. Rep. Online* **2005**, *E61*, o2691.
134. Wu, Q.; Liao, H.; Zhao, H.; Ye, G.; Jin, Y. *Zhongguo Yiyao Gongye Zazhi*, **2005**, *36*, 6.
135. Kaushik, V.K.; Handa, V.K.; Sivakumaran, M.S. IN 2005CH00036 (3/16/2007).
136. Szabadkai, I. HU 200301605 (11/28/2005).
137. Srinivasu, K.; Rajamannar, T. IN 2003MU00781 (4/1/2005).
138. Leone, M. IT 2001MI1529 (1/20/2003).
139. Gao, W.-G.; Ikemoto, T.; Iki, M. JP 2002114770 (4/16/2002).
140. (a) Bogeso, K.P. CA 1237147 (5/24/1988). (b) Bogeso, K.P. U.S. Patent 4,650,884 (3/17/1987). (c) Bogeso, K.P. EP 171943 (2/19/1986).
141. (a) Cotticelli, G.; Salvetti, R.; Bertoni, C. U.S. Patent 20080199922 (8/21/2008). (b) Cotticelli, G.; Salvetti, R.; Bertoni, C. WO 2006136521 (12/28/2006). (c) Cotticelli, G.; Salvetti, R.; Bertoni, C. EP 1736550 (12/27/2006).
142. Cotticelli, G.; di Lernia, G.; Silvia, M. EP 1506963 (4/13/2005).

143. Narahari Babu, A.; Srinivas Goud, V.; Gaonkar, S.L.; Manjunatha, S.G.; Kulkarni, A.K. WO 2005077927 (8/25/2005).

144. Gaonkar, S.L.; Das, P.P.; Narahari Babu, A.; Manjunatha, S. G.; Subramaniam, H. WO 2006106531 (10/12/2006).

145. Mei, R.; Guo, D.; Wang, S. U.S. Patent 7,435,838 (10/14/2008).

146. Patel, D.J.; Kumar, R.; Dwivedi, S.D. WO 2008059514 (5/22/2008).

147. (a) Humble, R.E.; Christensen, T.V.; Rock, M.H.; Nielsen, O.; Petersen, H.; Dancer, R. EP 1412341 (12/8/2004). (b) Humble, R.E.; Christensen, T.V.; Rock, M.H.; Nielsen, O.; Petersen, H.; Dancer, R. WO 2003000672 (1/3/2003).

148. Uttarwar, S.G.; Gawli, B.N. WO 2005012278 (2/10/2005).

149. (a) Boegesoe, K.P.; Perregaard, J.K. USRE 34712 (8/30/1994). (b) Boegesoe, K.P.; Perregaard, J. U.S. Patent 4943590 (7/24/1990). (c) Boegesoe, K.P.; Perregaard, J. EP 347066 (12/20/1989).

150. (a) Periyandi, N.; Kilaru, S.; Thennati, R. U.S. Patent 20080177096 (7/24/2008). (b) Periyandi, N.; Kilaru, S.; Thennati, R. WO 2006021971 (3/2/2006).

151. (a) Cotticelli, G.; Salvetti, R. U.S. Patent 7,482,477 (1/27/2009). (b) Cotticelli, G.; Salvetti, R. U.S. Patent 20070270599 (11/22/2007). (c) Cotticelli, G.; Salvetti, R. EP 1794140 (6/13/2007). (d) Cotticelli, G.; Salvetti, R. WO 2006037714 (4/13/2006).

152. (a) Hilden, L.; Huuhtanen, T.; Rummakko, P.; Grumann, A.; Pietikaeinen, P. U.S. Patent 6,753,433 (6/22/2004). (b) Hilden, L.; Huuhtanen, T.; Rummakko, P.; Grumann, A.; Pietikaeinen, P. U.S. Patent 6,579,993 (6/17/2003). (c) Hilden, L.; Huuhtanen, T.; Rummakko, P.; Grumann, A.; Pietikaeinen, P. WO 2002060886 (8/8/2002).

153. (a) Rock, M.H.; Petersen, H.; Ellegaard, P. U.S. Patent 6,566,540 (5/20/2003). (b) Rock, M.H.; Petersen, H.; Ellegaard, P. EP 1228056 (8/7/2002). (c) Rock, M.H.; Petersen, H.; Ellegaard, P. U.S. Patent 6,407,267 (6/18/2002). (d) Rock, M.H.; Petersen, H.; Ellegaard, P. WO 2000012044 (3/9/2000).

154. Zhu, X.; Ye, X.; Wu, X.; Yuan, Z.; Shi, H.; Wang, Q. CN 101265215 (9/17/2008).

155. Li, L.; Li, Q. CN 1510024 (7/7/2004).

156. Mital, A.; Kumar, R.; Ramachandran, U. *Org. Prep. Proced. Int.* **2006**, *38*, 423.

157. (a) De Faveri, C.; Huber, F.A.M.; Dancer, R.J. WO 2009033488 (3/19/2009). (b) De Faveri, C.; Huber, F.A.M.; Dancer, R.J. U. S. Patent 20090069582 (3/12/2009). (c) De Faveri, C.; Huber, F.A.M.; Dancer, R.J. EP 2061753 (3/19/2009).

158. Bech Sommer, M.; Nielsen, O.; Petersen, H.; Ahmadian, H.; Pedersen, H.; Brosen, P.; Geiser, F.; Lee, J.; Cox, G.; Dapremont, O.; Suteu, C.; Assenza, S.P.; Hariharan, S.; Nair, U. WO 2003006449 (1/23/2003).

159. Albert, M.; Berger, A. EP 1736462 (12/27/2006).

160. (a) Cotticelli, G.; Rocchietti, S.; Terreni, M.; Pregnolato, M.; Salvetti, R. U.S. Patent 7,470,526 (12/30/2008). (b) Cotticelli, G.; Rocchietti, S.; Terreni, M.; Pregnolato, M.; Salvetti, R. EP 1733042 (5/14/2008). (c) Cotticelli, G.; Rocchietti, S.; Terreni, M.; Pregnolato, M.; Salvetti, R. U.S. Patent 20070238887 (10/11/2007). (d) Cotticelli, G.; Rocchietti, S.; Terreni, M.; Pregnolato, M.; Salvetti, R. WO 2005098018 (10/20/2005).

161. Bayod Jasanada, M.; LLorente Garcia, I.; Gotor Santamaria, V.; Brieva Collado, M.R.; Fernandez Solares, L.; Quiros Alvarez, M. ES 2228274 (4/1/2005).

162. Wang, S.; Yang, L.; Wu, J.; Xu, G. CN 1974542 (6/6/2007).

163. (a) Taoka, N.; Kato, T.; Yamamoto, S.; Yoshida, T.; Takeda, T.; Ueda, Y.; Petersen, H.; Dancer, R.; Ahmadian, H.; Lyngso, L.O. WO 2004014821 (2/19/2004). (b) Taoka, N.; Kato, T.; Yamamoto, S.; Yoshida, T.; Takeda, T.; Ueda, Y.; Petersen, H.; Dancer, R.; Ahmadian, H.; Lyngso, L.O. EP 1534654 (2/19/2004).

164. Lyngso, LO. WO 2005077891 (8/25/2005).

165. Wang, S.-Z.; Wu, J.-P.; Xu, G.; Yang, L.-R. *Youji Huaxue* **2008**, *28*, 1584.

166. Albert, M.; Sturm, H.; Berger, A.; Kremminger, P. WO 2007082771 (7/26/2007).

167. (a) Petersen, H.; Felding, J. U.S. Patent 6,762,307 (7/13/2004). (b) Petersen, H.; Felding, J. U.S. Patent 20030060641 (3/27/2003).

168. The Aldrich 2007–2008 catalog price for 25 g is $213.50.

169. Paulon, A.; De Lucci, O.; Castellin, A.; Fabris, F.; Sbrogio, F.; Ceron, E.; Petersen, H.; Dancer, R. WO 2007054105 (5/18/2007).

170. The Sigma 2006–2007 catalog price (lipase acrylic resin from *Candida antarctica* from Novozyme Corporation) for 10 g is $109.50.

171. Petersen, H.; Dancer, R.; Christiansen, B.; Humble, R.E. WO 2004056754 (7/8/2004).

172. (a) Christensen, T.V.; Liljegren, K.; Elema, M.O.; Andresen, L.; Mahashabde, S.; Assenza, S.P. U.S. Patent 7,420,069 (9/2/2008). (b) Jensen, K.B.; Humble, R.E.; Liljegren, K.; Christensen, T.V. U.S. Patent 7,420,068 (9/2/2008). (c) Jensen, K.B.; Humble, R.E.; Liljegren, K.; Christensen, T.V. WO 2005084643 (9/15/2005). (d) Christensen, T.V.; Liljegren, K.; Elema, M.O.; Andresen, L.; Mahashabde, S.; Assenza, S.P. U.S. Patent 6,916,941 (7/12/2005). (e) Christensen, T.V.; Liljegren, K.; Elema, M.O.; Andresen, L.; Mahashabde, S.; Assenza, S.P. WO 2003011278 (2/13/2003).

173. Dancer, R.; Petersen, H.; Nielsen, O.; Rock, M.H.; Eliasen, H.; Liljegren, K. U.S. Patent 7,560,576 (7/14/2009).

# 4

# EFFEXOR XR® (VENLAFAXINE HYDROCHLORIDE)

## 4.1 EFFEXOR XR® AND THE ANTIDEPRESSANT MARKET

Effexor XR® (venlafaxine hydrochloride) belongs to the class of antidepressants known as SNRIs (serotonin–norepinephrine reuptake inhibitors).

Effexor XR® is approved for the treatment of major depressive disorder (MDD), generalized anxiety disorder (GAD), social anxiety disorder, and panic disorder. Effexor XR® is Wyeth's largest selling franchise, accounting for 17–18% of net revenue each year from 2006 ($3.7 billion) through 2008 ($3.9 billion). The sales figures for the most recent two quarters have dropped significantly primarily due to increasing generic competition in the antidepressant market. The total (domestic and international) net revenue for January–March 2009 was $819 million (down 17% from the same period in 2008). The total (domestic and international) net revenue for April–June 2009 was $772 million (down 22% from the same period in 2008).[1,2]

Depression affects about 121 million people worldwide. Major depression affects approximately 15 million Americans (about 8% of the 18- and older population) each year. Approximately 12% of women will experience clinical depression during their lifetime. One in 33 children and 1 in 8 adolescents experience depression. Most people (80%) experiencing depression are not currently receiving treatment despite the fact that depression is one of the most treatable illnesses. Most people (80%) treated for depression show some relief within 4–6 weeks of beginning medication, psychotherapy, participation in a support group, or some combination of these treatment options. In 2002, depression accounted for 4.5% of the worldwide total burden of disease in terms of disability-adjusted life years. Depression is expected to be the second most common health problem in the world by the year 2020 for all ages and both sexes.[3–6]

There are many antidepressants on the market and patients often go through several antidepressants before finding one that provides relief and has tolerable side effects. Patients are advised to continue with an effective antidepressant that is working rather than consider changing to a cheaper (perhaps generic) alternative. The impressive sales figures for Effexor® through 2008 suggest it is and continues to be an effective antidepressant for many people.

Antidepressants have some serious side effects. They may increase the potential for suicidal thinking in children. This is more likely when starting treatment and when the dose is increased or decreased. In 2003, the British Medicines and Healthcare Products Regulatory Agency, the equivalent of the FDA in Britain, ordered physicians to stop prescribing many antidepressants, including Effexor®, to children. In 2007, a review of 295 antidepressant trials with >77,000 patients also suggested a slight increase in suicidal thinking in young adults, aged 18–24. The FDA ordered that all antidepressants carry a black box warning to alert physicians and patients to these risks.

Generic venlafaxine hydrochloride is now available. Teva Pharmaceuticals USA launched venlafaxine hydrochloride immediate-release tablets in the US in August 2006 and will launch venlafaxine hydrochloride extended release capsules in July 2010. Osmotica Pharmaceutical Corp. launched

*Pharmaceutical Process Chemistry for Synthesis: Rethinking the Routes to Scale-Up*, By Peter J. Harrington

venlafaxine hydrochloride extended release tablets in October 2008. Impax Laboratories, Inc. was granted a license to launch extended release capsules in June 2011. Mylan Laboratories, Inc. received tentative approval from the FDA to manufacture immediate-release venlafaxine hydrochloride tablets in November 2008. Other manufacturers in the news include Anchen Pharmaceuticals, Alsa Corporation, Sandoz Inc., Barr Pharmaceuticals, Bentley Pharmaceuticals, and Zydus Pharma Inc.

Strategic planning for patent expiration and generic competition usually involves development of a second-generation drug. Since Effexor® is a racemate, a logical candidate for a second-generation drug would be the (+)- or (−)-enantiomer. However, both enantiomers have a role in the antidepressant activity with the (+)-enantiomer inhibiting serotonin reuptake and the (−)-enantiomer inhibiting norepinephrine reuptake. Wyeth's second-generation drug is a metabolite of venlafaxine. Pristiq® (*O*-desmethylvenlafaxine succinate (**2**)) was approved in February 2008 for the treatment of major depressive disorder and became available at retail pharmacies in May 2008. The total (domestic and international) net revenue for Pristiq® for January–March 2009 was \$49 million. For April–June 2009 that figure was \$62 million. Thomson Reuters quarterly review of the pharmaceutical pipeline for January–March 2008 lists Pristiq® among the top five most promising drugs launched or approved in January–March 2008. Analysts estimate annual sales may reach \$500 million in 2012.[1]

Pristiq® is awaiting approval as a nonhormone treatment for reducing hot flashes in menopause. Wyeth received an approvable letter from the FDA in July 2007. Key advantages of this nonhormone treatment would include rapid onset of action and ability to improve sleep quality and mood while not negatively affecting sexual function. The FDA requested a new clinical trial, lasting at least 1 year, to provide additional data on the potential for serious adverse cardiovascular and hepatic effects. Pristiq® is also in phase II development for the treatment of fibromyalgia.

In April 2009, Sepracor announced patent interference proceedings against Wyeth relating to Pristiq®. Sepracor asserts it has an earlier application date (April 6, 1999) than Wyeth (February 12, 2001).

A preferred method for manufacture of *O*-desmethylvenlafaxine (**2**) is by demethylation of venlafaxine (**1**) by reaction with nucleophile and a base in a polar aprotic solvent at high temperature (150–190°C). Suitable nucleophiles are alkanethiols and arenethiols (ethanethiol, dodecanethiol, 2-(dimethylamino)ethanethiol, thiophenol), sodium sulfide, diphenylphosphine, and L-Selectride®. Solvents are NMP, DMA, DMF, DMSO, and PEG-400.[7–10] With the demethylation link, the discussion of venlafaxine (**1**) manufacturing options in this chapter is relevant to manufacturing *O*-desmethylvenlafaxine (**2**) as well (Figures 4.1 and 4.2).

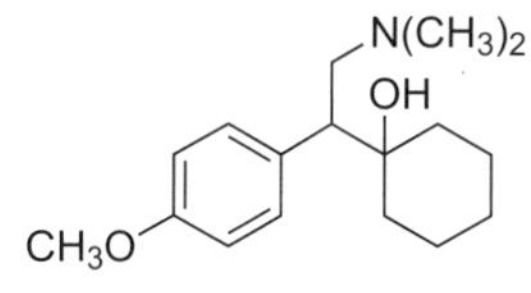


Venlafaxine
**1**

**FIGURE 4.1** Venlafaxine (**1**).

## 4.2 CYCLOHEXANONE CONDENSATIONS

The tertiary alcohol in venlafaxine (**1**) suggests a condensation with inexpensive cyclohexanone or addition of the Grignard reagent from inexpensive 1,5-dibromopentane to an ester. The starting material for both approaches is 4-methoxyphenylacetic acid or the ester, amide or nitrile. The cyclohexanone condensation would precede the conversion of the acid or acid derivative to the tertiary amine. The Grignard reagent addition to the ester would be the last step in the alternative sequence (Figure 4.3).

### 4.2.1 4-Methoxyphenylacetic Acid

The condensation of cyclohexanone with 4-methoxyphenylacetic acid was reported in the 1950s in a program targeting compounds with mydriatic activity similar to atropine but with a shorter duration (Scheme 4.1). A solution of 2-bromopropane (1.8 equivalents) in ethyl ether is added to a suspension of magnesium (1.8 equivalents) and sodium 4-methoxyphenylacetate[11] in ethyl ether. After the addition, the mixture is refluxed for 1 h to produce the enediolate. A solution of cyclohexanone[12] (1.0 equivalent) in ethyl ether is added, presumably at 25–35°C, and the mixture is again refluxed for 1 h. The mixture is quenched into dilute hydrochloric acid and extracted with ethyl ether. The extracts are dried and the ether is distilled. The crude condensation product **3** is crystallized from benzene. No yield is available.[13]

Moving ahead more than 50 years, this condensation is now utilized in the construction of a new generation of monoamine reuptake inhibitors that have functionalized piperazines on the side chain. In the updated version, the carboxylic acid is the starting material and the base is lithium diisopropylamide. The lithium diisopropylamide is prepared

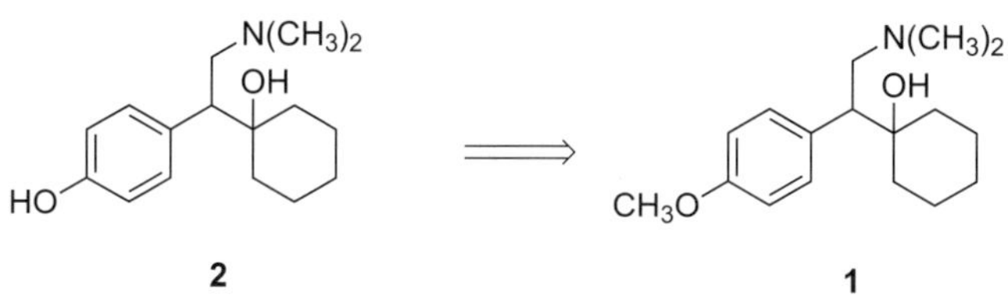


**FIGURE 4.2** *O*-Desmethylvenlafaxine (**2**) from venlafaxine (**1**).

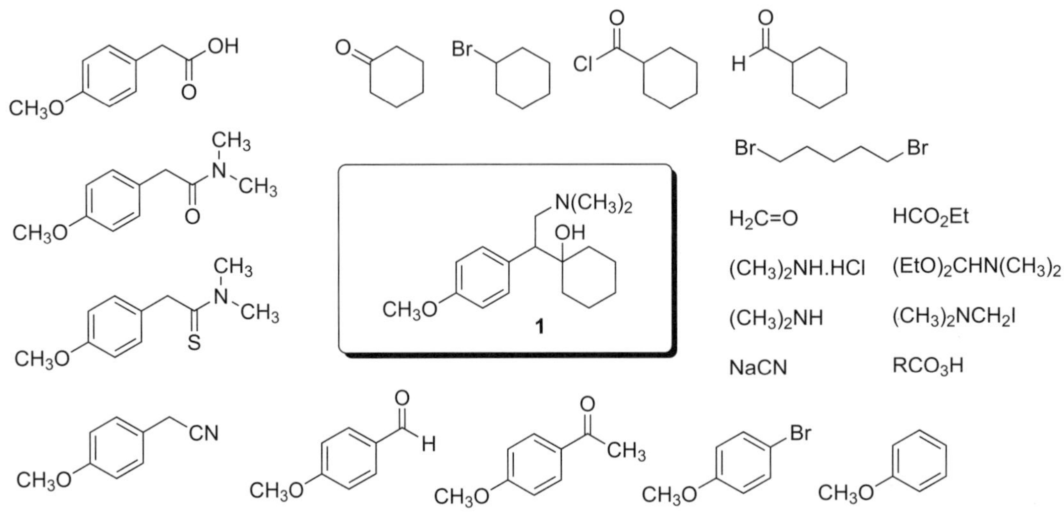


**FIGURE 4.3** Venlafaxine building blocks.

by addition of 2.5 M $n$-butyllithium in hexanes (2.4 equivalents) to diisopropylamine (2.4 equivalents) in THF at −78°C. The solution is warmed to 0°C, aged for 15 min, and cooled back to −78°C. A solution of 3-chlorophenylacetic acid in THF is added, presumably at −78°C, and the mixture is warmed to 25°C, aged at 25°C for 45 min, and cooled back to −78°C. A solution of cyclohexanone (1.5 equivalents) in THF is added and the mixture is aged at −78°C for 1.5 h. The reaction is quenched by adding saturated aqueous ammonium chloride. The THF is distilled at reduced pressure. The residue is dissolved in 2 M sodium hydroxide and washed with ethyl acetate. Hydrochloric acid (2 M) is added (to pH 1) and the mixture is extracted with ethyl acetate. The extracts are dried and concentrated at reduced pressure to afford 2-(1-hydroxycyclohexyl)-2-(3-chlorophenyl)acetic acid (**4**) (96%).[14]

### 4.2.2 2-(4-Methoxyphenyl)-*N,N*-dimethylacetamide (5)

2-(4-Methoxyphenyl)-*N,N*-dimethylacetamide (**5**) is likely prepared by reaction of the acid chloride **6** with 40% aqueous dimethylamine[15] or dimethylamine hydrochloride[16] and triethylamine. The acid chloride **6** is prepared from the acid using oxalyl chloride or thionyl chloride and a DMF catalyst. Available procedures for preparation of related 2-(4-substituted-phenyl)-*N,N*-dimethylacetamides suggest the yield of amide **5** from the acid is 90–97%. $n$-Butyllithium, lithium diisopropylamide, lithium hexamethyldisilazide, and isopropylmagnesium bromide are suitable bases for the amide–ketone condensation.[17,18]

Butyllithium is used as the base in the condensation of 2-(4-bromophenyl)-*N,N*-dimethylacetamide (**7**) with cyclohexanone. $n$-Butyllithium in hexane (1.4 M, 1.0 equivalent) is added to a solution of 2-(4-bromophenyl)-*N,N*-dimethylacetamide (**7**) in THF at <−70°C. The orange suspension is aged at −70°C for 20 min. Cyclohexanone (1.2 equivalents) is added and the mixture aged at −70°C for 50 min. The mixture is quenched into saturated aqueous ammonium chloride. The layers are separated and the aqueous layer is extracted with ethyl ether. The combined organic layers are washed with brine, dried, and concentrated at reduced pressure. The residue is crystallized from isopropanol to afford 2-(4-bromophenyl)-2-(1-hydroxycyclohexyl)-*N,N*-dimethylacetamide (**8**) (54%) (Scheme 4.2).[19,20]

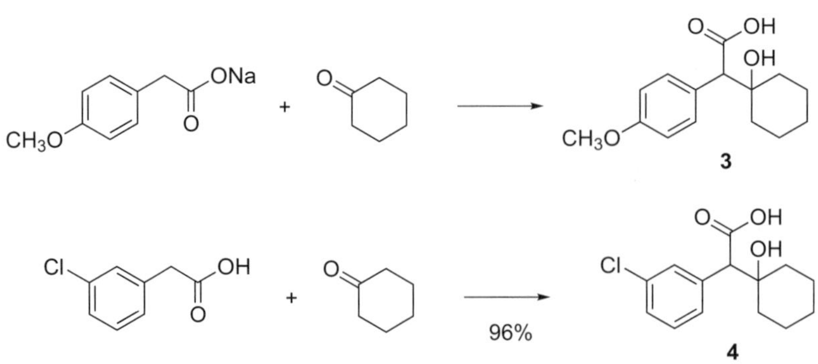


**SCHEME 4.1** Arylacetic acid–cyclohexanone condensations.

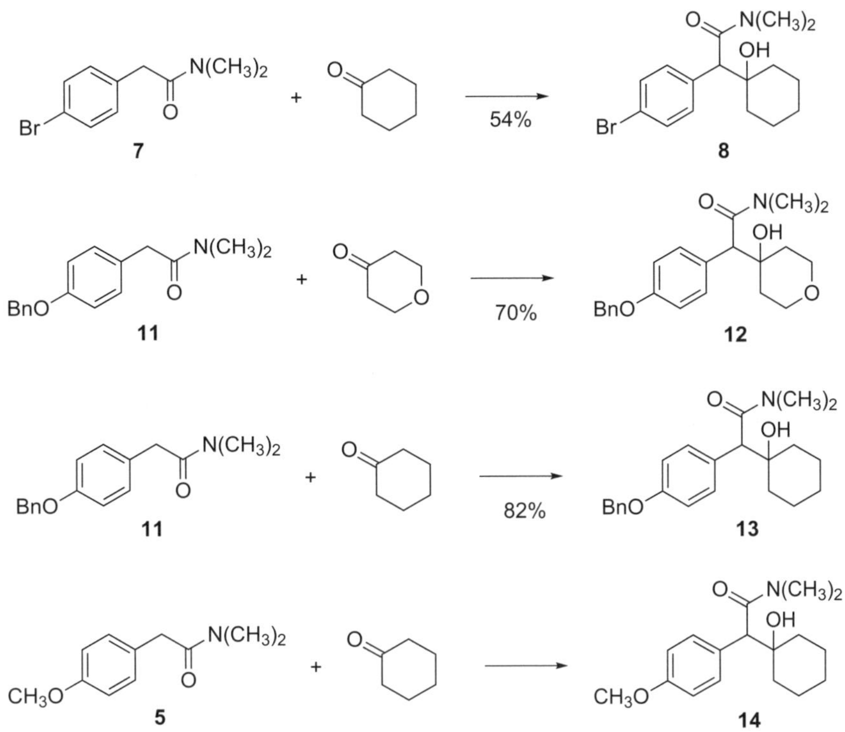

**SCHEME 4.2** 2-Aryl-*N*,*N*-dimethylacetamide–ketone condensations.

The analogous condensation of $d_{15}$-2-(4-methoxyphenyl)-*N*,*N*-dimethylacetamide (**9**) with $d_{10}$-cyclohexanone (1.2 equivalents) using *n*-butyllithium (1.0 equivalent) is quenched with deuterium oxide to afford $d_{24}$-2-(1-hydroxycyclohexyl)-2-(4-methoxyphenyl)-*N*,*N*-dimethylacetamide (**10**). No yield is available.[21]

The order of addition is amide-to-base in an amide–ketone condensation using lithium diisopropylamide as the base. A solution of 2-(4-(benzyloxy)phenyl)-*N*,*N*-dimethylacetamide (**11**) in THF is added to a 2 M solution of lithium diisopropylamide (1.6 equivalents) in THF at −78°C. The mixture is aged at −78°C for 1 h. A solution of tetrahydro-4*H*-pyran-4-one (1.4 equivalents) in THF is added and the mixture is aged, presumably at −78°C, for 2 h. The mixture is quenched into saturated aqueous ammonium chloride. The layers are separated and the aqueous layer extracted with ethyl acetate. The combined organic layers are concentrated at reduced pressure. The residue is separated between dichloromethane and water. The organic layer is dried and concentrated at reduced pressure. The residue is triturated with ethyl acetate–hexane to afford the crude 2-(4-hydroxytetrahydro-2*H*-pyran-4-yl)-*N*,*N*-dimethylacetamide (**12**) (70%, amide **12**: amide **11** is 98.6 : 1.4 by HPLC).[22]

A solution of lithium hexamethyldisilazide (2.5 equivalents) in THF is added to a solution of 2-(4-(benzyloxy)phenyl)-*N*,*N*-dimethylacetamide (**11**) in THF at −60 to −70°C. The mixture is aged for 15 min. Cyclohexanone (1.2 equivalents) is added at −60 to −70°C and the mixture is aged, presumably at −60 to −70°C, for an unspecified time. The mixture is presumably allowed to warm to 0°C and quenched by addition of saturated aqueous ammonium chloride. Dilute hydrochloric acid is added (to pH 4.0) and the mixture is extracted with ethyl acetate. The extracts are washed with water, dried, and concentrated at reduced pressure to afford crude 2-(4-(benzyloxy)phenyl)-2-(1-hydroxycyclohexyl)-*N*,*N*-dimethylacetamide (**13**) (82%). No purity data is provided. The crude product **13** is carried directly into the amide reduction.[23]

A solution of isopropylmagnesium bromide in THF is added to a solution of 2-(4-methoxyphenyl)-*N*,*N*-dimethylacetamide (**5**) at 10–30°C. The mixture is aged at 10–30°C for 4–8 h. Cyclohexanone is added and the mixture is aged at 40–45°C for 5–12 h. Quench and a standard workup procedure affords 2-(1-hydroxycyclohexyl)-2-(4-methoxyphenyl)-*N*,*N*-dimethylacetamide (**14**).[18]

### 4.2.3 2-(4-Methoxyphenyl)-*N*,*N*-dimethylethanethioamide (15)

Isopropylmagnesium bromide (1.45 M) (1.1 equivalents) in methyl *tert*-butyl ether is added to toluene and the solution is cooled to <10°C. A solution of 2-(4-methoxyphenyl)-*N*,*N*-dimethylethanethioamide (**14**) in toluene is added at <10°C and the mixture is aged, presumably at 0–10°C, for 2 h. A solution of cyclohexanone (1.0 equivalent) in toluene is added, presumably at <10°C, and the mixture is aged for

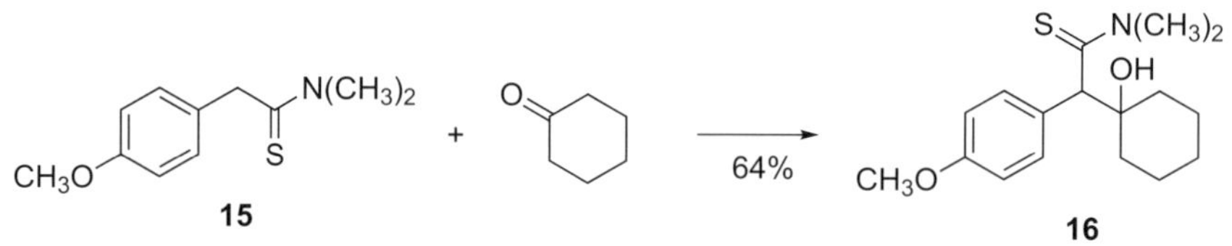


**SCHEME 4.3** Condensation of 2-(4-methoxyphenyl)-*N*,*N*-dimethylethanethioamide (**15**) with cyclohexanone.

15 min. The mixture is quenched into 1 M hydrochloric acid, presumably at 25°C. The layers are separated and the organic layer is washed with water, dried, and concentrated at reduced pressure. The residue is crystallized from diisopropyl ether to give 2-(1-hydroxycyclohexyl)-2-(4-methoxyphenyl)-*N*,*N*-dimethylethanethioamide (**16**) (64%) (Scheme 4.3).[24]

### 4.2.4 4-Methoxyphenylacetonitrile

The condensation of cyclohexanone with 4-methoxyphenylacetonitrile offers process flexibility not available with the acid or amide: *n*-butyllithium, lithium diisopropylamide, sodium hydride, sodium amide, sodium methoxide, potassium *tert*-butoxide, sodium hydroxide, and 1,8-diazabicyclo[5.4.0]undec-7-ene (DBU) are all effective bases. With alkoxide, hydroxide, and amine bases, only a catalytic amount of base is required, but the reversible condensation must be driven to completion by precipitation of the condensation product **17** from the reaction mixture (Scheme 4.4).

*4.2.4.1 **n-Butyllithium*** A solution of 4-methoxyphenylacetonitrile[25] in THF is cooled to −70°C and a 1.4 M solution of *n*-butyllithium in hexane (1.0 equivalent) is added at <−50°C. The resulting suspension (yellow precipitate) is aged at <−50°C for 30 min. Cyclohexanone (1.0 equivalent) is added at <−50°C and the mixture is aged at <−50°C for 45 min. The mixture is allowed to warm to 0°C and quenched by addition of saturated aqueous ammonium chloride. The layers are separated and the aqueous layer extracted with ethyl ether. The combined organic layers are dried and concentrated at reduced pressure to afford 2-(1-hydroxycyclohexyl)-2-(4-methoxyphenyl)acetonitrile (**17**) (30%). The same yield is observed using excess cyclohexanone (1.5 equivalents).[20,26]

The yield is greatly improved when the batch is kept at a lower temperature and quenched into 0–5°C aqueous ammonium chloride. A solution of 4-methoxyphenylacetonitrile in THF is cooled to −70°C and a 1.6 M solution of *n*-butyllithium in hexane (1.0 equivalent) is added, presumably at <−70°C. The resulting solution (perhaps this should read suspension) is aged for 30 min, presumably at <−70°C. Cyclohexanone (1.0 equivalent) is added, presumably at <−70°C, and the mixture is aged at −65°C for 2 h. The mixture is poured into cold aqueous ammonium chloride. The resulting suspension is filtered and the solid is washed with water and with ethyl ether and dried. Additional product is recovered from the THF layer in the liquors. The total yield is 83%.[19]

The yield is still higher when the anion suspension is prepared by adding the nitrile to *n*-butyllithium. A solution of 4-methoxyphenylacetonitrile in THF is added to a 1.6 M solution of *n*-butyllithium in hexane (1.0 equivalent) at −70 to −75°C. After aging, presumably at −70 to −75°C, for 30 min, cyclohexanone (1.0 equivalent) is added. The mixture is aged at −65 to −70°C for 4 h. The mixture is then poured into aqueous ammonium chloride at <0°C. Dilute hydrochloric acid is added (to pH 7). The resulting suspension is filtered and the solid is washed with hexane and dried to afford 2-(1-hydroxycyclohexyl)-2-(4-methoxyphenyl)acetonitrile (**17**) (89%, 99.8% pure by HPLC).[27]

A solution of 2-(4-(benzyloxy)phenyl)acetonitrile (**18**) in THF is cooled to −70°C and a 1.6 M solution of *n*-butyllithium in hexane (1.0 equivalent) is added, presumably at −70°C. After aging, presumably at −70°C, for 30 min, cyclohexanone (1.2 equivalents) is added at −70°C. The mixture is aged, presumably at −70°C, for 2.5 h. The mixture is then poured into aqueous ammonium chloride. The resulting suspension is filtered and the solid is dried to afford 2-(4-(benzyloxy)phenyl)-2-(1-hydroxycyclohexyl)acetonitrile (**19**) (93%). The yield is lower (72%) when the condensation is run at −30°C. No condensation product is isolated when the condensation is run at −5°C.[28,29]

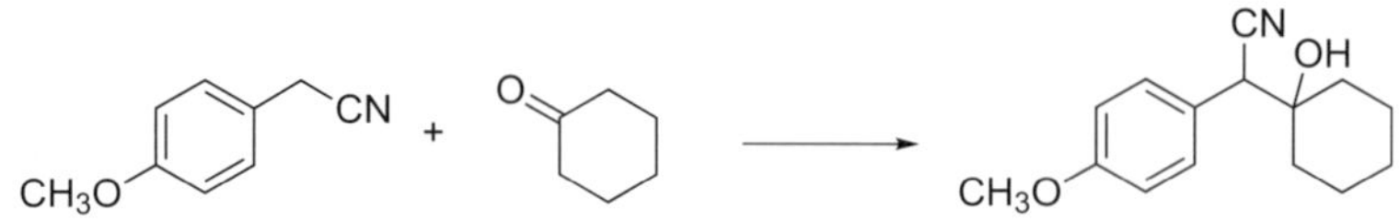


**SCHEME 4.4** Condensation of 4-methoxyphenylacetonitrile with cyclohexanone.

***4.2.4.2 Lithium Diisopropylamide*** A 2.0 M solution of lithium diisopropylamide (1.1 equivalents) in THF is added at <−65°C to a solution of 4-methoxyphenylacetonitrile in THF. The mixture is aged at −78°C for 30 min. Cyclohexanone (1.1 equivalents) is added at <−65°C and the mixture is aged at −78°C for 2 h. The mixture is quenched into cold aqueous ammonium chloride. After aging for 15 min, the mixture is extracted with ethyl acetate. The extracts are washed with water and with brine, dried, and concentrated at reduced pressure. The residue is suspended in hexane and the suspension is filtered. The solid is dried at reduced pressure to afford 2-(1-hydroxycyclohexyl)-2-(4-methoxyphenyl) acetonitrile (**17**) (81%).[30]

Addition of the nitrile in toluene to lithium diisopropylamide in toluene–hexane at <10°C gives similar results. A solution of 4-methoxyphenylacetonitrile (1.1 equivalents) in toluene is added to a solution of lithium diisopropylamide (1.1 equivalents) in 1 : 1 toluene–hexane at <10°C. The mixture is aged, presumably at <10°C, for 30 min. A solution of cyclohexanone in toluene is added at <10°C and the mixture is aged, presumably at <10°C, for 30 min. The solution is quenched into 1.1 M hydrochloric acid. The resulting suspension is filtered. The solid is dissolved in dichloromethane and the solution is washed with water. Dichloromethane is distilled and replaced by diisopropyl ether. The suspension is filtered and the solid dried to afford 2-(1-hydroxycyclohexyl)-2-(4-methoxyphenyl)acetonitrile (**17**) (79% based on cyclohexanone).[24,26]

The yield is lower when the nitrile anion is not generated before the cyclohexanone is added. A solution of lithium diisopropylamide in ethyl ether (from 1.4 equivalents plate-form lithium and 1.8 equivalents diisopropylamine) is cooled to <5°C. A toluene solution containing both 4-methoxyphenylacetonitrile and cyclohexanone (1.1 equivalents) is added, presumably at <5°C. The solution is quenched into ice-cold 4.5 M hydrochloric acid. The resulting suspension is filtered to afford solid crude product (50%). The liquors are separated and the organic layer is concentrated at reduced pressure to afford additional solid crude product (16%). The solids are combined and crystallized from toluene to afford 2-(1-hydroxycyclohexyl)-2-(4-methoxyphenyl)acetonitrile (**17**) (49%).[31]

***4.2.4.3 Sodium Hydride*** Sodium hydride, 60% oil dispersion (1.0 equivalent), is suspended in hexane–toluene (1 : 1) at 25°C. 4-Methoxyphenylacetonitrile is added at 25°C and the resulting suspension is aged at 25°C for 50 min. The mixture is cooled to −5°C and cyclohexanone (1.3 equivalents) is added. The mixture is aged, presumably at −5°C, for 5 h. The reaction is quenched by careful addition of 5% hydrochloric acid (to pH 6–7). The suspension is aged, presumably at 25°C, for 15 min and then filtered. The solid is washed with water, dried, and crystallized from toluene to afford 2-(1-hydroxycyclohexyl)-2-(4-methoxyphenyl)acetonitrile (**17**) (80%).[31]

***4.2.4.4 Sodium Amide*** 4-Methoxyphenylacetonitrile is added to sodium amide (1.0 equivalent) in THF at −78°C. Cyclohexanone (1.0 equivalent) is added at −78°C and the mixture is aged at −78°C for 3 h to produce 2-(1-hydroxycyclohexyl)-2-(4-methoxyphenyl)acetonitrile (**17**) (62%). The yield is lower (30%) when the additions and age are a higher temperature (−5°C).[29]

***4.2.4.5 Sodium Methoxide*** Four very similar procedures all indicate that the condensation using excess sodium methoxide (1.5–3.0 equivalents) in methanol at 0°C is quenched by simply adding water or dilute acetic acid to precipitate the product **17** in high (>90% yield). Volume throughputs can be as high as 175 g/L. Condensation product **17** is typically filtered and then triturated with a nonsolvent–solvent mixture or crystallized from toluene.

4-Methoxyphenylacetonitrile is added to a solution of sodium methoxide (2.5 equivalents) in methanol at −5 to 5°C. Cyclohexanone (1.3 equivalents) is added to the solution at −5 to 5°C. The mixture is aged at −5 to 5°C for 4–5 h. The reaction is quenched by addition of water at an unspecified temperature, perhaps at 0–5°C. The resulting suspension is filtered and the solid is suspended in hexane–ethyl acetate (10 : 1). The suspension is aged at 25°C for 3 h and then filtered. The solid is washed with hexane–ethyl acetate (presumably 10 : 1) and dried to afford 2-(1-hydroxycyclohexyl)-2-(4-methoxyphenyl)acetonitrile (**17**) (90%).[32]

4-Methoxyphenylacetonitrile is added to a solution of sodium methoxide (3.0 equivalents) in methanol at −3 to −5°C. The mixture is aged at −3 to −5°C for 2 h. Cyclohexanone (1.3 equivalents) is added to the solution at −5 to 5°C. The mixture is aged at −3 to −5°C for 10–12 h. The reaction is quenched by addition of water at 0–2°C. The resulting suspension is filtered and the solid is washed with water and dissolved in toluene at 50°C. The solution is filtered and cooled to 5–10°C, and the resulting suspension is aged at 5–10°C for 30 min and filtered. The solid is dried at 45–50°C and reduced pressure to afford 2-(1-hydroxycyclohexyl)-2-(4-methoxyphenyl)acetonitrile (**17**) (80%).[33]

4-Methoxyphenylacetonitrile is added to a solution of sodium methoxide (1.5 equivalents) in methanol at 0–5°C. The mixture is aged at 0–5°C for 15 min. Cyclohexanone (1.8 equivalents) is added to the solution at 0–5°C. The mixture is aged at 0–5°C for 2.5–3 h. The reaction is quenched by addition of dilute acetic acid (1.5 equivalents), presumably at 0–5°C. The resulting suspension is aged at 0–5°C for 30 min and filtered. The solid is crystallized from toluene and dried at 45–50°C and reduced pressure to afford 2-(1-hydroxycyclohexyl)-2-(4-methoxyphenyl)acetonitrile (**17**) (85–90%).[34]

4-Methoxyphenylacetonitrile is added to a solution of sodium methoxide (3.0 equivalents) in methanol at −2 to 5°C. The mixture is aged at 0–5°C for 2 h. Cyclohexanone (1.3 equivalents) is added to the solution, presumably at 0–5°C. The mixture is aged at 0–5°C for 4–5 h. The reaction is quenched by addition of water at 0–8°C. The resulting suspension is aged, presumably at 0–5°C, for 30–45 min and filtered. The solid is washed with water and crystallized from toluene. The isolated solid is washed with a small amount of toluene and dried at 45–50°C and reduced pressure to afford 2-(1-hydroxycyclohexyl)-2-(4-methoxyphenyl)acetonitrile (**17**) (86–89%).[35]

***4.2.4.6 Sodium or Potassium Butoxide*** In theory, the condensation requires only a catalytic amount of an alkoxide base. Since base also catalyzes the reverse reaction, a process using a catalytic amount of base must be driven to completion by precipitation of the product. It is not clear from the procedures just how much base is charged in three closely related processes using sodium *n*-butoxide, *iso*-butoxide, or *tert*-butoxide as the base. The amount specified in the patent claims is 0.1 equivalent or 10 mol%.

A 20 wt% solution of sodium 1-butoxide (presumably 10 mol% and presumably in 1-butanol) is added to a mixture of 4-methoxyphenylacetonitrile, cyclohexanone (1.0 equivalent), and 1-butanol (2.5 mL/g nitrile) at −10 to −5°C. The mixture is aged at −10 to −5°C for 2 h. Glacial acetic acid is added at −10 to −5°C (to pH 5.5). Water (3.0 mL/g nitrile) is added, presumably at −10 to −5°C, and the mixture is aged for 10 min. The layers are separated and the organic layer containing 2-(1-hydroxycyclohexyl)-2-(4-methoxyphenyl) acetonitrile (**17**) (>86% yield) in 1-butanol is used in a reduction of the nitrile. Comparable results (>83% yield) are achieved using isobutoxide in isobutanol or *tert*-butoxide in *tert*-butanol.[36]

A potassium *tert*-butoxide catalyst (0.2 mol%) is added to a mixture of 4-methoxyphenylacetonitrile, cyclohexanone (1.4 equivalents), and heptane (1.3 L/kg nitrile) at 25°C. The mixture is aged for 50 min with cooling to maintain the exothermic reaction at 25°C. The suspension is cooled to <10°C and a solution of acetic acid in heptane is added (to pH 3–4). The suspension is aged at <10°C for 30 min and filtered. The solid is washed with <10°C heptane and dried to afford 2-(1-hydroxycyclohexyl)-2-(4-methoxyphenyl) acetonitrile (**17**) (89%, 98.6% pure by HPLC). The yield is lower (82%) in the neat reaction. In this case, heptane is added during the workup after adding the acetic acid. The yield is also lower (68%) using toluene as the reaction solvent (0.71 L/kg nitrile) in place of heptane.[37]

***4.2.4.7 Sodium Hydroxide*** Both powdered sodium hydroxide and aqueous solutions are used. Powdered sodium hydroxide is difficult to work with and expensive. Aqueous solutions are preferred and 10% solutions give the best results. Phase transfer catalysts (tetrabutylammonium hydrogen sulfate, bromide, or hydroxide) or alcohol cosolvents (*tert*-butanol, methanol, or polyethylene glycol) are used. The order of mixing of nitrile, base, and catalyst is not critical. The anion is often generated before cyclohexanone is added but the results suggest this is not necessary. The condensation using sodium hydroxide is typically run at very high concentration to achieve nearly complete conversion in <24 h. The condensation is again driven to completion by precipitation of product **17**. The condensation product **17** has very limited solubility in water. The combination of high concentration and low product solubility suggests that *the mixture produced is likely to be a thick mass*. In most cases, mixing will be inadequate, the mixture will be difficult to transfer to a filter, and impurities (catalyst and 4-methoxyphenylacetonitrile) will be difficult to remove by simple washing of the solid on the filter. Crude **17** can be crystallized from toluene or ethyl acetate–hexanes.

A solution of 4-methoxyphenylacetonitrile in toluene is added to a mixture of sodium hydroxide (3.0 equivalents) (physical state not specified), *tert*-butanol (1.1 equivalents), and *n*-hexane at −5 to −10°C. The mixture is aged at −10°C for 45 min. A solution of cyclohexanone (1.1 equivalents) in toluene is added at −5 to −10°C and the mixture is aged at −5 to −10°C for 4 h. The reaction is quenched into cold dilute acetic acid (3.1 equivalents). After aging at 15–20°C for 30 min, the suspension is filtered. The product is washed with water and dried at 70°C and reduced pressure to afford 2-(1-hydroxycyclohexyl)-2-(4-methoxyphenyl)acetonitrile (**17**) (74%).[38]

A solution of sodium hydroxide (1.5 equivalents) and tetrabutylammonium bromide (4.6 mol%) in methanol–water (1 : 1) is added to a mixture of 4-methoxyphenylacetonitrile and cyclohexanone (1.5 equivalents). The mixture is aged at 25–30°C for 15 h. The suspension is filtered. The solid is washed with water and with hexane and dried to afford 2-(1-hydroxycyclohexyl)-2-(4-methoxyphenyl)acetonitrile (**17**) (96%). The volume throughput is 80 g/L.[39]

A mixture of 4-methoxyphenylacetonitrile, cyclohexanone (1.4 equivalents), and tetrabutylammonium hydroxide hydrate (0.2 mol%) is aged at 25°C (cooling required) for 15 min. The resulting thick white mass is broken up and dissolved by adding toluene and 0.1 M hydrochloric acid (amount not specified). The mixture is heated to 30°C and the layers are separated. The organic layer is washed with water and concentrated at reduced pressure. Heptane is added to the residue and the suspension is aged at 0°C for 30 min and filtered. The solid is presumably washed with heptane and dried to afford 2-(1-hydroxycyclohexyl)-2-(4-methoxyphenyl)acetonitrile (**17**) (74%).[37]

4-Methoxyphenylacetonitrile is mixed with 10% aqueous sodium hydroxide solution (25 mol%) at 0°C. Tetrabutylammonium hydrogen sulfate (2.1 mol%) is added at 0°C and the mixture is warmed to 15°C. Cyclohexanone (1.0 equivalent) is added rapidly, while maintaining the temperature at <15°C, to produce a thick suspension. The solid is broken up and the suspension is aged at 25°C for 1 h. The suspension is filtered and the solid is washed with water and dried. The crude product is crystallized from ethyl acetate–petroleum ether (10 : 7) to afford 2-(1-hydroxycyclohexyl)-2-(4-methoxyphenyl)acetonitrile (**17**) (97%). The same yield is achieved using 10 mol% tetrabutylammonium bromide. A lower yield (90%) is observed using the combination of powdered sodium hydroxide (50 mol%) and tetrabutylammonium iodide (10 mol%). The condensation also proceeds using powdered sodium hydroxide (25 mol%) when the phase transfer catalyst is omitted (78%).[40,41]

4-Methoxyphenylacetonitrile is mixed with 10% aqueous sodium hydroxide solution (1.1 equivalents) at 15°C. Tetrabutylammonium bromide (0.20 mol%) is added at 15°C and the mixture is aged at 15–18°C for 30 min. Cyclohexanone (1.3 equivalents) is added at 15–18°C and the mixture is aged at 15–18°C for 3 h. The suspension is filtered and the solid is washed with water and dried. The crude product is crystallized from ethyl acetate–petroleum ether (5 : 95) to afford 2-(1-hydroxycyclohexyl)-2-(4-methoxyphenyl)acetonitrile (**17**) (93%, >95% pure by HPLC). The same yield (92%) is achieved using 10% sodium hydroxide (0.46 equivalents), tetrabutylammonium bromide (0.20 mol%), and cyclohexanone (1.4 equivalents) at 27°C for 3 h. A lower yield (82%) is observed using 10% sodium hydroxide (0.46 equivalents), tetrabutylammonium bromide (0.20 mol%), and cyclohexanone (1.1 equivalents) at 27°C for 3–4 h.[29]

A solution of 10% sodium hydroxide (1.0 equivalent) and tetrabutylammonium bromide (0.2 mol%) is added to a mixture of 4-methoxyphenylacetonitrile and cyclohexanone (1.4 equivalents) at 27°C. The mixture is aged at 27°C for 2 h. The suspension is filtered and the solid is washed with water and dried to afford crude 2-(1-hydroxycyclohexyl)-2-(4-methoxyphenyl)acetonitrile (**17**) (95%). The same yield is observed using 10% sodium hydroxide (0.46 equivalents), tetrabutylammonium chloride (0.2 mol%), and cyclohexanone (1.4 equivalents) at 18°C for 6 h.[42]

4-Methoxyphenylacetonitrile is cooled to 15–18°C and added to 10% sodium hydroxide (1.0 equivalent). Tetrabutylammonium bromide (0.46 mol%) is added. Cyclohexanone (0.67 equivalents) is added, presumably at 15–18°C, followed by tetrabutylammonium borohydride (0.23 mol%). A second cyclohexanone charge (0.67 equivalents) is added, presumably at 15–18°C, and the mixture is aged at 15–18°C for 1.5 h. The suspension is filtered and the solid is washed with water until the washings are neutral. The solid is dissolved in toluene at 80–85°C and the solution cooled to 5–10°C. The suspension is filtered and the solid washed with cold toluene and dried at 60–65°C and reduced pressure to afford 2-(1-hydroxycyclohexyl)-2-(4-methoxyphenyl)acetonitrile (**17**) (86%). No rationale for the addition of tetrabutylammonium borohydride is offered.[43]

$d_3$-4-Methoxyphenylacetonitrile is cooled to 0°C and 2 M sodium hydroxide (0.42 equivalents) and tetrabutylammonium hydrogen sulfate (5.1 mol%) are added. The mixture is aged at 0°C for 30 min. Cyclohexanone (1.2 equivalents) is added at 0–5°C and the mixture is warmed to 25°C and aged for 1 h. The resulting suspension is filtered and the solid is washed with water and hexanes to afford $d_3$-2-(1-hydroxycyclohexyl)-2-(4-methoxyphenyl)acetonitrile (**20**) (91%). The analogous condensation of $d_3$-4-methoxyphenylacetonitrile with $d_{10}$-cyclohexanone and 2 M sodium deuteroxide afforded $d_{14}$-2-(1-hydroxycyclohexyl)-2-(4-methoxyphenyl)acetonitrile (**21**) (60%) after crystallization from ethyl acetate–hexanes.[21]

4-Methoxyphenylacetonitrile is cooled to 0–5°C and 10% sodium hydroxide (5.0 mol%) is added. Polyethylene glycol (PEG-400) (82 g/kg nitrile) is added. The mixture is aged at 0–5°C for 2 h. Cyclohexanone (0.70 equivalents) is added and the mixture is aged, presumably at 0–5°C, for 60–90 min. Additional charges of 10% sodium hydroxide (5.0 mol%) and cyclohexanone (0.70 equivalents) are added, presumably at 0–5°C. The mixture is warmed to 25°C and aged for 10–12 h. The suspension is diluted with hexane and filtered. The solid is washed with water and hexane and dried at 55–60°C and reduced pressure to afford 2-(1-hydroxycyclohexyl)-2-(4-methoxyphenyl)acetonitrile (**17**) (90%).[44]

***4.2.4.8 1,8-Diazabicyclo[5.4.0]undec-7-ene (DBU)*** A mixture of 4-methoxyphenylacetonitrile, cyclohexanone (1.5 equivalents) and DBU (10 mol%) is aged at 15–20°C for 144 h. Hydrochloric acid (1 M) is added (to pH <7). The suspension is aged for 1 h at 25°C and then filtered. The solid is washed with water and with ethyl acetate–hexane and then dried to afford 2-(1-hydroxycyclohexyl)-2-(4-methoxyphenyl)acetonitrile (**17**) (91%). Similar results (88%) are observed using DBU (50 mol%) and cyclohexanone (2.5 equivalents) after aging at 0°C for 60 h.

A mixture of 4-methoxyphenylacetonitrile, cyclohexanone (1.5 equivalents) and 1,5,7-triazabicyclo[4.4.0]dec-5-ene (TBD) (0.5 mol%) is aged at 20–25°C for 19 h. The mixture is dissolved in ethyl acetate. Water is added followed by 6 M hydrochloric acid (to pH 7). The mixture is warmed to 30–35°C and the layers are separated. The organic layer is concentrated at reduced pressure. The residue is probably suspended in ethyl acetate–hexane and the suspension is filtered. Results for the condensation of 4-nethoxy phenyl acetonitrile with cyclohexanone are summarized in Table 4.1. The solid is washed with water and with ethyl acetate–hexane and then dried to afford 2-(1-hydroxycyclohexyl)-2-(4-methoxyphenyl)acetonitrile (**17**) (89%).[26]

Results for the condensation of 4-methoxy phenyl acetonitrile with cyclohexanone are summarized in Table 4.1.

**TABLE 4.1 Base Equivalents and Yields in the Condensation of 4-Methoxyphenylacetonitrile with Cyclohexanone**

| Base | Base (equiv) | Yield (%) |
|---|---|---|
| *n*-BuLi | 1.0 | 89 |
| LDA | 1.1 | 81 |
| NaH | 1.0 | 80 |
| $NaNH_2$ | 1.0 | 62 |
| NaOMe | 2.5 | 90 |
| NaOBu | 0.1 | 86 |
| KO*t*Bu | 0.002 | 89 |
| NaOH | 1.5 | 96 |
| DBU | 0.1 | 91 |
| TBD | 0.005 | 89 |

> Many of the processes for 2-(1-hydroxycyclohexyl)-2-(4-methoxyphenyl)acetonitrile (**17**) are driven to near-completion by precipitation of the product. Published procedures rarely provide qualitative descriptions of or quantitative data on suspensions. Questions that will come up during scale-up of the lab process include as follows: (1) Is the mixing adequate? (2) What is the stir rate (rpm)? (3) What is the design of the agitator paddle? (4) What is the baffle arrangement (if any) in the reactor? (5) Can it be transferred (poured) onto a filter leaving little of the suspension in the reactor? (6) Can it be transferred through a Teflon transfer line with nitrogen pressure in the lab? If so, how much pressure (psi) is required and what is the diameter (mm) of the transfer line used? (7) How big are the crystals? (8) What is the crystal size distribution?

## 4.3 VENLAFAXINE (1) BY REDUCTION OF *N,N*-DIMETHYLACETAMIDE 14

While amides are generally not reduced by sodium borohydride, amide **14** is reduced by borane generated by reaction of potassium borohydride with a Lewis acid (boron trifluoride etherate or aluminum chloride) in THF or 1,2-dimethoxyethane.[17,18] Borane–THF complex is the preferred reagent in many amide reductions to produce venlafaxine analogs. Here, again, the reducing agent may be diborane released when borane–THF is heated at temperatures greater than 50°C. The borane–THF reduction is reliable on gram scale (54–66%).

Borane in THF (1.0 M) (2.2 equivalents) is added to (*R*)-2-(1-hydroxycyclohexyl)-2-(4-methoxyphenyl)-*N,N*-dimethylacetamide ((*R*)-**14**) in THF. The mixture is refluxed for 1.5 h. The mixture is then cooled to 0°C and excess hydride is quenched by the careful addition of 2 M hydrochloric acid. The mixture is aged at 75°C for 20 min and the THF is distilled at 25°C and reduced pressure. The remaining aqueous mixture is diluted with water and washed with ethyl acetate. Sodium hydroxide (2 M) is added (to pH 10) and the free base is extracted with dichoromethane. The extracts are dried and concentrated at reduced pressure. (*R*)-Venlafaxine ((*R*)-**1**) is isolated from the residual oil as the hydrochloride (88%) (Scheme 4.5).[45]

The reduction of the 4-bromophenyl analog **8** can be accomplished with aluminum hydride. Concentrated sulfuric acid (0.8 equivalents) is cautiously added to a suspension of lithium aluminum hydride (1.5 equivalents) in THF at 0°C. The mixture is aged, presumably at 0°C, for 1 h. A solution of the amide **8** in THF is added and the mixture is aged at 0°C for 1 h. Excess hydride is quenched by the careful addition of THF–water followed by 10% sodium hydroxide. The suspension is aged, presumably at 25°C, for 15 min and then filtered. The liquors are dried and concentrated at reduced pressure to afford 1-(1-(4-bromophenyl)-2-(dimethylamino)ethyl)cyclohexanol (**22**) (89%). $d_{26}$-Venlafaxine (**23**) is prepared from the $d_{24}$-amide **10** using lithium aluminium deuteride. No yield is available. The 4-amino analog **24** is prepared by a similar procedure using more sulfuric acid (5.2 equivalents). The yield is 80% after crystallization from isopropanol.[19–21]

The reduction of the 3,4-dichlorophenyl analog **25** is accomplished with borane–THF complex. A solution of amide **25** in THF is added to borane in THF (1.0 M) (1.5 equivalents) at 0°C. The mixture is refluxed for 2 h. The mixture is then cooled to 0°C and excess hydride is quenched by the careful addition of 2 M hydrochloric acid. The mixture is again refluxed for 90 min and then cooled to 25°C overnight. Solid potassium hydroxide is added (to pH 14) and the layers are separated. The organic layer is washed with brine, dried, and concentrated at reduced pressure. The solid residue is filtered, washed with petroleum ether, and dried to afford 1-(1-(3,4-dichlorophenyl)-2-(dimethylamino)ethyl)cyclohexanol (**26**) (54%).[19,20]

The 4-(benzyloxy)phenyl analog **13** is also reduced with borane–THF complex. Borane in THF (1.0 M) (5.0 equivalents) is added to a solution of amide **13** in THF at 0°C. The mixture is refluxed for 3 h. The mixture is then cooled to 10°C and excess hydride is quenched by the careful addition of 10% sodium hydroxide. The THF is distilled, water is added, and the mixture is extracted with toluene. The extracts are dried and concentrated. Hexane is added to the residue and the suspension is filtered. The solid is presumably washed with hexane and dried to afford 1-(1-(4-(benzyloxy)phenyl)-2-(dimethylamino)ethyl)cyclohexanol (**27**) (66%).[23]

Another 4-(benzyloxy)phenyl analog **12** is reduced with lithium aluminum hydride. A solution of amide **13** in THF is added to a solution of lithium aluminum hydride (3.5 equivalents) in THF at −78°C. The mixture is allowed to warm to 25°C and aged at 25°C overnight. Excess hydride is quenched by the careful addition of methanol followed by 1 M sodium hydroxide. The suspension is aged for 20 min, presumably at 25°C, and then filtered. The solid is washed

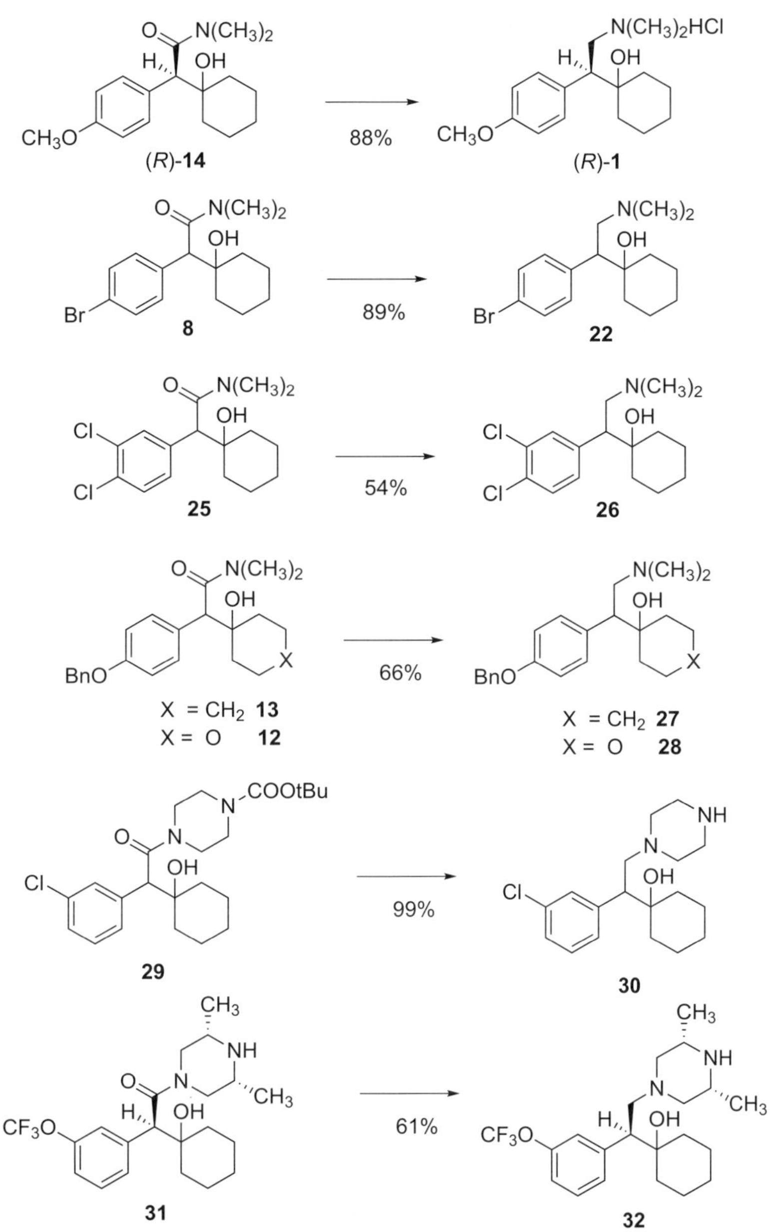


**SCHEME 4.5** Venlafaxine (**1**) and analogs by reduction of an amide.

with THF. The liquors are concentrated at reduced pressure. The residue is separated between dichloromethane and water. The organic layer is dried and concentrated at reduced pressure. Ethyl acetate–hexanes is added to the residue, the suspension is filtered, and the solid is dried to afford 4-(1-(4-(benzyloxy)phenyl)-2-(dimethylamino)ethyl)tetrahydro-2*H*-pyran-4-ol (**28**) (66%).[22]

The amide reduction with borane–THF complex is used to produce many new monoamine reuptake inhibitors that have piperazines on the side chain. In a 100–200 mg scale procedure, borane in THF (1.0 M) (3.5 equivalents) is added to a solution of *tert*-butyl 4-(2-(3-chlorophenyl)-2-(1-hydroxycyclohexyl) acetyl)piperazine-1-carboxylate (**29**) in THF. The mixture is refluxed for 2 h. The mixture is cooled to 0°C, carefully quenched by adding of 1 M hydrochloric acid, and aged at 70°C for 1 h. The resulting mixture is cooled, diluted with methanol, and concentrated at reduced pressure. The residue is dissolved in water and the solution is washed with ethyl acetate. Sodium hydroxide (2 M) is added (to pH 10) and the free base is extracted with ethyl acetate. The extracts are dried and concentrated at reduced pressure to afford 1-(1-(3-chlorophenyl)-2-piperazin-1-yl)ethyl)cyclohexanol (**30**) (99%).[14]

*The time required for complete conversion in the amide reduction with borane–THF increases with scale.* On a 25 g scale, borane in THF (1 M) (5.0 equivalents) is added to a solution of (*R*)-1-((3*R*,5*S*)-3,5-dimethylpiperazin-1-yl)-2-

(1-hydroxycyclohexyl)-2-(3-(trifluoromethoxy)phenyl) ethanone (**31**) in THF. The mixture is refluxed for 22 h. The mixture is then cooled, presumably to 0°C, and excess hydride is quenched by the careful addition of 2 M hydrochloric acid. The mixture is again refluxed for 1 h, cooled to 25°C, and washed with ethyl acetate. Sodium hydroxide (50%) is added to the aqueous layer (to pH 12) and the free base is extracted with ethyl acetate. The extracts are dried and concentrated at reduced pressure to afford 1-(*S*)-2-((3*R*,5*S*)-3,5-dimethylpiperazin-1-yl)-1-(3-(trifluoromethoxy)phenyl)ethyl)cyclohexanol (**32**) (61%).[45]

## 4.4 VENLAFAXINE (1) BY REDUCTION OF THE *N,N*-DIMETHYLETHANETHIOAMIDE 16

A solution of 2-(1-hydroxycyclohexyl)-2-(4-methoxyphenyl)-*N,N*-dimethylethanethioamide (**16**) in a 9 : 1 mixture (v/v) of 1,4-dioxane and acetic acid is added to a suspension of Raney nickel (20 g/g thioamide **16**) in a 9 : 1 mixture (v/v) of 1,4-dioxane and acetic acid at 25–50°C. The suspension is aged, presumably at 25–50°C, for up to 10 h and then filtered. The liquors are concentrated at reduced pressure. The residue is separated between dichloromethane and 2 M sodium hydroxide. The organic layer is dried and concentrated at reduced pressure to afford venlafaxine (**1**). No yield is available (Scheme 4.6).[24]

## 4.5 REDUCTION OF 2-(1-HYDROXYCYCLOHEXYL)-2-(4-METHOXYPHENYL)ACETONITRILE (17)

Since reversal of the nitrite-ketone condensation is catalyzed by base the nitrile reduction under basic conditions produces 2-(4-methoxyphenyl)ethanamine (**33**) as a side product. This is a significant problem in the reduction of 2-(1-hydroxycyclohexyl)-2-(4-methoxyphenyl)acetonitrile (**17**) with lithium aluminum hydride or lithium aluminum hydride–aluminum chloride.[29] The nitrile is more efficiently reduced with borane generated from sodium borohydride and a Lewis acid, tetrabutylammonium borohydride and iodomethane, or borane–dimethylsulfide at high (>90°C) temperature. Yields using *in situ*-generated borane can be >82%. However, as

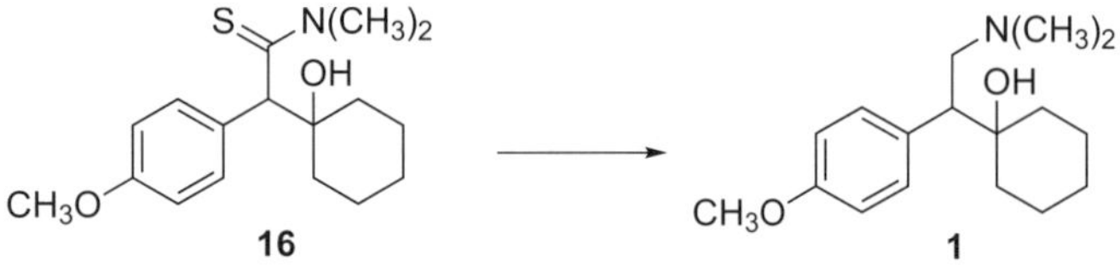


**SCHEME 4.6** Venlafaxine (**1**) from 2-(1-hydroxycyclohexyl)-2-(4-methoxyphenyl)-*N,N*-dimethylethanethioamide (**16**).

with the borane reduction of amide **14**, the utility of this method is likely limited to producing of gram-scale quantities in the laboratory. The nitrile is also reduced with sodium borohydride– cobalt boride, or by hydrogenation using a rhodium, nickel, cobalt, or palladium catalyst.

### 4.5.1 Lithium Aluminum Hydride and Aluminum Chloride

A solution of aluminum chloride (0.73 equivalents) in ethyl ether is added to a mixture of lithium aluminum hydride (2.1 equivalents) in ethyl ether with ice water cooling. The temperature maintained is not specified but is presumed to be <25°C. A solution of 2-(1-hydroxycyclohexyl)-2-(4-methoxyphenyl)acetonitrile (**17**) in THF is added with ice water cooling. The temperature maintained is not specified but is presumed to be <25°C. The mixture is aged at 25°C for 5 h. Excess hydride is quenched by careful addition of 1 : 1 THF–$H_2O$. Sodium hydroxide solution (50%) is added to produce a white precipitate. The liquid layer is decanted from the precipitate, the precipitate is washed with ethyl ether, and the wash is also decanted from the precipitate. The combined liquors are concentrated at reduced pressure. 1-(2-Amino-1-(4-methoxyphenyl)ethyl)cyclohexanol (**34**) hydrochloride is isolated from the residue. No yield is available.[31]

### 4.5.2 Sodium Borohydride and a Lewis Acid

Aluminum chloride (4.0 equivalents) is added in portions at 25–27°C to a solution of 2-(1-hydroxycyclohexyl)-2-(4-methoxyphenyl)acetonitrile (**17**) in THF. Sodium borohydride (4.0 equivalents) is then added in portions and the mixture is refluxed for 6–8 h. The mixture is cooled and quenched, presumably at 25–30°C, by careful addition of water (batch-to-water or water-to-batch is not specified). Hexanes is added and the layers are separated. Sodium hydroxide is added to the aqueous layer (presumably at 25–30°C to pH > 12) and the free base is extracted into toluene. The extracts are concentrated at 40°C and reduced pressure to afford 1-(2-amino-1-(4-methoxyphenyl)ethyl) cyclohexanol (**34**) (70%).[44]

Aluminum chloride (5.0 equivalents) is added in portions to THF at 25–30°C. Sodium borohydride (3.5 equivalents) and 2-(1-hydroxycyclohexyl)-2-(4-methoxyphenyl)acetonitrile (**17**) are then added in portions at 25–30°C. The mixture is aged at 40–50°C for 20–24 h. The mixture is cooled, quenched by careful addition of 50% sodium hydroxide at 25–30°C, and extracted with toluene. The extracts are concentrated at reduced pressure to afford crude 1-(2-amino-1-(4-methoxyphenyl)ethyl)cyclohexanol (**34**). No yield is available at this point but the yield for the three-step sequence (reduction–methylation–salt formation) to venlafaxine (**1**) hydrochloride is 47–53%. Similar results are

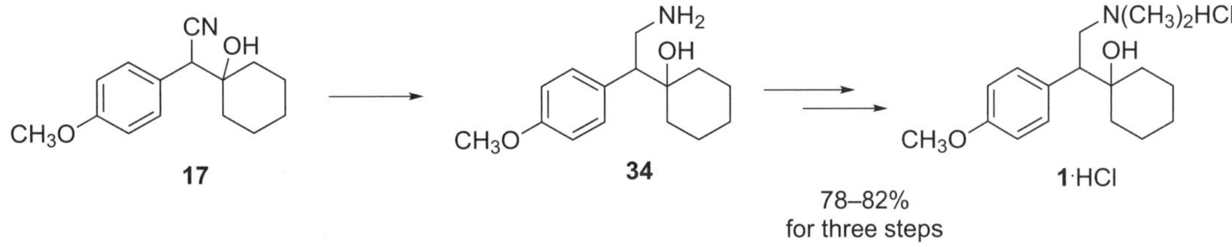


**SCHEME 4.7** Reduction of 2-(1-hydroxycyclohexyl)-2-(4-methoxyphenyl)acetonitrile (**17**) with sodium borohydride–boron trifluoride.

observed using zinc chloride or tin(II) chloride in place of aluminum chloride.[33]

Boron trifluoride etherate (2.0 equivalents) is added to sodium borohydride (2.3 equivalents) in THF at 0–5°C. A solution of 2-(1-hydroxycyclohexyl)-2-(4-methoxyphenyl) acetonitrile (**17**) in THF is then added at 0–5°C. The mixture is warmed to 25°C, aged at 25°C for 2.5 h, warmed to 40°C, and aged at 30–40°C for 2.5 h. The resulting mixture is cooled to 10°C and quenched by careful addition of 10% aqueous formic acid at 10–15°C. No yield is available at this point but the yield for the three-step sequence (reduction-–methylation–salt formation) to venlafaxine (**1**) hydrochloride is 78–82% (Scheme 4.7).[34]

### 4.5.3 Tetrabutylammonium Borohydride and Iodomethane

Tetrabutylammonium borohydride (1.5 equivalents) is added in portions to a solution of 2-(1-hydroxycyclohexyl)-2-(4-methoxyphenyl)acetonitrile (**17**) in dichloromethane at 25°C. The solution is cooled to 0–5°C and iodomethane (2.0 equivalents) is added. The mixture is aged at 0–5°C for 30 min, warmed to 25°C, and aged at 25–30°C for 90 min. Excess hydride is quenched by careful addition of ethanol. Dilute hydrochloric acid is added (to pH 2.0) and the layers are separated. The organic layer is extracted with water. Sodium hydroxide (48%) is added to the combined aqueous layers (to pH 10.5–11.0) and the free base is extracted with dichloromethane. The extracts are dried and filtered. Hydrogen chloride gas is added at 15–20°C and the mixture is aged at 20–25°C for 30 min and then concentrated at reduced pressure. Acetone is added to the residue and the suspension is filtered. The solid is washed with cold acetone and dried at 55–60°C and reduced pressure to afford 1-(2-amino-1-(4-methoxyphenyl)ethyl)cyclohexanol (**34**) hydrochloride (43%).[43]

### 4.5.4 Borane–Dimethyl Sulfide

Borane–dimethyl sulfide complex (1.6 equivalents) is added to a solution of 2-(1-hydroxycyclohexyl)-2-(4-methoxyphenyl)acetonitrile (**17**) in 1,4-dioxane at 25°C. The mixture is aged at 25°C for 15 min and then refluxed (92°C) until the reduction is complete. The resulting mixture is cooled to 25°C and the solution containing 1-(2-amino-1-(4-methoxyphenyl)ethyl)cyclohexanol (**34**) is carried directly into the methylation. The yield for the two-step (reduction–methylation) sequence is 65%.[44]

### 4.5.5 Sodium Borohydride and Cobalt Boride

The reaction of sodium borohydride with cobalt chloride produces a black precipitate of cobalt boride that can coordinate to and activate the nitrile toward reduction with additional borohydride. Cobalt chloride (2.0 equivalents) is added to a solution of 2-(1-hydroxycyclohexyl)-2-(4-methoxyphenyl)acetonitrile (**17**) in methanol at 25°C to produce a clear dark blue solution. Sodium borohydride (10.2 equivalents) is added in portions at <35°C. The resulting suspension (black precipitate) is aged at 25°C for 2 h. The suspension is cooled and quenched by careful addition of 3 M hydrochloric acid at <25°C. The suspension is aged at 25°C for 30 min. Methanol is distilled from the suspension (some black precipitate remains) at reduced pressure. The mixture is extracted with ethyl acetate. The extracts are washed with water and with brine and then dried and concentrated at reduced pressure to afford crude 1-(2-amino-1-(4-methoxyphenyl)ethyl)cyclohexanol (**34**) (84%).[30]

Cobalt chloride has been investigated as a tumorigen, mutagen, and reproductive effector. Cobalt and cobalt compounds have been shown to cause cancer in laboratory animals. Inhalation of cobalt dust and fumes is associated with an increased incidence of lung disease. The OSHA permissible exposure limit (PEL) is 0.1 mg/m$^3$ for cobalt metal dust. The ACGIH threshold limit value (TLV) for inorganic cobalt compounds is 0.02 mg/m$^3$ (TWA) as cobalt.

### 4.5.6 Catalytic Hydrogenation with Rhodium on Alumina

2-(1-Hydroxycyclohexyl)-2-(4-methoxyphenyl)acetonitrile (**17**) is dissolved in 4 : 1 (v/v) ethanol–ammonium hydroxide (28%) and hydrogenated over 5% rhodium on alumina (0.23 g/g **17**). (*Note*: Ammonia in the patent procedure is assumed to be 28% ammonium hydroxide.) The temperature, pressure, and time are not specified but may be 25°C, 30–60 psi, and 3 h. The suspension is filtered and the recovered catalyst is washed with ethanol. The combined liquors are concentrated at reduced pressure to afford the crude 1-(2-amino-1-(4-methoxyphenyl)ethyl)cyclohexanol (**34**) (98%). This is dissolved in toluene and hydrogen chloride in isopropanol is added. The suspension is diluted with ethyl ether and filtered. The solid is presumably washed with ethyl ether and dried to afford the hydrochloride salt (58% from crude free base). The temperature for the salt formation is not specified. Aqueous sodium hydroxide (10% or 0.1%) can be substituted for ammonium hydroxide.[19,20,26,29]

### 4.5.7 Catalytic Hydrogenation with Raney Nickel

Raney nickel is supplied as a suspension in water. It is presumed that standard procedures are used to wash the catalyst with water and charge it to the hydrogenation reactor. Since Raney nickel may be contaminated by basic impurities and cleavage of the β-hydroxynitrile **17** is catalyzed by base, the reduction may be more efficient when the catalyst is repeatedly washed with water or perhaps pretreated with dilute acid. The catalyst loadings are high, ranging from 150 to over 900 g of Raney nickel/kg of nitrile **17**. The catalyst can be recycled at least four times with no noticeable deactivation,[46] so the catalyst loading should be selected to maximize yield and selectivity at convenient temperatures (25–30°C) and hydrogen pressures (<100 psi). Alcohols (methanol, ethanol, isopropanol, 1-butanol, isobutanol, *tert*-butanol) are the most common solvents and ammonia, ammonium hydroxide, or dilute sodium hydroxide are usually added. The reduction can also be accomplished in a biphasic mixture (toluene–water) and in acetic acid at elevated temperatures and pressures. Volume throughputs range from 5.5 to 20 L/kg nitrile **17**. Crude yields are 85–90% and the crude product is likely to contain 10–15% 2-(4-methoxyphenyl)ethanamine (**33**). The purity is upgraded by precipitation of the acetate, formate, or hydrochloride salt from ethyl acetate.

Raney nickel (500 g/kg **17**) is added to a solution of 2-(1-hydroxycyclohexyl)-2-(4-methoxyphenyl)acetonitrile (**17**) in methanol (15.0 L/kg **17**). The suspension is hydrogenated at (presumably) 25°C and 60 psi hydrogen for 4 h. The suspension is filtered and the recovered catalyst is presumably washed with methanol. The combined liquors contain 1-(2-amino-1-(4-methoxyphenyl)ethyl)cyclohexanol (**34**) (99%, 93% pure by HPLC).[38]

Raney nickel (408 g/kg **17**) is added to a mixture of 2-(1-hydroxycyclohexyl)-2-(4-methoxyphenyl)acetonitrile (**17**) and methanol (12.2 L/kg **17**). The suspension is hydrogenated at 25°C and atmospheric pressure of hydrogen for 15 h. The suspension is filtered, the recovered catalyst is washed with methanol, and the liquors are concentrated at reduced pressure. 1-(2-Amino-1-(4-methoxyphenyl)ethyl)cyclohexanol (**34**) is isolated from the residue by chromatography (83%).[31]

Raney nickel (400 g/kg **17**) is added to a solution of 2-(1-hydroxycyclohexyl)-2-(4-methoxyphenyl)acetonitrile (**17**) and ammonia (5.8 equivalents) in methanol (20.0 L/kg **17**). The suspension is hydrogenated at 35–40°C and 140–150 psi hydrogen for 3 h. The contained yield of **34** is >83%.[39]

$d_3$-2-(1-Hydroxycyclohexyl)-2-(4-methoxyphenyl)acetonitrile (**20**) is reduced using a Raney nickel catalyst in a continuous flow hydrogenation reactor at 80°C and 1100–1200 psi. The eluent is 2.0 M ammonia in methanol. The yield of $d_3$-1-(2-amino-1-(4-methoxyphenyl)ethyl) cyclohexanol (**35**) is 69%. $d_{14}$-2-(1-Hydroxycyclohexyl)-2-(4-methoxyphenyl)acetonitrile (**21**) is reduced under the same conditions to $d_{14}$-1-(2-amino-1-(4-methoxyphenyl) ethyl)cyclohexanol (**36**) (92%).[21]

Raney nickel (720 g/kg theoretical **17**) and ammonia (7.0 equivalents based on theoretical **17**) in 1-butanol are added to a solution of 2-(1-hydroxycyclohexyl)-2-(4-methoxyphenyl)acetonitrile (**17**) in 1-butanol. The suspension is hydrogenated at 10–12°C and 120 psi hydrogen for 6–7 h. The suspension is filtered, the recovered catalyst is washed with 1-butanol, and the liquors are concentrated at reduced pressure. The residue is dissolved in ethyl acetate. Hydrogen chloride in isopropanol (20%) is added at 0–5°C (to pH 1–1.5) and the solution is aged at 0–5°C for 30 min. *n*-Hexane is added at 0–5°C. The suspension is aged at 0–5°C for 3 h and then filtered. The solid is presumably washed with cold ethyl acetate–hexane and dried at 50–52°C and reduced pressure to afford 1-(2-amino-1-(4-methoxyphenyl)ethyl) cyclohexanol (**34**) hydrochloride (90% for three steps from 4-methoxyphenylacetonitrile, 95% pure by HPLC). Similar yields and purities (88–89%, 94.1–94.5% pure by HPLC) are observed using isobutanol or *tert*-butanol in place of 1-butanol.[36]

Raney nickel (Kawaken, Grace 2400, 2724, Degussa B111W, 112 W) (500 g/kg **17**) is suspended in 10 : 1 (v/v) methanol–25% ammonium hydroxide (5.5 L/kg **17**). 2-(1-Hydroxycyclohexyl)-2-(4-methoxyphenyl)acetonitrile (**17**) is added followed by more methanol (to a total volume of 10.5 L/kg **17**). The suspension is hydrogenated at 10–20°C and 60 psi hydrogen for 20–30 h. The suspension is filtered, the recovered catalyst is presumably washed with methanol, and the combined liquors are concentrated at reduced pressure. The residue is dissolved in ethyl acetate–isopropanol (2 : 1). Acetic acid (0.85 equivalents) is added and the mixture is aged, presumably at 25°C, for 1–1.5 h. The resulting suspension is filtered and the solid is washed with ethyl acetate and dried to afford 1-(2-amino-1-(4-methoxyphenyl)ethyl)cyclohexanol (**34**) acetate (70%, 99% pure by HPLC). The same results (70%, 99% pure by HPLC) are observed using more acetic acid (1.3 equivalents). Lower yields (60–63%) are observed using less Raney nickel (300 g/kg **17**). Temperature plays an important role in the hydrogenation efficiency: hydrogenation at 5°C produces less side product 2-(4-methoxyphenyl)ethanamine (**33**) than hydrogenation at 25°C.[48]

Raney nickel (CORM III) (750 g/kg **17**) is added to a mixture of 2-(1-hydroxycyclohexyl)-2-(4-methoxyphenyl)acetonitrile (**17**) in 4 : 1 (v/v) methanol–28% ammonium hydroxide (16.7 L/kg **17**). (*Note*: The ammonium hydroxide is presumed to be 28–30%.) The suspension is hydrogenated at 27°C and 120 psi hydrogen for 10 h. The suspension is presumably filtered, the catalyst is washed with methanol, and combined liquors are concentrated at reduced pressure to afford crude 1-(2-amino-1-(4-methoxyphenyl)ethyl)cyclohexanol (**34**) (89%). Analysis of the crude product by HPLC reveals an 89 : 11 ratio of 1-(2-amino-1-(4-methoxyphenyl)ethyl)cyclohexanol (**34**) and 2-(4-methoxyphenyl)ethanamine (**33**) (Scheme 4.8).

In five experiments, the crude yields and product–side product ratios were 84–89% and 89 : 11 to 82 : 18. However, reproducibility is suspect. In another experiment, the crude yield is 90% but 2-(4-methoxyphenyl)ethanamine (**33**) is the major product. Nitrile reductions using Raney nickel type-B in place of Raney nickel type-CORM III are slow and do not go to completion. Attempted nitrile reduction using Raney nickel type-F afforded only 2-(4-methoxyphenyl)ethanamine (**33**).[29]

Raney nickel (170 g/kg **17**) is added to 2-(1-hydroxycyclohexyl)-2-(4-methoxyphenyl)acetonitrile (**17**) in 3 : 1 (v/v) isopropanol–15% ammonium hydroxide (6 L/kg **17**) at 10°C. The suspension is hydrogenated at 10–15°C and 28 psi hydrogen for 1 h and then at 35°C and 99 psi hydrogen for 6–8 h. The suspension is filtered, the recovered catalyst is presumably washed with isopropanol, and the combined liquors are concentrated at reduced pressure. The residue is dissolved in ethyl acetate and acetic acid (0.82 equivalents) is added. The resulting suspension is filtered and the solid is presumably washed with ethyl acetate and dried to afford 1-(2-amino-1-(4-methoxyphenyl)ethyl)cyclohexanol (**34**) acetate (80–85%, purity >99% by HPLC). Similar results are observed using *tert*-butanol in place of isopropanol (*Note*: The same weight% ammonium hydroxide is presumably used.) or using a lower Raney nickel loading (50 g/kg **17**). Similar results are also observed using 6 : 1 isopropanol–15% ammonium hydroxide (10.5 L/kg **17**) or 8 : 1 methanol–15% ammonium hydroxide (13.5 L/kg **17**). (*Note*: The same weight% ammonium hydroxide is presumably used.)[47]

Raney nickel (200 g/kg **17**) is added to a mixture of 2-(1-hydroxycyclohexyl)-2-(4-methoxyphenyl)acetonitrile (**17**), water, and toluene. Ammonium hydroxide (20%) is added to produce a 4 : 1 : 1 (v/v/v) mixture of water–toluene–20% ammonium hydroxide (6.0 L/kg **17**). The suspension is hydrogenated at <12°C and 60–85 psi hydrogen for 2 h and then at 45–50°C and 100–115 psi for 8 h. The suspension is diluted with toluene and filtered, the recovered catalyst is washed with toluene, and the layers of the combined liquors are separated. The organic layer is washed with 10% brine and then diluted with methanol. Acetic acid (0.86 equivalents) is added, presumably at 25°C, in two portions. The suspension is aged at 75–80°C for 15 min, cooled to 0–5°C, and filtered. The solid is washed with ethyl acetate and dried to afford 1-(2-amino-1-(4-methoxyphenyl)ethyl)cyclohexanol (**34**) acetate (66%, 99% pure by HPLC). Lower yield and purity (53%, 90% pure by HPLC) is observed when the hydrogenation is run in dilute (3.3%) ammonium hydroxide alone and the toluene is introduced later to follow the same workup procedure.[27]

Pretreatment of the Raney nickel with 5% acetic acid may remove traces of base that catalyze the cleavage of β-hydroxynitrile **17**. After the acid treatment, the Raney nickel is washed several times with water. The pretreated/washed Raney nickel (750 g/kg **17**) is then added to a mixture of 2-(1-hydroxycyclohexyl)-2-(4-methoxyphenyl)acetonitrile (**17**) in 4 : 1 methanol–25% ammonium hydroxide (16.7 L/kg **17**). The suspension is hydrogenated at 27–30°C and 120 psi hydrogen for 9–10 h. The suspension is filtered, the recovered catalyst is washed with methanol, and the liquors are concentrated at reduced pressure to afford crude 1-(2-amino-1-(4-methoxyphenyl)ethyl)cyclohexanol (**34**) (80%). The purity is not specified but is likely to be 85–90 wt% by HPLC. A comparable yield and purity (86% yield, 79 wt% by HPLC) is observed using more pretreated/washed catalyst (2.5 kg/kg **17**) and carrying out the hydrogenation in 4 : 1 ethanol–25% ammonium hydroxide (25 L/kg **17**) at higher temperature and pressure (60°C and 580 psi for 140 min). The purity increases (86% yield, 86 wt% by HPLC) when the catalyst charge is further increased (to 5.0 kg/kg **17**) under the same high temperature and pressure conditions.[46]

Pretreating Raney nickel with vanadium, tungsten, or molybdenum compounds is intended to minimize cleavage of β-hydroxynitrile **17**. Pretreatment with a saturated aqueous solution of vanadium(III) acetylacetonate is followed by

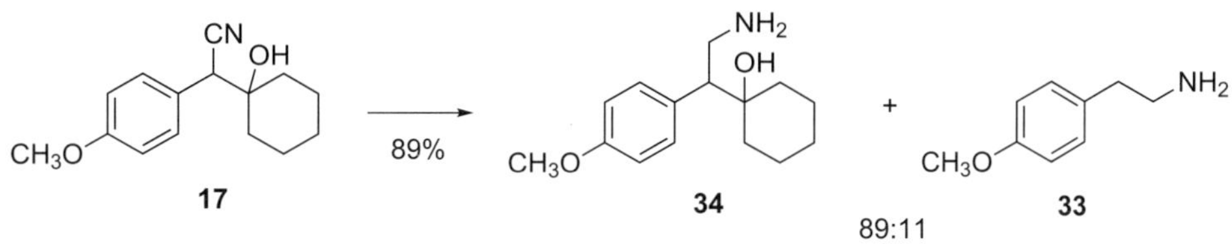


**SCHEME 4.8** Hydrogenation of 2-(1-hydroxycyclohexyl)-2-(4-methoxyphenyl)acetonitrile (**17**) catalyzed by Raney nickel.

several water washes. The pretreated Raney nickel (909 g/kg **17**) is then added to a mixture of 2-(1-hydroxycyclohexyl)-2-(4-methoxyphenyl)acetonitrile (**17**) in 4 : 1 ethanol–25% ammonium hydroxide (12.1 L/kg **17**). The suspension is hydrogenated at 60°C and 580 psi hydrogen for 11 h. The suspension is presumably cooled and filtered, the recovered catalyst is washed with ethanol, and the liquors are concentrated at reduced pressure to afford crude 1-(2-amino-1-(4-methoxyphenyl)ethyl)cyclohexanol (**34**) (86%, 93 wt% by HPLC). Vanadyl acetylacetonate and $H_3[P(W_3O_{10})_4]\cdot xH_2O$ are also used to pretreat the catalyst.[46]

1-(2-Amino-1-(4-methoxyphenyl)ethyl)cyclohexanol (**34**) can be isolated as a formate salt. The crude free base is suspended in ethyl acetate. Formic acid (1.5 equivalents) is added. The mixture is refluxed, cooled to 25°C, and filtered. The solid is washed with hexane and dried to afford the salt.[46]

Raney nickel (150 g/kg **17**) in acetic acid is added to a mixture of 2-(1-hydroxycyclohexyl)-2-(4-methoxyphenyl) acetonitrile (**17**) and acetic acid (total acetic acid is 6.0 L/kg **17**). The suspension is hydrogenated at 55°C and 140–210 psi hydrogen for 4–6 h. The suspension is cooled and filtered, the recovered catalyst is washed with acetic acid, and the liquors are concentrated at reduced pressure. The residue is separated between toluene and water. Ethyl acetate is added to the aqueous layer. Ammonium hydroxide (25%) is added to the biphasic mixture (presumably to pH > 10). After aging at 25°C for 30 min, the layers are separated and the aqueous layer is extracted with ethyl acetate. The combined organic layers are concentrated at reduce pressure. The residue is dissolved in ethyl acetate and acetic acid (1.5 equivalents) is added. The mixture is refluxed for 30 min and then cooled to 0°C. The suspension is filtered and the solid is washed with ethyl acetate and dried to afford 1-(2-amino-1-(4-methoxyphenyl)ethyl)cyclohexanol (**34**) acetate (70%).[32]

In the laboratory, a hydrogenation catalyst is usually filtered by suction using a filter aid (Celite or cellulose powder). The wet cake is quickly washed with fresh solvent. The liquors receiver is then changed and the wet cake is washed with water to deactivate the catalyst and wash off residual flammable solvent. The filtration and cake washing are carried out open to air. The entire operation takes <5 min. The water-wet cake is sent to a vendor for metal recovery. Similar operations, with a longer time allocation and under nitrogen, are feasible on pilot plant and plant scales.How can these procedures be modified to recycle an active catalyst directly back into the next hydrogenation? One attractive option is to *keep the catalyst in the reactor*. This eliminates mechanical losses and reduces the potential for catalyst deactivation by exposure to air. The solution is separated from the catalyst by decantation under nitrogen pressure.In the laboratory, a single decant line (a Teflon or stainless steel cannula) can be used. If the hydrogenation is run in a glass reactor, allow the suspension to settle. Decant off the top using nitrogen pressure. The decant line can be slowly lowered to just above the solid as the liquid level drops. If the hydrogenation is run in a reactor where you cannot see the solid, set the decant line at a predetermined depth. A filter screen can be used on the decant line inlet if the solid does not settle adequately.In the pilot plant or on production scale, it would be prudent to have two decant lines: nitrogen pressure is applied through one line and liquid is decanted through the other. In the event the filter screen for the liquid transfer line becomes plugged with catalyst, the roles of the two lines can be reversed.

### 4.5.8 Catalytic Hydrogenation with Raney Cobalt

The nitrile reduction can be accomplished using Raney cobalt. The Raney cobalt is pretreated with 5% acetic acid and then washed with water. The pretreated Raney cobalt (2.5 kg/kg **17**) is then added to a mixture of 2-(1-hydroxycyclohexyl)-2-(4-methoxyphenyl)acetonitrile (**17**) in 4 : 1 ethanol–25% ammonium hydroxide (25 L/kg **17**). The suspension is hydrogenated at 60°C and 580 psi hydrogen for 90 min. The suspension is presumably cooled and filtered, the recovered catalyst is washed with ethanol, and the liquors are concentrated at reduced pressure to afford crude 1-(2-amino-1-(4-methoxyphenyl)ethyl)cyclohexanol (**34**) (79%, 92 wt% by HPLC).[46]

### 4.5.9 Catalytic Hydrogenation with Palladium on Carbon

The nitrile is reduced using palladium on carbon under acidic conditions. The palladium on carbon catalyst charge ranges from 750 mg to 50 g palladium metal/kg of the nitrile **17**. The solvents are methanol, acetic acid, or formic acid. Methanesulfonic, sulfuric, or hydrochloric acid are added when the solvent is methanol. The nitrile reduction is accomplished at 15–55°C and 50–280 psi hydrogen pressure. The recovered palladium on carbon catalyst is presumably sent for reprocessing. 1-(2-Amino-1-(4-methoxyphenyl)ethyl)cyclohexanol (**34**) is isolated as the hydrochloride or acetate salt from ethyl acetate, ethyl acetate–isopropanol, isopropyl acetate–hexane, or isopropanol. An impressive 91% yield is achieved in one 300 kg scale nitrile reduction using methanesulfonic acid in methanol. In many cases, reduction using palladium under acidic conditions can be carried directly into a reductive amination of formaldehyde or run simultaneously with a reductive amination of formaldehyde.

Palladium on carbon (5% w/w, 50% water wet) (1.25 g Pd/kg **17**) is added to a mixture of 2-(1-hydroxycyclohexyl)-2-(4-methoxyphenyl)acetonitrile (**17**) and methanesulfonic acid (1.0 equivalent) in methanol (3.3 L/kg **17**) at 0–30°C. The suspension is hydrogenated at 15–25°C and 130–140 psi hydrogen for 4 h and then used directly in the next step. The yield of 1-(2-amino-1-(4-methoxyphenyl) ethyl)cyclohexanol (**34**) is >75%. The yield is lower (>71%) at lower hydrogen pressure (60–70 psi) and unchanged (>74%) at higher hydrogen pressure (580 psi). At the higher hydrogen pressure (580 psi), the catalyst charge can be reduced to 790 mg palladium/kg nitrile **17** (>74% yield) but the yield drops if the charge is reduced further (>67% yield at 450 mg Pd/kg **17**). The yield is >91% on 300 kg scale using 1.25 g Pd/kg nitrile **17** at 580 psi hydrogen pressure.[49]

Palladium on carbon (5% w/w) (10 g Pd/kg **17**) is added to a mixture of 2-(1-hydroxycyclohexyl)-2-(4-methoxyphenyl) acetonitrile (**17**) and 37% aqueous sulfuric acid (0.96 equivalents) in methanol (15 L/kg **17**), presumably at 25°C. The suspension is hydrogenated at 25°C and 50 psi hydrogen for 6 h. A completion check by HPLC reveals 97.8 area% **34** (area% residual **17** not provided). The suspension is filtered and the liquors are concentrated at reduced pressure. The residue is suspended in water, 3 M sodium hydroxide is added (to pH 12), and the mixture is extracted with isopropyl acetate. The extracts are concentrated to a smaller volume and acid (presumably hydrochloric acid) is added (to pH < 1). The mixture is dried by distillation of the isopropyl acetate–water azeotrope at reduced pressure. Hexane is added, the resulting suspension is aged at 25°C for 1 h and at 0–5°C for 1 h and then filtered. The solid is presumably washed with hexane and dried to afford 1-(2-amino-1-(4-methoxyphenyl)ethyl)cyclohexanol (**34**) hydrochloride (88%) (Scheme 4.9).[28]

Palladium on carbon (10% w/w) (31 g Pd/kg **17**) is added to a mixture of 2-(1-hydroxycyclohexyl)-2-(4-methoxyphenyl)acetonitrile (**17**) and concentrated hydrochloric acid (1.8 equivalents) in methanol (15 L/kg **17**), presumably at 25°C. The suspension is hydrogenated at 25°C and 50 psi hydrogen for 6 h. A completion check by HPLC reveals 1.1% nitrile **17** (97.3% of **34**). The suspension is filtered and the liquors are concentrated at reduced pressure. The residue is suspended in isopropanol. The suspension is concentrated at reduced pressure to a small volume and then filtered. The solid is dried to afford 1-(2-amino-1-(4-methoxyphenyl)ethyl)
cyclohexanol (**34**) hydrochloride (83%).[28]

Palladium on carbon (10% w/w) (31 g Pd/kg **17**) is added to a mixture of 2-(1-hydroxycyclohexyl)-2-(4-methoxyphenyl)acetonitrile (**17**) and hydrogen chloride (1.3 equivalents) in 7.5 : 1 (v/v) methanol–ethanol (17 L/kg **17**), presumably at 25°C. The suspension is hydrogenated at 25°C and 50 psi hydrogen for 6 h. A completion check by HPLC reveals 1.0 area% **17** (98.3 area% **34**). The suspension is filtered and the liquors are concentrated at reduced pressure. The residue is suspended in diisopropyl ether at 0–5°C and the suspension is filtered. The solid is dried to afford 1-(2-amino-1-(4-methoxyphenyl)ethyl)cyclohexanol (**34**) hydrochloride (91%).[28]

Palladium on carbon (10% w/w, 50% water wet) (50 g Pd/kg **17**) is added to a mixture of 2-(1-hydroxycyclohexyl)-2-(4-methoxyphenyl)acetonitrile (**17**) and hydrochloric acid in methanol (8 L/kg **17**), presumably at 25°C. (*Note*: The amount of hydrochloric acid charged is quoted as "1–3 moles," or 10–29 equivalents. Perhaps this is a typographical error.) The suspension is hydrogenated at 40–50°C and 210–280 psi hydrogen for 7–12 h. The suspension is cooled to 25–30°C and filtered. The filter cake is washed with methanol. Sodium hydroxide (50%) is added to the combined liquors (to pH 10.5–11.0) and the solution is concentrated at reduced pressure. The residue is separated between ethyl acetate and water. The organic layer is washed with brine and concentrated at reduced pressure to afford crude 1-(2-amino-1-(4-methoxyphenyl)ethyl)cyclohexanol (**34**) (86%). Hydrogen chloride (16%) in isopropanol is added to the crude free base in ethyl acetate at 5–10°C. The resulting suspension is aged at 5–10°C for 1 h and filtered. The solid is washed with ethyl acetate and dried to afford the hydrochloride salt (72%).[34]

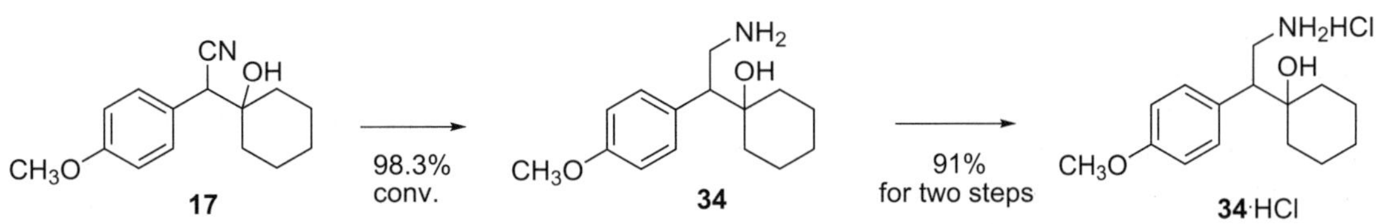


**SCHEME 4.9** Hydrogenation of 2-(1-hydroxycyclohexyl)-2-(4-methoxyphenyl)acetonitrile (**17**) catalyzed by palladium on carbon.

In the laboratory, the reagent and solvent charging operations required to get a gram-scale reaction up-and-running can usually be completed in short order. In a pilot plant or plant, the same charging operations may take much longer. Where will this difference be an issue? It is an issue when the mixtures produced along the way are not stable over time.An overriding concern using mixtures of sulfuric or sulfonic acids in methanol is the potential for generation of methyl sulfonates. Methyl methanesulfonate is likely formed during the time that elapses between charging the reactor and completing the nitrile reduction. The mixture is at 25°C, has no water cosolvent, and does not have sufficient base to neutralize the acid. The reaction mixture, including any methyl methanesulfonate produced, is carried into the final methylation step.[50]

Palladium on carbon (10% w/w, 50% water wet) (3 g Pd/kg **17**) is added to a mixture of 2-(1-hydroxycyclohexyl)-2-(4-methoxyphenyl)acetonitrile (**17**) and acetic acid (6 L/kg **17**), presumably at 25°C. The suspension is hydrogenated at 50–55°C and 210–240 psi hydrogen for 10–12 h. The suspension is cooled to 25–35°C and filtered and the liquors are concentrated at <70°C and reduced pressure. The residue is dissolved in dichloromethane and water. Ammonium hydroxide is added at 0–10°C (presumably to pH > 10). The layers are separated and the aqueous layer extracted with dichloromethane. The combined organic layers are concentrated at reduced pressure. The residue is suspended in 2 : 1 *n*-heptane–isopropanol. The suspension is aged at 0–5°C for 1–2 h and then filtered. The isolated free base is washed with *n*-heptane and then dissolved in ethyl acetate. The solution is cooled to 0–10°C and acetic acid (1.5 equivalents) is added. The suspension is aged at 0–10°C for 1–2 h and then filtered. The solid is washed with ethyl acetate and dried at 50–60°C and reduced pressure to afford 1-(2-amino-1-(4-methoxyphenyl)ethyl)cyclohexanol (**34**) acetate (45–55%). The yield is lower (42%) when the free base is isolated from isopropanol alone.[35]

Palladium on carbon (15% w/w, 50% water wet) (11.3 g Pd/kg **17**) in 98% formic acid is added to a mixture of 2-(1-hydroxycyclohexyl)-2-(4-methoxyphenyl)acetonitrile (**17**) and 98% formic acid (total formic acid 8.5 L/kg **17**) at 10–15°C. The suspension is hydrogenated at 10–15°C and 50–60 psi hydrogen for 4 h. The suspension is filtered and the liquors are concentrated at 40–45°C and reduced pressure. The residue is dissolved in water and the solution is washed with ethyl acetate. Sodium hydroxide (10%) is added at 15–20°C (to pH 9.5) and the free base is extracted with ethyl acetate. The extracts are concentrated at 45–50°C and reduced pressure. The residue is dissolved in ethyl acetate and hydrochloric acid is added at 20–30°C (to pH 1–2). The suspension is aged at 40–45°C for 2 h, cooled to 10–15°C, and filtered. The solid is suspended in ethyl acetate and the suspension is aged for 2 h and filtered. The solid is washed with cold ethyl acetate and dried at 40–45°C and reduced pressure to afford 1-(2-amino-1-(4-methoxyphenyl)ethyl)cyclohexanol (**34**) hydrochloride (56–61%, 99.0–99.5% pure by HPLC).[51]

Palladium on carbon (14% w/w, 50% water wet) (9.6 g Pd/kg **17**) is added to a mixture of 2-(1-hydroxycyclohexyl)-2-(4-methoxyphenyl)acetonitrile (**17**) and 85% formic acid (4.0 L/kg **17**) at 10–25°C. The suspension is hydrogenated at 10–25°C and 580 psi hydrogen for 2–4 h. The suspension is filtered and the liquors are concentrated at 60°C and reduced pressure. The mixture of 1-(2-amino-1-(4-methoxyphenyl)ethyl)cyclohexanol (**34**) formate in formic acid is used in the next step. The yield for the nitrile reduction is at least 65%.[49]

## 4.6 VENLAFAXINE (1) BY METHYLATION OF 1-(2-AMINO-1-(4-METHOXYPHENYL)ETHYL) CYCLOHEXANOL (34)

Eschweiler–Clarke methylation converts 1-(2-amino-1-(4-methoxyphenyl)ethyl)cyclohexanol (**34**) to venlafaxine (**1**). The methylation likely proceeds via 5-(4-methoxyphenyl)-1-oxa-3-aza-spiro[5.5]undecane (**37**) and 5-(4-methoxyphenyl)-3-methyl-1-oxa-3-aza-spiro[5.5]undecane (**38**). The hydrochloride, acetate, and formate salts and the free base of **34** are all suitable starting materials. A base (NaOH, $Et_3N$) is usually added when starting with the hydrochloride salt. Excesses of both formic acid and formaldehyde (usually 37% aqueous formaldehyde) are used. Charges range from 4.0 to 9.7 equivalents of formic acid and from 2.2 to 4.2 equivalents of formaldehyde. Mole ratios of formic acid to formaldehyde range from 2.6 : 1 to 1.3 : 1. There are a few processes that use higher charges, perhaps even use formic acid as the reaction solvent, and there are a few processes where the mole ratio of formic acid to formaldehyde is <1. Higher charges of formaldehyde increase the potential for side product formation by ring hydroxymethylation. The water charge varies from as low as 88 mL/kg of amine **34** to as high as 96 L/kg of amine **34**. Low water charges may reduce losses to the aqueous layer during the workup procedure. An extended reflux time (14–24 h) is required to complete the final conversion of 5-(4-methoxyphenyl)-3-methyl-1-oxa-3-aza-spiro[5.5]undecane (**38**) to venlafaxine (**1**). It is suggested that sodium formate accelerates the conversion to venlafaxine, reducing the total reflux time to 5–8 h. Long reflux times are associated with higher levels of side product formated by ring hydroxymethylation.

The workup procedure may include washing of an aqueous solution of venlafaxine hydrochloride with ethyl acetate to remove a pink-colored impurity. When this washing operation is omitted, the color may be removed by treating with carbon or perhaps removed during isolation of venlafaxine or venlafaxine hydrochloride. Venlafaxine can be isolated as the free base (mp 78.3–79.5°C) from hexane or heptane. Yields of the crude free base can be as high as 98% but purity data is usually not provided. Venlafaxine can also be isolated as the hydrochloride salt. The salt is produced by addition of hydrogen chloride in isopropanol to a solution of the free base in isopropanol, acetone, toluene, or ethyl acetate. Aqueous hydrochloric acid can be used when the water is subsequently removed by azeotropic distillation. Yields for venlafaxine hydrochloride are likely to be 84–87% but can be as high as 95%.

A one-pot reduction and methylation can be accomplished using palladium on carbon and formic or acetic acid.

> Formaldehyde is a suspected human carcinogen. The OSHA permissible exposure limit for formaldehyde is 0.75 ppm (TWA) and 2 ppm (STEL). Using an undesirable reagent, such as a suspected carcinogen, in a manufacturing process should always raise a "red flag." After exhausting all process options to eliminate the undesirable reagent, the next objective should be to *completely consume the reagent during the reaction.* Operator exposure is then limited to charging the reagent and there are many excellent methods for containment during charging operations. A 37% aqueous formaldehyde charge is likely to be easier to contain during charging than a solid paraformaldehyde charge. *The fate of any excess undesirable reagent should be defined.* The methylation to produce venlafaxine (**1**) is typically run at reflux under nitrogen. The nitrogen off-gas from the labscale reaction is vented in the hood. As the scale increases, the off-gas line is vented through a "scrubber" solution, perhaps aqueous hydroxide, to capture formaldehyde and formic acid. When the methylation is complete, excess formaldehyde in the reactor could be reduced to an acceptable level by distillation at atmospheric pressure. Charging aqueous hydrogen peroxide to the distillate and to the scrubber solution converts formaldehyde to sodium formate. At moderate temperatures (< 40°C) and pH 10–11, the hydrogen peroxide cost is $1.65 per kg formaldehyde. In the laboratory and pilot plant the scrubber could simply be a second reactor charged with aqueous hydroxide with a subsurface inlet line and above-surface outlet to the atmosphere. For production scale, there are packed bed air/gas scrubbers specifically designed to efficiently scrub formaldehyde.[52,53]

To emphasize the importance of minimizing the formaldehyde charge, the methylation procedures in each section are presented in order from most formaldehyde equivalents to least.

### 4.6.1 1-(2-Amino-1-(4-methoxyphenyl)ethyl) cyclohexanol (34)

The dimethylation likely proceeds via an intermediate 1-oxa-3-azaspiro[5.5]undecane. This intermediate is converted to venlafaxine (**1**) by extending the reflux time. A mixture of 1-(2-amino-1-(4-methoxyphenyl)ethyl)cyclohexanol (**34**), water (10 L/kg **34**), 99% formic acid (11.0 equivalents), and 37% aqueous formaldehyde (11.4 equivalents) is refluxed for 2 h. The mixture is concentrated to a smaller volume, diluted with water, and washed with ethyl acetate. Sodium hydroxide solution (50%) is added to the aqueous layer (to pH 10) and the free base is extracted with ethyl acetate. The extracts are dried and concentrated at reduced pressure. The main component of the residue is 5-(4-methoxyphenyl)-3-methyl-1-oxa-3-azaspiro[5.5]undecane (**38**). Reduction of the residue with lithium aluminum hydride in ethyl ether affords venlafaxine (**1**) (60%).[54]

A 3.3 : 1 mixture of venlafaxine (**1**) and the 1-oxa-3-azaspiro[5.5]undecane (**38**) is produced in the reaction of 1-(2-amino-1-(4-methoxyphenyl)ethyl)cyclohexanol (**34**), 88% formic acid (7.1 equivalents), aqueous formaldehyde (3.1 equivalents), and water (10.4 L/kg **34**) after 5 h at 100°C (Scheme 4.10). (*Note*: The formaldehyde concentration is not specified. Formaldehyde equivalents are calculated based on a 37% solution.)[20]

1-Oxa-3-azaspiro[5.5]undecane **37**, perhaps also an intermediate in the dimethylation, forms when 37% formaldehyde (16.5 equivalents) is added to the suspension at the completion of the nitrile reduction with Raney nickel. The suspension is aged, presumably at 25°C, for 3 h. The catalyst is filtered and presumably washed with methanol. The combined liquors are distilled to a small volume. The resulting suspension is cooled to 10–15°C and filtered. The solid is washed with cold hexane and dried to afford 5-(4-methoxyphenyl)-1-oxa-3-aza-spiro[5.5]undecane (**37**) (83% for the reduction and formaldehyde reaction). A mixture of 1-oxa-3-azaspiro[5.5]undecane **37** with 96% formic acid (10 equivalents) and 37% aqueous formaldehyde (5.0 equivalents) in water (5.0 L/kg **37**) is refluxed for 15 h. The mixture is cooled to 25°C, sodium hydroxide solution is added (to pH 12), and the free base is extracted with heptane. The extracts are concentrated at reduced pressure. The residue is dissolved in isopropanol and a solution of hydrogen chloride in isopropanol is added, presumably at 25°C (to pH 2). The suspension is filtered and the solid is washed with heptane and dried to afford venlafaxine (**1**) hydrochloride (85%, 99.7% pure by HPLC) (Scheme 4.11).[39]

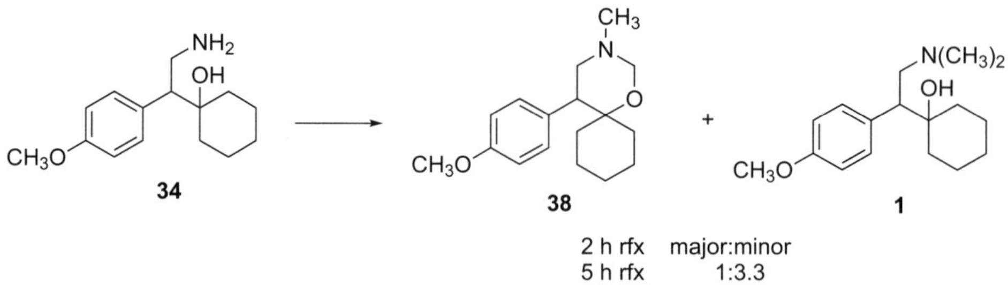


**SCHEME 4.10** Venlafaxine (**1**) from 5-(4-methoxyphenyl)-3-methyl-1-oxa-3-aza-spiro[5.5]undecane (**38**).

A mixture of 1-(2-amino-1-(4-methoxyphenyl)ethyl)cyclohexanol (**34**), 88% formic acid (11.0 equivalents), and 37% aqueous formaldehyde (11.4 equivalents) in water (10 L/kg **34**) is refluxed for 20 h. The mixture is concentrated to a smaller volume and cooled to 25°C. Hydrochloric acid (3 M) is added (to pH 2.0) and the mixture is washed several times with ethyl acetate to remove a pink-colored impurity. Sodium hydroxide solution (50%) is added to the aqueous layer (presumably to pH 10–11) and the free base is extracted with ethyl acetate. The extracts are dried and concentrated at reduced pressure to afford crude venlafaxine (**1**) (93%).[30]

A mixture of 1-(2-amino-1-(4-methoxyphenyl)ethyl)cyclohexanol (**34**), water (5.0 L/kg **34**), 88% formic acid (2.6 equivalents), and 40% aqueous formaldehyde (6.0 equivalents) is refluxed for 19 h. (*Note*: The formic acid concentration is not specified. Equivalents are based on 88% formic acid.) The mixture is cooled, presumably to 25°C, and washed with chloroform. The aqueous layer is cooled to 5°C, 48% sodium hydroxide is added (to pH > 7), and the free base is extracted with chloroform. The extracts are concentrated at reduced pressure. The residue is dissolved in isopropanol and a solution of hydrogen chloride in isopropanol is added (to pH 2), presumably at 25°C. The resulting suspension is filtered and the solid is washed with isopropanol and dried at 55–60°C and reduced pressure to afford venlafaxine (**1**) hydrochloride (63%).[35]

A mixture of 1-(2-amino-1-(4-methoxyphenyl)ethyl)cyclohexanol (**34**), 88% formic acid (15.1 equivalents), and 30% aqueous formaldehyde (5.7 equivalents) in water (2.7 L/kg **34**) is refluxed for 14 h. The solution is cooled to 15°C and 40% sodium hydroxide is added (presumably to pH 10–11). The mixture is extracted with chloroform and the extracts are concentrated at reduced pressure. The residual venlafaxine (**1**) (88% crude) is dissolved in acetone and a solution of hydrogen chloride in isopropanol is added at 5–10°C (to pH 2). The suspension is filtered and the solid is washed with isopropanol and dried to afford venlafaxine (**1**) hydrochloride (65%).[38]

A mixture of 1-(2-amino-1-(4-methoxyphenyl)ethyl)cyclohexanol (**34**), formic acid (2.1 equivalents), and 40% aqueous formaldehyde (5.3 equivalents) in water (5.0 L/kg **34**) is refluxed for 20–22 h. (*Note*: The formic acid concentration is not specified. Equivalents are based on 88% formic acid.) The mixture is cooled and washed with dichloromethane. Sodium hydroxide (5%) is added to the aqueous layer (to pH 9–10) and the free base is extracted with toluene. Hydrogen chloride in isopropanol is added to the combined extracts and the resulting suspension is cooled to 0°C, aged at 0°C for 30 min, and filtered. The solid is suspended in isopropanol. The suspension is refluxed for 30 min, cooled to 0°C, and filtered. The solid is washed with isopropanol and dried to afford venlafaxine (**1**) hydrochloride (65%).[32]

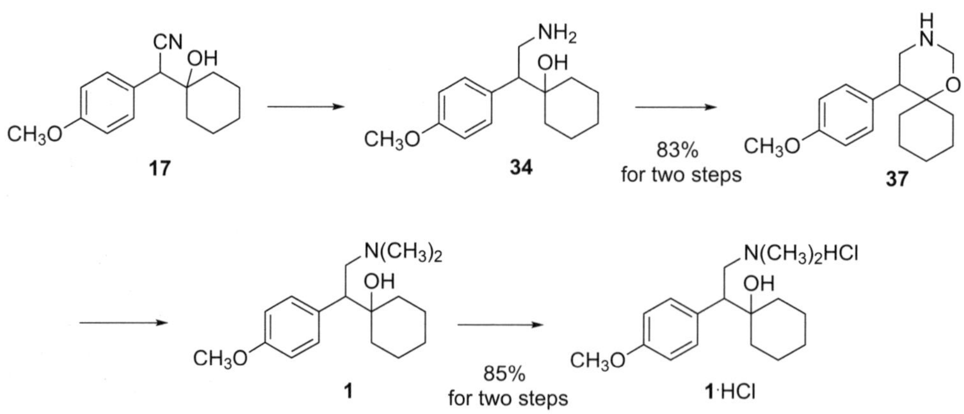


**SCHEME 4.11** Venlafaxine (**1**) from 5-(4-methoxyphenyl)-1-oxa-3-aza-spiro[5.5]undecane (**37**).

A mixture of 1-(2-amino-1-(4-methoxyphenyl)ethyl)cyclohexanol (**34**), formic acid (9.7 equivalents), and 37% aqueous formaldehyde (4.2 equivalents) in water (12.7 L/kg **34**) is refluxed for 16 h. (*Note*: The formic acid concentration is not specified. Equivalents are based on 88% formic acid.) After cooling to 30°C, hydrochloric acid is added (to pH < 7) and the mixture is presumably washed with ethyl acetate. (*Note*: The ethyl acetate wash was omitted from the procedure description.) Sodium hydroxide (50%) is added (to pH 10–11) and the free base is extracted with toluene. The extracts are dried and concentrated at reduced pressure. The residue is dissolved in isopropanol and a solution of 20% hydrogen chloride in isopropanol is added at 10°C (to pH 2.0). The suspension is aged, presumably at 10°C, for 1–2 h and then filtered. The solid is washed with 10°C isopropanol and dried to afford venlafaxine (**1**) hydrochloride (64%, >99.7% pure by HPLC).[43]

A mixture of 1-(2-amino-1-(4-methoxyphenyl)ethyl)cyclohexanol (**34**), formic acid (8.4 equivalents), and aqueous formaldehyde (3.8 equivalents) in water (88 mL/kg **34**) is refluxed for 14–15 h. (*Note*: The formic acid and formaldehyde concentrations are not specified. Equivalents are based on 88% formic acid and 37% formaldehyde.) The mixture is presumably cooled to 25°C and 20% sodium hydroxide is added (to pH 9–9.5). Ethyl acetate is added and the mixture is aged at 40°C for 30 min. The layers are separated and the organic layer is concentrated at reduced pressure. The residue is dissolved in ethyl acetate and 20% hydrogen chloride in isopropanol is added (to pH 1–1.5), presumably at 25°C. The resulting suspension is filtered and the solid dried at 50–55°C and reduced pressure to afford venlafaxine (**1**) hydrochloride (94–95%, 99.2–99.5% pure by HPLC).[36]

A mixture of 1-(2-amino-1-(4-methoxyphenyl)ethyl)cyclohexanol (**34**), 88% formic acid (8.0 equivalents), and 33% aqueous formaldehyde (3.1 equivalents) in water (9.6 L/kg **34**) is refluxed overnight. The mixture is concentrated to a smaller volume, cooled to 25°C, and diluted with water. Hydrochloric acid (12 M) is added (to pH 2.0) and the mixture is washed several times with ethyl acetate to remove a pink-colored impurity. Sodium hydroxide solution (50%) is added to the aqueous layer (presumably to pH 10–11) and the free base is extracted with ethyl acetate. The extracts are washed with brine, dried, and concentrated at reduced pressure. The residue is dissolved in ethyl acetate and a solution of hydrogen chloride in isopropanol is added. The resulting suspension is filtered and the solid is presumably washed with ethyl acetate and dried to afford venlafaxine (**1**) hydrochloride (80%).[19]

A mixture of 1-(2-amino-1-(4-methoxyphenyl)ethyl)cyclohexanol (**34**), 88% formic acid (8.0 equivalents), and 37% aqueous formaldehyde (3.1 equivalents) in water (9.6 L/kg **34**) is refluxed for 6 h. Some of the solvent is distilled and water is added. Hydrochloric acid (12 M) is added (to pH 2) and the mixture is washed with ethyl acetate. Sodium hydroxide (50%) is added (presumably to pH 10–11) and the free base is extracted with ethyl acetate. The extracts are washed with brine, dried, and concentrated at reduced pressure. The residue is dissolved in ethyl acetate and a solution of hydrogen chloride in isopropanol is added. The resulting suspension is filtered and the solid is presumably washed with ethyl acetate and dried to afford venlafaxine (**1**) hydrochloride (66%).[55]

A mixture of 1-(2-amino-1-(4-methoxyphenyl)ethyl)cyclohexanol (**34**), 88% formic acid (5.0 equivalents), and 36% aqueous formaldehyde (3.1 equivalents) in water (96 L/kg **34**) is refluxed for 21 h. After cooling to 25°C, 32% sodium hydroxide solution is added (to pH 11) and the mixture is extracted with toluene. The extracts are washed with water, dried, and concentrated at reduced pressure to afford crude venlafaxine (**1**) (98%, 99.5% pure by HPLC). Venlafaxine free base can be crystallized from hexane (mp 78.3–79.5°C).[57]

A mixture of 1-(2-amino-1-(4-methoxyphenyl)ethyl)cyclohexanol (**34**), formic acid (6.3 equivalents), and paraformaldehyde (2.9 equivalents) in water (7.9 L/kg theoretical **34**) is refluxed for 24–48 h. (*Note*: The formic acid concentration is not specified. Equivalents are based on 88% formic acid.) The mixture is cooled and washed with ethyl acetate. Sodium hydroxide (50%) is added to the aqueous layer (to pH > 12) and the free base is extracted into ethyl acetate. The extracts are dried and filtered. A 20% solution of hydrogen chloride in isopropanol is added at 25–30°C (to pH < 2.0). The resulting suspension is aged at 25–30°C for 30 min and filtered. The solid is washed with isopropanol and dried at 45–50°C and reduced pressure to afford venlafaxine (**1**) hydrochloride (47–53% for the nitrile reduction, methylation, and salt formation).[33]

### 4.6.2 1-(2-Amino-1-(4-methoxyphenyl)ethyl) cyclohexanol (34) Hydrochloride

***4.6.2.1 Sodium or Potassium Hydroxide*** Sodium hydroxide solution (50%) (3.7 equivalents) is added to a solution of 1-(2-amino-1-(4-methoxyphenyl)ethyl)cyclohexanol (**34**) hydrochloride in water (3.2 L/kg **34** free base) at 25°C. Formic acid (98%) (15.3 equivalents) and 37% aqueous formaldehyde (5.7 equivalents) are added, presumably at 25°C, and the mixture is refluxed for 5 h. After cooling to 25°C, 50% sodium hydroxide solution is added (to pH 12) and the mixture is extracted with isopropyl acetate. The extracts are washed with water and then dried by distillation of the isopropyl acetate–water azeotrope at atmospheric pressure. The solution is cooled, presumably to 25°C, and seeded with venlafaxine hydrochloride form I. Hydrochloric acid (2 M) (1.0 equivalent) is added. The resulting suspension is aged, presumably at 25°C, for 2 h and then filtered. The solid is washed with isopropyl acetate and dried to afford venlafaxine (**1**) hydrochloride polymorph form I (86%). Sodium formate may accelerate the conversion of 5-(4-methoxyphenyl)-3-methyl-1-oxa-3-aza-spiro [5.5]undecane (**38**) to venlafaxine (**1**).[56]

Sodium hydroxide solution (47%) (1.5 equivalents) is added to a solution of 1-(2-amino-1-(4-methoxyphenyl) ethyl)cyclohexanol (**34**) hydrochloride in water (4.9 L/kg **34** free base). Formic acid (98%) (4.5 equivalents) and paraformaldehyde (2.8 equivalents) are added, presumably at 25°C, and the mixture is refluxed for 20–24 h. The mixture is cooled to 60–70°C and 50% sodium hydroxide is added (to pH 9.5–11.5). Heptane and carbon are added and the biphasic suspension is aged at 60–70°C for 10–15 min. The suspension is filtered and the liquors layers are separated at 60–70°C. The organic layer is washed with water at 60–70°C and the heptane is distilled at atmospheric pressure. The residue is crystallized from heptane. The suspension, presumably at 25°C, is filtered and the solid is washed with heptane and dried at 40–50°C and reduced pressure to afford venlafaxine (**1**) (65–71%). The same results (65–71%) are achieved starting with the free base in place of the hydrochloride salt and hydroxide.[34]

Sodium hydroxide solution (32%) (1.1 equivalents) is added, presumably at 25°C, to a solution of 1-(2-amino-1-(4-methoxyphenyl)ethyl)cyclohexanol (**34**) hydrochloride in water (2.6 L/kg **34** free base). Formic acid (88.5%) (4.0 equivalents) and 35.8% aqueous formaldehyde (2.2 equivalents) are added, presumably at 25°C, and the mixture is refluxed for 8 h. After cooling to 25°C, 32% sodium hydroxide solution is added (to pH 11) and the mixture is extracted with heptane. The extract is washed with water, dried, and concentrated at reduced pressure to produce a suspension. The suspension is filtered and the solid is presumably washed with heptane and dried to afford venlafaxine (**1**). No yield is available.[57]

Sodium hydroxide solution (32%) (1.1 equivalents) is added, presumably at 25°C, to a solution of 1-(2-amino-1-(4-methoxyphenyl)ethyl)cyclohexanol (**34**) hydrochloride in water (2.6 L/kg **34** free base). Formic acid (88%) (4.0 equivalents) and 36% aqueous formaldehyde (2.2 equivalents) are added, presumably at 25°C, and the mixture is refluxed for 8 h. After cooling to 25°C, 32% sodium hydroxide solution is added (to pH 11) and the mixture is extracted with heptane. The extract is washed with water, dried, and concentrated at reduced pressure to produce a suspension. The suspension is filtered and the solid is presumably washed with heptane and dried to afford venlafaxine (**1**). No yield is available.[57]

Potassium hydroxide (1.1 equivalents) is dissolved in water (4.8 L/kg **34** free base) at 0–10°C. 1-(2-Amino-1-(4-methoxyphenyl)ethyl)cyclohexanol (**34**) hydrochloride is added and the mixture is aged at 25°C for 30 min. Formic acid (85%) (5.3 equivalents) and 37% aqueous formaldehyde (2.1 equivalents) are added, presumably at 25°C, and the mixture is refluxed for 15 h. After cooling to 5–10°C, ethyl acetate is added followed by 50% sodium hydroxide solution (to pH 11–12). The mixture is aged for 30 min, the layers are separated, and the aqueous layer is extracted with ethyl acetate. The combined organic layers are dried and filtered. Hydrogen chloride in isopropanol (5 M) (0.99 equivalents) is added at 5–10°C (to pH < 4). The suspension is aged at 0–5°C for 1 h and filtered. The solid is suspended in 1.3 : 1 (v/v) ethyl acetate–methanol and seed crystals of venlafaxine (**1**) hydrochloride form I are added. The suspension is refluxed for 30 min, cooled to 0–5°C, aged at 0–5°C for 1 h, and filtered. The solid is dried at 55–65°C and reduced pressure to afford venlafaxine (**1**) hydrochloride form I (87%, 99.97% pure by HPLC). This process is demonstrated on a 5 kg scale. Increasing the charges of formic acid and formaldehyde and extending the reflux time increases the potential for hydroxymethylation of the aromatic ring.[58]

***4.6.2.2 Triethylamine*** Triethylamine (4.9 equivalents) is added to 1-(2-amino-1-(4-methoxyphenyl)ethyl)cyclohexanol (**34**) hydrochloride in water (3.4 L/kg **34** free base). Formic acid (99%) (15.1 equivalents) and 34% aqueous formaldehyde (5.1 equivalents) are added at 25°C. The mixture is aged at 25°C for 15 min and then refluxed for 4–4.5 h. The mixture is cooled to 15–20°C, 20% sodium hydroxide is added (to pH 10–11), and the free base is extracted with ethyl acetate. The extracts are concentrated at reduced pressure. The residual oil is dissolved in 4 : 1 isopropanol–ethyl acetate, presumably at 25°C, and 20–22% hydrogen chloride in isopropanol is added (to pH 1–2). The resulting thick suspension is diluted with 4 : 1 isopropanol–ethyl acetate and the suspension is aged at 40–45°C for 2 h. The suspension is aged at 0–5°C for 2 h and then filtered. The solid is resuspended in 4 : 1 isopropanol–ethyl acetate, aged at 40–45°C for 1 h, aged at 0–5°C for 1 h, and filtered. The solid is washed with 4 : 1 isopropanol–ethyl acetate and dried at 45–50°C and reduced pressure to afford venlafaxine (**1**) hydrochloride (69–76%, 99.89–99.90% pure by HPLC).[51]

***4.6.2.3 No base*** 1-(2-Amino-1-(4-methoxyphenyl)ethyl)cyclohexanol (**34**) hydrochloride is isolated from the nitrile reduction with palladium on carbon and hydrochloric acid in methanol by filtering the catalyst and concentrating the liquors at reduced pressure. A mixture of this crude hydrochloride, 99% formic acid (6.0 equivalents), and formaldehyde (4.0 equivalents) in water (9.8 L/kg **34** free base) is refluxed for 20–24 h. (*Note*: The formaldehyde source is not specified. The quantity charged (quoted in grams and in moles) suggests paraformaldehyde is used.) The mixture is cooled to 25°C and 50% sodium hydroxide is added (to pH 9.5–11.5). The free base is extracted with ethyl acetate. The extracts are washed with water and with brine, dried, and concentrated at reduced pressure. The residue is suspended in heptane, presumably at 25°C, and the suspension is filtered. The solid is presumably washed with heptane and dried to afford venlafaxine (**1**) (77% for the reduction and methylation).[34]

### 4.6.3 1-(2-Amino-1-(4-methoxyphenyl)ethyl) cyclohexanol (34) Acetate

A mixture of 1-(2-amino-1-(4-methoxyphenyl)ethyl)cyclohexanol (**34**) acetate, formic acid (3.3 equivalents), and 40% aqueous formaldehyde (7.5 equivalents) in water (6.2 L/kg **34** free base) is refluxed for 19 h. (*Note*: The formic acid concentration is not specified. Equivalents are based on 88% formic acid.) The mixture is cooled and washed with chloroform. Sodium hydroxide (5%) is added to the aqueous layer at 5°C (to pH 9–10) and the free base is extracted with chloroform. The extracts are concentrated at reduced pressure. The residue is dissolved in isopropanol and hydrogen chloride in isopropanol is added (to pH 2). The resulting suspension is filtered and the solid is washed with isopropanol and dried at 60°C and reduced pressure to afford venlafaxine (**1**) hydrochloride (79%).[32]

A mixture of 1-(2-amino-1-(4-methoxyphenyl)ethyl)cyclohexanol (**34**) acetate, 88% formic acid (6.9 equivalents), and 40% aqueous formaldehyde (3.8 equivalents) in water (3.8 L/kg **34** free base) is aged at 98°C for 18 h. The mixture is cooled to 10°C and ethyl acetate is added. Aqueous sodium hydroxide is added (to pH 7) followed by ammonium hydroxide (to pH 10–10.5). The layers are separated and the aqueous layer is extracted with ethyl acetate. The combined organic layers are treated with carbon, filtered, and concentrated at reduced pressure. The residue is dissolved in isopropanol. A solution of hydrogen chloride in isopropanol is added (to pH 1–1.5). The mixture is aged at 60°C for 60–90 min, cooled to 10°C, aged at 10°C for 60 min, and then filtered. The solid is washed with isopropanol and dried at 60°C and reduced pressure to afford venlafaxine (**1**) hydrochloride (84%, 99.9% pure by HPLC) (Scheme 4.12). XRD data is provided.[27]

### 4.6.4 1-(2-Amino-1-(4-methoxyphenyl)ethyl) cyclohexanol (34) Formate

A mixture of 1-(2-amino-1-(4-methoxyphenyl)ethyl)cyclohexanol (**34**) formate, 98% formic acid (4.4 equivalents), and 37% aqueous formaldehyde (3.5 equivalents) in water (4.0 L/kg **34** free base) is refluxed for 20 h. After cooling to 25°C, 4 M hydrochloric acid is added (to pH < 1) and the mixture is washed several times with ethyl acetate to presumably remove a pink-colored impurity. Sodium hydroxide (30%) is added (to pH > 12), toluene is added, and the mixture is filtered. The liquor layers are separated and the aqueous layer is extracted with toluene. The combined organic layers are washed with water. Hydrogen chloride (4.2 M) (1.3 equivalents) in 1,4-dioxane is added, presumably at 25°C. The resulting suspension is aged, presumably at 25°C, for 1 h and then filtered. The solid is washed with hexane and dried at 40°C and reduced pressure to afford venlafaxine (**1**) hydrochloride (86%).[46]

### 4.6.5 Nonisolated 1-(2-Amino-1-(4-methoxyphenyl) ethyl)cyclohexanol (34)

The mixture produced on reduction of nitrile **17** with borane methylsulfide in 1,4-dioxane can be carried into the methylation. The mixture is cooled to 25°C and formic acid (7.4 equivalents), 37% aqueous formaldehyde (3.2 equivalents), and water (9.8 L/kg theoretical **34** free base) are added. (*Note*: The formic acid concentration is not specified. Equivalents are based on 88% formic acid.) The mixture is aged at 90–92°C until complete by TLC. After cooling to 25°C, the volatiles are distilled at <50°C and reduced pressure. The residual aqueous mixture is washed with ethyl acetate. Solid potassium hydroxide is added (to pH 9–10) and the free base is extracted with toluene. The extracts are washed with brine, dried, and concentrated at <50°C and reduced pressure to afford crude venlafaxine (**1**) (65% for the reduction and methylation).[44]

The mixture produced on reduction of nitrile **17** with sodium borohydride and boron trifluoride etherate in THF can also be carried directly into the methylation. The mixture is quenched by careful addition of 14% aqueous formic acid at 10–15°C. The mixture is aged at 20–25°C for 30–40 min and the organic solvents are then distilled at 45–95°C. The resulting mixture is presumably cooled to 25°C and 90% formic acid (4.4 equivalents, for a total of 6.8 equivalents), formaldehyde (2.8 equivalents), and water (total water is 7.6 L/kg theoretical **34** free base) are added. (*Note*: The formaldehyde source is not specified. The quantity charged (quoted in grams and in moles) suggests paraformaldehyde is used.) The mixture is refluxed for 8–16 h. The mixture is cooled to 20–25°C and 50% sodium hydroxide is added (to

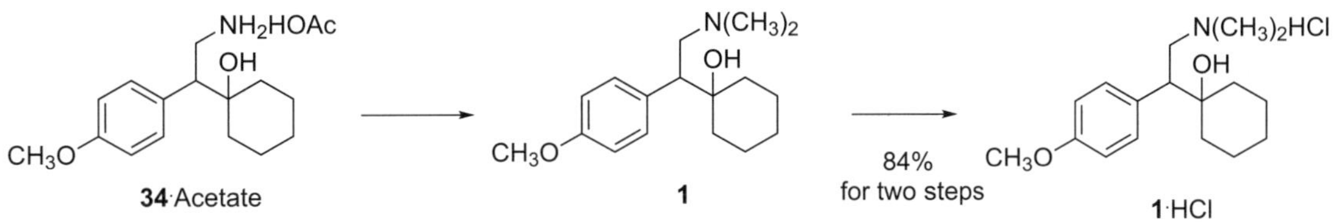


**SCHEME 4.12** Venlafaxine (**1**) by Eschweiler–Clarke methylation of 1-(2-amino-1-(4-methoxyphenyl)ethyl)cyclohexanol (**34**).

pH 9.5–11.5). The free base is extracted with ethyl acetate. The extracts are washed with water and with brine and then dried and concentrated at 45–50°C and reduced pressure. Hydrochloric acid in isopropanol (16%) (1.2 equivalents) is added to a solution of the free base in ethyl acetate at 20–45°C. The resulting suspension is aged at 5–10°C for 90–120 min and then filtered. The solid is washed with ethyl acetate and dried at 35–40°C and reduced pressure to afford venlafaxine (**1**) hydrochloride (78–82% for the reduction, methylation, and salt formation, 99.82% pure by HPLC).[34]

The mixture produced on hydrogenation of nitrile **17** with palladium on carbon (1.3 g Pd/kg **17**) and methanesulfonic acid (1.03 equivalents) in methanol can be carried directly into the methylation. The suspension is cooled and 30% sodium hydroxide (0.05 equivalents) and 36% formaldehyde (2.2 equivalents) are added. The mixture is aged at 20–90°C (*Note*: The temperature is gradually increased during the age time.) under 130–145 psi hydrogen for 4 h. The suspension is cooled to 20°C. The catalyst is filtered and then washed with methanol and with water. The combined liquors are concentrated at reduced pressure. The residue is dissolved in ethyl acetate and the solution is washed with 33% aqueous potassium carbonate and with 5% brine. The organic layer is concentrated at reduced pressure. The residual oil is crystallized from 1.6 : 1 (v/v) acetone–water. The suspension is filtered and the crystals are washed with 2 : 1 (v/v) acetone–water and dried at 40–60°C and reduced pressure to afford venlafaxine (**1**) (71–75%). The same results (74%) are achieved at 580 psi hydrogen pressure. The same results are achieved at 580 psi hydrogen pressure using less palladium (0.79 g Pd/kg **17**). The yield is lower (67%) at 580 psi hydrogen pressure when the catalyst charge is decreased further (0.45 g Pd/kg **17**). An impressive 91% yield is achieved in a 300 kg scale demonstration at 580 psi hydrogen pressure and higher catalyst and hydroxide charges (1.3 g Pd/kg **17**, 0.21 equivalents hydroxide).[49]

The mixture produced on hydrogenation of nitrile **17** using palladium on carbon (9.6 g Pd/kg **17**) in formic acid can be carried directly into the methylation. The suspension is filtered and the liquors are concentrated to afford a mixture of 1-(2-amino-1-(4-methoxyphenyl)ethyl)cyclohexanol **34** in formic acid (approximated as 3.5–4.0 equivalents). Sodium hydroxide (30%) (1.7 equivalents) is added at <40°C followed by 36% formaldehyde (2.4 equivalents). The solution is refluxed for >2 h. The mixture is cooled to <40°C and diluted with ethyl acetate. Sodium hydroxide (30%) is added (to pH > 12) and the layers are separated. The organic layer is washed with 5% brine and concentrated at reduced pressure. The residue is dissolved in heptane at 60°C. The solution is seeded with venlafaxine (**1**) at 60°C and then cooled to 0°C. The suspension is filtered and the solid is washed with heptane and dried at 40–60°C and reduced pressure to afford venlafaxine (**1**) (65%).[49]

### 4.6.6 One-Pot Nitrile Reduction and Amine Methylation

A mixture of 2-(1-hydroxycyclohexyl)-2-(4-methoxyphenyl)acetonitrile (**17**), palladium on carbon (10% w/w, 50% water wet) (50 g Pd/kg **17**), methanol (7.9 L/kg theoretical **34** free base), water (4.9 L/kg theoretical **34** free base), formic acid (3–5 equivalents), and formaldehyde (2–6 equivalents) is aged at 40–50°C and 140–280 psi hydrogen for 10–24 h. (*Notes*: The formic acid concentration is not specified. The formaldehyde source (aqueous solution or paraformaldehyde) is also not specified.) The mixture is cooled to 25–30°C and the suspension is filtered. The recovered catalyst is washed with methanol. The methanol in the combined liquors is distilled at reduced pressure. *n*-Heptane is added to the resulting aqueous mixture. Carbon and 50% sodium hydroxide are added (to pH 10–11). The biphasic mixture is aged at 60–70°C for 30 min and then filtered while hot. The solid is washed with hot water and with hot *n*-heptane. The liquors layers are separated and the organic layer is washed with hot water. *n*-Heptane is distilled from the organic layer at atmospheric pressure. The residue is crystallized from *n*-heptane by aging at 50–60°C and then at 0–5°C. The suspension is filtered and the solid washed with *n*-heptane and dried at 50–60°C and reduced pressure to afford venlafaxine (**1**) (76%). Similar results (78%) are achieved using platinum on carbon (5% w/w) (30 g Pt/kg **17**).[34]

The nitrile reduction and amine methylation are advantageously accomplished in separate stages in a one-pot process. A mixture of 2-(1-hydroxycyclohexyl)-2-(4-methoxyphenyl)acetonitrile (**17**), palladium on carbon (5% w/w, 50% water wet) (7.5 g Pd/kg **17**), acetic acid (5.9 L/kg theoretical **34** free base), water (3.9 L/kg theoretical **34** free base), and paraformaldehyde (2.2 equivalents) is aged at 5°C for 30 min. The mixture is aged at 5–15°C and 175 psi hydrogen pressure for 3.5 h to complete the nitrile reduction and then aged at 60°C and unspecified hydrogen pressure to complete the methylation. The time required for complete methylation is not specified. The suspension is cooled to 30–35°C and filtered. The liquors are concentrated at reduced pressure to afford a viscous but stirrable suspension. Water is added and the mixture is cooled to <40°C. Toluene is added and the mixture is cooled to <10°C. Sodium hydroxide (30%) is added (to pH 6.0–6.6). The aqueous phase is separated. The organic layer is extracted twice with water containing acetic acid (pH 6.0–6.6). The aqueous layers are combined and washed with heptane. More heptane is added, the mixture is cooled to 10°C, and 30% sodium hydroxide is added (to pH > 12). The resulting suspension is filtered and the solid is dried at 55°C and reduced pressure to afford venlafaxine (**1**) (94%, 97.5 wt% by HPLC).[59]

## 4.7 VENLAFAXINE (1) FROM ETHYL 3-(DIMETHYLAMINO)-2-(4-METHOXYPHENYL) PROPANOATE (39) AND PENTAMETHYLENEBIS (MAGNESIUM BROMIDE)

A cyclohexanol can be constructed by addition of the bis-Grignard reagent, pentamethylenebis(magnesium bromide), to an ester. The requisite ester, ethyl 3-(dimethylamino)-2-(4-methoxyphenyl)propanoate (**39**), is available from 4-methoxyphenylacetic acid. Fischer–Speier esterification of 4-methoxyphenylacetic acid with ethanol and a sulfuric acid catalyst (89%) affords ethyl 4-methoxyphenylacetate.[60]

Ethyl 3-(dimethylamino)-2-(4-methoxyphenyl)propanoate (**39**) can be produced in one, two or three steps (Scheme 4.13). In the one-step route, ethyl 4-methoxyphenylacetate is condensed with *N,N*-dimethylmethylene ammonium iodide. In the two-step route, ethyl 4-methoxyphenylacetate is condensed with dimethylformamide diethyl acetal and the resulting acrylate **40** is reduced. In the three-step route, ethyl 4-methoxyphenylacetate is condensed with ethyl formate, dimethylamine is introduced, and the acrylate **40** is reduced.

### 4.7.1 Ethyl 2-(4-Methoxyphenyl)-3-oxopropanoate (41)

Sodium hydride (60% oil dispersion) (2.0 equivalents) is added in portions to a mixture of ethyl 4-methoxyphenylacetate and ethyl formate (6.6 equivalents) at 0–5°C over 3–3.5 h. The mixture is aged at 25°C for 2.5 h and then cooled to 0–5°C and carefully quenched by adding water at <20°C. Hydrochloric acid (5 M) is added (to pH 5–6) followed by dichloromethane. The layers are separated and the aqueous layer is extracted with dichloromethane. The combined organic layers are washed with water, dried, and concentrated at reduced pressure to afford crude ethyl 2-(4-methoxyphenyl)-3-oxopropanoate (**41**) (98%).[60]

### 4.7.2 Ethyl 3-(Dimethylamino)-2-(4-methoxyphenyl) acrylate (40)

Potassium carbonate (0.57 equivalents) is added to a mixture of ethyl 2-(4-methoxyphenyl)-3-oxopropanoate (**41**), dimethylamine hydrochloride (4.0 equivalents), and ethanol. The resulting solution is aged at 25°C for 72 h. The solution is concentrated at reduced pressure and the residue is suspended in water. Hydrochloric acid (5 M) is added (to pH 5–6) and the mixture is extracted with dichloromethane. The extracts are dried and concentrated at reduced pressure to afford ethyl 3-(dimethylamino)-2-(4-methoxyphenyl)acrylate (**40**) (99%).

In the alternative preparation, a solution of ethyl 4-methoxyphenylacetate and dimethylformamide diethyl acetal[61] (6.0 equivalents) is refluxed (135°C) for 20 h. The remaining starting materials are then distilled under high vacuum to afford ethyl 3-(dimethylamino)-2-(4-methoxyphenyl)acrylate (**40**) (83%).[60]

### 4.7.3 Ethyl 3-(Dimethylamino)-2-(4-methoxyphenyl) propanoate (39)

The acrylate reduction can be accomplished with sodium borohydride (21%) or lithium aluminum hydride (68%) or by catalytic hydrogenation using a platinum catalyst (74–82%). Platinum on carbon (4.83%, 59.7% water wet)

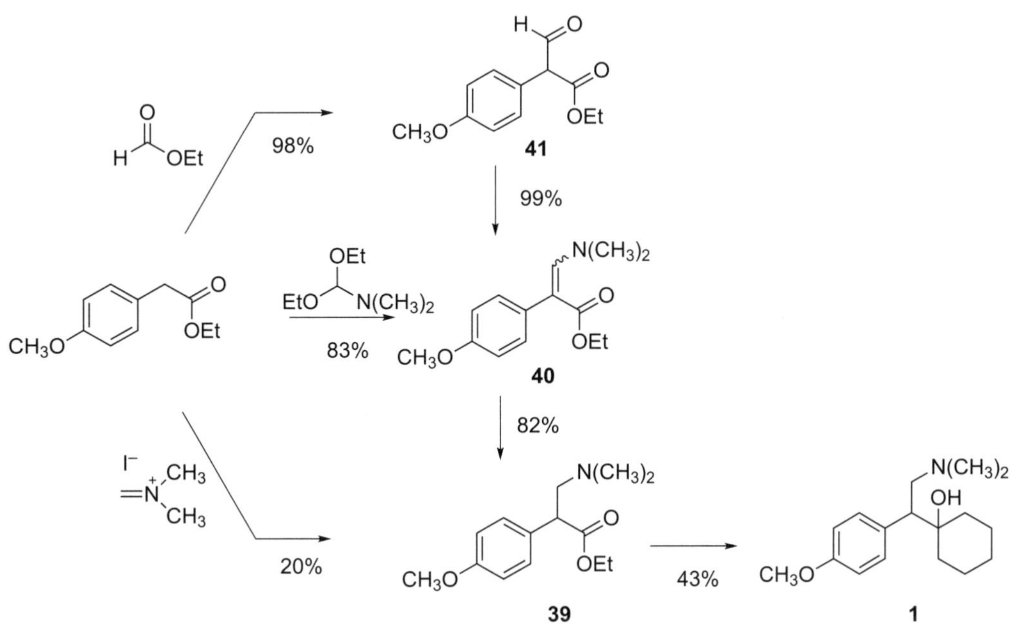


**SCHEME 4.13** Venlafaxine (**1**) from ethyl 3-(dimethylamino)-2-(4-methoxyphenyl)propanoate (**39**).

(31 g Pt/kg **40**) is added to ethyl 3-(dimethylamino)-2-(4-methoxyphenyl)propanoate (**40**) in ethanol (48 L/kg **40**). The suspension is aged at 25°C and 200 psi hydrogen pressure for 72 h. The suspension is filtered and the recovered catalyst is washed with ethanol. The combined liquors are concentrated at reduced pressure. The residue (*Note*: There is an error in the weight of the residue.) is suspended in 2 M hydrochloric acid and washed with dichloromethane. Sodium and potassium carbonates are added to the aqueous layer (to pH 11) and the free base is extracted with dichloromethane. The extracts are dried and concentrated at reduced pressure to afford ethyl 3-(dimethylamino)-2-(4-methoxyphenyl)propanoate (**39**) (82%). The yield is lower (74%) when the hydrogen pressure is increased and the age time decreased (600 psi hydrogen for 23 h).

The one-step preparation of ethyl 3-(dimethylamino)-2-(4-methoxyphenyl)propanoate (**39**) from ethyl 4-methoxyphenylacetate is not a viable option. *N,N*-Dimethylmethylene ammonium iodide[62] is expensive and the yield in the alkylation is just 20%.[60]

### 4.7.4 Venlafaxine (1) from Ethyl 3-(Dimethylamino)-2-(4-methoxyphenyl)propanoate (39)

In the key final step, a solution of ethyl 3-(dimethylamino)-2-(4-methoxyphenyl)propanoate (**39**) in THF is added at 10–20°C to a 0.5 M solution of pentamethylenebis(magnesium bromide)[63] (1.3 equivalents) in THF. The mixture is aged at 25°C for 3.5 h. The mixture is cooled to 0–10°C, and quenched by adding water, and concentrated at reduced pressure to remove THF. The resulting mixture is diluted with water and 50% sodium hydroxide is added (to pH 12–13). The free base is extracted into dichloromethane. Some insoluble solid (perhaps interface rag) is filtered and the layers are separated. The extract is concentrated at reduced pressure. The residue is dissolved in ethyl acetate and hydrogen chloride in ethanol (3.4 M) is added, presumably at 25°C. The suspension is filtered and the solid is presumably washed with ethyl acetate and dried to afford venlafaxine (**1**) hydrochloride (43%). The overall yield from 4-methoxyphenylacetic acid is at best 30%.[60]

## 4.8 VENLAFAXINE (1) FROM 2-(4-METHOXYPHENYL)-1-OXASPIRO[2.5]OCTANE-2-CARBONITRILE (42)

Hydrogenolysis of the benzylic epoxide in 2-(4-methoxyphenyl)-1-oxaspiro[2.5]octane-2-carbonitrile (**42**) is the key step in another novel approach to the venlafaxine skeleton (Scheme 4.14). There are three routes to epoxide **42**. The shortest route proceeds via dehydration of

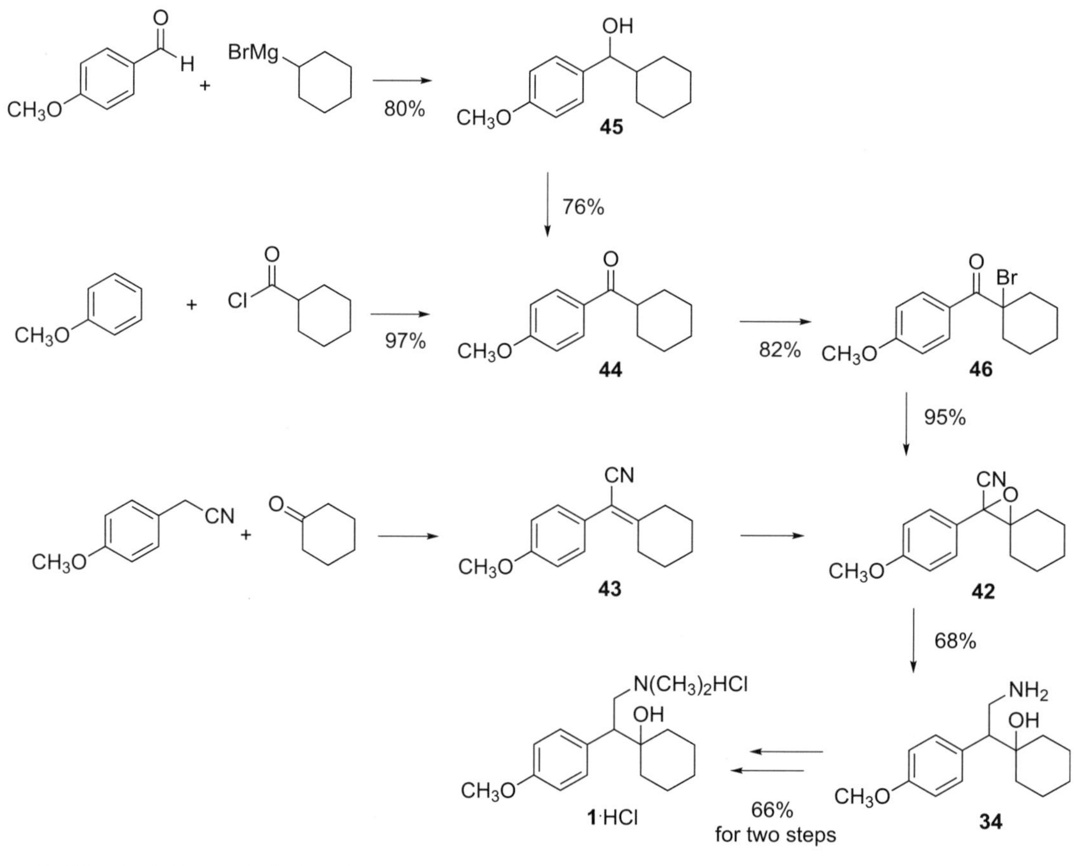


**SCHEME 4.14** Venlafaxine (**1**) from 2-(4-methoxyphenyl)-1-oxaspiro[2.5]octane-2-carbonitrile (**42**).

2-(1-hydroxycyclohexyl)-2-(4-methoxyphenyl)acetonitrile (**17**) and epoxidation of the resulting alkene **43**. Alternative three- and four-step routes generate the carbon skeleton from cyclohexyl(4-methoxyphenyl)methanone (**44**) and sodium cyanide.

### 4.8.1 2-(4-Methoxyphenyl)-1-oxaspiro[2.5]octane-2-carbonitrile (42) from 4-Methoxyphenylacetonitrile and Cyclohexanone

Reaction of 4-methoxyphenylacetonitrile, cyclohexanone, and sodium methoxide in methanol at 25°C for 6 h affords 2-cyclohexylidene-2-(4-methoxyphenyl)acetonitrile (**43**). Potassium *tert*-butoxide in *tert*-butanol can be used in place of methoxide in methanol. The experimental details and yields are not available. Epoxidation of alkene **43** with 3-chloroperbenzoic acid[64] (MCPBA) (1.0 equivalent) in dichloromethane requires 8 h at reflux. (*Note*: The purity of MCPBA is not provided. Equivalents are calculated based on 77% purity.) The suspension is cooled to 25°C and filtered to remove 3-chlorobenzoic acid. The liquors are washed with saturated sodium bicarbonate solution and then dried and concentrated at reduced pressure. 2-(4-Methoxyphenyl)-1-oxaspiro[2.5]octane-2-carbonitrile (**42**) is isolated from the residue by chromatography. The yield is not available.[55]

### 4.8.2 2-(4-Methoxyphenyl)-1-oxaspiro[2.5]octane-2-carbonitrile (42) from 4-Methoxybenzaldehyde, Cyclohexylmagnesium Bromide, and Cyanide

Cyclohexylmagnesium bromide is prepared by slow addition of bromocyclohexane[65] (1.5 equivalents) to a suspension of magnesium turnings (1.8 equivalents) in THF at reflux. The mixture is refluxed for 1 h and then cooled to 5–10°C. 4-Methoxybenzaldehyde[66] is added at 5–15°C. The mixture is allowed to warm to 25°C and aged at 25°C for 6 h. The mixture is quenched into aqueous ammonium chloride and the layers are separated. The aqueous layer is extracted with ethyl acetate. The combined organic layers are washed with brine, dried, and concentrated at reduced pressure to afford crude cyclohexyl(4-methoxyphenyl)methanol (**45**) (80%). Oxidation of the alcohol with sodium dichromate affords crude cyclohexyl(4-methoxyphenyl)methanone (**44**) (76%).

A mixture of ketone **44** and trimethylphenylammonium tribromide (1.0 equivalent) in THF is refluxed for 2–3 h. The mixture is presumably cooled and then poured into water. The resulting suspension is filtered and the solid presumably washed with water and dried to afford crude (1-bromocyclohexyl)(4-methoxyphenyl)methanone (**46**) (82%).

The α-bromoketone **46** is added to a solution of sodium cyanide (1.2 equivalents) in methanol at 25°C. The solution is aged at 25°C for 2 h and then poured into water. The mixture is extracted with dichloromethane. The extracts are washed three times with water and then dried and concentrated at reduced pressure to afford crude 2-(4-methoxyphenyl)-1-oxaspiro[2.5]octane-2-carbonitrile (**42**) (95%).[55]

### 4.8.3 2-(4-Methoxyphenyl)-1-oxaspiro[2.5]octane-2-carbonitrile (42) from Anisole, Cyclohexanecarbonyl Chloride, and Cyanide

The Friedel–Crafts reaction of anisole with cyclohexanecarbonyl chloride is a preferred approach to cyclohexyl(4-methoxyphenyl)methanone (**44**). Cyclohexanecarbonyl chloride[67] (1.1 equivalents) is added, presumably at 25°C, to a suspension of aluminum chloride (1.2 equivalents) in 1,2-dichloroethane. Anisole[68] is then added, again at 25°C, and the mixture is aged at 25°C overnight. The mixture is quenched into water at 0–5°C. The layers are separated and the aqueous phase is extracted with dichloromethane. The organic layers are washed with 1 M sodium hydroxide and with water and then dried and concentrated at reduced pressure to afford cyclohexyl(4-methoxyphenyl)methanone (**44**) (97%).[69]

### 4.8.4 1-(2-Amino-1-(4-methoxyphenyl)ethyl) cyclohexanol (34)

2-(4-Methoxyphenyl)-1-oxaspiro[2.5]octane-2-carbonitrile (**42**) is added to a suspension of Raney nickel (200 g/kg **42**) in ethanol–ammonia (equivalents of ammonia not specified) and the mixture is aged at 25°C and 70–75 psi hydrogen pressure for 7 h. The suspension is filtered and the liquors are concentrated at reduced pressure. The residue is dissolved in isopropanol and hydrogen chloride in isopropanol is added. The suspension is filtered. The solid is dissolved in water and the solution is washed with dichloromethane. (*Note*: In the patent procedure, the filtrate (liquors) is taken up in water.) Sodium hydroxide solution is then added to the aqueous layer (presumably to pH 10–11) and the free base is extracted with ethyl acetate. The extracts are concentrated at reduced pressure and the dry residue is dissolved in ethyl acetate. Dry hydrogen chloride gas is added with cooling (presumably at <25°C) and the suspension is filtered. The solid is suspended in ethyl ether and the suspension is filtered. The solid is dried to afford 1-(2-amino-1-(4-methoxyphenyl)ethyl)cyclohexanol (**34**) hydrochloride (68%).[55]

## 4.9 POLYMORPHS OF VENLAFAXINE (1) HYDROCHLORIDE

After many months at the bench developing and piloting a process that delivers the target molecule in acceptable overall yield and high purity, is it time for the process group to celebrate? Borrowing a quote from a colleague: *you may be wearing the coat but you have yet to meet the bear*. One critically important task remains: the search for polymorphs. You may be holding in your hands a bottle of the most beautiful needles you have ever produced and wondering why this is so important. If the polymorph you have prepared is not the most stable polymorph, rest assured that some day the most stable polymorph will be prepared. Perhaps it will be prepared and patented by a competitor. Or perhaps it will be found in one of your production batches 5 years later. You will receive the news from the formulation group. You will be assigned to "troubleshoot" the production process to stop making this new polymorph. The first option will be to rigorously clean the entire production train of every trace of this new polymorph. When this fails, the second option is to move the process to a production facility in another state or on another continent. Both options are time-consuming and expensive. Both options are undertaken under incredible time pressure to resume production. Both options are stop-gap measures that will only allow you to resume production until the new polymorph reappears.What methods do you use and what equipment do you need to search for polymorphs? The methods and equipment used to find and characterize new polymorphs of venlafaxine (**1**) hydrochloride serves as a case study.

There are many processes for producing polymorphs of racemic venlafaxine (**1**) hydrochloride and more are published every year. Polymorphs are often characterized by some combination of differential scanning calorimetry (DSC), X-ray diffraction (XRD), and infrared spectroscopy (IR). When a polymorph is described in the patent literature, the process for producing it and the characterization data are usually not compared and contrasted with processes and data for the known polymorphs.

To deconvolute a bewildering array of polymorph literature for a drug target, begin by attaching the patent number to the assigned name to give each polymorph a unique identification number. Form I in US20080167498 then becomes Form I-US20080167498 and Form A in US7045661 becomes Form A-US7045661. Next, assume all crystallization processes start from a polish-filtered solution so that the polymorph form going into a crystallization has no bearing on the polymorph coming out of a crystallization. Then sort processes by the solvent(s) used for crystallization or for aging a suspension. At this point, a head-to-head review of the procedures might suggest that two nearly identical procedures should afford the same polymorph. Compare and contrast the characterization data for both polymorphs to confirm or reject the hypothesis. In some cases, two nearly identical processes generate polymorphs that do not have the same XRD data. This leads to the conclusion that *the processes offered in the patents are not sufficiently defined to be differentiated or reproduced*. There are certainly many other variables beside solvent that may influence which polymorph is produced. These include the age time, maximum temperature during the age time, atmosphere over the mixture during the age time, temperature of the suspension when filtered, solid drying process (static or agitated), and drying temperature. After completing the head-to-head process review, sort the matching polymorphs and replace their polymorph designations with a new polymorph designation assigned to the matching set. Finally, review the procedures for consistent results and mine the results for useful process trends.

The available XRD data suggests there are five venlafaxine hydrochloride polymorphs. These are collected in Table 4.2. There are striking similarities in the XRD data for a mixture of Forms I and II-WO2002046140, Form I-WO2003050074, and Form II-US20080167498. These are assigned as mixtures of polymorphs I and II.

### 4.9.1 Solvents for Producing Venlafaxine (1) Hydrochloride-Polymorph not Specified

The processes for generating venlafaxine hydrochloride typically charge hydrogen chloride as a solution in isopropanol to a solution of venlafaxine free base. Suitable solvents for venlafaxine free base are isopropanol, acetone, toluene, ethyl acetate, and isopropyl acetate. Venlafaxine hydrochloride is often upgraded by crystallization from isopropanol or from ethyl acetate–methanol.

### 4.9.2 Solvents for Producing Specific Polymorphs of Venlafaxine (1) Hydrochloride

#### *4.9.2.1 Isopropanol*

*Salt Preparation with Hydrogen Chloride* A solution of hydrogen chloride in isopropanol is added to a solution of venlafaxine (**1**) in isopropanol at 30°C (to pH 1–1.5). The mixture is aged at 60°C for 60–90 min and then cooled to 10°C and aged for 60 min. The suspension is filtered and the solid is washed with isopropanol and dried to afford venlafaxine hydrochloride Form I-WO2006035047 (84% for methylation and salt formation). The XRD data provided is in a format that does not allow direct comparison to other XRD data.[27]

A solution of dry hydrogen chloride in isopropanol is added to a solution of venlafaxine (**1**) in isopropanol at 25–30°C (to pH 2). The suspension is heated to 64°C to

**TABLE 4.2 Polymorphs of Venlafaxine (1) Hydrochloride**

| Polymorph I | Polymorph II | Polymorph III | Polymorph IV | Polymorph V |
|---|---|---|---|---|
| II-WO2002046140 | I-WO2002046140 | III-WO2003050074 | hyd. US20030114536 | anh. US7030164 |
| II-WO2003042161 | I-WO2003042161 | II-US7030164 | 4-Ref.76 | A-7045661 |
| III-WO2003042161 | II-WO2003050074 | | D-7045661 | 6-Ref.75 |
| I-US7030164 | B-US7045661 | | I-US20080167498 | |
| C-US7045661 | III-US20080167498 | | | |

produce a solution. The solution is slowly cooled to 30°C. The resulting suspension is filtered and the solid is washed with isopropanol and dried at 60°C and reduced pressure to afford a mixture of venlafaxine hydrochloride polymorphs I and II (60%). XRD, IR, and solid-state $^{13}C$ NMR data are provided.[38]

A solution of dry hydrogen chloride in isopropanol is added to a solution of venlafaxine (**1**) in isopropanol, presumably at 25°C (to pH 7). The mixture is heated to produce a clear solution and then slowly cooled to 10°C. The suspension is filtered and the solid is washed with isopropanol and dried to afford venlafaxine hydrochloride polymorph IV. DSC and XRD data are provided. Other results suggest polymorph IV, the monohydrate, is formed when the solid was dried at 50–60°C using a rotary evaporator.[57]

A solution of dry hydrogen chloride in isopropanol is added to a solution of venlafaxine (**1**) in isopropanol at 0–5°C (to pH 2). The suspension is aged at 0–5°C for 1 h and filtered. The solid is washed with isopropanol and dried at 60°C and reduced pressure to afford a mixture of venlafaxine hydrochloride polymorphs I and II (80%). XRD, IR, and solid-state $^{13}C$ NMR data are provided.[38]

*Crystallization of Venlafaxine (1) Hydrochloride* Venlafaxine (**1**) hydrochloride is dissolved in isopropanol at reflux. The solution is aged at 25°C (air atmosphere) overnight and the suspension is filtered. The solid is polymorph II. (*Note*: The procedure reads Form II in the text but Form III in the title.) DSC and XRD data are provided.[57]

Venlafaxine (**1**) hydrochloride is dissolved in isopropanol at reflux. The solution is filtered hot and then cooled to 25°C. The resulting suspension is aged at 25°C for 26 h and then filtered. The solid is dried to afford venlafaxine hydrochloride polymorph II (75%). DSC, XRD, and IR data are provided.[72]

Venlafaxine (**1**) hydrochloride is dissolved in isopropanol at reflux. The solution is cooled to 0°C and the suspension is filtered. The solid is presumably washed with isopropanol and dried to afford polymorph I. XRD data is provided.[73]

Venlafaxine (**1**) hydrochloride is dissolved in isopropanol at 70°C. The solution is cooled to 63°C and seeded with polymorph II. The suspension is slowly cooled to 25°C and filtered. The solid is dried to afford polymorph II. XRD data is provided.[73]

Venlafaxine (**1**) hydrochloride is dissolved in isopropanol at reflux. The solution is quickly cooled until crystallization commences then slowly cooled to 2–4°C. The suspension is aged at 2–4°C for 30 min and filtered. The solid is washed with cold isopropanol and dried at 35–40°C and reduced pressure to afford polymorph II (80%). XRD data is provided.[74]

A mixture of venlafaxine (**1**) hydrochloride polymorph I in isopropanol is refluxed for 30 min. The suspension is cooled to 0–5°C, aged at 0–5°C for 1 h, and filtered. The solid is washed with cold isopropanol and dried at 35–40°C and reduced pressure to afford polymorph I (62%). XRD data is provided.[74]

Venlafaxine (**1**) hydrochloride is dissolved in isopropanol at reflux. The solution is cooled to 32°C and seeded with polymorph II. The suspension is cooled to 0–5°C, aged for 30 min at 0–5°C, and filtered. The solid is washed with cold isopropanol and dried at 35–40°C and reduced pressure to afford polymorph II (75%). XRD data is provided.[74]

Venlafaxine (**1**) Hydrochloride is dissolved in isopropanol at reflux. The solution is cooled to 25–30°C and the resulting suspension is aged at 25–30°C for 1 h and filtered. The solid is washed with isopropanol and dried at 60°C and reduced pressure to afford a mixture of polymorphs I and II (76%). XRD, IR, and solid-state $^{13}C$ NMR data are provided.[38]

Venlafaxine (**1**) hydrochloride is dissolved in isopropanol at reflux. The solution is cooled to 5–10°C and the suspension is filtered. The solid is washed with isopropanol and dried to afford a mixture of polymorphs I and II (83–86%). XRD, IR, and solid-state $^{13}C$ NMR data are provided.[38]

Venlafaxine (**1**) Hydrochloride is dissolved in isopropanol at reflux. The solution refluxed for 1 h then cooled in stages to 50°C and to 30°C. The resulting suspension is aged at 25°C for 1 h and filtered. The solid is washed with isopropanol and dried at 60°C and reduced pressure to afford a mixture of polymorphs I and II (91%). XRD, IR, and solid-state $^{13}C$ NMR data are provided.[38]

*Aging a Suspension at 25°C* A suspension of venlafaxine (**1**) hydrochloride in isopropanol is aged, presumably at 25°C, for 72 h. The suspension is filtered and the solid is dried to afford polymorph II. XRD data is provided.[73]

A suspension of venlafaxine (**1**) hydrochloride polymorph I in isopropanol (0.01% $H_2O$) is aged at 20–25°C for 110 h and then filtered. The solid is washed with cold isopropanol and dried at 35–40°C and reduced pressure to afford polymorph II (80%). The yield drops as the water content in the isopropanol increases (66% yield at 0.42% $H_2O$; 60% yield at 2.0% $H_2O$). XRD data is available for both polymorphs.[74]

***4.9.2.2 Ethanol*** Venlafaxine (**1**) hydrochloride is dissolved in ethanol at reflux. The solution is aged at 25°C (air atmosphere) overnight and the resulting suspension is filtered. The solid is polymorph II. Polymorph II is converted to the monohydrate, polymorph IV, when dried in a rotary evaporator at 60°C and reduced pressure. DSC and XRD data are provided.[57]

A suspension of venlafaxine (**1**) hydrochloride polymorph I in ethanol (0.12% $H_2O$) is aged at 20–25°C for 100 h and filtered. The solid is washed with cold ethanol and dried at 35–40°C and reduced pressure to afford polymorph II (44%). The yield drops as the water content in the ethanol increases (20% yield at 2.0% $H_2O$). XRD data is available for both polymorphs.[74]

***4.9.2.3 Ethyl Acetate*** Venlafaxine (**1**) is dissolved in ethyl acetate at 25–30°C. Dry hydrogen chloride gas is added (to pH 2). The suspension is refluxed for 1 h, cooled to 25–30°C, and filtered. The solid is washed with ethyl acetate and dried at 60°C and reduced pressure to afford venlafaxine hydrochloride polymorph II (89%). XRD, IR, and solid-state $^{13}C$ NMR data are provided.[38]

Venlafaxine (**1**) is dissolved in ethyl acetate at 25–30°C. The solution is cooled to 0–5°C and dry hydrogen chloride gas is added (to pH 2). The suspension is aged at 0–5°C for 1 h and filtered. The solid is washed with ethyl acetate and dried at 60°C and reduced pressure to afford a mixture of venlafaxine hydrochloride polymorphs I and II (93%). XRD, IR, and solid-state $^{13}C$ NMR data are provided.[38]

***4.9.2.4 Ethyl Acetate–Methanol*** Venlafaxine (**1**) hydrochloride is dissolved in methanol at 0–5°C. Ethyl acetate is added. The resulting suspension is aged for 30 min, presumably at 0–5°C, and filtered. The solid is a mixture of polymorphs I and II. The solid is dried in a rotary evaporator at 60°C and reduced pressure to afford a mixture of polymorphs I and II. DSC and XRD data are provided.[57]

Venlafaxine (**1**) hydrochloride is dissolved in methanol at 0°C. Ethyl acetate is added to produce a suspension with a 15 : 1 (v/v) ethyl acetate–methanol v/v ratio at 0–5°C. The suspension is aged, presumably at 0–5°C, for 24 h and then filtered. The fine needles are dried to afford polymorph I (78%). DSC, XRD, and IR data are provided.[72]

Venlafaxine hydrochloride is dissolved in methanol at reflux. Antisolvent (ethyl acetate, diisopropyl ether, or methyl *tert*-butyl ether) is added. The suspension is refluxed for 10 min, aged at 25°C (air atmosphere) overnight, and filtered. The solid is polymorph II. The solid is washed with methanol–antisolvent and dried in a rotary evaporator at 60°C and reduced pressure to afford the monohydrate, polymorph IV. DSC and XRD data are provided for both forms.[57]

Venlafaxine (**1**) hydrochloride polymorph II is presumably dissolved in 4 : 1 ethyl acetate–methanol at reflux for 30 min. The solution is cooled to 0–5°C. The suspension is aged at 0–5°C for 1 h and filtered. The solid is washed with cold 4 : 1 ethyl acetate–methanol and dried at 35–40°C and reduced pressure to afford polymorph I (42–60%). XRD data is available for both polymorphs.[74]

***4.9.2.5 Ethyl Acetate–Ethanol*** Hydrogen chloride (0.95 equivalents) is added to a solution of venlafaxine (**1**) in 1 : 1 (w/w) ethyl acetate–ethanol at 68–73°C and the resulting mixture is aged at 68–73°C for 15 min (pH 2.5). Half the weight of solvent is removed by azeotrope distillation at atmospheric pressure. The resulting suspension is diluted with ethyl acetate and the distillation is repeated. The dilution–distillation sequence is repeated a third time. The resulting suspension is cooled to 15°C, aged at 15°C for 15 h, and filtered. The solid is washed with 98 : 2 (w/w) ethyl acetate–ethanol and dried at 50°C and reduced pressure to afford venlafaxine hydrochloride Form I-WO2009080655 (95%, >99.8% pure). (*Note*: The polymorph is designated Form I to be consistent with the other polymorph names.) The Raman spectrum of the polymorph is provided.[49]

Anhydrous polymorph V is converted to polymorph I by crystallization from 4 : 1 ethyl acetate in ethanol. DSC and XRD data are provided.[70,71]

Venlafaxine (**1**) hydrochloride is dissolved in ethanol at reflux. Ethyl acetate is added. The suspension is refluxed for 10 min, aged at 25°C (air atmosphere) overnight, and filtered. The solid is polymorph II. DSC and XRD data are provided.[57]

***4.9.2.6 Ethyl Acetate–Water*** A suspension of venlafaxine (**1**) hydrochloride polymorph I in 97 : 3 (v/v) ethyl acetate-water is aged at 20–25°C for 224 h and filtered. The solid is washed with cold ethyl acetate and dried at 35–40°C and reduced pressure to afford polymorph I (85%). XRD data is provided.[74]

***4.9.2.7 Isopropyl Acetate*** Hydrochloric acid (2 M) is added to venlafaxine (**1**) in isopropyl acetate. Water is removed by distillation of the isopropyl acetate–water azeotrope at atmospheric pressure. A suspension forms as the water is removed. The suspension is refluxed for 90 min, cooled to 25°C, and filtered. The solid is presumably washed with isopropyl acetate and dried to afford venlafaxine (**1**) hydrochloride (88% for methylation and salt formation.) The polymorph is identified as Form I-US20090240082 but no DSC, XRD, or IR data are provided.[56]

***4.9.2.8 Isopropyl Acetate–Ethanol*** Venlafaxine (**1**) hydrochloride Form I-US20090240082 seed crystals are added to a solution of venlafaxine (**1**) in isopropyl acetate. Hydrogen chloride (2.5 M) in ethanol is added over 30 min, presumably at 25°C. The resulting suspension is aged at 25°C for 2 h and then filtered. The solid is washed with isopropyl acetate and dried to afford venlafaxine (**1**) hydrochloride (88% for methylation and salt formation.) The polymorph is identified as Form I-US20090240082 but no DSC, XRD or IR data are provided.[56]

***4.9.2.9 Toluene*** Venlafaxine (**1**) is dissolved in toluene at 30–35°C. Dry hydrogen chloride gas is added (to pH 2). The suspension is refluxed for 1 h, cooled to 30°C, aged at 30–35°C for 1 h, and filtered. The solid is washed with toluene and dried at 60°C and reduced pressure to afford venlafaxine hydrochloride polymorph II (94%). XRD, IR, and solid-state $^{13}C$ NMR data are provided.[38]

***4.9.2.10 Toluene–Isopropanol*** Venlafaxine (**1**) is dissolved in toluene at 30–35°C. A solution of dry hydrogen chloride in isopropanol is added (to pH 2). The suspension is refluxed for 1 h, cooled to 30–35°C, aged at 30–35°C for 1 h, and filtered. The solid is presumably washed with toluene and dried at 60°C and reduced pressure to afford venlafaxine hydrochloride polymorph III (83%). XRD, IR, and solid-state $^{13}C$ NMR data are provided.[38]

***4.9.2.11 Acetone*** Venlafaxine (**1**) is dissolved in acetone at 25–30°C. Dry hydrogen chloride gas is added (to pH 2). The suspension is refluxed for 5 min, cooled to 25–30°C, and filtered. The solid is washed with acetone and dried at 60°C and reduced pressure to afford venlafaxine hydrochloride polymorph II (82%). XRD, IR, and solid-state $^{13}C$ NMR data are provided.[38]

Venlafaxine (**1**) is dissolved in acetone at 25–30°C. The solution is cooled to 0–5°C and dry hydrogen chloride gas is added (to pH 2). The suspension is aged at 0–5°C for 1 h and filtered. The solid is washed with cold acetone and dried at 60°C and reduced pressure to afford venlafaxine hydrochloride polymorph II (87%). XRD, IR, and solid-state $^{13}C$ NMR data are provided.[38]

Venlafaxine (**1**) hydrochloride is suspended in acetone. The suspension is aged at 60°C for 1 h and at 0°C for 1 h and filtered. The solid is washed with cold acetone and dried with agitation, possibly in a rotary evaporator, at 50°C and reduced pressure to afford the monohydrate, polymorph IV (93%). DSC and XRD data are provided.[57]

Venlafaxine (**1**) hydrochloride is suspended in acetone. The suspension is aged at 60°C for 1 h and at 0°C for 1 h and filtered. The solid is washed with cold acetone and dried on a tray (no agitation) at 50°C and reduced pressure to afford a mixture of polymorphs I and II (93%). DSC and XRD data are provided.[57]

A suspension of venlafaxine (**1**) hydrochloride polymorph I in acetone (2.0% $H_2O$) is aged at 20–25°C for 72 h and filtered. The solid is washed with cold acetone and dried at 35–40°C and reduced pressure to afford polymorph II (87%). XRD data is provided.[74]

Venlafaxine (**1**) hydrochloride is dissolved in methanol and the solution is concentrated at reduced pressure. The residue is suspended in acetone and the suspension is concentrated at reduced pressure. The residue is again suspended in acetone. The suspension is refluxed for 30 min, aged at 30°C for 10 min, cooled to 5–10°C, and filtered. The solid is washed with acetone and dried to afford polymorph II (92%). XRD, IR, and solid-state $^{13}C$ NMR data are provided.[38]

***4.9.2.12 Acetone–Isopropanol*** Crude venlafaxine (**1**) is dissolved in acetone at 25–30°C. The solution is cooled to 5–10°C and a solution of hydrogen chloride in isopropanol is added at 5–10°C (to pH 2). The resulting suspension is aged at 5–10°C for 1 h and filtered. The solid is washed with acetone and dried at 60°C and reduced pressure to afford a mixture of venlafaxine (**1**) hydrochloride polymorphs I and II (65%). XRD, IR, and solid-state $^{13}C$ NMR data are provided.[38]

***4.9.2.13 Acetone–Water*** Venlafaxine (**1**) hydrochloride is dissolved in water at reflux. Acetone is added. The resulting suspension is refluxed for 10 min, aged at 25°C (air atmosphere) overnight, and filtered. The solid is a mixture of polymorphs I and II. The solid is washed with water–acetone and dried in a rotary evaporator at 60°C and reduced pressure to afford the monohydrate, polymorph IV. DSC and XRD data are provided for both forms.[57]

***4.9.2.14 Methyl Isobutyl Ketone*** Venlafaxine (**1**) is dissolved in methyl isobutyl ketone (MIBK) at 30–35°C. Dry hydrogen chloride gas is added (to pH 2). The resulting suspension is refluxed for 1 h, cooled to 30°C, aged at 30–35°C for 1 h, and filtered. The solid is washed with methyl isobutyl ketone and dried at 60°C and reduced pressure to afford venlafaxine hydrochloride polymorph II (92%). XRD, IR, and solid-state $^{13}C$ NMR data are provided.[38]

***4.9.2.15 Methyl Isobutyl Ketone–Methanol*** Venlafaxine (**1**) hydrochloride is dissolved in methanol. Methyl isobutyl ketone is added to produce a suspension with a 3.3 : 1 (v/v) MIBK–methanol ratio. The suspension is cooled to 0–5°C and filtered. The solid is dried to afford polymorph I (76%). DSC, XRD, and IR data are provided.[72]

***4.9.2.16 Methyl Isobutyl Ketone–Isopropanol*** Venlafaxine (**1**) is dissolved in methyl isobutyl ketone at 30–35°C.

The solution is cooled to 0–5°C and a solution of dry hydrogen chloride in isopropanol is added (to pH 2). The resulting suspension is aged at 0–5°C for 1 h and filtered. The solid is presumably washed with methyl isobutyl ketone and dried to afford a mixture of venlafaxine hydrochloride polymorphs I and II (83%). XRD, IR, and solid-state $^{13}C$ NMR data are provided.[38]

*4.9.2.17* ***Acetonitrile*** Venlafaxine (**1**) is dissolved in acetonitrile at 25–30°C. Dry hydrogen chloride gas is added (to pH 2). The resulting suspension is refluxed for 1 h, cooled to 25–30°C, and filtered. The solid is washed with acetonitrile and dried at 60°C and reduced pressure to afford venlafaxine hydrochloride polymorph II (51%). XRD, IR, and solid-state $^{13}C$ NMR data are provided.[38]

Venlafaxine (**1**) is dissolved in acetonitrile at 25–30°C. The solution is cooled to 5–10°C and dry hydrogen chloride gas is added (to pH 2). The resulting suspension is aged at 0–5°C for 1 h and filtered. The solid is washed with acetonitrile and dried at 60°C and reduced pressure to afford a mixture of venlafaxine hydrochloride polymorphs I and II (66%). XRD, IR, and solid-state $^{13}C$ NMR data are provided.[38]

A suspension of venlafaxine (**1**) hydrochloride polymorph I in acetonitrile (2.0% $H_2O$) is aged at 20–25°C for 48 h and filtered. The solid is washed with cold acetonitrile and dried at 35–40°C and reduced pressure to afford polymorph II (53–65%).[74]

*4.9.2.18* ***DMSO or DMF*** Venlafaxine (**1**) hydrochloride is dissolved in DMSO or DMF at reflux. The solution is aged at 25°C overnight and the suspension is filtered. The solid is Form IV-US20080167498. (*Note*: The procedure reads Form II in the text but Form IV in the title.) The solid is washed with the solvent and dried in a static oven at 160°C for 30 min to afford the monohydrate, polymorph IV. DSC and XRD data are provided for both forms.[57]

*4.9.2.19* ***DMSO–Water or DMF–Water*** Venlafaxine (**1**) hydrochloride is dissolved in water at reflux. Antisolvent (DMSO or DMF) is added. The resulting suspension is refluxed for 10 min, aged at 25°C overnight, and filtered. The solid is polymorph IV. The solid is washed with water–antisolvent and dried in a rotary evaporator at 60°C and reduced pressure to afford the monohydrate, polymorph IV. DSC and XRD data are provided.[57]

*4.9.2.20* ***Water*** Venlafaxine (**1**) hydrochloride is dissolved in water at 25°C. The water is then evaporated in air at 25°C to afford polymorph IV. XRD and Raman data are provided.[73]

Venlafaxine (**1**) hydrochloride is dissolved in water at reflux. The solution is cooled to 5°C and the resulting suspension is filtered. The crystals are dried in air for 8 h to afford polymorph IV. DSC, TGA, and XRD data are provided.[71]

*4.9.2.21* ***Other Solvents*** A mixture of venlafaxine (**1**) hydrochloride in 1,4-dioxane is refluxed until the salt dissolves. The solution is filtered hot and then slowly or rapidly cooled to 25°C. The resulting suspension is aged at 25°C for 24 h and then filtered. The solid is dried to afford polymorph II (82–83%, >99% pure). DSC, XRD, and IR data are provided.[72]

Venlafaxine (**1**) hydrochloride is dissolved in a 44 : 1 mixture (v/v) of 1,2-dichloroethane and water at 40°C. The cloudy solution is filtered and the liquors are aged at 20–25°C for 26 h. The resulting suspension is filtered and the solid is dried at 35–40°C and reduced pressure to afford polymorph I (62–73%).[74]

*4.9.2.22* ***Crystallization from the Melt*** Anhydrous polymorph V is produced by heating polymorph I or polymorph III at 200°C in glass under nitrogen for 1–2.5 h.[71]

Venlafaxine (**1**) hydrochloride polymorph I or II is held at 180–190°C to produce polymorph V by solid–solid phase transition. Polymorph V is the highest melting (219–220°C by DSC) and most stable polymorph of venlafaxine hydrochloride. Attempts to produce polymorph V by crystallization from isopropanol, ethyl acetate–methanol, ethyl acetate–ethanol, DMF, or DMSO afforded only polymorphs I and/or II.[73,75]

### 4.9.3 Polymorph Process Trends

Polymorph I is a kinetic product. It is best produced by rapid cooling of a solution and limiting the age time after producing the suspension. Polymorph II is a thermodynamic product. It is best produced by seeding a hot solution, controlled (slow) cooling to produce the suspension, and extending the age time after producing the suspension. A suspension of polymorphs I and II is converted to all-polymorph II at reflux temperatures (>60°C) over a few hours. A suspension of a mixture of polymorphs I and II is converted to all-polymorph II by aging at 5–25°C over a few days. These trends are consistent using all of the solvent systems discussed.

Polymorph III is only prepared using the toluene–isopropanol solvent system. Polymorph IV is a monohydrate produced when polymorph I or II or a mixture of polymorphs I and II is dried at 50–60°C using a rotary evaporator. Polymorph IV is also formed when venlafaxine hydrochloride is crystallized from water, DMF, or DMSO. Polymorph V is produced from the melt of polymorph I or polymorph III. Polymorph V is converted back to polymorph I (and II) by crystallization. The literature suggests venlafaxine (**1**) hydrochloride is marketed as polymorphs I and II. Our analysis suggests the target is polymorph II.[75,76]

## 4.10 THE BEST PROCESS AVAILABLE TODAY

There are just two conceptually unique methods for constructing the framework of venlafaxine: (1) by condensing cyclohexanone with 4-methoxyphenylacetic acid or its *N*,*N*-dimethylamide **5**, *N*,*N*-dimethylthioamide **15**, or nitrile and (2) by adding pentamethylenebis(magnesium bromide) to ethyl 3-(dimethylamino)-2-(4-methoxyphenyl)propanoate (**39**). The latter approach is not competitive because the Grignard reagent addition in the final step is inefficient (43%).

Assume that 4-methoxyphenylacetic acid, the *N*,*N*-dimethylamide **5**, *N*,*N*-dimethylthioamide **15**, or nitrile could all be sourced from a bulk chemical supplier. The condensation of cyclohexanone with 4-methoxyphenylacetic acid using lithium diisopropylamide (2.4 equivalents) is likely to be efficient (95%). Avoiding elimination of the tertiary alcohol during the two-step conversion of 2-(1-hydroxycyclohexyl)-2-(4-methoxyphenyl)acetic acid (**3**) to the *N*,*N*-dimethylamide **14** would be a challenge.

Results with the 4-benzyloxyphenyl analog suggest that the condensation of cyclohexanone with 2-(4-methoxyphenyl)-*N*,*N*-dimethylacetamide (**5**) could generate 2-(1-hydroxycyclohexyl)-2-(4-methoxyphenyl)-*N*,*N*-dimethylacetamide (**14**) in 82% crude yield. *n*-Butyllithium, lithium diisopropylamide, lithium hexamethyldisilazide, and isopropylmagnesium bromide are suitable bases, but the 82% yield is achieved using 2.5 equivalents of lithium hexamethyldisilazide. Reduction of the *N*,*N*-dimethylacetamide in **14** is accomplished using aluminum hydride or lithium aluminum hydride at low temperature (0–25°C) or borane–THF at reflux. Borane–THF is used in excess, with charges ranging from 1.5 to 5.0 equivalents. The results for producing (*R*)-venlafaxine (*R*)-**1**) and the new monoamine reuptake inhibitors suggest the borane–THF reduction is very efficient (>90% yield of crude **1**).

Borane–THF solutions are stable when stored at 0°C under nitrogen. *When borane–THF solutions are heated above 50°C, diborane is produced and THF is undergoes reductive cleavage*. The generation of diborane is a serious safety issue. The headspace and piping for the reactor must be purged with dry nitrogen to prevent accumulation of toxic and pyrophoric diborane. Special attention must be given to eliminating all "dead spots," areas in the headspace that are not efficiently purged. The OSHA permissible exposure limit and the ACGIH threshold limit value for diborane is 0.1 ppm (TWA). The $LC_{50}$ (human) by inhalation is 159 M ppm for 15 min. The lethal dose (human) by inhalation is 30–90 mg/$m^3$ for 4 h. The amide process is unattractive because of these and other issues associated with the use of borane–THF at reflux.

The thioamide process replaces undesirable borane–THF with Raney nickel and hydrogen. The product of the condensation of cyclohexanone with 2-(4-methoxyphenyl)-*N*,*N*-dimethylethanethioamide (**15**) is isolated by crystallization (64%). Assuming a 10–11% loss to mother liquors in the crystallization, the crude yield of **16** is 75%. Isopropylmagnesium bromide (1.1 equivalents) was the only base used. The very high Raney nickel charge (20 kg Raney nickel/kg thioamide) for the thioamide reduction is unattractive.

Bases used for the condensation of cyclohexanone with 4-methoxyphenylacetonitrile run the gamut from *n*-butyllithium to sodium hydroxide. What is the best base? The bases stronger than alkoxide (*n*-BuLi, LDA, NaH, $NaNH_2$) offer the advantage of complete conversion in the condensation but the disadvantages of high cost and/or significant storage and handling concerns. Alkoxide and hydroxide bases offer the advantages of lower cost and only routine storage and handling concerns but the disadvantage of incomplete conversion in the condensation. The alkoxide and hydroxide bases are preferred.

The condensation with cyclohexanone (1.3–1.8 equivalents) using sodium methoxide (1.5–3.0 equivalents) in methanol at 0°C requires 2–5 h for complete conversion. The reaction is quenched with water or dilute acetic acid. The suspension is filtered and the solid is then suspended in a solvent–nonsolvent mixture or crystallized from toluene (85–90%). The volume throughput can be as high as 175 g/L.

Condensation with cyclohexanone (1.4 equivalents) using a potassium *tert*-butoxide catalyst (0.2 mol%) in heptane at 25°C is run at very high concentration. High concentration offers the advantage of a fast reaction rate (<1 h for complete conversion). High concentration may also be associated with poor control of heat transfer for the fast and exothermic condensation and perhaps formation of a thick suspension that could be difficult to agitate and to transfer. The reaction is quenched with acetic acid in heptane. The suspension is filtered and the solid washed with heptane (89%, 98.6% pure by HPLC).

Condensation with cyclohexanone (1.2–1.5 equivalents) using 10% aqueous sodium hydroxide at 15–25°C requires a phase transfer catalyst. The condensation is typically run at high concentration. Again, high concentration offers the advantage of a fast reaction rate (<1 h for complete conversion). High concentration may also be associated with poor control of heat transfer for the fast and exothermic condensation and formation of a thick suspension that could be difficult to agitate and to transfer. Dilution with methanol–water is an attractive compromise. The time for complete conversion at 25°C is 15 h at a volume throughput of 80 g/L. The suspension is filtered and the solid washed with hexanes and dried (96%).

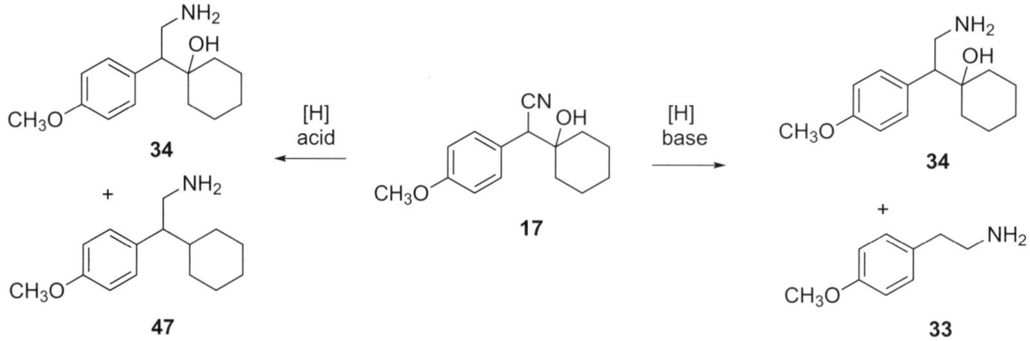


**SCHEME 4.15** Potential primary amine side products from reduction of nitrile **17** under basic and acidic conditions.

There are as many methods for the nitrile reduction as there are methods for the nitrile–cyclohexanone condensation. Reduction under basic conditions must avoid or minimize cleavage of the condensation product back to 4-methoxyphenylacetonitrile and cyclohexanone. Reduction of 4-methoxyphenylacetonitrile generates 2-(4-methoxyphenyl)ethanamine (**33**) that must be separated from the amine product **34**. Reduction under acidic conditions must minimize the formation of alkene by dehydration of 2-(1-hydroxycyclohexyl)-2-(4-methoxyphenyl)acetonitrile (**17**). Reduction of the alkene is likely to generate 2-cyclohexyl-2-(4-methoxyphenyl)ethanamine (**47**) that must be separated from the amine product **34** (Scheme 4.15).

Assuming the yields for hydride reduction and catalytic hydrogenation are comparable, none of the hydride reductions are competitive. The cleavage predominates in the attempted nitrile reduction with lithium aluminum hydride: 2-(4-methoxyphenyl)ethanamine (**33**) is the major product. Nitrile reductions with sodium borohydride and Lewis acids generate diborane. The nitrile reduction with borane–dimethyl sulfide at reflux may also generate diborane and will definitely generate odor complaints. The nitrile reduction with sodium borohydride–cobalt boride requires solid charging of cancer-causing cobalt chloride (2 equivalents) and isolation and handling of the spent cobalt boride catalyst.

Assuming the yields for the catalytic hydrogenation methods are all comparable, the methods catalyzed by rhodium on alumina and Raney cobalt are not competitive. The rhodium on alumina method is only described on 10–20 g scale using a high loading of expensive rhodium (12 g Rh/kg **17**). Rhodium metal is currently $1960 per oz or $63 per g.[77] Recycle of the rhodium catalyst is not addressed. The Raney cobalt method is only described on 1–2 g scale using a high loading of Raney cobalt.

The nitrile reduction and subsequent amine methylation can be combined in a "one-pot" process using palladium on carbon but is not known for Raney nickel. Could this be a deciding factor? The one-pot process offers the obvious advantage of one less process step. The one-pot process would also be expected to have one less isolation. In fact, the one-pot process has two isolations: venlafaxine (**1**) free base and venlafaxine (**1**) hydrochloride. Many of the two-step processes also have two isolations: the amine (**34**) hydrochloride salt and venlafaxine (**1**) hydrochloride. The known one-pot processes require hydrogen pressures (140–280 psi) not attainable in standard processing equipment. Lower hydrogen pressures (<100 psi) are used in some two-step processes. Finally, the recovery and recycle of a catalyst from a mixture containing aqueous formaldehyde is a decisive disadvantage of the one-pot process.

Nitrile **17** is reduced using a palladium on carbon catalyst using hydrochloric acid or hydrogen chloride in methanol. The palladium catalyst charge is high (31–50 g Pd/kg **17**) and there is no demonstrated recycle of the catalyst. Palladium metal is currently $348 per oz or $8 per g.[78] (Multiply the metal cost by 8 to factor in the cost of converting palladium metal to 10% palladium on carbon.[79]) The hydrogen pressure is high in one of the two processes (210–280 psi and 40–50°C). The compatibility of the acid with reactor materials of construction is an issue.[80]

Nitrile **17** is reduced using a palladium on carbon catalyst in acetic acid. The hydrogen pressures are high (210–240 psi). The palladium catalyst charge is high (11.3 g Pd/kg **17**) and there is no demonstrated recycle of the catalyst. The isolated yield of 1-(2-amino-1-(4-methoxyphenyl)ethylcyclohexanol (**34**) acetate is just 45–55%.

Nitrile **17** is reduced using a palladium on carbon catalyst in formic acid. The palladium catalyst charge is high (9.6–11.3 g Pd/kg **17**) and there is no demonstrated recycle of the catalyst. The isolated yield of 1-(2-amino-1-(4-methoxyphenyl)ethylcyclohexanol (**34**) is 65% as the formate and 61% as the hydrochloride.

Nitrile **17** is reduced using a palladium on carbon catalyst and methanesulfonic acid or sulfuric acid in methanol. In one outstanding example, the 300 kg scale reduction at 25°C and 580 psi hydrogen required just 1.25 g palladium/kg nitrile **17** and afforded 1-(2-amino-1-(4-methoxyphenyl)ethyl)cyclohexanol (**34**) in >91% yield. The yield may be 5–10% lower at 60–70 psi. The potential for generation of methyl methanesulfonate by reaction of methanol with methanesulfonic acid cannot be ignored.

Acid-catalyzed dehydration and reduction of the dehydration product to produce 2-cyclohexyl-2-(4-methoxyphenyl)ethanamine (**47**) are not specifically addressed in any of the palladium-catalyzed processes.

The results for nitrile reduction with Raney nickel suggest the virgin catalyst should be pretreated (perhaps by the vendor) to remove residual base that can catalyze cleavage back to 4-methoxyphenylacetonitrile. A series of water washes or an acid wash followed by water washes is preferred. Four recycles of the Raney nickel catalyst are demonstrated with no observed deactivation. In several cases, the yield and purity of the crude product increase with increasing Raney nickel charge. Considering the price of nickel metal, currently \$7.30/lb or \$0.16/g,[81] and the demonstrated recycle of the catalyst, even nitrile reductions with 400–500 g catalyst/kg nitrile **17** are viable. At these high catalyst loadings, the nitrile reduction in methanol at 25°C is complete in 15 h at atmospheric pressure or 4 h at 60 psi.

Most of the reductions are performed in a mixture of methanol and ammonium hydroxide. In one series of reductions in 4 : 1 (v/v) methanol–28% ammonium hydroxide at 27°C and 120 psi, analysis of the crude product revealed a mixture of 1-(2-amino-1-(4-methoxyphenyl)ethyl)cyclohexanol (**34**) and 2-(4-methoxyphenyl)ethanamine (**33**) in ratios ranging from 89 : 11 to 82 : 18. The specific Raney nickel catalyst, the virgin catalyst pretreatment, and the catalyst charge could all be factors in determining the ratio of product **34** to side product **33**. Another key to selectivity is the reaction temperature. The reduction with Raney nickel (500 g/kg **17**) in 10 : 1 (v/v) methanol–28% ammonium hydroxide at 5°C and 60 psi generates a 93 : 3 mixture of product **34** to side product **33** after 45 h. The same reaction at 25°C generates a 52 : 39 mixture after 22 h. Data suggests the hydrogen pressure may also be a factor. The ratios at 27°C and 120 psi (89 : 11 to 82 : 18) are significantly better that the 52 : 39 ratio at 25°C and 60 psi. Results in other publications suggest the dramatic improvement in selectivity is not due to a difference in catalyst loading (750 g RaNi/kg **17** at 27°C and 120 psi versus 500 g RaNi/kg **17** at 25°C and 60 psi). Taking into account all the factors influencing selectivity, an acceptable compromise would be to reduce nitrile **17** using Raney nickel (500 g/kg **17**) at 10°C and at the maximum hydrogen pressure attainable in standard equipment, perhaps 90–100 psi, to achieve complete conversion to a mixture with >90 : 10 ratio in <24 h.

The Raney nickel catalyst is kept in the reactor. The methanol–ammonia solution containing 1-(2-amino-1-(4-methoxyphenyl)ethyl)cyclohexanol (**34**) is decanted and concentrated by distillation at atmospheric pressure. The residual oil is dissolved in ethyl acetate, and acetic acid (1.0 equivalent) is added to produce the acetate salt (80–85%, >99% pure by HPLC). 1-(2-Amino-1-(4-methoxyphenyl)ethyl)cyclohexanol (**34**) could also be isolated as a hydrochloride salt (94.1–95% pure by HPLC). The purity data for the isolated salts suggests the side product 2-(4-methoxyphenyl)ethanamine (**33**) is more efficiently rejected in the acetate salt isolation.

1-(2-Amino-1-(4-methoxyphenyl)ethyl)cyclohexanol (**34**) acetate could be converted to its free base and then to venlafaxine (**1**) or it can be converted directly to venlafaxine (**1**). The direct conversion is preferred. The methylation reaction charges are: 88% formic acid (6.9 equivalents), 40% aqueous formaldehyde (3.8 equivalents), and water (3.8 L/kg **34** free base). After 18 h at reflux and 1 h cooling the mixture to 25°C, there are two processing options. The acidic aqueous solution can be washed with ethyl acetate to remove a pink impurity. Base is then added and venlafaxine is extracted into ethyl acetate. Alternatively, the ethyl acetate wash can be bypassed and the pink impurity rejected by carbon treatment and/or during the isolation of venlafaxine (**1**) hydrochloride. With very little information available on venlafaxine (**1**) hydrochloride color, the bypass option is selected.

Ethyl acetate is added. Sodium hydroxide is added (to pH 10–10.5) and the layers are separated. The aqueous layer is extracted with ethyl acetate and the combined extracts are treated with carbon. Since venlafaxine (**1**) hydrochloride has considerable water solubility and the target is polymorph II, the decolorized ethyl acetate extracts are dried by distillation of the ethyl acetate–water azeotrope (bp 70.3°C, 8.5 wt% $H_2O$). The bulk of the ethyl acetate is then recovered. The level of residual ethyl acetate can be reduced further by adding isopropanol and distilling a portion of the isopropanol. It is important to keep the concentrated mixture hot until isopropanol is added to prevent crystallization of venlafaxine (**1**) free base (mp 78.3–79.5°C).

The isopropanol solution is cooled and a solution of dry hydrogen chloride in isopropanol at 25–30°C (to pH 2). The suspension is heated to 64°C to produce a solution. The solution is slowly cooled to a target temperature and seeded with polymorph II. The resulting suspension is slowly cooled to 5°C, aged at 5°C, and filtered. The solid is washed with isopropanol and dried at 60°C and reduced pressure to afford venlafaxine (**1**) hydrochloride polymorph II (84%, 99.9% pure by HPLC). It is important to dry polymorph II under a dry nitrogen atmosphere to prevent conversion to the monohydrate, polymorph IV.

The overall process converts 4-methoxyphenylacetonitrile to venlafaxine (**1**) hydrochloride in four steps in a remarkable 65–69% overall yield. There are three solid isolations and three drying operations. There are just four process solvents (ethyl acetate, methanol, isopropanol, ammonium hydroxide) all common solvents in a pharmaceutical manufacturing plant. The most significant

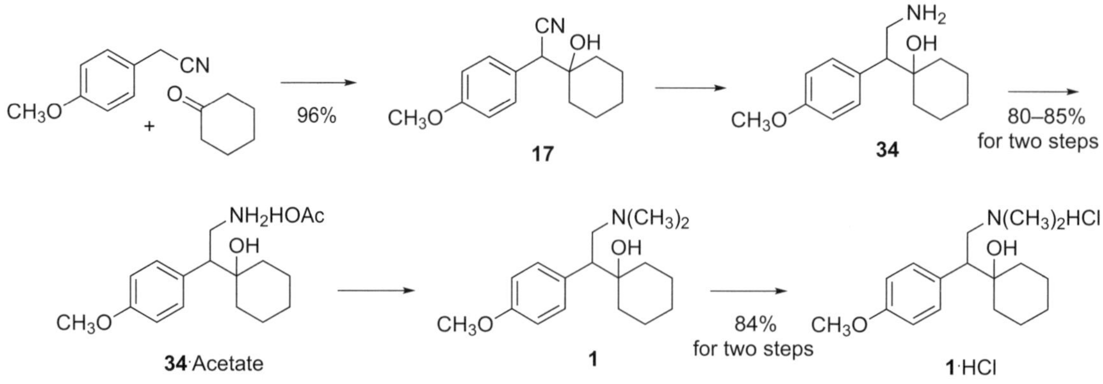


**SCHEME 4.16** The best venlafaxine (**1**) hydrochloride process available today.

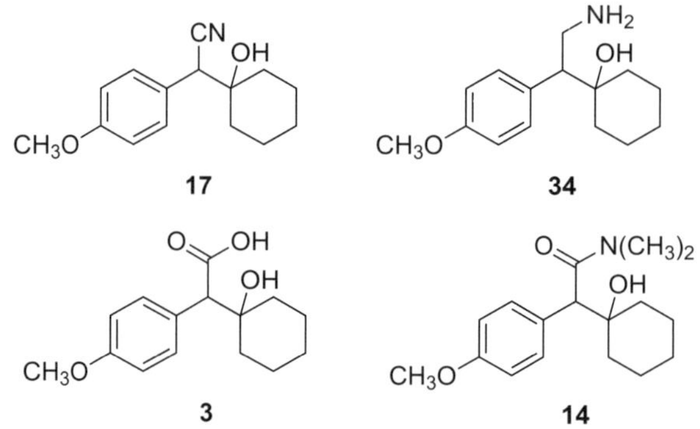


**FIGURE 4.4** Structures searched for venlafaxine (**1**) presentation.

process weakness is the use of aqueous formaldehyde (Scheme 4.16).

## 4.11 STRUCTURES SEARCHED

Four structure searches were used to generate all the information presented in this chapter (Figure 4.4).

## REFERENCES

1. (a) Wyeth News Release: *Wyeth Reports Earnings Results for 2009 First Quarter.* (b) Wyeth News Release: *Wyeth Reports Earnings Results for 2009 Second Quarter and First Half and Raises Full Year 2009 Guidance.*
2. Wyeth 2008 10K accessed at www.annualreports.com/Company/3968.
3. Accessed at www.pbs.org/wgbh/takeonestep/depression/pdf/dep_stats.pdf and www.pbs.org/wgbh/takeonestep/depression/pdf/dep_facts.pdf.
4. Accessed at www.who.int/mental_health/management/depression/definition.
5. Accessed at www.who.int/whosis/whostat2007.
6. Accessed at www.dbsalliance.org/site/PageServer?pagename=about_statistics_depression.
7. (a) Weber, B.T. U.S. Patent 7,435,854 (10/14/2008). (b) Weber, B.T. EP 1451143 (8/6/2008). (c) Weber, B.T. U.S. Patent 6,689,912 (2/10/2004). (d) Weber, B.T. WO 2003048104 (6/12/2003).
8. Niddam-Hildesheim, V.; Shenkar, N.; Shachan-Tov, S. WO 2008013995 (1/31/2008).
9. (a) Niddam-Hildesheim, V. WO 2007120923 (10/25/2007). (b) Niddam-Hildesheim, V. EP 2007708 (10/25/2007).
10. Bosch I Llado, J. WO 2008090465 (7/31/2008).
11. The Aldrich 2007–2008 catalog price for 500 g of the acid is $95.60.
12. The Aldrich 2007–2008 catalog price for 17 L is $163.50.
13. Treves, G.R. U.S. Patent 2,558,020 (6/26/1951).
14. (a) Mahaney, P.E.; Trybulski, E.J.; Gavrin, L.K.; Vu, A.T. U.S. Patent 7,531,543 (5/12/2009). (b) Mahaney, P.E. U.S. Patent 7,524,846 (4/28/2009). (c) Trybulski, E.J.; Mahaney, P.E.; Gavrin, L.K.; Moore, W.J.; Sabatucci, J.P. U.S. Patent 7,491,723 (2/17/2009). (d) Trybulski, E.J.; Sabatucci, J.P.; Vu, A.T.; Cohn, S.T.; Santilli, A.A.; Gavrin, L.K.; Ye, F.; Mahaney, P.E. U.S. Patent 7,365,076 (4/29/2008).
15. The Aldrich 2007–2008 catalog price for 2 L is $39.10.

16. The Aldrich 2007–2008 catalog price for 1 kg is $46.70.
17. Sheng, R.; Liu, T.; Hu, Y. *Zhejiang Daxue Xuebao, Yixueban* **2004**, *33*, 77.
18. Cheng, G.; Zhuo, Z.; Zhang, W.; Tang, H. CN 1240206 (1/5/2000).
19. Yardley, J.P.; Husbands, G.E.M.; Stack, G.; Butch, J.; Bicksler, J.; Moyer, J.A.; Muth, E.A.; Andree, T.; Fletcher, H.; James, M. N.G.; Sielecki, A.R. *J. Med. Chem.* **1990**, *33*, 2899.
20. (a) Husbands, G.E.M.; Yardley, J.P.; Muth, E.A. U.S. Patent 4,761,501 (8/2/1988). (b) Husbands, G.E.M.; Yardley, J.P.; Muth, E.A. U.S. Patent 4,611,078 (9/9/1986). (c) Husbands, G.E.M.; Yardley, J.P.; Muth, E.A. U.S. Patent 4,535,186 (8/13/1985).
21. (a) Gant, T.G.; Sarshar, S. U.S. Patent 7,456,317 (11/25/2008). (b) Gant, T.G.; Sarshar, S.; Woo, S.H. U.S. Patent 20080234257 (9/25/2008). (c) Gant, T.G.; Sarshar, S. WO 2007064697 (6/7/2007).
22. Shah, S.M.; Fawzi, M.B.; Ehrnsperger, E.C.; Raveendranath, P. U.S. Patent 7,595,340 (9/29/2009).
23. Gore, V.G.; Kulkarni, V.S.; Wakchaure, V.S.; Hublikar, M.G.; Wavhal, S.R. WO 2008093142 (8/7/2008).
24. Shepard, R.G. U.S. Patent 5,043,466 (8/27/1991).
25. The Aldrich 2007–2008 catalog price for 100 g is $75.60.
26. Kim, K.-S.; Kim, K.-I.; Lee, S.-W.; Park, J.-S.; Chai, K.-B. WO 2003000652 (1/3/2003).
27. Gokhale, U.B.; Parenky, C. WO 2006035457 (4/6/2006).
28. Zupancic, S.; Skrabanja, V. WO 2007147564 (12/27/2007).
29. (a) Saigal, J.C.; Gupta, R.P.; Pandit, V.V.; Desai, A.J.; Mehta, N.V.; Rane, S.H. EP 1490325 (11/5/2008). (b) Saigal, J.C.; Gupta, R.P.; Pandit, V.V.; Desai, A.J.; Mehta, N.V.; Rane, S.H. U.S. Patent 7,026,513 (4/11/2006). (c) Saigal, J.C.; Gupta, R.P.; Gorule, V.S.; Pandit, V.V.; Joshi, R.D. WO 2003080565 (10/2/2003). (d) Saigal, J.C.; Gupta, R.P.; Pandit, V.V.; Desai, A.J.; Mehta, N.V.; Rane, S.H. WO 2003080560 (10/2/2003).
30. (a) Jerussi, T.P.; Senanayake, C.H. U.S. Patent 20050197392 (9/8/2005). (b) Jerussi, T.P.; Senanayake, C.H.; Bhongle, N.N. U.S. Patent 6,911,479 (6/28/2005). (c) Jerussi, T.P.; Senanayake, C.H.; Bhongle, N.N. U.S. Patent 6.441.048 (8/27/2002). (d) Jerussi, T.P.; Senanayake, C.H. U.S. Patent 6,342,533 (1/29/2002). (e) Jerussi, T.P.; Senanayake, C.H. U.S. Patent 6,197,828 (3/6/2001). (f) Jerussi, T.P.; Senanayake, C.H. WO 2000059851 (10/12/2000). (g) Jerussi, T.P.; Senanayake, C.H. WO 2000032556 (6/8/2000). (h) Jerussi, T.P.; Senanayake, C.H. WO 2000032555 (6/8/2000).
31. Zhiyin, L.; Kaiyun, S.; Qizhuang, M.; Yulin, L. U.S. Patent 20040106818 (6/3/2004).
32. Devarakonda, S.N.; Yarraguntla, S.R.; Nalivela, V.; Thaimattam, R.; Peddireddy, S.; Raghupati, B.; Baig, M.A.; Gade, S.R.; Mallepalli, S.R.; Chitta, V.; Bollikonda, S.; Mohanarangam, S.; Kadaboina, R. WO 2009009665 (1/15/2009).
33. Soni, R.R.; Koftis, T.; Panagiotidis, T.; Georgopoulou, I. WO 2008083708 (7/17/2008).
34. Kansal, V.K.; Chaurasia, B.P.; Jaysing, M.S.; Shelke, S.H. WO 2007047972 (4/26/2007).
35. Reguri, B.R.; Kadaboina, R.; Gade. S.R.; Ireni, B. U.S. Patent 20050033088 (2/10/2005).
36. Desai, P.H.; Salvi, N.J.; Patravale, B.K.S.; Patil, S.T. WO 2007069277 (6/21/2007).
37. Marti, H.-R.; Trafelet, H. WO 2005035483 (4/21/2005).
38. Rameshchandra, S.B.; Patel, M.S.; Patel, G.B.; Ramakrishna, N.V.; Manakiwala, S.C.; Agarwal, V.K.; Pandita, K.; Patel, P.R. WO 2003050074 (6/19/2003).
39. Basappa; Kavitha, C.V.; Rangappa, K.S. *Bioorg. Med. Chem. Lett.* **2004**, *14*, 3279.
40. Chavan, S.P.; Khobragade, D.A.; Kamat, S.K.; Sivadasan, L.; Balakrishnan, K.; Ravindranathan, T.; Gurjar, M.K.; Kalkote, U.R. *Tetrahedron Lett.* **2004**, *45*, 7291.
41. (a) Chavan, S.P.; Kamat, S.K.; Sivadasan, L.; Balakrishnan, K.; Khobragade, D.A.; Thottapillil, R.; Gurjar, M.K.; Kalkote, U.R. U.S. Patent 6,504,044 (1/7/2003). (b) Chavan, S.P.; Kamat, S.K.; Balakrishnan, K.; Khobragade, D.A.; Thottapillil, R.; Gurjar, M.K.; Kalkote, U.R.; Sivadasan, L. EP 1238967 (9/11/2002). (c) Chavan, S.P.; Kamat, S.K.; Sivadasan, L.; Balakrishnan, K.; Khobragade, D.A.; Thottapillil, R.; Gurjar, M.K.; Kalkote, U.R. U.S. Patent 20020120164 (8/29/2002).
42. (a) Ekkundi, V.S.; Mumbaikar, V.N.; Paingankar, N.; van der Schaaf, P.A. EP 1313698 (11/3/2004). (b) Ekkundi, V.S.; Mumbaikar, V.N.; Paingankar, N.; van der Schaaf, P.A. WO 2002018325 (3/7/2002).
43. Rajnikant Vyas, J.; Satya Varma Nidadavolu, V. EP 1721889 (11/15/2006).
44. Ramakrishnan, A.; Subhash, V.D.; Ashwini, K.G.; Rushikesh, V.K. WO 2007094008 (8/23/2007).
45. Mahaney, P.E.; Antane, M.M.; Sun, J.S. WO 2007067501 (6/14/2007).
46. (a) Paingankar, N.; Mumbaikar, V.N.; Ekkundi, V.S.; Jalett, H.-P.; Siegrist, U.; van der Schaaf, P.A.; Bienewald, F.; Studer, M.; Burkhardt, S. U.S. Patent 7,141,697 (11/28/2006). (b) Paingankar, N.; Mumbaikar, V.N.; Ekkundi, V.S.; Jalett, H.-P.; Siegrist, U.; van der Schaaf, P.A.; Bienewald, F.; Studer, M.; Burkhardt, S. EP 1343750 (1/26/2005). (c) Paingankar, N.; Mumbaikar, V.N.; Ekkundi, V.S.; Jalett, H.-P.; Siegrist, U.; van der Schaaf, P.A.; Bienewald, F.; Studer, M.; Burkhardt, S. WO 2002050017 (6/27/2002).
47. Narayanaiyer, V.; Chaturvedi, R.; Sajikumari, B.; T-Pyati, R.S.; Yadav, S.K.; Yadav, P.K. WO 2006067808 (6/29/2006).
48. (a) Kim, K.-S.; Kim, K.-I.; Chai, K.-B. U.S. Patent 7,563,931 (7/21/2009). (b) Kim, K.-S.; Kim, K.-I.; Chai, K.-B. U.S. Patent 7,179,944 (2/20/2007). (c) Kim, K.-S.; Kim, K.-I.; Chai, K.-B. WO 2004080934 (9/23/2004). (d) Kim, K.-S.; Kim, K.-I.; Chai, K.-B. U.S. Patent 20040181093 (9/16/2004).
49. (a) Schlegel, J.; Demel, P. WO 2009080655 (7/2/2009).(b) Schlegel, J.; Demel, P. EP 2072495 (12/21/2007).
50. Teasdale, A.; Eyley, S.C.; Delaney, E.; Jacq, K.; Taylor-Worth, K.; Lipczynski, A.; Reif, V.; Elder, D.P.; Facchine, K.L.; Golec, S.; Oestrich, R.S.; Sandra, P.; David, F. *Org. Proc. Res. Dev.* **2009**, *13*, 429.
51. Srinivasan, C.V.; Siddiqui, I.H.; Wadhwa, L. WO 2007049302 (5/3/2007).

52. www.h2o2.com/applications/industrialwastewater/hcho.html.
53. www.monroeenvironmental.com/PDF/ME_Horizontal_Pack_Bed.pdf.
54. Turski, L.A.; Stoit, A.; Kruse, C.G.; Vader, S.; Tulp, M.T.M. WO 2009000797 (12/31/2008).
55. (a) Rathod, D.M.; Rengaraju, S.; Gharpure, M.M.; Patel, N.M.; Deoahar, M.M. EP 1249447 (10/22/2008). (b) Rathod, D.M.; Rengaraju, S.; Gharpure, M.M.; Patel, N.M.; Deoahar, M.M. U. S. Patent 6,756,502 (6/29/2004).
56. (a) Zupancic, S. U.S. Patent 20090240082 (9/24/2009). (b) Zupancic, S. WO 2005058796 (6/30/2005).
57. (a) Dolitzky, B.-Z.; Aronhime, J.; Wizel, S.; Nisnevich, G.A. U. S. Patent 20080167498 (7/10/2008). (b) Dolitzky, B.-Z.; Aronhime, J.; Wizel, S.; Nisnevich, G.A. U.S. Patent 6,924,393 (8/2/2005). (c) Dolitzky, B.-Z.; Aronhime, J.; Wizel, S.; Nisnevich, G. A. U.S. Patent 20040220278 (11/4/2004). (d) Dolitzky, B.-Z.; Aronhime, J.; Nisnevich, G.A. WO 2003048082 (6/12/2003). (e) Dolitzky, B.-Z.; Aronhime, J.; Wizel, S.; Nisnevich, G.A. U.S. Patent 20020183553 (12/5/2002). (f) Dolitzky, B.-Z.; Aronhime, J.; Wizel, S.; Nisnevich, G.A. U.S. Patent 20020143211 (10/3/2002). (g) Dolitzky, B.-Z.; Aronhime, J.; Wizel, S.; Nisnevich, G. A. WO 2002045658 (6/13/2002).
58. (a) Arnalot Aguilar, C.; Chamorro Gutierrez, I. WO 2008038146 (4/3/2008).(b) Arnalot Aguilar, C.; Chamorro Gutierrez, I. EP 2081885 (4/3/2008).
59. (a) Burgos, A.; Tonnel, J.; Dambrin, V.; Lucet, D.; Poirier, P. U. S. Patent 7,462,742 (12/9/2008). (b) Burgos, A.; Tonnel, J.; Dambrin, V.; Lucet, D.; Poirier, P. WO 2008062138 (5/29/2008).
60. (a) Arnalot I Aguilar, C.; Bosch I Llado, J.; Camps Garcia, P.; Onrubia Miguel, M.C.; Soldevilla Madrid, N. EP 1213279 (9/15/2004). (b) Arnalot I Aguilar, C.; Bosch I Llado, J.; Camps Garcia, P.; Onrubia Miguel, M.C.; Soldevilla Madrid, N. U.S. Patent 6,506,941 (1/14/2003). (c) Arnalot I Aguilar, C.; Bosch I Llado, J.; Camps Garcia, P.; Onrubia Miguel, M.C.; Soldevilla Madrid, N. WO 2001007397 (2/1/2001).
61. The Aldrich 2007–2008 catalog price for 50 g (95%) is $113.00.
62. The Aldrich 2007–2008 catalog price for 50 g is $188.00.
63. The Aldrich 2007–2008 catalog price for 100 mL (0.5 M) is $51.80. The Aldrich 2007–2008 catalog price for 1,5-dibromopentane is 500 g for $97.20.
64. The Aldrich 2007–2008 catalog price for 500 g ($\leq$77%) is $206.50.
65. The Aldrich 2007–2008 catalog price for 1 kg is $86.70.
66. The Aldrich 2007–2008 catalog price for 1 kg is $76.20.
67. The Aldrich 2007–2008 catalog price for 100 g is $79.70.
68. The Aldrich 2007–2008 catalog price for 20 L is $569.00.
69. Matzke, M.; Mohrs, K.-H.; Raddatz, S.; Fruchtmann, R.; Hatzelmann, A.; Kohlsdorfer, C.; Muller-Peddinghaus, R.; Theisen-Popp, P. U.S. Patent 5,399,698 (3/21/1995).
70. Sherman, D.M.; Clark, J.C.; Lamer, J.U.; White, S.A. U.S. Patent 6,274,171 (8/14/2001).
71. (a) Ali, K.A.N.; Han, J.; Lee, Y.J. U.S. Patent 7,030,164 (4/18/2006).(b) Han, J.; Lee, Y.J. U.S. Patent 20030114536 (6/19/2003).
72. Siripragada, M.R.; Krishnamurthi, V.; Arikatla, S.L.D.; Gaddam, O.R. WO 2002046140 (6/13/2002).
73. (a) van der Schaaf, P.A.; Marcolli, C.; Szelagiewicz, M.; Freiermuth, B. U.S. Patent 7,045,661 (5/16/2006). (b) van der Schaaf, P.A.; Marcolli, C.; Szelagiewicz, M.; Freiermuth, B. U. S. Patent 6,906,087 (6/14/2005). (c) van der Schaaf, P.A.; Marcolli, C.; Szelagiewicz, M.; Freiermuth, B. U.S. Patent 20030105359 (6/5/2003). (d) van der Schaaf, P.A.; Marcolli, C.; Szelagiewicz, M.; Freiermuth, B. WO 2002036542 (5/10/2002).
74. Barkoczy, J.; Kotay Nagy, P.; Simig, G.; Horvath, K.; Szent Kirallyi, Z.; Farkas, B.; Gregor, T.; Nagy, K. Krasznai, G. WO 2003042161 (5/22/2003).
75. Roy, S.; Bhatt, P.M.; Nangia, A.; Kruger, G.J. *Crystal Growth Design* **2007**, *7*, 476.
76. Roy, S.; Aitipamula, S.; Nangia, A. *Crystal Growth Design* **2005**, *5*, 2268.
77. Accessed on November 12, 2009 at www.rhodiumspotprice.com.
78. Accessed on November 12, 2009 at www.kitco.com/charts/livepalladium.html.
79. The Aldrich 2007–2008 catalog price for 50 g palladium on carbon (10% w/w, dry, Engelhard code C3645) is $314.00. The cost/g of palladium is $63.
80. www.hse.gov.uk/comah/sragtech/techmeasmaterial.htm.
81. Accessed on November 12, 2009 at www.kitco.com/charts/nickel_historical_large.html.

# 5

# SEROQUEL® (QUETIAPINE HEMIFUMARATE)

## 5.1 SEROQUEL® AND THE ANTIPSYCHOTIC MARKET

Quetiapine hemifumarate (Seroquel®) is an atypical antipsychotic. The atypical antipsychotics are a class of newer drugs thought to be free of the extrapyramidal side effects (EPS) associated with the older typical antipsychotic drugs (Figure 5.1). In fact, the atypical antipsychotics also have EPS, but to a lesser extent. Quetiapine hemifumarate is prescribed to control the symptoms of schizophrenia, bipolar depression, and bipolar mania, and for bipolar maintenance. The mechanism of action is not known. It is suggested that the efficacy in the treatment of schizophrenia and the mood stabilizing properties in both bipolar depression and mania are mediated through a combination of dopamine type 2 ($D_2$) and serotonin type 2 ($5HT_2$) antagonism. Quetiapine hemifumarate is distinguished from the other antipsychotics in that it is used to treat both the positive (hallucinations and delusions) and negative symptoms (emotional withdrawal and apathy) of psychosis and is associated with fewer neurological and endocrine-related side effects.[1]

Schizophrenia and bipolar depression are biological disorders of the brain. Schizophrenia is characterized by acute episodes of delusions, hallucinations, emotional withdrawal, apathy, and depressive signs and symptoms. The prevalence rate for schizophrenia is approximately 1.1% of the population over the age of 18. This translates to approximately 2.2 million people in the United States and 51 million people worldwide. The current combined direct and indirect costs of schizophrenia are estimated to be $63 billion. Bipolar I disorder is characterized by recurring episodes of mania and depression. Bipolar II disorder is characterized by one or more major depressive episodes accompanied by at least one episode of mild to moderate mania. It is estimated that 27 million people suffer from bipolar disorder worldwide.[2,3]

The global market for antipsychotics was $18.1 billion in 2006. AstraZeneca's Seroquel® remains the market leader in the U.S. antipsychotic market with a total prescription share of 31.6% in December 2008. Seroquel® has had phenomenal sales growth of no less than 27% per year through 2006. For the full year 2008, global sales figures were $4.452 billion, up 9% from the 2007 sales figure of $4.027 billion. Sales for the fourth quarter 2008 were $1.160 billion, up 7% from the fourth quarter 2007 figure of $1.086 billion. Global sales for 2006 were $3.4 billion and for 2005 $2.8 billion. It is estimated that Seroquel® has been used to treat more than 19 million people worldwide from launch through 2007.[4]

AstraZeneca has been working on several fronts to mitigate the effect of the quetiapine hemifumarate patent expiration in 2011. First, an extended release formulation (Seroquel XR™) was developed and AstraZeneca has exclusive marketing rights until 2017. Seroquel XR™ was launched in the United States in 2007 for the acute and maintenance treatment of schizophrenia. Second, clinical trials were designed and undertaken to demonstrate the efficacy of the extended release formulation for the treatment of bipolar mania and depression, major depressive disorder, and generalized anxiety disorder. Seroquel XR™ was launched in February 2009 for the treatment of depressive, manic, and mixed episodes of bipolar disorder as well as the maintenance treatment of bipolar I disorder as adjunctive therapy with lithium or divalproex. Third, at least 20

*Pharmaceutical Process Chemistry for Synthesis: Rethinking the Routes to Scale-Up*, By Peter J. Harrington

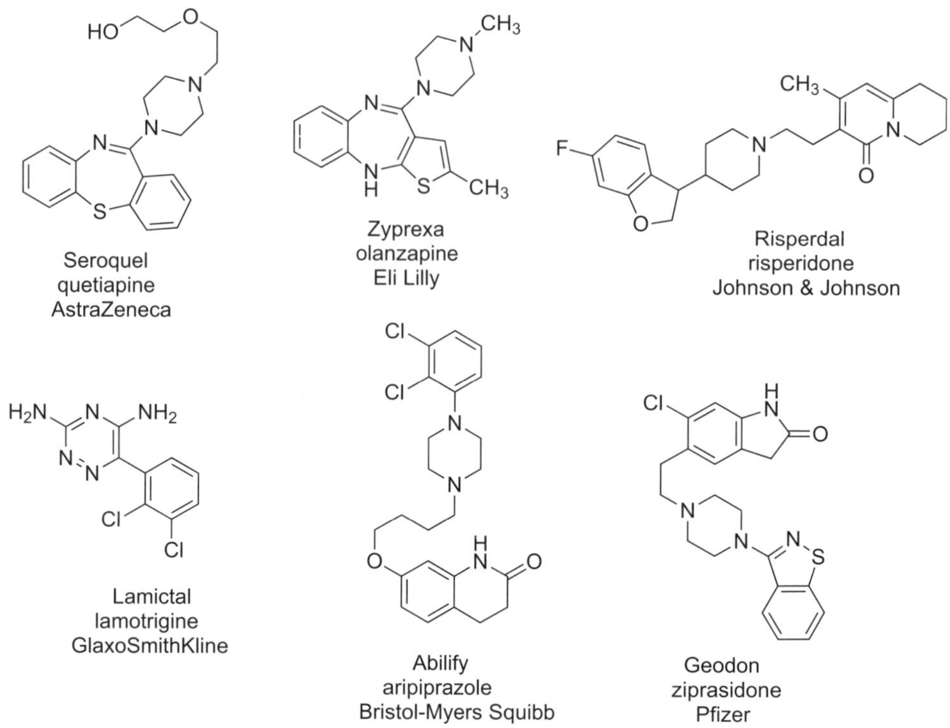


**FIGURE 5.1** Atypical antipsychotics.

metabolites of quetiapine have been identified. The metabolites are produced by conversion of the sulfide to sulfoxide, oxidation of the aromatic ring, and oxidation and cleavage reactions of the (2-hydroxyethoxy)ethyl side chain. There is little published information on the pharmacological activities of the metabolites, with one exception. 11-(Piperazin-1-yl)dibenzo[*b*,*f*][1,4]thiazepine, produced by cleavage of the side chain, may play an important role in the antidepressant activity of quetiapine through NET inhibition and partial 5-$HT_{1A}$ antagonism. AstraZeneca has recently patented several 11-(piperazin-1-yl)dibenzo[*b*,*f*][1,4]thiazepine derivatives, including carbamates, amides, and aminals, perhaps in search of a superior antidepressant.[5,6]

It would appear that the demand for quetiapine hemifumarate will continue to increase. However, there are three factors that recommend caution in predicting robust sales increases in the future. First, there are several other atypical antipsychotic agents on the market, including Zyprexa® (Eli Lilly), Risperdal® (Johnson & Johnson), Lamictal® (GlaxoSmithKline), Abilify® (Bristol-Myers Squibb), and Geodon® (Pfizer).

There is also one generic atypical antipsychotic agent. Johnson & Johnson lost patent exclusivity on Risperdal® in 2007. There is no one-size-fits-all in the antipsychotic market so there is no global initiative to reduce health care costs by switching patients from quetiapine hemifumarate to generic risperidone. Second, very high discontinuation rates are common with all the antipsychotic drugs. This suggests a new drug could quickly gain market share if it is better tolerated. Third, weight gain and hyperglycemia are clinically demonstrated side effects of quetiapine hemifumarate treatment. The proportion of patients in clinical trials meeting a weight gain criterion of >7% of body weight was 5–23% for quetiapine and 0–7% for placebo. In long-term trials of quetiapine, hyperglycemia was observed in 10.7% of patients receiving quetiapine (mean exposure 213 days) versus 4.6% in patients receiving placebo (mean exposure 152 days). Clinically significant increases in cholesterol (4–16% for quetiapine versus 2–7% for placebo) and triglycerides (8–23% for quetiapine versus 6–16% for placebo) are observed. There are 15,000 consumer claims that AstraZeneca failed to adequately inform the public about this clinical data and the weight gain side effect associated with Seroquel®. Many of the lawsuits claim a link between Seroquel® treatment and the development of type II diabetes.

## 5.2 SYNTHESIS OF DIBENZO[b,f][1,4] THIAZEPIN-11(10H)-ONE (2)

Disconnection at the C–N bond joining the piperazine and the dibenzo[*b*,*f*][1,4]thiazepine splits quetiapine (**1**) into two components of comparable complexity. The central ring of the dibenzothiazepine is created by C–N bond formation to produce an amide, imino chloride, or amidine, by C–C bond formation by acylation of an aryl thioether where the acylation reagent is generated from an isocyanate, urethane, or urea, or by C–S bond formation by biradical coupling. The

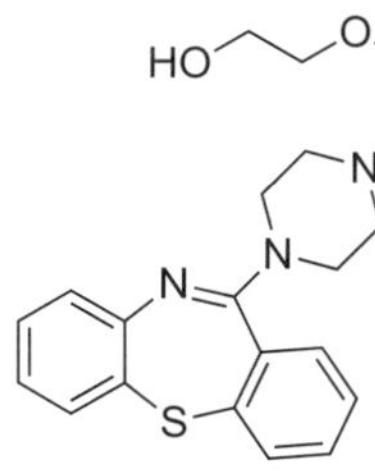

Quetiapine
1

**FIGURE 5.2** Quetiapine (**1**).

thioether for cyclization by C—N and C—C bond formation is constructed by halide displacement from a 1-halo-2-nitrobenzene by thiophenol or a thiosalicylate or from a 2-iodobenzoate or 2-fluorobenzonitrile by 2-aminothiophenol (Figures 5.2 and 5.3).

The hydroxyethoxy ethyl side chain can be introduced by a piperazine alkylation before or after creating the C—N bond joining the piperazine and the dibenzo[*b*,*f*][1,4]thiazepine and can be introduced in a single step using 2-chloroethoxyethanol or in two steps using ethylene oxide and ethylene glycol. Finally, rather than starting with an intact piperazine ring, the piperazine can be created by N-alkylation of a bis-(2-chloroethyl)amine derived from triethanolamine or diethanolamine. The symmetrical piperazine and the diethylene glycol components suggest that we will encounter statistical mixture problems somewhere along the way and that a protecting group strategy may be advantageous in certain cases. With many options introducing the piperazine and constructing the side chain, convergency will be a central theme in selecting a route for scale-up.

### 5.2.1 Constructing the Thiazepine Ring by C—C Bond Formation

Many of the methods for constructing the dibenzo[*b*,*f*][1,4]thiazepine ring were first described over 50 years ago. While it is often suggested that the heterocyclic ring can be generated by a C—S bond formation, in fact almost all of the available methods generate the ring by C—C or C—N bond formation. The C—C bond can be generated by activation of an isocyanate or carbamate followed by electrophilic aromatic substitution. The isocyanate and carbamate are produced from 2-(phenylthio)aniline (**3**), which is available in two steps from 1-chloro-2-nitrobenzene and thiophenol (Scheme 5.1). 2-Nitrodiphenylsulfide (**4**) precipitates in near quantitative yield from the reaction of 1-chloro-2-nitrobenzene,[7] thiophenol[8] (1.0 equivalent), and sodium hydroxide (1.1 equivalents) in ethanol–water at reflux.[9,10] Note that thiophenol has an objectionable garlic-like stench, is readily absorbed through the skin, and can cause cyanosis.

The nitro group is reduced with iron powder.[11] Reduction of 2-nitrodiphenylsulfide (**4**) with iron powder (3.3 equivalents) in aqueous ethanol is initiated by adding 36% hydrochloric acid (0.24 equivalents). After the exotherm subsides, the mixture is refluxed for 1 h. The iron salts are filtered and the cake washed with acetone. Hydrochloric acid 36% (1.2 equivalents) is added to the combined filtrates at 25°C to

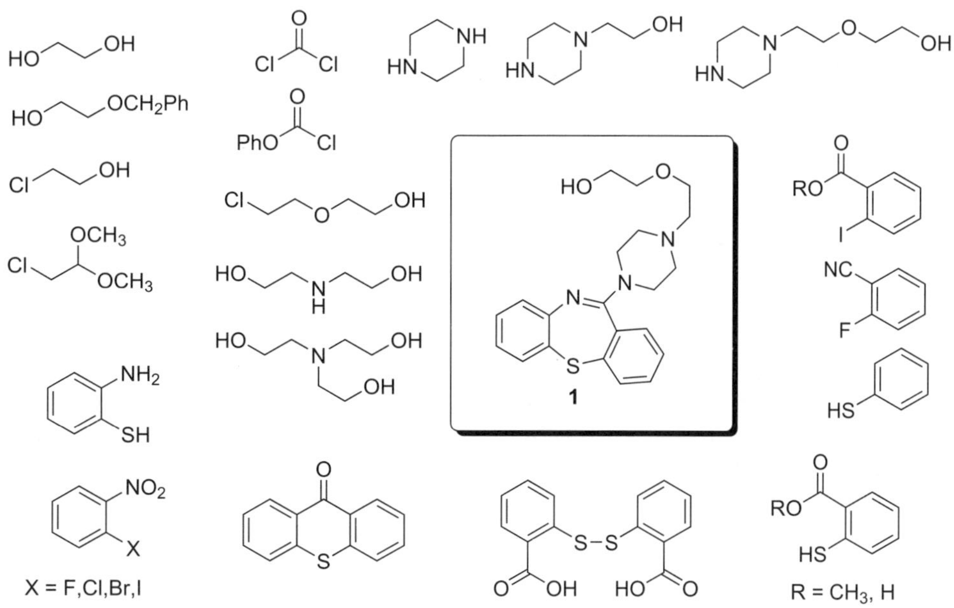

**FIGURE 5.3** Quetiapine (**1**) building blocks.

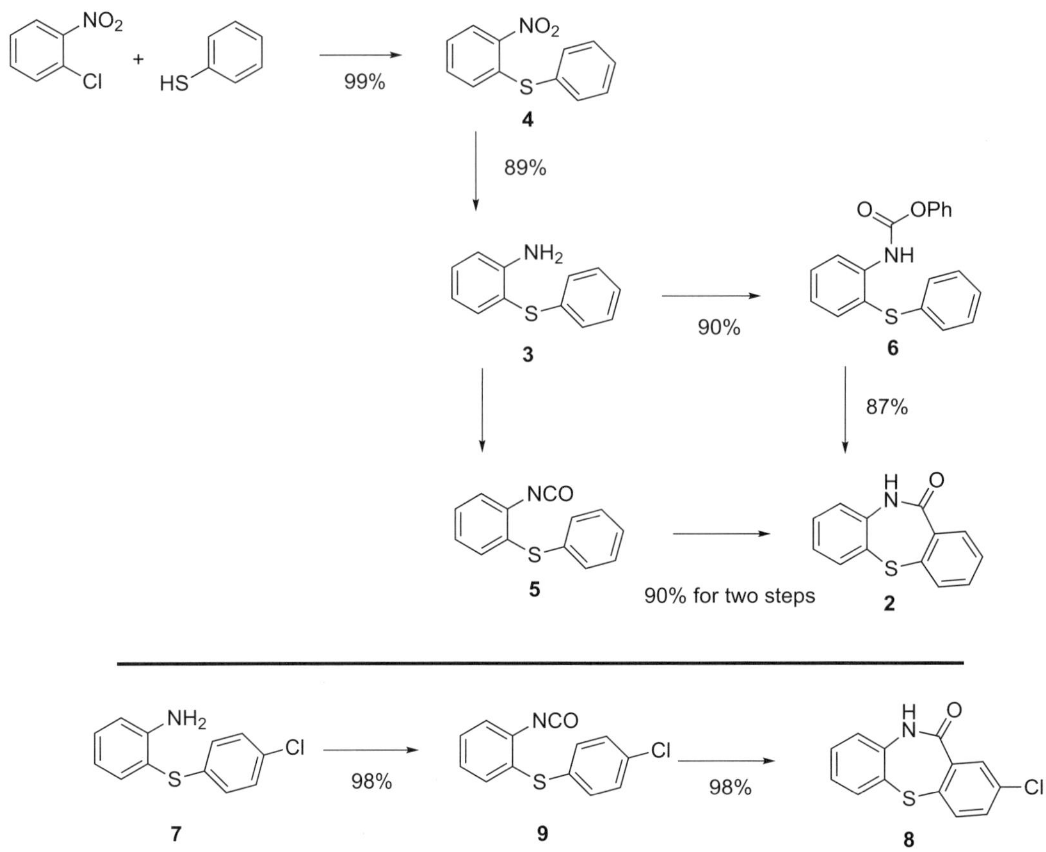


**SCHEME 5.1** 2-Chlorodibenzo[*b*,*f*][1,4]thiazepin-11(10*H*)-one (**8**) and dibenzo[*b*,*f*][1,4]thiazepin-11(10*H*)-one (**2**) via cyclization of an isocyanate.

precipitate 2-(phenylthio)aniline (**3**) hydrochloride that is filtered, washed with acetone, and dried (89% from thiophenol). There are at least two issues associated with scale-up of this process. First, with all the iron powder and 2-nitrodiphenylsulfide (**4**) in the pot at the start, how can the exotherm be controlled? Second, what is the cost associated with the treatment and disposal of the filtered iron salts and the aqueous liquors containing iron? The reduction uses 0.88 kg of iron powder at a cost of $55 to produce 1 kg of 2-(2-phenylthio)aniline (**3**) hydrochloride.[10]

Toxic phosgene or triphosgene are the reagents of choice for preparation of the isocyanate **5**. The isocyanate preparation and cyclization are described in detail for the conversion of 2-amino-4′-chlorodiphenylsulfide (**7**) to 2-chlorodibenzo[*b*,*f*][1,4]thiazepin-11(10*H*)-one (**8**). Slow addition of 2-amino-4′-chlorodiphenylsulfide (**7**) in toluene to phosgene in toluene at 0°C followed by reflux while bubbling in additional phosgene affords a toluene solution of the isocyanate **9**. Excess phosgene is eliminated by bubbling in nitrogen and the solvent is removed at reduced pressure to afford the isocyanate **9** (98%). The procedure for a Leuckart amide synthesis begins with slow addition of a solution of the isocyanate **9** in *o*-dichlorobenzene to anhydrous aluminum chloride (1.1 equivalents) in *o*-dichlorobenzene at 90–120°C. The mixture is aged at 150°C for 1 h then cooled and quenched into ice. Steam distillation yields a suspension. The solid is filtered and then crystallized from acetone to afford 2-chlorodibenzo[*b*,*f*][1,4]thiazepin-11(10*H*)-one (**8**) (98%). 2-(2-Phenylthio)aniline (**3**) is similarly converted to dibenzo[*b*,*f*][1,4]thiazepin-11(10*H*)-one (**2**) via isocyanate **5** (90% for two steps). The cyclization can also be promoted using polyphosphoric acid.[12,13]

A preferred carbamate is produced by reaction of 2-(phenylthio)aniline (**3**) with phenyl chloroformate[14] (1.2 equivalents), sodium hydroxide (0.75 equivalents), and sodium carbonate (0.88 equivalents) in toluene–water at 5°C or dichloromethane–water at 10–25°C. The phenyl carbamate **6** is isolated from the organic phase by concentration at reduced pressure and recrystallization from hexane (90%).[15,16] Activation of the carbamate **6** and cyclization are accomplished by reaction with polyphosphoric acid (7.5 equivalents) at 100°C for 6 h. The reaction mixture is cooled and quenched into water, The precipitate is isolated, probably washed with water, washed with acetone, and dried at an unspecified temperature to afford dibenzo[*b*,*f*][1,4]thiazepin-11(10*H*)-one (**2**) (87%).[15]

### 5.2.2 Constructing the Thiazepine Ring by C–N Bond Formation

***5.2.2.1 Assembly and Cyclization of Methyl 2-(2-Aminophenylthio)benzoate (10)*** The thiazepine ring is produced by C–N bond formation in an alternative sequence using methyl thiosalicylate,[17] thiosalicylic acid,[18] or the dimer 2,2′-dithiodibenzoic acid[19] in place of thiophenol (Scheme 5.2). A sequence starting with expensive methyl thiosalicylate utilizes cesium carbonate, sodium methoxide, or sodium hydroxide as the base and DMF, methanol, or methanol–water as the solvent for the chloride displacement. While a copper catalyst is sometimes added, there is no data to suggest that the copper catalyst increases the reaction rate or improves the efficiency of the chloride displacement.

The assembly and cyclization of a 2-(2-aminophenylthio) benzoic acid is described in detail for the preparation of 8-chlorodibenzo[*b,f*][1,4]thiazepin-11(10*H*)-one (**11**). The reaction of methyl thiosalicylate (2.0 equivalents) with 4-chloro-1-fluoro-2-nitrobenzene and cesium carbonate (2.0 equivalents) in DMF at 25°C is complete in 2 h. A routine dichloromethane–water workup followed by silica gel chromatography affords the thioether **12** (92%).[20] The reaction of methyl thiosalicylate with 1,4-dichloro-2-nitrobenzene (1.2 equivalents) and sodium hydroxide (1.0 equivalent) in methanol–water at reflux also affords thioether **12** (95%).[23]

The nitro group can be efficiently reduced by tin(II) or by catalytic reduction using Raney nickel or platinum oxide in methanol, ethanol, or ethyl acetate.[21–23] Methyl 2-(4-chloro-2-nitrophenylthio)benzoate (**12**) is reduced with tin(II) chloride dihydrate[24] (5.0 equivalents) in ethanol at reflux. The solvent is removed at reduced pressure and the residue is quenched with ice. Sodium carbonate is added (to pH 10). Ethyl acetate is added and the slurry is filtered through celite. The organic layer is washed with water and brine then dried and concentrated at reduced pressure to afford methyl 2-(2-amino-4-chlorophenylthio)benzoate (**13**) (70%). There are at least four issues associated with the scale-up of this procedure. First, the yield is low for this simple reduction. Second, the cost for treatment of the filtered tin salts, and the aqueous liquors containing tin must be considered. Third, there are two concentrations to dryness. Fourth, to prepare 1 kg of the dibenzo[*b,f*][1,4]thiazepin-11(10*H*)-one **11** by this procedure would require 5.6 kg of tin(II) chloride dihydrate at a cost of $950. While excess tin(II) may have been used because of the small scale (<1 g) of this reduction procedure, the cost would still be too high using a stoichiometric charge of the tin reagent.[20]

The catalytic reduction of methyl 2-(4-chloro-2-nitrophenylthio)benzoate (**12**) with Raney nickel (124 g Ni/kg **12**) in methanol at 25°C and 30 psi hydrogen is complete in 17 h. The catalyst is filtered and the solvent removed at reduced pressure to afford methyl 2-(2-amino-4-chlorophenylthio) benzoate (**13**) in quantitative yield.

The cyclization of methyl 2-(2-amino-4-chlorophenylthio)benzoate (**13**) is accomplished by reaction with

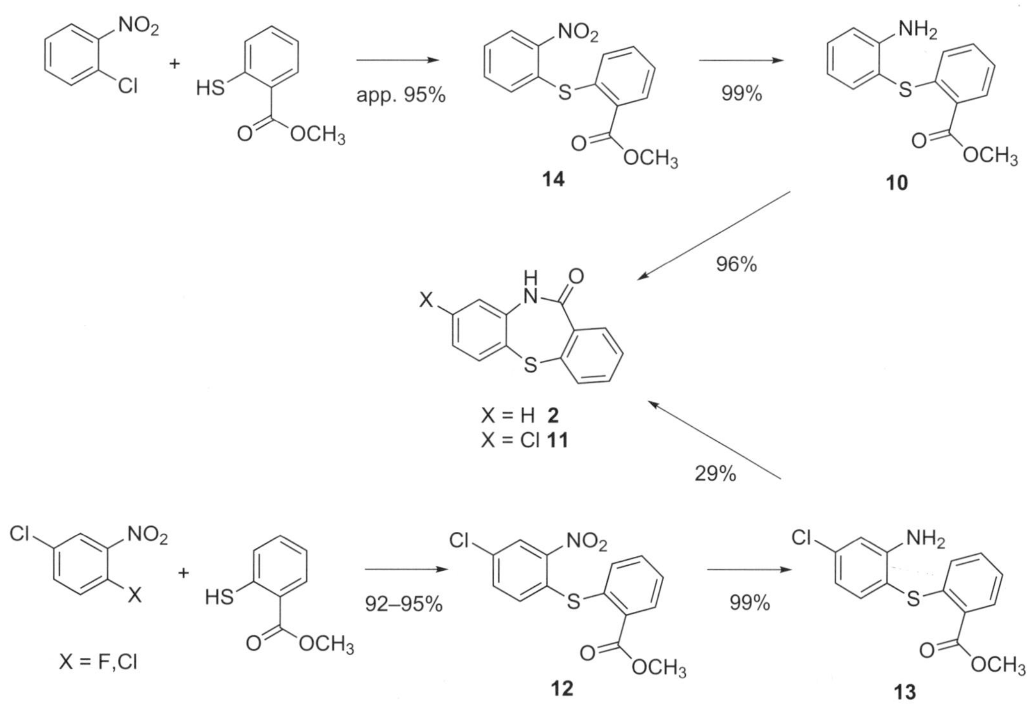


**SCHEME 5.2** 8-Chlorodibenzo[*b,f*][1,4]thiazepin-11(10*H*)-one (**11**) and dibenzo[*b,f*][1,4]thiazepin-11(10*H*)-one (**2**) via cyclization of a 2-(2-aminophenylthio)benzoate.

trimethylaluminum (1.4 equivalents) in toluene–dichloromethane at 25°C over 6 days. The workup involves careful addition of water to quench excess trimethylaluminum, addition of hydrochloric acid, and extraction with dichloromethane. The extracts are dried and concentrated at reduced pressure and the residue is chromatographed on silica to afford 8-chlorodibenzo[*b,f*][1,4]thiazepin-11(10*H*)-one (**11**) (29%).[20]

The reaction of methyl thiosalicylate with 1-chloro-2-nitrobenzene (1.0 equivalent) and sodium methoxide (1.4 equivalents) in methanol at 50°C affords the thioether **14** in 56% yield after recrystallization from methanol.[21,22] The reaction is likely to be more efficient using sodium hydroxide as the base in methanol–water. The cyclization of methyl 2-(2-aminophenylthio)benzoate (**10**) can be accomplished by simply heating the molten ester (mp 95–96°C) in an oil bath at 200–220°C for 7 h. Methanol is distilled from the reaction. The resulting solidified mass is crystallized from ethanol to afford dibenzo[*b,f*][1,4]thiazepin-11(10*H*)-one (**2**) (96%). The high-melting product **2** (mp 257–259°C) has low solubility in most organic solvents. This cyclization procedure is likely to produce a broken or at least a bound agitator even on a lab scale.

***5.2.2.2 Assembly and Cyclization of 2-(2-Aminophenylthio)benzoic Acid (15)*** The C–S bond of 2-(2-aminophenylthio)benzoic acid (**15**) is prepared by reaction of (1) a 1-halo-2-nitrobenzene and thiosalicylic acid, (2) 1-chloro-2-nitrobenzene, 2,2′-dithiodibenzoic acid and a reducing agent, (3) 2-aminothiophenol and 2-iodobenzoic acid, and (4) 2-aminothiophenol and 2-fluorobenzonitrile. The cyclization of 2-(2-aminophenylthio)benzoic acid (**15**) is accomplished by heating neat or in solution, with or without an acid catalyst, and with or without removal of the water by-product as an azeotrope with the reaction solvent.

*1-Halo-2-Nitrobenzene and Thiosalicylic Acid* The sequence starting with less expensive thiosalicylic acid utilizes sodium methoxide, sodium hydroxide, potassium hydroxide, or potassium carbonate as the base and methanol, isopropanol–water, or DMF as the solvent in the halide displacement. All of the 1-halo-2-nitrobenzenes are suitable reaction partners.

The reaction of thiosalicylic acid with 1-fluoro-2-nitrobenzene[25] (1.0 equivalent) and potassium carbonate (3.0 equivalents) in DMF is presumably complete in a few hours at 100–110°C. The suspension is cooled and filtered and the liquors are concentrated at reduced pressure. The residue is taken up in water and sodium hydroxide is added (to pH 11). The solution is washed with toluene then acidified (to pH 2) with hydrochloric acid. The precipitate is filtered, washed with water, and dried at 60°C and reduced pressure to afford 2-(2-nitrophenylthio)benzoic acid (**16**) (94–98%).[26,27]

The reaction of thiosalicylic acid with 1-chloro-2-nitrobenzene (1.2 equivalents) and potassium hydroxide (2.2 equivalents) in isopropanol–water is complete in 6 h at reflux. The reaction mixture is cooled and water and hydrochloric acid are added. The suspension is again heated to reflux and then allowed to slowly cool to 25°C. The precipitate is filtered, washed with water and toluene, and air dried to afford 2-(2-nitrophenylthio)benzoic acid (**16**) (90%).[28]

The reaction of thiosalicylic acid with 1-chloro-2-nitrobenzene (1.0 equivalent) and potassium carbonate (3.0 equivalents) in DMF at 100–110°C probably goes to completion in 2 h. The suspension is cooled and filtered and the liquors are concentrated at reduced pressure. The residue is taken up in aqueous sodium hydroxide, washed with toluene, and acidified (to pH 2) with hydrochloric acid. The precipitate is filtered, washed with water, and dried at 60°C and reduced pressure to afford 2-(2-nitrophenylthio)benzoic acid (**16**) (92–96%).[26]

The reaction of thiosalicylic acid with 1-chloro-2-nitrobenzene (1.2 equivalents) and potassium carbonate (2.3 equivalents) in DMF is complete in 6 h at 70°C. The mixture is cooled and water and ethyl acetate are added. The aqueous layer is separated and water and hydrochloric acid are added. The precipitate is filtered, washed with water, and dried to afford 2-(2-nitrophenylthio)benzoic acid (**16**) (98%).[29] Methanol can be used in place of DMF, and sodium methoxide or sodium hydroxide can be used in place of potassium carbonate when methanol is the solvent (94–96%). With these high yields and short reaction times, adding potassium iodide as a promoter to a methoxide–methanol procedure or tetrabutylammonium iodide as a phase transfer catalyst[30] to a potassium carbonate–methanol procedure does not provide a significant increase in the reaction rate or efficiency of the chloride displacement.

Here are two nearly identical processes using potassium carbonate in DMF where one has a filtration and a concentration at reduced pressure and the other does not. These operations improve the volume throughput of the process at the expense of time throughput. Which is preferred? As long as the volume throughput is acceptable (>100 g/L), eliminating the filtration and concentration makes sense. As a general rule, *Avoid adding manufacturing operations that are not absolutely necessary.* This is not being lazy, this is being forward thinking. A filtration may be a trivial operation in the lab ("a quick filtration") but this operation may become a problem (slow filtration/all filters in use) in the pilot plant or plant. A DMF distillation at reduced pressure may make it possible to generate more material in typical lab-scale equipment but that distillation will take much longer and be difficult to reproduce in the pilot plant or plant.

The reaction of thiosalicylic acid with 1-bromo-2-nitrobenzene[31] (1.2 equivalents) and potassium carbonate (2.3 equivalents) in DMF is complete in 6 h at 70°C. The mixture is cooled and water and ethyl acetate are added. The aqueous layer is separated and water and hydrochloric acid are added. The precipitate is filtered, washed with water, and dried to afford 2-(2-nitrophenylthio)benzoic acid (**16**) (98%).[29]

The reaction of thiosalicylic acid with 1-iodo-2-nitrobenzene[32] (1.0 equivalent) and copper(I) oxide (0.5 equivalents) in dry pyridine is stopped after 3 h at reflux. The mixture is cooled and transferred into 5% aqueous hydrochloric acid. The precipitate is filtered, washed with water, and dissolved in 2 M NaOH. The solution is polish filtered and then acidified with hydrochloric acid to precipitate 2-(2-nitrophenylthio)benzoic acid (**16**) that is filtered, washed, and dried (85%).[33] After reviewing the processes and the price and bulk availability for each 1-halo-2-nitrobenzene, 1-chloro-2-nitrobenzene is the preferred reaction partner for thiosalicylic acid.

The nitro group is efficiently reduced by iron or by catalytic reduction (Raney nickel, palladium on carbon, platinum oxide) in ethyl acetate, an alcohol (methanol, ethanol, or 1-butanol), or water. In the reduction with iron(II) sulfate heptahydrate, 2-(2-nitrophenylthio)benzoic acid (**16**) is dissolved in 28 wt% aqueous ammonia (14.5 mL/g **16**). A solution of ferrous sulfate heptahydrate[34] (7.9 g/g **16**) in water is added. After aging at 80°C for 10 min, the suspension is cooled and the metal by-product filtered. The liquors are concentrated at reduced pressure and ethyl acetate and acetic acid are added. The organic layer is dried and concentrated at reduced pressure to afford 2-(2-aminophenylthio) benzoic acid (**15**) (95%). There are at least four issues associated with scale-up of this process. First, the cost associated with the treatment and disposal of the iron by-product filtered and the aqueous phase containing iron must be considered. Second, the concentration of the aqueous ammonia solution at reduced pressure will require cryogenic cooling to capture ammonia from the vapor phase. Third, the volume throughput is low (21 g/L). Finally, this reduction uses 9.3 kg of iron(II) sulfate heptahydrate at a cost of $292 to produce 1 kg of 2-(2-aminophenylthio)benzoic acid (**15**). The cost for the reducing agent is higher ($360–370/kg of **15**) using iron powder and catalytic iron(III) chloride.[33] The discussion moving forward will be limited to catalytic reduction methods.[29]

> Good process design takes into account the alignment of economic and environmental incentives in downstream waste processing. When a metal is used as a stoichiometric reducing agent for a nitro group, an inexpensive metal or metal salt is usually chosen. The metal by-product is filtered and/or separated as an aqueous solution. The cost to recycle the metal is high relative to the cost of the metal. Thus, *there is an economic incentive to treat the metal waste and bury it.* The waste would be buried according to regulations in place in the year buried. But the story does not end at the gravesite. The company maintains responsibility for that site, that is, it has "cradle to grave" responsibility for materials used in the process. If the integrity of the gravesite is compromised by forces of nature (soil expansion, flooding, earthquake) at any time in the future, the company is responsible for waste leaking from the site. Changes in regulations and/or public perception may prompt improvements in or elimination of the gravesite in the future. The forward-looking alternative reduction using an expensive metal catalyst and stoichiometric hydrogen avoids future liabilities by burying nothing. With the catalytic reduction, *there is an economic incentive to be environmentally responsible*, to filter the heterogeneous catalyst using cellulose and recover the precious metal from the cellulose-metal cake.

The high catalyst loadings in available procedures suggest that the sulfide in substrate **16** may act as a catalyst poison. In most cases, the hydrogen pressures are also higher than the 30–60 psi range we would prefer for scale-up in a conventional plant reactor. A reduction at atmospheric pressure (by maintaining hydrogen atmosphere in the headspace by passing hydrogen through the reactor) is possible at elevated temperature. The reduction using Raney nickel (294 g Ni/kg **16**) in methanol at 25°C and 280 psi hydrogen requires 5 h. The catalyst is filtered and the solvent removed at reduced pressure to afford 2-(2-aminophenylthio)benzoic acid (**15**) (92%). The reduction using Raney nickel (200 g Ni/kg **16**) in methanol at 100°C and 450 psi hydrogen requires 6 h (98%). Complete reduction can be accomplished using Raney nickel (250 g Ni/kg **16** or 66 g Ni/kg **16**) in 1-butanol at 100°C and hydrogen at atmospheric pressure (91–92%). The reduction using Raney nickel (~200 g Ni/kg **16**) in an aqueous basic solution at 100–110°C and 450 psi hydrogen requires 3 h (100% conversion, 98% selectivity). After venting the hydrogen and cooling the suspension, the catalyst is filtered and the liquors neutralized with hydrochloric acid to precipitate 2-(2-aminophenylthio)benzoic acid (**15**). The reduction in aqueous base at 50°C and 145 psi hydrogen for 13.5 h is far less efficient (46% conversion, 69% selectivity) (Scheme 5.3).[29,35]

The reduction using water-wet 5% palladium on carbon (5 g Pd/kg **16**) in methanol at 25°C and 140 psi hydrogen requires 6 h (95–97%). The volume throughput can be as high as 240 g product **15**/L when the hydrogenation is performed in methanol at 50°C (97–98%). A 100 g scale reduction using 5% palladium on carbon (5 g Pd/kg **16**) in methanol at 30–35°C and 100 psi hydrogen requires 10–15 h. No yield is provided but the similarity to other processes

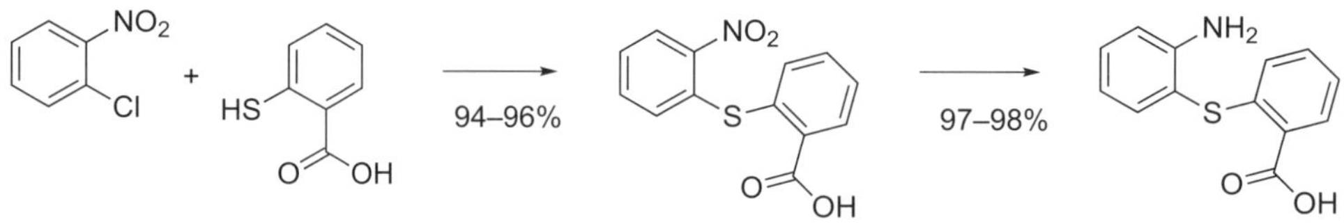


16 15

**SCHEME 5.3** 2-(2-Aminophenylthio)benzoic acid (**15**) from 1-chloro-2-nitrobenzene and thiosalicylic acid.

suggests the yield is likely to be >95%. The reduction using palladium or platinum on carbon in ethyl acetate at 50°C and 90 psi requires 5 h for complete conversion. No yield is available. The reduction can also be accomplished using platinum oxide (40 g Pt/kg **16**) in methanol at 50°C (88%).[28–30]

*1-Chloro-2-nitrobenzene and 2,2′-Dithiodibenzoic Acid* Thiosalicylic acid is produced from anthranilic acid. Diazotization and reaction with sodium sulfide and sulfur affords 2,2′-dithiodibenzoic acid. Reduction of this disulfide with zinc then affords thiosalicylic acid. Perhaps the raw material costs for producing 2-(2-aminophenylthio) benzoic acid (**15**) can be reduced by generating thiosalicylic acid when needed by reduction of inexpensive 2,2′-dithiodibenzoic acid.[36]

Zinc (10 equivalent) is added to a solution of 2,2′-dithiodibenzoic acid and sodium hydroxide (2.0 equivalents) in water. After aging 1 h at 40°C, 1-chloro-2-nitrobenzene (1.0 equivalent) is added and the mixture refluxed for 6 h. The mixture is cooled, washed with toluene, and used in the nitro group reduction. Reduction with Raney nickel (200 g Ni/kg **16**) in water at 110°C and 450 psi hydrogen is complete in 3 h. After venting the hydrogen and cooling the suspension, the catalyst is filtered and the liquors neutralized with hydrochloric acid to precipitate 2-(2-aminophenylthio)benzoic acid (**15**) hydrochloride (65% for three steps from 2,2′-dithiodibenzoic acid). A second procedure suggests that *the yield for the first two steps is 85–90%*! 1-Chloro-2-nitrobenzene (1.1 equivalents) is added to a solution of 2,2′-dithiodibenzoic acid and sodium hydroxide (2.0 equivalents) in water and the resulting mixture is refluxed for 5 h. (Note: The zinc is missing from the patent procedure.) The mixture is cooled and washed with ethyl acetate. The aqueous layer is neutralized with hydrochloric acid and extracted with ethyl acetate. The extracts are dried and ethyl acetate removed at reduced pressure to afford 2-(2-nitrophenylthio)benzoic acid (**16**) (87%). The nitro group reduction with Raney nickel (200 g Ni/kg **16**) in water at 100°C and 450 psi hydrogen is complete in 3 h (98%).[35]

While using the dimer does offer some raw material cost savings, we will be moving the disulfide reduction with zinc and the treatment and disposal of a zinc-containing aqueous waste stream from a commodity chemical manufacturer with presumably lower overhead costs to a pharmaceutical manufacturer with higher overhead costs. A better long-term approach to reduce raw material costs may be to improve the manufacturing process for thiosalicylic acid. For example, using a sodium–sulfur ratio of 1.64, the reaction of the diazonium salt from anthranilic acid with sodium sulfide and sulfur affords thiosalicylic acid (84%) directly.[37]

*2-Iodobenzoic Acid or 2-Fluorobenzonitrile and 2-Aminothiophenol* The nitro group reduction is taken out of the linear sequence when the alternative C–S bond is generated in the $S_NAr$ reaction. The halogen displaced is activated by copper catalysis or by an adjacent nitrile. The thiol partner is 2-aminothiophenol,[38] a low-melting solid (mp 19–21°C), which will likely have to be warmed to prevent line plugging during transfer as a liquid. The step-by-step procedure for 2-aminothiophenol transfer must be carefully designed to minimize the potential for release of vapors and the ensuing odor complaints. This transfer will probably be more challenging on pilot plant scale using multipurpose equipment than on plant scale where the equipment can be chosen for the task and remain dedicated to the task.

2-Aminothiophenol (1.1 equivalents) reacts with 2-iodobenzoic acid,[39] potassium hydroxide (4.6 equivalents), and copper bronze (0.8–1.0 equivalents) in water at reflux. The suspension is filtered and the liquors are cooled and acidified to precipitate 2-(2-aminophenylthio)benzoic acid (**15**). The solid is filtered, washed with ethanol and ethyl ether, and dried at reduced pressure (75%). 2-(2-Aminophenylthio) benzoic acid (**15**) can also be isolated as the hydrochloride salt (no yield provided) (Scheme 5.4).[40,41]

2-Aminothiophenol is converted to the sodium salt with sodium hydride (1.0 equivalent of 60% oil dispersion) in THF at 0°C. The mixture is then added portionwise to 2-fluorobenzonitrile[42] (1.0 equivalent) in THF at 25°C. After aging for 16 h, the solvent is removed at reduced pressure and the residue is chromatographed on silica to afford 2-(2-aminophenylthio)benzonitrile (**17**) (47%). The reaction of 2-aminothiophenol (1.1 equivalents) with 2-fluoro-4-trifluoromethylbenzonitrile (**18**) and sodium hydride (1.3 equivalents of 60% oil dispersion) in DMF at 0–10°C affords 2-(2-aminophenylthio)-4-trifluoromethylbenzonitrile (**19**) (69%) after a routine workup and chromatography.

2-(2-Aminophenylthio)benzonitrile (**17**) is hydrolyzed in a mixture of water, acetic acid, and sulfuric acid at reflux.

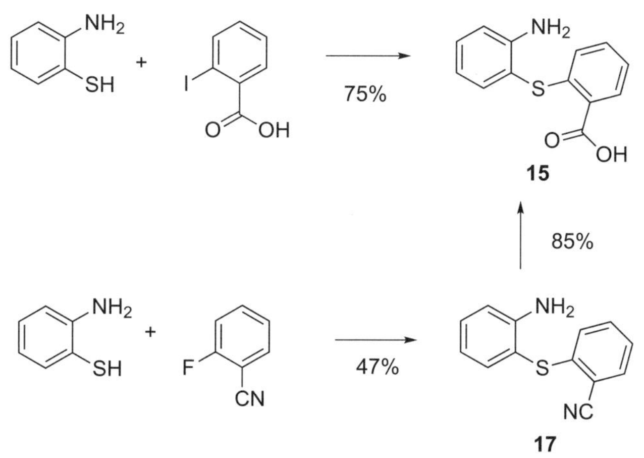

**SCHEME 5.4** 2-(2-Aminophenylthio)benzoic acid (**15**) from 2-aminothiophenol.

The mixture is cooled, diluted with water, pH adjusted to 6 with aqueous sodium hydroxide, and extracted with dichloromethane. The combined extracts are dried and concentrated at reduced pressure to afford 2-(2-aminophenylthio)benzoic acid (**15**). While no yield is provided at this point in the sequence, a well-established yield for the remaining step in the sequence to dibenzo[*b*,*f*][1,4]thiazepin-11(10*H*)-one (**2**) suggests the nitrile hydrolysis yield is 85%.[43]

*Cyclization of 2-(2-Aminophenylthio)benzoic Acid (15)* The pyrolysis of 2-(2-aminophenylthio)benzoic acid (**15**) at 200–230°C produces a solid mass that is crystallized from ethyl acetate to afford dibenzo[*b*,*f*][1,4]thiazepin-11(10*H*)-one (**2**) (23%) (Scheme 5.5). A bound or broken agitator is the likely outcome of this neat pyrolysis even on a lab scale.[40,41]

2-(2-Aminophenylthio)benzoic (**15**) acid can be converted to dibenzo[*b*,*f*][1,4]thiazepin-11(10*H*)-one (**2**) by

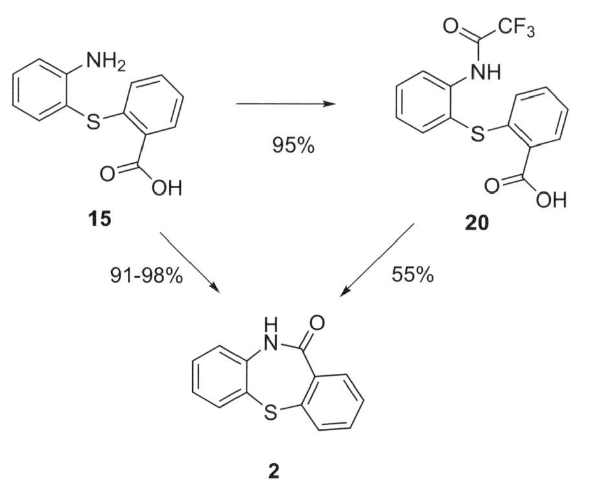

**SCHEME 5.5** Dibenzo[*b*,*f*][1,4]thiazepin-11(10*H*)-one (**2**) from 2-(2-aminophenylthio)benzoic acid (**15**) and the trifluoroacetamide **20**.

simply heating in toluene. The cyclization is slow, requiring more than 20 h for high conversion. The focus for process development has been to decrease the time required (time throughput) by (1) removing the water by-product by distillation as an azeotrope, (2) using an acid catalyst, and (3) replacing toluene with a higher boiling solvent. Sulfuric acid and *p*-toluenesulfonic acid are suitable catalysts and toluene (bp 110–111°C), xylenes (bp 137–140°C), and cumene (bp 152–154°C) are suitable solvents. Dibenzo[*b*,*f*][1,4]thiazepin-11(10*H*)-one (**2**) precipitates during the reaction using any of these solvents and is isolated by cooling and filtering the resulting suspension.

The suspension produced after 20 h in refluxing toluene is cooled to 25°C and filtered to afford dibenzo[*b*,*f*][1,4]thiazepin-11(10*H*)-one (**2**) (69%). When a Dean–Stark trap is used to collect the water by-product, the yield after 20 h is 80%.[29]

The suspension produced after 15 h in refluxing xylenes using the Dean–Stark trap is cooled to 25°C and filtered to afford dibenzo[*b*,*f*][1,4]thiazepin-11(10*H*)-one (**2**) (98%). Presumably to increase the purity of the crude product, saturated aqueous bicarbonate or dilute sodium hydroxide can be added to the suspension prior to the isolation (93–95%).[29,43]

The suspension produced after 10 h in refluxing cumene using the Dean–Stark trap is cooled to 25°C and filtered to afford dibenzo[*b*,*f*][1,4]thiazepin-11(10*H*)-one (**2**) (98%). Saturated aqueous bicarbonate can be added to the suspension prior to the isolation (97%).[29]

The suspension produced after just 10 h in refluxing xylenes using the Dean–Stark trap and a *p*-toluenesulfonic acid catalyst (0.15 mol%) is cooled to 55°C and filtered to afford dibenzo[*b*,*f*][1,4]thiazepin-11(10*H*)-one (**2**) (86%). A saturated bicarbonate wash would remove the catalyst.[28]

The suspension produced after just 6 h in refluxing xylenes using the Dean–Stark trap and a sulfuric acid catalyst (1.3 mol%) is cooled to 25°C and filtered to afford dibenzo[*b*,*f*][1,4]thiazepin-11(10*H*)-one (**2**) (91%). A saturated bicarbonate wash would remove the catalyst.[35]

The reaction is stopped after 30 min in refluxing xylenes when phosphorus pentoxide is added. No workup procedure or yield is provided.[44]

Finally, 2-(2-aminophenylthio)benzoic acid (**15**) hydrochloride is converted to the trifluoroacetamide **20** (95%) by reaction with trifluoroacetic anhydride (2.0 equivalents) at 25°C in dioxane. Dibenzo[*b*,*f*][1,4]thiazepin-11(10*H*)-one (**2**) is produced when the trifluoroacetamide **20** is reacted with excess trifluoroacetic anhydride (6.5 equivalents) in DMF at 100°C for 1 h. The reaction is cooled to 25°C and transferred into water. The suspension is filtered and the solid is washed with water and resuspended in dilute aqueous potassium bicarbonate. The suspension is filtered and the solid is presumably washed with water and dried (55%).[33]

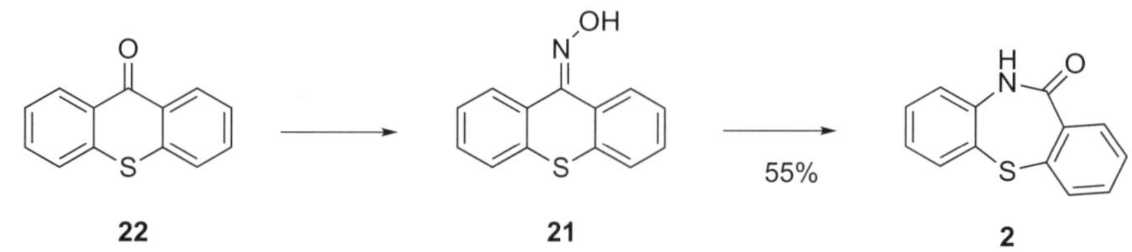


**SCHEME 5.6** Dibenzo[*b,f*][1,4]thiazepin-11(10*H*)-one (**2**) from thioxanthen-9-one (**22**).

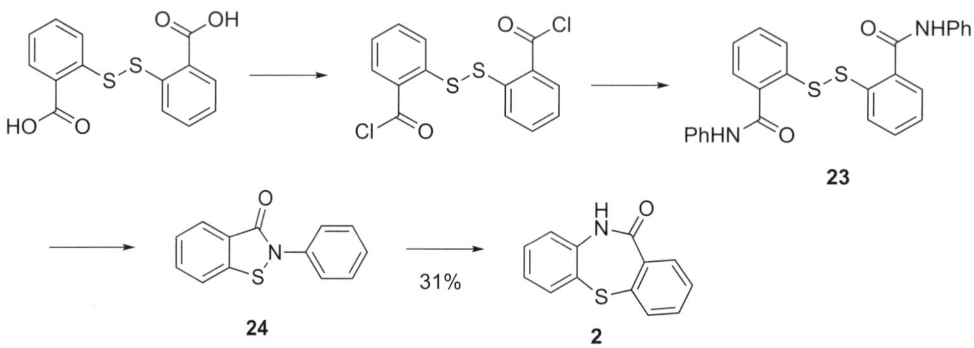


**SCHEME 5.7** Dibenzo[*b,f*][1,4]thiazepin-11(10*H*)-one (**2**) from 2,2′-dithiodibenzoic acid.

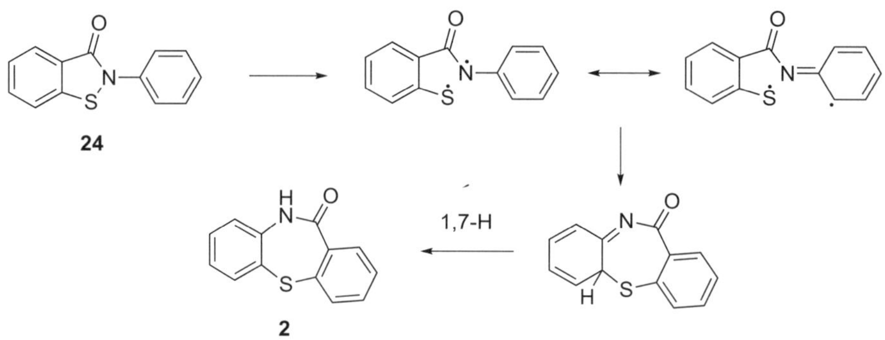


**SCHEME 5.8** Photoisomerization of 2-phenylbenzo[*d*]isothiazol-3(2*H*)-one (**24**).

***5.2.2.3 Beckmann Rearrangement of Thioxanthen-9-one Oxime (21)*** Thioxanthen-9-one[45] (**22**) is prepared by the reaction of diphenyl sulfide with phosgene and aluminum chloride. Beckmann rearrangement of the oxime **21** with polyphosphoric acid at 170°C for 30 min affords dibenzo[*b, f*][1,4]thiazepin-11(10*H*)-one (**2**) (55%). Lower yields are obtained in the rearrangement using polyphosphoric acid at 160–165°C for 1 h (38%) or phosphorus pentachloride in ethyl ether (52%) (Scheme 5.6).[46]

### 5.2.3 Constructing the Thiazepine Ring by C–S Bond Formation

2,2′-Dithiodibenzoic acid is converted to the acid chloride with thionyl chloride. Reaction of the acid chloride with aniline[47] affords the 2,2′-dithiodibenzamide **23**. This is converted to 2-phenylbenzo[*d*]isothiazol-3(2*H*)-one (**24**) by reaction with chlorine or bromine. Irradiation of a solution of 2-phenylbenzo[*d*]isothiazol-3(2*H*)-one (**24**) in *tert*-butanol with a 450 W medium pressure mercury lamp (Hanovia) through a Corex filter (>280 nm) under argon for 90 min results in conversion to dibenzo[*b,f*][1,4]thiazepin-11(10*H*)-one (**2**) (31%) (Scheme 5.7). Perhaps this conversion results from homolytic cleavage of the N–S bond to form a biradical, generation of the C–S bond by biradical recombination via a resonance form, and photochemical 1,7-hydrogen transfer (Scheme 5.8).[44,48]

## 5.3 QUETIAPINE HEMIFUMARATE FROM DIBENZO[b,f][1,4]THIAZEPIN-11(10H)-ONE (2)

### 5.3.1 11-Chlorodibenzo[*b, f*][1,4]thiazepine (25)

Dibenzo[*b,f*][1,4]thiazepin-11(10*H*)-one (**2**) is converted to quetiapine (**1**) by converting the oxygen at the 11-position to a leaving group and displacing the leaving group with a piperazine. Chloride, bromide, and methanethiolate are suitable leaving groups. To avoid the preparation[43] and methylation of a thione, recent processes almost exclusively use the chloride **25**. The chloride displacement is accomplished with piperazine, 1-(2-hydroxyethyl)piperazine, or 2-(2-(piperazin-1-yl)ethoxy)ethanol (**26**).

***5.3.1.1 Dibenzo[b,f][1,4]thiazepin-11(10H)-one (2) and Phosphorus Oxychloride*** Dibenzo[*b,f*][1,4]thiazepin-11(10*H*)-one (**2**) can be converted to 11-chlorodibenzo[*b,f*][1,4]thiazepine (**25**) with phosphorus oxychloride and *N,N*-dimethylaniline or triethylamine, with phosphorus pentachloride, or with oxalyl chloride and DMF. Suitable solvents for reaction with phosphorus oxychloride are high-boiling inert solvents such as toluene, xylenes, and chlorobenzene. The workup generally involves quench of the batch into water at <40°C, layer separation, and drying the organic layer by azeotropic distillation at reduced pressure and <60°C. The temperature limits on the quench and azeotropic distillation ensure minimal hydrolysis of chloride **25** to regenerate dibenzo[*b,f*][1,4]thiazepin-11(10*H*)-one (**2**). The dry solution can be carried directly into the next step.

The reaction of dibenzo[*b,f*][1,4]thiazepin-11(10*H*)-one (**2**) with phosphorus oxychloride (14.8 equivalents) (bp 105.8°C) and *N,N*-dimethylaniline (0.62 equivalents) is complete in 6 h at reflux. Excess phosphorus oxychloride is removed by distillation at reduced pressure. The residual syrup is dissolved in toluene and quenched with ice water. The layers are separated and the organic layer washed with water, dried, and concentrated at reduced pressure to afford 11-chlorodibenzo[*b,f*][1,4]thiazepine (**25**) (93%).[15]

The reaction of dibenzo[*b,f*][1,4]thiazepin-11(10*H*)-one (**2**) with phosphorus oxychloride (30 equivalents) and *N,N*-dimethylaniline (0.22 equivalents) is complete in 6 h at reflux. Excess phosphorus oxychloride is removed by distillation at reduced pressure. The residual syrup is dissolved in ethyl acetate and quenched with ice water. The layers are separated and the aqueous layer extracted with ethyl acetate. The combined organic layers are washed with water and concentrated at reduced pressure to afford 11-chlorodibenzo[*b,f*][1,4]thiazepine (**25**) (97%).[49–51]

The reaction of dibenzo[*b,f*][1,4]thiazepin-11(10*H*)-one (**2**) with phosphorus oxychloride (16.2 equivalents) and *N,N*-dimethylaniline (0.30 equivalents) is complete in 4 h at reflux. Excess phosphorus oxychloride is removed by distillation at reduced pressure. The residual syrup is dissolved in xylenes and quenched into ice water. The layers are separated and the organic layer is washed with dilute hydrochloric acid, water, and brine. The organic layer is then concentrated at reduced pressure to afford 11-chlorodibenzo[*b,f*][1,4]thiazepine (**25**) (no yield available).[15,52]

The water quench can be delayed until after reacting 11-chlorodibenzo[*b,f*][1,4]thiazepine (**25**) with a piperazine. For example, 8-chlorodibenzo[*b,f*][1,4]thiazepin-11(10*H*)-one (**11**) is reacted with phosphorus oxychloride (27.7 equivalents) and *N,N*-dimethylamine (4.1 equivalents) in toluene at reflux. The excess phosphorus oxychloride and toluene are removed by concentration at reduced pressure and the residue containing 8,11-dichlorodibenzo[*b,f*][1,4]thiazepine (**27**) dissolved in toluene and carried into the next step (no yield available).[23]

Phosphorus oxychloride is highly corrosive. It affects the respiratory system, central nervous system, and kidneys. It may be fatal if inhaled. The ACGIH threshold limit value (TLV) is 0.1 ppm (TWA). The quench of a phosphorus oxychloride reaction generates an aqueous phosphate waste stream. While phosphorus oxychloride may be the reagent of choice for this conversion, the potential for exposure and the waste issues provide incentive to develop a process that does not use phosphorus oxychloride as the solvent. The phosphorus oxychloride charge can be dramatically reduced using chlorobenzene or toluene as the solvent.

The reaction of dibenzo[*b,f*][1,4]thiazepin-11(10*H*)-one (**2**) with phosphorus oxychloride (4.4 equivalents) and *N,N*-dimethylaniline (0.46 equivalents) in chlorobenzene is complete in 3 h at 95–100°C. The mixture is cooled and quenched by addition to water and chlorobenzene at 10°C. The organic layer is washed with water and concentrated at reduced pressure to afford a solution of 11-chlorodibenzo[*b,f*][1,4]thiazepine (**25**) (>90%) for use in the next step.[53]

The reaction of dibenzo[*b,f*][1,4]thiazepin-11(10*H*)-one (**2**) with phosphorus oxychloride (3.1 equivalents) and *N,N*-dimethylaniline (0.59 equivalents) in toluene is complete in 4 h at reflux. Excess phosphorus oxychloride and toluene are removed by distillation at reduced pressure. The residue is dissolved in toluene and quenched by addition of cold aqueous potassium carbonate. The toluene layer containing 11-chlorodibenzo[*b,f*][1,4]thiazepine (**25**) (>90%) is separated and carried directly into the next step.[54]

The reaction of dibenzo[*b,f*][1,4]thiazepin-11(10*H*)-one (**2**) with phosphorus oxychloride (1.0 equivalent) and *N,N*-dimethylaniline (0.63 equivalents) in toluene is complete in 8 h at reflux. The mixture is cooled and quenched by addition of water at 30–35°C. The toluene layer is separated, washed with aqueous sodium bicarbonate and with water, and dried by azeotropic distillation. The toluene solution containing 11-chlorodibenzo[*b,f*][1,4]thiazepine (**25**) (HPLC purity 99.0%, no yield available) is carried directly into the next step.[55]

Headache, dizziness, and cyanosis result from acute short-tern inhalation exposure to *N,N*-dimethylaniline (bp 193°C). The OSHA permissible exposure limit (PEL) is 5 ppm (TWA) and the ACGIH threshold limit value is 5 ppm (TWA) and 10 ppm (STEL) with skin designation. The potential for exposure during charging and workup procedures provides some incentive to replace *N,N*-dimethylaniline with another amine.

The reaction of dibenzo[*b,f*][1,4]thiazepin-11(10*H*)-one (**2**) with phosphorus oxychloride (1.1 equivalents) and triethylamine (1.0 equivalent) in toluene is complete in 2 h at reflux. The mixture is cooled and quenched into water at <40°C. The toluene layer is separated, washed with water, and dried by azeotropic distillation. The toluene solution containing 11-chlorodibenzo[*b*,*f*][1,4]thiazepine (**25**) (no yield available) is carried directly into the next step.[56]

The reaction of 4 kg of dibenzo[*b*,*f*][1,4]thiazepin-11(10*H*)-one (**2**) with phosphorus oxychloride (0.75 equivalents) and triethylamine (0.62 equivalents) in xylenes is complete in 4 h at reflux. The mixture is cooled and quenched by addition of water at 25°C. Sodium hydroxide (50%) is added (to pH 2.5–3.5) and the suspension is filtered. The organic layer is separated, washed with water, and dried by azeotropic distillation at reduced pressure and <65°C. The toluene solution containing 11-chlorodibenzo[*b*,*f*][1,4]thiazepine (**25**) (no yield available) is carried directly into the next step.[57]

Does replacing *N*,*N*-dimethylaniline with triethylamine improve the safety profile for the process? Triethylamine (bp 90°C) has a strong ammonia-like odor, has a higher flammability rating, and is both toxic and corrosive. The OSHA permissible exposure limit is 25 ppm (TWA) but the ACGIH threshold limit value is just 1 ppm (TWA) and 3 ppm (STEL) with skin designation. The triethylamine process is not the answer, but time and money invested in developing a process with an alternative amine would be well spent.

Both phosphorus oxychloride and *N*,*N*-dimethylaniline can be replaced by phosphorus pentachloride. For example, 3-methyldibenzo[*b*,*f*][1,4]thiazepin-11(10*H*)-one (**28**) reacts with phosphorus pentachloride (1.2 equivalents) in refluxing chloroform. After 2 h, the volatiles are removed by distillation at reduced pressure and the crude product mixture containing 11-chloro-3-methyldibenzo[*b*,*f*][1,4]thiazepine (**29**) (no yield available) is used in the next step.[58]

Addition of phosphorus pentachloride (1.5 equivalents) in five lots to a solution of dibenzo[*b*,*f*][1,4]thiazepin-11(10*H*)-one (**2**) in toluene followed by 6 h at reflux results in complete conversion. The mixture is cooled and quenched into water at 10–15°C. The organic layer is separated, washed with brine, and dried by azeotropic distillation at reduced pressure and <50°C. The toluene solution containing 11-chlorodibenzo[*b*,*f*][1,4]thiazepine (**25**) (no yield available) can be used in the next step. Side products **30** and **31** are generated by a phosphorus pentachloride promoted substitution reaction on toluene (Figure 5.4).[59]

Dichloromethane is a superior solvent for the reaction with phosphorus pentachloride. Addition of phosphorus pentachloride (1.5 equivalents) in five lots to a solution of dibenzo[*b*,*f*][1,4]thiazepin-11(10*H*)-one (**2**) in dichloromethane at −15°C followed by 2–3 h aging at −15°C results in complete conversion. Most of the dichloromethane is distilled at reduced pressure, toluene is added, and the remaining dichloromethane and some toluene are distilled at reduced pressure. The mixture is cooled and quenched into water at 10–15°C. The organic layer is separated, washed with brine, and dried by azeotropic distillation at reduced pressure and <55°C. The toluene solution containing 11-chlorodibenzo[*b*,*f*][1,4]thiazepine (**25**) (no yield available) can be used in the next step. Comparable results are achieved when the dichloromethane is replaced by toluene after the water quench.[59]

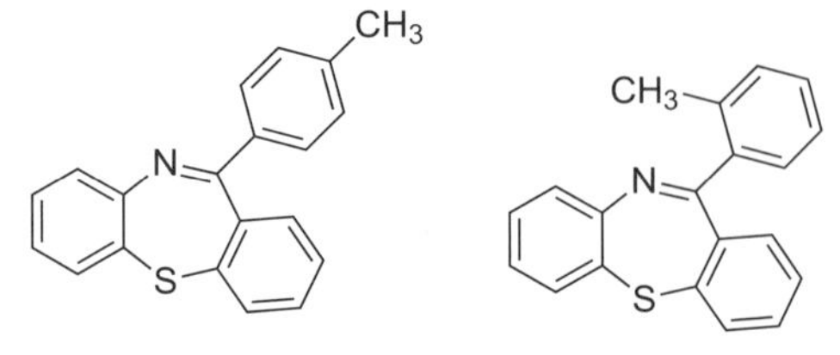


**30** **31**

**FIGURE 5.4** Side products from the reaction of dibenzo[*b*,*f*][1,4]thiazepin-11(10*H*)-one (**2**) with phosphorus pentachloride in toluene.

Replacing phosphorus oxychloride and *N*,*N*-dimethylaniline by phosphorus pentachloride does not improve the safety profile for the process. Phosphorus pentachloride is very hazardous in case of skin or eye contact (irritant and corrosive), ingestion, or inhalation. It is toxic to mucous membranes and may also be toxic to the kidneys and liver. The OSHA permissible exposure limit and the ACGIH threshold limit value are 0.1 ppm (TWA). The greater reactivity of phosphorus pentachloride necessitates a solvent change from toluene to dichloromethane and addition of the solid pentachloride in lots to control the temperature of the reaction. The addition in lots will have a much higher potential for operator exposure to phosphorus pentachloride dust and hydrogen chloride produced when the pentachloride contacts moist air. The addition in lots will also have a much higher potential for operator exposure to dichloromethane and hydrogen chloride from the reactor.

The reaction of dibenzo[*b*,*f*][1,4]thiazepin-11(10*H*)-one (**2**) with oxalyl chloride–DMF (1.1 equivalents) is stopped after 19–21 h at 95–110°C in toluene. Concentration at reduced pressure and workup with water and dichloromethane affords 11-chlorodibenzo[*b*,*f*][1,4]thiazepine (**25**) (66%, 91% pure by HPLC).[60] The reaction of dibenzo[*b*,*f*][1,4]thiazepin-11(10*H*)-one (**2**) with bromine and phosphorus tribromide affords 11-bromodibenzo[*b*,*f*][1,4]thiazepine (**32**) (no yield available).[61]

*5.3.1.2 2-(2-Aminophenylthio)benzoic Acid (15) and Phosphorus Oxychloride* The cyclization of 2-(2-aminophenylthio)benzoic acid (**15**) to produce dibenzo[*b*,*f*][1,4]thiazepin-11(10*H*)-one (**2**) is driven by dehydrating conditions. When phosphorus oxychloride or phosphorus pentachloride is used as the dehydrating agent, 2-(2-aminophenylthio)benzoic acid (**15**) can be converted directly to 11-chlorodibenzo[*b*,*f*][1,4]thiazepine (**25**) (Scheme 5.9). 2-(2-

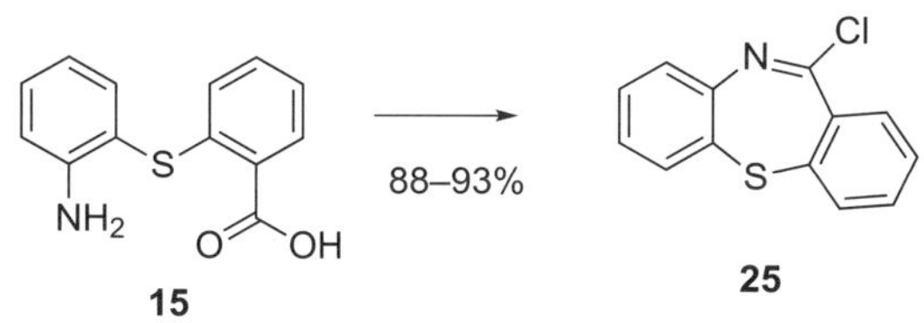

**SCHEME 5.9** 11-Chlorodibenzo[*b,f*][1,4]thiazepine (**25**) from 2-(2-aminophenylthio)benzoic acid (**15**).

Aminophenylthio)benzoic acid (**15**) is reacted with phosphorus oxychloride (13.2 equivalents) for 5–6 h at reflux. Excess phosphorus oxychloride is removed by distillation at reduced pressure. The residue containing crude 11-chlorodibenzo[*b,f*][1,4]thiazepine (**25**) (no yield available) is taken up in toluene and used in the next step.[30]

2-(2-Aminophenylthio)benzoic acid (**15**) hydrochloride is reacted with phosphorus oxychloride (10.0 equivalents) in toluene for 2 h at 90°C. *N,N*-Dimethylaniline (1.0 equivalent) and phosphorus oxychloride (1.1 equivalents) are then added and heating continued at reflux for 2 h. The mixture is cooled and quenched with water and the organic layer is carried into the next step. The yield is not available but is estimated to be 88–93%.[35]

### 5.3.2 11-(Piperazin-1-yl)dibenzo[*b,f*][1,4]thiazepine (33)

***5.3.2.1 11-Chlorodibenzo[b,f][1,4]thiazepine (25) and Piperazine*** The chloride displacement by piperazine probably begins with transfer of a toluene solution of 11-chlorodibenzo[*b,f*][1,4]thiazepine (**25**) into a reactor containing a piperazine–toluene suspension. The heat of the reaction will help to heat the mixture to the reaction temperature (110°C). The workup procedure may include operations to reduce the level of a dimer impurity (likely present at 1–2% when a sufficient excess of piperazine is used). The workup procedure may then concentrate a solution of 11-(piperazin-1-yl) dibenzo[*b,f*][1,4]thiazepine (**33**) to afford an oil. Possible crystallization of the oil should be avoided by concentration to remove some of the solvent followed by a solvent exchange. 11-(Piperazin-1-yl)dibenzo[*b,f*][1,4]thiazepine (**33**) can be crystallized from toluene–methyl *tert*-butyl ether or converted to a dihydrochloride salt.

The reaction of 11-chlorodibenzo[*b,f*][1,4]thiazepine (**25**) with piperazine[62] (equivalents not provided) in toluene is complete in 5 h at reflux. The suspension is cooled, piperazine hydrochloride is filtered, and the liquors are washed several times with water and concentrated at reduced pressure to afford an oil. The oil is dissolved in ethanol and hydrogen chloride is added to precipitate **33** as a dihydrochloride salt (88%).[15] It should be noted that piperazine is a corrosive solid and may cause severe eye and skin burns, dermatitis, and an allergic respiratory reaction.

The toluene solution of 11-chlorodibenzo[*b,f*][1,4]thiazepine (**25**) from an aqueous potassium carbonate wash can be used without an azeotropic distillation to remove trace water. The reaction with piperazine (2.6 equivalents) in toluene goes to completion in 24 h at reflux. The suspension is cooled, the solids are filtered, and the liquors are concentrated at reduced pressure to afford an oil. The oil is dissolved in isopropanol and 33% hydrogen chloride in ethanol is added to precipitate **33** as a dihydrochloride salt, which is filtered, washed with cold ethanol, and dried at 60–70°C (90% for two steps from **2**).[54]

The reaction of 11-chlorodibenzo[*b,f*][1,4]thiazepine (**25**) with piperazine (3.0 equivalents) in toluene is complete in 8 h at reflux. The suspension is cooled to 30–35°C and washed three times with water. The organic layer is dried by azeotropic distillation and the toluene solution of 11-(piperazin-1-yl)dibenzo[*b,f*][1,4]thiazepine (**33**) (no yield available, 99.0% pure by HPLC) is used in the next step.[55]

The reaction of 11-chlorodibenzo[*b,f*][1,4]thiazepine (**25**) with piperazine (5.1 equivalents) in xylenes is complete in <40 h at reflux. The suspension is cooled and concentrated at reduced pressure. The residue is separated between dilute sodium hydroxide and dichloromethane. The aqueous layer is extracted with additional dichloromethane and the combined extracts are dried and concentrated at reduced pressure. 11-(Piperazin-1-yl)dibenzo[*b,f*][1,4]thiazepine (**33**) is isolated as an amorphous foam by chromatography (78%). The foam is dissolved in methanol–diethyl ether. Addition of hydrogen chloride in ethyl ether produces a gummy precipitate. The solvent is decanted and ethanol is added to the gum to crystallize **33** as a dihydrochloride salt. The salt is filtered, washed with ethyl ether, and dried at 60°C (99%).[49–51]

The reaction of 11-chlorodibenzo[*b,f*][1,4]thiazepine (**25**) with piperazine (3.8 equivalents) in toluene goes to completion in 2–3 h at 70–80°C. The suspension is cooled to 30–35°C and washed twice with water. This process delivers a toluene solution of 11-(piperazin-1-yl)dibenzo[*b,f*][1,4] thiazepine (**33**) (no yield available, 97% pure by HPLC). The major impurity is identified as the dimer **34** resulting from reaction of 11-chlorodibenzo[*b,f*][1,4]thiazepine (**25**) with 11-(piperazin-1-yl)dibenzo[*b,f*][1,4]thiazepine (**33**). The dimer impurity level can be reduced by concentrating the toluene solution at reduced pressure, dissolving the residue in *n*-butanol or ethanol at 0–5°C, and filtering off the insoluble solids. The 11-(piperazin-1-yl)dibenzo[*b,f*][1,4]thiazepine (**33**) in solution is 98% pure by HPLC. The impurity is more efficiently removed by extracting 11-(piperazin-1-yl)dibenzo[*b,f*][1,4]thiazepine (**33**) from the toluene solution into aqueous (acetic or formic) acid and then recovering it from the aqueous solution by adding aqueous sodium carbonate and extracting with toluene. The purity of 11-(piperazin-1-yl)dibenzo[*b,f*][1,4]thiazepine (**33**) in solution is 99%.

Purities of >99.0% are possible when **33** free base is converted to the dihydrochloride salt. Hydrochloric acid (36%) is added to the toluene solution at 25–30°C and water is removed by distillation of the azeotrope. The resulting suspension is cooled and the solid filtered, washed with ethanol, and dried at 45–50°C to afford the dihydrochloride salt (89–95% from the free base). Based on results presented above and the HPLC purities provided, the overall yield from 11-chlorodibenzo[*b,f*][1,4]thiazepine (**25**) to **33** dihydrochloride salt is likely to be 89–95%.[59]

A very similar procedure uses hydrochloric acid in place of acetic or formic acid, sodium hydroxide in place of sodium carbonate, and methyl *tert*-butyl ether as the final extraction solvent. The reaction of 11-chlorodibenzo[*b,f*][1,4]thiazepine (**25**) with piperazine (3.6 equivalents) in toluene is complete in 1–3 h at reflux. The suspension is cooled, piperazine hydrochloride is filtered, and the liquors are washed with an aqueous solution of methyl *tert*-butyl ether and methanol (water:MTBE:methanol 2.5:1:0.6) and with water. The amount of dimer **34** is reduced by extracting 11-(piperazin-1-yl)dibenzo[*b,f*][1,4]thiazepine (**33**) from the toluene solution into aqueous hydrochloric acid (pH 3) and then recovering it from the aqueous solution by adding 30% sodium hydroxide and extracting with methyl *tert*-butyl ether. The solution is dried by distillation of the azeotrope and then diluted with isopropanol. Hydrochloric acid (36%) is added to precipitate the dihydrochloride salt. The suspension is filtered and the solid is washed with isopropanol and dried at 60–70°C. The overall yield from 11-chlorodibenzo[*b,f*][1,4]thiazepine (**25**) to **33** dihydrochloride salt is likely to be 85–90%.[56]

A yield is available from a similar process in xylenes. The reaction of 11-chlorodibenzo[*b,f*][1,4]thiazepine (**25**) with piperazine (1.4 equivalents) in xylenes is complete in <17 h at reflux. The suspension is cooled and diluted with ethyl ether and water. The layers are separated and the aqueous layer extracted with ethyl ether. The combined organic layers are washed with water. The amount of dimer **34** is reduced by extracting 11-(piperazin-1-yl)dibenzo[*b,f*][1,4]thiazepine (**33**) from the toluene–ether solution into aqueous hydrochloric acid and then recovering it from the aqueous solution by adding 5 N sodium hydroxide and extracting with ethyl ether. The extracts are concentrated at reduced pressure to afford 11-(piperazin-1-yl)dibenzo[*b,f*][1,4]thiazepine (**33**) (86%).[52]

A toluene solution of 11-chlorodibenzo[*b,f*][1,4]thiazepine (**25**) is produced by reaction of 2-(2-aminophenylthio)benzoic acid (**15**) with phosphorus oxychloride followed by concentration at reduced pressure and dissolution of the residue in toluene. Using this solution, the reaction with piperazine (6.0 equivalents) in toluene goes to completion in 6–8 h at reflux. The suspension is cooled, the solids are filtered, and the liquors are washed with water. The organic layer is dried by azeotropic distillation and the toluene solution of 11-(piperazin-1-yl)dibenzo[*b,f*][1,4]thiazepine (**33**) (no yield available) is used in the next step.[30]

A toluene solution of 11-chlorodibenzo[*b,f*][1,4]thiazepine (**25**) is produced by reaction of 2-(2-aminophenylthio)benzoic acid (**15**) with phosphorus oxychloride and *N,N*-dimethylaniline followed by a water wash. This solution can be used without an azeotropic distillation to remove the water. The reaction with piperazine (3.5 equivalents) in toluene goes to completion in just 1 h at reflux. The solution is cooled and washed with water. The organic layer is concentrated at reduced pressure to afford an oil. The oil is dissolved in ethanol and hydrogen chloride in ethanol is added to precipitate **33** as the dihydrochloride salt (79% from **15**).[35]

A long list of other crystalline salts of 11-(piperazin-1-yl)dibenzo[*b,f*][1,4]thiazepine (**33**) can be prepared in acetonitrile, acetone, or ethyl acetate. These include salts with L-tartaric acid, fumaric acid, methanesulfonic acid, sulfuric acid, and phosphoric acid.[63]

Despite the many processes that finish with isolation of 11-(piperazin-1-yl)dibenzo[*b,f*][1,4]thiazepine (**33**) as the dihydrochloride salt and the recent preparation and characterization of many other crystalline salts, perhaps the best option from a manufacturing perspective is to isolate **33** as the crystalline free base. A toluene solution from workup of the reaction with piperazine in toluene is extracted with aqueous hydrochloric acid at 70°C. The layers are separated and aqueous sodium hydroxide and toluene are added to the aqueous layer at 70°C to release the free base. The layers are separated and the toluene layer washed twice with water. The toluene layer is concentrated to a smaller volume and diluted with methyl *tert*-butyl ether. The solution is cooled to 25°C and seed crystals of polymorph A are added. The suspension is cooled to 10°C and the precipitate filtered, washed with methyl *tert*-butyl ether, and dried at 40°C to afford 11-(piperazin-1-yl)dibenzo[*b,f*][1,4]thiazepine (**33**) (86%, 99.9% pure by HPLC). Polymorph A can be recrystallized from isopropanol (83% recovery). Crash cooling in isopropanol produces another polymorph, Form I. Polymorph A is characterized by NMR, X-ray powder diffraction (XRD), differential scanning calorimetry (DSC), thermal gravimetric analysis (TGA), and DVS (dynamic vapor sorption). The DSC displays one endothermic event at 123.1°C corresponding to a melt event prior to degradation. DVS data reveals that polymorph A is nonhydroscopic.[50,51]

As part of a campaign to develop a more efficient quetiapine (**1**) manufacturing process, we are always looking to minimize manufacturing operations. Several of the processes demonstrate on a lab scale that good yields are achievable when the distillation of the toluene–water azeotrope is eliminated and the water-wet toluene solution is used in the chloride displacement step. Is this a good operation to target for elimination? In a manufacturing plant, inconsistency in the last phase split may result in varying levels of water in the chloride displacement step. Water in the toluene at elevated temperature will convert some 11-chlorodibenzo[*b,f*][1,4]thiazepine (**25**) back to dibenzo[*b,f*][1,4]thiazepin-11(10*H*)-one (**2**), resulting in a lower yield. The azeotrope distillation to dry the toluene solution is important for process reproducibility. Any process modifications that trade reproducibility ("robustness") for speed should be carefully scrutinized.

*5.3.2.2 Dibenzo[b,f][1,4]thiazepin-11(10H)-one (2) and Piperazine* 11-Chlorodibenzo[*b,f*][1,4]thiazepine (**25**) and the challenges of its preparation can be bypassed by reacting dibenzo[*b,f*][1,4]thiazepin-11(10*H*)-one (**2**) with piperazine and a titanium reagent. This single-step approach to 11-(piperazin-1-yl)dibenzo[*b,f*][1,4]thiazepine (**33**) is particularly attractive for quickly generating gram-quantity libraries of 11-aminodibenzo[*b,f*][1,4]thiazepines (Scheme 5.10). Addition of a solution of dibenzo[*b,f*][1,4]thiazepin-11(10*H*)-one (**2**) in dioxane to piperazine (5–10 equivalents) and titanium tetrachloride (1.1–2.2 equivalents) in dioxane–toluene at 50°C is followed by aging overnight at 100°C. The mixture is cooled to 25°C and excess titanium reagent is quenched with 2 M hydrochloric acid. The layers are separated, and the aqueous layer washed with ethyl acetate. Aqueous sodium hydroxide is added and the free base is extracted with ethyl acetate. The ethyl acetate extracts are concentrated at reduced pressure and the residue chromatographed. No yield for 11-(piperazin-1-yl)dibenzo[*b,f*][1,4]thiazepine (**33**) is available.[20]

The reaction of dibenzo[*b,f*][1,4]thiazepin-11(10*H*)-one (**2**) with piperazine (3.0 equivalents) and titanium tetraisopropoxide[64] (2.9 equivalents) is complete in 5 h at 170°C. The isopropanol by-product is distilled from the mixture as the reaction proceeds. The mixture is cooled to 100°C and toluene is added. Excess titanium reagent is quenched with water, the 25°C suspension is filtered, and the solid washed with toluene. The combined liquors are washed with water and concentrated at reduced pressure. 11-(Piperazin-1-yl)dibenzo[*b,f*][1,4]thiazepine (**33**) is isolated from the residue by chromatography (60%).[65]

While process streamlining by eliminating chemical steps is always a worthy development goal, the yield in this process is too low, maintaining 170°C for 5 h is not routine in a manufacturing plant, the water quench will be more challenging as the scale increases, and the solid titanium-containing by-product must be filtered and processed for disposal.

### 5.3.3 2-(4-(Dibenzo[*b,f*][1,4]thiazepin-11-yl)piperazin-1-yl)ethanol (35)

One strategy to avoid the formation of dimer **34** would be to use a monoprotected piperazine and deprotect after the chloride displacement. Chloride displacement with an *N*-monoalkylated piperazine where the alkyl group is all or a part of the quetiapine side chain could avoid the dimer formation without adding protection and deprotection steps. The challenging monofunctionalization of piperazine is moved from the longer linear sequence from thiosalicylic acid to quetiapine (**1**) to a parallel side chain sequence.

The reaction of 11-chlorodibenzo[*b,f*][1,4]thiazepine (**25**) with 1-(2-hydroxyethyl)piperazine[66] (1.5 equivalents) is complete in 12–18 h in refluxing xylenes. The mixture is cooled and diluted with ethyl ether and water. The layers are separated and the aqueous layer is extracted with ether. The extracts are washed with water and then extracted with dilute hydrochloric acid. Aqueous sodium hydroxide is added to the aqueous acid solution and the mixture is extracted into ethyl ether. The extracts are concentrated at reduced pressure to afford 2-(4-(dibenzo[*b,f*][1,4]thiazepin-11-yl)piperazin-1-yl)ethanol (**35**) (53% for the two steps from **2**) (Scheme 5.11).[52]

2-(4-(Dibenzo[*b,f*][1,4]thiazepin-11-yl)piperazin-1-yl)ethanol (**35**) can also be prepared directly from dibenzo[*b,f*][1,4]thiazepin-11(10*H*)-one (**2**) using a titanium reagent. The reaction of dibenzo[*b,f*][1,4]thiazepin-11(10*H*)-one (**2**) with 1-(2-hydroxyethyl)piperazine (2.0 equivalents) and titanium tetraisopropoxide (3.1 equivalents) is complete in 5.5 h at 170°C. The isopropanol by-product is distilled from the mixture as the reaction proceeds. The mixture is cooled

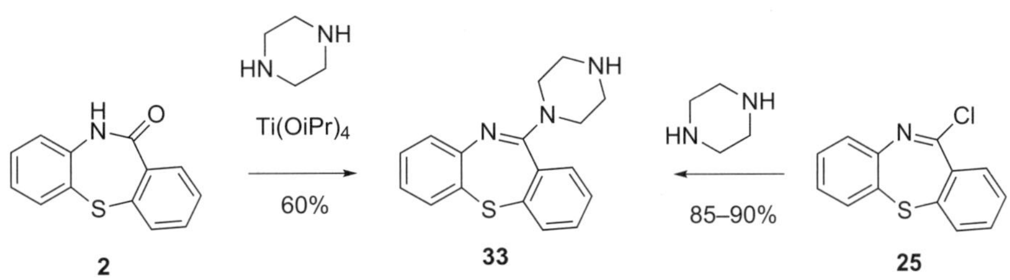


**SCHEME 5.10** 11-(Piperazin-1-yl)dibenzo[*b,f*][1,4]thiazepine (**33**) from 11-chlorodibenzo[*b,f*][1,4]thiazepine (**25**) or dibenzo[*b,f*][1,4]thiazepin-11(10*H*)-one (**2**).

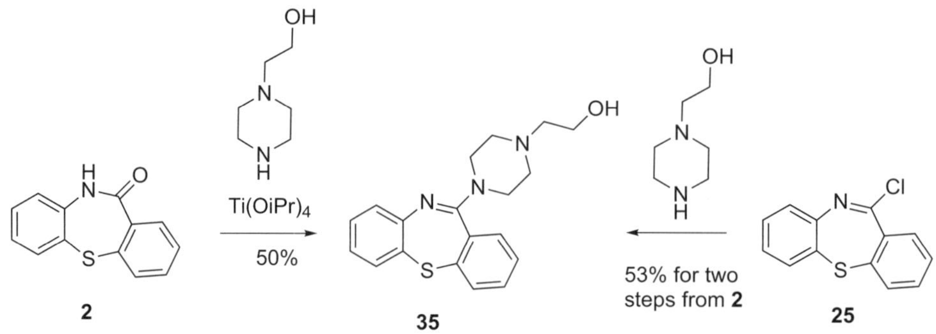

**SCHEME 5.11** 2-(4-(Dibenzo[*b,f*][1,4]thiazepin-11-yl)piperazin-1-yl)ethanol (**35**) from 11-chlorodibenzo[*b,f*][1,4]thiazepine (**25**) or dibenzo[*b,f*][1,4]thiazepin-11(10*H*)-one (**2**).

to 100°C and toluene is added. Excess titanium reagent is quenched with water, the 25°C suspension is filtered, and the solid washed with toluene. The combined toluene liquors are washed with water and concentrated at reduced pressure. 2-(4-(Dibenzo[*b,f*][1,4]thiazepin-11-yl)piperazin-1-yl)ethanol (**35**) is isolated from the residue by chromatography (50%).[65]

While process streamlining by eliminating chemical steps is always a worthy development goal, the yield in this process is too low, maintaining 170°C for 5.5 h is not routine in a manufacturing plant, the water quench will be more challenging as the scale increases, and the solid titanium-containing by-product must be filtered and processed for disposal.

### 5.3.4 Quetiapine (1) Hemifumarate

***5.3.4.1 11-Chlorodibenzo[b,f][1,4]thiazepine (25) and 2-(2-(Piperazin-1-yl)ethoxy)ethanol (26)*** Quetiapine (**1**) can be prepared in a single step from 11-chlorodibenzo[*b,f*][1,4]thiazepine (**25**) and 2-(2-(piperazin-1-yl)ethoxy)ethanol (**26**). The reaction partner, 2-(2-(piperazin-1-yl)ethoxy) ethanol (**26**), is commercially available but expensive.[67] It is likely produced from 2-(2-chloroethoxy)ethanol[68] and piperazine. 2-(2-Chloroethoxy)ethanol is a lacrymator that may cause eye inflammation and corneal injury. It is also a skin sensitizer and respiratory tract irritant. The high cost is a direct result of the handling and exposure issues associated with 2-(2-chloroethoxy)ethanol and piperazine, the high cost of 2-(2-chloroethoxy)ethanol, the low yield in the reaction of 2-(2-chloroethoxy)ethanol with excess piperazine in toluene (72%), and the final purification by high-vacuum distillation (bp 130°C at 0.7 mmHg and 168–172°C at 10 mmHg).[69] Perhaps the high-vacuum distillation could be replaced by preparation of a salt. The dihydrochloride, dihydrochloride monohydrate, and fumarate salts are known.

The conditions for the reaction of 11-chlorodibenzo[*b,f*][1,4]thiazepine (**25**) with 2-(2-(piperazin-1-yl)ethoxy)ethanol (**26**) are much like the conditions used to prepare 11-(piperazin-1-yl)dibenzo[*b,f*][1,4]thiazepine with one notable exception: an inorganic or organic base ($K_2CO_3$, $KHCO_3$, $Na_2CO_3$, $NaHCO_3$, *N,N*-dimethylaniline, triethylamine, *N,N*-diisopropylethylamine) is usually added so that a second equivalent of the expensive amine **26** is not consumed in salt formation. Toluene, xylenes, and *n*-butanol are suitable solvents. An iodide salt (KI, NaI), a phase transfer catalyst (tetrabutylammonium bromide), or a polar aprotic solvent (DMSO) is often added as a promoter. Reaction times are 6–30 h with no promoter, 12–20 h with an iodide catalyst, and 6–8 h with DMSO as cosolvent. Quetiapine (**1**) free base is usually upgraded by an acid–base extraction procedure. In some cases the free base yield and purity data are provided, but these may be misleading. The best indicators of process efficiency are yield and purity data for the hemifumarate. The hemifumarate is prepared in essentially quantitative yield from the free base (neat or as a concentrated solution in toluene or xylenes) and fumaric acid (0.55 equivalents) in ethanol, methanol, or ethyl acetate.[70,71]

The reaction of 11-chlorodibenzo[*b,f*][1,4]thiazepine (**25**) with 2-(2-(piperazin-1-yl)ethoxy)ethanol (**26**) (2.0 equivalents) in refluxing xylenes is complete in <30 h. A brown oil separates during the reaction. The suspension is cooled and dilute sodium hydroxide and ethyl ether are added. The layers are separated and the aqueous layer is extracted with ethyl ether. Quetiapine (**1**) is extracted from the combined organic layers with dilute hydrochloric acid. Aqueous sodium carbonate is added to the acidic extract to precipitate a brown oil that is extracted into dichloromethane. The organic extracts are dried and concentrated at reduced pressure and quetiapine (**1**) free base is isolated from the viscous amber oil by chromatography (78%). The free base is dissolved in ethanol and fumaric acid (1.1 equivalents) is added. The mixture is heated to reflux. The resulting suspension is cooled to 25°C and the solid is filtered and dried at an unspecified temperature to afford quetiapine (**1**) hemifumarate (99% from the free base). No purity data is provided.[15]

Another reaction of 11-chlorodibenzo[*b,f*][1,4]thiazepine (**25**) with 2-(2-(piperazin-1-yl)ethoxy)ethanol (**26**) (2.0

equivalents) in refluxing xylenes is complete in 6 h. The suspension is cooled and water and hydrochloric acid are added (to pH 4.5–5.5). The layers are separated and the aqueous acidic layer is washed with xylenes. Isopropyl acetate is added followed by 50% sodium hydroxide (to pH 9–10). The layers are separated and the aqueous layer is extracted with isopropyl acetate. The combined organic extracts containing quetiapine (**1**) crude (78% from **2**) are concentrated to a small volume at reduced pressure and <60°C. Methanol is added and the solution again concentrated to a small volume at reduced pressure and <60°C. Methanol is added and the solution heated to 50–55°C. A 50°C solution of fumaric acid (0.50 equivalents) in methanol is added while maintaining the temperature at 50–55°C. The resulting suspension is cooled to 25°C and the solid is filtered and dried at 60°C to afford quetiapine (**1**) hemifumarate (72% for three steps from **2**, 99.72% pure by HPLC). Higher yield and purity (77%, 99.91% pure by HPLC) are reported for a similar process where the quetiapine (**1**) hemifumarate is first isolated then recrystallized from methanol.[57]

The reaction of 11-chlorodibenzo[*b,f*][1,4]thiazepine (**25**) with 2-(2-(piperazin-1-yl)ethoxy)ethanol (**26**) (2.0 equivalents) in refluxing toluene is complete in 8 h. The mixture is cooled and washed twice with water. The organic layer is dried by azeotropic distillation and concentrated at reduced pressure to afford quetiapine (**1**) crude (98%). The crude free base is dissolved in ethanol and fumaric acid (0.60 equivalents) is added. The suspension is heated to reflux and then cooled to 30–35°C and aged for 2 h. The suspension is filtered and the solid is washed with ethanol and dried at 65°C to afford quetiapine (**1**) hemifumarate (94% for three steps from **2**).[55]

The reaction of 11-chlorodibenzo[*b,f*][1,4]thiazepine (**25**) with 2-(2-(piperazin-1-yl)ethoxy)ethanol (**26**) (1.1 equivalents) and potassium carbonate (1.0 equivalent) in refluxing toluene is complete in 6 h. The mixture is cooled and washed three times with water. After carbon treatment at 90–100°C, the suspension is filtered and the filtrate is concentrated at reduced pressure. The residual oil is dissolved in ethanol and fumaric acid (0.50 equivalents) is added at 25°C. The suspension is aged 30 min and filtered and the solid is washed with ethanol and dried at 50°C to afford quetiapine (**1**) hemifumarate (90%, 99.9% pure by HPLC). The yield and purity are not compromised when water is used as a cosolvent in the reaction.[60]

The reaction of 11-chlorodibenzo[*b,f*][1,4]thiazepine (**25**) with 2-(2-(piperazin-1-yl)ethoxy)ethanol (**26**) (1.1 equivalents), potassium carbonate (2.0 equivalents), and potassium iodide (0.5 equivalents) in refluxing toluene is complete in <12 h. The mixture is cooled and quetiapine (**1**) is extracted into dilute hydrochloric acid. The layers are separated and aqueous sodium hydroxide is added to the aqueous acid layer (to pH 9). The pH 9 aqueous layer is extracted with toluene. The toluene extracts are concentrated at reduced pressure to afford quetiapine (**1**) crude (78%). Quetiapine (**1**) (78%) results using xylenes as the reaction solvent. The yield is lower using *n*-butanol as the reaction solvent (62%). To prepare the hemifumarate salt, the toluene or xylenes solution from the workup is concentrated to 5% of the original volume. Ethanol is added and the solution is treated with carbon and filtered. Fumaric acid is added and the mixture is refluxed for 2 h. The resulting suspension is cooled to 25°C and the solid is filtered, washed with ethanol, and dried at 50–60°C to afford quetiapine (**1**) hemifumarate. The yield from 11-chlorodibenzo[*b,f*][1,4]thiazepine (**25**) is 57–63% using toluene or xylenes as reaction solvent and 45–51% using *n*-butanol. The yield using sodium carbonate and sodium iodide in toluene is 45–58%. Purity data for quetiapine (**1**) hemifumarate is not provided.[72]

The reaction of 11-chlorodibenzo[*b,f*][1,4]thiazepine (**25**) with 2-(2-(piperazin-1-yl)ethoxy)ethanol (**26**) dihydrochloride monohydrate (1.1 equivalents) and sodium carbonate (5.8 equivalents) in chlorobenzene–DMSO is complete in 6–8 h at 85–90°C. The solution is cooled and water and ethyl acetate are added. The layers are separated and the aqueous phase extracted with ethyl acetate. The combined organic layers containing quetiapine (**1**) crude are washed with water and brine. Fumaric acid (0.60 equivalents) is added in four portions at 58–62°C. The resulting suspension is aged at 58–62°C for 30 min then cooled to 0°C. The solid is filtered, washed with cold ethyl acetate, and dried at 50°C to afford quetiapine (**1**) hemifumarate (93% yield for three steps from **2**, 99.92% pure by HPLC).[53]

***5.3.4.2 Dibenzo[b,f][1,4]thiazepin-11(10H)-one (2) and 2-(2-(Piperazin-1-yl)ethoxy)ethanol (26)*** Quetiapine (**1**) can be prepared in a single step from dibenzo[*b,f*][1,4]thiazepin-11(10*H*)-one (**2**) using a titanium reagent. The reaction of dibenzo[*b,f*][1,4]thiazepin-11(10*H*)-one (**2**) with 2-(2-(piperazin-1-yl)ethoxy)ethanol (**26**) (1.8–3.0 equivalents) and titanium tetraisopropoxide (1.7–3.0 equivalents) is complete in 6 h at 170°C. The isopropanol by-product is distilled from the mixture as the reaction proceeds. The mixture is cooled to 100°C and toluene is added. Excess titanium reagent is quenched with water, the 25°C suspension is filtered, and the solid washed with toluene. The combined liquors are washed with water and concentrated at reduced pressure to afford quetiapine (**1**) crude (86–94%). Fumaric acid (0.50 equivalents) is added to quetiapine (**1**) crude in ethanol. The mixture is refluxed for 2 h. The resulting suspension is cooled to 0°C and the solid filtered, washed with ethanol at 0°C, and dried at an unspecified temperature to afford quetiapine (**1**) hemifumarate (70–76% from **2**). No purity data is provided (Scheme 5.12).[65]

While the case for a single-step process from dibenzo[*b,f*][1,4]thiazepin-11(10*H*)-one (**2**) is compelling, there are

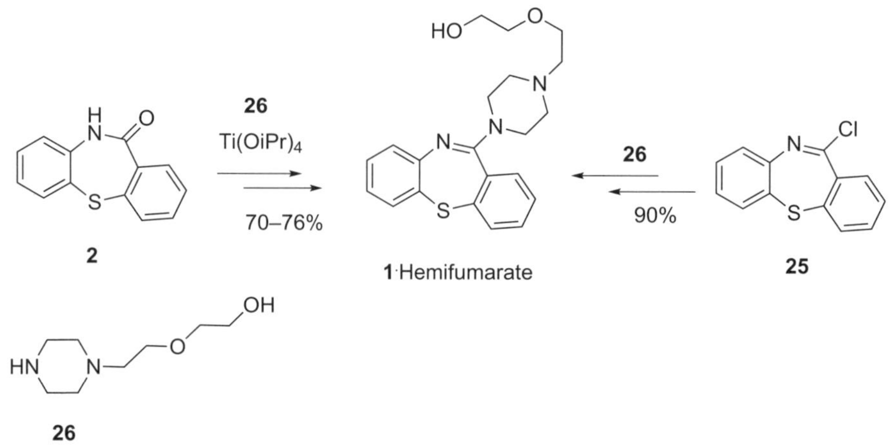


**SCHEME 5.12** Quetiapine (**1**) hemifumarate from 11-chlorodibenzo[*b*,*f*][1,4]thiazepine (**25**) and dibenzo[*b*,*f*][1,4]thiazepin-11(10*H*)-one (**2**).

scale-up concerns. Excess (2–3 equivalents) of the expensive amine **26** is required, maintaining 170°C for 6 h is not routine in a manufacturing plant, the water quench will be more challenging as the scale increases, and the solid titanium-containing by-product must be filtered and processed for disposal.

***5.3.4.3 11-(Piperazin-1-yl)dibenzo[b,f][1,4]thiazepine (33) and 2-(2-Chloroethoxy)ethanol*** Quetiapine (**1**) is most often produced by alkylation of 11-(piperazin-1-yl) dibenzo[*b*,*f*][1,4]thiazepine (**33**) with 2-(2-chloroethoxy) ethanol. 11-(Piperazin-1-yl)dibenzo[*b*,*f*][1,4]thiazepine (**33**) is prepared from the dihydrochloride salt in a separate step or generated *in situ*. An inorganic or organic base ($K_2CO_3$, $Na_2CO_3$, $Et_3N$) is added so that the amine **33** is not consumed by salt formation. The alkylation reaction can be run neat or in *n*-butanol, *n*-propanol, DMF, toluene, or water. Sodium iodide, a phase transfer catalyst (tetrabutylammonium bromide), and polar aprotic solvents (DMF, NMP), perhaps in combination, are added as promoters. Reaction times are 6–24 h. The workup in some cases includes an acid–base extraction procedure, although the procedure cannot separate the starting material **33** from the product **1** since they are both amines. Since secondary amine **33** is alkylated to produce tertiary amine **1**, competitive quaternization of the tertiary amine is a concern, especially as the alkylation approaches complete conversion. Free base yield and purity data are provided in some cases, but these may be misleading. The best indicators of process efficiency are yield and purity data for quetiapine (**1**) hemifumarate. The hemifumarate is prepared in essentially quantitative yield from the free base (as a solution in butanol or as a concentrated solution in toluene) and fumaric acid in an alcohol (methanol, ethanol, isopropanol, *n*-propanol, *n*-butanol), ethyl acetate, or acetone.

*Alkylation in n-Butanol* The reaction of 11-(piperazin-1-yl)dibenzo[*b*,*f*][1,4]thiazepine (**33**) with 2-(2-chloroethoxy) ethanol (1.1 equivalents) and sodium carbonate (1.5 equivalents) in *n*-butanol is refluxed for 10–12 h. The mixture is cooled and concentrated at reduced pressure. The residue is separated between ethyl acetate and water. The organic layer is extracted with dilute hydrochloric acid. Aqueous ammonia is added to the acidic extracts and the mixture is extracted with ethyl acetate. The organic extracts are concentrated at reduced pressure to afford quetiapine (**1**) crude. No yield or purity data is available.[30]

11-(Piperazin-1-yl)dibenzo[*b*,*f*][1,4]thiazepine (**33**) is prepared from the dihydrochloride salt by reaction with 50% sodium hydroxide (1.1 equivalents) in *n*-butanol–water at 20–30°C. The layers are separated, the organic layer is dried by distillation of the *n*-butanol–water azeotrope at reduced pressure (maximum temperature 55–60°C), and the *n*-butanol solution is cooled and carried forward. The reaction with 2-(2-chloroethoxy)ethanol (1.2 equivalents), extra fine potassium carbonate (1.0 equivalent), and sodium iodide (0.45 equivalents) in *n*-butanol is complete in 12–16 h at 102°C. The suspension is cooled to 80°C and water is added. The layers are separated at 40–50°C and the organic layer washed with water at 40–50°C. The organic layer is dried by azeotropic distillation at 55–65°C. Isopropanol is added and the solution heated to 80–85°C. A solution of fumaric acid in isopropanol at 80–85°C is added at 80°C. The resulting suspension is aged at 80°C for 1.5 h then cooled to 5°C. The solid is filtered, washed with cold isopropanol, and dried at an unspecified temperature. This quetiapine (**1**) hemifumarate crude is recrystallized from water and from isopropanol to remove impurities. After one recrystallization from water, the solid is filtered and dried at 70–80°C (74% from **33**).[56]

There is data to suggest that addition of a phase transfer catalyst increases the reaction rate in *n*-butanol. The reaction of 11-(piperazin-1-yl)dibenzo[*b,f*][1,4]thiazepine (**33**) dihydrochloride with 2-(2-chloroethoxy)ethanol (1.3 equivalents), sodium carbonate (6.0 equivalents), and sodium iodide (4.4 mol%) in *n*-butanol still contains 4.3% 11-(piperazin-1-yl)dibenzo[*b,f*][1,4]thiazepine (**33**) after 17 h at reflux. The same reaction with tetrabutylammonium bromide (0.21 equivalents) added contains 0.79% 11-(piperazin-1-yl)dibenzo[*b,f*][1,4]thiazepine (**33**) after 17 h at reflux.[73]

The reaction of 11-(piperazin-1-yl)dibenzo[*b,f*][1,4]thiazepine (**33**) with 2-(2-chloroethoxy)ethanol (1.2 equivalents), sodium carbonate (1.8 equivalents), and tetrabutylammonium bromide (0.23 equivalents) is complete in 6–7 h in refluxing *n*-butanol. The suspension is cooled and filtered and the solid washed with *n*-butanol. Fumaric acid (0.50 equivalents) is added to the combined liquors. The resulting suspension is filtered and the solid recrystallized from ethanol and dried at 50–55°C to afford quetiapine (**1**) hemifumarate (62–67%). The same yield achieved with 11-(piperazin-1-yl)dibenzo[*b,f*][1,4]thiazepine (**33**) dihydrochloride salt, 2-(2-chloroethoxy)ethanol (1.2 eqvivalents) sodium carbonate (2.0 equivalents), and tetrabutylammonium bromide (0.23 equivalents) in 18–20 h in refluxing *n*-butanol.[59]

Purity data (HPLC) is sometimes reported as area percent, percent purity or, rarely, as weight percent. A weight percent (wt%) is determined by analysis of a sample relative to a reference sample. The reference sample is usually produced by multiple recrystallizations, perhaps recrystallizations from different solvents. The reference sample is rigorously analyzed by several methods and determined to be free of impurities. The sample area count and reference sample area count for same-weight samples are compared to arrive at a weight percent.

An area percent is determined by analysis of a sample without regard to a reference sample. Thus, it is possible that a sample is 98 area% by HPLC but only 80 wt% if the impurity is not detected in the HPLC analysis. Probably the most common impurity not detected is residual solvent. We must assume the authors are reporting a weight percent, not an area percent, when purity is reported as percent purity.

The reaction of 11-(piperazin-1-yl)dibenzo[*b,f*][1,4]thiazepine (**33**) dihydrochloride with 2-(2-chloroethoxy)ethanol (1.3 equivalents), sodium carbonate (6.0 equivalents), sodium iodide (4.0 mol%) and tetrabutylammonium bromide (0.21 equivalents) in *n*-butanol is refluxed for 24 h. The mixture is cooled and the solid is filtered and washed with *n*-butanol. Fumaric acid (0.50 equivalents) is added to the liquors and the solution heated at 100°C. The solution is cooled and the resulting precipitate is filtered, recrystallized from ethanol or *n*-butanol, and dried at an unspecified temperature to afford quetiapine (**1**) hemifumarate (60–71%). No purity data is provided.[73]

*Alkylation in n-Propanol* The reaction of 11-(piperazin-1-yl)dibenzo[*b,f*][1,4]thiazepine (**33**) dihydrochloride salt with 2-(2-chloroethoxy)ethanol (2.0 equivalents), sodium carbonate (6.0 equivalents), and sodium iodide (5.0 mol%) is complete in 24 h in refluxing *n*-propanol–DMF. The suspension is cooled and the solids are filtered. The liquors are concentrated at reduced pressure. The residue is separated between dichloromethane and water. The layers are separated and the organic layer is concentrated at reduced pressure. The residue is dissolved in ethanol and a solution of fumaric acid (0.55 equivalents) in ethanol is added. The resulting suspension is filtered and the solid dried at 60–70°C to afford quetiapine (**1**) hemifumarate crude. The solid is recrystallized from methanol and dried at 60–70°C to afford quetiapine (**1**) hemifumarate (81%, 98.8% pure).[54]

The reaction of 11-(piperazin-1-yl)dibenzo[*b,f*][1,4]thiazepine (**33**) dihydrochloride salt with 2-(2-chloroethoxy)ethanol (1.1 equivalents), sodium carbonate (6.0 equivalents), and sodium iodide (4.0 mol%) is complete in 24 h in refluxing *n*-propanol–NMP. The mixture is cooled, diluted with ethyl acetate, and washed twice with water. The organic layer is concentrated at reduced pressure. The residual oil is dissolved in ethanol and fumaric acid (0.55 equivalents) is added. The resulting suspension is filtered and the solid is presumably washed with ethanol and dried at an unspecified temperature to afford quetiapine (**1**) hemifumarate (71%). No purity data is available.[35]

*Alkylation in Toluene* There is data to suggest that addition of a phase transfer catalyst increases the alkylation reaction rate in toluene. The reaction of 11-(piperazin-1-yl)dibenzo[*b,f*][1,4]thiazepine (**33**) dihydrochloride with 2-(2-chloroethoxy)ethanol (1.3 equivalents), sodium carbonate (6.0 equivalents), and sodium iodide (4.4 mol%) in toluene still contains 7.1% 11-(piperazin-1-yl)dibenzo[*b,f*][1,4]thiazepine (**33**) after 17 h at reflux. The same reaction with tetrabutylammonium bromide (0.21 equivalents) added contains 0.45% 11-(piperazin-1-yl)dibenzo[*b,f*][1,4]thiazepine (**33**) after 17 h at reflux.[73]

The reaction of 11-(piperazin-1-yl)dibenzo[*b,f*][1,4]thiazepine (**33**) with 2-(2-chloroethoxy)ethanol (1.2 equivalents), sodium carbonate (1.8 equivalents), and tetrabutylammonium bromide (0.23 equivalents) is complete in 10–12 h in refluxing toluene. The mixture is cooled and washed with water. The water wash is back-extracted with toluene. Quetiapine (**1**) is extracted from the combined organic layers with dilute hydrochloric acid (pH 2–3). The aqueous acidic extract is washed with toluene. Toluene and sodium carbonate are added (to pH 8–10). The layers are separated and the aqueous layer extracted with toluene. The combined organic extracts are washed with water and dried by distillation of the toluene–water azeotrope at reduced pressure and <50°C. The concentrated toluene solution of quetiapine (**1**) free base is diluted with ethanol, refluxed with activated carbon, and filtered. Fumaric acid (0.48 equivalents) is added at 50°C. The suspension is refluxed for 2 h then cooled to 25°C. The solid is filtered, washed with ethanol, and dried at 50–55°C to afford quetiapine (**1**) hemifumarate (62–67%).[59]

The reaction of 11-(piperazin-1-yl)dibenzo[*b*,*f*][1,4]thiazepine (**33**) dihydrochloride with 2-(2-chloroethoxy)ethanol (1.3 equivalents), sodium carbonate (6.2 equivalents), sodium iodide (4.1 mol%), and tetrabutylammonium bromide (0.21 equivalents) in toluene is refluxed for 24–40 h. The mixture is dried by distillation of the toluene–water azeotrope. The suspension is cooled and the solid filtered and washed with toluene. Fumaric acid (0.50 equivalents) is added to the toluene liquors and the solution heated at 100°C. The solution is cooled and the resulting precipitate is filtered, recrystallized from ethanol, and dried at an unspecified temperature to afford quetiapine (**1**) hemifumarate (73%). No purity data is provided. An alternative workup, similar to one used in other processes, gives a lower yield. The toluene liquors from the filtration are washed with water and concentrated to a small volume. Ethanol is added followed by fumaric acid (0.50 equivalents). The solution is refluxed. The resulting suspension is cooled and the solid is filtered, recrystallized from ethanol, and dried at an unspecified temperature to afford quetiapine (**1**) hemifumarate (66%). Again, no purity data is provided.[73]

The reaction of 11-(piperazin-1-yl)dibenzo[*b*,*f*][1,4]thiazepine (**33**) dihydrochloride salt with 2-(2-chloroethoxy)ethanol (1.1 equivalents), sodium carbonate (1.8 equivalents), and sodium iodide (0.16 equivalents) in toluene–NMP is refluxed for 24 h. The mixture is cooled, washed with water, and dried by distillation of the toluene–water azeotrope. A solution of fumaric acid (0.60 equivalents) in ethanol is added at 60–70°C. The resulting suspension is cooled and the solid is filtered, probably washed with ethanol, and dried at an unspecified temperature to afford quetiapine (**1**) hemifumarate (78%). No purity data is available.[15]

Starting with a toluene solution of 11-(piperazin-1-yl)dibenzo[*b*,*f*][1,4]thiazepine (**33**) prepared from dibenzo[*b*,*f*][1,4]thiazepin-11(10*H*)-one (**2**), the reaction with 2-(2-chloroethoxy)ethanol (0.82 equivalents based on **2**), sodium carbonate (2.1 equivalents), and sodium iodide (0.10 equivalents) in toluene–NMP is refluxed for 8 h. The mixture is cooled and washed with water twice. The organic layer is dried by distillation of the toluene–water azeotrope to afford a concentrated toluene solution of quetiapine (**1**) crude (80% yield from **2**). The solution is diluted with ethanol and fumaric acid (0.60 equivalents) is added. The resulting solution is aged at 80–85°C for 30 min, then cooled to 30°C. The solid is filtered, washed with ethanol, and dried at 65°C to afford quetiapine (**1**) hemifumarate (77% for the four steps from **2**) (Scheme 5.13).[55,74]

*Alkylation in DMF* A completion check for the reaction of 11-(piperazin-1-yl)dibenzo[*b*,*f*][1,4]thiazepine (**33**) dihydrochloride with 2-(2-chloroethoxy)ethanol (1.3 equivalents), sodium carbonate (6.0 equivalents), and sodium iodide (4.5 mol%) in DMF at 103°C after 18 h reveals 95.4% quetiapine (**1**) and 1.1% 11-(piperazin-1-yl)dibenzo[*b*,*f*][1,4]thiazepine (**33**). A nearly identical mixture is produced at 103°C after 18 h when tetrabutylammonium bromide (0.21 equivalents) is added.[73]

*Alkylation in Water* The reaction of 11-(piperazin-1-yl)dibenzo[*b*,*f*][1,4]thiazepine (**33**) with 2-(2-chloroethoxy)ethanol (1.2–1.5 equivalents), sodium carbonate (4.8–6.0 equivalents), and sodium iodide (3.2–3.9 mol%) is complete in 9 h in water at 100–105°C (<0.1% **33** remains by HPLC). The mixture is cooled to 30°C and extracted with dichloromethane. The organic layer is washed with water and concentrated at reduced pressure. The residue is suspended in ethanol and fumaric acid (0.85–1.0 equivalents) is added. After aging at 25°C for 6 h, the suspension is cooled to 10°C and the solid is filtered and dried at an unspecified temperature to afford quetiapine (**1**) hemifumarate (80%, 99.5–99.6% pure by HPLC). The time to complete conversion is 7 h when tetrabutylammonium bromide (9.1 mol%) is also added (80%, 99.8% pure by HPLC).[75]

*Alkylation Neat* The reaction can be run with no solvent when triethylamine is the base. The reaction of 11-(piperazin-1-yl)dibenzo[*b*,*f*][1,4]thiazepine (**33**) with 2-(2-chloroethoxy)ethanol (1.5 equivalents), triethylamine (1.6 equivalents), and sodium iodide (3.9 mol%) is complete in 5 h at 80–85°C (<0.1% **33** remains by HPLC). The mixture is cooled to 30°C and separated between water and dichloromethane. The organic layer is washed with water and concentrated at reduced pressure. The residue is suspended in ethanol and fumaric acid (0.85 equivalents) is added. After aging at 25°C for 6 h, the suspension is cooled to 15°C and the solid is filtered and dried at an unspecified temperature to afford quetiapine (**1**) hemifumarate (80%, 99.8% pure by HPLC).[75]

***5.3.4.4 Reductive Alkylation*** Some quaternary ammonium salt is likely produced when the alkylation of a secondary amine with a primary alkyl chloride is pushed to completion. Perhaps the reaction of 11-(piperazin-1-yl)dibenzo[*b*,*f*][1,4]thiazepine (**33**) with 2-(2-chloroethoxy)ethanol generates quaternary ammonium salt as a side product. A tertiary amine can usually be produced in higher yield and purity by a reductive alkylation of a secondary amine with an aldehyde. For a reductive alkylation to produce quetiapine (**1**), the

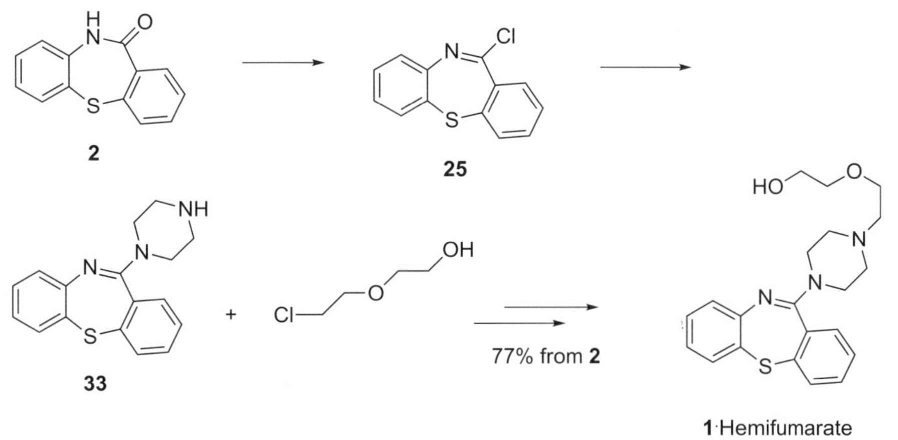


**SCHEME 5.13** Quetiapine (**1**) hemifumarate by alkylation of 11-(piperazin-1-yl)dibenzo[*b*,*f*][1,4]thiazepine (**33**).

required aldehyde is 2-(2-hydroxyethoxy)acetaldehyde (**36**). Aldehyde **36** is a specialty chemical and is likely produced from chloroacetaldehyde dimethyl acetal and ethylene glycol. Sodium borohydride (38 equivalents) is added in portions to a mixture of 11-(piperazin-1-yl)dibenzo[*b*,*f*][1,4]thiazepine (**33**), (2-hydroxyethoxy)acetaldehyde (**36**) (1.0 equivalent), and sodium acetate (2.9 equivalents) in acetic acid at 0°C. After aging the mixture at 10°C for 1 h, aqueous sodium hydroxide is added (to pH 8–9) and the mixture is extracted with ethyl ether. The combined extracts are dried and concentrated at reduced pressure to afford quetiapine (**1**) free base (90%, about 98% pure by HPLC). The free base is not converted to quetiapine (**1**) hemifumarate. Quetiapine (**1**) (84%, 90% pure by HPLC) is also produced using magnesium borohydride in place of sodium borohydride. The reductive alkylation of 11-(piperazin-1-yl)dibenzo[*b*,*f*][1,4]thiazepine (**33**) with 2-(2-oxoethoxy)ethyl acetate (**37**) (1.0 equivalent), an ester-protected form of (2-hydroxyethoxy)acetaldehyde (**26**), also affords quetiapine (**1**) (77%, 82% pure by HPLC) (Scheme 5.14).[76]

***5.3.4.5 2-(4-(Dibenzo[b,f][1,4]thiazepin-11-yl)piperazin-1-yl)ethanol (35) and Ethylene Glycol*** 2-(4-(Dibenzo[*b*,*f*][1,4]thiazepin-11-yl)piperazin-1-yl)ethanol (**35**) is converted to quetiapine (**1**) hemifumarate by Williamson ether synthesis (Figure 5.5). Methanesulfonyl chloride (1.5 equivalents) is added dropwise to 2-(4-(dibenzo[*b*,*f*][1,4]thiazepin-11-yl)piperazin-1-yl)ethanol (**35**) and triethylamine (2.0 equivalents) in dichloromethane at 0–5°C. The mixture is aged at 25°C for 2 h then quenched with water. The layers are separated and the organic layer is washed with aqueous bicarbonate and concentrated at reduced pressure. A toluene solution of the crude methanesulfonate **38** is added to a solution prepared from sodium hydride (0.99 equivalents) and ethylene glycol. The suspension is refluxed (110–120°C) for 10–12 h, cooled to 25°C, and diluted with water. The layers are separated and the organic layer is extracted with dilute hydrochloric acid. Ammonia is added to the aqueous acidic solution and the mixture is extracted with ethyl acetate. The organic extract is concentrated at reduced pressure. The residue is dissolved in acetone and fumaric

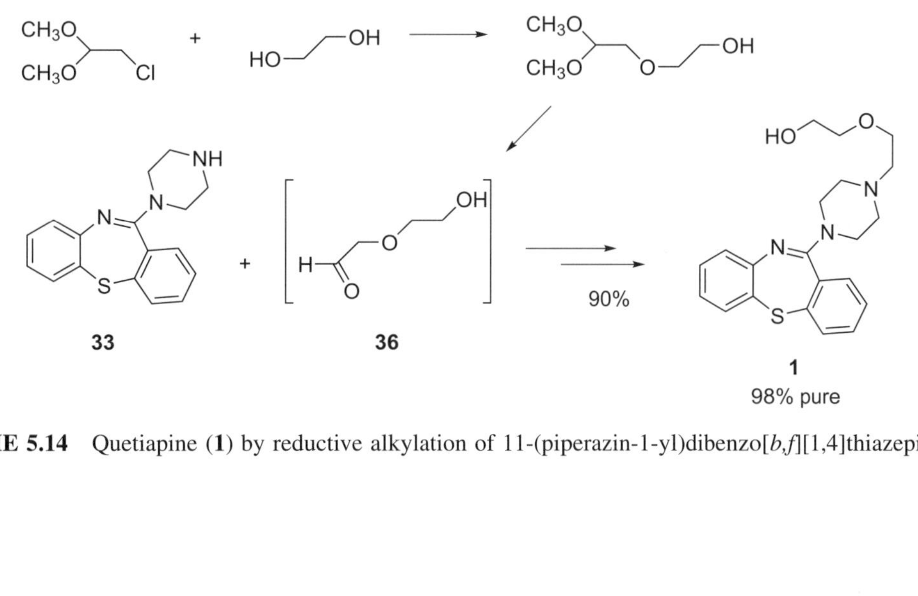


**SCHEME 5.14** Quetiapine (**1**) by reductive alkylation of 11-(piperazin-1-yl)dibenzo[*b*,*f*][1,4]thiazepine (**33**).

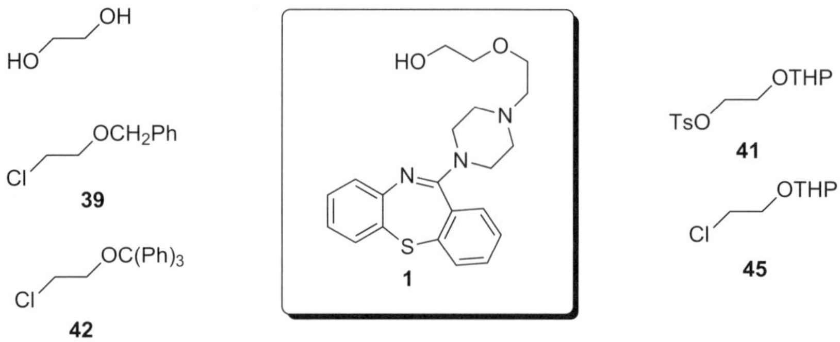


**FIGURE 5.5** Reagents for introducing two carbons of the quetiapine side chain by Williamson ether synthesis.

acid (0.55 equivalents) is added. After aging at 25°C, the suspension is filtered and the solid is washed with acetone and dried at an unspecified temperature to afford quetiapine (**1**) hemifumarate. No yield or purity data is provided.[30]

The intermediate β-aminoethyl methanesulfonate **38** is capable of cyclization to produce an aziridinium salt, a potent alkylating agent. Intermediate **38** can be avoided by reversing the reactivity of the partners in the Williamson ether synthesis. Reversing the reactivity requires a protection–deprotection strategy for the alcohol of the reaction partner. The known reaction partners have the alcohol protected as a benzyl, tetrahydropyranyl, or triphenylmethyl ether.

((2-Chloroethoxy)methyl)benzene (**39**) is prepared in a single step from commercially available ethylene glycol monobenzyl ether[77] by reaction with thionyl chloride in chloroform (60%).[78] The reaction of excess ethylene glycol with dihydropyran and catalytic *p*-toluenesulfonic acid affords 2-(tetrahydro-2*H*-pyran-2-yloxy)ethanol (**40**) (80%). The *p*-toluenesulfonate **41** is then prepared by reaction of **40** with *p*-toluenesulfonyl chloride and triethylamine in dichloromethane (99%).[79] The reaction of 2-chloroethanol[80] with triphenylmethyl chloride and triethylamine in acetonitrile affords the triphenylmethyl ether (**42**) (86%).[81]

The reaction of 2-(4-(dibenzo[*b*,*f*][1,4]thiazepin-11-yl) piperazin-1-yl)ethanol (**35**) with ((2-chloroethoxy)methyl) benzene (**39**) (4.3 equivalents), 50% aqueous sodium hydroxide (34 equivalents), and tetrabutylammonium hydrogen sulfate (9.8 mol%) is complete in 9 h at 60°C. The mixture is cooled and separated between toluene and water. The organic layer is washed with water and then extracted with dilute hydrochloric acid. Ammonium hydroxide (25%) is added to the aqueous acidic solution and the mixture is extracted with toluene. The toluene extract is concentrated at reduced pressure to afford quetiapine benzyl ether (**43**) (93%). Reaction with hydrogen bromide in acetic acid affords acetate-protected quetiapine **44** (89%). Acetate-protected quetiapine **44** is hydrolyzed with powdered potassium hydroxide in methanol to afford quetiapine (**1**) free base (94%). No purity data is available and the free base is not converted to the hemifumarate.[82]

The reaction of 2-(4-(dibenzo[*b*,*f*][1,4]thiazepin-11-yl) piperazin-1-yl)ethanol (**35**) with 2-(tetrahydro-2*H*-pyran-2-yloxy)ethyl 4-methylbenzenesulfonate (**41**) (4.3 equivalents), 50% aqueous sodium hydroxide (34 equivalents), and tetrabutylammonium hydrogen sulfate (9.8 mol%) is complete in 8 h at 60–65°C. The mixture is cooled and separated between toluene and water. The organic layer is washed twice with water and then extracted with dilute hydrochloric acid. Deprotection of the THP ether is accomplished by aging the toluene–aqueous acid mixture for 3 h at 25°C. The layers are separated and the aqueous acidic extract is washed with *n*-butanol and with toluene. Toluene and aqueous potassium carbonate are added to the aqueous acidic extract (to pH 10). The layers are separated and the aqueous layer is extracted with toluene. The combined toluene extracts are concentrated at reduced pressure to afford quetiapine (**1**) free base (90%). No purity data is available and the free base is not converted to the hemifumarate. Lower yields are observed using powdered potassium hydroxide and 18-crown-6 or powdered potassium hydroxide and Aliquat® 336 in place of 50% aqueous sodium hydroxide–tetrabutylammonium hydrogen sulfate. Lower yields are also observed using 2-(2-chloroethoxy)tetrahydro-2*H*-pyran (**45**) in place of the *p*-toluenesulfonate.[82,52]

The only Williamson ether synthesis in this series completed through to quetiapine (**1**) hemifumarate starts with the triphenylmethyl ether **42**. The reaction of 2-(4-(dibenzo[*b*,*f*][1,4]thiazepin-11-yl)piperazin-1-yl)ethanol (**35**) with the triphenylmethyl ether **42** (1.2 equivalents), powdered potassium hydroxide (7.4 equivalents), and 18-crown-6 (5.0 mol %) is heated at 100–115°C for 2 h. Toluene and water are added, the mixture is allowed to cool to 25°C, and the layers are separated. The organic layer is washed with dilute brine and concentrated at reduced pressure. The residue is triturated with methanol–toluene and recrystallized from methanol–methyl ethyl ketone. The solid is filtered, washed with cold methanol–methyl ethyl ketone and with cold methanol, and dried at 45°C to afford triphenylmethyl-protected quetiapine **46** (82%). Reaction with *p*-toluenesulfonic acid monohydrate (1.54 equivalents) in toluene–methanol in 4 h at reflux affords quetiapine (**1**) free base (95%). The

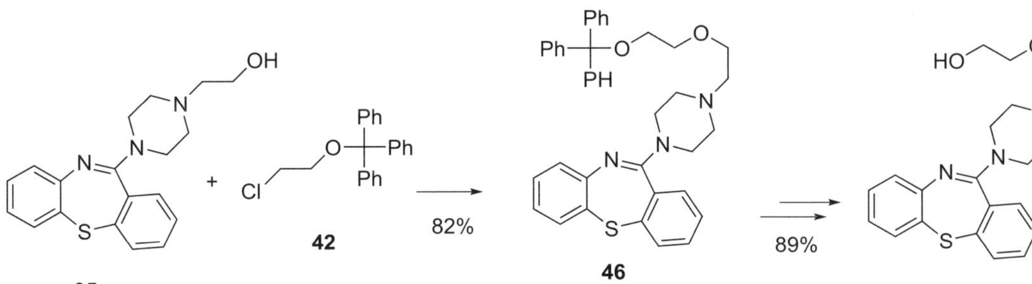


35

**SCHEME 5.15** Quetiapine (**1**) hemifumarate from 2-(4-(dibenzo[*b,f*][1,4]thiazepin-11-yl)piperazin-1-yl)ethanol (**35**) by a Williamson ether synthesis.

free base is dissolved in methanol and fumaric acid (0.55 equivalents) is added. The resulting suspension is refluxed for 5 h then cooled to 10–15°C. The solid is filtered, washed with cold methanol, and dried at 45°C to afford quetiapine (**1**) hemifumarate (94%, >99.7% pure) (Scheme 5.15).[82]

## 5.4 ALTERNATIVE METHODS FOR CONSTRUCTING THE AMIDINE LINK OF QUETIAPINE (1) HEMIFUMARATE

Consider the amidine carbon as the focal point for construction of quetiapine (**1**) (Figure 5.6). The methods presented thus far construct the amidine from dibenzo[*b,f*][1,4]thiazepin-11(10*H*)-one (**2**) by (a) activation and oxygen replacement by piperazine, 1-(2-hydroxyethyl)piperazine, or 2-(2-(piperazin-1-yl)ethoxy)ethanol (**26**). The amidine can also be constructed from a urea by (b) activation and oxygen replacement by aromatic carbon or (c) from a benzamide or by activation and oxygen replacement by an aniline nitrogen.

### 5.4.1 Thiazepine Construction by Assembly and Cyclization of a Urea

***5.4.1.1 Phenyl 2-(phenylthio)phenylcarbamate (6) and Piperazine*** The reaction of phenyl 2-(phenylthio)phenylcarbamate (**6**) with piperazine (2.0 equivalents) is complete in 1 h in refluxing toluene. The mixture is cooled and washed with aqueous sodium carbonate, and the solid is filtered and washed with toluene. The two phases of the liquors are separated and the organic layer is washed with water and with brine and concentrated at reduced pressure to afford the urea **47**. Urea **47** is reacted with phosphorus pentoxide (2.9 equivalents) and phosphorus oxychloride (8.8 equivalents) over 12 h at reflux. After cooling to 25°C, dichloromethane is added. The solid is filtered, washed with dichloromethane,

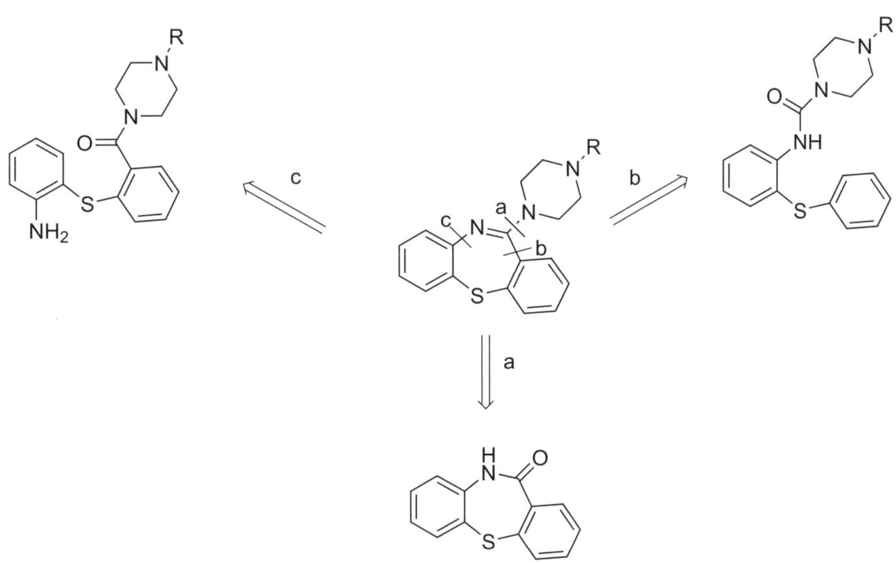


**FIGURE 5.6** The amidine carbon as the focal point for construction of quetiapine (**1**).

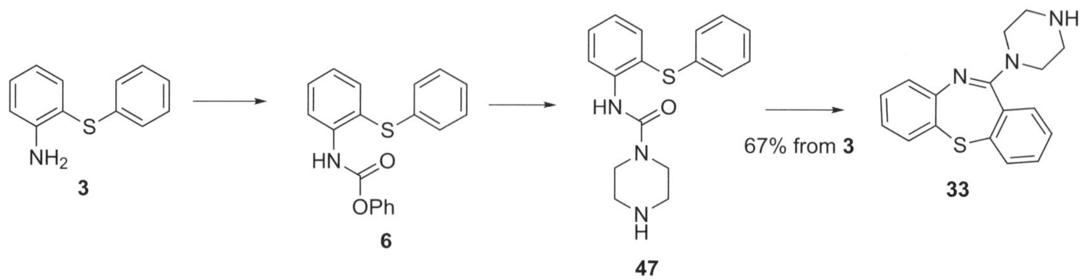


**SCHEME 5.16** 11-(Piperazin-1-yl)dibenzo[*b,f*][1,4]thiazepine (**33**) from 2-(phenylthio)aniline (**3**) via urea **47**.

and quenched (exothermic) into water. The aqueous acidic solution is washed with dichloromethane. Aqueous sodium hydroxide is added and the mixture is extracted with dichloromethane. The extracts are dried and concentrated at reduced pressure to afford 11-(piperazin-1-yl)dibenzo[*b,f*][1,4]thiazepine (**33**) as a thick colorless oil (67%). No purity data is available (Scheme 5.16).[10]

***5.4.1.2 Phenyl 2-(phenylthio)phenylcarbamate (6) and 1-(2-Hydroxyethyl)piperazine*** The reaction of phenyl 2-(phenylthio)phenylcarbamate (**6**) with 1-(2-hydroxyethyl) piperazine (1.0 equivalent) is complete in 2 h in refluxing toluene. The mixture is cooled, washed with dilute sodium hydroxide and with water, dried, and concentrated at reduced pressure. The residue is recrystallized from hexane–ethyl acetate to afford the urea **48** (95%). A mixture of urea **48**, phosphorus pentoxide (2.2 equivalents) and phosphorus oxychloride (22 equivalents) is refluxed for 7 h. After cooling to 25°C, excess phosphorus oxychloride is removed by distillation at reduced pressure. The residue is quenched with ice water. Aqueous ammonia is added and the mixture is extracted with dichloromethane. The extract is dried and concentrated at reduced pressure. The residue is recrystallized from diisopropyl ether to afford 11-(4-(2-chloroethyl) piperazin-1-yl)dibenzo[*b,f*][1,4]thiazepine (**49**) (80%).[16]

The side chain ether linkage is then created by a Williamson ether synthesis. A solution of 11-(4-(2-chloroethyl)piperazin-1-yl)dibenzo[*b,f*][1,4]thiazepine (**49**) in toluene is added to a solution prepared from sodium (1.7 equivalents) and ethylene glycol. The mixture is refluxed for 9 h. After cooling to 25°C, the mixture is washed with water and extracted with dilute hydrochloric acid. Aqueous ammonia is added and the mixture is extracted with dichloromethane. The extracts are dried and concentrated at reduced pressure to afford quetiapine (**1**) free base (98%). The free base is dissolved in ethanol and fumaric acid (1.1 equivalents) is added. The resulting suspension is refluxed for 25 min, aged overnight at 0°C, and filtered. The solid is washed with cold ethanol and dried at an unspecified temperature to afford quetiapine (**1**) hemifumarate (85%). No purity data is available (Scheme 5.17).[16]

***5.4.1.3 Phenyl 2-(phenylthio)phenylcarbamate (6) and 2-(2-(Piperazin-1-yl)ethoxy)ethanol (26)*** The reaction of 2-(phenylthio)aniline (**3**) with phosgene and 2-(2-(piperazin-1-yl)ethoxy)ethanol (**26**) affords the urea **51** possessing the full quetiapine side chain (80%). Perhaps phenyl 2-(phenylthio) phenylcarbamate (**6**) could be used in place of the carbamyl chloride **52** to avoid the use of phosgene. The alcohol is protected as a benzoate ester (97%). A mixture of the

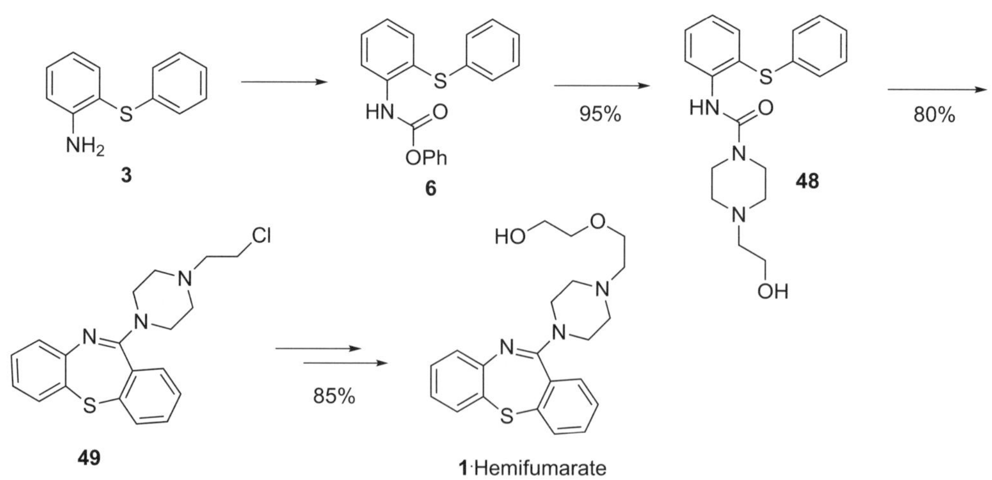


**SCHEME 5.17** Quetiapine (**1**) hemifumarate from 2-(phenylthio)aniline (**3**) via urea **48**.

benzoate-protected urea **53**, phosphorus pentoxide (3.6 equivalents), and phosphorus oxychloride (41 equivalents) is heated at 90°C for 19 h. After cooling to 25°C, excess phosphorus oxychloride is removed by distillation at reduced pressure. The residue is quenched with ice water. Sodium bicarbonate is added (to pH 7–8) and the mixture is extracted with dichloromethane. The extract is dried and concentrated at reduced pressure to afford benzoate-protected quetiapine **54**. The benzoate ester is hydrolyzed with 50% aqueous sodium hydroxide in ethanol to afford quetiapine (**1**) crude (82%) (Scheme 5.18).[83]

### 5.4.2 Thiazepine Construction by Assembly and Cyclization of a Benzamide

Amide **55** is generated by the mixed anhydride method (pivaloyl chloride, $Et_3N$, NMM) from 2-(2-nitrophenylthio) benzoic acid (**16**) and 2-(2-(piperazin-1-yl)ethoxy)ethanol (**26**) (0.98 equivalents) in dichloromethane. After a routine workup, the crude amide **55** is isolated as a viscous liquid (82–85%, 97–99% pure by HPLC). The alcohol is protected as an acetate ester by reaction with acetic anhydride and pyridine (90–92%, 97–99% pure by HPLC). The nitro group is reduced using 10% palladium on carbon and hydrogen at 140 psi and 50°C. The suspension is cooled to 25°C, the catalyst is filtered, and the liquors are concentrated at reduced pressure to afford the aniline **57** as a viscous liquid (83–87%, 97–99% pure by HPLC). The amidine is produced by reaction of the aniline **57** with phosphorus oxychloride (4.8 equivalents) over 5–6 h in toluene at reflux. The mixture is cooled and the excess phosphorus oxychloride and toluene are distilled at reduced pressure. The residue is quenched with water. Toluene and aqueous sodium hydroxide are added (to pH 8–12). The layers are separated and the aqueous phase is extracted with toluene. The toluene extract is treated with carbon and then concentrated at reduced pressure to afford acetate-protected quetiapine **44**. The acetate ester is hydrolyzed with sodium hydroxide in methanol to afford quetiapine (**1**) free base. The free base is dissolved in isopropanol and fumaric acid (0.80 equivalents) is added. The resulting suspension is refluxed for 1 h. The suspension is cooled to 30°C and the solid is filtered and dried at 60°C to afford quetiapine (**1**) hemifumarate, 66–74% from **57**). No purity data is available (Scheme 5.19).[26]

Another sequence leading to acetate-protected quetiapine **44** begins with 2-iodobenzoic acid. Amide **58** is produced by the reaction of the acid chloride, generated using thionyl chloride, with 2-(2-(piperazin-1-yl)ethoxy)ethanol (**26**) (1.0 equivalent) and triethylamine (1.1 equivalents) in THF–water at 0–5°C (95%). The iodide is displaced by reaction with 2-aminothiophenol (1.1 equivalents) and copper iodide (5 mol%) in isopropanol–ethylene glycol at reflux over 12–18 h (89%). Reaction with acetic anhydride and triethylamine results in acetylation of the alcohol and the amino group to afford **60** (84%). The reaction of **60** with phosphorus oxychloride (13 equivalents) requires 3 h at reflux. The mixture is cooled and quenched by careful addition of water and methanol. Sodium hydroxide is added (to pH 11) and the mixture is extracted with toluene. The toluene extracts are concentrated at reduced pressure to afford acetate-protected quetiapine **44** (88%). The acetate ester is hydrolyzed with 1 M sodium hydroxide in ethanol to afford quetiapine (**1**) free base (98%). No purity data is provided for the free base and the free base is not converted to the hemifumarate (Scheme 5.20).[84,85]

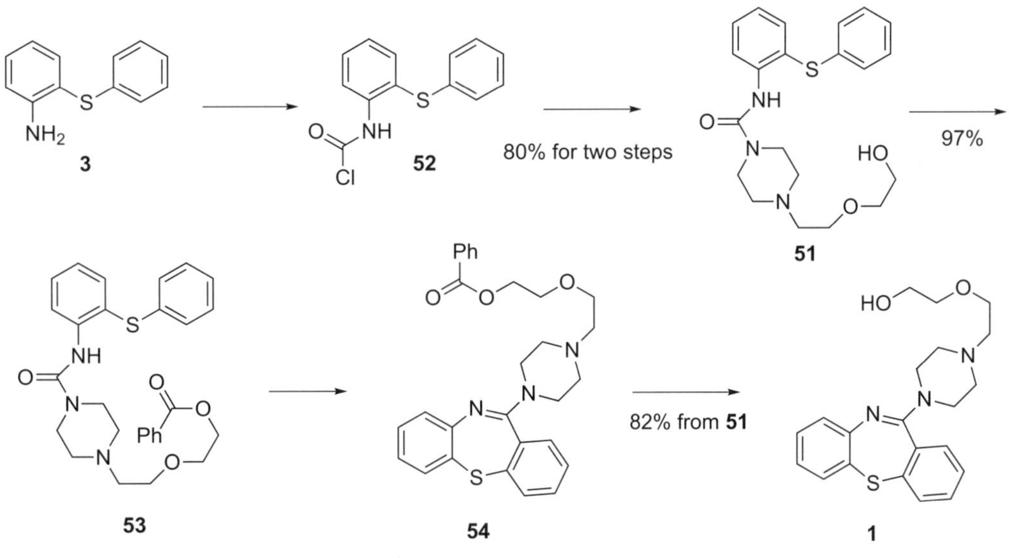


**SCHEME 5.18** Quetiapine (**1**) from 2-(phenylthio)aniline (**3**) via urea **51**.

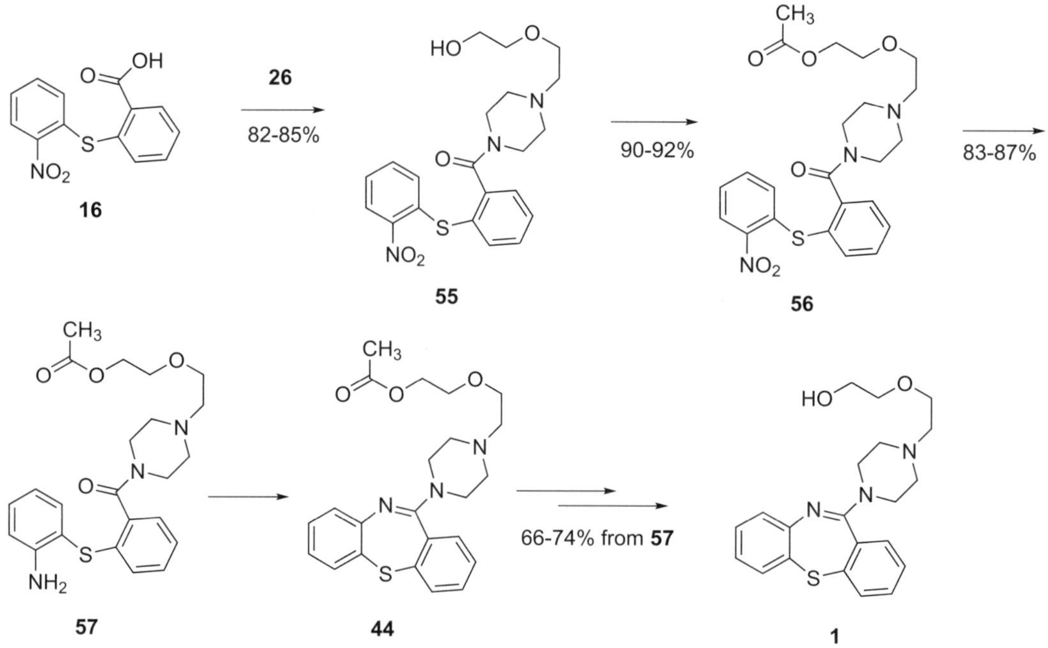


**SCHEME 5.19** Quetiapine (**1**) hemifumarate from 2-(2-nitrophenylthio)benzoic acid (**16**) via benzamide **57**.

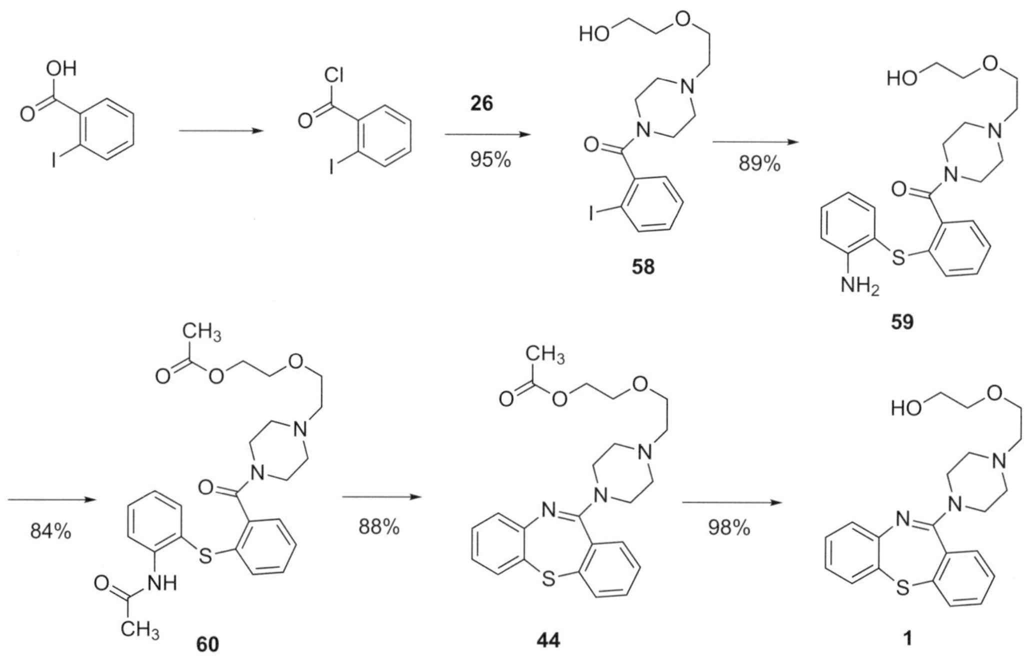


**SCHEME 5.20** Quetiapine (**1**) from 2-iodobenzoic acid via amide **60**.

## 5.5 CONSTRUCTING THE PIPERAZINE OF QUETIAPINE (1) HEMIFUMARATE

Piperazine is almost universally used to construct quetiapine (**1**) hemifumarate. Two alternative processes use inexpensive triethanolamine or diethanolamine to construct the piperazine (Scheme 5.21). A protection–deprotection strategy for the side chain alcohol is required and the preferred protecting groups are benzyl and tetrahydropyranyl. The other component is dibenzo[*b,f*][1,4]thiazepin-11-amine (**61**), prepared in two steps from 2-aminothiophenol and 2-fluorobenzonitrile.

The reaction of 2-fluorobenzonitrile with 2-aminothiophenol (1.1 equivalents) and sodium hydride (1.1 equivalents) in DMF is complete in 12 h at 25°C. Water is carefully added and the solvents are distilled at reduced pressure. The residue is taken up in chloroform, dried, concentrated at reduced pressure, and chromatographed to afford 2-(2-

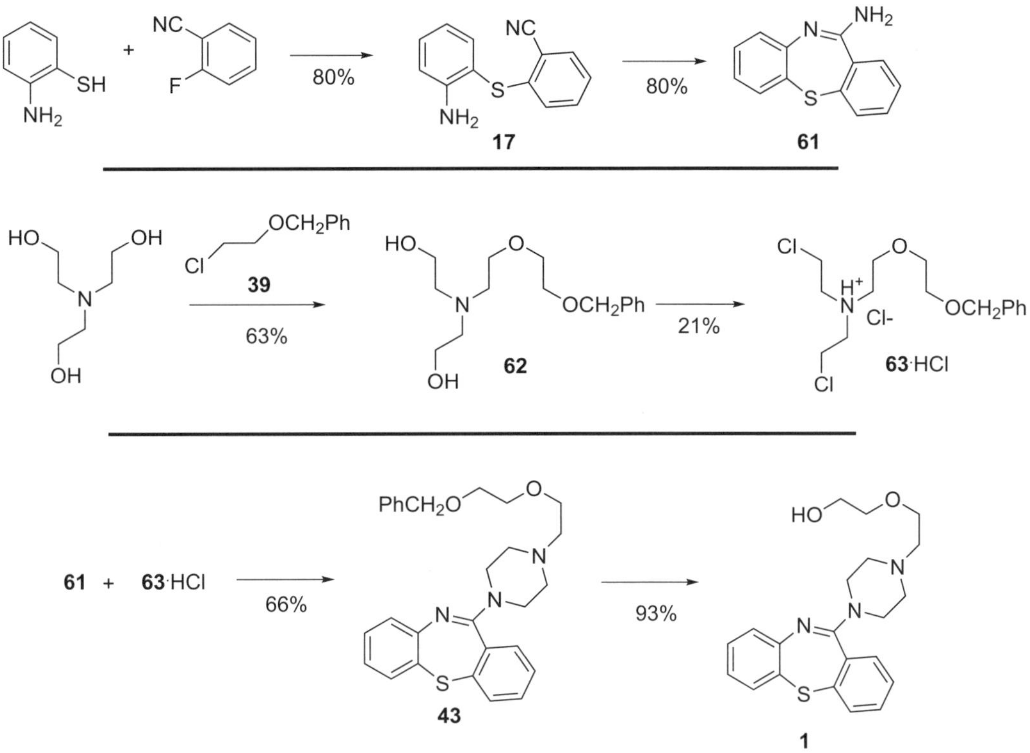

**SCHEME 5.21** Quetiapine (**1**) by construction of the piperazine ring from dibenzo[*b*,*f*][1,4]thiazepin-11-amine (**61**) and triethanolamine.

aminophenylthio)benzonitrile (**17**) (80%). Cyclization is accomplished by reaction with sodium hydride (1.5 equivalents) in THF over 2 h at 70°C. The mixture is cooled, water is carefully added, and the suspension is filtered. The solid is dried and chromatographed to afford dibenzo[*b*,*f*][1,4]thiazepin-11-amine (**61**) (80%). When the water quench and chromatography in the workup of 2-(2-aminophenylthio) benzonitrile (**17**) are omitted, the yield of dibenzo[*b*,*f*] [1,4]thiazepin-11-amine (**61**) is 54%.[86]

Again, the ether linkage in the side chain is prepared by a Williamson ether synthesis. The process begins with the reaction of triethanolamine[87] (2.9 equivalents) with sodium hydride (1.1 equivalents). ((2-Chloroethoxy)methyl)benzene (**39**) and potassium iodide (0.2 mol%) are added and the mixture aged at 140°C for 6 h to afford the diol **62** (63%). The reaction of diol **62** with thionyl chloride (6.5 equivalents) and catalytic DMF (12 mol%) in refluxing toluene affords 2-(2-(benzyloxy)ethoxy)-*N*,*N*-bis(2-chloroethyl) ethanamine (**63**) as the hydrochloride salt (21%). In the key step, sodium hydride (60%) (2.9 equivalents) is added in small portions to a solution of dibenzo[*b*,*f*][1,4]thiazepin-11-amine (**61**) in DMF at 20°C. After aging at 40°C for 15 min, the solution is cooled to 20°C and 2-(2-(benzyloxy)ethoxy)-*N*,*N*-bis-(2-chloroethyl)ethanamine (**63**) hydrochloride (1.0 equivalent) is added. The mixture is aged at 80°C for 5 h and then cooled to 25°C. DMF is removed by distillation at reduced pressure. The residue is separated between dichloromethane and water. The organic layer is dried and concentrated at reduced pressure. The residue is chromatographed to afford quetiapine benzyl ether (**43**) (66%). Deprotection is accomplished with boron trichloride (2.4 equivalents) in xylenes (93%). No purity data is available for the quetiapine (**1**) free base produced and it is not converted to the hemifumarate.[88]

The route using diethanolamine to construct the piperazine begins with the protection of 2-(2-chloroethoxy)ethanol by reaction with 3,4-dihydro-2*H*-pyran (1.5 equivalents) and Amberlyst H-15 catalyst in dichloromethane. After 17 h at 25°C, the catalyst is filtered and the solvent removed by distillation at reduced pressure. The THP ether **64** is reacted with diethanolamine[89] (2.4 equivalents), sodium carbonate (0.95 equivalents), tetrabutylammonium bromide (0.2 mol %), and sodium iodide (3.3 mol%) in DMF at 140°C. The mixture is cooled and the solvent distilled at reduced pressure. A routine workup of the residue affords the crude bis-β-(2-hydroxyethyl)amine **65** (79%). This is converted to the bis-(2-chloroethyl)amine **66** with *N*-chlorosuccinimide (2.3 equivalents) and triphenylphosphine (2.3 equivalents) in dichloromethane at 25°C (82%). In the key step, sodium hydride (60%) (5.0 equivalents) is added in small portions to a solution of dibenzo[*b*,*f*][1,4]thiazepin-11-amine (**61**) and the bis-(2-chloroethyl)amine **66** (1.9 equivalents) in DMF. The mixture is aged at 105°C for 3.5 h and then cooled and quenched with water. The water is decanted and the solid

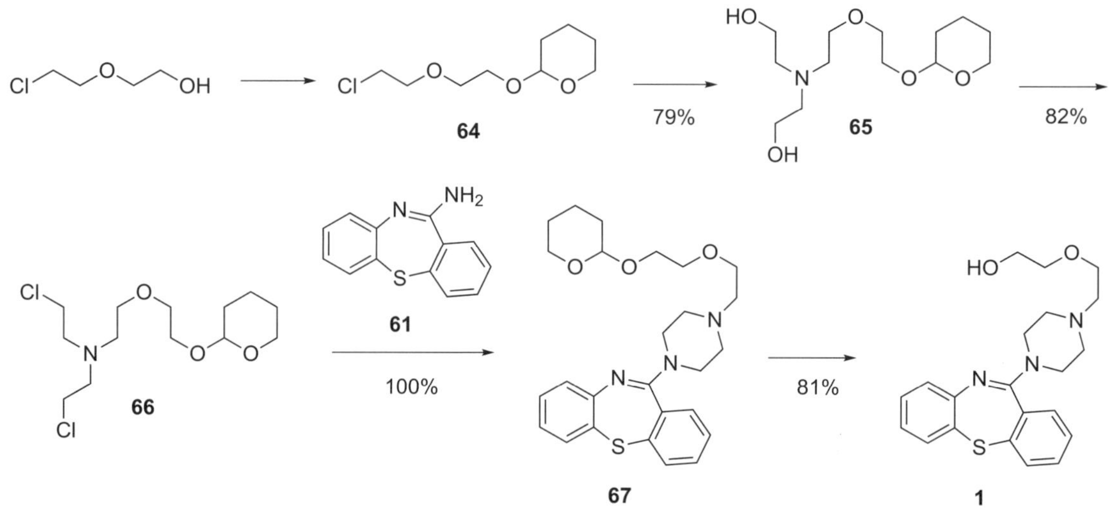

**SCHEME 5.22** Quetiapine (**1**) by construction of the piperazine ring from dibenzo[*b,f*][1,4]thiazepin-11-amine (**61**) and diethanolamine.

residue is dissolved in ethyl acetate. The solution is washed with water and the solvent distilled at reduced pressure to afford tetrahydropyranyl-protected quetiapine **67** (100%). Deprotection is accomplished with hydrochloric acid in aqueous methanol (81%). No purity data is available for the quetiapine (**1**) free base and it is not converted to the hemifumarate (Scheme 5.22).[88]

## 5.6 MEETING SPECIFICATIONS: RECRYSTALLIZATON OF QUETIAPINE (1) FREE BASE AND POLYMORPHS OF QUETIAPINE (1) HEMIFUMARATE

More than 20 processes for converting quetiapine (**1**) free base to quetiapine (**1**) hemifumarate have been presented. The process can be run in toluene, ethyl acetate, methanol, ethanol, isopropanol, *n*-butanol, and mixtures of these solvents. The preferred solvent is ethanol. The preferred fumaric acid charge is 0.50–0.60 equivalents. The mixture is usually refluxed to produce a solution and then allowed to cool to crystallize the salt. The salt is dried at 65°C and reduced pressure with no reports of decomposition during drying. The yield of quetiapine hemifumarate from chromatographed (and presumably >99% pure) quetiapine free base is 99%. The reported HPLC purities range from 98.8% to 99.9%. Quetiapine hemifumarate can be recrystallized from water, methanol, ethanol, isopropanol, or *n*-butanol. There is no information available on removal of specific impurities by recrystallization of quetiapine hemifumarate.

While there is every indication that the salt formation will produce quetiapine hemifumarate that meets specifications, a rework procedure for the penultimate intermediate, quetiapine free base, is an important feature of the manufacturing process. The free base can be recrystallized from non-aromatic solvents (ethyl acetate, isobutyl acetate, methyl isobutyl ketone, and methyl *tert*-butyl ether) in the absence of water. A preferred rework procedure begins with dissolving the free base in toluene. The free base is extracted into dilute aqueous acid, released with aqueous sodium hydroxide, and extracted into methyl *tert*-butyl ether at 45°C. The organic extract is washed with water at 45°C, dried by distillation of the water–methyl *tert*-butyl ether azeotrope at 55°C, and cooled to 25°C and seeded. The resulting suspension is cooled to 0°C and the solid filtered, washed with methyl *tert*-butyl ether, and dried at 50°C to afford quetiapine (79–86%).[90]

Crystallization of quetiapine hemifumarate from alcohol, acetone, or ethyl acetate affords polymorph I. Polymorph I is also produced when quetiapine hemifumarate is dissolved in a polar aprotic solvent (DMF, DMA, DMSO, NMP) at 80°C and precipitated with water, isobutanol, acetone, acetonitrile, ethyl acetate, isopropyl acetate, methyl *tert*-butyl ether, or toluene. Polymorph II chloroform solvate is produced when quetiapine hemifumarate is refluxed in chloroform for 6 h. Polymorph III chloroform solvate is produced when quetiapine hemifumarate is dissolved in a polar aprotic solvent at 80°C and precipitated by adding chloroform. Polymorph II dichloromethane solvate is produced by dissolving quetiapine hemifumarate in a polar aprotic solvent at 80°C and adding dichloromethane. Polymorph II and III chloroform solvates and polymorph II dichloromethane solvate are dried at reduced pressure and 65°C. The oven drying must be carefully monitored to avoid loss of the bound solvent. Polymorphs I, and II and III solvates are characterized by XRD, Fourier transform IR spectroscopy (FTIR), DSC, and TGA.[91]

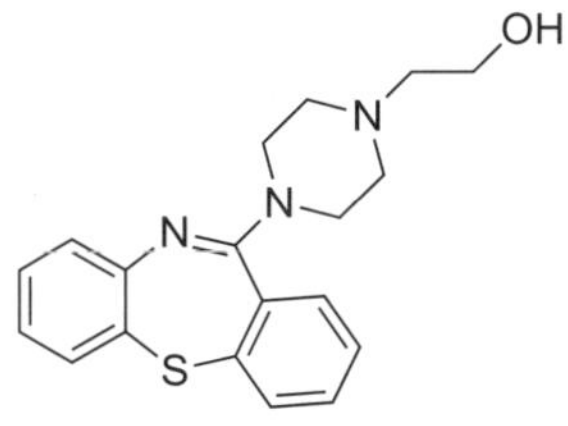

35

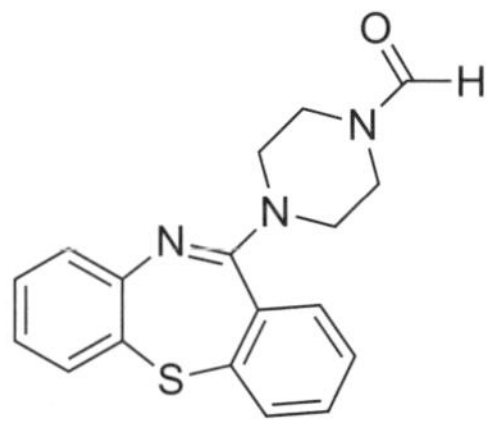

68

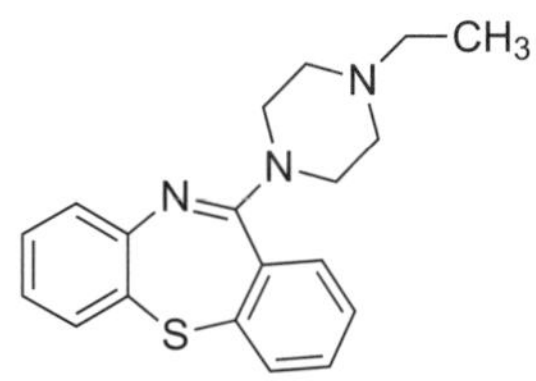

69

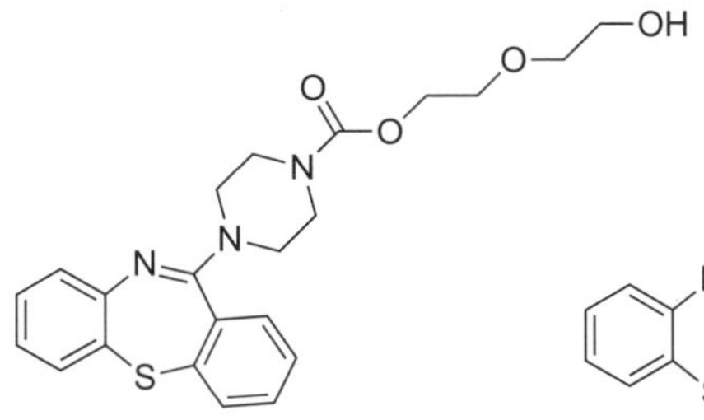

70

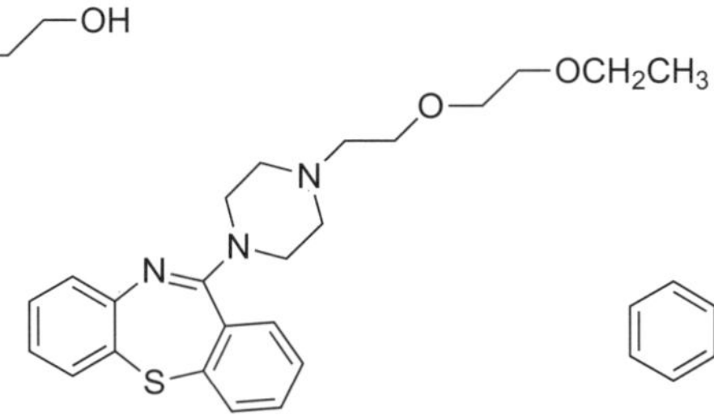

71

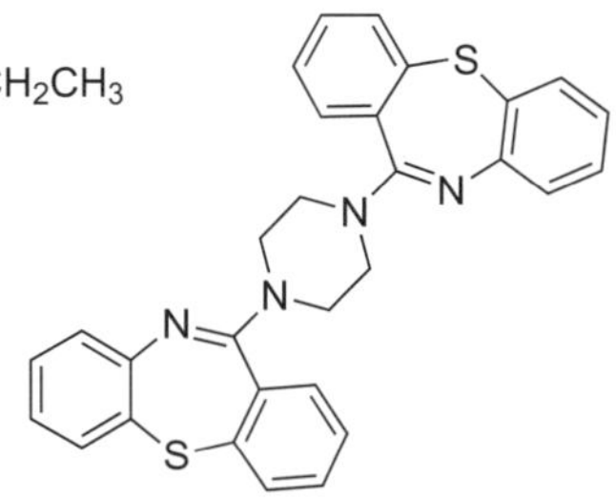

34

**FIGURE 5.7** Impurities in quetiapine (**1**) hemifumarate.

Another polymorph, also designated polymorph II-US7238686 is produced when fumaric acid is added to a solution of quetiapine in refluxing methyl *tert*-butyl ether. The resulting suspension is cooled to 25°C and the solid filtered, washed with methyl *tert*-butyl ether, and dried at an unspecified temperature. An amorphous form is isolated when quetiapine hemifumarate is dissolved in chloroform–methanol at 45°C and the solution is vacuum dried. Polymorph II-US7238686 and the amorphous form are characterized by XRD.[92]

A second amorphous form of quetiapine hemifumarate is produced when quetiapine hemifumarate is completely melted at 170°C and then cooled to 25°C. The melting point is lower (110°C) at reduced pressure. This amorphous form is characterized by XRD and FTIR.[93]

## 5.7 TRADE SECRETS AND THE CONTINUING REFINEMENT OF ANALYTICAL METHODS

An impurity in quetiapine (**1**) hemifumarate from Huihai Inc. (Taizhou, China) is identified by comparison of the multistage mass spectrometry for quetiapine hemifumarate and the impurity (positive ESI/MS$^n$) and an accurate mass for the impurity obtained by Fourier transform ion cyclotron resonance (FTICR). The impurity is the dimer **34** likely resulting from reaction of 11-(piperazin-1-yl)dibenzo[*b,f*][1,4]thiazepine (**33**) with 11-chlorodibenzo[*b,f*][1,4]thiazepine (**25**). This suggests that the Huihai manufacturing process proceeds via alkylation of 11-(piperazin-1-yl)dibenzo[*b,f*][1,4]thiazepine (**33**) by 2-(2-chloroethoxy)ethanol and highlights the challenge to reduce the dimer **34** to acceptable levels in the API.[94]

Six previously unknown impurities and the dimer **34** are present at the 0.05–0.15% level in quetiapine (**1**) hemifumarate from an undisclosed source (Figure 5.7). The impurities are isolated by preparative reverse-phase HPLC, characterized by $^1$H and $^{13}$C NMR, IR, and MS, and independently synthesized. Dimer **34** again suggests the manufacturing process proceeds via alkylation of 11-(piperazin-1-yl)dibenzo[*b,f*][1,4]thiazepine (**33**) by 2-(2-chloroethoxy)ethanol. Formamide **68** and carbamate **70** impurities suggest the solvent and base for the alkylation are DMF and carbonate. The identification of 2-(4-(dibenzo[*b,f*][1,4]thiazepin-11-yl)piperazin-1-yl)ethanol (**35**) as an impurity highlights the importance of setting specifications for 2-chloroethanol in 2-(2-chloroethoxy)ethanol.[95]

## 5.8 THE BEST PROCESS AVAILABLE TODAY

Four criteria will be used narrow the field of potential quetiapine hemifumarate manufacturing processes. The process must use raw materials that are commercially available in bulk, have a minimum number of steps in the linear sequence, have no β-chloroamine (nitrogen mustard) intermediates, and utilize a minimal charge of phosphorus reagent(s). While the phosphorus reagents are all inexpensive, the phosphorus reagent charge should be minimized to minimize phosphate waste and avoid a reduced pressure distillation to remove excess phosphorus reagent during the workup procedure.

Constructing the side chain by Williamson ether synthesis requires two additional steps in the linear sequence. One approach uses ethylene glycol and the β-aminoethyl methanesulfonate **38**, a potent alkylating agent. The other approaches use alcohol-protected 2-chloroethanol partners **39**, **42**, and **45** that must be prepared in-house or outsourced from a custom chemical manufacturer.

There are five routes constructing the thiazepine by cyclization of a urea or benzamide. Urea **48** produced from 1-(2-hydroxyethyl)piperazine is converted to the β-chloroamine 11-(4-(2-chloroethyl)piperazin-1-yl)dibenzo[*b,f*][1,4]thiazepine (**49**). The sequences via urea **51** and benzamides **57** and **60** produced from 2-(2-(piperazin-1-yl)ethoxy)ethanol (**26**) have two additional steps, protection of the side chain alcohol before cyclization and deprotection after. Processes via **51**, **57**, and **60** also incorporate expensive 2-(2-(piperazin-1-yl)ethoxy)ethanol (**26**) early in the manufacturing sequence. Finally, the routes via **48**, **51**, **57**, and **60** and the route via urea **47** require more phosphorus oxychloride (and in most cases phosphorus pentoxide) than is required to convert dibenzo[*b,f*][1,4]thiazepin-11(10*H*)-one (**2**) to 11-chlorodibenzo[*b,f*][1,4]thiazepine (**25**).

There are two clear disadvantages with the piperazine construction approaches. First, while dibenzo[*b,f*][1,4]thiazepin-11-amine (**61**) can be prepared in two steps from two commercially available materials, one of these (2-fluorobenzonitrile) is significantly more expensive than the raw materials used in other processes for constructing the thiazepine ring. Second, the other component in the key piperazine ring construction step is bis-β-chloroamine **63** or **66**. Confirming that these nitrogen mustard reagents used near the end of the manufacturing process are not present even at low levels in quetiapine (**1**) will be difficult to track by HPLC or GC since the reagents contain only a weak chromophore (**63**, benzyl protected) or no chromophore (**66**, THP-protected) and they are thermally unstable.

Having eliminated the alternative approaches for constructing the thiazepine and piperazine rings, 11-chlorodibenzo[*b,f*][1,4]thiazepine (**25**) is likely a key intermediate for manufacture of quetiapine (**1**) hemifumarate. Six criteria will be used to select a manufacturing process to this intermediate. The manufacturing process should require only standard processing equipment, have a high overall yield, use raw materials that are commercially available in bulk, have minimal exposure and odor issues, generate no heavy metal waste sludge by-product, and utilize a minimal charge of phosphorus reagent(s).

11-Chlorodibenzo[*b,f*][1,4]thiazepine (**25**) is produced from either 2-(2-aminophenylthio)benzoic acid (**15**) or dibenzo[*b,f*][1,4]thiazepin-11(10*H*)-one (**2**). Dibenzo[*b,f*][1,4]thiazepin-11(10*H*)-one (**2**) is the preferred precursor since conversion to 11-chlorodibenzo[*b,f*][1,4]thiazepine (**25**) can be accomplished with less phosphorus oxychloride.

There are many routes to the key starting material dibenzo[*b,f*][1,4]thiazepin-11(10*H*)-one (**2**). The process featuring photoisomerization of 2-phenylbenzo[*d*]isothiazol-3(2*H*)-one (**24**) has the lowest cost pair of raw materials (2,2′-dithiodibenzoic and aniline) but the photoisomerization step requires nonstandard equipment and the yield is 31%. The process featuring the Beckmann rearrangement starts with expensive thioxanthen-9-one (**22**) and the yield in the rearrangement step is just 55%. The displacement of fluoride from 2-fluorobenzonitrile by 2-aminothiophenol is inefficient (47%), 2-fluorobenzonitrile is too expensive, and there are exposure and odor issues associated with handling 2-aminothiophenol. The displacement of iodide from 2-iodobenzoic acid by 2-aminothiophenol is inefficient (75%), there are exposure and odor issues associated with handling 2-aminothiophenol, and the displacement process also produces a heavy metal (copper) waste stream. While the displacement of chloride from 1-chloro-2-nitrobenzene by thiophenol is efficient, there are exposure and odor issues associated with handling thiophenol and the sequence from thiophenol via the phenyl carbamate **6** to dibenzo[*b,f*][1,4]thiazepin-11(10*H*)-one (**2**) is too long (four steps, 70% overall yield). Methyl thiosalicylate, thiosalicylic acid, or 2,2′-dithiodibenzoic acid can all be used to displace chloride from 1-chloro-2-nitrobenzene. Methyl thiosalicylate is too expensive and the use of 2,2′-dithiodibenzoic acid is accompanied by generation of a heavy metal (zinc) waste stream. The best process converts thiosalicylic acid and 1-chloro-2-nitrobenzene to dibenzo[*b,f*][1,4]thiazepin-11(10*H*)-one (**2**) in three steps in 83–92% yield.

The reaction of thiosalicylic acid and 1-chloro-2-nitrobenzene in methanol using sodium hydroxide as the base affords 2-(2-nitrophenylthio)benzoic acid (**16**), which is isolated. The nitro group reduction is run using a water-wet palladium catalyst in aqueous methanol at elevated temperature and pressures accessible using standard equipment. The methanol solution of 2-(2-aminophenylthio)benzoic acid (**15**) is carried directly into the cyclization by concentration at reduced pressure and solvent exchange with xylenes. The cyclization is run in xylenes using a Dean–Stark water separator and an acid catalyst. After cooling, aqueous sodium bicarbonate is added to wash out the catalyst and improve the quality of the dibenzo[*b,f*][1,4]thiazepin-11(10*H*)-one (**2**) isolated.

What is the best process for converting dibenzo[*b,f*][1,4]thiazepin-11(10*H*)-one (**2**) to 11-chlorodibenzo[*b,f*][1,4]thiazepine (**25**)? There are three reagent options. Of these, oxalyl chloride is too expensive and we would prefer to handle liquid phosphorus oxychloride over solid phosphorus pentachloride. To convert dibenzo[*b,f*][1,4]thiazepin-11(10*H*)-one (**2**) to 11-chlorodibenzo[*b,f*][1,4]thiazepine (**25**) with a minimum amount of phosphorus oxychloride (2–4 equivalents), a solvent (toluene) and a promoter (*N,N*-dimethylaniline, 0.5–0.6 equivalents) are used. Refluxing for 4–6 h will produce a clear solution containing <2% starting material **2**. The solution is cooled to 30°C and quenched into water while maintaining the temperature at 20–40°C. This temperature range is selected based on two considerations. First, some hydrolysis back to dibenzo[*b,f*][1,4]thiazepin-11

(10*H*)-one (**2**) is possible during a quench over several hours at higher temperatures. Second, slow layer separation after the quench is likely at quench temperatures below 20°C. A filtration to remove an interface rag may be helpful.[56] The organic layer is washed with water and dried by azeotropic distillation at reduced pressure and <60°C. The toluene solution of 11-chlorodibenzo[*b,f*][1,4]thiazepine (**25**) (>90% contained yield) is carried into the next step.

11-Chlorodibenzo[*b,f*][1,4]thiazepine (**25**) can be converted to quetiapine (**1**) in a single step by reaction with 2-(2-(piperazin-1-yl)ethoxy)ethanol (**26**). 11-Chlorodibenzo[*b,f*][1,4]thiazepine (**25**) can also be converted to quetiapine (**1**) in two steps via 11-(piperazin-1-yl)dibenzo[*b,f*][1,4]thiazepine (**33**). All indications are that the two-step sequence via **33** is the current manufacturing process. However, the efficiency of the 11-(piperazin-1-yl)dibenzo[*b,f*][1,4]thiazepine (**33**) alkylation is limited (75–80% yield of quetiapine (**1**) hemifumarate for two steps from **33**). On the other hand, the reaction of 11-chlorodibenzo[*b,f*][1,4]thiazepine with 2-(2-(piperazin-1-yl)ethoxy)ethanol is highly efficient: it "looks more like" a late step in a pharmaceutical manufacturing process (>90% yield of quetiapine (**1**) hemifumarate for three steps from **2**). The challenges associated with manufacturing 2-(2-(piperazin-1-yl)ethoxy)ethanol (**26**) are clear. It is also clear that a supplier able to overcome these challenges and produce 2-(2-(piperazin-1-yl)ethoxy)ethanol (**26**) in bulk at a competitive price could play a pivotal role in the manufacturing of quetiapine (**1**) hemifumarate in the future.

The toluene solution of 11-chlorodibenzo[*b,f*][1,4]thiazepine (**25**) (>90% contained yield) is transferred into a reactor already charged with piperazine (3–5 equivalents) and toluene and the mixture is refluxed for 3–8 h. The suspension is cooled, water is added, and the layers are separated. The dimer level is reduced by extracting 11-(piperazin-1-yl)dibenzo[*b,f*][1,4]thiazepine (**33**) from the toluene solution into aqueous hydrochloric acid at 70°C and then recovering it from the aqueous solution by adding aqueous sodium hydroxide and extracting with toluene at 70°C. The toluene layer is washed with water, concentrated to a smaller volume, and diluted with methyl *tert*-butyl ether. The solution is cooled to 25°C and seed crystals of 11-(piperazin-1-yl)dibenzo[*b,f*][1,4]thiazepine (**33**) polymorph A are added. The suspension is cooled to 10°C and the precipitate filtered, washed with methyl *tert*-butyl ether, and dried at 40°C (85–90%, 99.9% pure by HPLC). If necessary, 11-(piperazin-1-yl)dibenzo[*b,f*][1,4]thiazepine (**33**) can be upgraded by recrystallization from isopropanol.

Since 11-(piperazin-1-yl)dibenzo[*b,f*][1,4]thiazepine (**33**) will be available as the crystalline free base (>99.8% pure by HPLC), the reaction solvent can be selected based on yield and purity of quetiapine (**1**) hemifumarate, time and volume throughput, operational simplicity of the transition from the alkylation into the salt formation, operational simplicity of the salt formation, and overall process robustness. The available data suggests the yield from 11-(piperazin-1-yl)dibenzo[*b,f*][1,4]thiazepine (**33**) to quetiapine (**1**) hemifumarate can be 75–80% using any one of the solvents. With the right combination of phase transfer catalyst, iodide salt, and perhaps a polar aprotic cosolvent, the reaction time can be less than 12 h.

While the neat alkylation, by definition, has superior volume throughput *for the reaction*, the mixture must be taken up in dichloromethane and water before leaving the reactor. Thus, the volume throughput *for the process* is similar to the processes that use a solvent for the reaction. The neat reaction process is less robust because the heat transfer will be poor from the time agitation is started until a stable reaction temperature is reached and because the cooled reaction mixture may solidify prior to adding the workup solvents. The neat reaction is less desirable from an environmental health and safety perspective since it uses dichloromethane rather than toluene, *n*-butanol, or ethyl acetate as the workup solvent.

While *the alkylation reaction* in water would be preferred from an environmental perspective, *the process* incorporating this alkylation is less desirable from an environmental perspective since it uses dichloromethane rather than toluene, *n*-butanol, or ethyl acetate as the other workup solvent. Assuming 11-(piperazin-1-yl)dibenzo[*b,f*][1,4]thiazepine (**33**) and quetiapine (**1**) free base have low solubility in water, the process is less robust because the heat transfer will be poor from the time agitation is started until a stable reaction temperature is reached and because quetiapine free base may partially solidify from the cooled reaction mixture prior to adding dichloromethane.

In processes using *n*-propanol, the high solubility of *n*-propanol in water necessitates use of a water–immiscible solvent (dichloromethane or ethyl acetate) in the workup of the alkylation. No additional solvent is required in the workup of the alkylation in *n*-butanol or toluene. In processes using *n*-butanol, the transition from the alkylation into the salt formation does not require a solvent change since *n*-butanol is a suitable solvent for the hemifumarate salt formation. However, even without a solvent change, the transition from the alkylation into the salt formation still requires distillation of the *n*-butanol–water azeotrope. In addition, the hemifumarate salt produced in and isolated from *n*-butanol must be upgraded by recrystallization from ethanol. In processes using toluene, the transition from the alkylation into the salt formation involves cooling the suspension, washing with water, distillation of the toluene–water azeotrope to dry the solution, then distillation at reduced pressure to produce a concentrated toluene solution that can be carried directly into the salt formation in ethanol. The toluene–ethanol ratio in the salt formation can be precisely controlled by distillation of the toluene–ethanol azeotrope prior to adding fumaric acid. The quetiapine (**1**) hemifumarate isolated from toluene–ethanol does not require an additional purity upgrade. Finally, toluene is already used in the preparation of 11-chlorodibenzo[*b*,*f*][1,4]thiazepine (**25**) and 11-(piperazin-1-yl)dibenzo[*b*,*f*][1,4]thiazepine (**33**). For these reasons, toluene is the preferred solvent for the alkylation reaction.

The reaction of 11-(piperazin-1-yl)dibenzo[*b*,*f*][1,4]thiazepine (**33**) with 2-(2-chloroethoxy)ethanol (1.2 equivalents), sodium carbonate (1.5–2.0 equivalents), and a phase transfer catalyst (0.10–0.25 equivalents) is complete in 10–12 h in refluxing toluene. The mixture is cooled and washed with water. The organic layer is dried by distillation of the toluene–water azeotrope and then distilled at reduced pressure to afford a concentrated toluene solution of quetiapine (**1**). The solution is diluted with ethanol and fumaric acid (0.55–0.60 equivalents) is added as a solution in ethanol at 80°C. The resulting solution is aged at 80–85°C for 30 min and then cooled to 30°C. The solid is filtered, washed with ethanol, and dried at 65°C to afford quetiapine (**1**) hemifumarate (85–90% from **33**, approximately 77% for four steps from **2**). An investment of development time to define the design space for this transformation would pay big dividends.

The FDA defines design space as "The multidimensional combination and interaction of input variables (e.g., material attributes) and process parameters that have been demonstrated to provide assurance of quality. Working within the design space is not considered as a change. Movement outside the design space is considered to be a change and would normally initiate a regulatory post-approval change process. Design space is proposed by the applicant and is subject to regulatory assessment and approval."[96] The first step is listing all the variables for the process and then prioritizing from most likely to least likely to impact yield and quality. The variables for the process converting 11-(piperazin-1-yl)dibenzo[*b*,*f*][1,4]thiazepine (**33**) to quetiapine (**1**) hemifumarate include specific impurities in 2-(2-chloroethoxy)ethanol, equivalents of 2-(2-chloroethoxy)ethanol, equivalents of carbonate base, counterion for the carbonate base, equivalents of the phase transfer catalyst, quaternary ammonium ion and halide of the catalyst, concentration of 11-(piperazin-1-yl)dibenzo[*b*,*f*][1,4]thiazepine (**33**) in toluene, reaction time, number of water washes, volume of water, temperature of the batch during the water washes, toluene–ethanol ratio during the salt formation, temperature of the batch during the fumaric acid addition, equivalents of fumaric acid, weight of quetiapine (**1**) hemifumarate per liter of solvent during the age time, age time after the fumaric acid addition, temperature of the batch during isolation, isolation equipment and associated filtering and washing protocol, and temperature and volume of the ethanol wash.

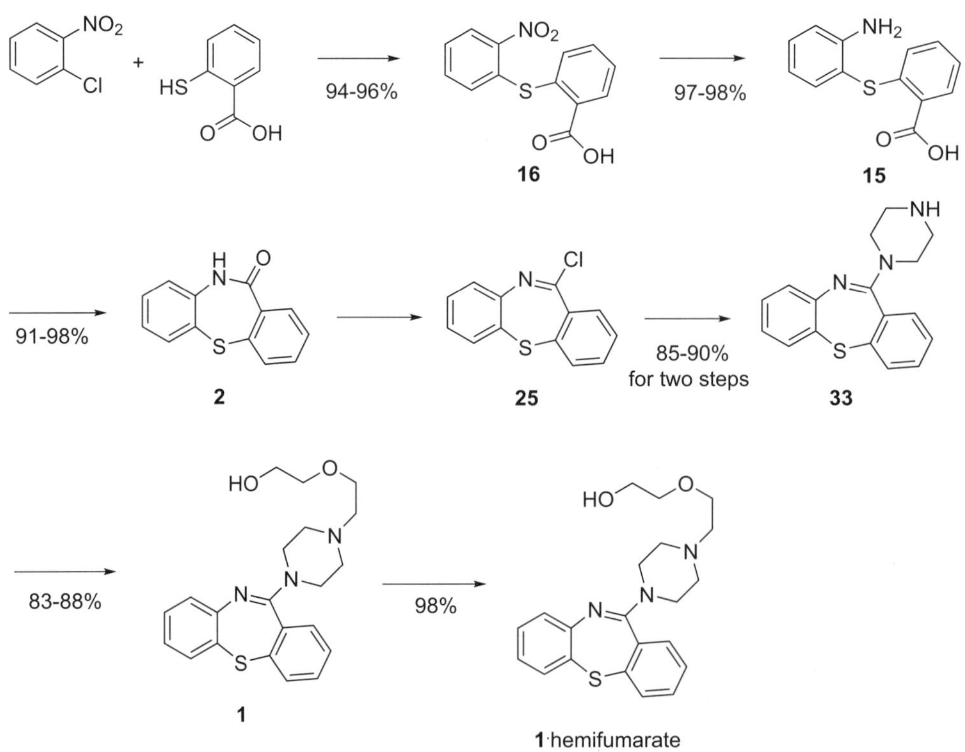


**SCHEME 5.23** The best quetiapine (**1**) hemifumarate process available today.

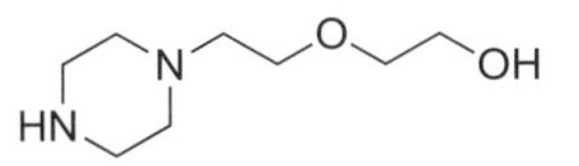
26

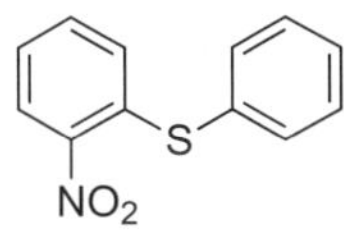
4

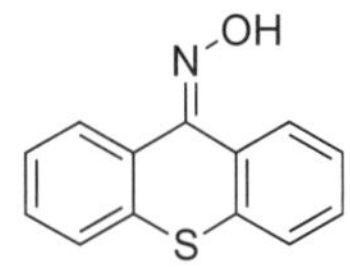
21

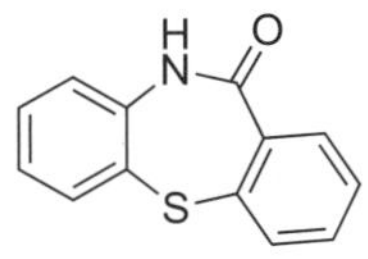
2

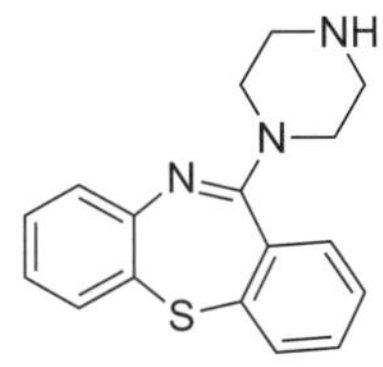
33

**FIGURE 5.8** Structures searched for quetiapine (**1**) hemifumarate presentation.

The process work is not complete until yield and purity specifications are met. If in-process analysis of the concentrated toluene solution of the free base indicates the hemifumarate would not meet specifications, the free base can be upgraded by crystallization from methyl *tert*-butyl ether.

To summarize, thiosalicylic acid is converted into quetiapine (**1**) hemifumarate in seven steps in an overall yield of 57–72% (Scheme 5.23). There are four solid isolations. The raw materials are all commercially available in bulk. One of the raw materials, 2-(2-chloroethoxy)ethanol, is a lacrymator. The process can be run in standard equipment with one vessel dedicated for hydrogenation at elevated temperature and pressure. The five solvents are all from the list of process solvents commonly found in a pharmaceutical manufacturing plant: methanol, ethanol, methyl *tert*-butyl ether, xylenes, and toluene.

Eliminating one late-stage lower yielding step can have a tremendous impact. The process using 2-(2-(piperazin-1-yl) ethoxy)ethanol (**26**) as a raw material would have just three isolations, eliminate the lacrymator 2-(2-chloroethoxy)ethanol from the reagent list, and convert thiosalicylic acid into quetiapine (**1**) hemifumarate in six steps in an overall yield of 75–80%.

## 5.9 STRUCTURES SEARCHED

Five structure searches were used to generate all the information presented in this chapter (Figure 5.8).

## REFERENCES


1. Accessed at www1.astrazeneca-us.com/pi/Seroquel.pdf.
2. Accessed at www.schizophrenia.com/szfacts.htm.
3. Accessed at www.who.int/mip/2003/other_documents/en/globalburdenofdisease.pdf.
4. AstraZeneca PLC Fourth Quarter and Full Year Results 2008 (1/29/2009).
5. Jensen, N.H.; Rodriguiz, R.M.; Caron, M.G.; Wetsel, W.C.; Rothman, R.B.; Roth, B.L. *Neuropsychopharmacology* **2008**, *33*, 2303.
6. (a) Bernstein, P.; Campbell, J.B. WO 2008079848 (7/3/2008). (b) Bernstein, P.; Campbell, J.B. WO 2008079847 (7/3/2008). (c) Bernstein, P.; Campbell, J.B. WO 2008079839 (7/3/2008). (d) Bernstein, P.; Campbell, J.B. WO 2008079838 (7/3/2008).
7. The Aldrich 2007–2008 catalog price for 1 kg is $51.30 ($8.10/mol).
8. The Aldrich 2007–2008 catalog price for 1 kg (97%) is $93.20.
9. (a) Sowinski, F.A.; Yale, H.L.U.S. Patent 3,188,320 (6/8/1965). (b) Yale, H.L.; Sowinski, F.A.U.S. Patent 3,050,524 (11/21/1962).
10. (a) Kaczmarek, L.; Badowska-Roslonek, K.; Stolarczyk, E.; Szelejewski, W. EP 1660469 (5/31/2006).(b) Kaczmarek, L.; Badowska-Roslonek, K.; Stolarczyk, E.; Szelejewski, W. WO 2005012274 (2/10/2005).
11. The Aldrich 2007–2008 catalog price for 2 kg (97%, 325 mesh) is $125.00.
12. (A) Schmutz, J.; Hunziker, F.; Schindler, O.; Kunzle, F.M.U.S. Patent 3,367,930 (2/6/1968).(b) Schmutz, J.; Kuenzle, F.; Hunziker, F.; Buerki, A. *Helv. Chim. Acta* **1965**, *48*, 336.
13. Liang, W.; Liu, T.; Chen, Y.; Chen, X. *Zh. Yiyao Gongye Zazhi* **2004**, *35*, 705.
14. The Aldrich 2007–2008 catalog price for 100 g (97%) is $25.00.
15. (a) Warawa, E.J.; Migler, B.M.; Ohnmacht, C.J.; Needles, A.L.; Gatos, G.C.; McLaren, F.M.; Nelson, C.L.; Kirkland, K.M. *J. Med. Chem.* **2001**, *44*, 372.(b) Barker, A.C.; Copeland, R.J. EP 282236 (9/14/1988).(c) Warawa, E.J.; Migler, B.M.U.S. Patent 4,879,288 (11/7/1989).(d) Warawa, E.J.; Migler, B.M. EP 240288 (10/7/1987).
16. (a) Bozsing, D.; Kovanyine Lax, G.; Simig, G.; Rakoczy, G.; Tompe, P.; Krasznai, G.; Vereczkeyne Donath, G.; Nagy, K. EP 1252151 (10/30/2002).(b) Bozsing, D.; Kovanyine Lax, G.; Simig, G.; Rakoczy, G.; Tompe, P.; Krasznai, G.; Vereczkeyne Donath, G.; Nagy, K. WO 2001055125 (8/2/2001).
17. The Aldrich 2007–2008 catalog price for 25 g is $209.50.
18. The Aldrich 2007–2008 catalog price for 500 g is $129.00.

19. The Aldrich 2007–2008 catalog price for 500 g is $76.60.
20. Ek, F.; Olsson, R.; Ohlsson, J.U.S. Patent 7,491,715 (2/17/2009).
21. (a) Hach, V.; Protiva, M. *Chem. Listy 1957, 51*, **1909**. (b) Protiva, M.; Borovicka, M.; Hach, V.; Votava, Z.; Sramkova, J.; Horakova, Z. *Experientia* **1957**, *13*, 291.
22. (a) Jaques, R.; Rossi, A.; Urech, E.; Bein, H.J.; Hoffmann, K. *Helv. Chim. Acta* **1959**, *42*, 1265. (b) Hoffmann, K.; Urech, E. DE1062246 (7/30/1959).(c) Hoffmann, K.; Urech, E.U.S. Patent 2,852,510 (9/16/1958).
23. Kapur, S.; McClelland, R.U.S. Patent 6,890,919 (5/10/2005).
24. The Aldrich 2007–2008 catalog price for 500 g (98%) is $84.60.
25. The Aldrich 2007–2008 catalog price for 250 g is $150.50 ($85.00/mol).
26. Deshpande, P.B.; Holkar, A.G.; Gudaparthi, O.; Kumar, J.D.U.S. Patent 20050080072 (4/15/2005).
27. Tan, Y.; Shao, J.-Y.; Zhao, M.; Wu, F.-H. *Zh. Xinyao Zazhi* **2007**, *16*, 867.
28. Murray, P.M.; Vaz, L.-M.; Ainge, D.; Harada, K.; Nishino, S.; Yoshii, K.U.S. Patent 20070203336 (8/30/2007).
29. (a) Harada, K.; Nishino, S.; Yoshii, K.U.S. Patent 7,314,930 (1/1/2008). (b) Harada, K.; Nishino, S.; Yoshii, K.U.S. Patent 7,045,621 (5/16/2006). (c) Harada, K.; Nishino, S.; Yoshii, K. EP 1201663 (7/9/1999). (d) Harada, K.; Nishino, S.; Yoshii, K. JP 11199574 (7/27/1999).
30. Pathak, S.; Sharma, J.; Kaushik, G.; Thaper, R.K.; Dubey, S.K. WO 2006027789 (3/16/2006).
31. The Aldrich 2007–2008 catalog price for 100 g is $104.50 ($211/mol).
32. The Aldrich 2007–2008 catalog price for 25 g is $62.60 ($624/mol).
33. Kuti, M.; Rabai, J.; Kapovits, I. *Phosphorus, Sulfur, Silicon* **1993**, *85*, 119.
34. The Aldrich 2007–2008 catalog price for 12 kg is $382.00.
35. (a) Kwak, B.S.; Lee, S.I.; Hwang, H.J.; Lim, J.H. WO 2006001619 (1/5/2006). (b) Kwak, B.-S.; Chung, K.-N.; Koh, K.-H.; Hwang, H.-J. WO 2004047722 (6/10/2004).
36. Allen, C.F.H.; MacKay, D.D. *Org. Syn. Coll. Vol. 2* **1943**, 580.
37. Yokota, K.U.S. Patent 6,768,022 (7/27/2004).
38. The Aldrich 2007–2008 catalog price for 500 g (90% technical grade) is $85.60.
39. The Aldrich 2007–2008 catalog price for 100 g is $43.50 (or $108/mol).
40. (a) Hargrave, K.D.; Schmidt, G.; Engel, W.; Schromm, K. EP 419861 (4/3/1991).(b) Schromm, K.; Engel, W.; Schmidt, G.; Hargrave, K.D. CA2024040 (3/1/1991).
41. Belanger, P.C.; Rokach, J.; Scheigetz, J.U.S. Patent 4,728,735 (3/1/1988).
42. The Aldrich 2007–2008 catalog price for 100 g is $188.00.
43. (a) Nicol, R.H.; Slater, M.J.; Hodgson, S.T.U.S. Patent 5,607,929 (3/4/1997).(b) Nicol, R.H.; Slater, M.J.; Hodgson, S.T. WO 1992019607 (11/12/1992).(c) Nicol, R.H.; Slater, M. J.; Hodgson, S.T. WO 1992019277 (11/12/1997).
44. McClelland, E.W.; Peters, R.H. *J. Chem. Soc.* **1947**, 1229.
45. The Aldrich 2007–2008 catalog price for 100 g is $91.80.
46. (a) Nagarajan, K.; Kulkarni, C.L.; Venkateswarlu, A. *Ind. J. Chem.* **1974**, *12*, 247. (b) Nagarajan, K.; Kulkarni, C.L.; Venkateswarlu, A. *Ind. J. Chem.* **1968**, *6*, 225. (c) Catsoulacos, P. *Chim. Therapeut.* **1970**, *5*, 401. (d) Rasanu, N. *Revis. Chim.* **1969**, *20*, 175.
47. The Aldrich 2007–2008 catalog price for 18 kg is $304.50.
48. (a) Kamigata, N.; Hashimoto, S.; Kobayashi, M.; Hiroshi, N. *Bull. Chem. Soc. Jpn.* **1985**, *58*, 3131. (b) Kamigata, N.; Hashimoto, S.; Kobayashi, M. *Sulfur Lett.* **1984**, *2*, 17. (c) Kamigata, N.; Hashimoto, S.; Fujie, S.; Kobayashi, M. *J. Chem. Soc., Chem. Commun.* **1983**, 765.
49. Goldstein, J. WO 2005002586 (1/13/2005).
50. Christoph, G.; Grimm, S.W.; Widzowski, D. WO 2008116144 (9/25/2008).
51. (a) Black, S.N.; Broady, S.N.; Kirschner, A.S.; Osborn, J.A.; Jolly, S.D. WO 2007062337 (5/31/2007). (b) Korey, D.; Main, K.B.; Creekmore, R.J.R.; Surian, J.M.; Bi, V.Y. WO 2007062339 (5/31/2007). (c) Brown, D.B.; Korey, D.; Main, K.B.; Creekmore, R.J.R. WO 2007062338 (5/31/2007). (d) Davis, P.C.; Goldstein, J.M.; Grimm, S.W.; Winter, H.R.; Suckow, R.F.U.S. Patent 20060217367 (9/28/2006). (e) Davis, P.C.; Goldstein, J.M.; Grimm, S.W.; Winter, H.R.; Suckow, R. F.U.S. Patent 20060217366 (9/28/2006). (f) Davis, P.C.; Goldstein, J.M.; Grimm, S.W.; Winter, H.R.; Suckow, R.F.U.S. Patent 20060217365 (9/28/2006). (g) Davis, P.C.; Goldstein, J.M.; Grimm, S.W.; Suckow, R.F.; Winter, H.R. WO 2006073360 (7/13/2006).
52. Tung, R.; Harbeson, S. WO 2008066620 (6/5/2008).
53. Czibula, L.; Dobay, L.; Feher, K.; Werkne Papp, E. WO 2008152434 (12/18/2008).
54. Tarur, V.R.; Sathe, D.G.; Naidu, A.V.; Aher, U.P.; Patil, S.S. WO 2007004234 (1/11/2007).
55. Holkar, A.G.; Pise, A.C. WO 2007020011 (2/22/2007).
56. Buenger, G.S.; Alexander, A. WO 2006135544 (12/21/2006).
57. Bosch Llado, J.; Burgarolas Montero, M.C.; Chamorro Gutierrez, I. WO 2006117700 (11/9/2006).
58. Schmutz, J.; Hunziker, F.U.S. Patent 3,539,573 (11/10/1970).
59. Kansal, V.K.; Ahmad, S.; Lal, K.; Patil, B.T.U.S. Patent 20080241949 (10/2/2008).
60. Etlin, O.; Brand, M.; Ditkovich, J.; Davidi, G.; Shookrun, M.; Arad, O.; Kaspi, J.U.S. Patent 20060063927 (3/23/2006).
61. Kvapil, L.; Hradil, P.; Grepl, M.; Simunkova, P.; Melnicky, R.; Slezar, P.; Obadalek, R. CZ 295017 (5/18/2005).
62. The Aldrich 2007–2008 catalog price for 500 g is $34.50.
63. Kirschner, A.S.; Osborn, J.A.; Hulsizer, J. WO 2007062336 (5/31/2007).
64. The Aldrich 2007–2008 catalog price is 2 L ($d$=0.96 g/mL) for $120.00.
65. Comely, A.C.; Verdaguer Espaulella, F.X.; Rafecas Jane, L.; Domingo Coto, A. WO 2006094549 (9/14/2006).
66. The Aldrich 2007–2008 catalog price for 500 g is $92.40.
67. The Aldrich 2007–2008 catalog price for 5 g is $31.30.

68. The Aldrich 2007–2008 catalog price for 250 g is $102.00.
69. (a) Yu, I.-C.; Lui, C.-M.; Lu, C.-L. *Yaoxue Xuebao* **1966**, *13*, 514.(b) Morren, H. BE556239 (9/30/1957).(c) Morren, H.G. BE523902 (2/16/1954).(d) Morren, H.G.; Denayer, R.; Trolin, S.; Grivsky, E.; Strubbe, H.; Linz, R.; Dony, G.; Maricq, J. *Ins. Chim. Belg.* **1954**, *19*, 1176.
70. Wu, N.; Tang, H. CN101190902 (6/4/2008).
71. Lu, L.; Ren, H.; Wan, X.; Ji, Y.; Huang, G.; Deng, K.; Qu, D.; Gu, Z.; Liang, J.; Ning, Y. CN101293879 (10/29/2008).
72. Kansal, V.K.; Lal, K.; Ahmad, S.; Leonov, D. WO 2006113425 (10/26/2006).
73. (a) Diller, D.; Dolitzky, B.-Z.U.S. Patent 7,071,331 (7/4/2006). (b) Diller, D.; Dolitzky, B.-Z.U.S. Patent 20,040,220,400 (11/4/2004).(c) Diller, D.; Dolitzky, B.-Z. WO 2004076431 (9/10/2004).
74. (a) Anon. *IP.com J.* **2006**, *6*, 12.(b) Anon. *IP.com J.* **2006**, *6*, 9.
75. (a) Mahale, G.D.; Kumar, A.; Singh, D.; Ramaswamy, A.V.; Waghmode, S.B. *J. Serb. Chem. Soc.* **2008**, *73*, 385. (b) Kumar, A.; Singh, D.; Patil, S.H.; Mahale, G.D.; Sawant, U.A.; Jadhav, B.G.; Rana, R.U.S. Patent 20,070,293,471 (12/20/2007). (c) Kumar, A.; Singh, D.; Patil, S.H.; Mahale, G.D.; Sawant, U.A.; Jadhav, B.G.; Rana, R. WO 2006077602 (7/27/2006).
76. Michalski, K.; Wieczorek, M. EP 1602650 (12/7/2005).
77. The TCI America 2006–2007 catalog price for 250 mL ($d$ = 1.07 g/mL) is $94.25.
78. Ludeman, S.M.; Bartlett, D.L.; Zon, G. *J. Org. Chem.* **1979**, *44*, 1163.
79. de Vries, E.F.J.; Steenwinkel, P.; Brussee, J.; Kruse, C.G.; van der Gen, A. *J. Org. Chem.* **1993**, *58*, 4315.
80. The Aldrich 2007–2008 catalog price for 1 kg is $39.10.
81. (a) ("Shell" Research Ltd.) NL294635 (4/26/1965).(b) ("Shell" Research Ltd.) FR1370678 (8/28/1964).
82. (a) Puig Torres, S.; Herbera Espinal, R.; Dalmases Barjoan, P. EP 1660468 (7/27/2004).(b) Puig Torres, S.; Herbera Espinal, R.; Dalmases Barjoan, P. WO 2005014590 (2/17/2005).
83. Hilden, L.; Grumann, A.; Huhta, S.; Rummakko, P. WO 2005028459 (3/31/2005).
84. Deshpande, P.B. IN2003CH00804 (11/18/2005).
85. Rummakko, P.; Hutha, S.; Grumann, A. WO 2005028458 (3/31/2005).
86. Mettey, Y.; Lehuede, J.; Vierfond, J.-M. *J. Het. Chem.* **1997**, *34*, 465.
87. The Aldrich 2007–2008 catalog price is 20 kg for $238.50.
88. Grumann, A.; Huhta, S.; Rummakko, P. WO 2005028457 (3/31/2005).
89. The Aldrich 2007–2008 catalog price for 3 kg is $42.20.
90. (a) Snape. E.W.U.S. Patent 6,372,734 (4/16/2002).(b) Snape, E.W. WO 1999006381 (2/11/1999).
91. Lifshitz-Liron, R.; Kovalevski-ishai, E.; Dolitzky, B.-Z.; Wizel, S.; Lidor-Hadas, R.U.S. Patent 20030216376 (11/20/2003).
92. (a) Parthasaradhi, R.B.; Rathnakar, R.K.; Raji, R.R.; Muralidhara, R.D.; Subash Chander, R.K.U.S. Patent 7,238,686 (7/3/2007).(b) Parthasaradhi, R.B.; Rathnakar, R.K.; Raji, R.R.; Muralidhara, R.D.; Subash Chander, R.K.U.S. Patent 20040242562 (12/2/2004).(c) Parthasaradhi, R.B.; Rathnakar, R.K.; Raji, R.R.; Muralidhara, R.D.; Subash Chander, R.K. WO 2004078735 (9/16/2004).
93. Rao, M.A.; Kuduva, S.S.; Shreenivas, M.T.U.S. Patent 7,488,821 (2/10/2009).
94. (a) Xu, H.; Wang, D.; Sun, C.; Pan, Y.; Zhou, M. *J. Pharm. Biomed. Anal.* **2008**, *47*, 658.(b) Xu, H.; Wang, D.; Sun, C.; Pan, Y.; Zhou, M. *J. Pharm. Biomed. Anal.* **2007**, *44*, 414.
95. Bharathi, C.; Prabahar, K.J.; Prasad, C.S.; Srinivasa Rao, M.; Trinadhachary, G.N.; Handa, V.K.; Dandala, R.; Naidu, A. *Pharmazie* **2008**, *63*, 14.
96. Accessed at www.fda.gov/CbER/gdlns/ichq8pharm.htm#IIC.

# 6

# SINGULAIR® (MONTELUKAST SODIUM)

## 6.1 SINGULAIR® AND THE ASTHMA MARKET

Asthma is defined as "a chronic inflammatory disorder of the airways in which many cells and cellular elements play a role. The chronic inflammation is associated with airway hyperresponsiveness that leads to recurrent episodes of wheezing, breathlessness, chest tightness, and coughing, particularly at night or early in the morning. These episodes are usually associated with widespread, but variable, airflow obstruction within the lung that is often reversible either spontaneously or with treatment."[1]

There is no question that asthma is one of the most prevalent chronic diseases. For the 3-year period 2001–2003, an average of 20 million Americans had asthma.[2] The prevalence of asthma in the United States across all age groups increased 85% from 1982 to 1996.[3] It is estimated that 300 million people have asthma worldwide.[4] Some of the highest prevalences are found in the United Kingdom (18.4–15.3%), New Zealand (15.1%), Australia (14.7%), the Republic of Ireland (14.6%), Canada (14.1%), and the United States (10.9%).[5] There is a correlation between asthma prevalence and degree of urbanization in developing regions of Africa, Asia, Central and South America, and the Pacific.[5] It is estimated that with increasing urbanization there will be an additional 100 million asthmatics by 2025.[4]

What causes asthma and why is asthma becoming more prevalent? A radio quote from Winston Churchill from 1939 comes to mind: "It is a riddle wrapped in a mystery inside an enigma." We could offer statistics that point to a genetic factor. For example, approximately 40% of children in the United States who have asthmatic parents will develop asthma.[6] But the genetic component is complex. Multiple genes are involved and different genes may be involved with different ethnic subgroups.[7,8] We could discuss key environmental factors such as allergen exposure and sensitization in children, the protective effect of certain childhood respiratory infections,[9,10] exposure to occupational sensitizers,[11] and indoor and outdoor air pollution. The bottom line is this: we are many years away from solving asthma but we do have small-molecule drugs that can dramatically improve the quality of life for asthmatics right now.

The combined market for asthma treatment in the United States and top five E.U. countries was $9.7 billion in 1999 and $11.8 billion in 2003. The current U.S./E.U. market is estimated to be as high as $33 billion. Perhaps just as important to understanding the market potential for asthma treatment as the increase in asthma prevalence is our changing demographics. It is estimated that the proportion of elderly asthmatics will increase 18% by 2015.[12,13]

Drugs for asthma treatment can be grouped into three categories. The first- and second-line therapies are beta-2-agonists, inhaled corticosteriods (ICS), and combination inhaled corticosteroids/long acting beta-agonists (ICS/LABA) drugs. The third-line therapies are the leukotriene antagonists and anticholinergics. Singulair® (montelukast sodium) (Figure 6.1), is an orally active selective leukotriene

*Pharmaceutical Process Chemistry for Synthesis: Rethinking the Routes to Scale-Up*, By Peter J. Harrington

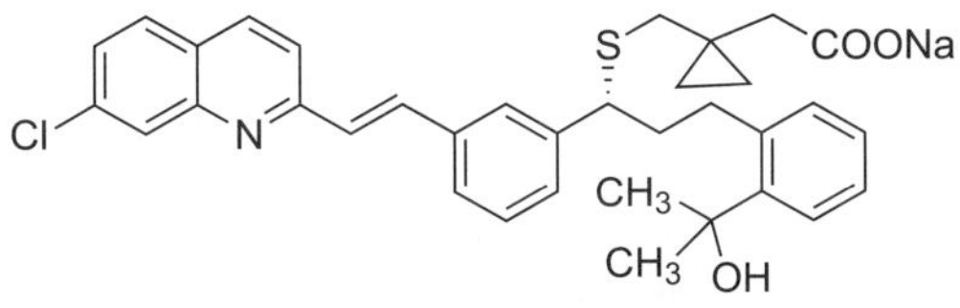


**FIGURE 6.1** Montelukast (**1**) sodium.

receptor antagonist that inhibits the cysteinyl leukotriene $CysLT_1$ receptor. This receptor is found in the human airway and on other proinflammatory cells. Singulair® is the most prescribed allergy drug in the United States and Merck's biggest seller. Singulair® sales from 2005 through the first quarter of 2008 show double-digit growth. Sales were $4.266 billion for 2007, up 19.2% from 2006. Singulair® sales were $1.1 billion for the first quarter of 2008, 10% higher than the first quarter 2007 sales figures.

While the asthma market potential and sales figures for Singulair® are certainly impressive, there are postmarketing reports linking Singulair® use with an increase in suicidal thoughts and behavior. The prescribing information is now updated to alert doctors and patients to a number of adverse reactions reported in postmarketing use including aggressive behavior, anxiousness, dream abnormalities and hallucinations, depression, insomnia, irritability, restlessness, suicidal thinking and behavior (including suicide), and tremor.[14] An FDA investigation is currently underway. Merck's Dr. R. Alan B. Ezekowitz concludes "In over 40 placebo-controlled trials, no reports of suicide in Singulair® treated groups have been found." An independent review of three trials conducted by the American Lung Association recently concluded there is no evidence of a negative effect of Singulair® on emotional well being.[15] In a joint statement from the American Academy of Allergy Asthma & Immunology and the American College of Allergy, Asthma, & Immunology on May 28, 2008, it is recommended that patients taking Singulair® should continue to take it provided (1) the patient and physician feel the medication is effective and (2) the patient does not experience any suicidal behavior or thoughts. An update on the investigation of a link between use of leukotriene modifying medications (Accolate®, Zyflo® and ZyfloCR®, and Singulair®) and suicidal thinking can be expected from the FDA in the next few months.

## 6.2 BACKBONE SYNTHESIS LEFT TO RIGHT

### 6.2.1 (E)-Methyl 2-(3-(3-(2-(7-Chloroquinolin-2-yl)vinyl)phenyl)-3-oxopropyl)benzoate (25)

At first glance, montelukast (**1**) sodium has four components. Three of the components contain heteroaromatic or aromatic rings (quinoline, *meta*-substituted phenyl, *ortho*-substituted phenyl) and are linked by two-carbon and three-carbon chains to form a backbone. The fourth component, 2-(1-mercaptomethyl)cyclopropyl)acetic acid (**2**), is linked to the backbone three-carbon chain to create the chiral center. Retrosynthetic analysis leads us back to some building blocks (Figure 6.2).

The heteroaromatic component in many montelukast processes is introduced by the condensation of 7-chloroquinaldine (**3**) (7-chloro-2-methylquinoline) with an aldehyde. 7-Chloroquinaldine is commercially available but expensive.[16] Many quinolines, including 7-chloroquinaldine (**3**), can be produced by the acid-catalyzed condensation of an aniline with an α,β-unsaturated aldehyde or ketone in the

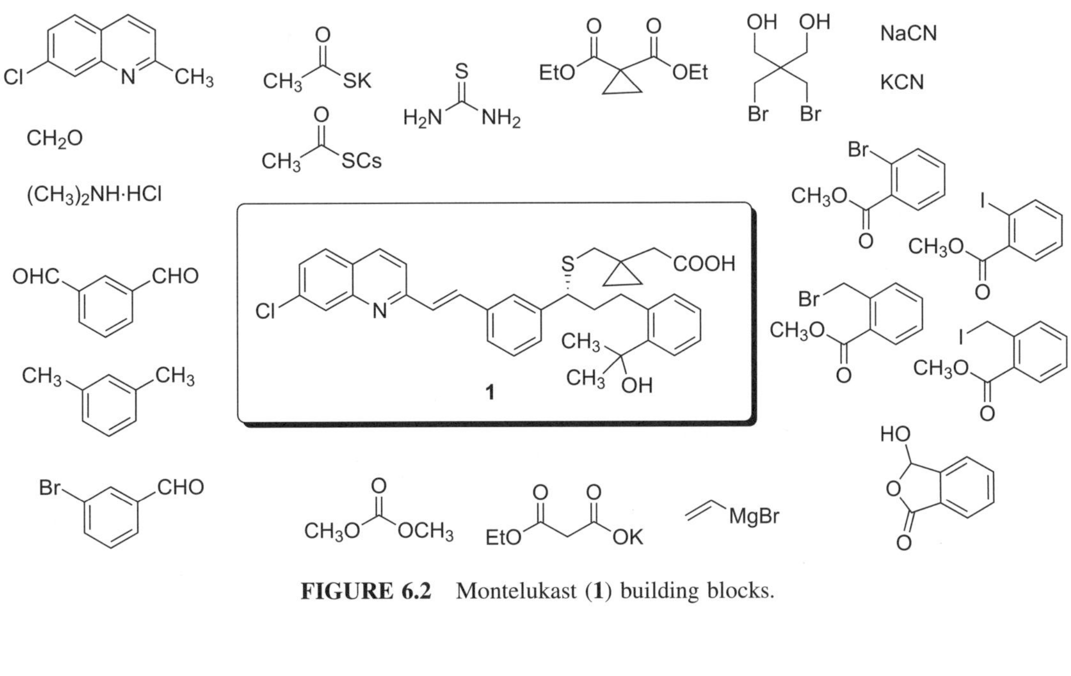

**FIGURE 6.2** Montelukast (**1**) building blocks.

presence of an acid and an oxidizing agent (Skraup's synthesis/Doebner–von Miller reaction). The starting materials, 3-chloroaniline[17] and crotonaldehyde,[18] are both inexpensive but the condensation using aqueous hydrochloric acid or aqueous hydrochloric acid in ethanol is low yielding and produces a complex mixture of at least eight components including 7-chloroquinaldine (**3**) and 5-chloroquinaldine (**4**). 7-Chloroquinaldine (**3**) is "fished out" of this mixture by precipitation as a zinc complex. 7-Chloroquinaldine (**3**) is then released from the zinc complex with ammonium hydroxide (35–42% from 3-chloroaniline). The 5:1 ratio of 7-chloroquinaldine (**3**) to 5-chloroquinaldine (**4**) observed in ethanol increases to 7.5:1 when the reaction is run in 2-butanol (Scheme 6.1).[19]

The other components in the mixture in both cases are reduction products, suggesting that adding an oxidant may increase the yield of 7-chloroquinaldine (**3**). In fact, the reaction of 3-chloroaniline with crotonaldehyde (1.2 equivalents), hydrogen chloride (1.6 equivalents), and *p*-chloranil (1.0 equivalent) in 2-butanol at reflux is more efficient. Crude 7-chloroquinaldine (**3**) hydrochloride is isolated by concentration at reduced pressure and resuspension of the residue in THF. Recrystallization of the crude salt from methanol–THF affords 7-chloroquinaldine (**3**) hydrochloride in 61% yield.[19] While this process does have a higher yield and eliminates the zinc waste stream, *p*-chloranil[20] is too expensive, more expensive than either of the starting materials.

1-Chloro-3-nitrobenzene[21] can serve as an inexpensive oxidant. The reaction of 3-chloroaniline, crotonaldehyde (1.7 equivalents), and 1-chloro-3-nitrobenzene (0.42 equivalents) in 64% sulfuric acid requires several h at 110°C. The reaction mixture is cooled, quenched into water, neutralized, extracted into chloroform, and concentrated at reduced pressure. The residual oil is distilled at 140–185°C and 15 mmHg to afford a 2.2:1 mixture of 7-chloroquinaldine (**3**) and 5-chloroquinaldine (**4**) (65% yield of the mixture and 45% yield of **3**). 7-Chloroquinaldine (**3**) is efficiently

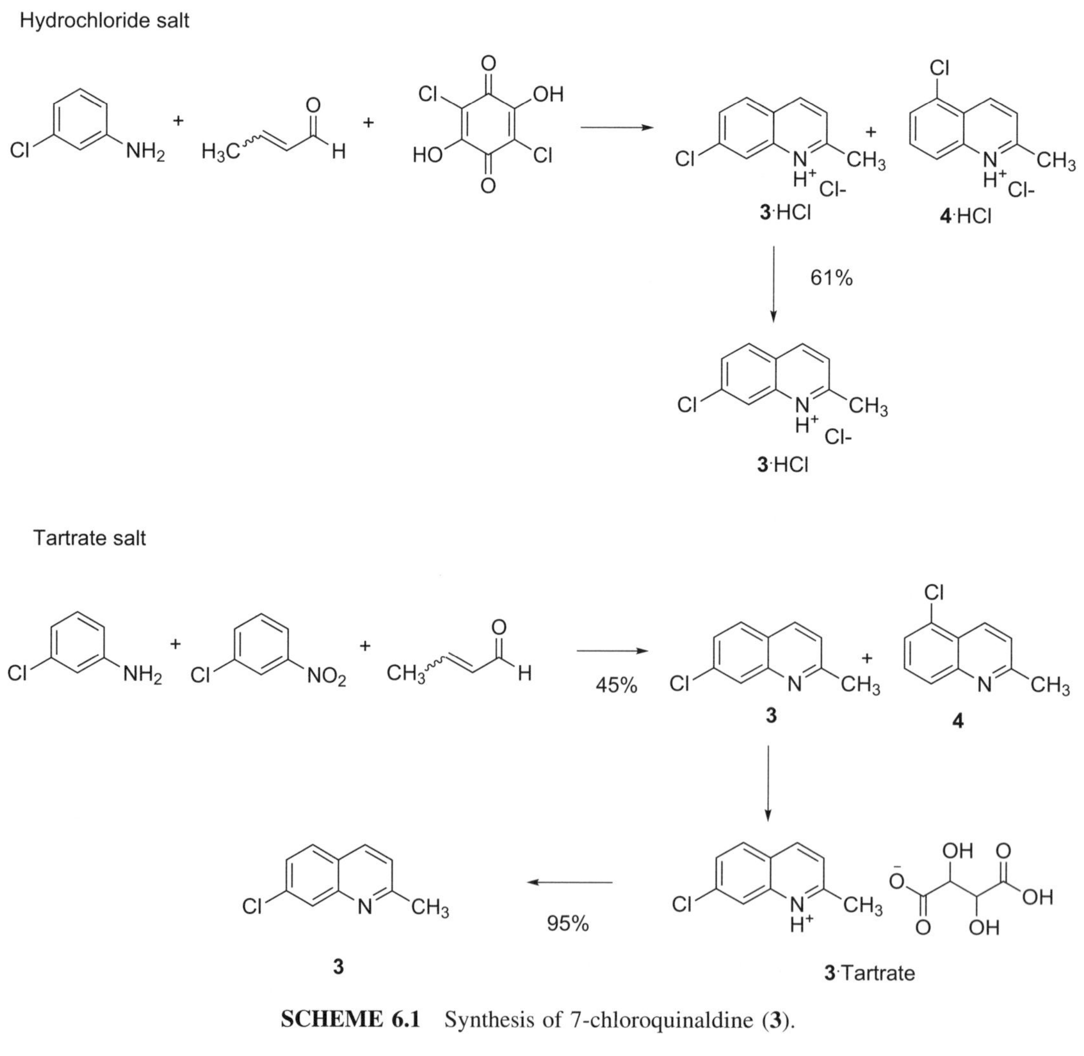


**SCHEME 6.1** Synthesis of 7-chloroquinaldine (**3**).

separated from the mixture by salt formation with L-(+)-tartaric acid (0.80 equivalents) in acetone at 50–60°C. 7-Chloroquinaldine (**3**) precipitates from an aqueous solution of the tartrate salt when sodium hydroxide is added. The yield of 99% pure 7-chloroquinaldine (**3**) from the 2.2:1 mixture is 95%.[22]

> Now that the challenges of producing 7-chloroquinaldine (**3**) are understood, a specification for 5-chloroquinaldine (**4**) in the starting material must be set and the fate of the side products from 5-chloroquinaldine (**4**) produced in the following step(s) must be considered. Our first inclination, as synthetic chemists, is to demand high-purity starting material. However, it would be prudent to invest time to demonstrate efficient rejection of the side products from 5-chloroquinaldine (**4**) downstream. This data will empower us to use a lower grade of 7-chloroquinaldine (**3**) that will be available at a better price.

A left-to-right construction of the montelukast backbone begins by condensing 7-chloroquinaldine (**3**) with a *meta*-substituted benzaldehyde. The quinaldine methyl group is activated enough to react with isophthalaldehyde and acetic anhydride.[23,24] The crude product obtained from reaction with isophthalaldehyde (1.5 equivalents) and acetic anhydride (2.9 equivalents) in xylenes at reflux is typically a 4:1 mixture of the desired monocondensation product **5** and the double-condensation side product **6**. Crude product (**5**) is upgraded by suspending in hot ethyl acetate, filtering the very insoluble double-condensation product **6**, and cooling the mother liquors to crystallize **5** (65% from **3**). The statistical problem associated with the dialdehyde is addressed by using a larger excess (3.0 equivalents) of dialdehyde and designing a solvent system, 3:1 heptane–toluene (6.4 mL/g **3**), in which the condensation proceeds at a reasonable rate (at 100°C) and the product **5** precipitates. Using this approach, the yield and the **5:6** ratio are much improved (82% of **5:6** at 9–10:1). The mother liquors containing isophthalaldehyde are recycled. Similar results are achieved in the condensation using isophthalaldehyde (1.2 equivalents) and acetic anhydride (1.8 equivalents) in toluene alone. After precipitation of the crude product **5** with hexanes and the ethyl acetate crystallization, the monocondensation product **5** is obtained in 75% yield (98.9% purity by HPLC).[25] However, the yield and purity are lower when this procedure is run on 200 kg scale (Scheme 6.2).[26]

There is no statistical problem associated with condensing 7-chloroquinaldine (**3**) with methyl 3-formylbenzoate. Refluxing a mixture of 7-chloroquinaldine (**3**), methyl 3-formylbenzoate (1.1 equivalents), acetic acid and pyridine in toluene affords (*E*)-methyl 3-(2-(7-chloroquinolin-2-yl)vinyl)benzoate (**7**) (89%).[27]

The (*E*)-alkene can be produced from a benzaldehyde with more complex functionality at the *meta*-position by Wittig reaction with a stabilized ylid. The ylid is derived

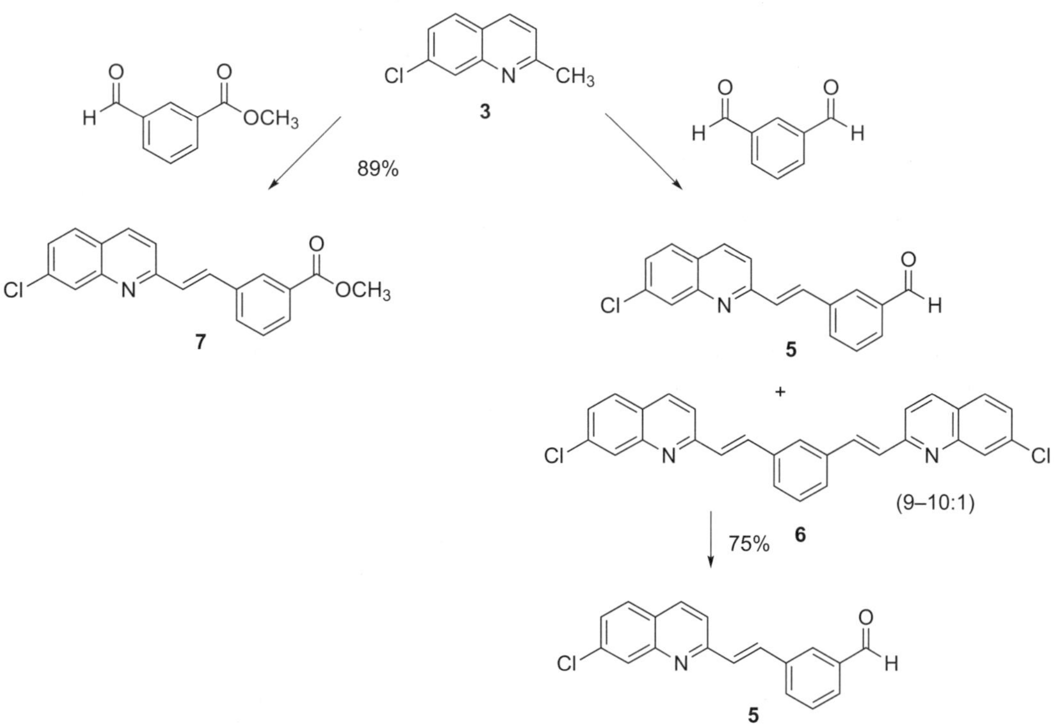


**SCHEME 6.2** 7-Chloroquinaldine (**3**) condensations.

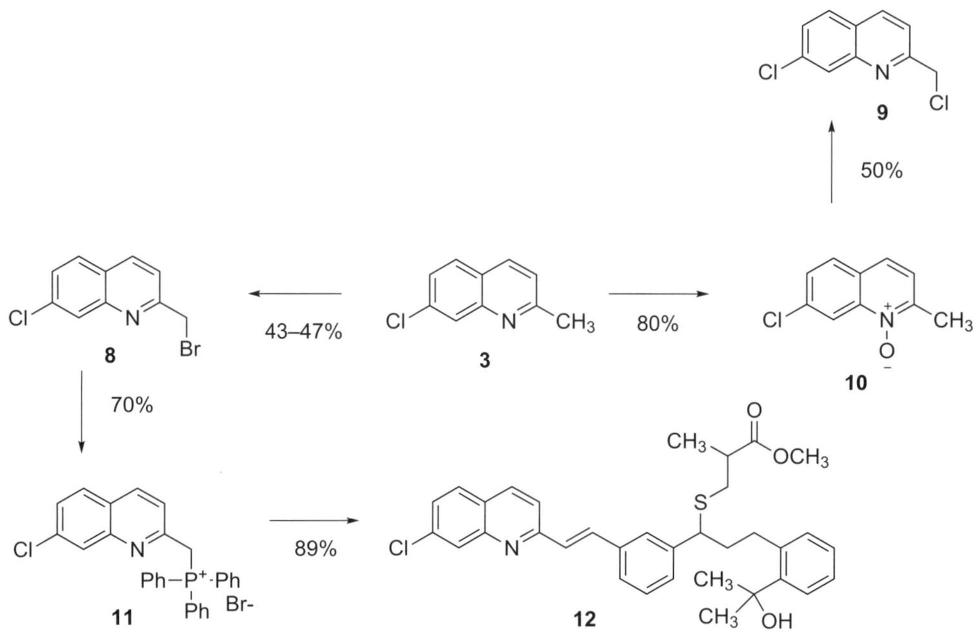


**SCHEME 6.3** 7-Chloroquinaldine (**3**) to a phosphonium salt (**11**) for Wittig reaction.

from 7-chloroquinaldine (**8**) (Scheme 6.3). Free radical bromination of 7-chloroquinaldine (**3**) with *N*-bromosuccinimide (1.0 equivalent) and catalytic benzoyl peroxide in carbon tetrachloride at reflux affords 2-(bromomethyl)-7-chloroquinoline (**8**) as the major component in a statistical mixture. After a tedious workup procedure including two chromatographic separations, the yield of 2-bromomethyl-7-chloroquinoline (**8**) is low (43–47%).[28] Of course, we would not use carbon tetrachloride and chromatography is out of the question.

Reaction of 7-chloroquinaldine (**3**) with lithium diisopropylamide (LDA) (1.1 equivalents) and chlorotrimethylsilane (1.5 equivalents) activates the methyl group for an electrophilic attack in THF–hexanes at −10 to 0°C. Reaction with bromine (1.0 equivalent) at −29°C is quenched into water. After a routine workup, 2-(bromomethyl)-7-chloroquinoline (**8**) is isolated by chromatography (76%).[29]

As an alternative approach to functionalizing the methyl group, *N*-oxide chemistry can be utilized to convert 7-chloroquinaldine (**3**) to 2-(chloromethyl)-7-chloroquinoline (**9**). The *N*-oxide (**10**) is produced by reaction of 7-chloroquinaldine (**3**) with MCPBA (80%) in dichloromethane at 5°C. Reaction of the *N*-oxide (**10**) with benzenesulfonyl chloride affords 2-(chloromethyl)-7-chloroquinoline (**9**) in toluene at 50°C (50%).[30]

A more efficient approach to 2-(chloromethyl)-7-chloroquinoline (**9**) utilizes the novel reactivity of *tert*-butyl hypochlorite to produce **9** in one step. The mixture from the reaction of 7-chloroquinaldine (**3**) with *tert*-butyl hypochlorite (2.0 equivalents) in chlorobenzene at 40–45°C is washed with aqueous hydrochloric acid to remove unreacted 7-chloroquinaldine (**3**). Drying by concentration and addition of a solution of hydrogen chloride in ethyl ether results in precipitation of 2-(chloromethyl)-7-chloroquinoline (**9**) hydrochloride (88%).[31]

Reaction of 2-(bromomethyl)-7-chloroquinoline (**8**) with triphenylphosphine (1.4 equivalents) in acetonitrile at 60°C affords the phosphonium salt **11** (70–80%).[28,29,32] While a Wittig reaction with a fully elaborated aldehyde to produce montelukast has not been published, a Wittig reaction to generate the 3-mercapto-2-methylpropanoate analog **12** is known. The ylid is generated with *n*-butyllithium (0.94 equivalents) in THF–hexanes at −78°C. Benzaldehyde **13** is then added and the reaction mixture aged at −78°C for 30 min. After workup and chromatography to separate the triphenylphosphine oxide, the (*E*)-alkene **12** is isolated in 89% yield.[32]

The construction of fully elaborated benzaldehydes such as **13** will be discussed later. What about isophthalaldehyde[33] and methyl 3-formylbenzoate?[34] Both are available but expensive. Isophthalaldehyde can be produced from *m*-xylenes by conversion to the tetrachloro derivative (**14**) and hydrolysis,[35] from isophthaloyl chloride by Rosenmund reduction[36] or reduction with lithium tri-*t*-butoxyaluminum hydride,[37] and from 1,3-bis(bromomethyl)benzene (**15**) or 1,3-phenylenedimethanamine (**16**) by the Sommelet reaction.[38,39] The Rosenmund reduction route is most promising since isophthaloyl chloride is commercially available and inexpensive (Scheme 6.4).[40]

Inexpensive *m*-toluic acid[41] is converted to methyl 3-formylbenzoate in three steps: esterification, free radical bromination, and Sommelet reaction to convert the

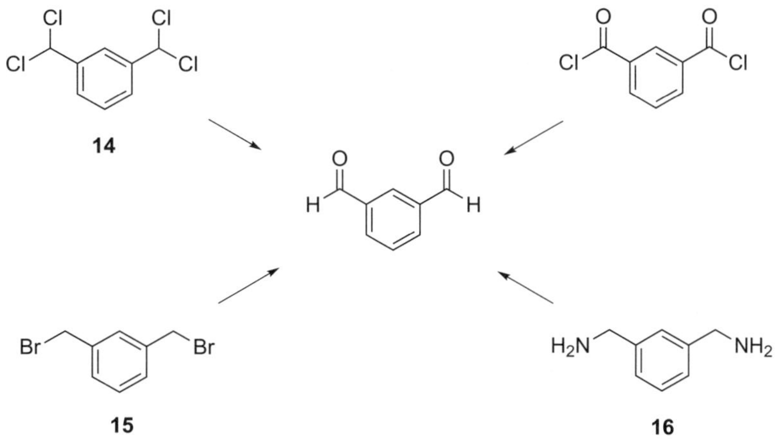


**SCHEME 6.4** Routes to isophthalaldehyde.

bromomethyl benzoate to the formyl benzoate.[27] The free radical bromination with 1,3-dibromo-5,5-dimethylhydantion and catalytic 4,4′-azobis(isobutyronitrile) (AIBN) in dichloromethane at reflux affords a statistical mixture of products (96%). Reaction of the mixture with excess hexamethylenetetramine in acetic acid at 50°C followed by quench into ice affords methyl 3-formylbenzoate in remarkably good yield (83%). 4,4′-Azobis(isobutyronitrile), which is no longer available, would have to be replaced by another initiator. The throughput at the time of the quench into ice is just 34 g/L. The ice quench produces an aqueous waste stream containing acetic acid, ammonia, and formaldehyde (Scheme 6.5).

There are a many options for elaborating (*E*)-3-(2-(7-chloroquinolin-2-yl)vinyl)benzaldehyde (**5**) to montelukast (**1**). Addition of methylmagnesium bromide in THF at 0°C affords the alcohol (**17**) (89%).[25] The yield is lower on 100 kg scale in THF–toluene at −5°C (78%).[26] The alcohol is then oxidized to the ketone (**18**) with activated manganese dioxide in ethyl acetate or dichloromethane (69%)[25,26,32] or by Swern oxidation using DMSO, oxalyl chloride, and triethylamine (92%) in dichloromethane at −60°C.[25] These oxidations, and in fact most oxidations of secondary alcohols to ketones, are not good candidates for scale-up. Consider scale-up of the specific cases at hand. Manganese dioxide can be prepared by a number of methods. The amount of manganese dioxide required to complete the oxidation will depend on the method of preparation and the particle size. A typical ratio of manganese dioxide to substrate is 5:1 to 50:1 by weight. Using the 2.5:1 ratio (69% yield of ketone **18**) we have to filter and treat 3.4 kg of solid waste for each kilogram of product. It should also be noted that the procedure calls for adding the solid manganese dioxide in portions to the alcohol in dichloromethane. While this may be only inconvenient in the lab, solid additions on scale may expose operators to solvent vapors, allow air into a reactor with a flammable solvent, and be difficult to complete due to wetting of the powder in the powder addition funnel by condensing solvent vapors. The throughput for this oxidation is 16 g/L.

The Swern oxidation presents its own unique challenges. First and foremost, the by-product of the Swern oxidation is volatile and malodorous dimethylsulfide. It will be very challenging to efficiently recover both dimethylsulfide and dichloromethane during the subsequent concentration at

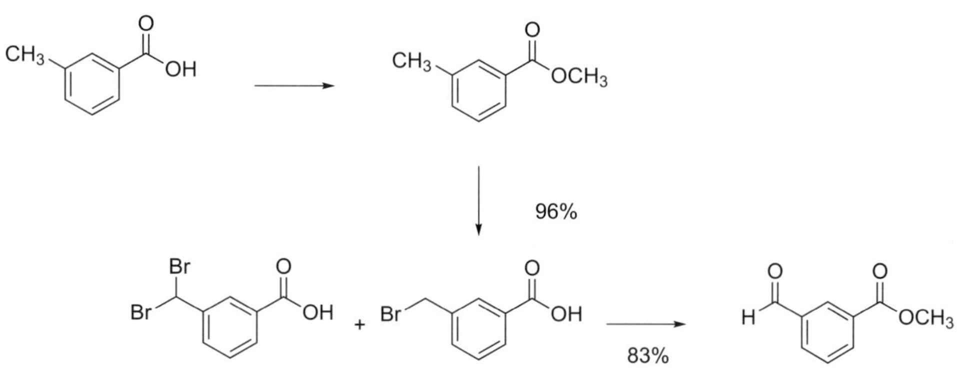


**SCHEME 6.5** Route to methyl 3-formylbenzoate.

reduced pressure. The odor threshold for dimethylsulfide is less than 1 ppm and dimethylsulfide vapor is heavier than air and may travel a considerable distance. Assuming we could engineer a satisfactory solution to the odor problem, the low reaction temperature (−50 to −60°C), the expense of oxalyl chloride (2.2 equivalents),[42] and the low throughput (14 g/L) of this oxidation are additional concerns.

> Oxidations of secondary alcohols to ketones can be separated into three categories: those that are dangerous, those that generate large volumes of often hazardous by-products, or those that are dangerous *and* generate large volumes of often hazardous by-products. When confronted with the scale-up of an alcohol oxidation it is advisable to invest time in redesigning the route to avoid the oxidation.

Reaction of (*E*)-3-(2-(7-chloroquinolin-2-yl)vinyl)benzaldehyde (**5**) with vinylmagnesium bromide in THF at 0–5°C affords the allylic alcohol **19** contaminated with 30–40% of the benzyl alcohol (**20**) from aldehyde reduction. The reaction of (*E*)-3-(2-(7-chloroquinolin-2-yl)vinyl)benzaldehyde (**5**) with vinylmagnesium bromide (1.06 equivalents) is much more efficient in a mixture of toluene–THF (2:1) at 0–5°C. After workup including quench with ammonium acetate, layer separation, and concentration at reduced pressure, addition of hexanes affords the crude allylic alcohol **19** contaminated with just 3% of the benzyl alcohol **20** (93%).[32,43] Vinylmagnesium bromide is expensive and is only available as a 1 M solution in THF.[44] The first step in evaluating the suitability of a vinylmagnesium bromide reaction for scale-up should be a throughput calculation. In this case, the throughput is 52 g/L.

Claisen condensation of the methyl ketone, (*E*)-1-(3-(2-(7-chloroquinolin-2-yl)vinyl)phenyl)ethanone (**18**), with dimethyl carbonate (2.5 equivalents) using 80% sodium hydride (3.0 equivalents) and a methanol initiator in THF at 70°C affords the β-keto ester, (*E*)-methyl 3-(3-(2-(7-chloroquinolin-2-yl)vinyl)phenyl)-3-oxopropanoate (**21**) (89%).[32] The Claisen condensation with dimethyl carbonate (2.7–3.7 equivalents) in carcinogenic 1,4-dioxane[45] at 75–80°C is also efficient using sodium methoxide (2.1–3.2 equivalents) in place of sodium hydride (81–87%) (Scheme 6.6).[25,26]

The β-keto ethyl ester **22** is produced from (*E*)-methyl 3-(2-(7-chloroquinolin-2-yl)vinyl)benzoate (**7**), in four steps: ester hydrolysis in methanol–water at reflux (95%), conversion of the acid **23** to the acid chloride **24** with thionyl chloride (10 equivalents), and condensation of the acid chloride **24** with potassium ethyl malonate followed by decarboxylation (90% from acid **23**).[27] With process development to eliminate three concentrations to dryness, address the transfer of the acid chloride as a thick slurry, and improve the throughput (57 g/L) this procedure could be attractive for scale-up (Scheme 6.7).

The downstream processing of allylic alcohol **19**, methyl ketone **18**, and β-keto esters **21** and **22** all converge

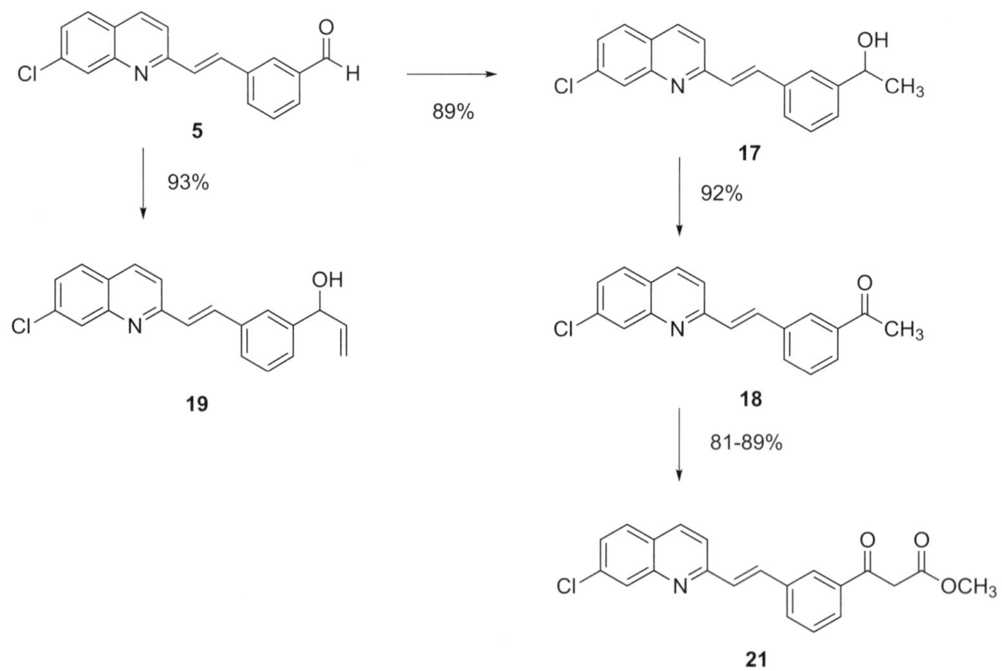


**SCHEME 6.6** Allylic alcohol **19**, methyl ketone **18**, and β-keto ester **21** from (*E*)-3-(2-(7-chloroquinolin-2-yl)vinyl)benzaldehyde (**5**).

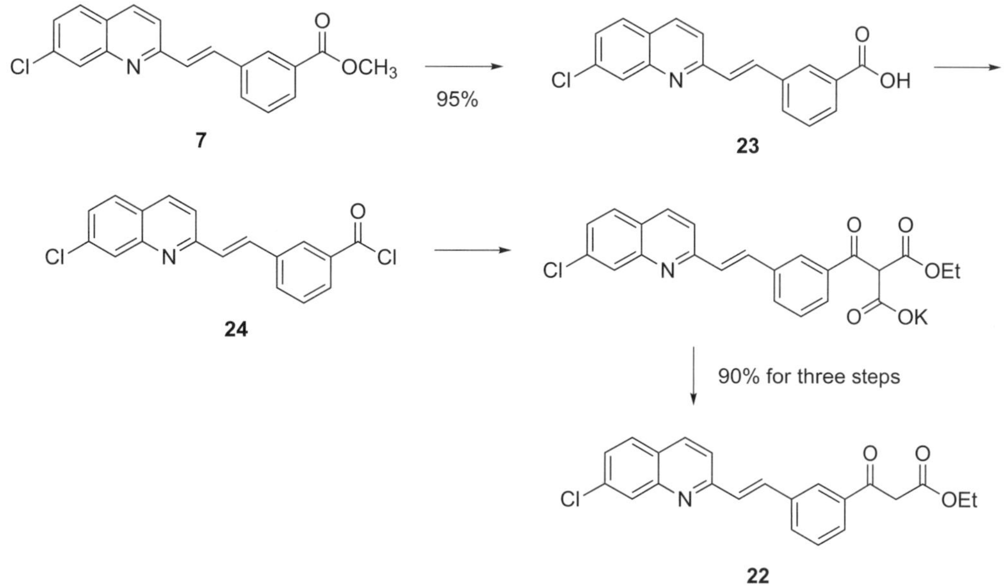


**SCHEME 6.7** The β-keto ester **22** from (*E*)-methyl 3-(2-(7-chloroquinolin-2-yl)vinyl)benzoate (**7**).

at (*E*)-methyl 2-(3-(3-(2-(7-chloroquinolin-2-yl)vinyl)phenyl)-3-oxopropyl)benzoate (**25**) (Scheme 6.8). The partner for the allylic alcohol route is methyl 2-iodobenzoate or methyl 2-bromobenzoate. Both are likely produced by the Sandmeyer reaction of anthranilic acid.[46] The methyl ketone and β-keto ester routes utilize methyl (2-iodomethyl)benzoate (**26**) or methyl (2-bromomethyl)benzoate (**27**). Methyl (2-bromomethyl)benzoate (**27**) is produced from inexpensive phthalide[47] by ring opening with hydrogen bromide in acetic acid (73%)[48] and esterification with oxalyl chloride and methanol. Finkelstein reaction with sodium iodide in acetone then affords methyl (2-iodomethyl)benzoate (**26**) (100%).[32]

Elaboration the allylic alcohol beautifully illustrates the power of the Heck arylation of alkenes in synthesis. Reaction of crude allylic alcohol **19** with methyl 2-iodobenzoate (1.0 equivalent), triethylamine (1.5 equivalents), and palladium acetate (0.5 mol%) in acetonitrile at reflux affords (*E*)-methyl 2-(3-(3-(2-(7-chloroquinolin-2-yl)vinyl)phenyl)-3-oxopropyl)benzoate (**25**), which simply precipitates on cooling the reaction mixture (82%).[43] The reaction of the crude allylic alcohol **19** with methyl 2-bromobenzoate (1.0 equivalent) is accomplished with palladium acetate (2.9 mol%), lithium acetate dihydrate (2.6 equivalents), and lithium chloride (1.0 equivalent) in DMF at 95°C. After aqueous workup, (*E*)-methyl 2-(3-(3-(2-(7-chloroquinolin-2-yl)vinyl)phenyl)-3-oxopropyl)benzoate (**25**) is isolated by recrystallization from ethyl acetate–hexanes (86%).[32]

(*E*)-Methyl 2-(3-(3-(2-(7-chloroquinolin-2-yl)vinyl)phenyl)-3-oxopropyl)benzoate (**25**) is isolated by chromatography (55%) from the crude mixture produced in the alkylation of methyl ketone **18** by methyl (2-iodomethyl)benzoate (**26**) (1.5 equivalents) using lithium diisopropylamide (0.95 equivalents) as the base in DMPU–THF at −60 to 10°C.[32] Alkylation of β-keto ester **21** with methyl (2-iodomethyl) benzoate (**26**) (1.1 equivalents) using sodium hydride (1.1 equivalents) in DMF at 0°C is much more efficient (97%). Alkylation of the β-keto ester **21** can also be accomplished with methyl (2-bromomethyl)benzoate (**27**) (1.25 equivalents) and crushed potassium carbonate (1.3 equivalents) in DMF at 60°C (88%). The yield is lower using methyl (2-bromomethyl)benzoate (**27**) (1.2 equivalents) and potassium carbonate (1.1 equivalents) in DMF at 55°C on 200 kg scale (78%).[26] Selective hydrolysis the β-keto ester in **28** or **29** in the presence of the benzoate ester is not possible. Both esters are hydrolyzed and the β-keto acid is decarboxylated by reaction with acetic acid–hydrochloric acid at 70°C.[26] Alkylation with iodomethane and potassium carbonate in acetone at 50–55°C converts benzoic acid **30** to the methyl benzoate **25** (53–63%).[25,26,32] While detouring through β-keto ester **22** nearly doubles the efficiency of the alkylation step and eliminates the chromatography, the detour cost is three additional steps and addition of undesirable iodomethane, a potential occupational carcinogen[49] with OSHA permissible exposure limit of 5 ppm (TWA) with skin designation, to the list of reagents.

### 6.2.2 Ketone Reduction to the (*S*)-Alcohol 32

The left-to-right routes converge on (*E*)-methyl 2-(3-(3-(2-(7-chloroquinolin-2-yl)vinyl)phenyl)-3-oxopropyl)benzoate (**25**) leading up an asymmetric reduction of the ketone to create the chiral center. There are five options for the asymmetric reduction: microbial reduction to (*R*)-alcohol

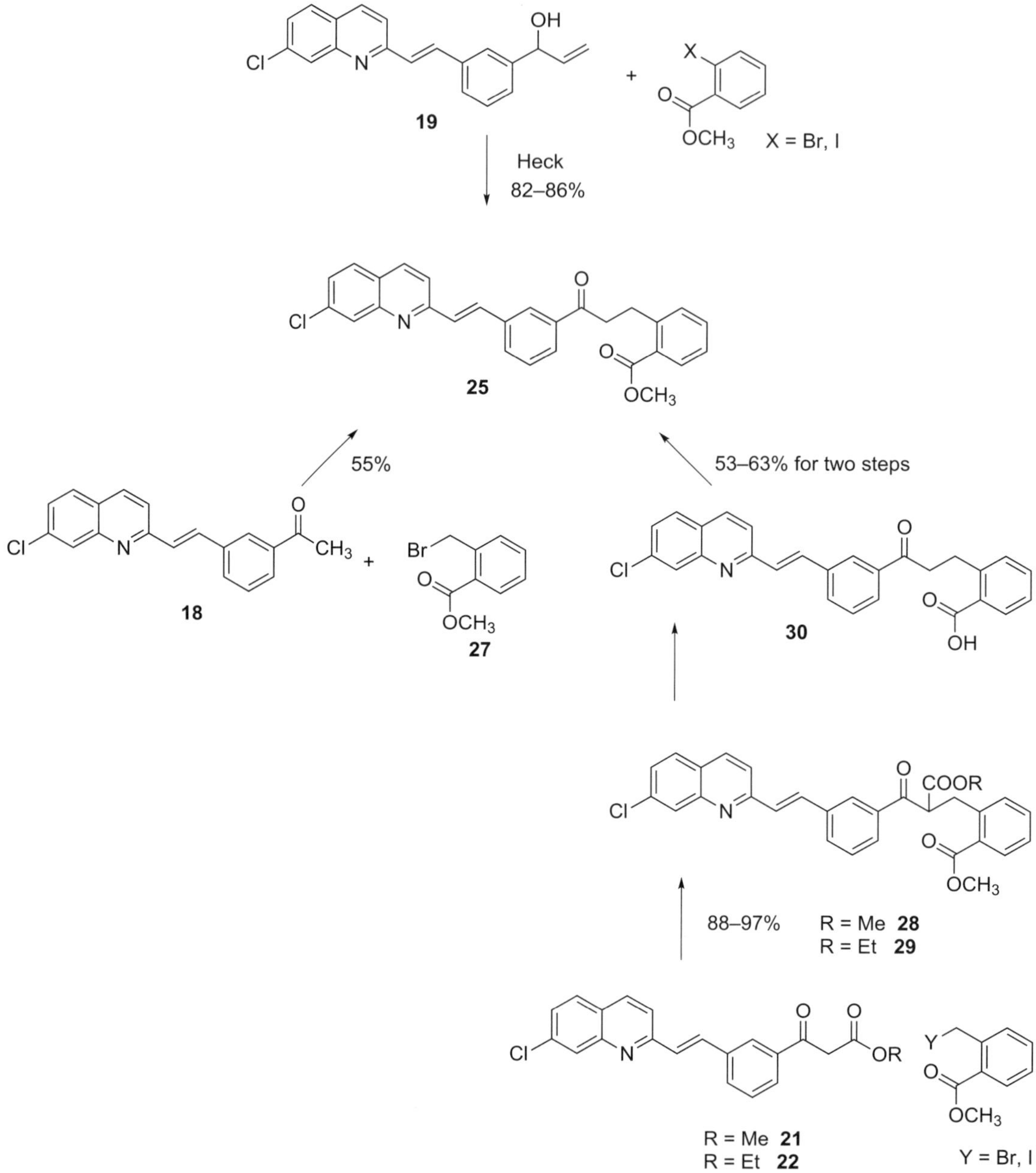


**SCHEME 6.8** (*E*)-Methyl 2-(3-(3-(2-(7-chloroquinolin-2-yl)vinyl)phenyl)-3-oxopropyl)benzoate (**25**) from allylic alcohol **19**, methyl ketone **18**, and β-keto esters **21** and **22**.

**31** with the novel microorganism Microbacterium MB5614 (ATCC 55557) and a Mitsunobu inversion,[50,32] microbial reduction with *Mucor hiemalis* IFO 5834 to the (*S*)-alcohol **32**,[51] reduction to (*S*)-alcohol **32** with borane–THF catalyzed by an oxazaborolidine,[32] reduction to (*S*)-alcohol **32** with diisopinocampheylchloroborane,[43] and ruthenium-catalyzed transfer hydrogenation to produce (*S*)-alcohol **32**.[52] Since the microbial reduction patents provide only milligram-scale procedures and are more than 10 years old, we will focus on the chemical methods. The chemical reductions are all capable of producing (*S*)-alcohol **32** in high enantiomeric excess. Which is best suited for a manufacturing process?

***6.2.2.1 (R)-Oxazaborolidine-Catalyzed Ketone Reduction by Borane–THF*** The chiral amino alcohol component of the (*S*)-oxazaborolidine, (*S*)-diphenyl(pyrrolidin-2-yl)methanol, is efficiently prepared from (*S*)-proline.[53] But the (*R*)-oxazaborolidine is required. The much higher cost of (*R*)-proline[54] prompted the development of an efficient route from pyroglutamic acid to the racemic diphenyl(pyrrolidin-2-yl)methanol that can be resolved. Resolution with (*S*)-(+)-*O*-acetylmandelic acid separates (*S*)-diphenyl(pyrrolidin-2-yl)methanol. (*R*)-Diphenyl(pyrrolidin-2-yl)methanol is recovered from the crystallization liquors with (*R*)-(+)-*O*-acetylmandelic acid.[55] The oxazaborolidine is prepared from diphenyl(pyrrolidin-2-yl)methanol by reaction with methylboronic acid or trimethylboroxine. In a preferred process, trimethylboroxine is added to a toluene solution of the amino alcohol. Aging the mixture at 25°C for 30 min ensures complete conversion of the amino alcohol to the oxazaborolidine–methylboronic acid adduct. This adduct is cleanly converted to the oxazaborolidine by distillation of toluene, excess methylboronic acid (as trimethylboroxine), and the toluene–water azeotrope. Toluene is added and the distillation repeated to ensure complete removal of any residual methylboronic acid and water.[56] It is critically important to remove the water by-product since the hydrated catalyst disproportionates back to the amino alcohol and oxazaborolidine–methylboronic acid adduct. In reductions using 10 mol% of the oxazaborolidine catalyst, just 1 mol% of the amino alcohol or methylboronic acid or trace water causes a significant decrease in the enantioselection. The (*S*)- and (*R*)-oxazaborolidine catalysts are commercially available for the same price and are conveniently handled as toluene or THF solutions (Scheme 6.9).[57]

Optimal conditions for the oxazaborolidine-catalyzed reduction of a ketone are 0.6 equivalents of borane–THF complex and 5 mol% of the oxazaborolidine in THF at 25°C. Under these conditions propiophenone is converted by the (*S*)-oxazaborolidine to (*R*)-1-phenylpropanol with 90% ee.[58] The reduction of propiophenone using borane–THF (0.6 equivalents) and 10 mol% of the (*S*)-oxazaborolidine catalyst at −10°C is even more selective (96.7% ee).[55] Using higher amounts of borane–THF often results in lower selectivity due to increased competition from the noncatalyzed reduction. Complexation of the borane by a pyridine ring nitrogen increases the borane–THF requirement for complete reduction of pyridine-containing ketones. Thus, 1.7 equivalents of borane–THF may be required for complete reduction of our quinoline-containing ketone.

(*E*)-Methyl 2-(3-(3-(2-(7-chloroquinolin-2-yl)vinyl)phenyl)-3-oxopropyl)benzoate (**25**) produced by the Heck arylation is contaminated with trace palladium. This amount is sufficient to catalyze a competing reduction of the ethylene bridge during the oxazaborolidine-catalyzed ketone reduction. Formation of the bridge-reduced alcohol side product **33** (3–10%) not only reduces the yield, it necessitates a chromatography or recrystallization to separate side product **33** from the product. Increasing the oxazaborolidine catalyst charge to 20 mol% favors the ketone reduction. Thus, reduction of ketone **25** with borane–THF[59] (1.7 equivalents)

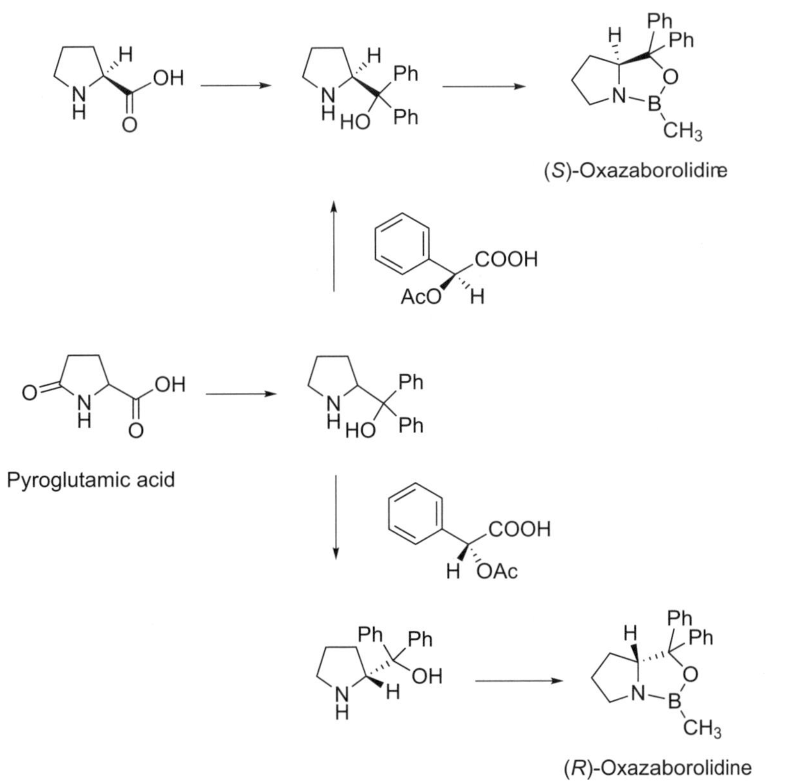


**SCHEME 6.9** Synthesis of (*R*)-1-methyl-3,3-diphenylhexahydropyrrolo[1,2-*c*][1,3,2]oxazaborole.

and the (*S*)-oxazaborolidine catalyst (20 mol%) in THF at 25°C affords (*R*)-alcohol **31** with 98.5% ee. In the workup, the reaction is quenched with 2 M HCl then (*R*)-alcohol **31** is extracted with ethyl acetate. The organic layer is washed with aqueous ammonium acetate and brine then concentrated at reduced pressure. (*R*)-Alcohol **31** is isolated from the residual oil by flash chromatography. Assuming a 95% yield and distillation of some of the THF before adding ethyl acetate, the throughput for this process is just 31 g/L.[32,43]

With the data for the oxazaborolidine-catalyzed reduction now on the table, what are the scale-up issues and is there any upside potential? The borane reagent itself is a concern. Commercial solutions of borane–THF are stable for extended times at 0°C. However, maintaining any bulk chemical at low temperature during shipping and on site is inconvenient and expensive. Commercial borane–THF now contains *N*-isopropyl-*N*-methyl *tert*-butylamine (0.005 M) as a stabilizer. There is significant upside potential. The reduction of ketone **25** containing no palladium, perhaps produced by the β-keto ester route, would not suffer from the competing ethylene bridge reduction. The oxazaborolidine charge could be reduced to 5 mol% or less and the flash chromatography could be eliminated.

> As a catalyst charge decreases to <1 mol%, the contribution of catalyst cost to the raw material cost for a catalyzed reaction becomes vanishingly small. Development time planned for catalyst recovery and recycle might be better spent determining the minimum catalyst charge and identifying catalyst poisons in the substrate.

***6.2.2.2 Ketone Reduction by (−)-Diisopinocampheylchloroborane*** The chiral component of diisopinocampheylchloroborane is derived from α-pinene. Hydroboration of α-pinene (3.8 equivalents) with borane–dimethylsulfide (1.7 equivalents) in THF at 0°C produces diisopinocampheylborane. Rather than isolate this air- and moisture-sensitive intermediate, slow addition of a solution of dry hydrogen chloride (1.7 equivalents) in THF at 0°C affords a solution of diisopinocampheylchloroborane (Scheme 6.10).[60] Diisopinocampheylchloroborane can be prepared in a single step by hydroboration of α-pinene (4.0 equivalents) with chloroborane–dimethylsulfide (1.8 equivalents) in hexanes at 25–30°C.[43] With either of these methods we would have to deal with the dimethylsulfide odor during the reduction workup procedure. There is no dimethylsulfide odor associated with the preparation of diisopinocampheylchloroborane using boron trichloride and sodium borohydride. A 1.0 M solution of boron trichloride in heptane (1.0 equivalent) is added at 0°C to a mixture of sodium borohydride (0.9 equivalents) and α-pinene (3.6 equivalents) in 1,2-dimethoxyethane. After aging at 0, 25, and 40°C, the slurry of diisopinocampheylchloroborane is ready to use.[61] While a solution of diisopinocampheylchloroborane in THF can be stored for up to a year at 0°C, to maximize plant efficiency the slurry could be generated and used the same day. (+)-α-Pinene is converted to (−)-diisopinocampheylchloroborane that affords the desired (*S*)-alcohol **32**. (−)–αPinene is converted to (+)-diisopinocampheylchloroborane that affords the (*R*)-alcohol **31**.

Are raw material costs for producing (−)-diisopinocampheylchloroborane by these three methods an important consideration? The three methods use approximately the same amount of (+)-α-pinene (91% ee)[62] at $42/mol. A relatively large quantity (14.4 kg) of borane dimethylsulfide complex[63] is available for $24/mol. A small quantity (50 g) of chloroborane–dimethylsulfide complex[64] is $223/mol. Boron trichloride[65] (454 g) is $87/mol and reagent grade sodium borohydride[66] (2 kg) is $9/mol. What stands out in this analysis is the "specialty chemical" quantity and price of the chloroborane–dimethylsulfide complex.

(+)-α-Pinene is available in several grades (e.g., 97% or 91% ee) and in the analysis above the lower price for the 91% grade is quoted. This grade is selected because asymmetric

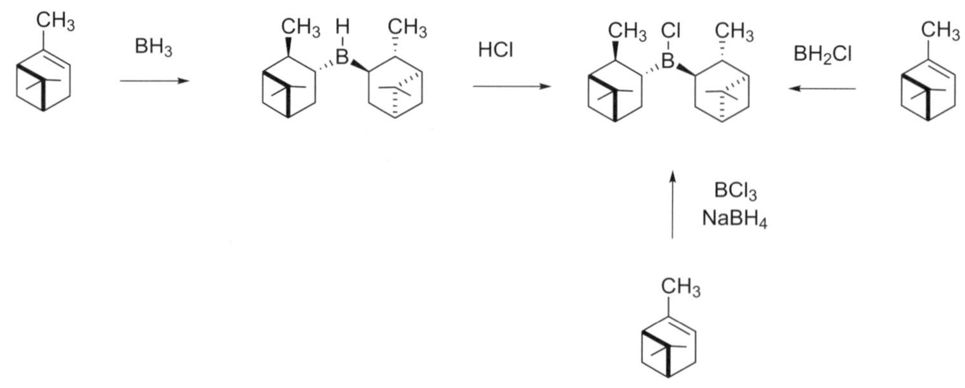


**SCHEME 6.10** Routes to (−)-diisopinocampheylchloroborane.

amplification is observed in the ketone reduction with (−)-diisopinocampheylchloroborane. Consider that each borane reagent contains two α-pinene units. Starting with a mixture of (+)-α-pinene and (−)-α-pinene we can generate the (+,+), (−,−), and (+,−) boranes. Apparently formation of the (+,−) borane is disfavored or the rate of ketone reduction by the (+,−) borane is much slower. For example, reduction of (*E*)-methyl 2-(3-(3-(2-(7-chloroquinolin-2-yl)vinyl)phenyl)-3-oxopropyl)benzoate (**25**) with diisopinocampheylchloroborane (1.8 equivalents) generated from 70% ee (−)-α-pinene and chloroborane–dimethylsulfide in hexanes–THF at −20°C affords the (*R*)-alcohol **31** with 99.5% ee (87%). With three methods for generating diisopinocampheylchloroborane and many options for quenching the excess reducing agent, releasing the alcohol from the borate ester, and isolating the alcohol there can be an infinite number of variations of the ketone reduction process. In this case, excess reducing agent is quenched by adding acetone. The alcohol is released with a 20% solution of sodium potassium tartrate. After separating the aqueous layer, brine wash, and solvent exchange, the (*R*)-alcohol **31** monohydrate is crystallized from isopropyl acetate–hexanes. The throughput for this process is 41 g/L.[43]

Reduction of (*E*)-methyl 2-(3-(3-(2-(7-chloroquinolin-2-yl)vinyl)phenyl)-3-oxopropyl)benzoate (**25**) with (−)-diisopinocampheylchloroborane (1.5 equivalents) in THF at 15°C affords the (*S*)-alcohol **32**. Here the reaction is quenched into ice water and the precipitate is isolated. (*S*)-Alcohol **32** is released from the borate ester using diethanolamine. After separating the aqueous layer, brine wash and solvent exchange, (*S*)-alcohol **32** monohydrate is crystallized from 10:1 methanol–water (65%). The throughput is just 12 g/L. The enantiomeric purities of the (+)-α-pinene and the (*S*)-alcohol **32** are not provided.[32]

Reduction of (*E*)-methyl 2-(3-(3-(2-(7-chloroquinolin-2-yl)vinyl)phenyl)-3-oxopropyl)benzoate (**25**) with (−)-diisopinocampheylchloroborane in dichloromethane at −5 to 0°C affords the (*S*)-alcohol **32** (95%). The reaction is quenched and the alcohol released using triethanolamine. After a water workup, the dichloromethane solution is concentrated at reduced pressure. The residue is dissolved in methanol and (*S*)-alcohol **32** is precipitated with water. The solid is filtered and then slurried in heptane and the suspension is filtered. The equivalents and optical purity of the borane reagent and the optical purity of (*S*)-alcohol **32** are not provided.[67]

Reduction of (*E*)-methyl 2-(3-(3-(2-(7-chloroquinolin-2-yl)vinyl)phenyl)-3-oxopropyl)benzoate (**25**) with (−)-diisopinocampheylchloroborane in THF–heptane at 15°C affords the (*S*)-alcohol **32** (98%). The reaction is quenched and the alcohol released using aqueous 1% diethanolamine. Adding hexanes results in precipitation of the (*S*)-alcohol **32** monohydrate. While this simple "one-pot" procedure requires no layer separations or distillations, the throughput is just 22 g/L. The enantiomeric purities of the (+)-α-pinene and (*S*)-alcohol **32** are not provided.[25]

> While "one-pot" procedures are intellectually satisfying accomplishments in the laboratory and can facilitate process development by quickly supplying material for downstream development efforts, these procedures often have low throughput. One-pot procedures are also less efficient in the manufacturing plant because, by definition, the operations are performed in a linear mode: each step is completed before the next begins. A multipot process in a manufacturing facility may have several process operations running simultaneously.

Reduction of (*E*)-methyl 2-(3-(3-(2-(7-chloroquinolin-2-yl)vinyl)phenyl)-3-oxopropyl)benzoate (**25**) with diisopinocampheylchloroborane (1.8 equivalents) from 85% ee (+)-α-pinene and boron trichloride–sodium borohydride in THF–DME–heptane at −20°C affords (*S*)-alcohol **32** in >99.0% ee (95%). The reaction is quenched with benzaldehyde. The alcohol is released with 30 wt% aqueous potassium carbonate. After separating the aqueous layer and concentration to 1/3 of the original volume, adding heptane and water induces crystallization of (*S*)-alcohol **32** monohydrate. The throughput for this process is 59 g/L.[61]

***6.2.2.3 Ketone Reduction by Asymmetric Transfer Hydrogenation*** The asymmetric transfer hydrogenation uses a ruthenium precatalyst and a chiral diamine ligand. Reduction with bis($\eta^6$-mesitylene)ruthenium(II) chloride (2.1 mol%), (1*S*,2*S*)-*N*-piperidylsulfamoyl-1,2-diphenylethylenedianine (2.1 mol%), formic acid and triethylamine affords (*S*)-alcohol **32** in 99.8% ee (83%) in DMF at 40°C. The reaction is quenched with water and extracted with dichloromethane. The combined extracts are washed with water and aqueous bicarbonate and then concentrated at reduced pressure. The residue is taken up in isopropyl acetate and water. Dilution with hexanes results in precipitation of (*S*)-alcohol **32**. The throughput is 29 g/L. The ligand is prepared from (1*S*,2*S*)-(−)-1,2-diphenylethylenediamine[68] and *N*-chlorosulfonylpiperidine (55%).[52,69]

Using a 1 mol% catalyst charge in DMF at 40°C, (*S*)-alcohol **32** is produced in 99.5% ee (75%). Analysis of the reaction mixture reveals (*S*)-alcohol **32** (77%), starting ketone **25** (21%), and the bridge-reduced side product **33** (2%). Ethylene bridge reduction is much more prevalent using a rhodium catalyst.[52]

***6.2.2.4 A Cost Comparison of the Ketone Reduction Methods*** There are three chemical methods for the reduction of (*E*)-methyl 2-(3-(3-(2-(7-chloroquinolin-2-yl)vinyl)phenyl)-3-oxopropyl)benzoate (**25**) (Scheme 6.11). The

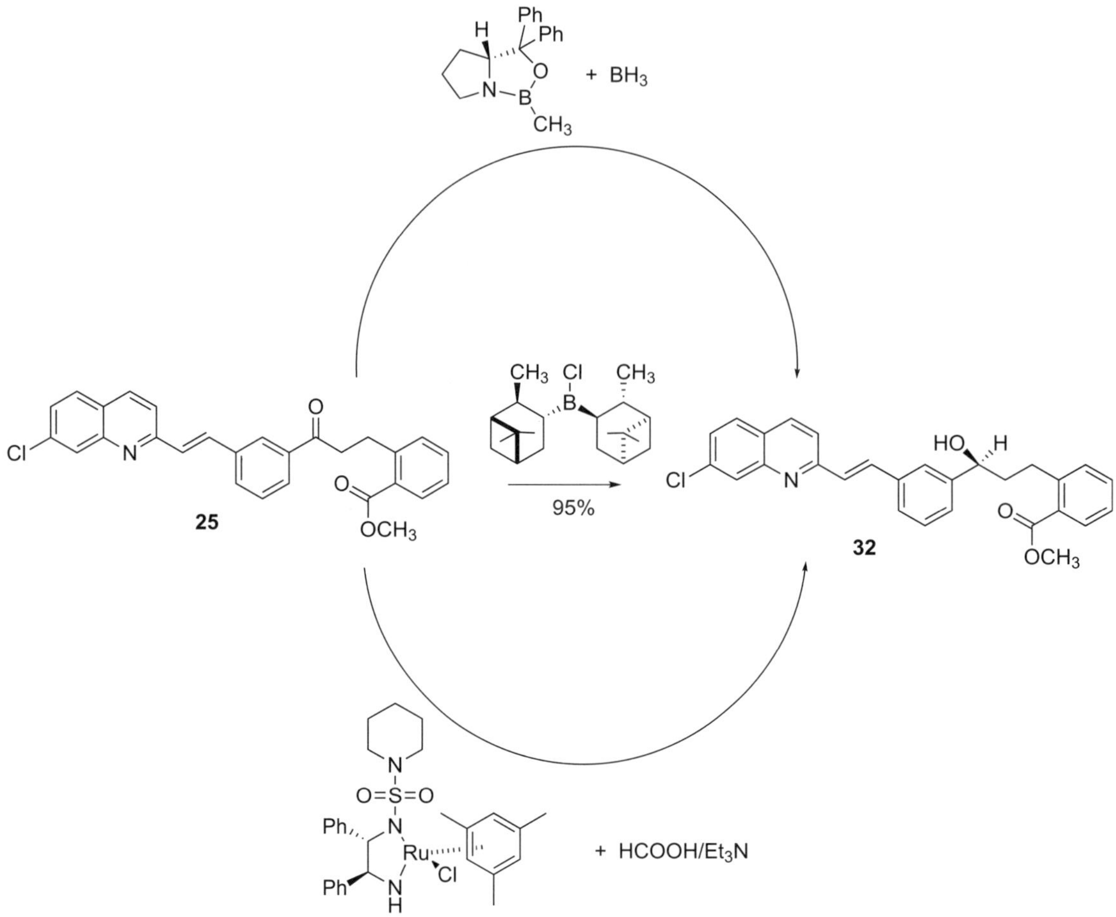


**SCHEME 6.11** (*S*,*E*)-Methyl 2-(3-(3-(2-(7-chloroquinolin-2-yl)vinyl)phenyl)-3-hydroxypropyl)benzoate (**32**) by asymmetric reduction of ketone **25**.

yields and optical purities are comparably high for all three processes. Because the oxazaborolidine catalyst is available in toluene or THF solution, catalyst decomposition by contact with air and water is no longer an issue. The catalyst loading is high and the catalyst is expensive. There may be an opportunity to reduce the catalyst loading using palladium-free ketone **25**. (+)-α-Pinene is inexpensive especially since (+)-α-pinene with a low optical purity can be used.

How low would the catalyst loading have to be to make the raw material cost for the oxazaborolidine-catalyzed process competitive with the diisopinocampheylchloroborane process? With some reasonable assumptions, calculating relative costs for the two processes is straightforward. First, assume the yields and optical purities are the same for products from both processes (95% yield and >99% ee). Second, assume borane dimethylsulfide (1.7 equivalents) can be used in both processes. Using the small-quantity prices provided, the oxazaborolidine catalyst, at 30 mol% loading, contributes \$5400 to the per kilogram cost of producing (*S*)-alcohol **32**. (+)-α-Pinene (3.8 equivalents) contributes just \$360. If the catalyst loading can be reduced to 2 mol%, the cost contributions from the catalyst and (+)-α-pinene would be equivalent.

How does the cost for the transfer hydrogenation compare? The cost drivers are ruthenium metal and the chiral ligand. Approximately 7.0 g of ruthenium and 21.5 g of (1*S*,2*S*)-(−)-1,2-diphenylethylenediamine are required to produce 1 kg of (*S*)-alcohol **32**. At \$300/oz, the ruthenium contribution per kilogram could be as low as \$75. The diamine contribution per kilogram could be as high as \$3300.[70]

### 6.2.3 Elaboration of the Backbone Methyl Ester

After following the more-or-less linear path from 7-chloroquinaldine **3** to the (*S*)-alcohol **32**, we reach an intersection where we have many roads to choose from and all "lead to Rome." Which path will produce montelukast that meets purity specifications for the lowest cost? We can convert the secondary alcohol to a methanesulfonate, displace the methanesulfonate, and convert the benzoate ester to the tertiary alcohol. We can convert the benzoate ester to the methyl ketone, convert the secondary alcohol to the methanesulfonate, displace the methanesulfonate, and convert the methyl ketone to the tertiary alcohol. We can convert the benzoate ester to the tertiary alcohol, convert the secondary alcohol of the resulting diol to a methanesulfonate, and displace the methanesulfonate. Should the methanesulfonate be displaced by a thiolate with the fully elaborated side chain, with a partially elaborated side chain, or with no side chain? Will we run into selectivity issues necessitating additional protection and deprotection steps? Where can we overlay similar operations to achieve maximum efficiency? The possibilities are almost limitless.

We will begin with conversion of the methyl benzoate to the tertiary alcohol (Scheme 6.12). While addition of methylmagnesium halide to a benzoate ester is usually a reliable method for generating a 2-aryl-2-propanol, abnormal reactions including enolization, reduction, conjugate addition, and pinacol coupling may reduce the efficiency of the conversion and necessitate purification of the product by chromatography or crystallization.

(*S*,*E*)-Methyl 2-(3-(3-(2-(7-chloroquinolin-2-yl)vinyl) phenyl)-3-hydroxypropyl)benzoate (**32**) is isolated from ketone reduction as a hydrate. The water of hydration is removed at the start of the next step as the water–toluene azeotrope. This is most convenient when toluene can be used as the solvent or cosolvent in the next step. The addition of 3.0 M methylmagnesium chloride (2.2 equivalents) in THF to a solution of (*S*,*E*)-methyl 2-(3-(3-(2-(7-chloroquinolin-2-yl)vinyl)phenyl)-3-hydroxypropyl)benzoate (**32**) in toluene at 0°C does produces tertiary alcohol **34**. Workup involves quench with 25% ammonium acetate, extraction with ethyl acetate, concentration, and chromatography. No yield or purity data is provided.[32] Adding the toluene solution of (*S*,*E*)-methyl 2-(3-(3-(2-(7-chloroquinolin-2-yl)vinyl)phenyl)-3-hydroxypropyl)benzoate (**32**) to 1.5 M methylmagnesium chloride in THF at 10°C affords tertiary alcohol **34** in 85% yield.[71] A very similar procedure with (*E*)-methyl 3-((3-(2-(7-chloroquinolin-2-yl)vinyl)phenyl)(hydroxyl)methyl)benzoate (**35**) affords the tertiary alcohol **36** in 86% yield.[32]

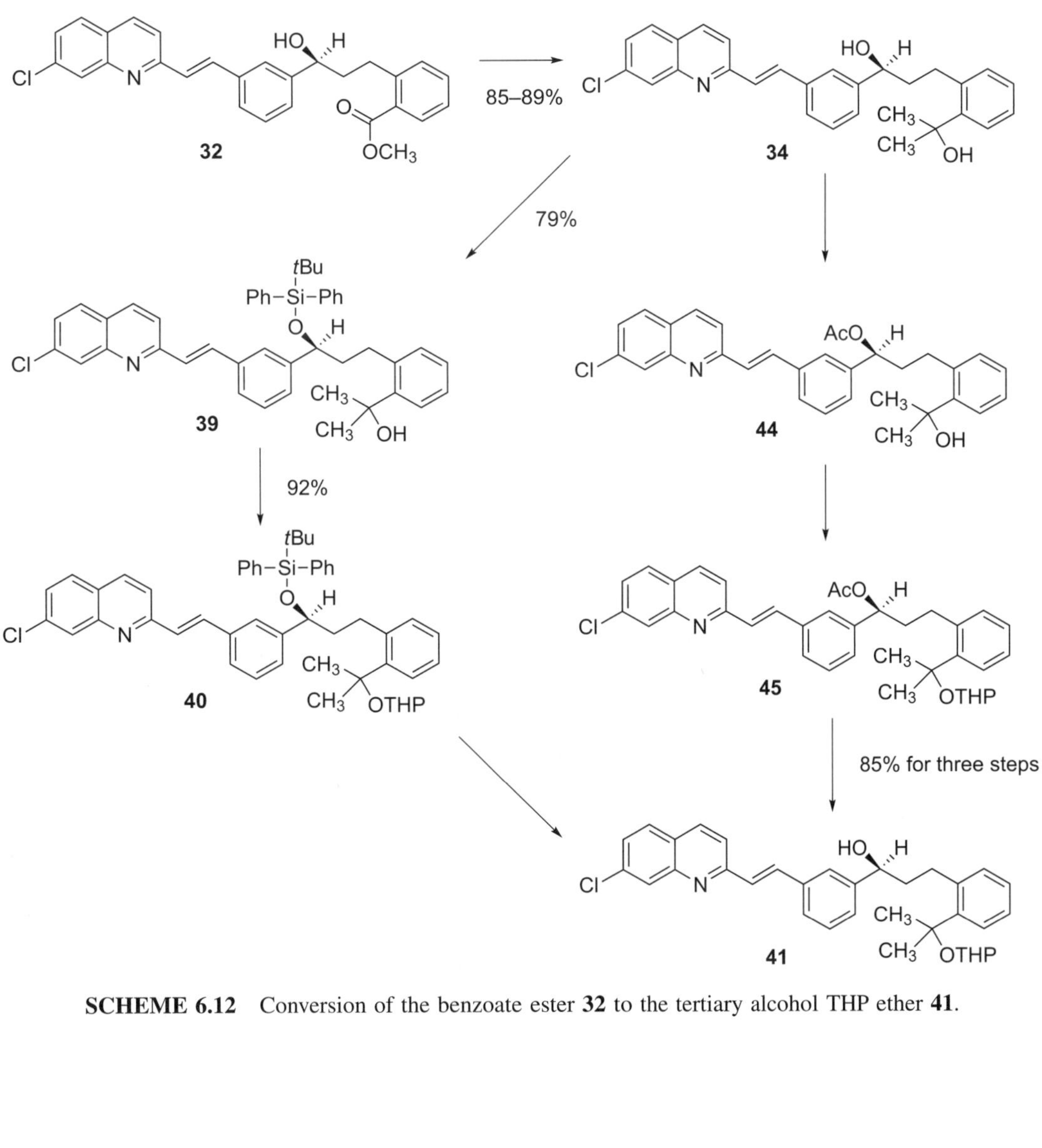


**SCHEME 6.12** Conversion of the benzoate ester **32** to the tertiary alcohol THP ether **41**.

The toluene cosolvent is critically important in achieving these high yields. The addition of 2–3 M methylmagnesium chloride to a dichloromethane solution of (*S*,*E*)-methyl 2-(3-(3-(2-(7-chloroquinolin-2-yl)vinyl)phenyl)-3-hydroxypropyl)benzoate (**32**) at 5°C results in incomplete conversion. After quench with aqueous acetic acid and a routine workup, the crude product mixture is again reacted with 2–3 M methylmagnesium chloride in THF–dichloromethane at 5°C. Aqueous acetic acid quench, extraction with dichloromethane, and concentration of the organic layers at reduced pressure affords crude diol **34**. Diol **34** is crystallized from toluene–hexanes (65%, 98.1% pure by HPLC). Recrystallization of diol **34** from toluene–hexanes affords higher purity diol (90% recovery, 99.8% pure by HPLC).[72]

At this late stage, with this now quite valuable benzoate ester **32**, some development time should be allocated to improve upon the yield and the impurity profile of crude product **34**. The addition of methylmagnesium chloride first produces the methyl ketone **37**. Perhaps the addition stalls at the methyl ketone stage if enolization competes with the second methylmagnesium chloride addition. In fact, the reaction can be redesigned to produce methyl ketone **37** by adding a base such as methoxide[73] or hexamethyldisilazide (HMDS). Methyl ketone **38** is the major product of the reaction of (*R*,*E*)-methyl 2-(3-(3-(2-(7-chloroquinolin-2-yl)vinyl)phenyl)-3-hydroxypropyl)benzoate (**31**) with 1:2 methylmagnesium chloride–lithium hexamethyldisilazide (3.3 equivalents) in toluene–THF at 0°C (78–80%, 80% pure by HPLC).[43,67]

Enolization of methyl ketone **37** may be minimized by converting the strongly basic methylmagnesium chloride to an organocerium reagent.[74] Cerium chloride is available as anhydrous beads and as a heptahydrate.[75] The large price differential between anhydrous and heptahydrate forms and the challenge of weighing and transferring the very hygroscopic anhydrous form have compelled most groups working with cerium chloride in the lab to use the heptahydrate and remove the water in the reaction flask at the start of the procedure. The water of hydration is removed by stirring the solid heptahydrate in the reaction flask at 150°C and high vacuum (0.2 mmHg) for 2 h. The flask is then cooled and filled with dry nitrogen. THF is added with stirring to produce a suspension of the cerium chloride–THF complex. This simple dehydration procedure likely results in some hydrolysis to a cerium oxychloride and hydrogen chloride. While this dehydration procedure is unusual but feasible in a lab, anhydrous cerium chloride,[76] whether dried by a vendor or dried in-house, would be charged in a manufacturing plant. The addition of THF to the anhydrous cerium chloride will be challenging. If agitation is not started in time, a hard cake will form that may bind the agitator.[77]

The ratio of cerium chloride to benzoate ester **32** is typically 0.5:1 to 1:1. For example, reaction of the benzoate ester **32** with the reagent generated from anhydrous cerium chloride (1.0 equivalent) and methylmagnesium chloride (5.2 equivalents) is complete in 30 min in toluene–THF at 0°C. After quenching with dilute acetic acid and a routine workup, diol **34** is isolated by crystallization from toluene–hexanes (89%, 99.6 wt% purity).[78] Reaction of benzoate ester **32** with the reagent generated from anhydrous cerium chloride (0.45 equivalents) and methylmagnesium chloride (5.4 equivalents) affords diol **34** (85%, 98.1% purity). Reaction of benzoate ester **32** with the reagent generated from anhydrous cerium chloride (0.54 equivalents) and methylmagnesium chloride (5.2 equivalents) affords diol **34** (103%, 94.5% purity).[52] Increasing the amount of cerium chloride from 0.5 to 1.0 equivalents appears to have little effect.

With diol **34** in hand, consider protecting the tertiary alcohol. The secondary alcohol of diol **34** is selectively protected as a silyl ether using a bulky silylating agent. Diol **34** is monoprotected by reaction with *tert*-butylchlorodiphenylsilane (2.0 equivalents), *N*,*N*-dimethylaminopyridine (1.7 equivalents), and triethylamine (3.0 equivalents) in dichloromethane at 25°C to reflux. After a routine workup procedure, isolation of silyl ether **39** requires chromatography (79%). The silyl ether survives protection of the tertiary alcohol by reaction with 3,4-dihydro-2*H*-pyran (3.3 equivalents) and triphenylphosphine hydrobromide (5 mol%) in dichloromethane at 25°C. Doubly protected **40** is again isolated by chromatography (92%). The silyl ether is then removed with tetrabutylammonium fluoride in THF at 8–25°C. No yield is provided for the chromatography of THP ether **41**.[32]

A similar sequence to THP ether **41** begins with monoprotection of diol **34** with *tert*-butyldimethylsilyl chloride (1.5 equivalents) and imidazole (3.3 equivalents) in toluene at 70°C. After a routine workup, addition of hexanes precipitates silyl ether **42** from toluene (65%). The silyl ether survives protection of the tertiary alcohol by reaction with 3,4-dihydro-2*H*-pyran (10 equivalents) and triphenylphosphine hydrobromide (1.0 equivalent) in dichloromethane at 40°C to produce **43** (104% crude). After a routine workup, the silyl ether is removed with tetrabutylammonium fluoride trihydrate (2.8 equivalents) in dichloromethane at 40°C (103% crude). While this sequence to THP ether **41** suggests that the chromatographic upgrades of the previous sequence can be eliminated, the proof is in the link to montelukast (**1**) yield and purity data, which is not available.[52]

A less expensive and more efficient route to THP ether **41** begins with protection of the secondary alcohol as an acetate using acetic anhydride (1.5 equivalents), pyridine (1.5 equivalents) and catalytic *N*,*N*-dimethylaminopyridine in dichloromethane at 25°C. The tertiary alcohol is converted to the THP ether with 3,4-dihydro-2*H*-pyran (7.5 equivalents) and triphenylphosphine hydrobromide (25 mol%) in dichloromethane at reflux. The acetate is then hydrolyzed with sodium hydroxide in THF–methanol at 25°C. After a routine workup, THP ether **41** is isolated by chromatography (85% from **34**).[79]

There is no reported Heck arylation of the allylic alcohol, (*E*)-1-(3-(2-(7-chloroquinolin-2-yl)vinyl)phenyl)prop-2-en-1-ol (**19**) with the THP ether of 2-(2-iodophenyl)propan-2-ol (**46**). Heck arylation of the homoallylic alcohol (**47**) with this THP ether is described but no yield is reported. This alternative Heck arylation would take the Grignard addition and THP protection out of the linear sequence from 7-chloroquinaldine (**3**) to montelukast (**1**).[32]

### 6.2.4 Methanesulfonylation of (*S*)-Alcohols

The next step is to activate the secondary alcohol as a leaving group and then displace the leaving group with a thiolate with clean inversion of stereochemistry. Acetate, *p*-toluenesulfonate, benzenesulfonate, trifluoromethanesulfonate, methanesulfonate, chloride, bromide, and iodide have all been used as the leaving group *but the optical purity of the thioether produced is only provided in the procedures using the methanesulfonate*. There are four secondary alcohols that differ only in the functionality on the *ortho*-substituted aromatic ring of the backbone. The methanesulfonylation is typically run in toluene, dichloromethane, or THF at low temperature using triethylamine or *N,N*-diisopropylethylamine as the base. The stability of these benzylic methanesulfonates should be a concern as we consider the workup details, especially drying by azeotropic distillation and concentration at reduced pressure (Figure 6.3).

Methanesulfonaton of the secondary alcohol of **41** with methanesulfonyl chloride (1.3 equivalents) and triethylamine (1.5 equivalents) or methanesulfonyl chloride (2.0 equivalents) and *N,N*-diisopropylethylamine (2.3 equivalents) in dichloromethane requires 1 h at −10 to −40°C. The workup procedure involves quench with aqueous sodium bicarbonate, layer separation, and concentration of the organic layer. No yield is provided for methanesulfonate **48**.[32,52,80]

Recall that secondary alcohol **32** is isolated as a monohydrate. The water of hydration is first removed as the azeotrope with toluene or **2**-butanol. (The **2**-butanol is then replaced by toluene.) Reaction with methanesulfonyl chloride (1.3 equivalents) and triethylamine (1.5 equivalents) in toluene requires 45 min at −40°C and 45 min at −10 or 0°C. The workup procedure involves an aqueous quench, layer separation, and concentration of the organic layer. No yield is provided.[79,81,82] Methanesulfonylation of alcohol **32** can also be run in dichloromethane. After removing the water of hydration as the toluene azeotrope, the toluene is removed at reduced pressure. Methanesulfonylation in dichloromethane with methanesulfonyl chloride (1.3 equivalents) and triethylamine (1.5 equivalents) requires 30 min at −40°C and 1 h at 0°C. Methanesulfonylation in dichloromethane using methanesulfonyl chloride (5.0 equivalents) and diisopropylethylamine (8.0 equivalents) is run at 25–35°C. The workup procedure in both cases involves quench with water, layer separation, and concentration of the organic layer. No yield is provided for methanesulfonate **49**.[32,82,83]

Secondary alcohol **37** is also isolated as a monohydrate. The water of hydration is first removed as the azeotrope with toluene. Reaction with methanesulfonyl chloride (1.4 equivalents) and triethylamine (1.8 equivalents) in toluene requires 1 h at −5°C. The workup procedure

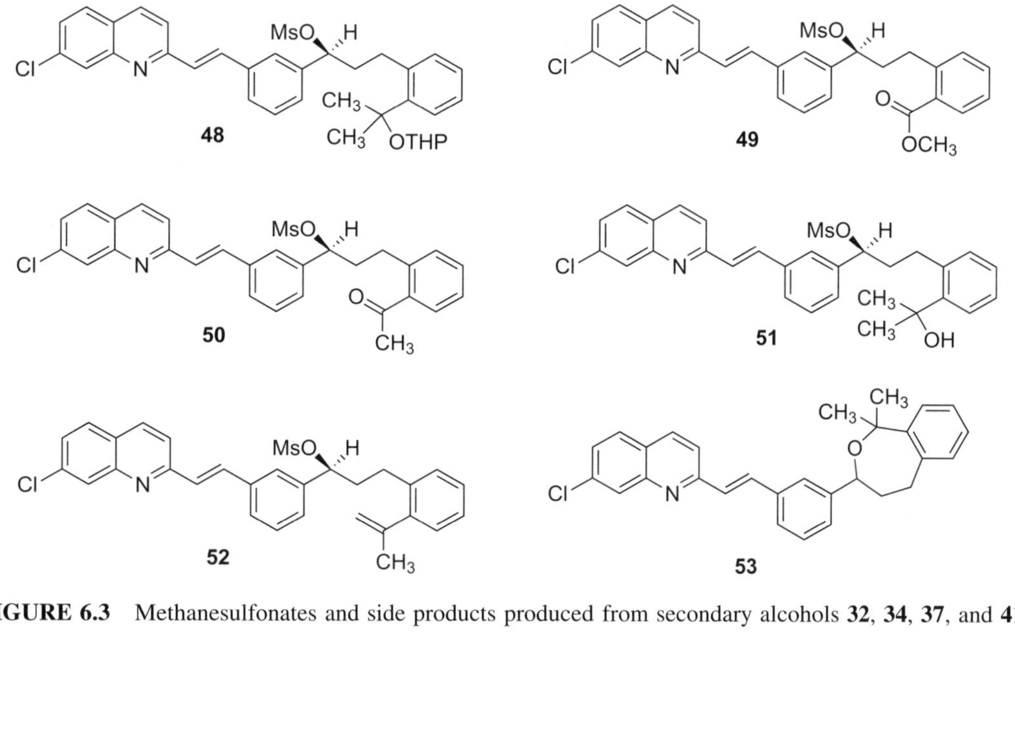


**FIGURE 6.3** Methanesulfonates and side products produced from secondary alcohols **32**, **34**, **37**, and **41**.

involves quench into cold saturated aqueous bicarbonate, layer separation, and drying over anhydrous sodium carbonate. No yield is provided for methanesulfonate **50**.[43]

A selective methanesulfonylation of the secondary alcohol of diol **34** would make the protection–protection–deprotection sequence to protect the tertiary alcohol unnecessary. Diol **34** is also isolated as a monohydrate. The methanesulfonylation procedure typically begins with removing the water as the toluene–water azeotrope. Several procedures are available from this point, many carrying the crude methanesulfonate **51** into the displacement step. The stoichiometry is critically important when preparing a solution of methanesulfonate **51** in THF. Reaction of diol **34** with methanesulfonyl chloride (1.22 equivalents) and *N,N*-diisopropylethylamine (1.42 equivalents) in THF at 0°C results in high conversion to methanesulfonate **51** containing manageable levels of methanesulfonate **52** produced by elimination of the tertiary alcohol.[84]

Methanesulfonate **51** is isolated in a preferred procedure. The dry solution of diol **34** in toluene is diluted with acetonitrile. With an acetonitrile–toluene ratio of 3:1, methanesulfonate **51** crystallizes from the reaction with methanesulfonyl chloride (1.1 equivalents) and *N,N*-diisopropylethylamine (1.2 equivalents) at −25°C. Methanesulfonate **51** is filtered, washed with cold acetonitrile and cold hexanes, dried by pulling nitrogen through the cake at 5°C for 20 h, then packaged cold and stored cold (81–89%).[78,80,85] In one case, methanesulfonylation of diol **34** is efficiently accomplished in a dipolar aprotic solvent. The mixture produced by reaction of diol **34** with methanesulfonyl chloride (1.2 equivalents) and diisopropylamine (1.6 equivalents) at −15°C in DMF is diluted with acetonitrile and then aged at −20°C to precipitate methanesulfonate **51** (95%).[86] The low reaction temperature is critically important in achieving good yields of the isolated methanesulfonate. Cyclization to ether **53** is competitive at higher temperatures in the presence of the amine hydrochloride salt. We should be able to prepare and hold a solution of isolated methanesulfonate **51** in dry THF for >24 h at 25°C.[87]

> Why review these simple methanesulfonylation procedures in such detail? These benzylic methanesulfonates are unstable in solution. Holding any of the methanesulfonate solutions or the crude oils residues for extended times will result in some decomposition. Yield is lost, new impurities may be created, and levels of impurities already present may increase. The methanesulfonylation procedures for secondary alcohols **32**, **37**, and **41** have no *hold point* after adding methanesulfonyl chloride. In contrast, methanesulfonylation of diol **34** can be held at the isolated methanesulfonate **51**.
>
> While two steps can certainly be run back-to-back in the lab, the choreography for multiple operations becomes increasingly more challenging as the scale increases. Thus, hold points become increasingly valuable as the scale increases. Allocating lab development time to identifying and demonstrating all the potential hold points for a process will provide flexibility in designing the process fit as it transitions to the pilot plant and plant.

## 6.3 SIDE CHAIN SYNTHESIS

### 6.3.1 Synthesis from Cyclopropane-1,1-diyldimethanol (54)

Many side chain approaches begin with cyclopropane-1,1-diyldimethanol (**54**) that is produced from the commercially available diethyl 1,1-cyclopropanedicarboxylate[88] by reduction with lithium aluminum hydride (84%) in THF at 15°C[32,74,89] or sodium borohydride in methanol at 75°C.[26] The objective is to convert one of two identical hydroxyl groups to a nitrile (*en route* to the carboxylic ester or acid) and then convert the other to the thiol (Scheme 6.13).

Monoprotection of diol **54** with benzoyl chloride (1.3 equivalents) and pyridine (1.5 equivalents) in dichloromethane at 0–25°C produces a nearly 1:1 mixture of the dibenzoate **55** and monobenzoate **56**. Monobenzoate **56** is separated by chromatography (55%). Reaction of monobenzoate **56** with methanesulfonyl chloride (1.3 equivalents) and triethylamine (3.0 equivalents) in dichloromethane at −40 to −10°C affords the methanesulfonate **57**. Methanesulfonate displacement is accomplished with sodium cyanide (4.5 equivalents) in DMSO at 25°C. Yields for these two steps are not available. Hydrolysis of the benzoate ester and nitrile of **58** with potassium hydroxide in ethanol–water at reflux affords 2-(1-(hydroxymethyl)cyclopropyl)acetic acid (**59**). The carboxylic acid is converted to methyl ester **60** with diazomethane (50% from **58**).[32,73]

The diol **54** can be converted to diacetate **61** by reaction with acetic anhydride and pyridine. Hydrobromic acid (33%) in glacial acetic acid then cleaves one ester to give (1-(bromomethyl)cyclopropyl)methyl acetate (**62**). No procedure for separation of the statistical mixture produced or contained yield for monoacetate **62** in the mixture is available. The statistical mixture containing monoacetate **62** is also produced by reaction of diol **54** with hydrogen bromide (3.0 equivalents) and acetic acid in dichloromethane at 10–22°C. Bromide displacement with potassium cyanide in DMF is complete in 1 h at 80°C. Selective hydrolysis of the ester with potassium carbonate in methanol at 50°C affords 1-(1-(hydroxymethyl)cyclopropyl)acetonitrile (**67**). Yields are not available for the cyanide displacement or ester hydrolysis.[90]

The diol **54** is converted to cyclic acetal **64** by reaction with *para*-anisaldehyde (1.0 equivalent) and catalytic *p*-toluenesulfonic acid in cyclohexane at reflux. The reaction is

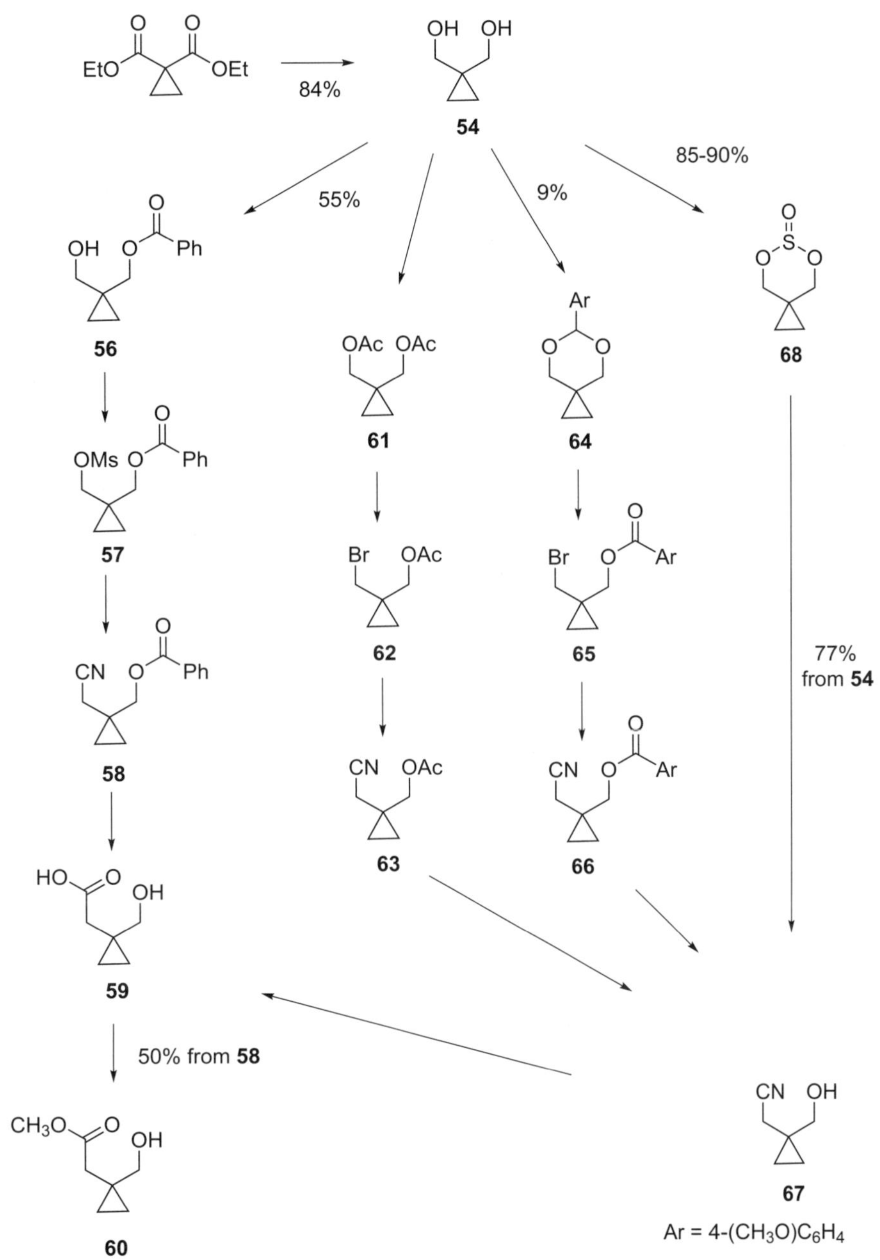


**SCHEME 6.13** Replacing one hydroxyl group of 1,1-cyclopropanediyldimethanol (**54**) by cyanide.

driven by removal of water as the cyclohexane–water azeotrope. The yield of cyclic acetal **64** is just 9%. Cleavage of the cyclic acetal with copper bromide (1.0 equivalent), tetramethylammonium bromide (1.0 equivalent) and an oxidant (DDQ or chloranil) (1–2 equivalents) in 1,2-dichloroethane at 25°C affords (1-(bromomethyl)cyclopropyl) methyl 4-methoxybenzoate (**65**). Bromide is displaced using potassium cyanide in ethanol at 35°C and the ester is cleaved with potassium hydroxide in ethanol–water at 25°C. No yields are available for the three-step sequence from the cyclic acetal **64** to 1-(1-(hydroxymethyl)cyclopropyl)acetonitrile (**67**).[79,82]

An elegant and scaleable solution to the statistical problem associated with elaborating the diol **54** begins with conversion to cyclic sulfite **68**. There are three procedures available for preparing the cyclic sulfite. Reaction of 1,1-cyclopropanediyldimethanol **54** with thionyl chloride (1.0 equivalent) and *N*,*N*-diisopropylethylamine (2.0 equivalents) in dichloromethane at 0°C or toluene at 40°C followed by quench with cold phosphate buffer (pH 7.2), layer separation, brine wash, and distillation of the dichloromethane–water azeotrope affords a solution of the crude cyclic sulfite **68** (85–90%).[78] The preferred procedure using diisopropyl sulfite delivers superior quality cyclic sulfite by avoiding contact of the reactive sulfite with water. The diisopropyl sulfite is prepared by reaction of thionyl chloride and isopropanol (95%) in toluene at 15–25°C. Hydrogen chloride, isopropanol, and some toluene are removed at reduced pressure to afford a concentrated solution of diisopropyl sulfite in toluene. Triethylamine is added to the solution as a stabilizer. After a solvent exchange to DMF, reaction of 1,1-cyclopropanediyldimethanol **54** with diisopropyl sulfite (1.2 equivalents) and catalytic sodium *tert*-butoxide is driven by a reduced pressure distillation of the isopropanol by-product. Since *tert*-butoxide slowly decomposes during the reaction, "kicker charges" of the THF solution of sodium *tert*-butoxide are added at intervals. The bulk of the isopropanol (99%) is removed using a total catalyst charge of 2.8 mol%. The resulting solution of cyclic sulfite **68** in DMF (89% yield) is carried directly into the cyanide displacement. The reaction with sodium cyanide (1.1 equivalents) and sodium iodide (0.2 equivalents) at 70°C requires 40 h to reach complete conversion. Workup by separation between toluene and water and concentration at reduced pressure affords a toluene solution of 1-(1-(hydroxymethyl)cyclopropyl)acetonitrile (**67**) (77% from **54**).[91]

There are two options for converting 1-(1-(hydroxymethyl)cyclopropyl)acetonitrile (**67**) to 2-(1-(mercaptomethyl)cyclopropyl)acetic acid (**2**): hydrolyze the nitrile then introduce the thiol or introduce the thiol then hydrolyze the nitrile. Hydrolysis of 1-(1-(hydroxymethyl)cyclopropyl) acetonitrile (**67**) with potassium hydroxide in ethanol–water at reflux affords 2-(1-(hydroxymethyl)cyclopropyl)acetic acid (**59**). This can be converted to the methyl ester **60** with methanol and sulfuric acid. No yield is available.[79,82] Do acid **59** or methyl ester **60** form a γ-lactone? This question leads us into a second synthetic manifold based on building the side chain from 2,2-bis(bromomethyl)-1,3-propanediol.

### 6.3.2 Synthesis from 2,2-Bis(bromomethyl)-1,3-propanediol

Our first inclination in designing an expeditious route to the side chain was to find a commercially available starting material with the correct carbon skeleton and then manipulate the functional groups to arrive at 2-(1-(hydroxymethyl)cyclopropyl)acetic acid (**59**). This approach locked in expensive diethyl 1,1-cyclopropanedicarboxylate as a starting material and a hazardous lithium aluminum hydride reduction as the first step. To expand the scope to include less expensive starting materials and avoid the hydride reduction, we must consider building the cyclopropane as part of the synthetic sequence. This expansion opens up routes too numerous to present even in the focused context of synthesis of the montelukast side chain.[32] The discussion here will be limited to one highly successful approach (Scheme 6.14).

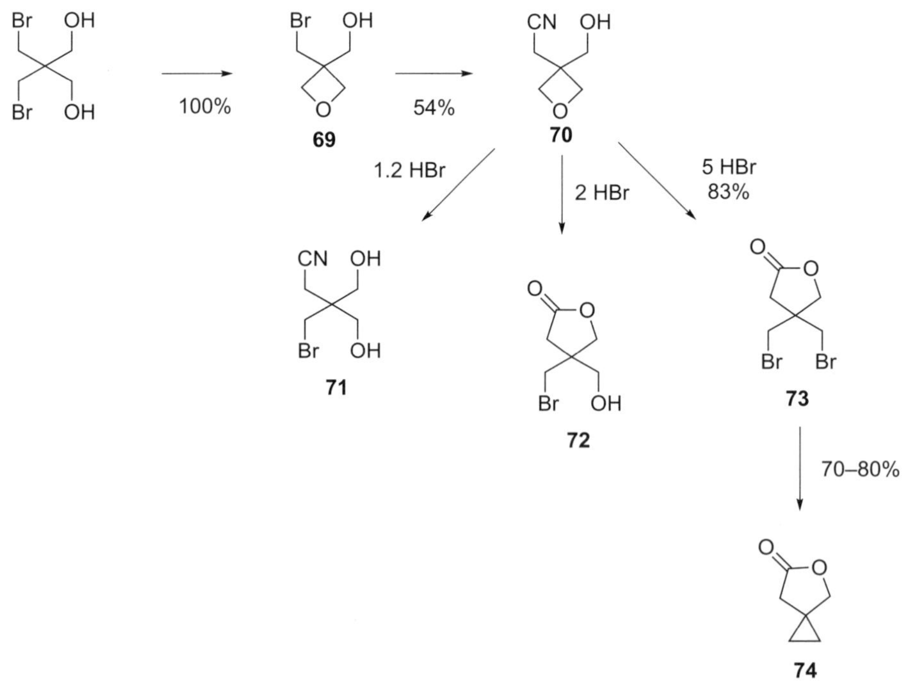


**SCHEME 6.14** 5-Oxaspiro[2.4]heptane-6-one (**74**) from 2,2-bis(bromomethyl)-1,3-propanediol.

2,2-bis(Bromomethyl)-1,3-propanediol is commercially available and inexpensive.[92] It is an industrial chemical used as a flame retardant in polyester resins and polyurethane foams. The statistical problem associated with manipulating this symmetrical substrate is cleverly addressed in the very first step. Reaction of 2,2-bis(bromomethyl)-1,3-propanediol with potassium hydroxide (1.0 equivalent) affords (3-(bromomethyl)oxetan-3-yl)methanol (**69**) in quantitative yield in ethanol at 25°C. The bromide is displaced using sodium cyanide (1.1 equivalents) (54%) in ethanol at reflux. In the next step, reaction with hydrobromic acid will certainly cleave the oxetane but hydrolysis of the nitrile, cyclization of the γ-hydroxy acid, and cleavage of the γ-lactone are also possible. In fact, the product isolated from reaction of 2-(3-(hydroxymethyl)oxetan-3-yl)acetonitrile (**70**) with hydrobromic acid depends on the equivalents of hydrobromic acid used and the reaction temperature. Oxetane cleavage with 1.2 equivalents of 48% hydrobromic acid in *N,N*-dimethylacetamide at 10°C affords **71**. Oxetane cleavage, nitrile hydrolysis, and lactone formation using 2.0 equivalents of 48% hydrobromic acid in *N,N*-dimethylacetamide at 90°C affords **72**. 4,4-bis(Bromomethyl)dihydrofuran-2(3*H*)-one (**73**), perhaps formed by oxetane cleavage, nitrile hydrolysis, lactone formation, lactone cleavage, and relactonization, is isolated from the reaction with 5.0 equivalents of hydrogen bromide in acetic acid at 100°C (83%). The cyclopropane ring is produced by reaction of dibromide **73** with zinc (2.0 equivalents) in *N,N*-dimethylacetamide at 120°C (70–80%).[93]

### 6.3.3 Introduction of the Side Chain Thiol

The three available side chain intermediates are 2-(1-(hydroxymethyl)cyclopropyl)acetonitrile (**67**), methyl 2-(1-(hydroxymethyl)cyclopropyl)acetate (**60**), and 5-oxaspiro[2.4]heptane-6-one (**74**). Routes from these intermediates to 2-(1-(mercaptomethyl)cyclopropyl)acetic acid (**2**) proceed through the methanesulfonate or bromide and use thioacetic acid[94] or thiourea[95] as the thiol source (Scheme 6.15). Hydrolysis of the thioacetate or isothiouronium salt is usually accomplished under basic conditions. The conditions for release of the thiol must be designed to either avoid the oxidative dimerization to the disulfide or to purposely produce the disulfide, which is then reduced in a separate step. The disulfide is produced using hydrogen peroxide or iodine–potassium iodide and it is reduced with zinc. In most cases, 2-(1-(mercaptomethyl)cyclopropyl)acetic acid (**2**) is crystallized from deoxygenated hexanes or heptane. The isolated solid is best stored at low temperature under nitrogen to minimize oxidative dimerization. In the event that some disulfide **75** is formed, perhaps during storage, the disulfide level can be reduced by dissolving the sulfide–disulfide mixture in toluene at 35°C and cooling to precipitate disulfide **75**. Filtration and concentration of the filtrate at reduced pressure then affords the upgraded sulfide **2**. For example, starting with 10 kg of an 87:13 sulfide–disulfide mixture, 8.5 kg of sulfide **2** (96.6% by GC) are isolated.[72]

Hydrolysis of the nitrile and conversion of the hydroxyl to a leaving group are linked when 2-(1-(hydroxymethyl)cyclopropyl)acetonitrile (**67**) is reacted with lithium bromide (1.1 equivalents) and sulfuric acid (0.57 equivalents) in isopropyl acetate at 0–25°C. Two products form : 2-(1-(bromomethyl)cyclopropyl)acetamide (**76**) and 5-oxaspiro[2.4]heptane-6-imine (**77**) hydrobromide salt. After concentration of the reaction mixture at reduced pressure to remove hydrogen bromide and the bulk of the solvent, reaction of the residue with thiourea (1.0 equivalent) in acetone at reflux results in precipitation of the isothiouronium salt **78** (83% from **67**). Hydrolysis of the isothiouronium salt and amide with aqueous sodium hydroxide requires 12 h at reflux. After cooling and filtering inorganic salts, disulfide **75** is generated by addition of dilute hydrogen peroxide (0.55 equivalents) at 5°C. Disulfide **75** precipitates when formic acid is added to lower the pH to 3.5–4. The disulfide is filtered, upgraded by a carbon treatment in hot isopropyl acetate–water, and precipitated by adding heptane (81% from **78**). The disulfide bond is reduced with zinc (1.1 equivalents) in aqueous ammonium hydroxide at 30–45°C. A precipitate forms when citric acid is added to lower the pH to 3.5. The precipitate is extracted with isopropyl acetate and the isopropyl acetate removed at reduced pressure to give an oil. The oil is then taken up in heptane and crystallized to afford 2-(1-(mercaptomethyl)cyclopropyl)acetic acid (**2**) (88%). The overall yield from 2-(1-(hydroxymethyl)cyclopropyl)acetonitrile (**67**) is 59%. The cost of appropriate disposal of the zinc hydroxide by-product (430 g/kg **67**) is an important consideration.[96]

2-(1-(Hydroxymethyl)cyclopropyl)acetonitrile (**67**) can be cleanly converted to 2-(1-(bromomethyl)cyclopropyl)acetonitrile (**79**) with bromine and triphenylphosphine in acetonitrile at 60°C. The suspension is filtered and the filtrate concentrated at reduced pressure. 2-(1-(Bromomethyl)cyclopropyl)acetonitrile (**79**) is isolated from the residue using methyl *tert*-butyl ether (87%). Reaction of 2-(1-(bromomethyl)cyclopropyl)acetonitrile (**79**) with thiourea (1.0 equivalent) in acetone at reflux affords a precipitate of the isothiouronium salt (**80**) (91%). Complete hydrolysis of the salt with 20% aqueous sodium hydroxide solution requires 14 h at reflux. After cooling, the solution is acidified to pH 3.5–4.0 with 85% formic acid and extracted with ethyl acetate. The extracts are washed with water and concentrated at reduced pressure. The residue is recrystallized from hexanes to afford 2-(1-(mercaptomethyl)cyclopropyl)acetic acid (**2**) (77%, 97.9% pure by HPLC). The cost of appropriate disposal of the triphenylphosphine oxide by-product (2.7 kg/kg of product) is an important consideration.[97]

From 2-(1-(hydroxymethyl)cyclopropyl)acetonitrile (**67**), the process can be streamlined by using the same solvent for

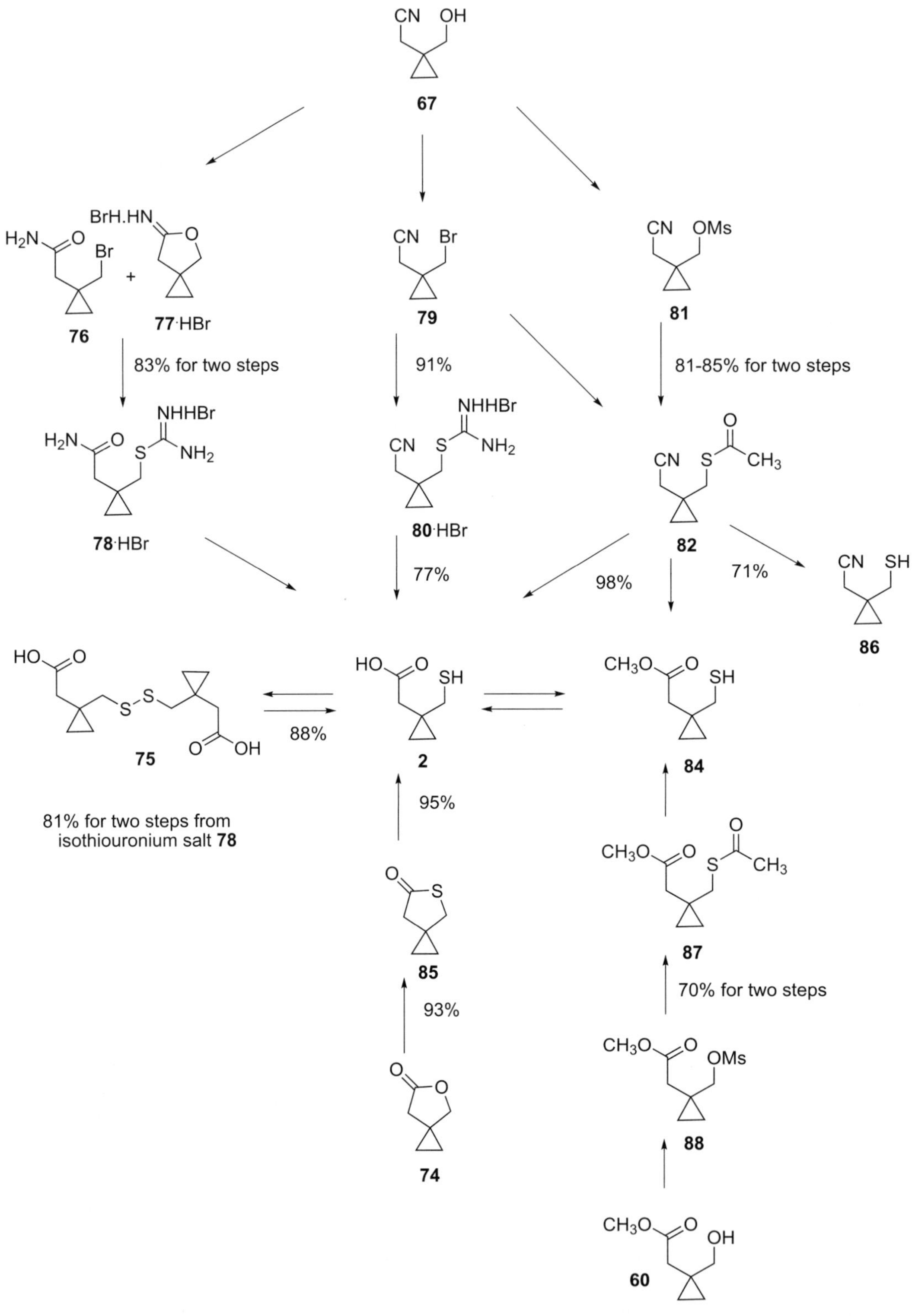


**SCHEME 6.15** Thiols and thioesters for the montelukast (**1**) side chain.

methanesulfonylation and methanesulfonate displacement. Suitable solvents are isopropyl acetate, toluene, and 3:1 toluene–DMF. For example, reaction of **67** with methanesulfonyl chloride (1.2 equivalents) and triethylamine (1.3 equivalents) in isopropyl acetate at 0°C is followed by a water wash and addition of potassium thioacetate (1.3 equivalents) at 20°C. Water wash and concentration at reduced pressure affords crude 2-(1-acetylthiomethyl)cyclopropyl)acetonitrile

(**82**) (81%).[78] Even the water wash between the two steps can be eliminated. The reaction with methanesulfonyl chloride and triethylamine in toluene at 5°C is followed by reaction with thioacetic acid and triethylamine at 30°C. Water wash and concentration at reduced pressure affords crude 2-(1-acetylthiomethyl)cyclopropyl)acetonitrile (**82**), which is upgraded by distillation at reduced pressure (87–88% purity by GC).[90]

Using this same streamlining approach, 2-(1-(mercaptomethyl)cyclopropyl)acetic acid (**2**) is prepared from 2-(1-(hydroxymethyl)cyclopropyl)acetonitrile (**67**) with just one solid isolation. A solution of methanesulfonate **81** is prepared by reaction of with methanesulfonyl chloride (1.25 equivalents) and triethylamine (1.3 equivalents) in 3:1 toluene–DMF at 5°C. The methanesulfonate is then displaced with thioacetic acid (1.2 equivalents) and triethylamine (1.5 equivalents) at 35°C. After quench with water, the toluene solution of thioacetate ester **82** (93%) is deoxygenated and carried on. Hydrolyses of the nitrile and the thioacetate with deoxygenated sodium hydroxide (5.0 equivalents) in the two-phase mixture are nearly complete in 6–10 h at 25°C. At this point, the phases are separated and the aqueous layer aged at 90°C for an additional 12–16 h to hydrolyze the intermediate amide **83**. The solution is cooled, deoxygenated heptane is added, and the pH is adjusted to 3.5–4.0 with sulfuric acid. The layers are separated and the aqueous is extracted with deoxygenated heptane. The combined heptane layers are concentrated then cooled to crystallize 2-(1-(mercaptomethyl)cyclopropyl)acetic acid (**2**) (83% from **67**).[78,91]

Reaction of 2-(1-(hydroxymethyl)cyclopropyl)acetonitrile (**67**) with phosphorus tribromide in dichloromethane at 10°C followed by bromide displacement with thioacetic acid and potassium carbonate in DMF at 10°C is another route to the thioacetate ester **82**. In this process, the hydrolyis of the nitrile is performed in two stages. Reaction with 32% hydrochloric acid in methanol at reflux affords methyl 2-(1-mercaptomethyl)cyclopropyl)acetate (**84**) that is isolated as an oil by fractional distillation at reduced pressure. The distilled ester contains some thiolactone **85**. Hydrolysis of the ester–thiolactone mixture with sodium hydroxide in methanol at reflux affords 2-(1-(mercaptomethyl)cyclopropyl)acetic acid (**2**). Yields are not available. This two-stage hydrolysis appears to offer no advantage since the nitrile hydrolysis, distillation, and ester hydroysis must all be carefully deoxygenated to minimize oxidative dimerization of the thiol.[90] Alkyl 2-(1-mercaptomethyl)cyclopropyl)acetates may also be prepared by Fisher esterification of 2-(1-(mercaptomethyl)cyclopropyl)acetic acid (**2**). The methyl ester **84** distills at 75–76°C and 4 mmHg.[98]

In some processes the nitrile is hydrolyzed after the thioether link is created. 2-(1-Mercaptomethyl)cyclopropyl)acetonitrile (**86**) is produced from 2-(1-acetylthiomethyl)cyclopropyl)acetonitrile (**82**) by a selective cleavage of the thioacetate ester in the presence of the nitrile. Reaction with potassium hydroxide (3.0 equivalents) in methanol–water at 50°C is not selective, affording a 2:3 mixture of 2-(1-mercaptomethyl)cyclopropyl)acetonitrile (**86**) and of 2-(1-mercaptomethyl)cyclopropyl)acetamide (**83**). Selective cleavage is accomplished with sodium methoxide (1.5 equivalents) in methanol at −12°C. After a water quench and extractions with heptane and dichloromethane, the combined organic layers are concentrated at reduced pressure. 2-(1-Mercaptomethyl)cyclopropyl)acetonitrile (**86**) is distilled, perhaps using an agitated thin film evaporator at reduced pressure, from the residual oil (71%).[26,84]

Product isolation and storage at low temperature under nitrogen is not necessary until the thiol is released. To streamline processing, one synthesis of the side chain is stopped at methyl 2-(1-acetylthiomethyl)cyclopropyl)acetate (**87**). The thiol will be released *in situ* at the start of the thioether formation procedure by reaction of the thioacetate ester with hydrazine or sodium methoxide. From methyl 2-(1-(hydroxymethyl)cyclopropyl)acetate (**60**), routine workup of the reaction with methanesulfonyl chloride (1.5 equivalents) and triethylamine (3.0 equivalents) in dichloromethane at −40 to −10°C affords a solution of the methanesulfonate **88**.[32,73,79,82] This solution must be dried ($Na_2SO_4$) and concentrated before dilution with DMF and reaction with cesium thioacetate (2.0 equivalents) at 25°C. Methyl 2-(1-acetylthiomethyl)cyclopropyl)acetate (**87**) is isolated by chromatography (70%).[32,73]

5-Oxaspiro[2.4]heptane-6-one (**74**) is converted to 5-thiaspiro[2.4]heptane-6-one (**85**) by reaction with potassium thioacetate[99] (1.2 equivalents) and hydroquinone (1 mol%) in DMA for 5 h at 155°C. After a routine aqueous workup, thiolactone **85** and DMA are separated by distillation. Thiolactone **85** distills at 76–78°C at 4 mmHg (93%). The thiolactone is opened with sodium hydroxide (13 equivalents) in water at reflux. The solution is cooled, hydrochloric acid is added, and the precipitate extracted with methyl *tert*-butyl ether. The extracts are then concentrated at reduced pressure to afford 2-(1-(mercaptomethyl)cyclopropyl)acetic acid (**2**) (95%).[100]

## 6.4 METHANESULFONATE DISPLACEMENT TO MONTELUKAST (1) SODIUM

The methanesulfonate displacements are run using the lithium, sodium, potassium, or cesium thiolate in a polar aprotic solvent such as DMF, DMSO, NMP, acetonitrile, or THF at 0–25°C. With three thiol side chains for the methanesulfonate displacement (the acid, ester, and nitrile) and four methanesulfonates that differ only in the *ortho*-substituted aromatic ring of the backbone, we have 12 potential

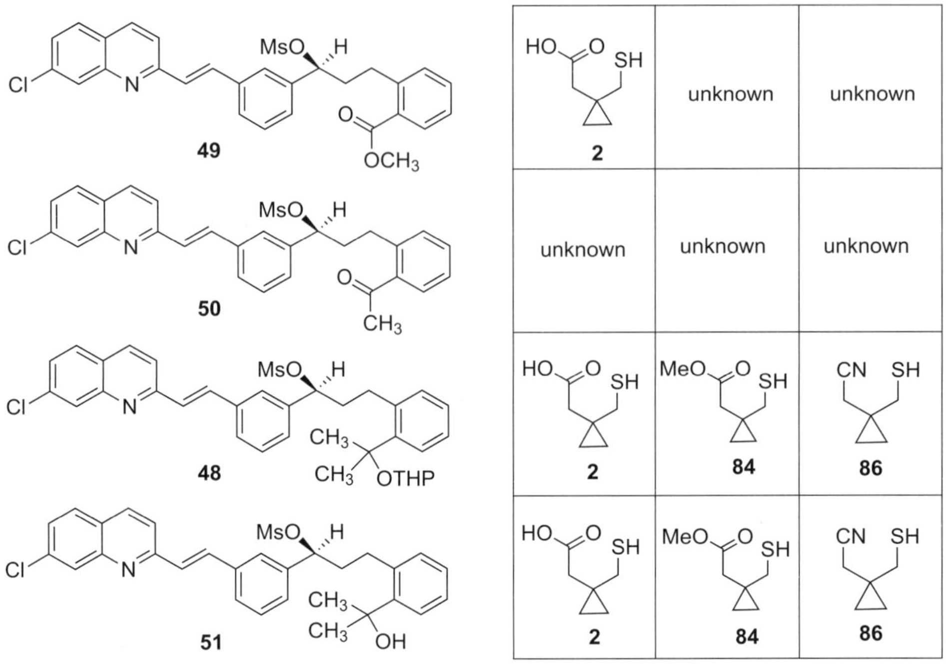


**FIGURE 6.4** Possible methanesulfonate displacements in syntheses of montelukast (**1**) with the available side chain thiols and backbone methanesulfonates.

combinations. How many combinations are known and which are preferred? Seven of the combinations are known. The combination of the thiol–acid **2** and methanesulfonate **49** is known. Reaction of the thioether **92** with methylmagnesium chloride affords montelukast (**1**). The combination of the thiol–acid **2** and the methanesulfonate **48** is known. Deprotection converts montelukast THP ether **89** to montelukast (**1**). The combinations of thiol–ester **84** or thiol–nitrile **86** and methanesulfonate **48** are known. Deprotection of the THP ether and hydrolysis of the ester in **90** or nitrile in **91** affords montelukast (**1**). The combinations of the thiol–ester **84** or thiol–nitrile **86** with methanesulfonate **51** are known. Hydrolysis of the ester in **93** or nitrile in **94** affords montelukast (**1**). Finally, the combination of the thiol–acid **2** with methanesulfonate **51** leads directly to montelukast (**1**) (Figure 6.4).

Crude montelukast (**1**) can be upgraded by recrystallization[101] or by conversion to an isopropylamine, *n*-butylamine, (+)-*sec*-butylamine, isobutylamine, *tert*-butylamine, *tert*-octylamine, dipropylamine, cyclopentylamine, cyclohexylamine, cycloheptylamine, cyclooctylamine, cyclododecylamine, *trans*-4-aminocyclohexanol, *N,N*-dicyclohexylamine, adamantylamine,[102] *N*-methylmorpholine (NMM), arginine, (*R*)-(+)-α–methylbenzylamine (MBA), cyclohexylethylamine, tris(hydroxymethyl)aminomethane,[103] (1*S*,2*S*)-(+)-2-amino-1-phenyl-1,3-propanediol,[103] (*S*)-(+)-2-phenylglycinol,[103] cinchonidine, quinine, strychnine, (+)-phenylpropanolamine, benzhydrylamine, or calcium or magnesium salt prior to conversion to montelukast sodium.

Most of the montelukast amine salts are crystalline.[104]

> A crystallization at or near the end of the process must be extraordinarily efficient or the yield losses will be unacceptable from a cost perspective. When the material to be upgraded possesses an acidic or basic functionality, a purity upgrade procedure involving salt formation, salt isolation, and salt release is often more efficient than a simple crystallization.

### 6.4.1 Displacement of Methanesulfonate 48 (Tertiary Alcohol THP Ether)

***6.4.1.1 Methyl 2-(1-Mercaptomethyl)cyclopropyl)acetate (84)*** THP-protected montelukast methyl ester **90** results from the reaction of methyl 2-(1-(mercaptomethyl)cyclopropyl)acetate (**84**) (1.5 equivalents) and sodium methoxide (2.5 equivalents) with the crude methanesulfonate **48** in DMF–methanol at 40°C. The ester is hydrolyzed during the workup procedure. THP-protected montelukast **89** is then deprotected by reaction with pyridinium *p*-toluenesulfonate[105] (30 mol%) in 3:1 methanol–THF at 40°C. Yields are not available.[52]

The thiol can be generated *in situ* from methyl 2-(1-acetylthiomethyl)cyclopropyl)acetate (**87**) (1.5 equivalents) by reaction with hydrazine (1.75 equivalents) in deoxygenated acetonitrile at 0°C. Reaction with crude methanesulfonate **48** and cesium carbonate (2.4 equivalents) in acetonitrile at 0–25°C affords THP-protected montelukast methyl ester **90**, which is isolated by chromatography. The THP ether is removed by reaction with pyridinium *p*-toluenesulfonate (3.0 equivalents) in 3:1 methanol–THF at 60°C. After removing the solvent at reduced pressure, montelukast methyl ester (**93**) is isolated by chromatography. The ester is hydrolyzed by reaction with sodium hydroxide in methanol–THF–water at 25°C. The methanol and THF are distilled at reduced pressure and the resulting aqueous solution acidified (pH = 5) with acetic acid and extracted with ethyl acetate. The extracts are concentrated at reduced pressure and the residue chromatographed to give montelukast (**1**). Yields are not available but probably average close to 90% per step. This conservative approach, using the THP ether protecting group to avoid selectivity issues in the methanesulfonylation, using the side chain methyl ester to avoid any involvement of the carboxylate during the methanesulfonate displacement, and isolating the products of the methanesulfonate displacement, THP ether deprotection, and ester hydrolysis by chromatography certainly leaves nothing to chance (Scheme 6.16).[32]

***6.4.1.2 2-(1-Mercaptomethyl)cyclopropyl)acetonitrile (86)*** Thioether **91** is produced by reaction of 2-(1-mercaptomethyl)cyclopropyl)acetonitrile (**86**) (1.5 equivalents) and sodium methoxide (2.0 equivalents) with crude methanesulfonate **48** in DMF–methanol at 25°C. The THP ether is removed by reaction with pyridinium *p*-toluenesulfonate (30 mol%) in 3:1 methanol–THF at 40°C. The nitrile is then hydrolyzed with sodium hydroxide to give montelukast (**1**) in ethanol–water at reflux. Yields are not available.[52]

***6.4.1.3 2-(1-(Mercaptomethyl)cyclopropyl)acetic Acid (2)*** THP-protected montelukast **89** results from the reaction of 2-(1-(mercaptomethyl)cyclopropyl)acetic acid (**2**) and sodium methoxide with crude methanesulfonate **48** in DMF–methanol at 25°C. The THP ether is removed by reaction with pyridinium *p*-toluenesulfonate (30 mol%) in 3:1 methanol–THF at 40°C. After a routine workup and solvent change to ethyl acetate, addition of adamantylamine affords an amorphous montelukast (**1**) adamantylamine salt (ee > 99.5%), which is isolated after diluting the suspension with hexane. Yields are not available.[52]

THP-protected montelukast **89** results from the reaction of 2-(1-(mercaptomethyl)cyclopropyl)acetic acid (**2**) (1.2 equivalents) and butyllithium with crude methanesulfonate **48** in THF at 0°C. The THP ether is removed by reaction with pyridinium *p*-toluenesulfonate (3.0 equivalents) in 3:1 methanol–THF at 60°C. After a routine workup and solvent change to toluene, addition of *N,N*-dicyclohexylamine affords montelukast (**1**) *N,N*-dicyclohexylamine salt (41% from alcohol **41**) that is isolated after diluting the resulting suspension with hexane. The salt is cleaved using acetic acid in toluene–water. After layer separation and concentration at reduced pressure, the residue is taken up in dichloromethane and montelukast (**1**) is precipitated by adding hexanes (89%).[80]

### 6.4.2 Displacement of Methanesulfonate 49 (Benzoate Ester)

Methanesulfonate **49** has only been displaced with 2-(1-(mercaptomethyl)cyclopropyl)acetic acid (**2**). This sets the

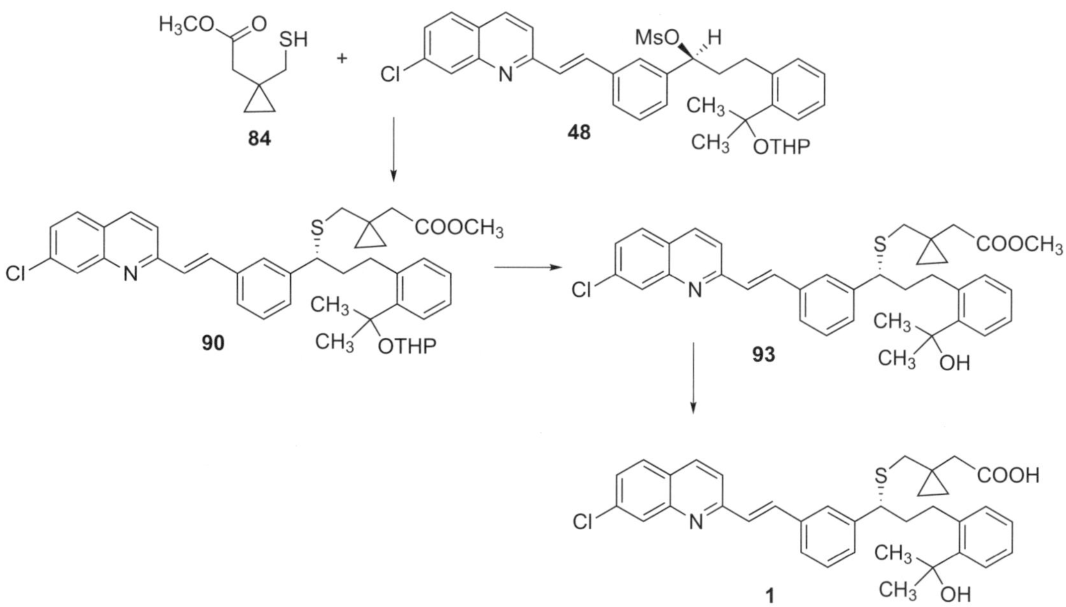


**SCHEME 6.16** Montelukast (**1**) from methanesulfonate **48** and methyl 2-(1-(mercaptomethyl)cyclopropyl)acetate (**84**).

stage for easy differentiation of the side chain acid and backbone ester in the Grignard reaction to follow.

One reaction of 2-(1-(mercaptomethyl)cyclopropyl) acetic acid (**2**) with a backbone methanesulfonate is reported to be successful with no added base. We suggest the base may have been omitted while preparing the patent. 2-(1-(Mercaptomethyl)cyclopropyl)acetic acid (**2**) (1.5 equivalents) reacts with crude methanesulfonate **49** in DMF–methanol–dichloromethane at reflux. After a routine aqueous workup and removal of the solvents at reduced pressure, the thioether **92** is isolated as the *N,N*-dicyclohexylamine (1.2 equivalents) salt from acetone at 25–35°C (30% yield for the methanesulfonylation, methanesulfonate displacement, and salt formation). A toluene solution of the free acid **92** is produced by adding acetic acid to the salt suspension and washing with water. The solution is dried and concentrated at reduced pressure. The residue is taken up in toluene–THF and reacted with methylmagnesium chloride (6.0 eqiuivalents) in THF at 0–5°C. After quench with dilute acetic acid and routine workup the conversion of the ester to the tertiary alcohol is not complete. The methylmagnesium chloride addition and workup is repeated several times to achieve complete conversion. At least part of the problem may be the competing enolization of the methyl ketone. Perhaps the enolization and the resulting reprocessing to achieve complete conversion can be avoided by using the cerium chloride procedure. Crude montelukast (**1**) is isolated from toluene (23% for this step and just 7% for the four steps from (*S*)-alcohol **32**) (Scheme 6.17).[83]

Reaction of 2-(1-(mercaptomethyl)cyclopropyl)acetic acid (**2**) (1.7 equivalents) and hexamethyldisilazide (3.4 equivalents) with crude methanesulfonate **49** in DMF at 0°C affords the crude thioether **92** (80%). Thioether **92** is upgraded by conversion to a hydrochloride salt (80%) in toluene–acetone at 5–20°C. Now the cerium chloride suggestion is implemented. Reaction of thioether **92** hydrochloride salt with anhydrous cerium chloride (1.1 equivalents) and methylmagnesium chloride (4.3 equivalents) in THF at 10°C is quenched into aqueous sodium acetate–acetic acid. Layer separation, extraction with isopropyl acetate, and concentration of the extracts at reduced pressure affords a solution of montelukast (**1**) (80% for this step and a much-improved 51% for the four steps from (*S*)-alcohol **32** (Scheme 6.18).[83]

### 6.4.3 Displacement of the Methanesulfonate 50 (Methyl Ketone)

There is limited information available about the conversion of the methanesulfonate with backbone methyl ketone to montelukast since a process that proceeds through the ketone has two steps using methylmagnesium chloride. Why not combine them and shorten the sequence? We can discuss only a likely outcome of the displacement with 2-(1-(mercaptomethyl)cyclopropyl)acetic acid (**2**).

Reaction of 2-(1-(mercaptomethyl)cyclopropyl)acetic acid (**2**) with the crude methanesulfonate **50** is not known. The reaction of 2-(1-(mercaptomethyl)cyclopropyl)acetic acid (**2**) with the *p*-toluenesulfonate **95** is known. The

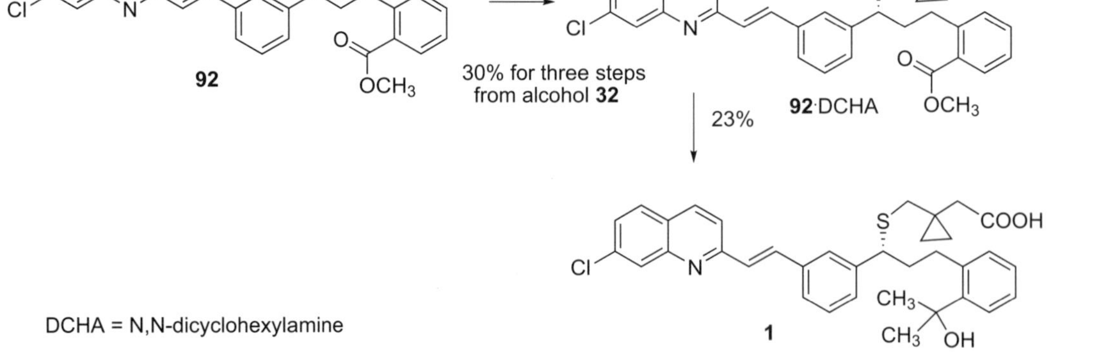


**SCHEME 6.17** Montelukast (**1**) from methanesulfonate **49** and 2-(1-(mercaptomethyl)cyclopropyl)acetic acid (**2**).

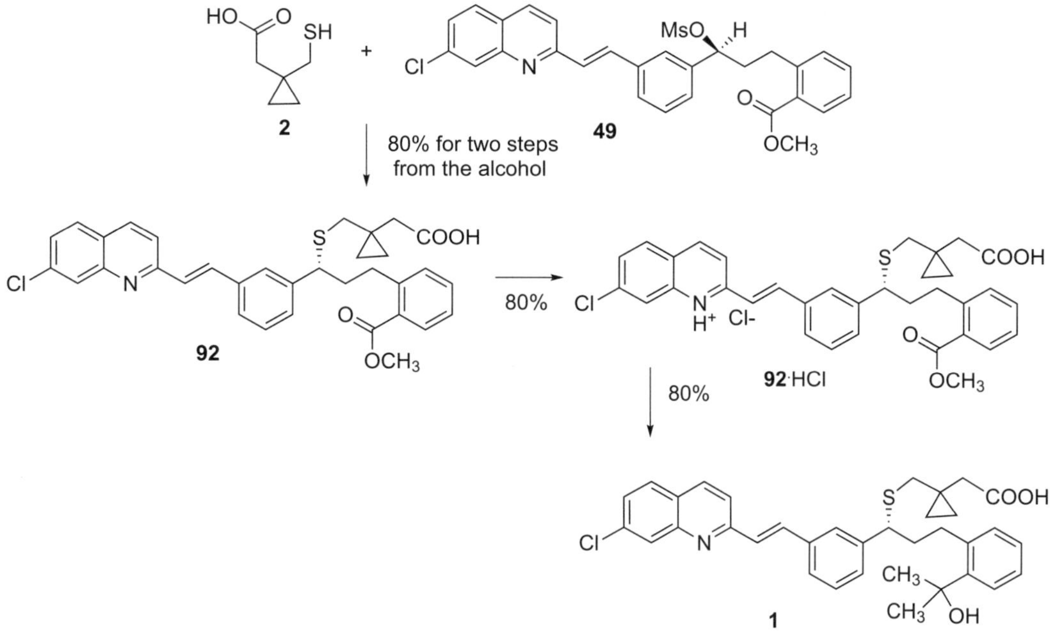


**SCHEME 6.18** Montelukast (**1**) from methanesulfonate **49**, 2-(1-(mercaptomethyl)cyclopropyl)acetic acid (**2**), and hexamethyldisilazide.

reaction of (*S*)-3-mercapto-2-methylpropanoic acid with the methanesulfonate **50** is also known. Reaction of the crude *p*-toluenesulfonate **95** with 2-(1-(mercaptomethyl)cyclopropyl)acetic acid (**2**) (1.4 equivalents) and cesium carbonate (5.2 equivalents) in DMF at 35–40°C affords the thioether **96**, which is isolated as a *N,N*-dicyclohexylamine salt from ethyl acetate (30%, 95% pure by HPLC). Salts of thioether **96** with isopropylamine and cyclohexylamine are also candidates for isolation. The yield from the reaction of dilithium salt from (*S*)-3-mercapto-2-methylpropanoic acid (**97**) (1.2 equivalents) with the methanesulfonate **50** may be more representative of what is achievable (82%).[43] A toluene solution of thioether **96** is produced by cleavage of the thioether *N,N*-dicyclohexylamine salt with acetic acid. Reaction with methylmagnesium chloride (8.1 equivalents) and cerium chloride (1.1 equivalents) in THF–toluene at −5 to 0°C is quenched in ice water and extracted into ethyl acetate. Montelukast (**1**) is precipitated from ethyl acetate as an isopropylamine, cyclohexylamine, or morpholine salt (67–70% yield, >98% pure by HPLC). Montelukast (**1**) is released from the salt with acetic acid in dichloromethane–water. The dichloromethane is distilled at reduced pressure and the residual montelukast (**1**) converted to montelukast (**1**) sodium (90%). Assuming a 99% yield in the methanesulfonylation and an 82% yield in the methanesulfonate displacement, the yield of montelukast (**1**) sodium from (*S*)-alcohol **37** could be as high as 51% (Scheme 6.19).[67]

### 6.4.4 Displacement of Methanesulfonate 51 (Tertiary Alcohol)

Methanesulfonate **51** can be isolated or carried on in solution. The hold point advantage of using isolated methanesulfonate has already been discussed. Using nonisolated methanesulfonate may be the best available option in a manufacturing plant, which does not have low-temperature isolation, drying, and storage capabilities. Many bases, including butyllithium, sodium hydride, sodium *tert*-butoxide, sodium methoxide, and sodium hydroxide, can be used to generate the thiolate. Montelukast amine salt yields of 75–85% from diol **34** are achievable.

#### *6.4.4.1 Methyl 2-(1-Mercaptomethyl)cyclopropyl)acetate (84)*

*Isolated Methanesulfonate* Several procedures describe the methanesulfonate displacement with the thiolate from methyl 2-(1-mercaptomethyl)cyclopropyl)acetate (**84**). These illustrate that many bases, in particular sodium hydride, cesium carbonate, potassium carbonate, sodium methoxide, and sodium hydroxide, can be used to generate the thiolate. Reaction of methyl 2-(1-mercaptomethyl)cyclopropyl) acetate (**84**) (1.0 equivalent) with isolated methanesulfonate **51** uses cesium carbonate (2.0 equivalents) in acetonitrile at 25°C. The methyl ester is hydrolyzed and montelukast (**1**) is isolated as the *N,N*-dicyclohexylamine salt (35% from **51**). A

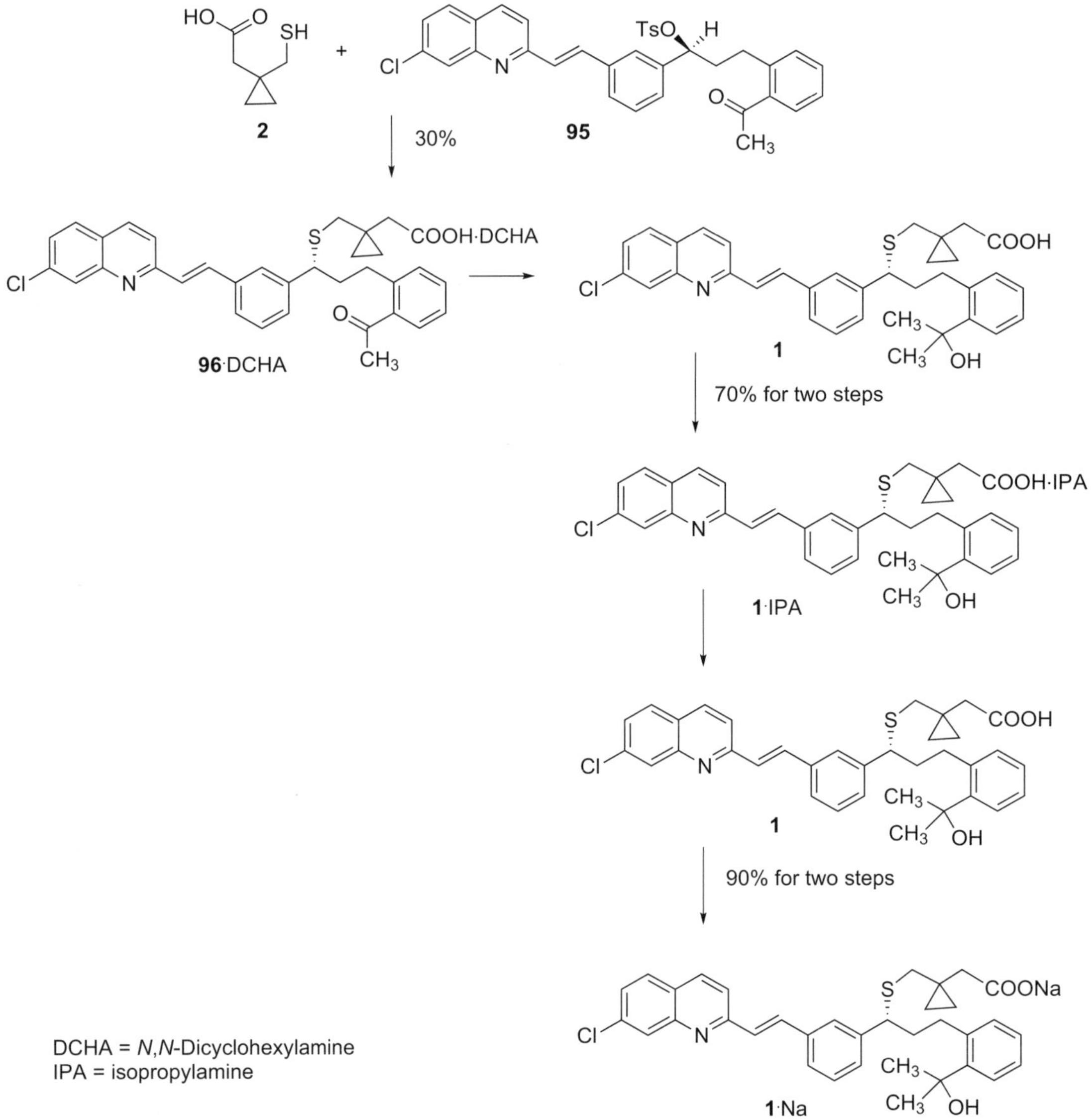


**SCHEME 6.19** Montelukast (**1**) from *p*-toluenesulfonate **95** and 2-(1-(mercaptomethyl)cyclopropyl)acetic acid (**2**).

better yield of montelukast *N,N*-dicyclohexylamine salt is obtained using potassium carbonate (2.5 equivalents) as the base (59% from **51**).[80]

Reaction of methyl 2-(1-mercaptomethyl)cyclopropyl) acetate (**84**) (1.1 equivalents) with isolated methanesulfonate **51** uses sodium methoxide in DMSO–methanol at 0°C. The reaction is quenched into with water and toluene. The ester is hydrolyzed with added aqueous sodium hydroxide (at pH 13.4). The layers are separated and the aqueous layer washed with toluene. The aqueous layer pH is reduced to 6.6 with acetic acid, the mixture with ethyl acetate, and the extracts concentrated at reduced pressure. The residue is dissolved in ethyl acetate and *tert*-butylamine is added to precipitate the montelukast (**1**) salt (49% from diol **34**). Isobutylamine, *n*-butylamine, and (+)-*sec*-butylamine also afford montelukast (**1**) salts (45–49%). These montelukast amine salts can be directly converted to montelukast (**1**) sodium using sodium hydroxide in methanol–toluene. Concentration at reduced pressure removes the amine, methanol, and water. The residue is dissolved in toluene then added to heptane to precipitate montelukast (**1**) sodium (87% from the *tert*-butylamine salt). Volatile amines such as *tert*-butylamine (bp 46°C) are much more easily removed by distillation than *N,N*-dicyclohexylamine (bp 256°C).[72]

*Nonisolated Methanesulfonate* The sodium thiolate is also generated by reaction of methyl 2-(1-mercaptomethyl) cyclopropyl)acetate (**84**) (1.4 equivalents) with using sodium hydride (2.4 equivalents) in DMF–THF at 25°C. Reaction with nonisolated methanesulfonate **51** in

DMF–THF at 25°C affords montelukast methyl ester (**93**) (69% from **51**, 66% pure by HPLC).[98]

Reaction of methyl 2-(1-mercaptomethyl)cyclopropyl) acetate (**84**) (1.2 equivalents) with nonisolated methanesulfonate **51** is also accomplished with 47% aqueous sodium hydroxide (1.5 equivalents) as base in DMF–THF–water at −10°C. The reaction mixture is then concentrated at reduced pressure and 45°C to remove the bulk of the DMF. (This distillation of DMF at reduced pressure would be very difficult to achieve at the pilot plant scale!) After a routine water–THF workup, the ester is hydrolyzed to afford montelukast (**1**) (74% yield, 81% pure by HPLC). Montelukast (**1**) can be crystallized from many solvents, including acetonitrile, toluene, ethyl acetate, and acetone–water. No data on any purity upgrade obtained by recrystallization is available.[98]

Reaction of methyl 2-(1-mercaptomethyl)cyclopropyl) acetate (**84**) (1.4 equivalents) with nonisolated methanesulfonate **51** uses 47% aqueous sodium hydroxide (2.5 equivalents) as base in DMA–THF at −6 to 18°C. After reaction completion, heating at 38°C with additional sodium hydroxide then hydrolyzes the ester. The reaction is quenched with brine then the aqueous layer is separated. The organic layer is washed with brine and aqueous tartaric acid. Isopropylamine is added and the solvent removed by concentration at reduced pressure. The residual crude montelukast (**1**) isopropylamine salt is crystallized from 2-butanone (68% from diol **34**, 98.4% pure by HPLC). The (*S*)-enantiomer is not detected. Montelukast (**1**) can also be isolated as the di-*n*-propylamine salt (Scheme 6.20).[98]

#### *6.4.4.2 2-(1-Mercaptomethyl)cyclopropyl)acetonitrile (86)*

Only the isolated methanesulfonate **51** has been used with 2-(1-mercaptomethyl)cyclopropyl)acetonitrile (**86**). The reaction of the thiolate generated from 2-(1-mercaptomethyl)cyclopropyl)acetonitrile (**86**) (1.5 equivalents) and butyllithium (6.4 equivalents) with isolated methanesulfonate **51** in DMF–hexanes is complete in 6–8 h at −10°C. Quench into brine, extraction with toluene, addition of acetic acid (to pH 5), and concentration of the organic layer at reduced pressure affords the crude thioether **94** (95%). Nitrile hydrolysis with sodium hydroxide in water at 120°C is worked up by extraction of the sodium salt into toluene and concentration of the toluene solution at reduced pressure. The residue is taken up in dichloromethane, washed with aqueous acetic acid, and concentrated at reduced pressure to give montelukast (**1**) crude. The residue is taken up in acetone and *tert*-butylamine is added to precipitate a salt. The montelukast (**1**) *tert*-butylamine salt is isolated but no yield is available. The salt can be upgraded by crystallization from acetone (79% recovery). The salt is taken up in dichloromethane–water and converted to montelukast (**1**) by addition of acetic acid. The layers are separated and the dichloromethane solution concentrated at reduced pressure. Montelukast (**1**) crude is taken up in methanol and sodium hydroxide in methanol is added. Again, the solution is concentrated at reduced pressure to replace methanol with toluene then *N*-heptane is added to precipitate montelukast (**1**) sodium (95% from montelukast *tert*-butylamine salt) (Scheme 6.21). A similar sequence converts a mixture of 2-(1-mercaptomethyl)cyclopropyl)acetonitrile (**86**) and 2-(1-mercaptomethyl)cyclopropyl)acetamide (**83**) to montelukast (**1**).[85]

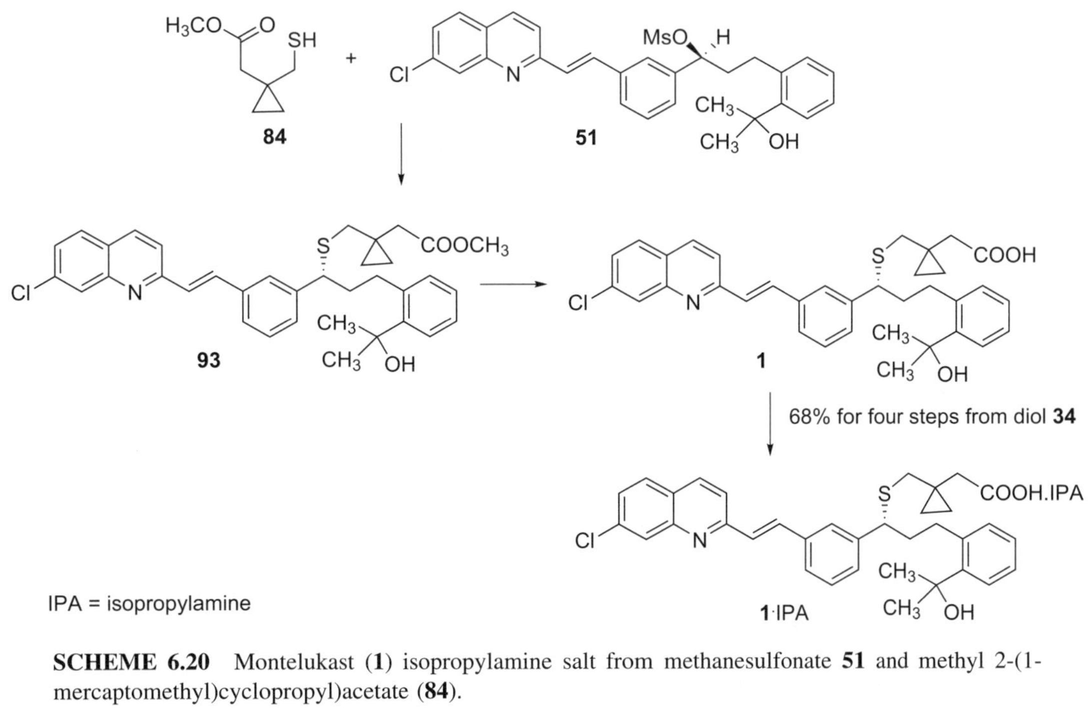


**SCHEME 6.20** Montelukast (**1**) isopropylamine salt from methanesulfonate **51** and methyl 2-(1-mercaptomethyl)cyclopropyl)acetate (**84**).

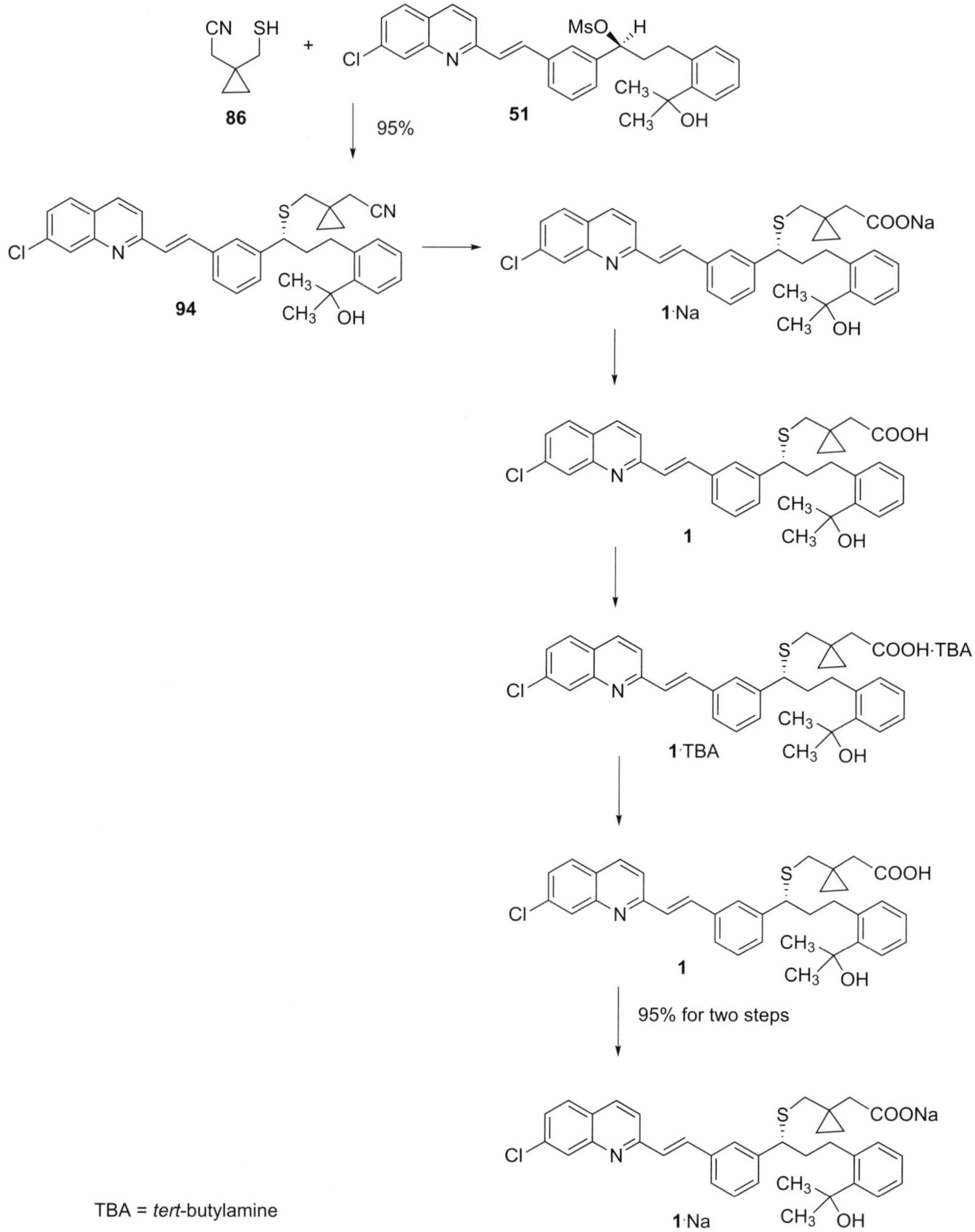


**SCHEME 6.21** Montelukast (**1**) from the methanesulfonate **51** and 2-(1-mercaptomethyl)cyclopropyl)acetonitrile (**86**).

The aqueous solution of montelukast (**1**) sodium is produced by similar procedure on a 10–20 kg scale. The montelukast sodium in water is acidified with acetic acid then extracted with toluene. The toluene extract is concentrated at reduced pressure. The residue containing crude montelukast is suspended in toluene and the suspension filtered. The crude montelukast (**1**) is digested in methanol at 25–35°C and the suspension cooled and filtered. Digestion is methanol is repeated to give montelukast (**1**) (99% pure by HPLC) (40% from diol **34**).[26]

Are these viable manufacturing processes? Consider the small-scale process. First, only 1 equivalent of butyllithium should be consumed in generating the lithium thiolate. Is some additional butyllithium consumed in reaction with the solvent DMF? If so, an alternative solvent should be considered. Second, what is the purity of crude thioether **94** and what is the yield in the nitrile hydrolysis step? If we consider that butyllithium may react with the nitrile before the methanesulfonate is added, these two missing data points may be linked. Third, while concentration at reduced pressure to a solid residue is a common lab procedure, it is not possible in a pilot plant or manufacturing plant (and this process has five such operations).

The 10–20 kg scale process still has two concentrations to dryness. The *tert*-butylamine salt procedure is replaced by two methanol digestions. All told there are three isolations of montelukast (**1**) and, if you include converting the final montelukast to montelukast sodium, each of these isolations has an associated solids handling as the material is returned to a reaction vessel for the next operation. The overall recovery (40%) is unacceptable at this late stage of the process. Missing yield data for these processes makes a comparison of the efficiencies of upgrade via the *tert*-butylamine salt and via methanol digestion impossible.

Why is it not possible to concentrate at reduced pressure to a solid residue in a pilot plant or plant? In a lab, this operation is performed every day. It is typically performed on a rotary evaporator using a water bath as a heat source. The flask is rapidly rotating to provide adequate heat transfer from the water bath to the solution to prevent bumping. The vacuum is adjusted to allow rapid but controlled distillation of the solvent with the water bath at <50°C. In a pilot plant or plant, a circulating reactor jacket fluid provides the heat and the solution is stirred with a fixed agitator. There is a space between the bottom of the agitator and the bottom of the reactor. As the distillation proceeds and the volume decreases it will reach a point where the solution is no longer being agitated. We no longer have efficient heat transfer. We may see bumping and caking of the solid up onto the reactor walls if we attempt to push the distillation by increasing the vacuum. We will risk thermal decomposition of the solid if we attempt to push the distillation by increasing the jacket temperature. Whenever you encounter a solvent change by concentration at reduced pressure to a residue you should consider how you might accomplish this by chasing the first solvent with the second. In some cases, this is easy (methanol to toluene), in others it is just not feasible (DMF to ethyl acetate).

The reaction of the thiolate generated from 2-(1-mercaptomethyl)cyclopropyl)acetonitrile (**86**) (1.1 equivalents) and sodium methoxide with isolated methanesulfonate **51** in DMSO–methanol is complete in 10 h at 5°C. After quench into water and extraction with toluene, the nitrile is hydrolyzed in a mixture of caustic lye solution and toluene at 125°C. The layers are separated. The aqueous layer is acidified with acetic acid and extracted with ethyl acetate. The extracts are concentrated at reduced pressure. The residue is dissolved in acetone and *N,N*-dicycloheylamine is added to precipitate montelukast (**1**) *N,N*-dicycloheylamine salt (44% from **51**). The salt is directly converted to montelukast sodium with sodium methoxide in methanol. The methanol is removed at reduced pressure and the residue dissolved in toluene. The toluene solution is added to heptane to precipitate montelukast sodium (89%).[106]

#### *6.4.4.3 2-(1-Mercaptomethyl)cyclopropyl)acetic Acid (2)*

*Isolated Methanesulfonate* The reaction of 2-(1-(mercaptomethyl)cyclopropyl)acetic acid (**2**) (1.5 equivalents) and sodium hydroxide (3.0 equivalents) with isolated methanesulfonate **51** in DMF at 0°C is quenched into brine and isopropyl acetate. The layers are separated and organic phase is extracted with water. Aqueous tartaric acid is added to a two-phase mixture of the combined aqueous phases and isopropyl acetate, the layers are separated, and the organic phase is concentrated to a small volume at reduced pressure to precipitate crude montelukast (**1**) (51%, 96.9% pure by HPLC). The crude montelukast (**1**) can be upgraded by slurry washing with isopropyl acetate (83% recovery, 98.1% pure by HPLC). A better yield is observed when montelukast (**1**) is precipitated from the combined aqueous phases by adding aqueous tartaric acid (71%, 95% pure by HPLC).[107]

The reaction of 2-(1-(mercaptomethyl)cyclopropyl) acetic acid (**2**) (2.2 equivalents) and 47% sodium hydroxide (5.7 equivalents) with isolated methanesulfonate **51** in DMF–THF–water at 25°C is quenched into brine and ethyl acetate. The layers are separated and the organic phase washed with aqueous tartaric acid and with water. The organic phase is then concentrated at reduced pressure. The residue is taken up in ethyl acetate and cyclohexylamine is added to precipitate montelukast cyclohexylamine salt (70%, 99% pure by HPLC).[104]

The reaction of 2-(1-(mercaptomethyl)cyclopropyl) acetic acid (**2**) (1.2 equivalents) and sodium hydride (3.8 equivalents) with isolated methanesulfonate **51** in 2:1 THF–DMSO at 0°C is quenched with acetic acid and water. Extraction with ethyl acetate, followed by brine and water washes and concentration at reduced pressure, affords a solution of montelukast (**1**) crude in ethyl acetate. Addition of a solution of magnesium chloride hexahydrate in ethyl acetate–methanol results in precipitation of a magnesium salt of montelukast (**1**). The methanesulfonate displacement is also run using butyllithium as the base in THF and the montelukast (**1**) is also precipitated as a calcium salt. The magnesium and calcium salts are converted to montelukast sodium via montelukast. Yields and purities are not available. These magnesium and calcium salts of montelukast are not hygroscopic and are more easily handled than the amine salts (Scheme 6.22).[108]

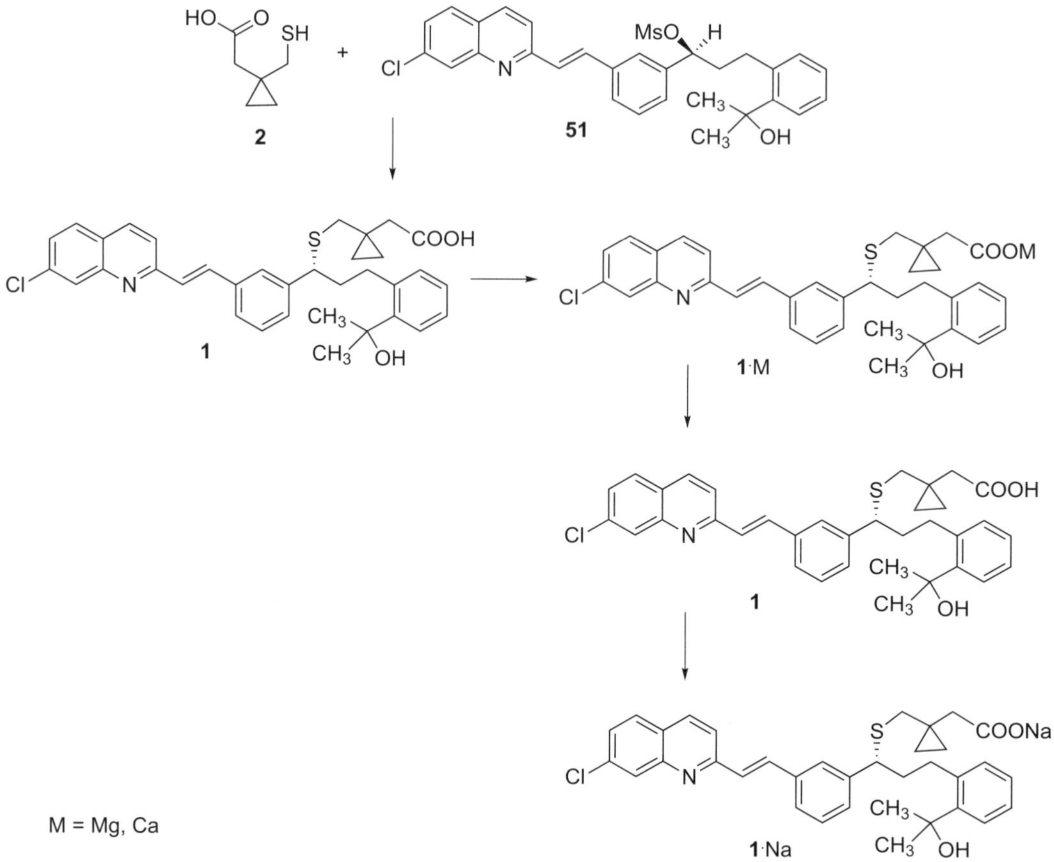


**SCHEME 6.22** Montelukast (**1**) sodium from methanesulfonate **51** and 2-(1-(mercaptomethyl)cyclopropyl)acetic acid (**2**) via a magnesium or calcium salt.

The reaction of 2-(1-(mercaptomethyl)cyclopropyl) acetic acid (**2**) (0.98 equivalents) and sodium *tert*-butoxide (2.0 equivalents) with isolated methanesulfonate **51** in NMP at 10°C is quenched with brine and toluene. Washing the toluene solution with aqueous tartaric acid and with water and concentration at reduced pressure affords crude montelukast (**1**). Addition of *tert*-butylamine (2.6 equivalents) to a toluene solution of crude montelukast (**1**) results in precipitation of the *tert*-butylamine salt (57%). Additional montelukast (**1**) *tert*-butylamine salt (20%) is recovered by concentration of the liquors at reduced pressure. Montelukast *tert*-butylamine salt is cleaved in toluene–0.5 M tartaric acid. Layer separation and water wash produced a thick emulsion. Montelukast is recovered from the emulsion and crystallized from methanol (75%). Montelukast sodium is produced by reaction with sodium hydroxide in toluene–methanol then precipitated by concentration at reduced pressure and addition of the toluene solution to hexanes (93%) (Scheme 6.23).[86]

The reaction of the thiolate generated from 2-(1-(mercaptomethyl)cyclopropyl)acetic acid (**2**) (1.3 equivalents) and cesium carbonate (2.0 equivalents) with the isolated methanesulfonate **51** in DMSO–methanol is complete in 10 h at 5°C. After quench into water, sodium hydroxide and toluene are added and the layers are separated. The aqueous layer is acidified with acetic acid and extracted with ethyl acetate. The extracts are washed with water and concentrated at reduced pressure. The residue is dissolved in acetone and *N,N*-dicyclohexylamine is added to precipitate the montelukast (**1**) *N,N*-dicyclohexylamine salt (70% from **51**). A lower yield of the montelukast *N,N*-dicyclohexylamine salt (56%) is observed using sodium methoxide in place of cesium carbonate. Ethyl acetate can be used in place of acetone for the salt formation step and *tert*-butyl amine can be used in place of *N,N*-dicyclohexylamine. Montelukast *N,N*-dicyclohexylamine salt is directly converted to montelukast sodium with sodium methoxide in methanol. The methanol is removed at reduced pressure and the residue dissolved in toluene. The toluene solution is added to heptane to precipitate montelukast sodium (89%).[106]

The reaction of 2-(1-(mercaptomethyl)cyclopropyl) acetic acid (**2**) (1.1 equivalents) and butyllithium (2.2 equivalents) with isolated methanesulfonate **51** in THF–hexanes at 0°C is quenched into ethyl acetate and brine. Layer separation, wash with 0.5 M tartaric acid, and concentration at reduced pressure affords crude montelukast (**1**). This is

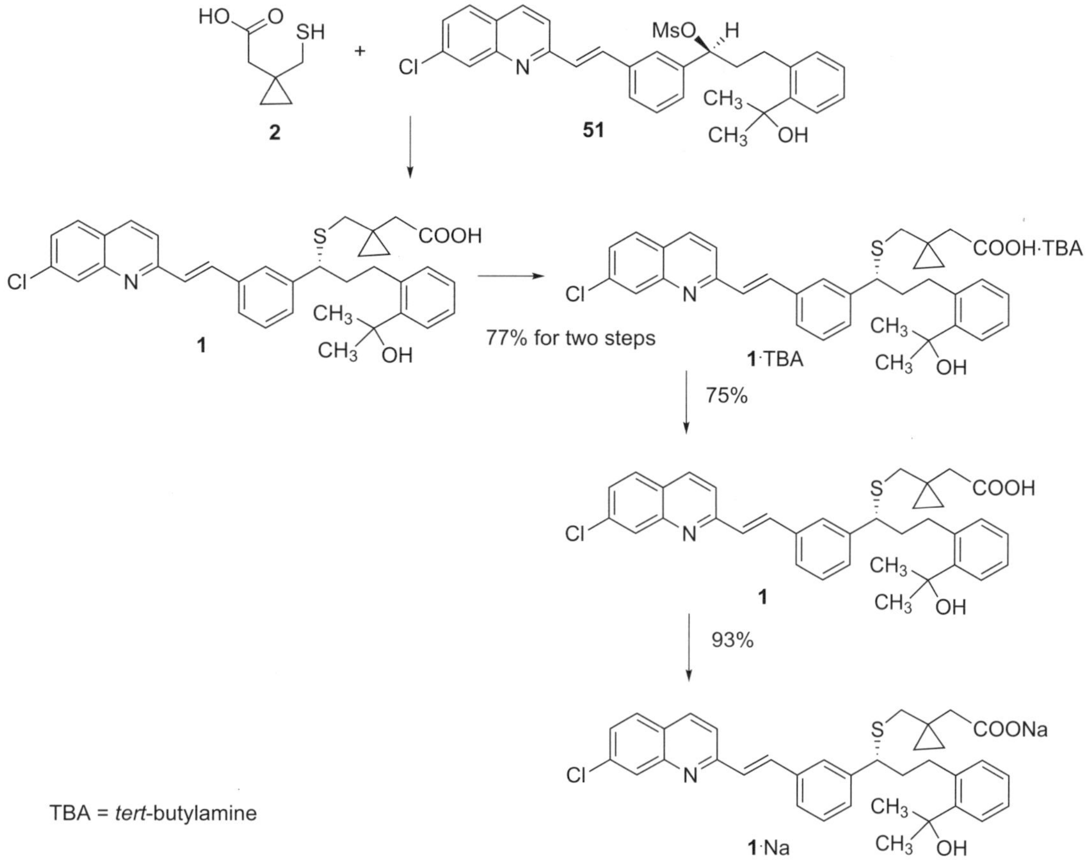


**SCHEME 6.23** Montelukast (**1**) sodium from methanesulfonate **51** and 2-(1-(mercaptomethyl)cyclopropyl)acetic acid (**2**) via the *tert*-butylamine salt.

taken up in ethyl acetate and precipitated as the *N,N*-dicyclohexylamine salt. The salt is isolated from the ethyl acetate suspension after dilution with hexanes (79%, 96 wt%, 99.8% ee). A second polymorph of the salt is produced in an alternative workup and salt isolation procedure. The reaction is quenched into brine alone. Layer separation and wash with 0.5 M tartaric acid affords a wet THF solution of crude montelukast (**1**). This is diluted with toluene and *N,N*-dicyclohexylamine is added. The resulting solution is carbon treated then concentrated at reduced pressure to remove water and THF. Toluene and seed crystals are added and the suspension aged at 25°C to produce a thick slurry of montelukast (**1**) *N,N*-dicyclohexylamine salt. The salt is isolated after diluting this suspension with heptane (81%, 99.8% ee). If the salt does not meet purity specifications it can be reslurried in toluene–heptane. Which workup and salt isolation is superior? Apparently there is no need for a carbon treatment or salt rework with the ethyl acetate–hexanes procedure. The answer may come from a comparison of the impurity profiles of montelukast sodium produced from the two montelukast *N,N*-dicyclohexylamine salts.[78,91]

Both polymorphs of the montelukast *N,N*-dicyclohexylamine salt are converted to montelukast sodium. Two procedures are offered for this conversion. In the first procedure, the amine salt is taken up in toluene–water and acetic acid (1.5 equivalents) is added. The layers are separated and the organic layer washed with water. A titrated solution of sodium hydroxide in 1% aqueous ethanol (1.00 equivalent) is added to produce a solution of the sodium salt. The solution is clarified by a polishing filtration then concentrated at reduced pressure. Acetonitrile is added at 40°C to produce a 1:1 acetonitrile–toluene solution of the sodium salt. The solution is seeded and aged at 40°C. After adding acetonitrile to increase the acetonitrile–toluene ratio to 2:1 and then 3:1 at 40°C, the suspension is cooled to 25°C and montelukast sodium is isolated (98%). A continuous crystallization of montelukast sodium is described in the second procedure. The amine salt is taken up in toluene–THF–water. Aqueous acetic acid (1.5 equivalents) is added. Essentially the same workup procedure leads to the concentrated toluene solution of the sodium salt. A suspension of seed crystals in 2:1 acetonitrile–toluene is prepared then the concentrated toluene solution of the sodium salt and acetonitrile are added as separate streams simultaneously while maintaining a 2:1 acetonitrile–toluene ratio and 20°C. This procedure affords montelukast sodium in comparable yield and

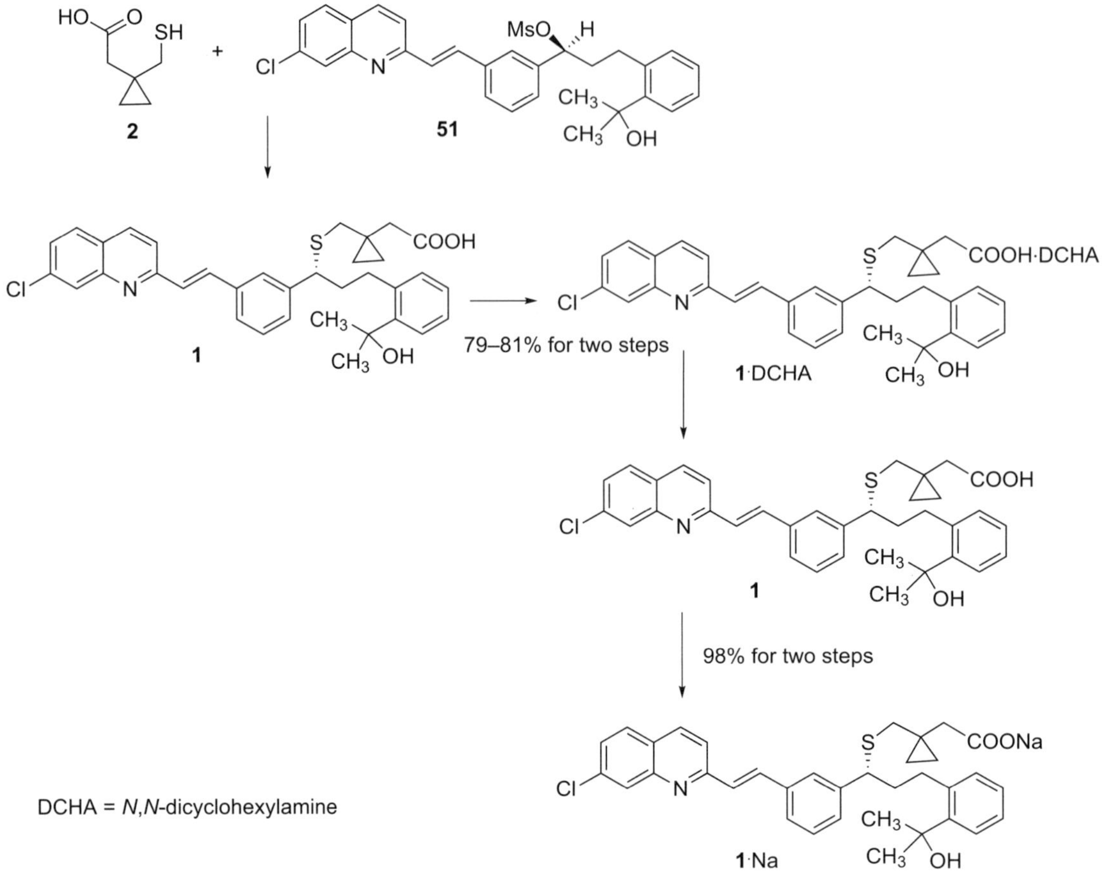


**SCHEME 6.24** Montelukast (**1**) sodium from isolated methanesulfonate **51** and 2-(1-(mercaptomethyl)cyclopropyl)acetic acid (**2**) via a the *N,N*-dicyclohexylamine salt.

quality. The overall yield from diol **34** through the isolated methanesulfonate **51** to montelukast (**1**) sodium is 62% (Scheme 6.24).[78,91]

In almost every procedure with methanesulfonate **51** the reaction is quenched to produce crude montelukast (**1**). This is converted to an amine salt. The amine salt is then converted back to pure montelukast with an organic acid (acetic acid, tartaric acid) and the pure montelukast is converted to the sodium salt. The sequence of operations can be streamlined by adding the amine, for example *N*-methylmorpholine, to the methanesulfonate displacement reaction mixture and by converting montelukast NMM salt directly to montelukast sodium with sodium methoxide. The reaction of 2-(1-(mercaptomethyl)cyclopropyl)acetic acid (**2**) (1.5 equivalents), butyllithium (1.7 equivalents), and *N*-methylmorpholine (1.7 equivalents) with the isolated methanesulfonate **51** in THF at −10°C is not quenched. The volatiles are removed by concentration at reduced pressure and the residue taken up in toluene. The suspension is filtered and the filtrate diluted with hexane to precipitate montelukast *N*-methylmorpholine salt (83%, 98.7% pure by HPLC). The amine salt is dissolved in toluene and sodium methoxide is added. After carbon treatment and filtration, the toluene solution is added to heptane to precipitate montelukast (**1**) sodium (90%) (Scheme 6.25).[109]

*Nonisolated Methanesulfonate* The reaction of 2-(1-(mercaptomethyl)cyclopropyl)acetic acid (**2**) (3.2 equivalents) and sodium methoxide (12.7 equivalents) with nonisolated methanesulfonate **51** in DMF–THF–methanol at 0°C is quenched with water. Extraction with ethyl acetate, washing of the extracts with aqueous bicarbonate, aqueous tartaric acid, and water, and concentration at reduced pressure affords montelukast (**1**) crude. Reaction with *N,N*-dicyclohexylamine in ethyl acetate is followed by dilution with acetonitrile to precipitate the salt (66% from diol **34**). Montelukast (**1**) salts are also prepared using *trans*-4-aminocyclohexanol (59%) and cyclohexylamine (64%). The nonisolated methanesulfonate **51** is generated with methanesulfonyl chloride (1.5 equivalents) and diisopropylethylamine (2.0 equivalents) in THF at 0°C.[110]

The reaction of 2-(1-(mercaptomethyl)cyclopropyl) acetic acid (**2**) (2.0 equivalents) and sodium hydroxide (5.7 equivalents) with nonisolated methanesulfonate **51** in

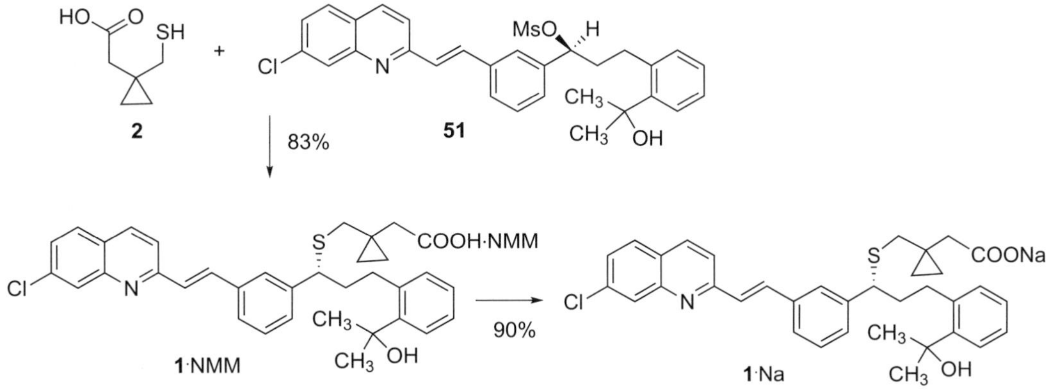

**SCHEME 6.25** Montelukast (**1**) sodium from isolated methanesulfonate **51** and 2-(1-(mercaptomethyl)cyclopropyl)acetic acid (**2**) via the *N*-methylmorpholine salt.

THF–NMP–water at 25°C is quenched with brine. Extraction with ethyl acetate, washing of the extracts with brine, aqueous tartaric acid, and water, and concentration at reduced pressure affords montelukast (**1**) crude. Reaction with cyclooctylamine in toluene precipitates the salt (68% yield from diol **34**). The analogous kilogram scale reaction of 2-(1-(mercaptomethyl)cyclopropyl)acetic acid (**2**) (1.2 equivalents) and 46% sodium hydroxide (2.4 equivalents) with methanesulfonate **51** in THF–NMP–water at 25°C is quenched with brine. A similar workup using toluene in place of ethyl acetate affords montelukast (**1**) cyclooctylamine salt (70% from diol (**34**), 97.7% pure by HPLC). The montelukast (**1**) cyclooctylamine salt contains just 0.25% of impurity **98** resulting from elimination of the tertiary alcohol. With the addition of two recrystallizations of the salt from toluene, this process is capable of converting diol **34** containing up to 8.0% by weight of the (*R*)-isomer **99** to montelukast (**1**) cyclooctylamine salt containing less than 0.1% by weight of the (*S*)-isomer **100** (41% from the diol **34**). DMF can be used in place of NMP and montelukast (**1**) can also be isolated as the cyclohexylamine and cycloheptylamine salts. Montelukast cyclooctylamine salt is cleaved using ethyl acetate and aqueous tartaric acid. The layers are separated and the ethyl acetate layer is washed with water and concentrated to a small volume at reduced pressure. Extraction of the concentrated ethyl acetate solution with aqueous sodium hydroxide affords an aqueous solution suitable for spray drying.[84]

Sodium 2-(1-(mercaptomethyl)cyclopropyl)acetate (**101**) can be isolated from the reaction of methyl 2-(1-mercaptomethyl)cyclopropyl)acetate (**84**) with 24% aqueous sodium hydroxide (5.0 equivalents) by adjusting the pH to 4.0 and extracting with toluene. The extracts are concentrated at reduced pressure and the residue isolated from cyclohexane (85%). Using carboxylate salt **101** reduces the base charge necessary to achieve high yields in the methanesulfonate displacement. The reaction of sodium 2-(1-(mercaptomethyl)cyclopropyl)acetate (**101**) (1.4 equivalents) and butyllithium (1.4 equivalents) with nonisolated methanesulfonate **51** in THF at −10°C is quenched into water, ethyl acetate, and acetic acid. The organic layer is separated and washed with water and brine. Adding (*R*)-(+)-α–methylbenzylamine precipitates the montelukast (**1**) salt (86% from diol **34**, 98.4% pure by HPLC). Cinchonidine, quinine, strychnine, (+)-phenylpropanolamine, and benzhydrylamine also afford montelukast (**1**) salts (83–86% from diol **34**, 98.1–98.5% pure by HPLC). The montelukast (**1**) salt with (*R*)-(+)-α–methylbenzylamine is converted directly to montelukast (**1**) sodium using sodium methoxide (93%) (Scheme 6.26).[111]

Methanesulfonate displacements with the dilithium salt from 2-(1-(mercaptomethyl)cyclopropyl)acetic acid (**2**) are typically accomplished in THF–hexanes or THF–hexanes–toluene. In every displacement of methanesulfonate **51** with the disodium salt from 2-(1-(mercaptomethyl)cyclopropyl) acetic acid (**2**), dipolar aprotic solvents (DMF, NMP, DMSO, acetonitrile) are used to achieve results that are comparable to those observed with the dilithium salt. In the absence of the dipolar aprotic solvent, the poor solubility of the disodium salt reduces the displacement rate. At the slower displacement rate, side reactions, particularly cyclization to ether **53**, become competitive. The disodium salt solubility in THF–toluene can be increased using a polyethylene glycol (PEG-600 and PEG-1500) or crown ether (18-crown-6 and 15-crown-5). Thus, the reaction of methanesulfonate **51** with 2-(1-(mercaptomethyl)cyclopropyl)acetic acid (**2**) (1.0 equivalent), sodium *tert*-butoxide (2.0 equivalents), and 18-crown-6 (1.0 equivalent) has a higher selectivity for montelukast (**1**) (79%) than the reaction with 2-(1-(mercaptomethyl)cyclopropyl)acetic acid (**2**) (1.0 equivalent) and lithium *tert*-butoxide (2.1 equivalents) in THF–toluene at −10 to 21°C (61%). Lower selectivities are observed using the dipotassium salt (9.6% with no crown ether and 52% with 18-crown-6).[112]

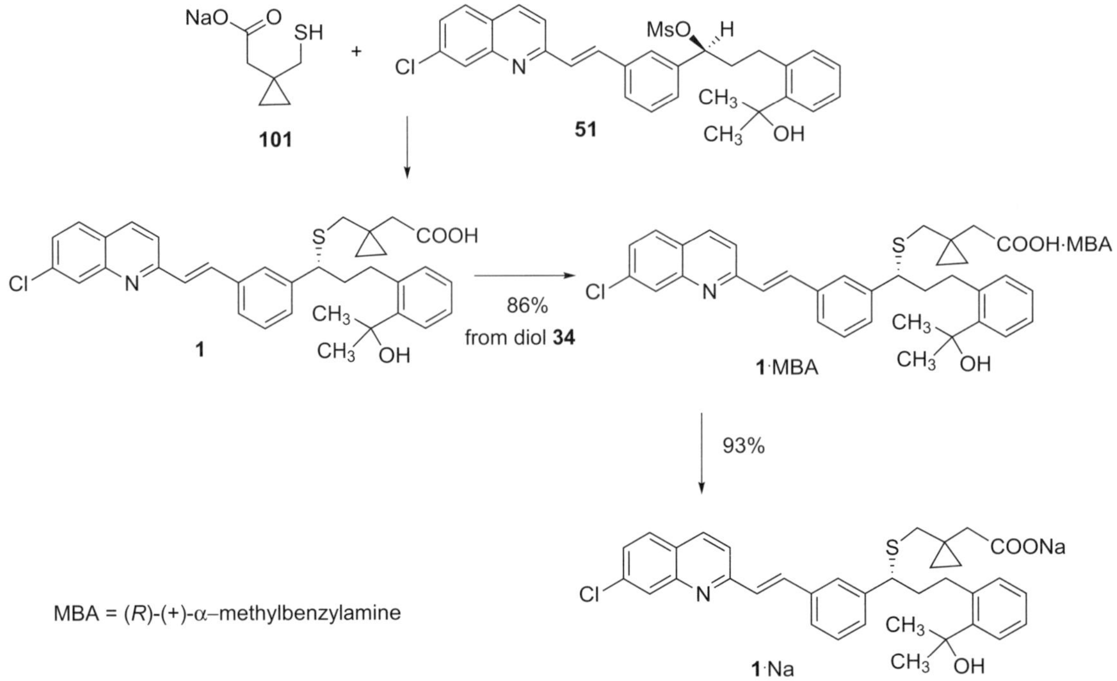


**SCHEME 6.26** Montelukast (**1**) sodium from nonisolated methanesulfonate **51** and sodium 2-(1-(mercaptomethyl)cyclopropyl)acetate (**101**) via the (*R*)-(+)-*α*-methylbenzylamine salt.

## 6.5 CHLORIDE DISPLACEMENT TO MONTELUKAST (1) SODIUM

The reaction of (*S*)-alcohol **32** with thionyl chloride affords (*S*)-chloride **102**. Water is removed from (*S*)-alcohol **32** monohydrate by distillation of the toluene–water azeotrope. The toluene is then removed at reduced pressure and the residue dissolved in dichloromethane. Reaction with thionyl chloride (3.1 equivalents) and DMF (3.1 equivalents) in dichloromethane at 25°C affords (*S*)-chloride **102**. The workup involves concentration at reduced pressure and digestion of the residue in acetonitrile at 45°C. The mixture is cooled to 25°C and (*S*)-chloride **102** is filtered (70%). The (*S*)-bromide **103** can be prepared from (*S*)-alcohol **32** in two steps. The reaction with methanesulfonyl chloride (2.0 equivalents) and triethylamine (3.1 equivalents) in dichloromethane at 25°C is quenched in water. The organic layer is concentrated at reduced pressure and the residue taken up in acetonitrile. Reaction with lithium bromide (2.1 equivalents) in acetonitrile at reflux, a routine workup and chromatography affords (*S*)-bromide **103** (37%). The (*S*)-iodide **104** can also be prepared from (*S*)-alcohol **32**. (*S*)-Iodide **104** precipitates from the reaction of **32** with iodotrimethylsilane generated from sodium iodide (2.4 equivalents) and chlorotrimethylsilane (3.1 equivalents) in acetonitrile at 45°C. The precipitate is filtered, dissolved in dichloromethane–methanol, washed with aqueous bicarbonate and with water, and concentrated at reduced pressure. Chromatography of the residue affords (*S*)-iodide **104** (35%). Optical purities of the starting (*S*)-alcohol and the product (*S*)-halides are not provided.[113]

Should we expect clean retention of configuration in the conversion of the alcohol to the chloride? The mechanism for the conversion of an alcohol to a chloride may be $S_N1$, $S_N2$, or $S_Ni$. In the absence of a base, the $S_Ni$ mechanism normally operates and retention of configuration is expected. It is known that addition of a base such as pyridine results in a shift to an $S_N2$ mechanism and inversion is observed. While no pyridine is added, the quinoline component of the (*S*)-alcohol might function as the base. We might also expect some inversion of configuration since this chlorination is catalyzed by DMF.

Some procedures using (*S*)-chloride **102** probably begin with drying by distillation of the toluene–water azeotrope. The source of the water is not specified. Only 2-(1-(mercaptomethyl)cyclopropyl)acetic acid (**2**) has been used for chloride displacements to date.

Reaction of (*S*)-chloride **102** with 2-(1-(mercaptomethyl)cyclopropyl)acetic acid (**2**) (1.2 equivalents) and cesium carbonate (3.0 equivalents) in DMF at 55°C affords the thioether **92**. After routine workup, thioether **92** is isolated as the *N,N*-dicyclohexylamine salt from ethyl acetate–hexanes at 25°C (62%). The salt can be upgraded by crystallization from 1:1 acetonitrile–toluene at 25°C (50% recovery). Salts are also produced using (*R*)-

α-methylbenzylamine and cyclohexylethylamine. The *N,N*-dicyclohexylamine salt is released with acetic acid in dichloromethane–water. The layers are separated and the organic layer is concentrated at reduced pressure. Thioether **92** in toluene is reacted with methylmagnesium chloride (5.8 equivalents) and cerium chloride (1.2 equivalents) in toluene–THF at 0°C to produce montelukast (**1**). After quench with aqueous acetic acid and routine workup, montelukast (**1**) is recovered as the (*R*)-α-methylbenzylamine salt from ethanol solution by addition of the amine and dilution with diisopropyl ether (76%). The salt can be upgraded by crystallization from 3:1 acetonitrile–toluene at 25°C (81% recovery). Montelukast (**1**) can also be isolated as the cyclohexylethylamine salt.[114]

Reaction of the (*S*)-chloride **102** with 2-(1-(mercaptomethyl)cyclopropyl)acetic acid (**2**) (1.5 equivalents) and sodium hydride (3.3 equivalents) in DMF at 0°C affords the thioether **92**. After routine workup, thioether **92** is isolated as the isopropylamine salt from ethyl acetate–hexanes at 25°C (74%). The isopropylamine salt is released with acetic acid in dichloromethane–water. The layers are separated and the organic layer is concentrated at reduced pressure. The residue in toluene is reacted with methylmagnesium chloride (6.9 equivalents) and cerium chloride (0.94 equivalents) in toluene–THF at 0°C to produce montelukast (**1**). After quench with aqueous acetic acid and routine workup, montelukast (**1**) is recovered as the cyclohexylamine salt from ethyl acetate solution by addition of the amine and dilution with heptane (67%). The yield of the thioether **92** *N,N*-dicyclohexylamine salt is lower (53%). This salt is converted to montelukast (**1**) *N,N*-dicyclohexylamine salt (65%), montelukast (**1**) dipropylamine salt (78%), and montelukast (**1**) (52%). Optical purities of the (*S*)-chloride **102**, amine salts, and montelukast (**1**) are not provided (Scheme 6.27).[114]

Perhaps (*S*)-chloride **102** is also prepared by reaction of the (*R*)-alcohol **31** with methanesulfonyl chloride (1.1 equivalents) and *N,N*-diisopropylethylamine (1.3 equivalents) in dichloromethane at 35°C. Reaction of crude (*S*)-chloride **102** with 2-(1-(mercaptomethyl)cyclopropyl)acetic acid (**2**) (1.5 equivalents) and sodium methoxide (3.7 equivalents) in DMF–dichloromethane at reflux affords thioether **92**. Thioether **92** is isolated as the *N,N*-dicyclohexylamine salt from acetone (75% from the (*R*)-alcohol **31**). The salt is cleaved with acetic acid in toluene–water. The layers are separated and the organic layer is concentrated at reduced

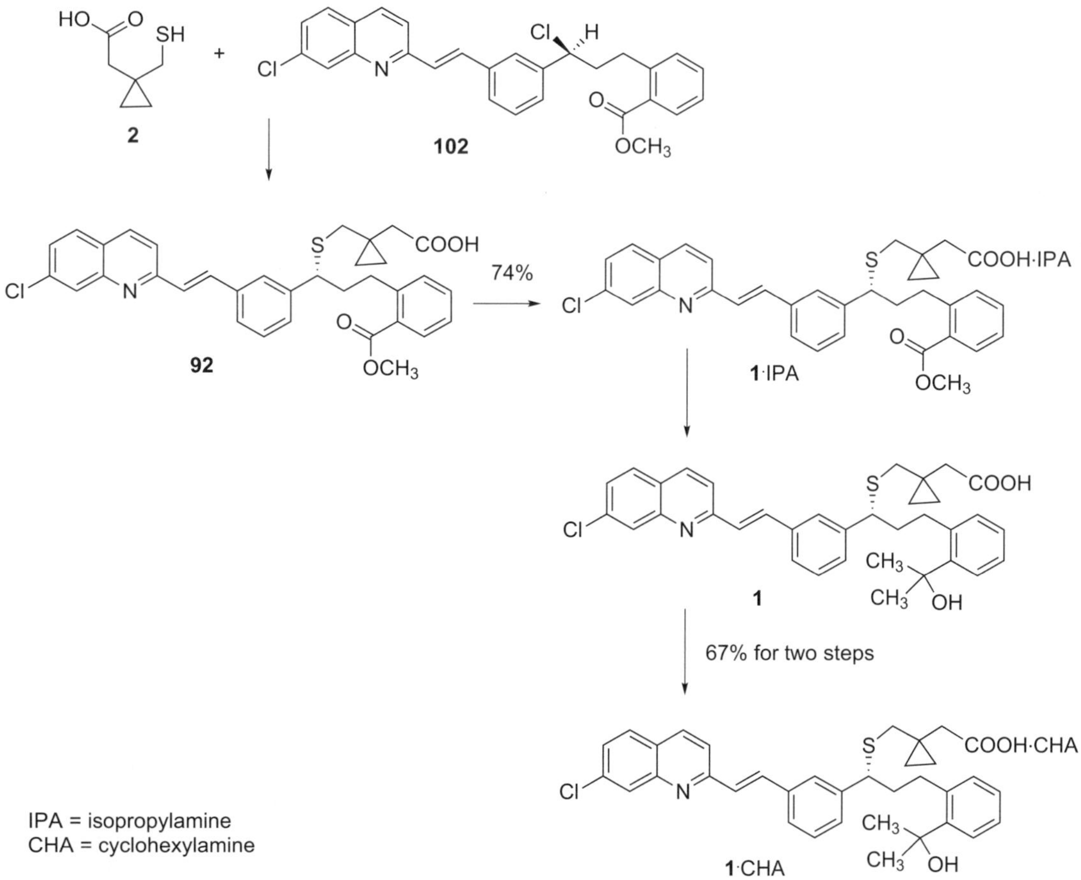


**SCHEME 6.27** Montelukast (**1**) cyclohexylamine salt from (*S*)-chloride **102** and 2-(1-(mercaptomethyl)cyclopropyl)acetic acid (**2**).

pressure. The residue in toluene is reacted with methylmagnesium chloride (6.0 equivalents) in toluene–THF at 5°C to afford montelukast (**1**) (70%).[115]

The reaction of (*S*)-chloride **102** with methylmagnesium chloride (7.1 equivalents) and cerium chloride (1.0 equivalent) in THF at 0°C affords (*S*)-chloride **105** (65%). Reaction of (*S*)-chloride **105** with 2-(1-(mercaptomethyl)cyclopropyl) acetic acid (**2**) (1.1–1.2 equivalents) and cesium carbonate (3.0 equivalents) in DMF at 35°C affords montelukast (**1**). After a routine workup, montelukast (**1**) is recovered as the isopropylamine salt from ethyl acetate solution by addition of the amine and dilution with heptane (71%) (Scheme 6.28). A similar procedure can be used to produce montelukast (**1**) (*R*)-α–methylbenzylamine salt (54%).[114]

Reaction of (*S*)-chloride **105** with 2-(1-(mercaptomethyl)cyclopropyl)acetic acid (**2**) (1.5 equivalents) and sodium hydride (3.3 equivalents) in DMF at 0°C also affords montelukast (**1**). After a routine workup, montelukast (**1**) is recovered as the *N*,*N*-dicyclohexylamine salt from ethyl acetate solution by addition of the amine and dilution with hexane (41%). The same procedure can be used to produce montelukast (**1**) (*R*)-α–methylbenzylamine salt (47%) and montelukast (**1**) (70%). Montelukast (*R*)-α-methylbenzylamine and isopropylamine salts are efficiently converted to montelukast sodium (83–90%). Optical purities of the (*S*)-chloride **105**, amine salts, montelukast, and montelukast sodium are not provided.[114]

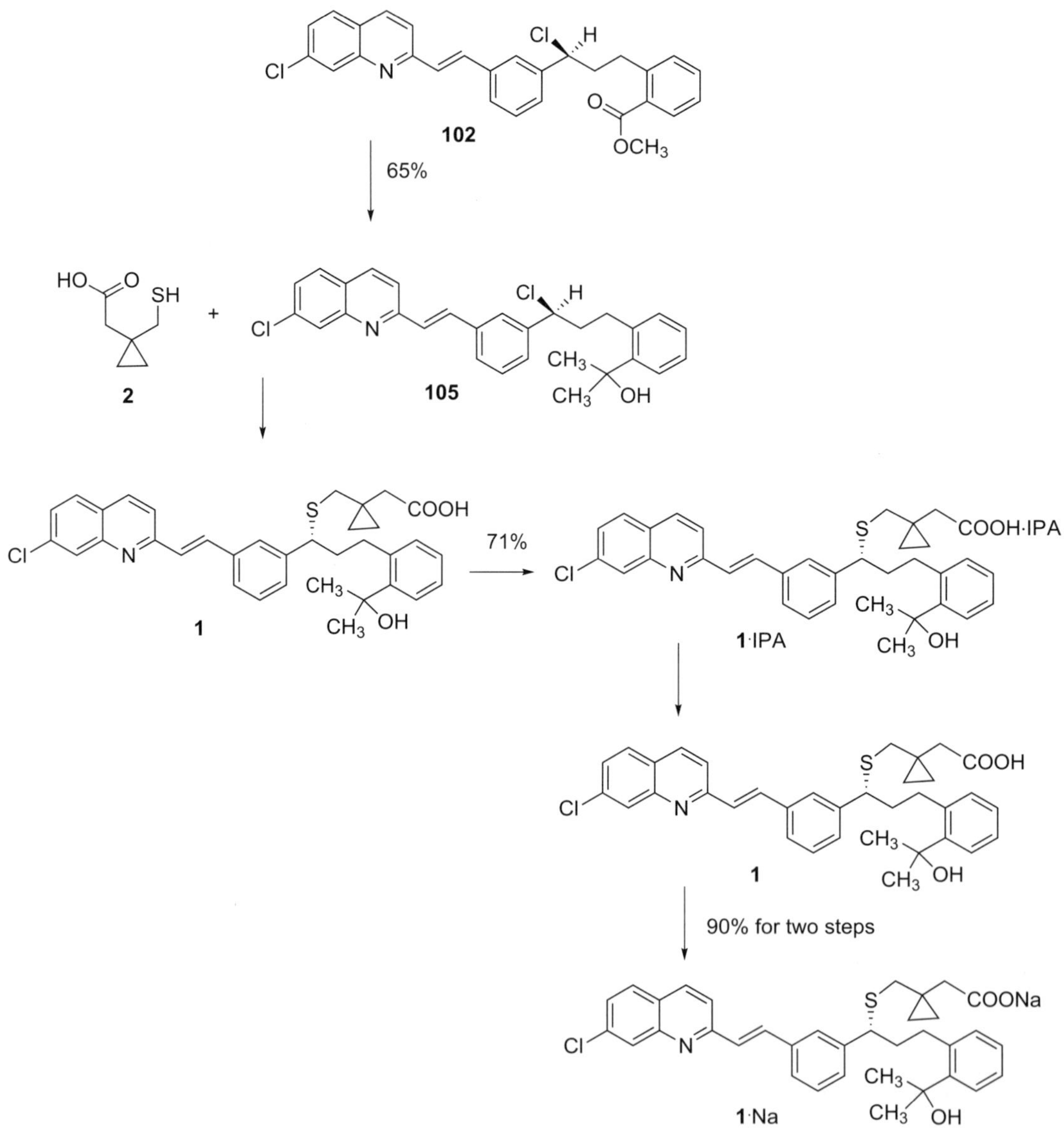


**SCHEME 6.28** Montelukast (**1**) sodium from (S)-chloride **105** and 2-(1-(mercaptomethyl)cyclopropyl)acetic acid (**2**) via the isopropylamine salt.

## 6.6 REACTION OF A BACKBONE THIOLATE AND SIDE CHAIN METHANESULFONATE OR BROMIDE

What might we hope to gain by adding a thioacetate displacement to the already long backbone construction sequence? Recall that the methanesulfonates are unstable in solution. In the one case where a methanesulfonate can be isolated, it must then be stored cold. A thioacetate ester would be much more stable in solution and the isolated thioacetate ester would not require cold storage. The thioester would be a good hold point, a campaignable intermediate.

Methanesulfonate **49** prepared in toluene is reacted with potassium thioacetate (1.2 equivalents) in 3:1 toluene–DMF at 25°C. Separation between ethyl acetate and water, wash with water, and concentration at reduced pressure affords a brown oil. The thioacetate ester **106** is isolated from the oil by chromatography (68%). The yield is lower when methanesulfonate **49** is prepared in dichloromethane (57%). Adding lithium hydroxide (1.1 equivalents) to a mixture of thioacetate ester **106** and methyl 2-(1-(bromomethyl)cyclopropyl)acetate (**107**) in acetonitrile–methanol–water at 25°C results in cleavage of the thioacetate ester and formation of the thioether **108**. The side chain methyl ester is hydrolyzed with sodium hydroxide in THF–methanol–water at 25°C. Methyllithium–lithium bromide complex is added to the backbone methyl benzoate to give montelukast (**1**) in THF–ethyl ether at −30°C. No yields are available. The hydrolysis of the side chain methyl ester is not likely to be selective and methyllithium and the ethyl ether cosolvent are undesirable (Scheme 6.29).[82]

The poorly selective ester hydrolysis can be avoided by changing the order of steps. The thioacetate ester **106** is reacted with methyllithium (4.0 equivalents) in THF–ethyl ether at −78°C. After quench by adding solid ammonium chloride, routine workup affords an oil. Thioacetate ester **109** is isolated from the oil by chromatography in just 12% yield. The low yield may be due to cleavage of the thioacetate ester, which is observed in the reaction with a Grignard reagent.[79,116]

The poorly selective ester hydrolysis and the use of methyllithium and ethyl ether can all be avoided by starting with methanesulfonate **48**. Methanesulfonate **48** is reacted with potassium thioacetate (1.2 equivalents) in 3:1 toluene–DMF at 25°C. The THP ether is then removed with

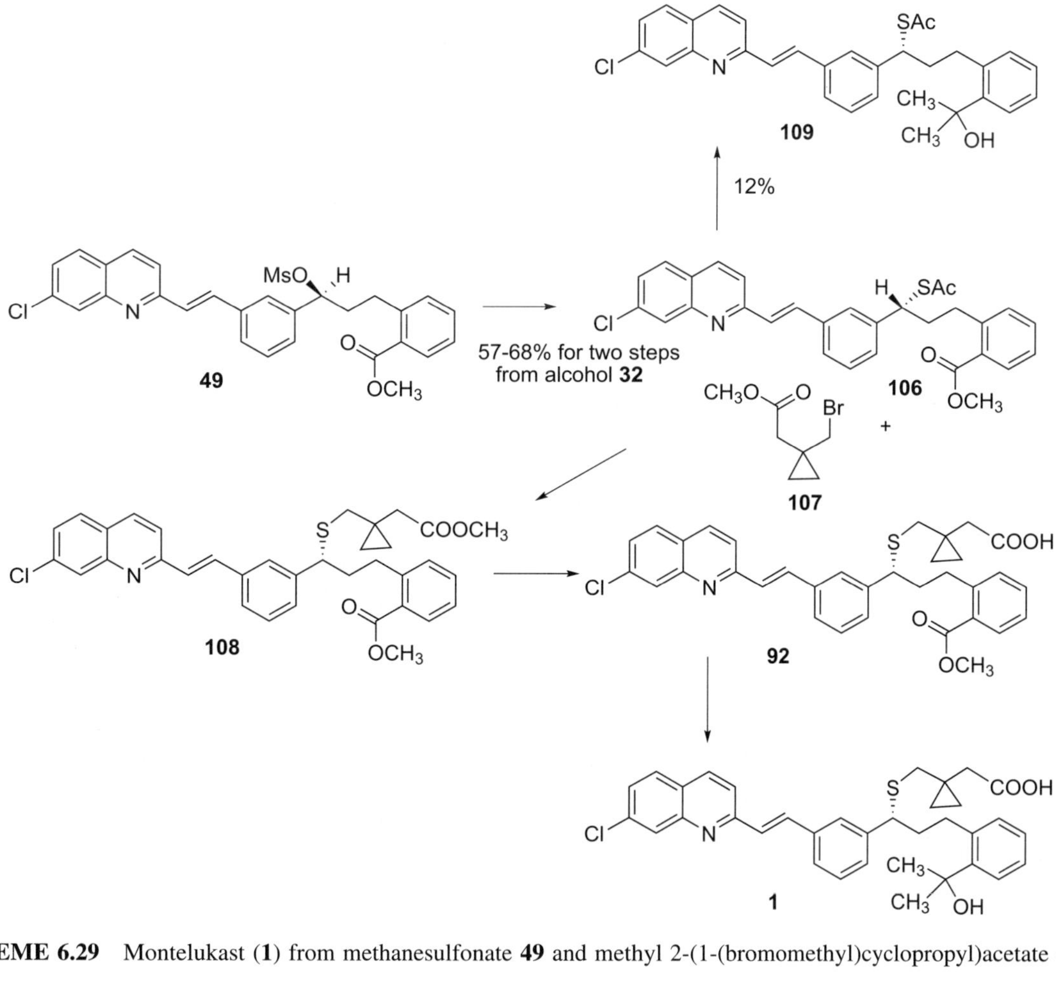


**SCHEME 6.29** Montelukast (**1**) from methanesulfonate **49** and methyl 2-(1-(bromomethyl)cyclopropyl)acetate (**107**).

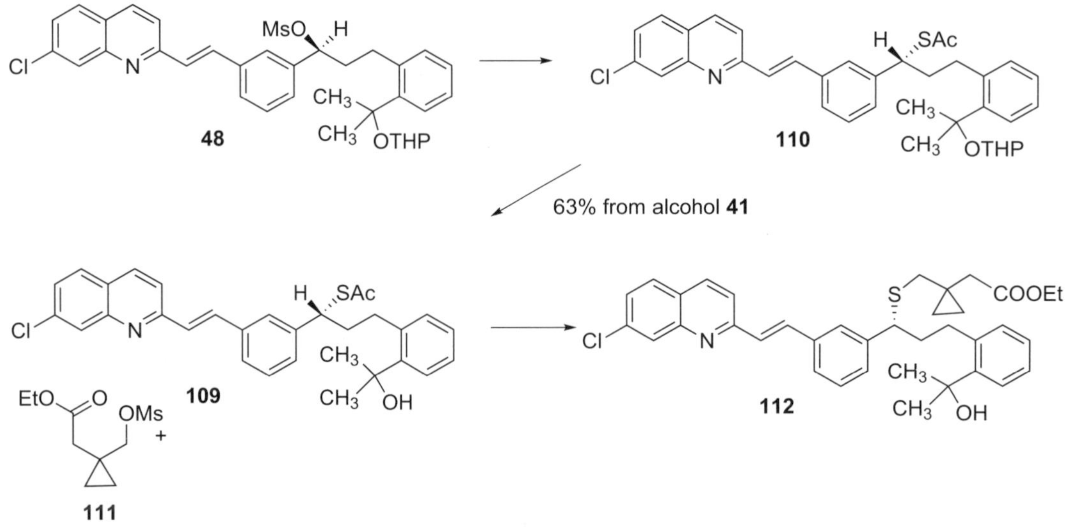


**SCHEME 6.30** Montelukast ethyl ester (**112**) from methanesulfonate **48** and ethyl 2-(1-(methanesulfonyloxy)methyl)cyclopropyl)acetate (**111**).

*p*-toluenesulfonic acid monohydrate (0.60 equivalents) in THF–methanol at 25°C. After separation between ethyl acetate and water, the organic layer is washed with brine and concentrated at reduced pressure. Thioester **109** is isolated from the residue by chromatography (63% from (*S*)-alcohol **41**).

> Any mixture of *p*-toluenesulfonic acid (benzenesulfonic acid, methanesulfonic acid) and methanol may generate some methyl *p*-toluenesulfonate (methyl benzenesulfonate, methyl methanesulfonate). The workup procedure should be designed to ensure that the sulfonate ester is destroyed.

The thiolate is released *in situ* by reaction with hydrazine hydrate (1.2 equivalents) in acetontirile at 0°C. Reaction with ethyl 2-(1-(methanesulfonyloxy)methyl)cyclopropyl) acetate (**111**) and cesium carbonate (2.0 equivalents) in acetonitrile at 0–25°C affords montelukast ethyl ester (**112**). No yield is provided (Scheme 6.30).[79]

## 6.7 BACKBONE SYNTHESIS RIGHT TO LEFT

While the majority of routes build the montelukast backbone from left to right, there are a growing number of routes in which backbone construction progresses right to left. These routes utilize the methodology already established in the left to right routes to create the three-carbon link and introduce chirality then create the ethylene bridge by 7-chloroquinaldine condensation with an aldehyde or by Heck arylation of 7-chloro-2-vinylquinoline (**113**).

### 6.7.1 Finishing with Condensation of 7-Chloroquinaldine (3) with an Aldehyde

The aldehyde can be carried through the sequence as a THP-protected benzylic alcohol (Scheme 6.31). 3-Bromobenzaldehyde[117] is reduced with sodium borohydride in ethanol at 25°C. The alcohol is then protected by reaction with dihydropyran and triphenylphosphine hydrobromide in dichloromethane at 25°C. The *ortho*-substituted aromatic is produced from α-tetralone.[118] Reaction with isopropenyl acetate converts the ketone to the vinyl acetate that is then ozonolyzed to give 2-(3-oxopropyl)benzoic acid (**117**) (75% from α-tetralone). While we would not consider doing an ozololysis in a manufacturing process, it is important to continue the discussion on the assumption that a suitable process for making **117** could be developed if we are motivated by promising results downstream. 2-((3-Bromophenyl)methoxy)tetrahydropyran (**115**) (2.4 equivalents) is converted to a Grignard reagent that is then added to 2-(3-oxopropyl)benzoic acid (**117**) in THF at 25°C to create the first link in the backbone. Two equivalents of the Grignard reagent are required since the first equivalent is consumed in deprotonation of the carboxylic acid. The product **118** is isolated by chromatography. No attempt is made to set the correct stereochemistry at the secondary alcohol center. The carboxylic acid is converted to the methyl ketone with methyllithium (4.2 equivalents) in THF–ethyl ether at 0°C. Methanesulfonate **120** is produced from alcohol **119** with methanesulfonyl chloride (1.3 equivalents) and triethylamine (1.5 equivalents) in dichloromethane at −40 to −10°C. While this prototype process uses a different thiol for the methanesulfonate displacement, it is instructive to finish construction of the backbone. The methyl ketone in **122** is converted to the tertiary alcohol with methylmagnesium

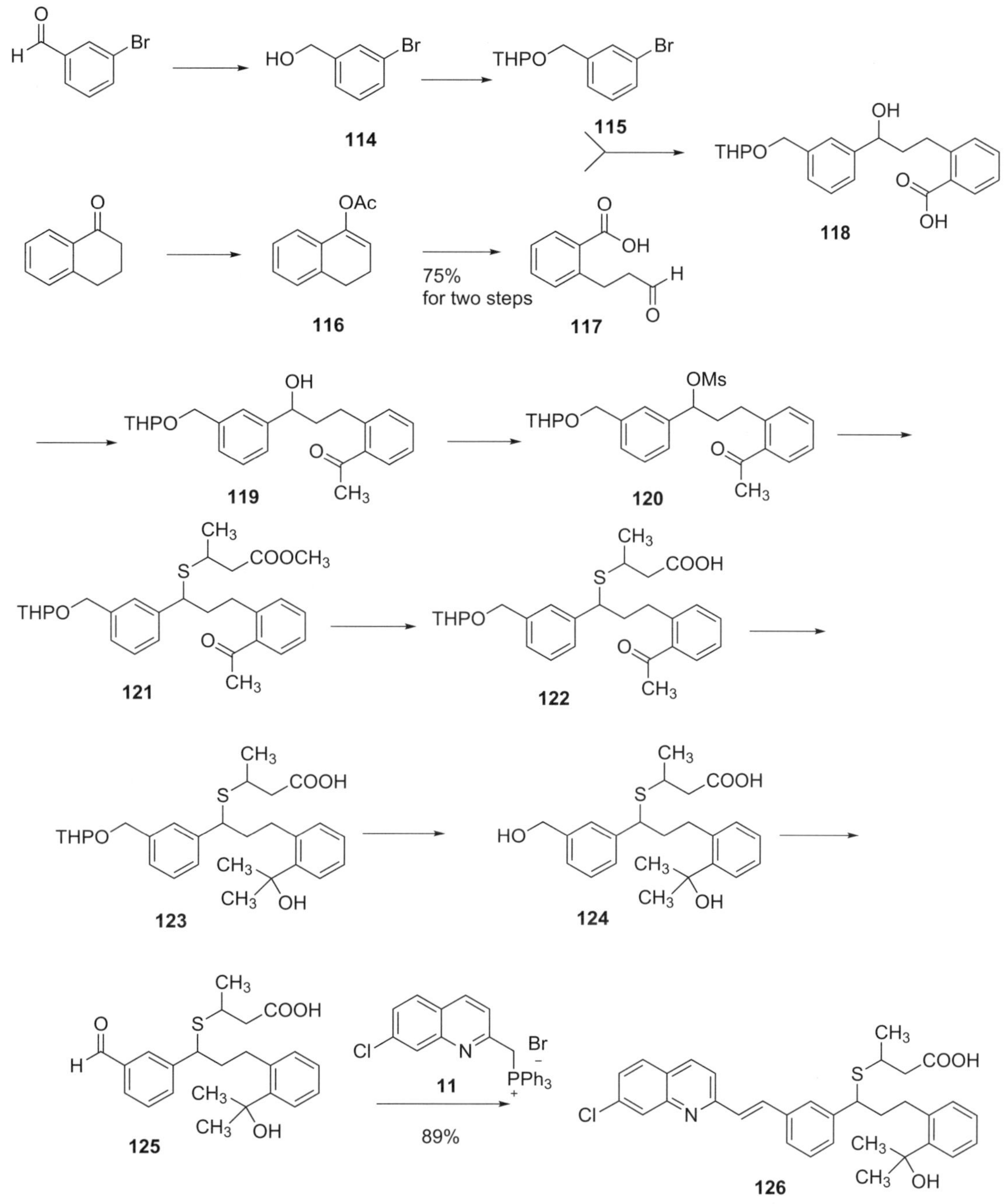


**SCHEME 6.31** A prototype process for montelukast by right-to-left backbone construction.

bromide (2.5 equivalents) in toluene at 0°C. Now the THP ether is released by reaction with pyridinium *p*-toluenesulfonate (0.4 equivalents) in methanol at 25°C and the alcohol oxidized with manganese dioxide (18 equivalents) in ethyl acetate at 25°C. All products up to this point are isolated by chromatography and no yields are provided. In the final step, the backbone is completed by Wittig condensation of the aldehyde **125** with the ylid produced from ((7-chloroquinolin-2-yl)methyl)triphenylphosphonium bromide (**11**) (3.0 equivalents) and butyllithium (2.8 equivalents) in THF at −78°C (89%).[32]

What can be learned from this prototype process and what improvements can be made? First, why start with an aldehyde, reduce it to an alcohol, and then oxidize the alcohol back to the aldehyde? Use an aldehyde protecting group that can be carried through the sequence and released when the aldehyde is needed. Second, the secondary alcohol stereochemistry must be set. Third, why convert the carboxylic

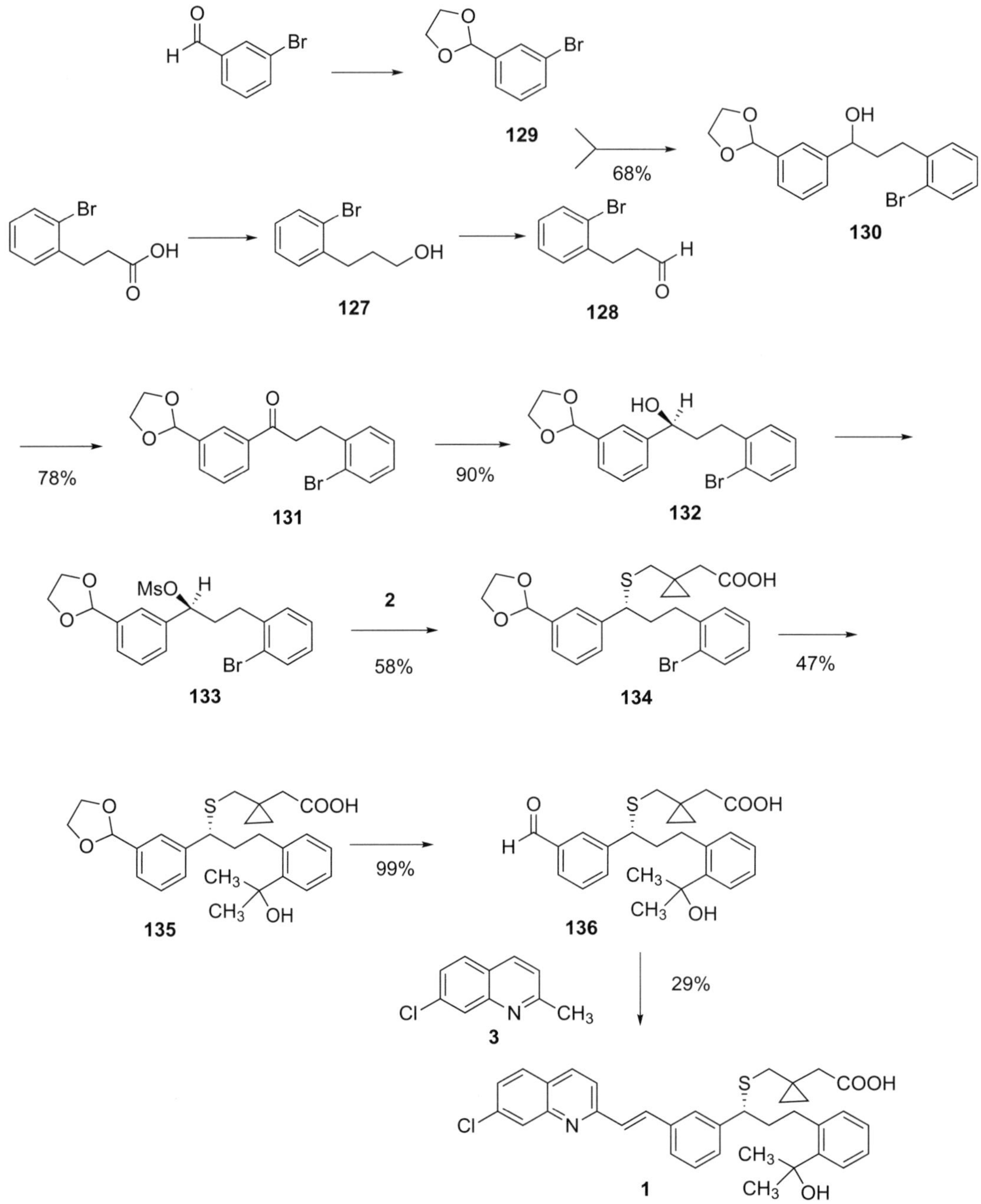


**SCHEME 6.32** Montelukast (**1**) by right-to-left backbone construction (alcohol oxidation) with a late quinaldine–aldehyde condensation.

acid to the tertiary alcohol in two steps (and why use methyllithium and ethyl ether?). Convert the methyl carboxylate to the tertiary alcohol in a single step using methylmagnesium chloride. Fourth, the conditions for deprotection of the THP ether are likely to cause some elimination of the tertiary alcohol. Choose an aldehyde protecting group that can be selectively removed in the presence of the tertiary alcohol. Finally, from the perspective of the valuable aldehyde, the Wittig condensation is very efficient.

Perhaps the latent aldehyde can be carried through the sequence as an acetal (Scheme 6.32). 3-(2-Bromophenyl) propionic acid[119] is reduced to the alcohol **127** with borane dimethylsulfide and the alcohol is oxidized to give 3-(2-bromophenyl)propanal (**128**).[120,121] While this sequence would not be suitable for manufacturing the aldehyde **128**, let us continue the discussion on the assumption that a suitable process could be developed if we are motivated by promising results downstream. 3-Bromobenzaldehyde **128** is converted to an acetal with ethylene glycol. The Grignard reagent derived from 2-(3-bromophenyl)-1,3-dioxolane (**129**) (1.0 equivalent) is condensed with 3-(2-bromophenyl)propanal (**128**) in THF at reflux (68%). The alcohol is oxidized to the ketone with pyridinium chlorochromate (78%) then reduced with borane dimethylsulfide (1.5 equivalents) using the (*R*)-oxazaborolidine catalyst (20 mol%) in THF at 25°C to set the correct stereochemistry (90%). Methanesulfonate **133** is produced by reaction of alcohol **132** with methanesulfonyl chloride (1.3 equivalents) and triethylamine (1.8 equivalents) in dichloromethane at −20°C. After quench with saturated sodium hydrogen carbonate, the organic layer is dried and concentrated at reduced pressure to replace dichloromethane by THF. Methanesulfonate displacement by the dilithium salt from 2-(1-mercaptomethyl)cyclopropyl)acetic acid (**2**) (1.05 equivalents) produced using butyllithium afforded the crude thioether **134** (58%). The bromide is converted to the tertiary alcohol by lithium-halogen exchange with butyllithium (2.1 equivalents) in THF at −94°C and quench with acetone (47%). The aldehyde is deprotected with *p*-toluenesulfonic acid (7 mol%) in 1:1 THF–water at 50°C (99%). The crude aldehyde **136** is then condensed with 7-chloroquinaldine (**3**) (1.0 equivalent) using a piperidine catalyst (0.5 equivalents) in toluene at reflux. Montelukast (**1**) is isolated by chromatography (29% from acetal **135**).[122]

What can be learned from this process and what improvements can still be made? The acetal protecting group is maintained throughout this sequence. The stereochemistry can be set by reducing the ketone but the alcohol oxidation to make the ketone is not suitable for scale-up. The tertiary alcohol can be introduced in a single step by converting a bromide to an aryllithium and adding acetone but the lithium-halogen exchange requires very low (−94°C) temperature and the yield is low. The conditions for acetal deprotection may cause some elimination of the tertiary alcohol. This question is left unanswered since no yield is provided for the release of the aldehyde and the yield of montelukast (**1**) from the aldehyde condensation with 7-chloroquinaldine (**3**) is low.

An alternative route to aldehyde **136** also starts with the aldehyde protected as an acetal (Scheme 6.33). Reaction of the Grignard reagent derived from 2-(3-bromophenyl)-1,3-dioxolane (**129**) with acetic anhydride (1.9 equivalents) in THF at −10°C affords 1-(3-(1,3-dioxolan-2-yl)phenyl)ethanone (**137**) (86%). The ketone is then elaborated using methodology already available from the left-to-right sequences. The β-ketoester **138** is produced by reaction with dimethyl carbonate (1.1 equivalents) and sodium hydride (1.3 equivalents) in DMF at 0–25°C (92%). The β-ketoester is alkylated with methyl (2-bromomethyl)benzoate (**27**) (1.0 equivalent) using sodium hydride (1.1 equivalents) as base in DMF at 25–40°C. No yield is provided but similar alkylations in the left-to-right routes were efficient (88%). Hydrolysis of the β-ketoester, decarboxylation, and hydrolysis of the acetal with hydrochloric acid in water–dioxane at reflux affords methyl 2-(3-(3-formylphenyl)-3-oxopropyl)benzoate (**140**) (80%). Since selective hydrolysis of the β-ketoester in the presence of the benzoate ester was not possible in the analogous left-to-right routes, it is difficult to rationalize why the benzoate ester would be retained in this process. The aldehyde is again protected, this time as the acetal with neopentyl glycol in toluene at reflux. Acetal **141** is isolated by chromatography (54%). The ketone is reduced with borane dimethylsulfide (1.6 equivalents) using the (*R*)-oxazaborolidine catalyst (7.2 mol%) to give the (*S*)-alcohol **142** (100% yield, 93.7% ee). Methanesulfonate **143** is produced by reaction of alcohol **142** with methanesulfonyl chloride (1.2 equivalents) and triethylamine (1.6 equivalents) in dichloromethane at 0–10°C. After a quench with water and a wash with saturated sodium hydrogen carbonate, the organic layer is dried and concentrated at reduced pressure to replace dichloromethane with THF. Reaction with the dilithium salt from 2-(1-mercaptomethyl)cyclopropyl) acetic acid (**2**) (1.05 equivalents) produced using lithium hexamethyldisilazide (2.1 equivalents) in THF at 5°C afforded crude thioether **144** (100%). The benzoate ester is converted to the tertiary alcohol by reaction with cerium chloride (0.5 equivalents) and methylmagnesium chloride (5.1 equivalents) in THF at 10–25°C (88%). The aldehyde is released by reaction with maleic acid (1.0 equivalent) in 1:1 acetone–water at 50°C for 11 h. After a routine workup, the release procedure and workup are repeated and aldehyde **136** is then isolated by chromatography (50%). The need to repeat the release procedure suggests the conditions are designed to minimize dehydration of the tertiary alcohol. The condensation of aldehyde **136** with 7-chloroquinaldine (**3**) (1.0 equivalent) catalyzed by piperidine (0.5 equivalents) in isobutyl alcohol at reflux did not go to completion. Citing the right-to-left prototype process already presented, we can assume a yield of 89% in a Wittig reaction of this aldehyde with the ylid from ((7-chloroquinolin-2-yl)methyl)triphenylphosphonium bromide (**11**).[122]

This process also falls short for several reasons. First, the acetal is lost during the ester hydrolysis and decarboxylation. The aldehyde must be reprotected and the yield in that step is low (54%). The final acetal deprotection is slow and inefficient (50%). The piperidine-catalyzed condensation is apparently also slow and inefficient. If we are able to develop an efficient route to the fully elaborated aldehyde **136**, it appears the only option is to use a Wittig condensation to finish the backbone construction. The key to success of the three related construction strategies that finish with the 7-chloroquinaldine condensation with an aldehyde is to find an aldehyde protecting group that will survive elaboration of the backbone and be selectively removed in the presence of a tertiary alcohol. This protecting group has yet to be identified.

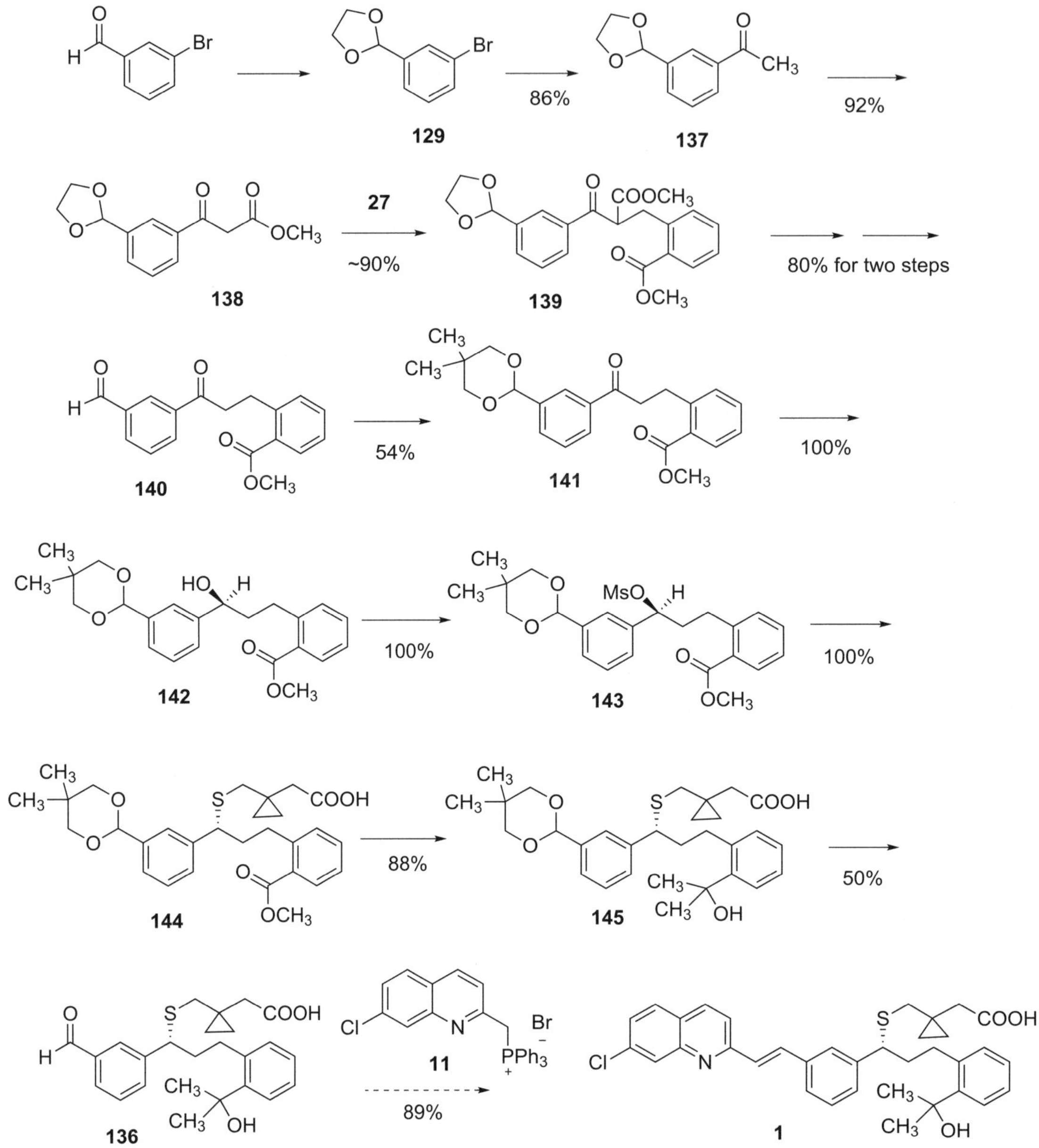

**SCHEME 6.33** Montelukast (**1**) by right-to-left backbone construction (β-keto ester) with a proposed late Wittig condensation.

### 6.7.2 Finishing with Heck Arylation of 7-Chloro-2-vinylquinoline (113)

Many methods for construction of montelukast (**1**) have a common strategy. They all create the ethylene bridge by condensation of 7-chloroquinaldine (or the related ylid) with a benzaldehyde. A Heck arylation of 7-chloro-2-vinylquinoline (**113**) is the only alternative strategy currently available for constructing the ethylene bridge. 7-Chloro-2-vinylquinoline (**113**) can be prepared by Hofmann elimination of a quaternary salt. Mannich reaction of 7-chloroquinaldine (**3**) with 37% formalin (0.98 equivalents), triethylamine (7.2 mol %), and diethylamine hydrochloride (1.0 equivalent) in ethanol–water at 60°C affords the crude tertiary amine **146**. Reaction of crude amine **146** with iodomethane (2.4 equivalents) in ethanol at 25°C followed by elimination with triethylamine (0.93 equivalents) in ethanol–water at reflux affords 7-chloro-2-vinylquinoline (**113**) (45% from **3**).[31] There are very few suppliers listed for this key raw material. For Heck arylation of **113** to be competitive, the price and availability of 7-chloro-2-vinylquinoline (**113**) should be comparable to the price and availability for 7-chloroquinaldine (**3**) (Scheme 6.34).

Right-to-left synthesis utilizing a Heck arylation begins with the construction of a 2-(3-(3-bromophenyl)-3-

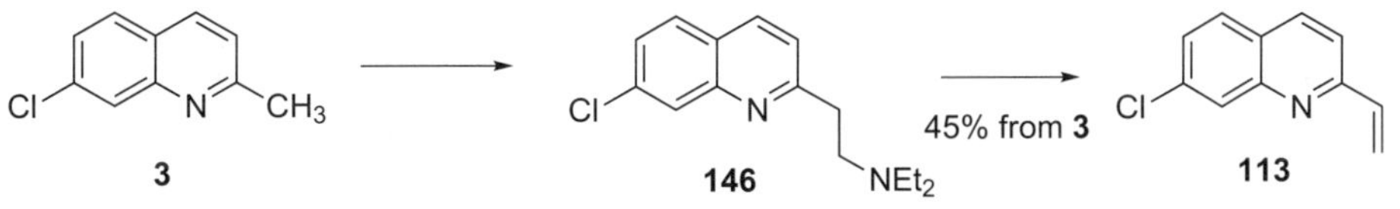


**SCHEME 6.34** 7-Chloro-2-vinylquinoline (**113**) from 7-chloroquinaldine (**3**).

oxopropyl)benzoate (Scheme 6.35). 3′-Bromoacetophenone[123] and 2-carboxybenzaldehyde[124] (1.1 equivalents) are condensed by Claisen–Schmidt condensation using sodium hydroxide (1.1 equivalents) in ethanol–water at 5°C. Addition of sulfuric acid and age at 60°C generates the lactone **148**, which precipitates from solution (82%). Elimination with sodium hydroxide (0.95 equivalents) and double bond reduction with Wilkinson's catalyst (1.2 mol%) in ethanol–water at 25°C and 40 psi hydrogen generates the benzoic acid **149** (96%). Esterification with methanol and sulfuric acid affords methyl 2-(3-(3-bromophenyl)-3-oxopropyl)benzoate (**150**) (92%).[31]

In an alternative process for methyl 2-(3-(3-bromophenyl)-3-oxopropyl)benzoate (**150**), the sodium salt of (*E*)-2-(3-(3-bromophenyl)-3-oxoprop-1-enyl)benzoic acid **151** precipitates when the mixture produced by condensation of 3′-bromoacetophenone and 2-carboxybenzaldehyde (1.1 equivalents) in ethanol–water at 0–25°C is diluted with brine (93%). The enone is reduced with zinc (1.5 equivalents) in ethanol–water at 25°C. After filtration, concentration to remove ethanol, acidification, and extraction with dichloromethane, the extracts are concentrated at reduced pressure to afford 2-(3-(3-bromophenyl)-3-oxopropyl)benzoic acid (**152**) (70%). This is converted to methyl ester **150** with methanol and sulfuric acid (89%).[71]

Three methods are available for the asymmetric reduction of methyl 2-(3-(3-bromophenyl)-3-oxopropyl)benzoate (**150**). Two methods are familiar from routes already presented. Reduction with (−)-diisopinocampheylchloroborane (1.5 equivalents) in THF at −20 to 30°C is quenched with acetone. After routine workup, the (*S*)-alcohol **153** is isolated by chromatography (90%, 96% ee). Asymmetric transfer hydrogenation of the ketone with bis($\eta^6$-mesitylene)ruthenium(II) chloride, (1*S*,2*S*)-*N*-piperidylsulfamoyl-1,2-diphenylethylenediamine, formic acid and triethylamine is also known (86%). In the new method, a catalyst solution is generated from 3-nitrophenylboronic acid (2.0 equivalents), (D)-tartaric acid (1.9 equivalents), and calcium hydride

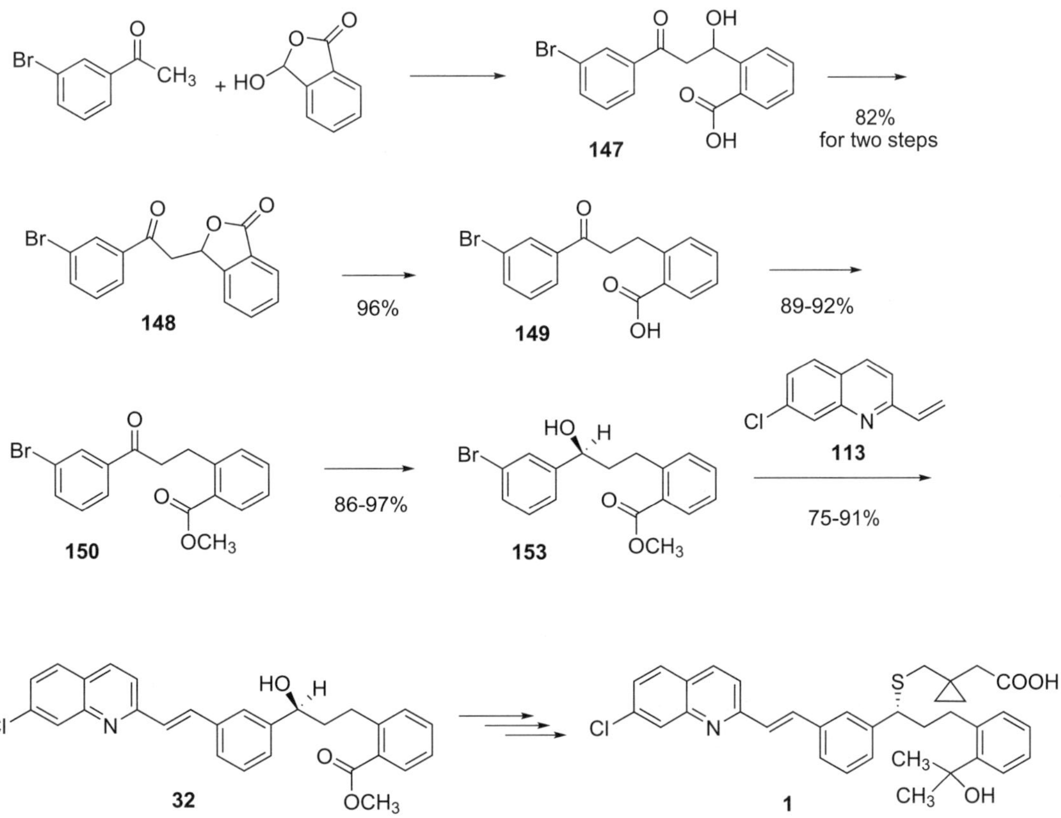


**SCHEME 6.35** Montelukast (**1**) by right-to-left backbone construction with a Heck arylation before introducing the side chain.

(4.1 equivalents) in THF at reflux. Reduction of the ketone using this catalyst solution and sodium borohydride (2.0 equivalents) in THF at 25°C is very efficient (97%).[71,125,126]

(*S*)-Alcohol **153** is also available by an aerobic oxidative kinetic resolution of the racemic alcohol **154**. Racemic alcohol **154** is produced by sodium borohydride reduction of ketone **150** in ethanol–dichloromethane (91%). Reaction of racemic alcohol **154** with palladium (nbd) dichloride (5 mol%), (−)-sparteine (20 mol%), cesium carbonate (0.5 equivalents), *tert*-butanol (1.5 equivalents), 3A molecular sieves, and oxygen (1 atm) in toluene at 60°C affords ketone **150** (63% conversion) and (*S*)-alcohol **153** (92.9% ee). The contact of toluene with oxygen at elevated temperature would not be attractive for scale-up.[127]

Moving ahead with (*S*)-alcohol **153**, the Heck arylation can be inserted before introduction of the side chain, after introduction of the side chain, or as the last step. The Heck arylation is perhaps best run before introducing the side chain. Reaction of (*S*)-alcohol **153** with 7-chloro-2-vinylquinoline (**113**) (1.2 equivalent), triethylamine (1.6 equivalents), palladium acetate (5.3 mol%), and tri(*o*-tolyl)phosphine (21 mol%) in DMF at 100°C affords (*S*)-alcohol **32** monohydrate (75%, >95% pure by HPLC, ee > 99%). The reaction of (*S*)-alcohol **153** with 7-chloro-2-vinylquinoline (**113**) (1.1 equivalents), triethylamine (1.5 equivalents), palladium acetate (3.0 mol%), and tri(*o*-tolyl)phosphine (9.4 mol%) in DMF at 100°C affords (*S*)-alcohol **32** monohydrate in still higher yield (91%).[31,71] Heck arylation of the (*S*)-alcohol **155** already possessing the tertiary alcohol is also possible (90%).[127]

The (*S*)-alcohol **153** is converted to the (*S*)-chloride **156** using thionyl chloride (1.7 equivalents) in dichloromethane at −10 to 0°C (92%). Since (*S*)-alcohol **153** has no quinoline component and no DMF catalyst is used, it is reasonable to expect retention of configuration by an $S_Ni$ mechanism. The (*S*)-chloride is displaced with 2-(1-(mercaptomethyl)cyclopropyl)acetic acid (**2**) (1.3 equivalents) using sodium *tert*-butoxide (2.7 equivalents) as the base in DMF at 25°C. The crude product **157** is precipitated as the *N*,*N*-dicyclohexylamine salt from toluene–hexane (54%). The final backbone link is created by Heck arylation with 7-chloro-2-vinylquinoline (**113**) (1.1 equivalents), triethylamine (2.7 equivalents), palladium acetate (5.9 mol%), and tri(*o*-tolyl)phosphine (18 mol%) in DMF at 100°C. After a routine workup, the (*E*)-alkene **92** is isolated by chromatography (62%). Reaction with methylmagnesium iodide (6.0 equivalents) in toluene–ethyl ether at 25°C affords montelukast (**1**) (72% chromatographed). The preferred conditions for this conversion (MeMgCl and $CeCl_3$ in THF) would give a higher yield (80%) (Scheme 6.36).[83,125]

The methyl benzoate conversion to the tertiary alcohol can be moved ahead in the sequence, either after making the (*S*)-chloride or after introducing the side chain (Scheme 6.37). Reaction of (*S*)-chloride **156** with methylmagnesium iodide (3.9 equivalents) in toluene–ethyl ether at 25°C affords the tertiary alcohol **158** (74%). The chloride is displaced with 2-(1-(mercaptomethyl)cyclopropyl)acetic acid (**2**) (1.4 equivalents) using sodium *tert*-butoxide (2.9 equivalents) as the base in DMF at 25°C. The crude thioether **159** is chromatographed (68%). Heck arylation

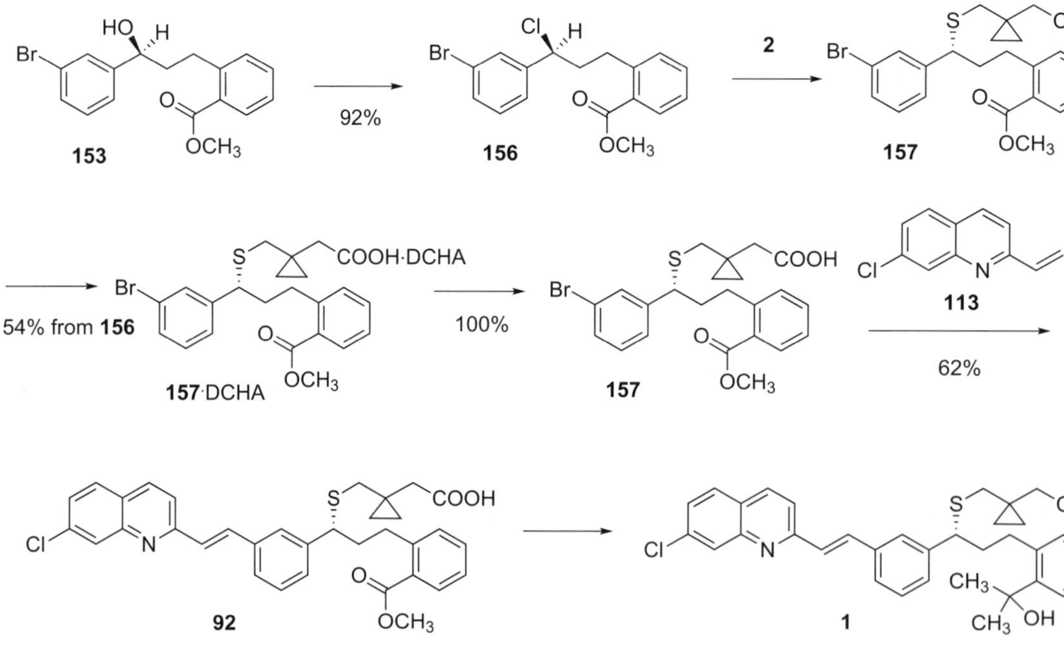


**SCHEME 6.36** Montelukast (**1**) by right-to-left backbone construction with Heck arylation after introducing the side chain.

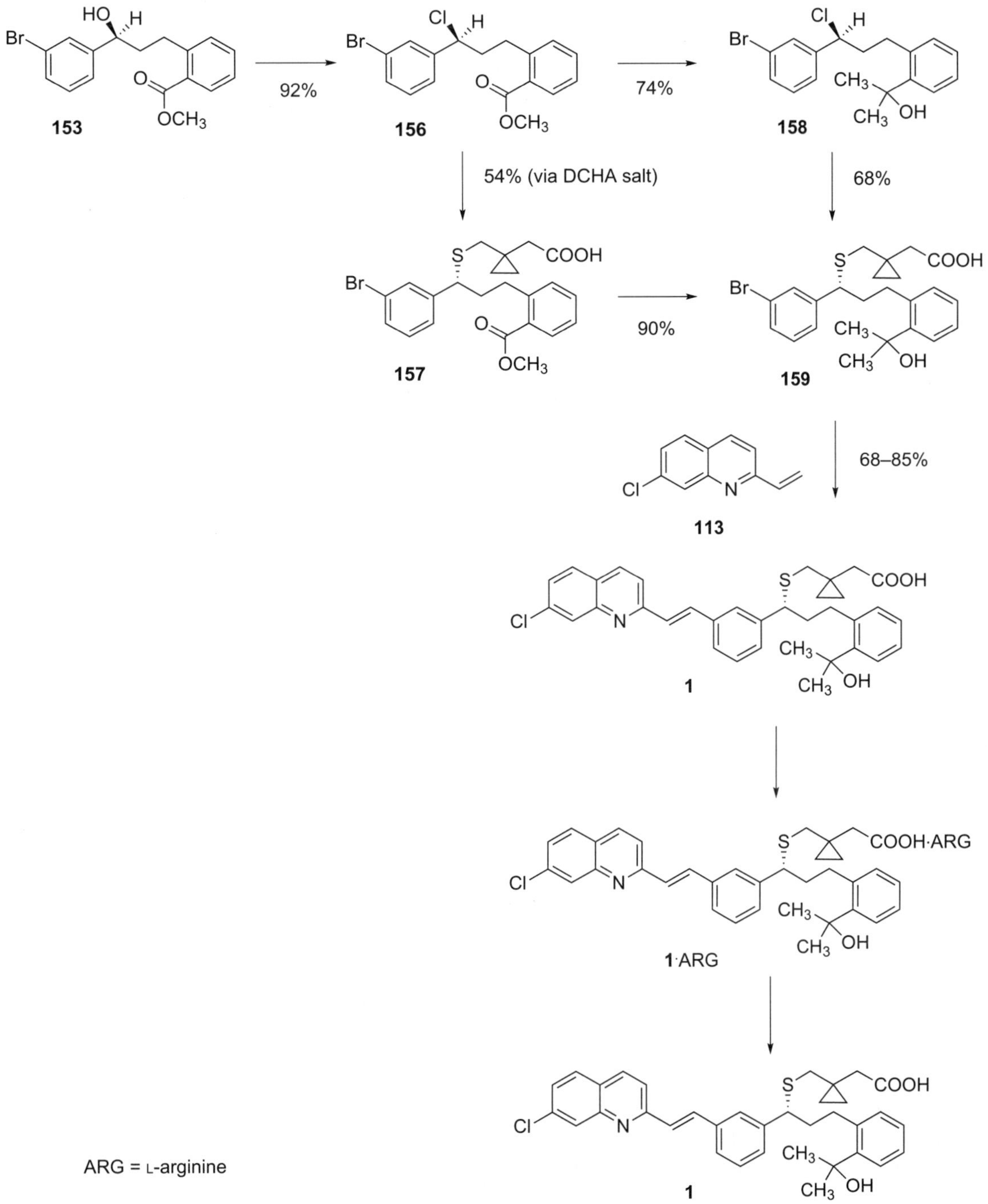


ARG = L-arginine

**SCHEME 6.37** Montelukast (**1**) by right-to-left backbone construction with Heck arylation as the last step.

with 7-chloro-2-vinylquinoline (**113**) (1.0 equivalent), triethylamine (2.5 equivalents), palladium acetate (6.7 mol %), and tri(*o*-tolyl)phosphine (30 mol%) in DMF at 100°C affords montelukast (**1**) (68% chromatographed). The methyl benzoate conversion to the tertiary alcohol after introducing the side chain is accomplished with methyllithium (3.8 equivalents) in toluene–ethyl ether at 0°C. The product **159** is chromatographed (90%). Montelukast (**1**) produced by this sequence can be converted to an amorphous L-arginine salt and the L-arginine salt can be converted to montelukast and montelukast sodium (Scheme 6.37).[125]

Optical purity is monitored by chiral HPLC in an alternative sequence, which introduces the side chain via the backbone thiolate (Scheme 6.38). Starting with (*S*)-alcohol **153**, reaction with methanesulfonyl chloride (1.5 equivalents) and diisopropylethylamine (1.7 equivalents) in

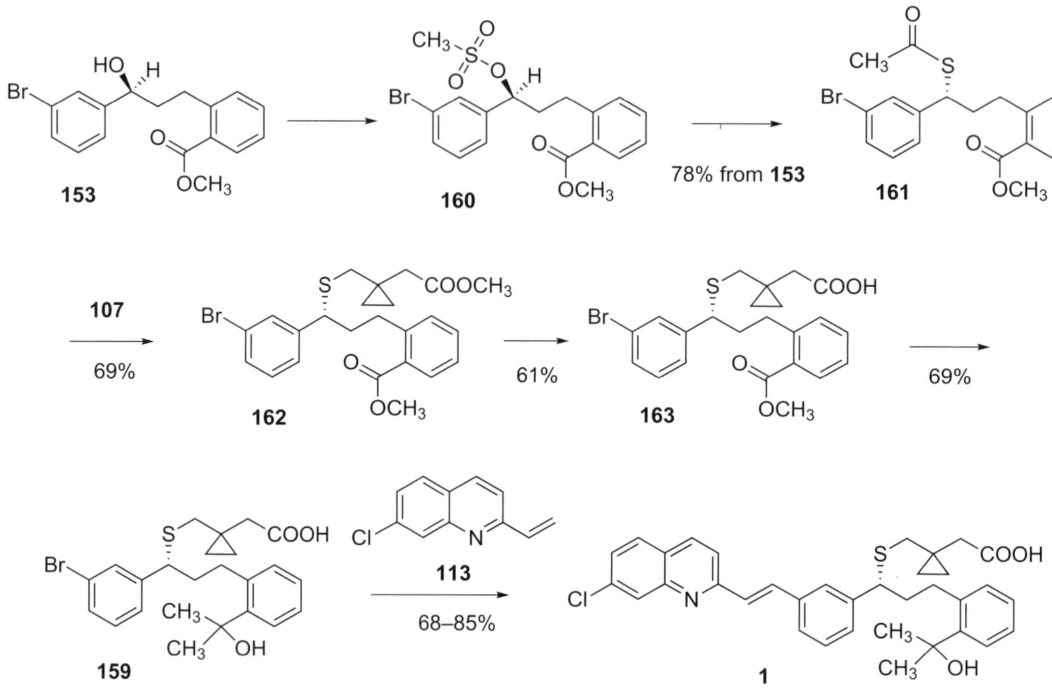


**SCHEME 6.38** An alternative route to montelukast (**1**) by right-to-left backbone construction with Heck arylation as the last step.

dichloromethane at −10°C affords methanesulfonate **160**. After a routine workup, the methanesulfonate is displaced with potassium thioacetate (1.1 equivalents) in toluene–DMF at 25°C. The (*R*)-thioacetate **161** is isolated by chromatography (78%, 99.6% ee). The side chain is introduced by hydrolysis of the thioacetate with lithium hydroxide monohydrate (1.0 equivalent) and alkylation of the resulting lithium thiolate with methyl 2-(1-bromomethyl) cyclopropyl)acetate (**107**) (1.5 equivalents) in acetonitrile–methanol–water at 20°C. The (*R*)-thioether **162** is isolated by chromatography (69%, 98.6% ee). Here, again, we encounter the problem of selective hydrolysis of the side chain methyl ester in the presence of the methyl benzoate. The hydrolysis to **163** is accomplished with sodium hydroxide (1.5 equivalents) in THF–methanol–water at 20°C (61% chromatographed, 99.5% ee). Methyl benzoate conversion to the tertiary alcohol is then accomplished with methyllithium (4.2 equivalents) in THF–ethyl ether at 20°C (69% chromatographed, 99.3% ee). Here, the Heck arylation yield is increased by doubling the catalyst charge. Reaction with 7-chloro-2-vinylquinoline (**113**) (1.0 equivalent), triethylamine (2.9 equivalents), palladium acetate (12 mol%), and tri (*o*-tolyl)phosphine (47 mol%) in DMF at 100°C affords montelukast (**1**), which is isolated by chromatography (85%, 99.4% ee).[126]

While the lack of availability of a key raw material, the long linear sequences, the multiple chromatographic isolations, and the use of ethyl ether as a reaction cosolvent certainly tell us that the Heck arylation routes are not ready for scale-up, they do offer fertile new ground for development. Production by Heck arylation at or near the end of the sequence will require careful attention to the level of residual palladium in the montelukast (**1**) produced.

## 6.8 TRADE SECRETS AND THE CONTINUING REFINEMENT OF ANALYTICAL METHODS

Two impurities in montelukast (**1**) sodium are the tertiary alcohol elimination product **98** and the sulfoxide **164** (Figure 6.5). These are readily separated by HPLC from montelukast sodium. These impurities can be reduced to <0.1% by weight by isolation and optional upgrade of a montelukast (**1**) amine salt.[98]

Useful analytical methods are available for the (*S*)-alcohol **32**[52] and for following the conversion of diol **34** to the methanesulfonate **51**.[87]

Almost every route to montelukast (**1**) includes an upgrade via precipitation of an amine salt or isolation of crude montelukast and crystallization. A more efficient direct crystallization of montelukast (**1**) from the toluene workup solvent may be possible when impurity levels are reduced. This can be accomplished using an incremental pH adjustment during the acetic acid addition. A toluene wash of a pH 12–13 aqueous quench solution removes impurities that do not possess a carboxylate group (diol **34**, cyclic ether **53**, secondary alcohol elimination product **165**, side chain nitrile **94**, and side chain amide **166**). After adding acetic acid to pH

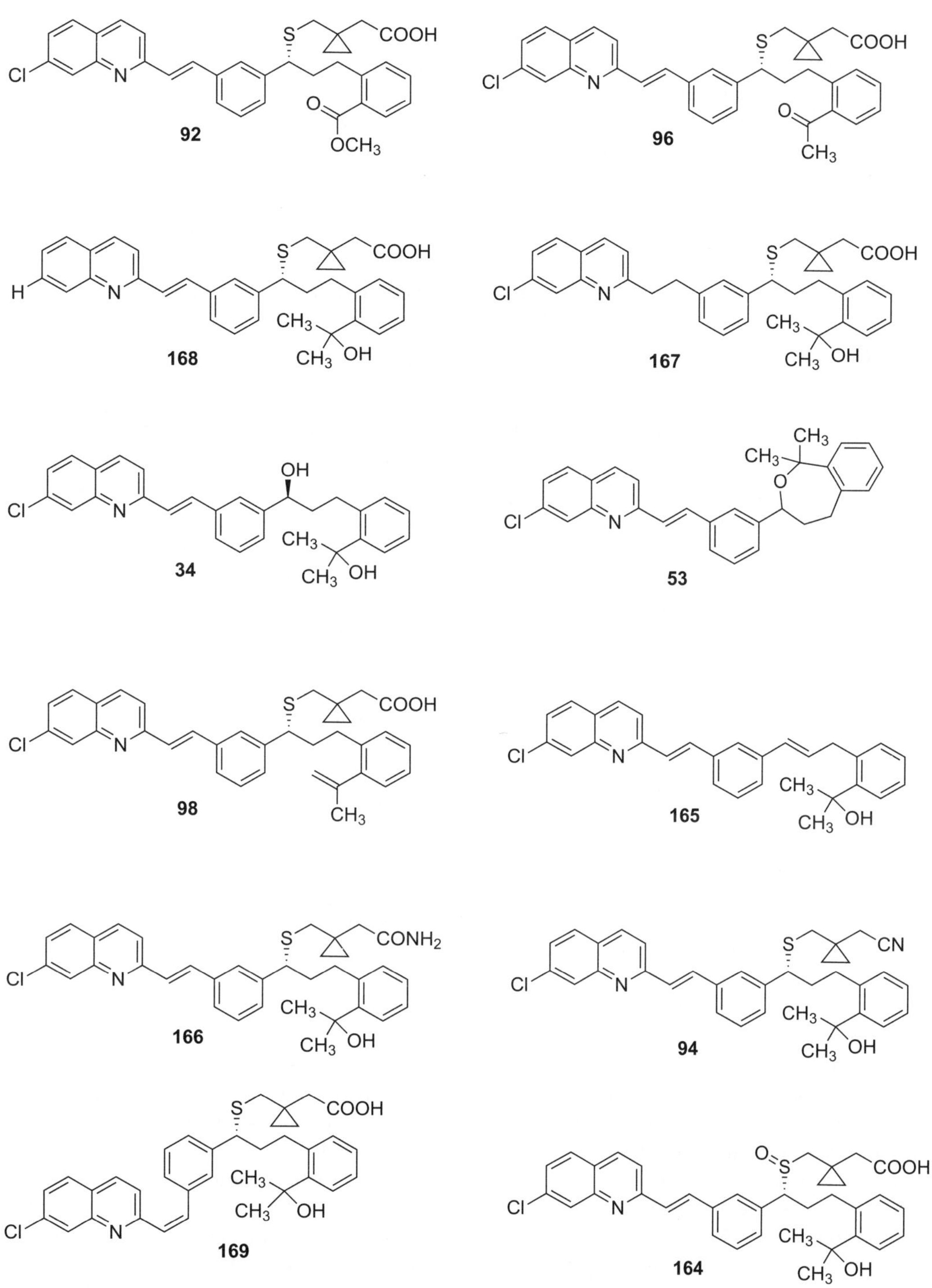


**FIGURE 6.5** Potential impurities in montelukast (**1**).

9.3, a toluene wash of the aqueous solution removes two additional impurities (ethylene bridge reduction product **167** and tertiary alcohol elimination product **98**). After adding acetic acid to pH 5–6, montelukast (**1**) is extracted into an organic solvent. The remaining impurities (including the (*S*)-enantiomer **100**) can be reduced to acceptable levels by isolation of montelukast (**1**) as an amine salt.[104]

## 6.9 THE BEST PROCESS AVAILABLE TODAY

With the wealth of information available on processes for the manufacture of montelukast sodium, narrowing the field of options is perhaps the best place to start. It is clear that the right-to-left routes offer no shorter linear sequence for the backbone construction. Two critical improvements are needed to make right-to-left routes that create the ethylene bridge by an aldehyde condensation competitive. First, these routes must have an aldehyde protecting group that can survive all the transformations required for backbone construction and be released without causing elimination of the tertiary alcohol. Second, these routes must find conditions that generate the ethylene bridge in high yield (90%) by direct condensation of aldehyde **136** with 7-chloroquinaldine (**3**). While the yield in the Wittig condensation is high, the condensation required excess ylid (3.0 equivalents) and, using the available routes to phosphonium salt **11**, the ylid is too expensive. The right-to-left routes that create the ethylene bridge by Heck arylation are more promising. The critical improvement required is a manufacturing process that will make 7-chloro-2-vinylquinoline (**113**) as available and inexpensive as 7-chloroquinaldine (**3**).

The carbon–sulfur bond at the chiral center can be set by reaction of a backbone methanesulfonate or a backbone chloride with a side chain thiol. While a backbone chloride might have greater storage stability, a methanesulfonate displacement is preferred for several reasons. First, the preparation and isolation of (*S*)-chloride **102** from the (*S*)-alcohol **32** is less efficient than the preparation and isolation of methanesulfonate **51** (70% versus 81–89%). Second, the approaches from (*S*)-chloride **102** to montelukast (**1**) are not more efficient than the many approaches to montelukast that proceed via a methanesulfonate displacement. Finally, there is no optical purity data available on montelukast (**1**) or any montelukast process intermediate produced via a chloride displacement.

There are two options for creating the thioether link: displacement of a backbone methanesulfonate with a side chain thiol or displacement of a side chain methanesulfonate or bromide with a backbone thiol. While a backbone thioacetate would have greater storage stability than a backbone methanesulfonate, displacement of a backbone methanesulfonate by a side chain thiol is preferred. The shortest sequence from the value-added methanesulfonate **51** to montelukast (**1**) sodium proceeds via displacement of the backbone methanesulfonate with the side chain thiol.

With the decision to proceed via left-to-right backbone construction to a methanesulfonate, the next step is to select one of four methanesulfonates (**48–51**) for the displacement reaction. Methanesulfonate **48** can be eliminated. The THP ether protection strategy adds four steps to the linear sequence from (*S*)-alcohol **41** to montelukast (**1**) sodium. Methanesulfonate **50** can also be eliminated since it adds one step to the linear sequence from (*S*)-alcohol **32** to montelukast (**1**) sodium and there is no evidence to suggest this strategy offers any yield improvement. Methanesulfonate **51** is the preferred intermediate for five reasons. First, there is no data to suggest that any process from methanesulfonate **49** to montelukast (**1**) sodium is more efficient than the available processes from methanesulfonate **51** to montelukast (**1**) sodium. Second, while the side chain is certainly less expensive than any of the backbone methanesulfonates, it is still a value-added reagent. Displacement of methanesulfonate **51** offers the shortest sequence from the side chain to montelukast (**1**) sodium. Third, the yields for the Grignard addition to convert (*S*)-alcohol **32** to diol **34** (85–89%) are higher than the yield for the Grignard addition converting thioether **92** to montelukast (**1**) (80%). Fourth, the Grignard addition (*S*)-alcohol **32** can be accomplished without using expensive cerium chloride (85–86%). Finally, diol **34** is an excellent cleanup point. It can be recrystallized from toluene–hexanes (90% recovery, 99.8% purity by HPLC).

There are two remaining decisions on backbone construction: the route to ketone **25** and the conditions for ketone reduction to (*S*)-alcohol **32**. There are currently three approaches to the ketone: Heck arylation of the allylic alcohol **19**, alkylation of the methyl ketone **18**, and alkylation of the β-ketoester **21**. The Heck arylation route produces the ketone **25** from 7-chloroquinaldine (**3**) in the shortest sequence (three steps) and highest yield (60%). The closest competitor is the β-ketoester route via (*E*)-methyl 3-(2-(7-chloroquinolin-2-yl)vinyl)benzoate (**7**). This route produces ketone **25** from 7-chloroquinaldine (**3**) and methyl 3-formylbenzoate in eight steps (47% overall yield). From a cost perspective, the ketone reduction by (−)-diisopinocampheylchloroborane is the best option but an (*R*)-oxazaborolidine-catalyzed reduction of palladium-free ketone by borane dimethylsulfide could be competitive.

Selecting 2-(1-mercaptomethyl)cyclopropyl)acetic acid (**2**) as the reagent for introducing the side chain gives the shortest sequence from methanesulfonate **51** to montelukast (**1**). The preferred starting material is 2,2-bis(bromomethyl)-1,3-propanediol. This starting material is less expensive than diethyl 1,1-cyclopropanedicarboxylate and avoids the hazardous lithium aluminum hydride reduction. The process from 2,2-bis(bromomethyl)-1,3-propanediol to 2-(1-mercaptomethyl)cyclopropyl)acetic acid (**2**) certainly has room for improvement. The process is long and linear, uses sodium cyanide and potassium thioacetate, and generates a zinc waste stream.

Should methanesulfonate **51** be isolated? While the isolation reduces the levels of several impurities, the available data suggests the methanesulfonate **51** need not be isolated if the montelukast (**1**) is isolated as an amine salt. Low-temperature isolation equipment will still be necessary even when methanesulfonate **51** is not isolated since the preferred procedure for preparation of a solution of methanesulfonate **51** in THF requires a filtration of the amine hydrochloride salt from a 0°C suspension.

The methanesulfonate displacement with 2-(1-(mercaptomethyl)cyclopropyl)acetic acid (**2**) can be accomplished with butyllithium, sodium hydride, sodium *tert*-butoxide, sodium methoxide, sodium hydroxide, or cesium carbonate. We can quickly eliminate two of the bases. Sodium hydride

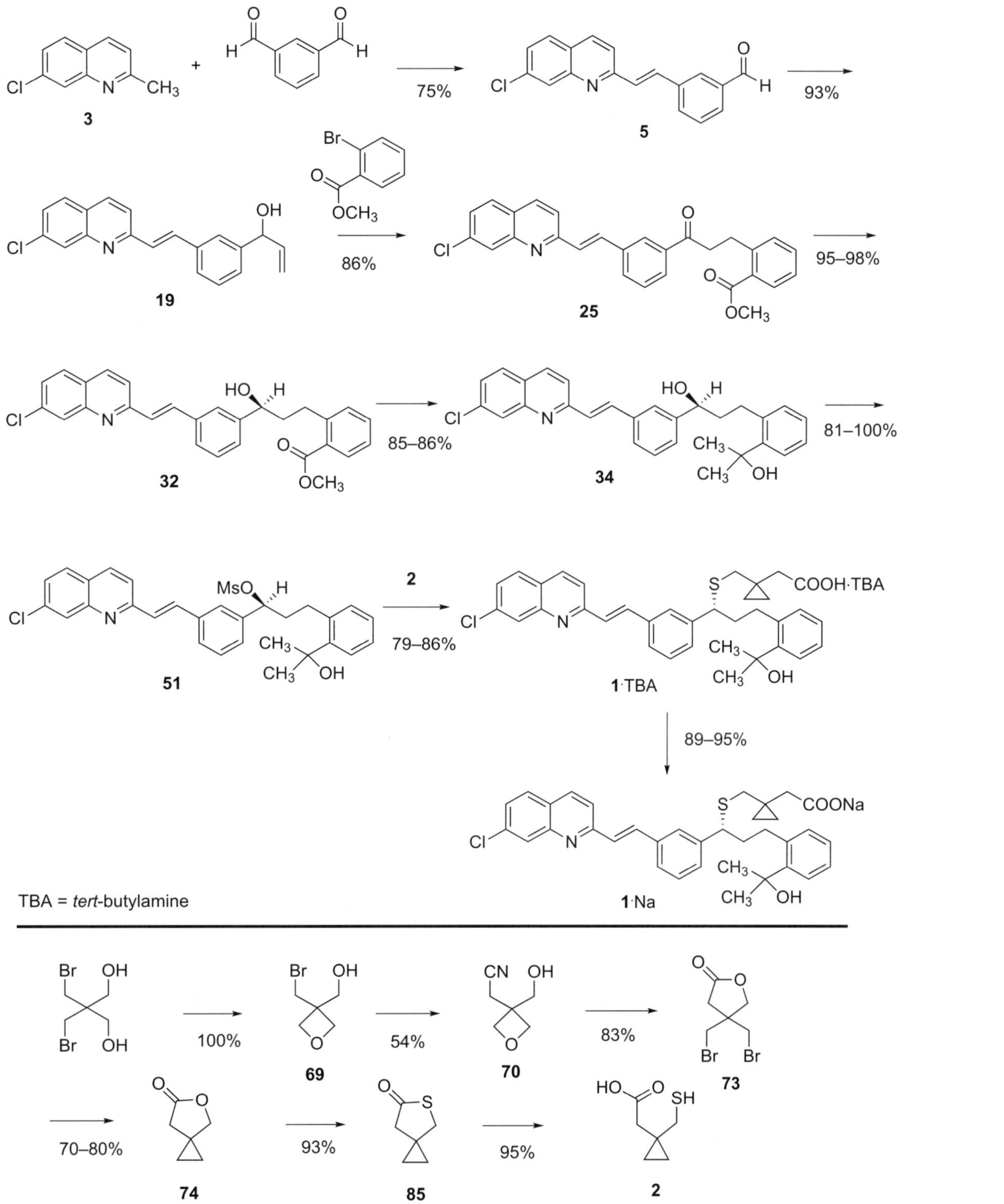


**SCHEME 6.39** The best process for manufacture of montelukast (**1**) sodium in 2008.

is not desirable for a large-scale plant operation. It is typically handled as a solid dispersion in oil. The oil must be removed by slurry washes with a dry nonpolar solvent such as hexanes or perhaps THF. The residual solid after the wash liquid is decanted can decompose rapidly on exposure to moist air and the heat evolved may cause spontaneous ignition. Cesium carbonate is far more expensive than sodium hydroxide and offers no higher yield of the

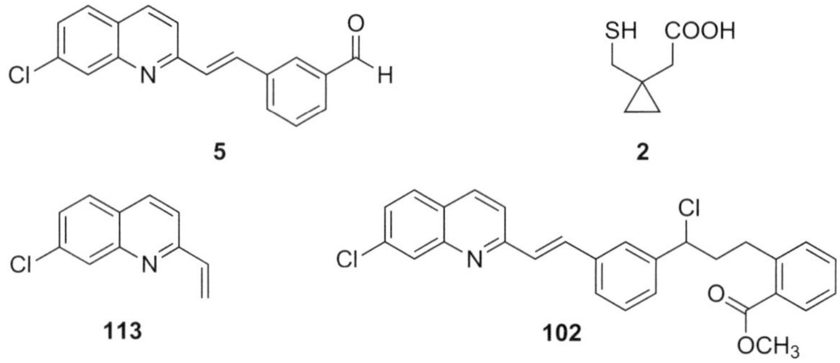

**FIGURE 6.6** Structures searched for montelukast (**1**) sodium presentation (November–December 2008).

montelukast (**1**) amine salt (70%). The remaining bases afford montelukast (**1**) amine salts in yields of 65–86% from diol **34** or 70–81% from isolated methanesulfonate **51**.

There is sufficient data available for the crude montelukast (**1**) *N,N*-dicyclohexylamine salt produced by displacement using butyllithium (79% yield, 96 wt%, 99.8% ee). There is very limited optical purity data for any crude montelukast (**1**) amine salt produced using sodium *tert*-butoxide, sodium methoxide, or sodium hydroxide. The best data comes from montelukast (**1**) isopropylamine salt isolated from the reaction of methyl 2-(1-mercaptomethyl) cyclopropyl)acetate (**84**) (1.4 equivalents) with nonisolated methanesulfonate **51** and 47% aqueous sodium hydroxide. The montelukast (**1**) isopropylamine salt contains no (*S*)-enantiomer (68% from diol **34**, 98.4% pure by HPLC). The reaction of sodium 2-(1-(mercaptomethyl)cyclopropyl)acetate (**101**) (1.4 equivalents) and butyllithium (1.4 equivalents) with nonisolated methanesulfonate **51** in THF at −10°C appears to be the most efficient, affording the montelukast (**1**) (*R*)-(+)-α–methylbenzylamine salt in 86% yield from the diol (98.4% pure by HPLC). In a plant with the requisite storage and handling equipment, butyllithium will be the best base option. A plant that does not have butyllithium capabilities can produce a montelukast (**1**) amine salt of comparable purity but most likely at a lower yield.

The majority of processes include isolation of a montelukast (**1**) amine salt. This penultimate intermediate can be upgraded by recrystallization to ensure that the montelukast sodium will meet purity specifications. Can we select a single montelukast amine salt, from the twenty-seven-and-counting list, as a preferred salt? This question leads to the final question. Should montelukast sodium be prepared from the montelukast amine salt in two steps via montelukast or in a single step by reaction with sodium methoxide in toluene? A low boiling amine such as *tert*-butylamine would be preferred for the direct conversion process. Should montelukast sodium be isolated from toluene–acetonitrile or from toluene–hexanes? These critical process decisions cannot be made based solely on economics or environmental impact. These decisions can only be made working in close collaboration with the analytical and formulation groups.

The best process affords montelukast (**1**) sodium in eight steps in overall 41% yield from 7-chloroquinaldine (**3**) and in seven steps in overall 55% yield from (*E*)-3-(2-(7-chloroquinolin-2-yl)vinyl)benzaldehyde (**5**). The process solvents are DMF, THF, ethyl acetate, toluene, heptane, hexanes, perhaps acetone and water, all solvents you will find in a pharmaceutical manufacturing plant. While every intermediate can be isolated as a solid, some of the early isolations might not be necessary. Key upgrade points are (*E*)-3-(2-(7-chloroquinolin-2-yl)vinyl)benzaldehyde (**5**), diol **34**, and the montelukast (**1**) amine salt (Scheme 6.39).

A superior process will come from analysis of the weaknesses of this preferred process. The weaknesses that have not been adequately addressed to date include the long linear sequences for construction of both the backbone and the side chain, the statistical problem encountered in the 7-chloroquinaldine (**3**) condensation with isophthalaldehyde, the poor throughput associated with the vinylmagnesium bromide reaction, and the instability of methanesulfonate **51** at elevated temperatures.

## 6.10 STRUCTURES SEARCHED

Four structure searches were used to generate all the information presented in this chapter (Figure 6.6).

## REFERENCES


1. Global Initiative for Asthma. *GINA Workshop Report 2007*, p. 2.
2. Moorman, J.E.; Rudd, R.A.; Johnson, C.A.; King, M.; Minor, P.; Bailey, C.; Scalia, M.R.; Akinbami, L.J. *National Surveillance for Asthma-United States, 1980–2004*. Accessed at www.cdc.gov.
3. *Trends in Asthma Morbidity and Mortality*, November **2007**. Accessed at www.lungusa.org.

4. Braman, S.S. *Chest* **2006**, *130*, 4S–12S.
5. Masoli, M.; Fabian, D.; Holt, S.; Beasley, R. *Allergy* **2004**, *59*, 469.
6. (a) Accessed at www.aaaai.org/media/resources/media_kit/asthma_statistics.(b) Martinez, F.D.; Wright, A.L.; Taussig, L.M.; Holberg, C.J.; Halonen, M.; Morgan, W.J. *New Engl. J. Med.* **1995**, *332*, 133.
7. Holloway, J.W.; Beghe, B.; Holgate, S.T. *Clin. Exp. Allergy* **1999**, *29*, 1023.
8. Wiesch, D.G.; Meyers, D.A.; Bleecker, E.R. *J. Allergy Clin. Immunol.* **1999**, *104*, 895.
9. Stein, R.T.; Sherrill, D.; Morgan, W.J.; Holberg, C.J.; Halonen, M.; Taussig, L.M.; Wright, A.L.; Martinez, F.D. *Lancet* **1999**, *354*, 541.
10. Shaheen, S.O.; Aaby, P.; Hall, A.J.; Barker, D.J.; Heyes, C.B.; Shiell, A.W.; Goudiaby, A. *Lancet* **1996** *347*, 1792.
11. A list of known sensitizers can be accessed at www.bohrf.org.uk.
12. Accessed at www.iii.co.uk.
13. Accessed at www.globalbusinessinsights.com.
14. The prescribing information is available at www.singulair.com.
15. Holbrook, J.T.; Harik-Khan, R. *J. Allergy Clin. Immunol.* **2008**, *122*, 828.
16. The Aldrich 2007–2008 catalog price for 5 g is $137.00.
17. The Aldrich 2007–2008 catalog price for 500 g is $60.40.
18. The Aldrich 2007–2008 catalog price for 1 L ($d$ = 0.846 g/mL) is $78.70 (predominantly *trans*, 98%).
19. (a) Leir, C.M. *J. Org. Chem.* **1977**, *42*, 911.(b) Song, Z.; Mertzman, M.; Hughes, D.L. *J. Heterocycl. Chem.* **1993**, *30*, 17.(c) Song, Z.; Hughes, D.L.U.S. Patent 5,126,456 (6/30/1992).(d) Song, Z.; Hughes, D.L. EP 537847 (10/9/1992).
20. The Aldrich 2007–2008 catalog price for 100 g is $78.50.
21. The Aldrich 2007–2008 catalog price for 1 kg is $91.40.
22. Pastorek, E.; Orth, W.; Weiss, W.; Kleffner, H.W.U.S. Patent 5,066,806 (11/19/1991).
23. Tung, H.H.; Hobbs, D.M.U.S. Patent 5,869,673 (2/9/1999).
24. McNamara, J.M.; Leazer, J.L.; Bhupathy, M.; Amato, J.S.; Reamer, R.A.; Reider, P.J.; Grabowski, E.J.J. *J. Org. Chem.* **1989**, *54*, 3718.
25. Suri, S.; Sarin, G.S.; Mahendru, M. WO 2006021974 (3/2/2006).
26. Padi, P.R.; Bollikonda, S.; Naidu, N.; Ghosh, D.; Venkata Kandirelli, K.K.; Jinna, R.R.; Kasturi, R.K.; Mohanarangam, S. WO 2008058118 (5/15/2008).
27. Gottumukkala, V.S.; Chava, S.; Casukhela, S.R. WO 2008035379 (3/27/2008).
28. (a) Young, R.N.; Zamboni, R.; Leger, S.U.S. Patent 4851409 (7/25/1989).(b) Zamboni, R.; Belley, M.; Champion, E.; Charette, L.; DeHaven, R.; Frenette, R.; Gauthier, J.Y.; Jones, T.R.; Leger, S.; Masson, P.; McFarlane, C.S.; Metters, K.; Pong, S.S.; Piechuta, H.; Rokach, J.; Therien, M.; Williams, H.W.R.; Young, R.N. *J. Med. Chem.* **1992**, *35*, 3832.
29. Bach, N.J.; Baker, S.R.; Gilmore, J.; Lewthwaite, R.A.; McKillop, A.; Sawyer, J.S.; Stephenson, G.R.; Urquhart, M.W.J. EP 775694 (11/18/1996).
30. von Sprecher, A.; Gerspacher, M.; Beck, A.; Kimmel, S.; Wiestner, H.; Anderson, G.P.; Niederhauser, U.; Subramanian, N.; Bray, M.A. *Biorg. Med. Chem. Lett.* **1998**, *8*, 965.
31. Larsen, R.D.; Corley, E.G.; King, A.O.; Carroll, J.D.; Davis, P.; Verhoeven, T.R.; Reider, P.J. *J. Org. Chem.* **1996**, *61*, 3398.
32. (a) Belley, M.L.; Leger, S.; Labelle, M.; Roy, P.; Xiang, Y.B.; Guay, D.U.S. Patent 5,565,473 (10/15/1996).(b) Belley, M.L.; Leger, S.; Labelle, M.; Roy, P.; Xiang, Y.B.; Guay, D.U.S. Patent 5,270,324 (12/14/1993).(c) Belley, M.L.; Leger, S.; Labelle, M.; Roy, P.; Xiang, Y.B.; Guay, D. EP 480717 (10/10/1991).
33. The Aldrich 2007–2008 catalog price for 25 g is $141.00.
34. The Aldrich 2007–2008 catalog price for 5 g is $179.50.
35. Colson, A.; Gautier, H. *Bull. Soc. Chim. Fr.* **1886**, *45*, 509.
36. Rosenmund, K.W.; Zetzsche, F.; Flutsch, C. *Chem. Ber.* **1921**, *54*, 2888.
37. Brown, H.C.; Subba Rao, B.C. *J. Am. Chem. Soc.* **1958**, *80*, 5377.
38. Jennings, K.F. *J. Chem. Soc.* **1957**, 1172.
39. Ackerman, J.H.; Surrey, A.R. *Org. Syn. Coll. Vol. 5*, **1973**, 668.
40. The Aldrich 2007–2008 catalog price for 500 g is $53.20.
41. The Aldrich 2007–2008 catalog price for 1 kg is $47.40.
42. The Aldrich 2007–2008 catalog price for 1 kg (98% reagent grade) is $320.50.
43. King, A.O.; Corley, E.G.; Anderson, R.K.; Larsen, R.D.; Verhoeven, T.R.; Reider, P.J.; Xiang, Y.B.; Belley, M.; Leblanc, Y.; Labelle, M.; Prasit, P.; Zamboni, R.J. *J. Org. Chem.* **1993**, *58*, 3731.
44. The Aldrich 2007–2008 catalog price for 18 L of 1 M is $1105.00.
45. National Toxicology Program. *Report on Carcinogens*, 11th edition, Public Health Service, U.S. Department of Health and Human Services, Atlanta, GA.
46. The Aldrich 2007–2008 catalog price for 2 kg is $81.80.
47. The Aldrich 2007–2008 catalog price for 500 g is $78.00.
48. Pifferi, G.; Testa, E. *Tetrahedron*, **1966**, *22*, 2107.
49. Chemicals Known to the State to Cause Cancer or Reproductive Toxicity, State of California, March 21, 2008.
50. Chartrain, M.M.; Shieh-Shung, T.C.; Garrity, G.M.; Heimbuch, B.; Roberge, C.; Shafiee, A.U.S. Patent 5,491,077 (2/13/1996).
51. Chen, S.-S.T.; Shafiee, A.U.S. Patent 5,427,933 (6/27/1995).
52. (a) Avdagic, A.; Jerkovic, J.; Pavlicic, D. WO 2008009970 (1/24/2008).(b) Avdagic, A.; Mohar, B.; Sterk, D.; Stephan, M. WO 2006008562 (1/26/2006).(c) Fujii, A.; Hashiguchi, S.; Uematsu, N.; Ikariya, T.; Noyori, R. *J. Am. Chem. Soc.* **1996**, *118*, 2521.
53. The Aldrich 2007–2008 catalog price for 5 kg is $1610.00.
54. The Aldrich 2007–2008 catalog price for 5 g is $104.50.
55. Corey, E.J.; Bakshi, R.K.; Shibata, S.; Chen, C.-P.; Singh, V. K. *J. Am. Chem. Soc.* **1987**, *109*, 7925.

56. Mathre, D.J.; Jones, T.K.; Xavier, L.C.; Blacklock, T.J.; Reamer, R.A.; Mohan, J.J.; Turner-Jones, E.T.; Hoogsteen, K.; Baum, M.W.; Grabowski, E.J.J. *J. Org. Chem.* **1991**, *56*, 751.
57. The Aldrich 2007–2008 catalog price for 25 mL of 1 M solution in toluene is $196.00 and for 25 mL of 1 M solution in THF is $312.00.
58. Corey, E.J.; Bakshi, R.K.; Shibata, S. *J. Am. Chem. Soc.* **1987**, *109*, 5551.
59. The Aldrich 2007–2008 catalog price for 18 L of 1.0 M in THF is $1980.00.
60. (a) Simpson, P.; Tschaen, D.; Verhoeven, T.R. *Synth. Commun.* **1991**, *21*, 1705.(b) Simpson, P.M.; Tschaen, D.M.; Verhoeven, T.R.U.S. Patent 5,292,946 (3/8/1994).
61. (a) King, A.O.; Larsen, R.D.; Verhoeven, T.R.; Zhao, M.U.S. Patent 5,693,816 (12/2/1997). (b) King, A.O.; Larsen, R.D.; Verhoeven, T.R.; Zhao, M.U.S. Patent 5,545,758 (8/13/1996). (c) King, A.O.; Larsen, R.D.; Verhoeven, T.R.; Zhao, M. WO 9605206 (2/22/1996). (d) King, A.O.; Larsen, R.D.; Verhoeven, T.R.; Zhao, M. EP 775146 (6/12/2002).
62. The Aldrich 2007–2008 catalog price for 500 mL ($d = 0.858$ g/mL) is $131.50.
63. The Aldrich 2007–2008 catalog price for 18 L (0.801 g/mL) is $4480.00.
64. The Aldrich 2007–2008 catalog price for 50 g is $101.00.
65. The Aldrich 2007–2008 catalog price for 454 g is $337.00.
66. The Aldrich 2007–2008 catalog price for 2 kg of reagent grade powder is $495.00.
67. Pathi, S.L.; Kankan, R.N.; Rao, D.R.; Phull, M.S. WO 2008032099 (3/20/2008).
68. The Aldrich 2007–2008 catalog price for 500 mg is $76.40.
69. Sterk, D.; Stephan, M.S.; Mohar, B. *Tetrahedron Asym.* **2002**, *13*, 2605.
70. *The Platinum Metals Report*, August 2008. Accessed at www.platinum.matthey.com.
71. Wiffen, J.; Wilson, I. WO 2008035086 (3/27/2008).
72. Kishore Kumar, M.; Rama Subba Reddy, K.; Shyla Prasad, D. WO 2008062478 (5/29/2008).
73. Belley, M.L.U.S. Patent 5,212,180 (5/18/1993).
74. Imamoto, T.; Takiyama, N.; Nakamura, K.; Hatajima, T.; Kamiya, Y. *J. Am. Chem. Soc.* **1999**, *111*, 4392.
75. The Aldrich 2007–2008 catalog price for 500 g of heptahydrate is $145.50. The Aldrich 2007–2008 catalog price for 50 g of anhydrous beads is $194.00.
76. Bunnelle, W.H.; Narayanan, B.A. *Org. Syn., Coll. Vol. 8*, **1993**, 602.
77. Refluxing the heptahydrate with thionyl chloride is the preferred method for producing anhydrous cerium chloride, in: Edelmann, F.T., ed., *Synthetic Methods of Organometallic and Inorganic Chemistry*, Vol. 6, Georg Thieme Verlag, Stuttgart, **1997**, pp. 32–33.
78. (a) Bhupathy, M.; Sidler, D.R.; McNamara, J.M.; Volante, R. P.; Bergan, J.U.S. Patent 6,320,052 (11/20/2001). (b) Bhupathy, M.; Sidler, D.R.; McNamara, J.M.; Volante, R.P.; Bergan, J.U.S. Patent 5,614,632 (3/25/1997). (c) Bhupathy, M.; McNamara, J.M.; Sidler, D.R.; Volante, R.P.; Bergan, J. WO 9518107 (7/6/1995).
79. Overeem, A.; Gieling, R.G.; Zhu, J.; Lambertus, T. WO 2005105749 (11/10/2005).
80. Ray, U.K.; Boju, S.; Pathuri, S.R.; Meenakshisunderam, S. WO 2008001213 (1/3/2008).
81. (a) Turchetta, S.; Tuozzi, A.; Ullucci, E.; de Ferra, L.U.S. Patent 7,417,149 (8/26/2008). (b) Turchetta, S.; Tuozzi, A.; Ullucci, E.; de Ferra, L. EP 1693368 (8/23/2006).
82. Overeem, A.; Gieling, R.G.; Zhu, J.; Lambertus, T. WO 2005105750 (11/10/2005).
83. Reguri, B.R.; Bollikonda, S.; Naga, V.V.; Bulusu, C.S.; Kasturi, R.K.; Aavula, S.K.U.S. Patent 20050107612 (5/19/2005).
84. (a) Noiman, M.; Salman, A.; Levin, I.; Arad, O.; Kaspi, J.; Weisman, A.; Gafni, Y. WO 2008126075 (10/23/2008). (b) Brand, M.; Weisman, A.; Gafni, Y.; Zelikovitch, L.; Davidi, G.; Manoff, E. WO 2008087628 (7/24/2008). (c) Brand, M.; Shookrun, M.; Arad, O.; Kaspi, J. WO 2007096889 (8/30/2007).
85. (a) Sundaram, V.; Rajan, S.T.; Naga, V.V.; Bulusu, C.S.; Srivastav, A.K.; Kasturi, R.K.; Aavula, S.K.U.S. Patent 7,189,853 (3/13/2007). (b) Sundaram, V.; Rajan, S.T.; Naga, V.V.; Bulusu, C.S.; Srivastav, A.K.; Kasturi, R.K.; Aavula, S. K.U.S. Patent 20050234241 (10/20/2005).
86. Achmatowicz, O.; Wisniewski, K.; Ramza, J.; Szelejewski, W.; Szechner, B. WO 2006043846 (4/27/2006).
87. Egekeze, J.O.; Danielski, M.C.; Grinberg, N.; Smith, G.B.; Sidler, D.R.; Perpall, H.J.; Bicker, G.R.; Tway, P.C. *Anal. Chem.* **1995**, *67*, 2292.
88. The Aldrich 2007–2008 catalog price for 25 g is $139.50. The Aldrich 2007–2008 catalog price for 50 mL ($d = 1.147$ g/mL) of dimethyl 1,1-cyclopropanedicarboxylate is $41.90.
89. Labelle, M.; Belley, M.; Gareau, Y.; Gauthier, J.Y.; Guay, D.; Gordon, R.; Grossman, S.G.; Jones, T.R.; Leblanc, Y.; McAuliffe, M.; McFarlane, C.; Masson, P.; Metters, K.M.; Ouimet, N.; Patrick, D.H.; Piechuta, H.; Rochette, C.; Sawyer, N.; Xiang, Y.B.; Pickett, C.B.; Ford-Hutchinson, A.W.; Zamboni, R.J.; Young, R.N. *Bioorg. Med. Chem. Lett.* **1995**, *5*, 283.
90. Suen, R.-T.; Liu, Y.-L.; Wei, C.-P.U.S. Patent 7,271,268 (9/18/2007).
91. (a) King, S.; Pipik, B.; Conlon, D.A. WO 9622987 (8/1/1996). (b) King, S.; Pipik, B.; Conlon, D.A.U.S. Patent 5,523,477 (6/4/1996).
92. The Aldrich 2007–2008 catalog price for 3 kg is $126.50.
93. (a) Hanselmann, P.U.S. Patent 5,486,622 (1/23/1996). b) Hanselmann, P. EP 678515 (4/19/1995).
94. The Aldrich 2007–2008 catalog price for 500 g is $107.50.
95. The Aldrich 2007–2008 catalog price for 1 kg is $37.20.
96. Liu, L.W.; Wang, Y.; Marrugo, H.; Harper, S.; Quan, D.D.C.; Zhou, Z.; Bydlinski, G.U.S. Patent 6,512,140 (1/28/2003).
97. Wang, Y.; Wang, Y.; Brand, M.; Kaspi, J. WO 2007088545 (8/9/2007).

98. (a) Sterimbaum, G.; Shapiro, E.; Chen, K. WO 2007005965 (1/11/2007). (b) Shapiro, E.; Yahalomi, R.; Niddam-Hildesheim, V.; Sterimbaun, G.; Chen, K.U.S. Patent 20050256156 (11/17/2005). (c) Shapiro, E.; Yahalomi, R.; Niddam-Hildesheim, V.; Sterimbaun, G.; Chen, K. WO 2005105751 (11/10/2005).
99. The Aldrich 2007–2008 catalog price for 100 g is $97.10.
100. (a) Quittmann, W.; McGarrity, J.U.S. Patent 5,534,651 (7/9/1996). (b) Quittmann, W.; McGarrity, J. EP 641775 (9/2/1994).
101. Niddam-Hildesheim, V.; Aronhime, J.; Chen, K. WO 2005074935 (8/18/2005).
102. (a) Bartl, J.; Picha, F.U.S. Patent 7,446,116 (11/4/2008). (b) Bartl, J.; Picha, F. WO 2007107297 (9/27/2007).
103. (a) Gasanz Guillen, Y.; Montsalvatje Llagostera, M.; Talavera Escasany, P. WO 2008017669 (2/14/2008). (b) Gasanz Guillen, Y.; Montsalvatje Llagostera, M.; Talavera Escasany, P. EP 1886998 (2/13/2008).
104. Adin, I.; Deutsch, Z.; Brand, M.; Shookrun, M.; Arad, O.; Kaspi, J. WO 2007096875 (8/30/2007).
105. The Aldrich 2007–2008 catalog price for 100 g is $134.00.
106. Satyanarayana, R.M.; Kihsore, K.M.; Thirumalai, R.S.; Ramasubba, R.K. WO 2007069261 (6/21/2007).
107. Chamorro Gutierrez, I.; Bosch I Llado, J.; Molins I Grau, E. WO 2006058545 (6/8/2006).
108. Srinivas, P.L.; Rao, D.R.; Kankan, R.N.; Relekar, J.P. WO 2006064269 (6/22/2006).
109. Chawla, H.P.S.; Chowdhary, A.S.; Patel, A.M.; Joshi, V.N.; Patel, M.P. WO 2007072114 (6/28/2007).
110. Bendre, D.S.; Holkar, A.G. WO 2008023044 (2/28/2008).
111. Chawla, H.P.S.; Chowdhary, A.S.; Patel, A.M.; Joshi, V.N.; Patel, M.P. WO 2007116240 (10/18/2007).
112. Halama, A.; Jirman, J. WO 2008083635 (7/17/2008).
113. Chava, S.; Gorantla, S.R.; Indukuri, V.S.K.; Simhadri, S. WO 2006008750 (1/26/2006).
114. (a) Gorantla, S.R.; Bandari, M.; Indukuri, V.S.K.; Simhadri, S.; Ketavarapu, N.R.; Jammulaveera, V.K.K.; Mohammed, Y. WO 2008015703 (2/7/2008). (b) Chava, S.; Gorantla, S.R.; Indukuri, V.S.K. WO 2007004237 (1/11/2007). (c) Chava, S.; Gorantla, S.R.; Indukuri, V.S.K.; Simhadri, S.; Jammula, V.V.K.K. WO 2006008751 (1/26/2006).
115. Satyanarayana, B.; Reddy, P.P. *Synth. Commun.* **2007**, *37*, 547.
116. (a) Wang, D.; Zhang, Y.; Li, J. CN 1428335 (7/9/2003). (b) Wang, D.; Zhang, Y.; Li, J. CN 1420113 (5/28/2003).
117. The Aldrich 2007–2008 catalog price for 100 g is $75.90.
118. The Aldrich 2007–2008 catalog price for 500 g is $142.50.
119. The Aldrich 2007–2008 catalog price for 25 g is $75.80.
120. Stambuli, J.P.; Stauffer, S.R.; Shaughnessy, K.H.; Hartwig, J.F. *J. Am. Chem. Soc.* **2001**, *123*, 2677.
121. Cooke, M.P.; Widener, R.K. *J. Org. Chem.* **1987**, *52*, 1381.
122. Bessa Bellmunt, J.; Rafecas Jane, L.; Pasto Aguila, M.; Verdaguer Espaulella, X.; Gordo Campon, E. WO 2007101841 (9/13/2007).
123. The Aldrich 2007–2008 catalog price for 100 g is $99.10.
124. The Aldrich 2007–2008 catalog price for 100 g is $62.50.
125. Kidemet, D.; Kroselj, V.; Osolnik, R. WO 2008049922 (5/2/2008).
126. (a) Perez Andres, J.A.; Huguet Clotet, J.; Dalmases Barjoan, P. EP 1958936 (8/20/2008). (b) Perez Andres, J.A.; Huguet Clotet, J.; Dalmases Barjoan, P.U.S. Patent 20080194825 (8/14/2008).
127. Caspi, D.D.; Ebner, D.C.; Bagdanoff, J.T.; Stoltz, B.M. *Adv. Synth. Catal.* **2004**, *346*, 185.

# 7

# PREVACID® (LANSOPRAZOLE)

## 7.1 PREVACID® AND THE PROTON PUMP INHIBITORS

Lansoprazole belongs to a class of antisecretory substituted benzimidazoles, which suppress gastric acid secretion by inhibiting the ($H^+$, $K^+$)-ATPase enzyme system at the secretory surface of the gastric parietal cell. This enzyme is referred to as the proton pump and lansoprazole is a proton pump inhibitor (PPI) (Figure 7.1). Lansoprazole (Figure 7.2) is used for the treatment and maintenance of healing of duodenal or gastric ulcers, NSAID-induced ulcers, erosive and reflux esophagitis, Barrett's esophagus, and Zollinger–Ellison syndrome.

The U.S. sales for lansoprazole were \$3.3 billion in 2007. U.S. sales figures for the period 2002–2007 were a steady \$3–4 billion per year.[1] The total PPI market was estimated to be worth \$13 billion in 2006. Other PPIs appearing on the list of top 200 drugs by U.S. sales in 2007 are Nexium® (esomeprazole), Protonix® (pantoprazole), and Aciphex® (rabeprazole).

Three recent developments signal a change in the steady annual U.S. sales of lansoprazole in the next few years. TAP Pharmaceuticals, a 30-year joint venture between Takeda and Abbott Laboratories, was recently split, with Takeda regaining the commercial rights to lansoprazole in the U.S. and Abbott getting the oncology drug Lupron.

On another front, in 2007 Takeda and Teva were in the U.S. District Court for the District of Delaware, Takeda claiming that Teva's abbreviated new drug application (ANDA) for a generic version of Prevacid capsules infringes their patents, and Teva contending that key Takeda patents (US4628098 and US5045321) are invalid for obviousness and unenforceable due to inequitable conduct. The court ruled in favor of Takeda. Teva will delay launch of their generic Prevacid until the US4628098 patent expires in 2009.[2] There is an ever-increasing pressure to reduce health care costs and a financial incentive to demonstrate the interchangeability of PPIs. When generic lansoprazole does hit the market it will likely be priced to compete with other PPIs already on the market.[3]

Takeda has been preparing for the generic competition. On January 31, 2009, Takeda gained FDA approval for Kapidex® dual delayed release (DDR) capsules for the once-daily treatment of heartburn associated with symptomatic nonerosive gastroesophageal reflux disease (GERD), for the healing of erosive esophagitis (EE), and for the maintenance of healed EE. Kapidex® is dexlansoprazole, the dextrorotatory (+) (*R*)-enantiomer of lansoprazole. Kapidex® has a unique formulation, containing two different types of enteric-coated granules. The concentration–time profile for Kapidex® has two peaks, the first in 1–2 h and the second in 4–5 h. This delayed release formulation enables a once-daily treatment for heartburn. In patients with EE, Kapidex® (60 mg) produced higher overall healing rates at week 8 when compared to lansoprazole (30 mg) (87% versus 85% and 85% versus 79% in two studies).[4]

In recent clinical news, a study reported that patients 66 years and older taking the ADP-induced platelet aggregation inhibitor Plavix® (clopidogrel) following a heart

*Pharmaceutical Process Chemistry for Synthesis: Rethinking the Routes to Scale-Up*, By Peter J. Harrington

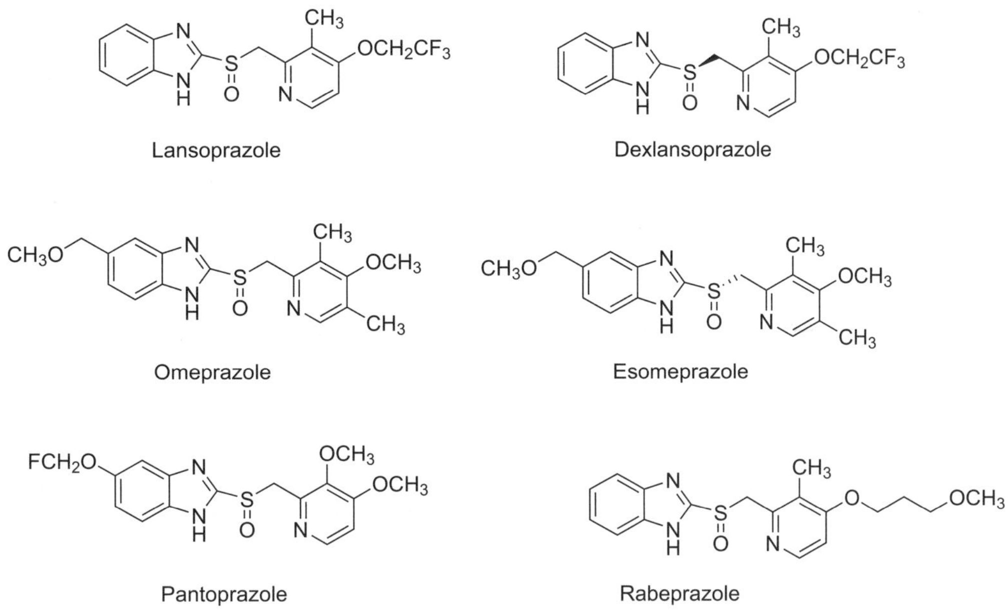


**FIGURE 7.1** Proton pump inhibitors.

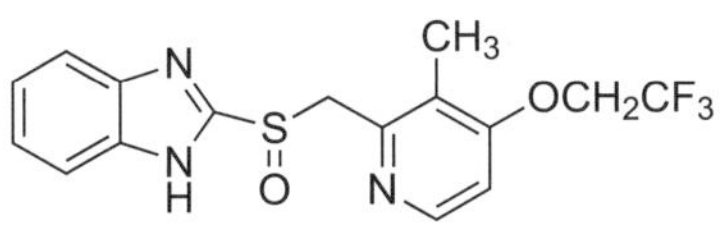

1

**FIGURE 7.2** Lansoprazole (**1**).

attack are at significantly higher risk of a recurrence if they are also taking the PPI's lansoprazole, omeprazole, or rabeprazole. There was no increased risk associated with pantoprazole or with $H_2$ receptor antagonists. The PPIs other than pantoprazole can block the conversion of clopidogrel to its active form in the liver.[5]

## 7.2 SYNTHESIS OF LANSOPRAZOLE SULFIDE (2)

Lansoprazole (**1**) can be produced by oxidation of the sulfide **2**. Lansoprazole sulfide **2** is best disconnected near the center into two components, 2-mercaptobenzimidazole and 2-(chloromethyl)-3-methyl-4-(2,2,2-trifluoroethoxy)pyridine (**3**). 2-Mercaptobenzimidazole is produced by condensation of potassium ethyl xanthate with 1,2-phenylenediamine.[6] It is commercially available and inexpensive.[7] Disconnection of the other C–S bond suggests expensive 2-chlorobenzimidazole[8] and the 2-(methylthio)pyridine (**4**) as alternative components (Figure 7.3).

### 7.2.1 Synthesis of 2-(Chloromethyl)-3-methyl-4-(2,2,2-trifluoroethoxy)pyridine (3)

Construction of 2-(chloromethyl)-3-methyl-4-(2,2,2-trifluoroethoxy)pyridine (**3**) begins with a commercially available pyridine and utilizes three unique features of pyridine *N*-oxide chemistry. First, while pyridines undergo electrophilic substitution only with difficulty at the 3- and 5-position, pyridine *N*-oxides undergo electrophilic substitution at the 4-position. Second, the nitro group of a 4-nitropyridine *N*-oxide can be displaced by a nucleophile. Third, reaction of a 2-methylpyridine *N*-oxide with acetic anhydride results in acetylation on oxygen and migration of the acetoxy substituent to produce a 2-(acetoxymethyl)pyridine (the Boekelheide reaction).

The synthesis of the appropriately functionalized pyridine **5** begins with commercially available 2,3-lutidine.[9] Slow addition of magnesium monoperoxyphthalate hexahydrate[10] to 2,3-lutidine and potassium carbonate in water at 25°C is followed by pH adjustment to 7.5–8.0 and extraction with dichloromethane (98% combined). Dilution of the organic extracts with 98% sulfuric acid and distillation of the dichloromethane affords a solution of the *N*-oxide **6** in sulfuric acid ready for use in the nitration step. Alternatively, reaction of 2,3-lutidine with 35 wt% hydrogen peroxide[11] (1.3 equivalents) in glacial acetic acid at 105°C, quench with paraformaldehyde, dilution with 98% sulfuric acid, and distillation at reduced pressure also affords a solution of the *N*-oxide **6** in sulfuric acid ready for use in the nitration. The nitration is accomplished by simultaneous dropwise addition of 98% sulfuric acid and 98% nitric acid to this

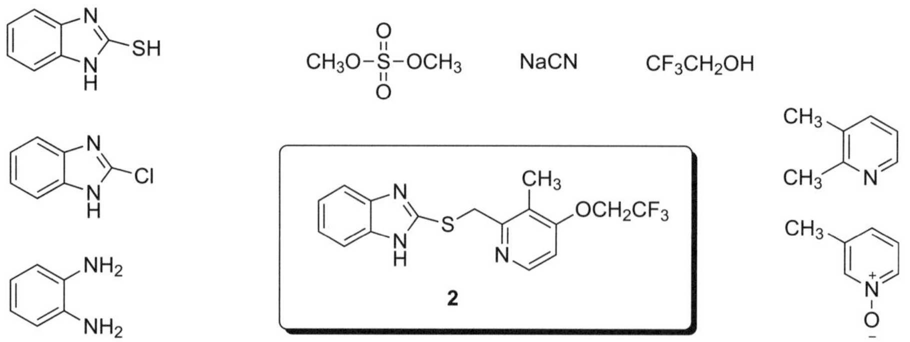


**FIGURE 7.3** Lansoprazole (**1**) building blocks.

solution at 80°C over 4 h followed by aging at 80°C for 5 h. The reaction mixture is cooled, poured into cold water, neutralized with 30% sodium hydroxide, and extracted with dichloromethane to afford a solution of 2,3-dimethyl-4-nitropyridine 1-oxide (**7**) (90% from 2,3-lutidine) (Scheme 7.1).[12,13]

Phase transfer catalysts, dipolar aprotic cosolvents, or palladium catalysts are used to facilitate nitro group displacements by 2,2,2-trifluoroethanol, 2,2,3,3,3-pentafluoropropanol, or 2,2,3,3,4,4,4,-heptafluorobutanol. The base is typically potassium carbonate or potassium *tert*-butoxide. In cases where a slow addition of potassium *tert*-butoxide is recommended, the base and 2,2,2-trifluoroethanol can be premixed and the resulting solution slowly added.

The reaction with 2,2,3,3,4,4,4-heptafluorobutanol (1.5 equivalents) and potassium carbonate (2.9 equivalents) in acetonitrile for 41 h at reflux affords the 4-(heptafluorobutoxy)pyridine 1-oxide **8** (73%) after a routine water–chloroform workup, chromatography, and recrystallization.[14] The reaction with 2,2,2-trifluoroethanol[15] (3.3 equivalents), potassium carbonate (1.9 equivalents), and benzyltributylammonium chloride (1.7 mol%) in 70% aqueous acetonitrile for 25 h at reflux affords 2,3-dimethyl-4-(2,2,2-trifluoroethoxy) pyridine 1-oxide (**9**) (78%) after a water–dichloromethane workup.[13]

The reaction with 2,2,3,3,3-pentafluoropropanol (2.6 equivalents) and potassium carbonate (2.0 equivalents) in methyl ethyl ketone–HMPA (12:1) requires 4.5 days at reflux. A routine workup followed by chromatography and recrystallization affords the 4-(2,2,3,3,3-pentafluoropropyl) pyridine 1-oxide (**10**) (74%).[14,16–19] The reaction with 2,2,3,3,3-pentafluoropropanol (3.0 equivalents) and potassium *tert*-butoxide (1.5 equivalents) can be accomplished at room temperature in pyridine. After removing the solvent at reduced pressure, chromatography and recrystallization affords **10** in lower yield (51%).[14]

Optimal results are achieved using the fluorinated alcohol as the solvent. The reaction using potassium *tert*-butoxide (1.7 equivalents) in 2,2,2-trifluoroethanol requires 42 h at 50–60°C. The suspension is filtered and the filtrate concentrated at reduced pressure. Chromatography and recrystallization of the residue affords 2,3-dimethyl-4-(2,2,2-trifluoroethoxy)pyridine 1-oxide (**9**) (94%).[14] Alternatively, the nitro group displacement using palladium

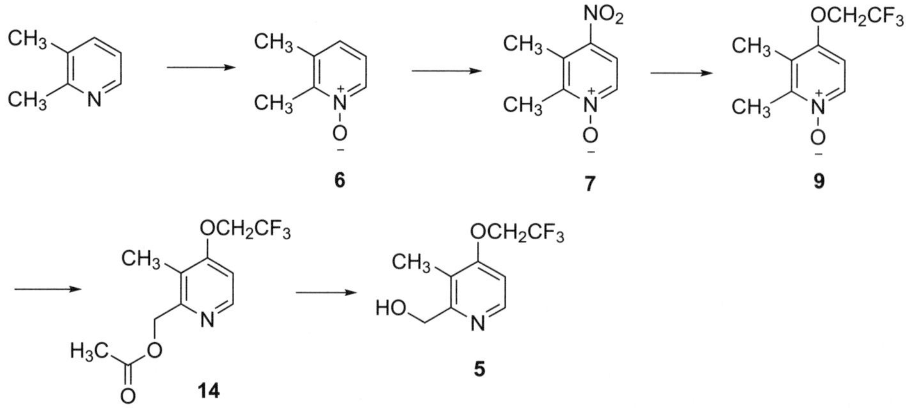


**SCHEME 7.1** Synthesis of appropriately functionalized pyridine **5** from 2,3-lutidine via pyridine *N*-oxide chemistry.

chloride (0.24 mol%) and potassium *tert*-butoxide (1.7 equivalents) in 2,2,2-trifluoroethanol is complete in just 6–8 h at 83–88°C. Distillation of the excess 2,2,2-trifluoroethanol and separation of the residue between water and dichloromethane affords a dichloromethane solution of 2,3-dimethyl-4-(2,2,2-trifluoroethoxy)pyridine 1-oxide (**9**) (95–97%).[12]

The acetylation and rearrangement are accomplished by reaction of the pyridine *N*-oxide with acetic anhydride at 90–120°C over 2–6 h. The lower reaction temperatures (90–110°C) are accessible using sulfuric acid or 4-dimethylaminopyridine (DMAP) as catalyst. The 4-(2,2,3,3-tetrafluoropropoxy)pyridine 1-oxide **11** undergoes acetylation and rearrangement in acetic anhydride using a sulfuric acid catalyst at 110°C and 4 h. After removing volatiles at reduced pressure, the acetate ester is hydrolyzed with sodium hydroxide in methanol–water at 25°C. Methanol is removed at reduced pressure and the residue extracted with ethyl acetate. The extracts are dried and concentrated at reduced pressure and the crude product purified by both chromatography and recrystallization to afford (3-methyl-4-(2,2,3,3-tetrafluoropropoxy)pyridin-2-yl)methanol (**12**) (61%).[16,17] The 4-(2,2,3,3,3-pentafluoropropoxy)pyridine 1-oxide **10** undergoes acetylation and rearrangement in acetic anhydride using a sulfuric acid catalyst at 110°C over 2 h. After removing volatiles at reduced pressure, the acetate ester is hydrolyzed with sodium hydroxide in methanol–water at 25°C. Methanol is removed at reduced pressure and the residue extracted with ethyl acetate. The extracts are dried and concentrated at reduced pressure and the crude product purified by both chromatography and recrystallization to afford (3-methyl-4-(2,2,3,3,3-pentafluoropropoxy)pyridin-2-yl)methanol **13** (64%).[18,19] The 4-(2,2,2-trifluoroethoxy)pyridine 1-oxide **9** undergoes acetylation and rearrangement in acetic anhydride using a sulfuric acid catalyst at 100–120°C over 5 h. After removing volatiles at reduced pressure, the acetate ester is hydrolyzed with sodium hydroxide in methanol–water at 25°C. Methanol is removed at reduced pressure and the residue extracted with ethyl acetate. The extracts are dried and concentrated at reduced pressure and the crude product purified by both chromatography and recrystallization to afford (3-methyl-4-(2,2,2-trifluoroethoxy)pyridin-2-yl) methanol **5** (66%).[14]

The 4-(2,2,2-trifluoroethoxy)pyridine 1-oxide **9** also undergoes acetylation and rearrangement in acetic anhydride–acetic acid (1:1) at 115°C over 6 h. The mixture is cooled, water is added, and the volatiles are removed at reduced pressure. A superior workup procedure eliminates the need for the chromatography and recrystallization. The acetate ester is hydrolyzed with sodium hydroxide in water–methanol (17:3) at 35°C. The supernatant liquid is removed, perhaps by a decantation.

> Replacing a filtration with a decantation eliminates two solid transfers from a process. Considering the time and potential operator exposure associated with solid transfers, decantation is underutilized as a solid isolation technique in pharmaceutical processes. To assess the viability of decanting a suspension in the lab, stop agitating the suspension and allow the solid to settle. If it settles quickly (<5 min) and leaves a clear supernatant, the suspension is a good candidate for decantation. Particle size is often a factor in determining how quickly a solid will settle. A large particle size solid will settle more quickly and will be less likely to blind a dipleg filter. In the lab it is convenient to use a Teflon dipleg and do a vacuum transfer. Lower the dipleg to the surface of the solid as the volume of supernatant decreases. A gas dispersion tube with porous glass frit can be used on the end of the Teflon dipleg if a clear decantate is desired. Gas dispersion tubes with glass frits with a range of porosities are available from Ace Glass.

The residual solid is dissolved in methanol and water is added to produce a precipitate. The precipitate is filtered, washed with water (decantation might be an option here as well), and dissolved in aqueous hydrochloric acid. After a pH adjustment to 3, some insoluble solids are filtered using diatomaceous earth. The filtrate is washed several times with dichloromethane, treated with carbon, diluted with ethanol, and neutralized with aqueous sodium hydroxide to precipitate (3-methyl-4-(2,2,2-trifluoroethoxy)pyridin-2-yl)methanol (5) as colorless crystals (66%).[13] While this is certainly an improvement over the previous procedure, the low yield, the reduced pressure distillation of acetic acid, and at least two solid isolations should prompt a search for a more efficient process.

An alternative process begins with addition of a dichloromethane solution of 2,3-dimethyl-4-(2,2,2-trifluoroethoxy)pyridine 1-oxide (**9**) to acetic anhydride and DMAP (1.9 mol%). The dichloromethane is distilled off and the resulting solution heated at 90–95°C for 4–5 h. The mixture is cooled and water is added, and the volatiles are removed at reduced pressure. The residual crude acetate ester **14** is an oil. Ester **14** is hydrolyzed with sodium hydroxide in methanol–water at 25–30°C. The mixture is neutralized with hydrochloric acid and (3-methyl-4-(2,2,2-trifluoroethoxy)pyridin-2-yl)methanol (**5**) is extracted with dichloromethane (89–93% contained yield from **9** by HPLC).[12]

It is informative to track the fate of the excess acetic anhydride and acetic acid in three of the processes. In the first process, excess acetic anhydride (bp 140°C, 44°C at 15 mmHg) is distilled at reduced pressure to produce a residue. Some acetic anhydride should be left in the residue to ensure it remains fluid and to keep the distillation time to a minimum. The anhydride left in the residue is then hydrolyzed during the acetate ester hydrolysis by sodium hydroxide in methanol–water. It is important to charge sufficient sodium hydroxide to convert all the acetate ester **14** to alcohol **5** and all the anhydride in the residue to acetate. After distillation at reduced pressure to remove the methanol, the product is separated from sodium acetate by extraction with ethyl acetate. In the second process, excess acetic anhydride is hydrolyzed by adding water. The resulting acetic acid (bp 118°C, 17°C at 10 mmHg) is distilled at reduced pressure. The OSHA permissible exposure limit (PEL) for acetic acid is 10 ppm (TWA) and the odor threshold is 0.21–1.0 ppm.[20] Some acetic acid should be left in the residue to ensure that it remains fluid and to keep the distillation time to a minimum. The remaining acetic acid is then converted to sodium acetate during the ester hydrolysis with sodium hydroxide in methanol–water. Again, it is important to charge sufficient sodium hydroxide to convert all acetate ester **14** to alcohol **5** and all the acetic acid to sodium acetate. The product is then separated from sodium acetate by the decantation. In the third process, excess acetic anhydride is hydrolyzed by adding water. Acetic acid is distilled at reduced pressure. Some acetic acid is left in the residue to ensure that it remains fluid and to keep the distillation time to a minimum. The remaining acetic acid is then converted to sodium acetate during the ester hydrolysis with sodium hydroxide in methanol–water. Again, it is important to charge sufficient sodium hydroxide to convert all acetate ester **14** to alcohol **5** and all the acetic acid to sodium acetate. Neutralization with hydrochloric acid converts some sodium acetate back to acetic acid. The product is separated from the neutralized mixture by a dichloromethane extraction. It is important to define the maximum amount of acetic acid (or anhydride) in the distillation residue that can be carried forward.

The reaction of a (4-alkoxypyridin-2-yl)methanol with thionyl chloride (1.1–2.1 equivalents) can be carried out in dichloromethane, chloroform, dioxane, or toluene at 25–62°C. The product, typically the hydrochloride salt, may be isolated or carried directly into the next step. The reaction of (3-methyl-4-(2,2,2-trifluoroethoxy)pyridin-2-yl)methanol (**5**) with thionyl chloride (2.1 equivalents) is complete in 20 min in refluxing chloroform. Workup involves concentration at reduced pressure to a solid residue of **3** as the hydrochloride salt (estimate 99% yield).[14,16–19] The reaction of (3-methyl-4-(2,2,2-trifluoroethoxy)pyridin-2-yl)methanol (**5**) with thionyl chloride (1.3 equivalents) is complete in 3 h in dioxane at 50°C. Workup by dilution with dichloromethane, washing with aqueous bicarbonate, and concentration at reduced pressure affords 2-(chloromethyl)-3-methyl-4-(2,2,2-trifluoroethoxy)pyridine (**3**) as the free base (91%).[21] While these procedures might be suitable for preparing small quantities in the lab, on large scale we would not want to use the suspected carcinogens chloroform or dioxane[22], to isolate a solid by concentration at reduced pressure to a solid residue, or to isolate the unstable free base.

Starting with a dichloromethane solution of (3-methyl-4-(2,2,2-trifluoroethoxy)pyridin-2-yl)methanol (**5**), a solvent exchange by distillation at atmospheric pressure affords a toluene solution. After a carbon treatment, thionyl chloride (1.1 equivalents) is added and the mixture aged for 3 h at 25°C. The hydrogen chloride and sulfur dioxide by-products are removed at reduced pressure. (The by-products might be more efficiently removed by also sparging the mixture with dry nitrogen gas.) The resulting suspension is cooled to 0°C and **4** hydrochloride salt filtered and dried (84%).[12]

Perhaps the dichloromethane–toluene exchange (and the separation of the gaseous by-products) can follow the reaction of **5** with thionyl chloride in dichloromethane at 25°C.[23]

The reaction of (3-methyl-4-(2,2,2-trifluoroethoxy)pyridin-2-yl)methanol (**5**) with thionyl chloride (1.5 equivalents) in dichloromethane is complete in 1 h at 25°C. Rather than exchange the solvents to drive off by-products and precipitate the salt, water is added to quench excess thionyl chloride. Dichloromethane is then distilled from the two-phase mixture at reduced pressure and the aqueous solution of the hydrochloride salt (estimate 99% yield) carried directly into the next step.[13] This procedure is suitable for scale-up to produce 100 kg of 2-(chloromethyl)-3-methyl-4-(2,2,2-trifluoroethoxy)pyridine (**3**) hydrochloride salt (Scheme 7.2).[24]

> The addition of water to the batch to quench excess thionyl chloride is perhaps easily monitored and controlled in the lab but adding the batch to water would be preferred in a pilot plant or plant. The two primary safety concerns are controlling the generation of heat and containing the toxic gas by-products of the quench at the start of the quench. The heat capacity $C_p$ for dichloromethane is 1.19 J/(g K). If water is added too rapidly to the batch, the batch could reach reflux temperature and sulfur dioxide, hydrogen chloride, and some dichloromethane could be carried over to the scrubber. The heat capacity $C_p$ for water is 4.18 J/(g K). The speed at which the batch can be added to water is limited only by the size of the water reservoir used.

### 7.2.2 Conversion of 2-(Chloromethyl)-3-methyl-4-(2,2,2-trifluoroethoxy)pyridine (3) to Lansoprazole Sulfide (2)

The reaction of 2-(chloromethyl)-3-methyl-4-(2,2,2-trifluoroethoxy)pyridine (**3**) with 2-mercaptobenzimidazole (1.0–1.3 equivalents) can be carried out in methanol, aqueous methanol, or acetone–*N*,*N*-dimethylacetamide. A base ($NaHCO_3$, NaOH, KOH, $NaOCH_3$, or a borohydride

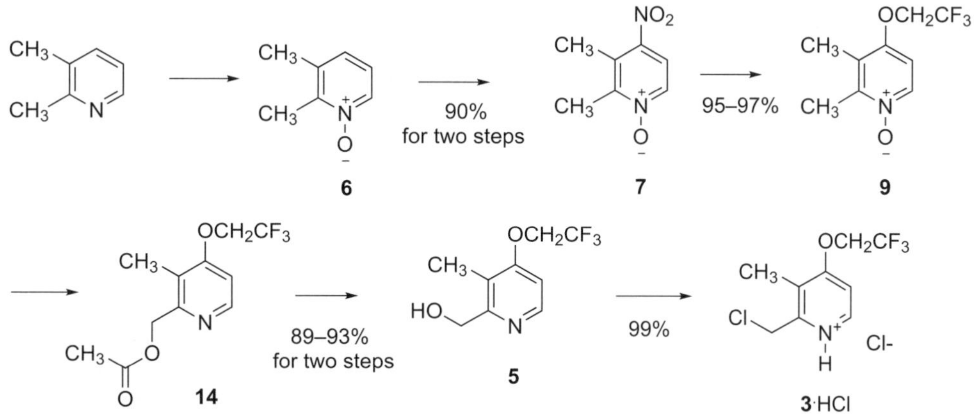

**SCHEME 7.2** Process for 2-(chloromethyl)-3-methyl-4-(2,2,2-trifluoroethoxy)pyridine (**3**).

exchange resin) is added either during the reaction or the workup. After contact with water during the reaction or the workup procedure, lansoprazole sulfide (**2**) is isolated as a monohydrate in high yield (>94%).

Reaction of 2-(chloromethyl)-3-methyl-4-(2,2,2-trifluoroethoxy)pyridine (**3**) with 2-mercaptobenzimidazole (1.1 equivalents) in methanol is complete in 1.5 h at reflux. Methanol is removed at reduced pressure. The residue is triturated with acetone–diisopropyl ether (1:5). The solid is then separated between chloroform–methanol (10:1) and aqueous sodium bicarbonate. The organic layer is dried and concentrated at reduced pressure to afford lansoprazole sulfide (**2**) (96%).[21]

Reaction of 2-(chloromethyl)-3-methyl-4-(2,2,2-trifluoroethoxy)pyridine (**3**) hydrochloride with 2-mercaptobenzimidazole (1.0 equivalent) and sodium methoxide (3.8 equivalents) is complete in 1 h in refluxing methanol. The methanol is removed at reduced pressure. The residue is separated between water and ethyl acetate and the organic layer is washed with aqueous hydroxide and concentrated at reduced pressure. Lansoprazole sulfide (**2**) is isolated from the residue by chromatography (96%).[14,16–19]

Additional 2-mercaptobenzimidazole can be used *in lieu* of an external base. The reaction of 2-(chloromethyl)-3-methyl-4-(2,2,2-trifluoroethoxy)pyridine (**3**) hydrochloride with 2-mercaptobenzimidazole (1.3 equivalents) and *N,N*-dimethylacetamide (DMA) (mol%) in acetone is complete after 5 h at reflux. The suspension is cooled and filtered. The solid, lansoprazole sulfide (**2**) hydrochloride, is suspended in aqueous acetone. Disodium EDTA (chelating agent used to sequester trace $Ca^{2+}$ and $Mg^{2+}$ ions) is added and the suspension is neutralized with aqueous potassium hydroxide (to pH 7.0–7.5) and filtered. The solid is suspended in water at 25°C. The suspension is filtered and the solid is dried to afford lansoprazole sulfide (**2**) (94%).[12]

Reaction of 2-(chloromethyl)-3-methyl-4-(2,2,2-trifluoroethoxy)pyridine (**3**) hydrochloride with 2-mercaptobenzimidazole (1.0 equivalent) and sodium hydroxide (2.6 equivalents) in methanol–water is complete in 4 h at 25°C. After diluting with water, concentrated hydrochloric acid is added (to pH 9) and the suspension is cooled to 5°C and filtered. The solid is washed, presumably with water, and dried to afford lansoprazole sulfide (**2**) (97%, >99% pure by HPLC).[25]

An aqueous solution of 2-(chloromethyl)-3-methyl-4-(2,2,2-trifluoroethoxy)pyridine (**3**) hydrochloride is added to 2-mercaptobenzimidazole (1.0 equivalent) in methanol. Aqueous 30% sodium hydroxide is added at 25°C over 1 h (to pH 11.0–11.5). The solution is aged for 30 min. After diluting with water, concentrated hydrochloric acid is added (to pH 8.5–10) and the precipitated lansoprazole sulfide (**2**) is filtered, washed, and dried (97%). This process is suitable for scale-up to produce 150 kg of lansoprazole sulfide (**2**) (Scheme 7.3).[13,24]

An alternative nonaqueous workup procedure is featured in the reaction of 2-(chloromethyl)-3-methyl-4-(2,2,2-

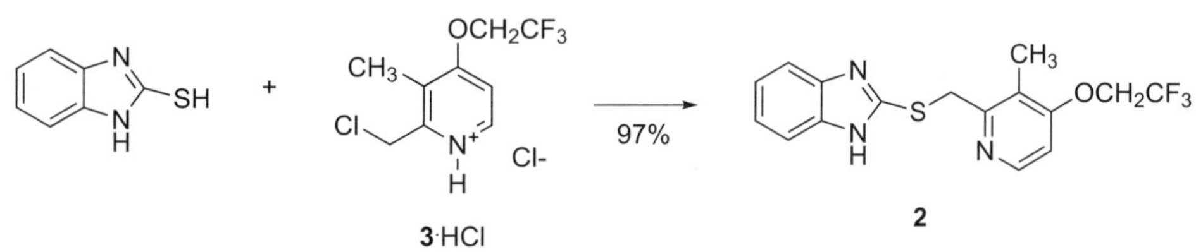

**SCHEME 7.3** Construction of lansoprazole sulfide (**2**) from 2-mercaptobenzimidazole and 2-(chloromethyl)-3-methyl-4-(2,2,2-trifluoroethoxy)pyridine (**3**).

trifluoroethoxy)pyridine (**3**) hydrochloride with 2-mercaptobenzimidazole (1.0 equivalent) and a borohydride exchange resin (1.2 equivalents) in methanol at reflux. The resin is filtered and the filtrate concentrated at reduced pressure. The residue is crystallized from ethyl acetate–hexanes to afford lansoprazole sulfide (**2**) (97%). The borohydride exchange resin is prepared by reacting an X-type resin (X = Cl, Br, CN, OAc, etc.) with an aqueous solution of sodium borohydride. Considering the same high yield (97%) is observed in the process using aqueous sodium hydroxide as base and the requirement for physical manipulation of the resin in its preparation, recovery, and recycle, this method offers no advantages.[26]

Reaction times on the order of 1 h at 25°C and routinely high yields of lansoprazole sulfide (**2**) suggest that no advantage will be gained by adding a surfactant or phase transfer catalyst.[27]

### 7.2.3 Alternative Starting Materials for Synthesis of 2-(Chloromethyl)-3-methyl-4-(2,2,2-trifluoroethoxy) pyridine (3)

While an efficient large-scale lansoprazole sulfide (**2**) process is available, perhaps it will be informative to pinpoint some of the weaknesses of the process and focus on addressing these issues in the search for a superior process. The process begins with relatively expensive 2,3-lutidine. Is there an alternative less expensive starting material?

One possible approach starts with 3-picoline[28] and introduces the 2-substituent by hydroxymethylation. In an analogous sequence, the nitration of 3,5-dimethylpyridine-*N*-oxide affords 3,5-dimethyl-4-nitropyridine 1-oxide (**15**). Nitro group displacement using potassium *tert*-butoxide (1.5 equivalents) in 2,2,3,3,3-pentafluoropropanol is complete in 18 h at 60°C. Workup by chromatography affords the 4-(2,2,3,3,3-pentafluoropropoxy)pyridine 1-oxide (**16**) (81%). *O*-Methylation with dimethyl sulfate (1.1 equivalents) followed by dilution with methanol and slow addition of ammonium persulfate at reflux affords crude (3,5-dimethyl-4-(2,2,3,3,3-pentafluoropropoxy)pyridine-2-yl)methanol (**17**) (84%).[16,17] Starting with 3,5-dimethyl-4-(2,2,2-trifluoroethoxy)pyridine 1-oxide (**18**), the same dimethyl sulfate–ammonium persulfate hydroxymethylation sequence affords crude (3,5-dimethyl-4-(2,2,2-trifluoroethoxy)pyridin-2-yl)methanol (**19**) (Scheme 7.4). After chromatography and recrystallization, the yield is just 18%. While hydroxymethylation of these symmetrical substrates can give only one product, hydroxymethylation of unsymmetrical 3-methyl-4-(trifluoroethoxy)pyridine 1-oxide (**20**), derived from 3-picoline would likely produce a mixture.[14]

3-Picoline is in fact a viable starting material. 3-Picoline is converted to lansoprazole sulfide (**2**) via 2-cyano-3-methyl-4-nitropyridine (**21**). Nitration of 3-picoline-*N*-oxide[29] affords 3-methyl-4-nitropyridine 1-oxide (**20**).[30] *O*-Methylation with dimethyl sulfate followed by Reissert– Kaufmann reaction with sodium cyanide affords 2-cyano-3-methyl-4-nitropyridine (**21**) (81%).[31] Nitro group displacement with sodium trifluoroethoxide (1.3 equivalents) (from the reaction of sodium metal with 2,2,2-trifluoroethanol) is complete in 1 h at 0°C. After a routine water–dichloromethane workup, the organic extracts are concentrated at reduced pressure to afford crude 2-cyano-3-methyl-4-(2,2,2-trifluoroethoxy) pyridine (**22**) (82%). The nitrile is hydrolyzed to the amide with sulfuric acid at 120°C and the amide hydrolyzed *in situ* by adding sodium nitrite at 25°C. After quench with ice, the pH is increased to 1.2 to precipitate the acid **23** (92%). Fisher esterification using a sulfuric acid catalyst in methanol at reflux (90%) is followed by reduction of ester **24** with sodium borohydride in methanol at 0°C. Water quench of the excess borohydride and routine water– dichloromethane workup affords 3-methyl-4-(2,2,2-trifluoroethoxy)pyridin-2-yl)methanol (**5**) (97%) (Scheme 7.5).[21]

This is a good point to discuss two important issues associated with this route and with the front-running route. During one synthesis of 3-methyl-4-nitropyridine 1-oxide (**20**) in 1989, an explosion occurred. It is suggested that the explosion resulted from carry over of some peracetic acid, used to produce the *N*-oxide, into the nitration step. We should be concerned about the potential for a similar incident in the preparation of 2,3-dimethyl-4-nitropyridine 1-oxide (**7**). Special care must also be taken in solids handling of these carcinogenic and mutagenic 4-nitropyridine 1-oxides.[32,33]

At first glance, 2,3-dimethyl-4-hydroxypyridine (**25**) might appear to be an attractive alternative starting material. Chlorination, *N*-oxide formation, and chloride displacement by 2,2,2-trifluoroethanol could lead to 2,3-dimethyl-4-(2,2,2-trifluoroethoxy)pyridine 1-oxide (**9**). The chloride

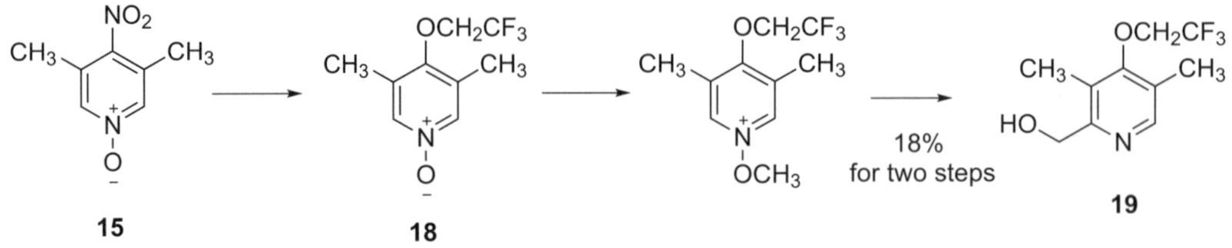


**SCHEME 7.4** (3,5-Dimethyl-4-(2,2,2-trifluoroethoxy)pyridin-2-yl)methanol (**19**) by hydroxymethylation.

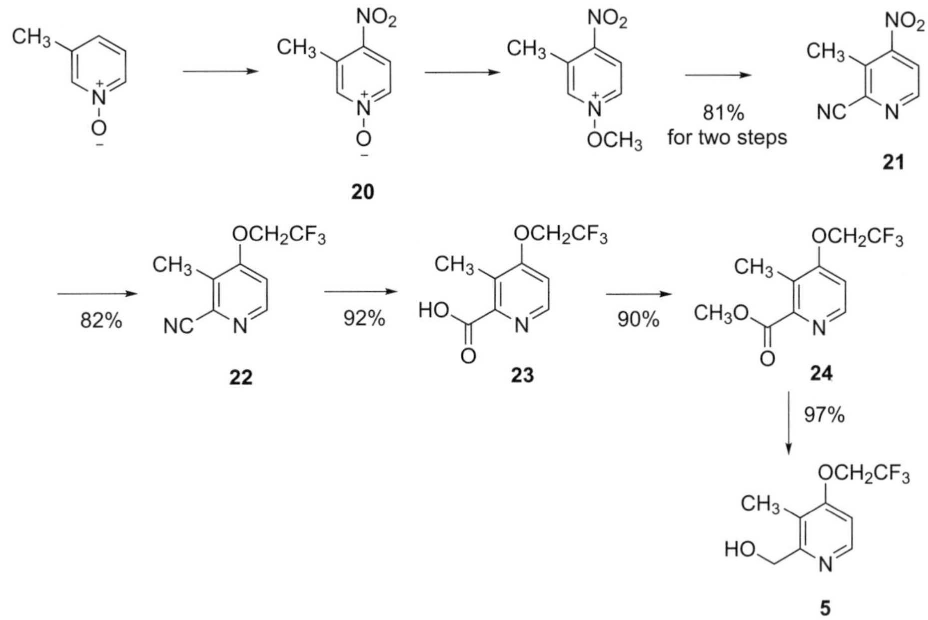


**SCHEME 7.5** (3-Methyl-4-(2,2,2-trifluoroethoxy)pyridin-2-yl)methanol (**5**) from 3-picoline-*N*-oxide.

displacement could also be moved to the end of the sequence. These approaches are not competitive since the best synthesis of the 2,3-dimethyl-4-hydroxypyridine (**25**) is from 2,3-dimethyl-4-nitropyridine (**26**) (Scheme 7.6).[18,34] [Note added in proofs.[108]]

### 7.2.4 Alternative Processes to Lansoprazole Sulfide (2) from 3-Methyl-4-(2,2,2-trifluoroethoxy)pyridin-2-yl) methanol (5) and 2-Mercaptobenzimidazole

The conversion of 3-methyl-4-(2,2,2-trifluoroethoxy)pyridin-2-yl)methanol (**5**) to the 2-(chloromethyl)pyridine **3** is accomplished in three steps, the last step using thionyl chloride. Thionyl chloride is very corrosive. Reaction with water liberates toxic hydrogen chloride and sulfur dioxide gases. Vapor contact with the skin can be severely irritating. Vapor contact with the eyes can cause severe burns and permanent eye damage. The ACGIH threshold limit value (TLV) is 1 ppm.

There are alternative processes for creating an electrophile for the reaction with 2-mercaptobenzimidazole. Processes that do not use thionyl chloride use trifluoroacetic anhydride to convert a 2-methylpyridine 1-oxide to a 2-(trifluoroacetoxymethyl)pyridine, methanesulfonic anhydride to convert a 2-methylpyridine *N*-oxide to a 2-(methanesulfonyloxymethyl)pyridine, *p*-toluenesulfonyl chloride

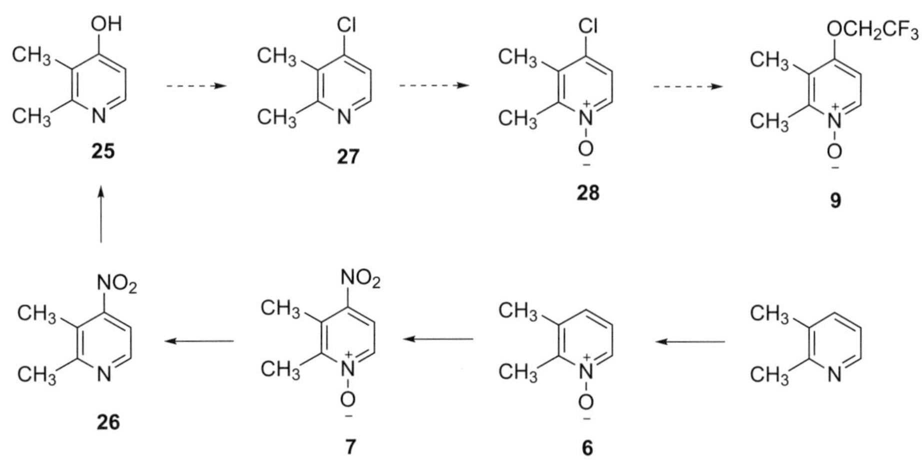


**SCHEME 7.6** A proposed route to 2,3-dimethyl-4-(2,2,2-trifluoroethoxy)pyridine 1-oxide (**9**) from 2,3-dimethyl-4-hydroxypyridine (**25**).

and triethylamine to convert a 2-methylpyridine *N*-oxide to a 2-(chloromethyl)pyridine, phosphorus tribromide to convert the pyridin-2-yl methanol to the 2-(bromomethyl)pyridine, and Mitsunobu conditions to convert 3-methyl-4-(2,2,2-trifluoroethoxy)pyridin-2-yl)methanol (**5**) directly to lansoprazole sulfide (**2**). In some of these alternative strategies, the 4-nitropyridine 1-oxide **6** is a better substrate for the Boekelheide rearrangement than the 4-(2,2,2-trifluoroethoxy) pyridine-1 oxide **9**. The nitro group displacement by 2,2,2-trifluoroethanol can be moved later in the sequence.

Lansoprazole sulfide (**2**) is produced in just 38% yield from 2,3-dimethyl-4-(2,2,2-trifluoroethoxy)pyridine 1-oxide (**9**) via the methanesulfonate. Reaction of **9** with methanesulfonic anhydride (2.1 equivalents) in 1,2-dichloroethane at reflux is followed by reaction of the methanesulfonate **30** with 2-mercaptobenzimidazole (1.0 equivalent) and triethylamine (3.3 equivalents) at 25°C. The solvent is removed at reduced pressure and lansoprazole sulfide (**2**) crystallized from the residue using ethanol–water.[35]

Reaction of 2,3-dimethyl-4-nitropyridine 1-oxide (**6**) with trifluoroacetic anhydride (1.5 equivalents) in dichloromethane is complete in 4 h at reflux. The excess trifluoroacetic anhydride (bp 40°C) and some dichloromethane are distilled at atmospheric pressure and the residue containing the trifluoromethanesulfonate **31** is reacted with 2-mercaptobenzimidazole (1.0 equivalent) and triethylamine (2.2 equivalents) in dichloromethane at 25°C. The dichloromethane is distilled at reduced pressure and the sulfide **32** crystallized from the residue using ethanol–water (43%). The nitro group of sulfide **32** is displaced with 2,2,2-trifluoroethanol (6.0 equivalents) and potassium carbonate (5.0 equivalents) in 12–15 h in refluxing acetonitrile. After filtering off the salts and concentration at reduced pressure, lansoprazole sulfide (**2**) is crystallized from the residue using acetone–water (86%).[35]

Another route to lansoprazole sulfide (**2**) begins with the reaction of 2,3-dimethyl-4-nitropyridine 1-oxide (**6**) with methanesulfonic anhydride (1.5 equivalents) in chloroforrm at reflux. The resulting suspension is cooled to 10°C and methanesulfonate ester **33** is filtered (94%). The displacement of the methanesulfonate with 2-mercaptobenzimidazole (0.72 equivalents) and triethylamine (1.61 equivalents) affords the sulfide **32** (82% yield based on 2-mercaptobenzimidazole but only 59% yield based on methanesulfonate **33**). The nitro group displacement then affords lansoprazole sulfide (**2**) (86%) (Scheme 7.7).[35]

A preferred process utilizes methanesulfonic anhydride and eliminates the isolation of the methanesulfonate ester. When the reaction of 2,3-dimethyl-4-nitropyridine 1-oxide (**6**) with methanesulfonic anhydride (1.5 equivalents) and triethylamine (0.75 equivalents) in 1,2-dichloroethane at 60–70°C is complete, the mixture is cooled and 2-mercaptobenzimidazole (1.0 equivalent) and triethylamine (2.2 equivalents) are added. The mixture is aged for 2 h at 25°C and the solvent is distilled at reduced pressure. The residue is dissolved in ethanol–water and aqueous sodium hydroxide is added (to pH 10.5–11) to precipitate sulfide **32** (72%). Sulfide **32** is then converted to lansoprazole sulfide (86%) (Scheme 7.7).[35] When the nitro group displacement follows formation of the sulfide bridge, there is the potential for a competing nitro group displacement by 2-mercaptobenzimidazole. This question cannot be addressed since no purity data for the sulfides is provided. Despite the shorter sequence and promising yields, the large-scale manufacturing potential of an approach based on methanesulfonic anhydride[36] is limited by the high cost for the anhydride.

The reaction of 2,3-dimethyl-4-(2,2,2-trifluoroethoxy)pyridine 1-oxide (**9**) with *p*-toluenesulfonyl chloride (0.95 equivalents) and triethylamine (1.15 equivalents) in toluene at 65°C is worked up by washing with aqueous sodium bicarbonate and concentration at reduced pressure. Reaction of the crude *p*-toluenesulfonate **34** with 2-mercaptobenzimidazole (0.73 equivalents) is complete in 4 h in refluxing methanol. The methanol is removed by distillation at reduced

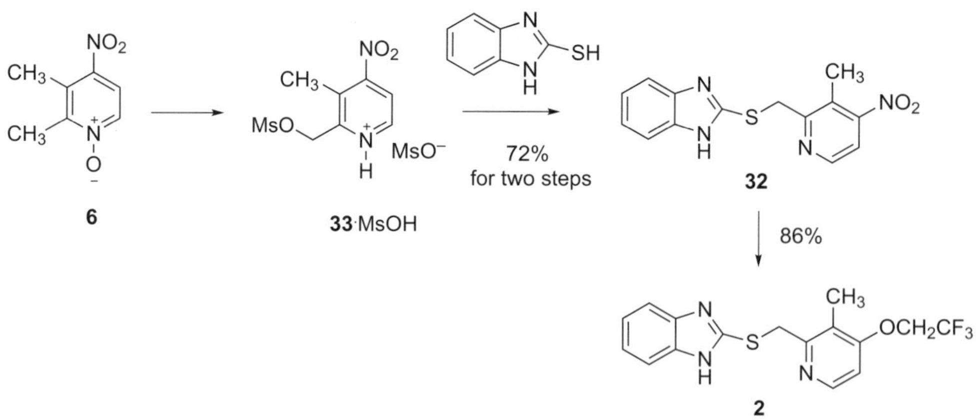


**SCHEME 7.7** Lansoprazole sulfide (**2**) from 2,3-dimethyl-4-nitropyridine 1-oxide (**6**) using methanesulfonic anhydride.

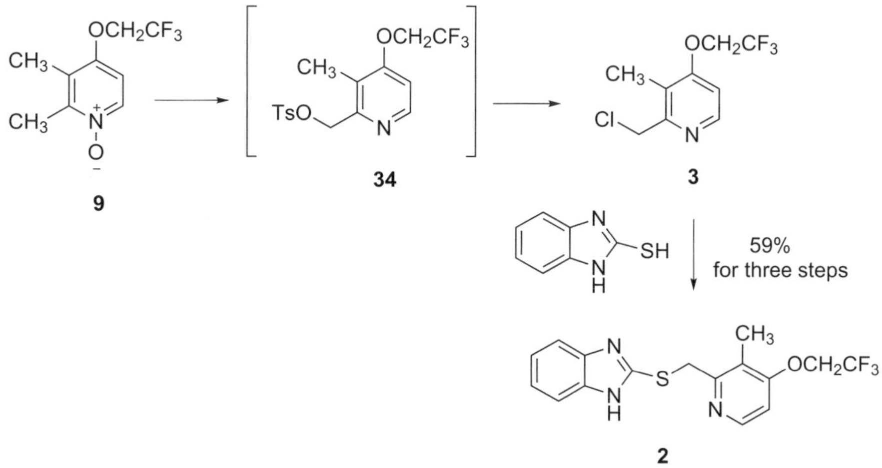


**SCHEME 7.8** Lansoprazole sulfide (**2**) from 2,3-dimethyl-4-(2,2,2-trifluoroethoxy)pyridine 1-oxide (**9**) using *p*-toluenesulfonyl chloride.

pressure. The residue is dissolved in ethyl acetate and washed with aqueous 5% sodium hydroxide. The ethyl acetate is distilled at reduced pressure and the residue chromatographed and recrystallized to afford lansoprazole sulfide (**2**) (59%) (Scheme 7.8). With the knowledge that the chloride displacement by 2-mercaptobenzimidazole in methanol is very efficient (>94%), the Boekelheide rearrangement using *p*-toluenesulfonyl chloride–triethylamine must be less efficient than the rearrangement using acetic anhydride (89–93%). A hybrid process starting with 2,3-dimethyl-4-nitropyridine 1-oxide (**6**) and *p*-toluenesulfonyl chloride–triethylamine might be more efficient and eliminate the expense associated with methanesulfonic anhydride.[37]

Thionyl chloride can also be replaced by phosphorus tribromide. Phosphorus tribromide is particularly hazardous in case of skin contact (a permeator, corrosive, and an irritant). This fuming liquid has a higher boiling point that thionyl chloride (175°C) but a vapor pressure of 10 mmHg at 48°C. Reaction of phosphorus tribromide with water liberates hydrogen bromide with an exposure limit in the UK-STEL of 3 ppm. While we are trading one "bad actor," thionyl chloride, for another, the reaction with phosphorus tribromide does offer one unique advantage: the introduction and displacement of the bromide can be carried out simultaneously. Thus, (3-methyl-4-(2,2,2-trifluoroethoxy) pyridin-2-yl)methanol (**5**) is reacted with phosphorus tribromide (1.0 equivalent), 2-mercaptobenzimidazole (1.0 equivalent), and sodium thiosulfate (0.2 equivalents) in dichloromethane at reflux. The mixture is cooled and quenched with 4 M sodium hydroxide to pH 13.5–14. (A quench into cold water followed by a pH adjustment would be preferable at scale.) The layers are separated and the organic layer dried and concentrated at reduced pressure. The residue is dissolved in ethyl acetate and precipitated with hexanes to afford lansoprazole sulfide (**2**) (93%) (Scheme 7.9).[38]

In a Mitsunobu process, diethyl azodicarboxylate (1.1 equivalents) in THF is added dropwise at 25°C to a THF solution of (3-methyl-4-(2,2,2-trifluoroethoxy)pyridin-2-yl) methanol (**5**), 2-mercaptobenzimidazole (1.0 equivalent), and triphenylphosphine (1.1 equivalents). The solvent is distilled at reduced pressure and the residue is separated between ethyl acetate and dilute hydrochloric acid. The organic layer is extracted with dilute hydrochloric acid. The combined aqueous layers are washed with ethyl ether and neutralized with aqueous sodium hydroxide to precipitate lansoprazole sulfide (**2**) (95%). While the separation of lansoprazole sulfide (**2**) from the by-products is straightforward with this workup procedure, the heat- and light-sensitivity and high cost for diethyl azodicarboxylate[39] and the cost associated with the disposal of the by-products, triphenylphosphine oxide (0.91 kg/kg of **2**) and diethyl 1,2-hydrazine dicarboxylate (0.58 kg/kg of **2**), make this Mitsunobu process unattractive for scale-up.[40]

### 7.2.5 Alternative Routes to Lansoprazole Sulfide (2) from 2-(Chloromethyl)-3-methyl-4-(2,2,2-trifluoroethoxy)pyridine (3)

The benzimidazole ring could be formed at the end of the sequence in another approach to lansoprazole sulfide (**2**). This approach would have a longer linear sequence and is likely to be less efficient. The 2-(chloromethyl)pyridine **3** is converted to the thioformic acid derivative **36** and this is condensed with a 1,2-phenylenediamine in 4 M hydrochloric acid at reflux. This approach has been demonstrated for at least 30 sulfides with a range of substitution patterns on both rings but is not described for lansoprazole sulfide (**2**). The yield in the benzimidazole ring formation is just 19% in one procedure (Scheme 7.10).[41]

The most unusual approach to lansoprazole sulfide (**2**) serves as an introduction to the unique reactivity of

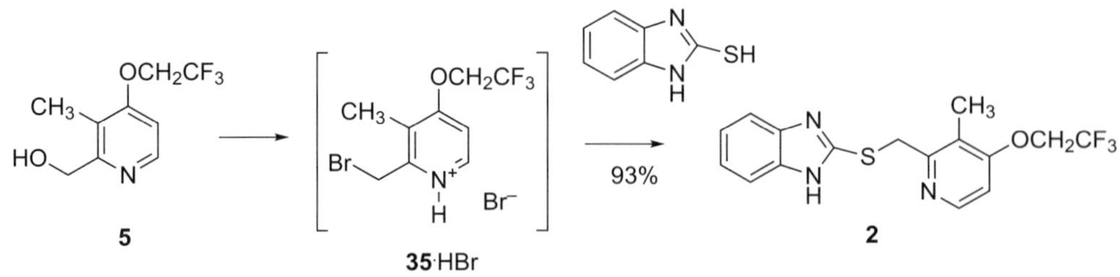

SCHEME 7.9 Lansoprazole sulfide (**2**) from (3-methyl-4-(2,2,2-trifluoroethoxy)pyridin-2-yl)methanol (**5**) using phosphorus tribromide.

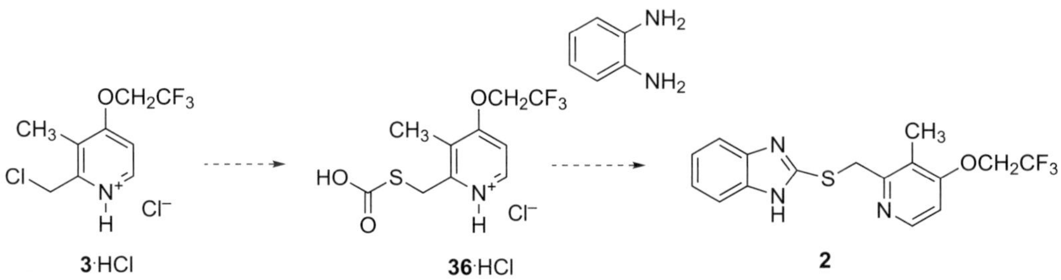

SCHEME 7.10 A potential route to lansoprazole sulfide (**2**) by benzimidazole ring formation at the end of the sequence.

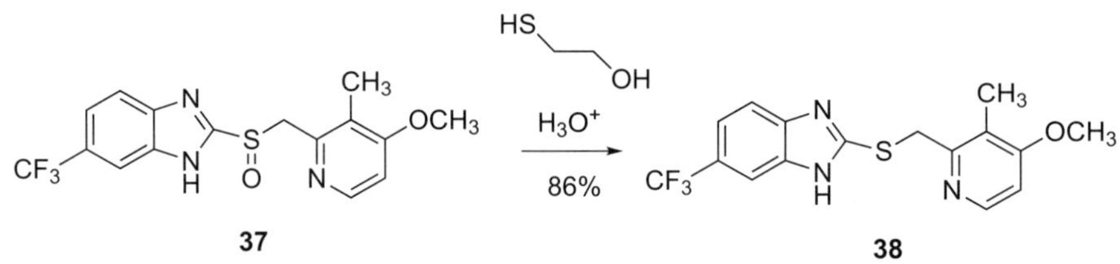

SCHEME 7.11 2-((4-Methoxy-3-methylpyridin-2-yl)methylthio)-6-(trifluoromethyl)-1*H*-benzo[*d*]imidazole (**38**) from the sulfoxide **37**.

lansoprazole (**1**) itself. The reaction of lansoprazole (**1**) with 2-mercaptoethanol in dilute hydrochloric acid is expected to produce lansoprazole sulfide (**2**) not by a simple sulfoxide reduction but rather by an acid-catalyzed rearrangement-disulfide formation and a disulfide cleavage and rearrangement back to the original skeleton. We will discuss the rearrangement chemistry further in the context of lansoprazole stability. The reaction might be worked up by neutralization and extraction with ethyl acetate. The combined extracts are then concentrated at reduced pressure and the residue crystallized from acetonitrile. Under these reaction and workup conditions, 2-((4-methoxy-3-methylpyridin-2-yl) methylthio)-6-(trifluoromethyl)-1*H*-benzo[d]imidazole (**37**) is produced from the sulfoxide **38** (86%) (Scheme 7.11).[23]

## 7.3 SYNTHESIS OF LANSOPRAZOLE (1) BY OXIDATION OF LANSOPRAZOLE SULFIDE (2)

While the selective oxidation of a sulfide to a sulfoxide is challenging,[42] there is no question that lansoprazole sulfide (**2**) is the preferred penultimate intermediate in the manufacture of lansoprazole (**1**). Methods for the oxidation have evolved over the past 10 years, with *m*-chloroperoxybenzoic acid (MCPBA) the preferred oxidant in the early work. Many new methods are now available that provide increased purity and stability of the isolated lansoprazole (**1**), reduce the oxidant cost, improve the process safety profile, and streamline the workup procedure. The more recent oxidants and oxidation methods are *m*-chloroperoxybenzoic acid, peracetic acid, sodium perborate, sodium hypochlorite, *N*-chlorosuccinimide (NCS) and *N*-bromosuccinimide (NBS), *tert*-butyl hydroperoxide (TBHP) with a vanadium catalyst, cumene hydroperoxide with a titanium catalyst, hydrogen peroxide with a vanadium, molybdenum, rhenium, or tungsten catalyst, oxaziridines, and sulfoxidation by biotransformation.

As we start the oxidation methods discussion, we should be aware that the lansoprazole sulfide (**2**) is typically charged as a monohydrate, that lansoprazole (**1**) is difficult to dry in many cases, that water-wet lansoprazole (**1**) undergoes decomposition during drying at elevated (>40°C) temperatures, and that lansoprazole (**1**) must be dried because it is unstable when stored in the presence of water. The drying and storage stability issues will be addressed later. Conditions for the oxidation are limited to slightly to strongly basic pH since lansoprazole (**1**) undergoes an acid-catalyzed rearrangement. When oxidation in aqueous solution is complete, lansoprazole (**1**) is often precipitated by adding acid to reduce the pH to 8–9.

*Safety should be of the utmost importance when designing an oxidation process.* It is advisable to minimize the oxidant inventory in the reaction vessel. This is best accomplished by selecting reaction conditions (temperature, catalyst charge, sulfide and catalyst concentrations, and solvent) that provide a rapid rate of oxidation and by metering in the oxidant at a rate matching the rate of oxidation. The best indication of a low oxidant inventory we have from published procedures will be minimal or no age time after the oxidant addition is complete. It is also advisable to quench the oxidation with an aqueous solution of a reducing agent such as sodium thiosulfate ($Na_2S_2O_3$), sodium hydrosulfite ($Na_2S_2O_4$), sodium metabisulfite ($Na_2S_2O_5$), or sodium sulfite ($Na_2SO_3$) and then test for residual oxidant before continuing the workup procedure. This is not always done in the published oxidation procedures in this section.

The decomposition of a peroxide is exothermic. When the energy released is not dissipated fast enough, the temperature increases and so does the rate of decomposition. The self-accelerating decomposition temperature, or SADT, is the storage temperature under which self-accelerating decomposition does not occur. The SADT is a function of the specific formulation of the peroxide (e.g., Luperox CU90 from Arkema, Inc. contains acetophenone (1.5%), cumene (5%), α-cumyl alcohol (6%), and cumene hydroperoxide (88%)) but also the size and shape of the packaging. The SADT for a peroxide is determined in a Heat Accumulation Storage Test. It should be emphasized that as the quantity of a peroxide stored increases, the ratio of heat transfer area to volume of peroxide and the SADT of the peroxide both decrease. While the oxidants we will discuss are available from many suppliers, we will quote an SADT from just one supplier's material safety data sheet (MSDS). The size and shape of the packaging are also quoted if provided.

### 7.3.1 *m*-Chloroperoxybenzoic Acid

Pure *m*-chloroperoxybenzoic acid may explode from heat, shock, friction, or contamination. It may ignite combustibles such as clothing, paper, and oil.[43] MCPBA is commercially available (77%) as a solid mixture with *m*-chlorobenzoic acid and water.[44] The SADT for Perkadox® MCPBA (70–75%) from AkzoNobel is 55°C.

The reaction of lansoprazole sulfide (**2**) with MCPBA can be run in dichloromethane, chloroform, or xylene–ethanol. The MCPBA is typically added as a solution at −30 to 0°C. The amount of MCPBA is 1.0–1.5 equivalents to maximize sulfide conversion and minimize sulfone formation.

The reaction of lansoprazole sulfide (**2**) with MCPBA (1.3 equivalents) is rapid at 0°C in chloroform. The mixture is washed with saturated sodium bicarbonate, dried, and concentrated at reduced pressure. The residue is chromatographed then recrystallized from acetone–ether–hexanes to afford lansoprazole (**1**) (45%).[14] This same procedure applied to the 2,2,3,3,3-pentafluoropropoxy analog **39** affords the sulfoxide **40** after crystallization from acetone–isopropyl ether (78%).[16–19]

The reaction of lansoprazole sulfide (**2**) with MCPBA (1.5 equivalents) is rapid at 5°C in chloroform. The mixture is washed with saturated sodium bicarbonate, dried, and concentrated at reduced pressure. The residue is suspended in cold ethanol. The solid is filtered and washed then dissolved in ethanol–water (9:1) at 65–70°C. The solution is polish filtered and the filtrate cooled to 5°C. The crystals produced are filtered and dried at an unspecified temperature to afford lansoprazole (**1**) (75%, 98.9% pure by HPLC) containing 0.6% of lansoprazole *N*-oxide (**41**).[45]

On a 40 kg scale, the reaction of lansoprazole sulfide (**2**) monohydrate with MCPBA (1.7 equivalents, purity not available) in chloroform at −20°C affords a mixture of lansoprazole (**1**), lansoprazole sulfide (**2**) (0.8%) and lansoprazole sulfone (**42**) (3.3%). A workup including separation of sulfone **42** by reaction with potassium carbonate and a recrystallization from ethyl acetate affords lansoprazole pure (**1**) (57%).[46]

The oxidation of lansoprazole sulfide (**2**) with MCPBA (1.0–1.3 equivalents) is complete in 1 h at −30°C in xylenes–ethanol (7:3). Aqueous potassium carbonate is added, the mixture is stirred at 35–40°C, the mixture is cooled to 0°C, and lansoprazole (**1**) is filtered, washed with cold xylene–ethanol, and dried at 35–45°C (83%). The crude lansoprazole (**1**) contains 0.27% lansoprazole *N*-oxide (**41**), 0.78% lansoprazole sulfone (**42**), and 0.09% lansoprazole sulfide (**2**). A purity upgrade is accomplished by resuspension of the crude lansoprazole (**1**) in xylenes–ethanol (7:3), addition of tetra-*n*-butylammonium hydroxide (1.0 equivalent) as a 40 wt% solution in water or methanol, age at 45–50°C, dilution with water, cooling to 0°C, and filtration and washing of the solid. The solid is dried at 40–50°C to

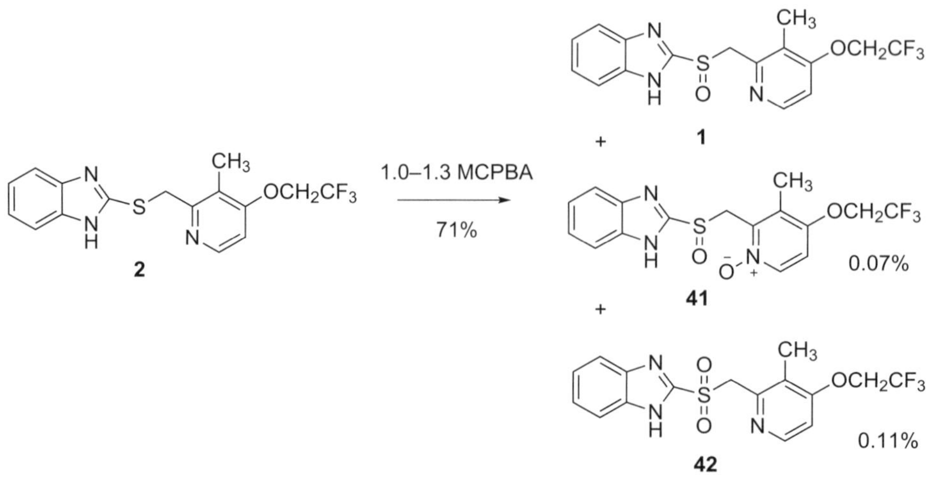


**SCHEME 7.12** Lansoprazole (**1**) by oxidation of lansoprazole sulfide (**2**) with *m*-chloroperoxybenzoic acid.

afford lansoprazole (71% from **2**) containing 0.07% lansoprazole *N*-oxide (**41**) and 0.11% lansoprazole sulfone (**42**) (Scheme 7.12).[47]

### 7.3.2 Peracetic Acid

Peracetic acid is produced by sulfuric acid-catalyzed reaction of hydrogen peroxide with glacial acetic acid. Concentrated hydrogen peroxide (70–90%) is required to produce the more concentrated solutions (>10 wt%) of peracetic acid. Peracetic acid is potentially explosive on contact with metals or in the presence of trace contaminants. The SADT for peracetic acid is 55°C.[48]

A solution of peracetic acid (1.7 equivalents) in isopropanol is added over 90 min to lansoprazole sulfide (**2**), sodium carbonate (0.49 equivalents) and DMSO (1.0 equivalent) in isopropanol at −5 to 0°C. After 3 h age at −5 to 0°C, excess peracetic acid is quenched by adding aqueous sodium thiosulfate and the pH is increased to 8–9 to precipitate lansoprazole (**1**) (57%).[49]

Alternatively, a 22.5 wt% solution of peracetic acid (unspecified equivalents) in an unspecified solvent (probably isopropanol) is added over 4–6 h to lansoprazole sulfide (**2**) in dichloromethane–isopropanol at −10 to 5°C. During the reaction, the pH is maintained at 4–7 by adding acetic acid or sodium acetate. Excess peracetic acid is quenched with aqueous sodium thiosulfate and the pH is adjusted to 8–9 with aqueous sodium hydroxide. The organic layer is separated and extracted with aqueous hydroxide. Acetic acid is added to the aqueous extract to lower the pH to 8–9 and precipitate crude lansoprazole (**1**). The solid is filtered and then dissolved in ethanol–water (9:1) containing a small amount of sodium carbonate. The mixture is heated to 60–70°C and filtered and the filtrate gradually cooled to 5°C to precipitate lansoprazole (**1**). The suspension is filtered and the solid is suspended in water and then dissolved by adding 8 wt% sodium hydroxide solution. After a carbon treatment, the pH is lowered to 8–9 by bubbling in carbon dioxide gas. The suspension is filtered and the solid is washed and dried to afford lansoprazole (**1**) (48%, 99.2% pure by HPLC, <50 ppm residual solvent).[50]

### 7.3.3 Magnesium Monoperoxyphthalate

Magnesium monoperoxyphthalate is commercially available but relatively expensive.[51] It is irritating to eyes, respiratory system, and skin. The SADT for magnesium monoperoxyphthalate is 90°C.[52]

Magnesium monoperoxyphthalate (1.1 equivalents) is added to lansoprazole sulfide (**2**) in 95% ethanol at −20°C. The mixture is aged 4 h at −20°C. Water is added and the suspension aged at 0°C. The solid is filtered, washed with cold isopropanol–water (1:1), and dried at an unspecified temperature to afford lansoprazole (**1**) (95%).[53]

### 7.3.4 Sodium Hypochlorite

The reaction of lansoprazole sulfide (**2**) with sodium hypochlorite is typically run in water–organic solvent (acetonitrile, chloroform, or dichloromethane) mixtures at 0–10°C. Hypochlorite solution concentrations ranging from 1% to 12% can be used. The more concentrated solutions (11–12%) tend to decrease in potency during storage. It would be preferable to generate these when needed or at least minimize their storage time and check potency just prior to charging. Oxidation in a two-phase mixture is accelerated by addition of a dipolar aprotic solvent or a phase transfer catalyst. In some cases, the excess oxidant is quenched with sodium metabisulfite or sodium thiosulfate to start the workup procedure.

Sodium hypochlorite solution (4.2%) (1.2 equivalents) is added over 4 h to a suspension of lansoprazole sulfide (**2**) and sodium hydroxide (unspecified equivalents) in acetonitrile–water (7:3) at 5–10°C. Excess hypochlorite is decomposed with 3% aqueous sodium metabisulfite. Acetone is added and the pH adjusted to 7.5–8.5 with acetic acid to precipitate a solid. The solid is filtered, washed with water, and dried (temperature not specified) to afford crude lansoprazole (**1**). The purity is upgraded by dissolving the solid in acetone–aqueous sodium hydroxide and then adding acetic acid (to pH 7–8). The resulting suspension is filtered and the solid is suspended in water. The suspension is filtered and the solid is dried at an unspecified temperature to afford lansoprazole (**1**) (64%).[54]

In a related procedure, a solution of sodium hydroxide (1.2 equivalents) and sodium hypochlorite (1.1 equivalents) in water is added over an unspecified amount of time to lansoprazole sulfide (**2**) in acetonitrile at 25–35°C. The mixture is then aged at 25–35°C for 2 h. Water is added, acetic acid is added (to pH 9–9.5), and the suspension is cooled to 0°C and filtered. The solid is washed with water and dried at 50°C to afford crude lansoprazole (**1**). The crude lansoprazole (**1**) is upgraded by three slurry washing and isolation procedures, one with THF–water and two with ethyl acetate, to afford lansoprazole (**1**) (35%, 99.7% pure by HPLC).[55]

A solution of sodium hypochlorite (~1 wt%) (~1.1 equivalents) is added to lansoprazole sulfide (**2**), sodium hydroxide (0.35 equivalents), and *N,N*-dimethylacetamide in chloroform at 5°C. Excess hypochlorite is quenched with sodium thiosulfate solution. After warming the mixture to 25°C, aqueous ammonium sulfate (50 wt%) is added and the layers separated. The organic layer is washed with aqueous bicarbonate. Some DMA is added and the solution cooled to 0°C to precipitate lansoprazole (**1**) (82%, 99.8% pure by HPLC).[56]

A solution of sodium hypochlorite (3.1 wt%) (0.57 equivalents) is added over 1–2 h to lansoprazole sulfide (**2**) and sodium hydroxide (2.2 equivalents) in 1:1 dichloromethane–water at 0–5°C. After aging at 10–15°C for 1–2 h, excess hypochlorite is quenched by addition of aqueous sodium thiosulfate. The layers are separated and the aqueous layer washed twice with dichloromethane to extract residual lansoprazole sulfide (**2**). The combined organic layers are washed with water. The combined aqueous layers are treated with carbon, the pH adjusted to 7.5–8.0 with acetic acid, and some acetone added to produce a suspension. The solid is filtered and washed with water. The solid is dried at 40–45°C to afford lansoprazole (**1**) (43–47%, 95.0–95.7% pure by HPLC) containing 2.2% lansoprazole sulfone (**42**). When this same procedure is repeated with a tetrabutylammonium bromide catalyst (2.2 mol%) added before the hypochlorite, lansoprazole (**1**) is isolated in higher yield (64–67%) and with higher purity (99.6% pure by HPLC with only 0.06–0.08% **42**) (Scheme 7.13). The higher yield here suggests the titer of the hypochlorite solution may have been at least 3.5 wt%. Less than stoichiometric oxidant charges suggest that the oxidation is difficult to stop at the sulfoxide. This low conversion process will only be a viable manufacturing process if the recycle of lansoprazole sulfide (**2**) is simple and there are no problematic impurities that build up in the recycle stream.[57]

### 7.3.5 *N*-Chlorosuccinimide

*N*-Chlorosuccinimide, *N*-bromosuccinimide, and other *N*-halogenated derivatives oxidize sulfides to sulfoxides with reasonable selectivity and can be used to produce lansoprazole (**1**).[42,58] A solution of NCS (1.2 equivalents) in DMF is added to lansoprazole sulfide (**2**) and sodium hydroxide (2.5 equivalents) in acetonitrile–water (7:3) at −4 to 3°C. After aging at 0°C for 1.5 h, excess oxidant is quenched with aqueous sodium thiosulfate. Water is added and the pH adjusted to 8.5 with acetic acid. Ethyl acetate and some sodium chloride are added and the layers separated. The aqueous layer is extracted with ethyl acetate. The combined organic layers are washed with dilute brine and concentrated at reduced pressure. The residue is suspended in a mixture of ethyl acetate, hexane, and toluene (volume ratio 3:7:2.5) and the suspension is filtered. The solid is washed with ethyl acetate–hexane and dried at an unspecified temperature to afford lansoprazole (**1**) (84%).[59]

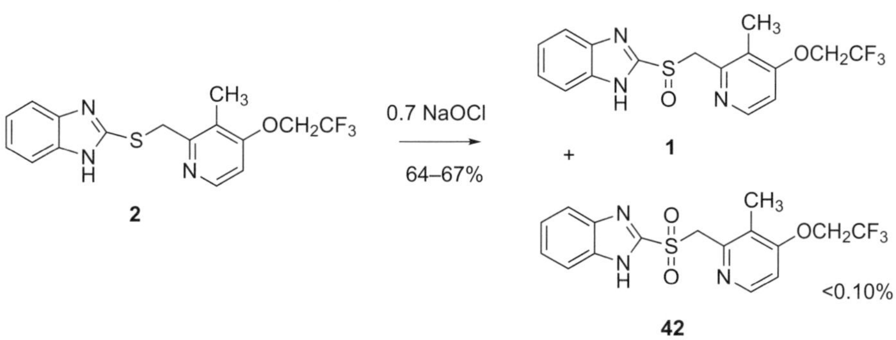


**SCHEME 7.13** Lansoprazole (**1**) by oxidation of lansoprazole sulfide (**2**) with sodium hypochlorite.

### 7.3.6 Sodium Perborate

Sodium perborate tetrahydrate is an inexpensive and relatively safe oxidant. It is a moderate irritant and has been investigated as a mutagen and reproductive effector. The SADT for sodium perborate monohydrate from Henan Hongye Chemical is 55°C (50 kg container) with oxygen release starting at 50°C.[60] Lansoprazole sulfide (**2**) can be oxidized by sodium perborate in aqueous methanol at reflux or in toluene–methanol–water at low temperature.

A solution of sodium perborate tetrahydrate (1.8 equivalents) and sodium hydroxide (1.8 equivalents) in water is added over 2 h to a refluxing solution of lansoprazole sulfide (**2**) in methanol. The methanol and water are removed at reduced pressure and the residue suspended in saturated sodium bicarbonate solution and extracted with dichloromethane. The extracts are concentrated at reduced pressure to afford lansoprazole (**1**) (92%, 90.3% pure by HPLC).[61]

A solution prepared by dissolving sodium perborate tetrahydrate (1.3 equivalents) and acetic anhydride (1.3 equivalents) in water is added dropwise over 16 min to a solution of lansoprazole sulfide (**2**) in toluene–methanol (5:1) at 0°C. After aging for 1.5 h at 0°C and 4 h at 10°C, the mixture is cooled to −15°C to produce a suspension. The suspension is filtered and the solid is washed with toluene and dried at an unspecified temperature to afford lansoprazole (**1**) (78%). With these fast reaction times at low temperature, a sodium perborate oxidation catalyzed by vanadyl acetylacetonate offers little advantage.[59]

### 7.3.7 Potassium Peroxymonosulfate

Potassium peroxymonosulfate (Oxone®) is a triple salt with the formula $2KHSO_5 \cdot KHSO_4 \cdot K_2SO_4$. It is the first neutralization salt of peroxymonosulfuric acid (Caro's acid). Since it is both acidic and an oxidizer, potassium peroxymonosulfate is very hazardous in case of skin or eye contact (irritant and corrosive). Slow decomposition liberating oxygen and releasing heat must be taken into account when designing storage facilities for large quantities. Decomposition associated with high temperature may generate sulfur dioxide, sulfur trioxide, and sulfuric acid.[62] The SADT for potassium monoperoxysulfate from Hansa is 80°C. Even small amounts of moisture can significantly lower the SADT.[63]

Potassium monoperoxysulfate (0.67 equivalents) is added to lansoprazole sulfide (**2**) and sodium bicarbonate (4.2 equivalents) in aqueous methanol at 0°C. After aging 4 h at 0°C, more potassium monoperoxysulfate (0.19 equivalents for a total of 0.86 equivalents) is added and the mixture aged 1.5 h at 0°C. Any excess persulfate is quenched with aqueous sodium metabisulfite. The solid is filtered, washed with water and methanol–water (1:1), and dried at an unspecified temperature to afford lansoprazole (**1**) (yield not available but probably 80–85%, 98.1% pure). To produce one 1 kg of lansoprazole (**1**) we will generate an aqueous waste stream containing at least 1.3 kg of sulfate/bisulfate.[64]

### 7.3.8 Sodium Percarbonate with a Molybdenum Catalyst

Sodium percarbonate is produced by reaction of hydrogen peroxide with sodium carbonate. It is very hazardous in case of skin contact (an irritant and perhaps a sensitizer) or eye contact. The SADT for Oxyper® sodium percarbonate from Solvay Chemicals is 50°C. Sulfide oxidation with sodium percarbonate is catalyzed by ammonium molybdate ($(NH_4)_2MoO_4$).

Sodium percarbonate (1.1 equivalents) is added to lansoprazole sulfide (**2**) and catalytic ammonium molybdate (5.4 mol%) in methanol at 10°C. After aging for 15 h at 10°C, water is added and the pH is adjusted to 10 with acetic acid. The suspension is filtered and the solid washed with water and dried at 40°C to afford lansoprazole (**1**) (90%). The long age time suggests there is a large inventory of percarbonate in the reaction vessel in the first few h at 10°C. Slow addition of a liquid oxidant or oxidant solution would be more easily controlled and have less potential for operator exposure compared to slow addition of solid sodium percarbonate or any other solid oxidant.[65]

> The solid oxidants (MCPBA, monoperoxyphthalate, percarbonate, perborate, monoperoxysulfate) are added slowly to control the reaction temperature and minimize the oxidant inventory in the reactor. Solid additions in the lab on small scale might be done by simply opening the reactor (in a well-ventilated hood) and adding the oxidant in several equal portions. Solid additions in the lab can also be done in a continuous mode using a powder addition funnel with a pressure equilibration side arm and an auger. The physical properties of the solid oxidant will be critical in a continuous feed. If the solid sticks to the walls, a significant amount of the oxidant will be left in the charging funnel. If the solid packs at the bottleneck, the flow may stop altogether. If the solid is hygroscopic, the acceptable flow properties at the start of the feed may deteriorate as the feed progresses. In a pilot plant or plant, charging solids in portions increases the risk of operator exposure and a continuous solid feed will likely encounter the same sticking and packing problems seen in the lab.

### 7.3.9 *tert*-Butyl Hydroperoxide with a Vanadium Catalyst

*tert*-Butyl hydroperoxide is available as a solution in a hydrocarbon solvent (decane or nonane) or as a 70% solution in water. It is hazardous in case of skin contact (corrosive and irritant), eye contact (corrosive), or if ingested or inhaled (lung corrosive). The SADT for T-Hydro® (70 wt% *tert*-butylhydroperoxide in water) from Lyondell Chemical Company is 88°C.

A solution of vanadium(V) oxytriisopropoxide is prepared from vanadium(V) oxychloride and isopropanol. A solution of 70% TBHP in water (1.5 equivalents) is added to lansoprazole sulfide (**2**) and vanadium(V) oxytriisopropoxide (1.0 mol%) in isopropanol at 10°C. After aging at 10°C for 6 h, excess oxidant is quenched with aqueous sodium sulfite (0.7 equivalents) over 1 h. The suspension is filtered and the solid is washed with water and dried at 40°C and 10 mmHg to afford lansoprazole (**1**) (83% based on purity of 86%). The low purity may result from a rearrangement of lansoprazole (**1**) catalyzed by hydrogen chloride, the by-product of vanadium(V) oxytriisopropoxide preparation. When this acid is neutralized with diethylamine before adding the oxidant, lansoprazole (**1**) is isolated in 90% yield. The oxidant inventory is high throughout since excess TBHP is used (hence the large amount of quench reagent) and a long age time is required.[66]

Vanadyl acetylacetonate ($VO(acac)_2$) is a preferred catalyst for lansoprazole sulfide (**2**) oxidation by TBHP. A solution of 70% TBHP in water (1.9 equivalents) is added to lansoprazole sulfide (**2**) and vanadyl acetylacetonate (0.6 mol%) in ethanol at 16°C. After aging for 3 h, excess oxidant is quenched with aqueous sodium metabisulfite. The solid is filtered, washed with cold ethyl acetate, and dried at an unspecified temperature to afford lansoprazole (**1**) (79%). In a 200 g scale lab procedure, a solution of 70% TBHP in water (1.5 equivalents) is added to lansoprazole sulfide (**2**) and vanadyl acetylacetonate (2.0 mol%) in 95% ethanol at 5°C. After aging for 6 h, excess oxidant is quenched with aqueous sodium sulfite (0.56 equivalents) over 17 h at 25°C. The suspension is cooled to 5°C and the solid filtered and dried at an unspecified temperature to afford lansoprazole (**1**) (85%) containing 0.15 wt% lansoprazole sulfone (**42**) and 0.3 wt% lansoprazole sulfide (**2**).[64,67]

In the procedures with a vanadyl acetylacetonate or vanadium(V) oxytriisopropoxide catalyst, the oxidant inventory is high throughout since excess TBHP is used and several hours of age time is required. After the lower selectivities in oxidations with peracids or hypochlorite, the selectivities seen here are encouraging. The inherently high selectivity is perhaps enhanced by precipitation of lansoprazole (**1**) as the oxidation proceeds. Concerns about handling highly toxic vanadium catalysts and disposing of aqueous waste streams containing vanadium will be addressed later.

### 7.3.10 Cumene Hydroperoxide with a Titanium Catalyst

In the context of manufacturing lansoprazole (**1**), presenting the asymmetric oxidation of lansoprazole sulfide might be considered going off on a tangent. In this case, the tangent is directly relevant to manufacture of the "next-generation" proton pump inhibitor Kapidex® (dexlanasoprazole) (**43**) and may provide new insights that are relevant to the problem at hand. While there are many promising methods for asymmetric sulfide oxidation,[68] the available lansoprazole literature offers four: oxidation with cumene hydroperoxide and chiral titanium alcoholate catalyst, oxidation with a hydrogen peroxide and a chiral tungsten alkaloid catalyst, oxidation with a cholesterol-derived oxaziridinium salt, and microbiological oxidation. We will discuss each of these oxidation methods up to the isolation of the crude (*R*)-lansoprazole **43**.

Cumene hydroperoxide, produced as an intermediate in the cumene process for converting benzene and propylene to phenol and acetone, is commercially available and inexpensive. Cumene hydroperoxide has a skin contact effect and is designated as a skin sensitizer. With a WEEL skin exposure limit of just 1 ppm (TWA), cutaneous absorption should be carefully monitored. The SADT for Luperox CU90 (88% cumene hydroperoxide) from Arkema is 82°C for a 5 gal container.[69,70]

A modified Sharpless reagent for asymmetric epoxidation catalyzes the asymmetric oxidation of prochiral sulfides to sulfoxides. The reagent can be produced from titanium(IV) isopropoxide, diethyl tartrate, water, and *tert*-butyl hydroperoxide (1:2:1:1). Higher enantioselectivities are observed using cumene hydroperoxide in place of *tert*-butyl hydroperoxide (Scheme 7.14).[71,72]

The asymmetric oxidation of lansoprazole sulfide (**2**) by cumene hydroperoxide can be accomplished with a full

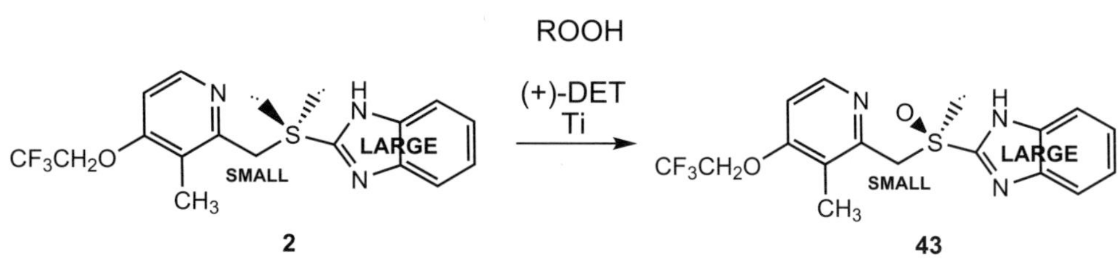


**SCHEME 7.14** Enantioselectivity observed in the oxidation of prochiral sulfide **2** by cumene hydroperoxide using a water-modified titanium (+)-diethyl tartrate catalyst.

equivalent of catalyst and near-stoichiometric quantities of oxidant or with 10–20 mol% catalyst and an excess (2–5 equivalents) of oxidant. Oxidation at lower temperatures (0°C rather than 25°C) results in less conversion to lansoprazole sulfone (**42**) and higher enantioselectivity. A solution of lansoprazole sulfide (**2**), tertiary amine, and the sulfoxidation catalyst is prepared by heating a solution of the sulfide, toluene, water (60 mol%), (−)-diethyl tartrate (250 mol%), and titanium(IV) isopropoxide (100 mol%) at 50°C for 1 h. The solution is cooled to 25°C and diisopropylethylamine (100 mol%) and cumene hydroperoxide (1.0 equivalent) are added. The mixture is stirred at 25°C for 16 h. The HPLC completion check reveals 11% lansoprazole sulfide (**2**), 7% lansoprazole sulfone (**42**), and 78% lansoprazole (**1**). Toluene is added and the mixture is extracted with 12% aqueous ammonia three times. The combined extracts are neutralized with acetic acid. Extraction with an unspecified organic solvent, concentration of the extracts at reduced pressure, and chromatography of the residue affords (*S*)-lansoprazole (**44**) (52%, 99.9% pure by HPLC) with 55% ee. The parallel procedure using (+)-diethyl tartrate affords (*R*)-lansoprazole (**43**) (37%, 99.9% pure by HPLC) with 46% ee.[73]

Superior chemoselectivity and enantioselectivity are achieved by reducing the catalyst charge, increasing the oxidant charge, and lowering the reaction temperature. A solution of lansoprazole sulfide (**2**), tertiary amine, and the sulfoxidation catalyst is prepared by heating sulfide **2** (50 g), toluene, water (12 mol%) and (+)-diethyl tartrate (44 mol %) at 50–55°C for 30 min, adding titanium(IV) isopropoxide (20 mol%), aging at 50–55°C for 1 h, and then adding diisopropylethylamine (34 mol%). The mixture is cooled to −10°C and cumene hydroperoxide (3.1 equivalents) is added. The mixture is stirred at −10 to 10°C for 4.5 h. A completion check by HPLC shows 0.74% lansoprazole sulfide (**2**), 1.46% lansoprazole sulfone (**42**), and (*R*)-lansoprazole (**43**) with 96.5% ee. Excess oxidant is quenched with aqueous sodium thiosulfate (30%) (2.7 equivalents) and the layers are separated. Water, heptane–diisopropyl ether (1:2) and heptane are added to the organic layer and the suspension is aged for 2 h and filtered. The solid is washed with toluene–diisopropyl ether (1:4) and dried at an unspecified temperature to afford (*R*)-lansoprazole (**43**) (98%, 100% ee) containing no detectable lansoprazole sulfide (**2**) and 1.5% lansoprazole sulfone (**42**). Comparable results are observed in the oxidation using a lower catalyst charge (5 mol% water, 22 mol% diethyl tartrate, and 10 mol% titanium(IV) isopropoxide). Comparable results are observed using 1.9 or 5.0 equivalents of cumene hydroperoxide or using ethyl acetate in place of toluene (Scheme 7.15).[74–76]

A 5 kg scale oxidation using 5.9 mol% water, 22 mol% (+)-diethyl tartrate, 10 mol% titanium(IV) isopropoxide and 3.0 equivalents of cumene hydroperoxide is nearly complete in 3 h at −8 to 2°C (1.0% of **2**, 1.7% of **42**, and **43** with 96.9% ee in the completion check). Excess oxidant is quenched with aqueous sodium thiosulfate (30%) and the layers are separated. The organic layer is concentrated to a smaller volume by distillation at reduced pressure then diluted with heptane–methyl *tert*-butyl ether (1:1) and additional heptane at 0°C. The suspension is filtered and the solid is washed with toluene–methyl *tert*-butyl ether (1:4) to afford (*R*)-lansoprazole (**43**) (no yield available, 98.3% ee) containing 0.45% lansoprazole sulfide (**2**) and 1.8% lansoprazole sulfone (**42**). A similar 4.5 kg scale run using 12 mol% water, 44 mol% (+)-diethyl tartrate, 20 mol% titanium(IV) isopropoxide and 3.0 equivalents of cumene hydroperoxide affords (*R*)-lansoprazole (**43**) (100% ee) containing no detectable lansoprazole sulfide (**2**) and 0.9% lansoprazole sulfone (**42**). After several crystallizations, (*R*)-lansoprazole (**43**) is recovered in 69% yield with 100% optical purity.[75–77]

The oxidant inventory in the reactor is typically high throughout since an excess (2.0–5.0 equivalents) of cumene hydroperoxide is used and several hours of age time are required after all the cumene hydroperoxide is added. The inherently high selectivity at 0°C makes it possible to achieve complete conversion with <2% lansoprazole sulfone (**42**) formation.

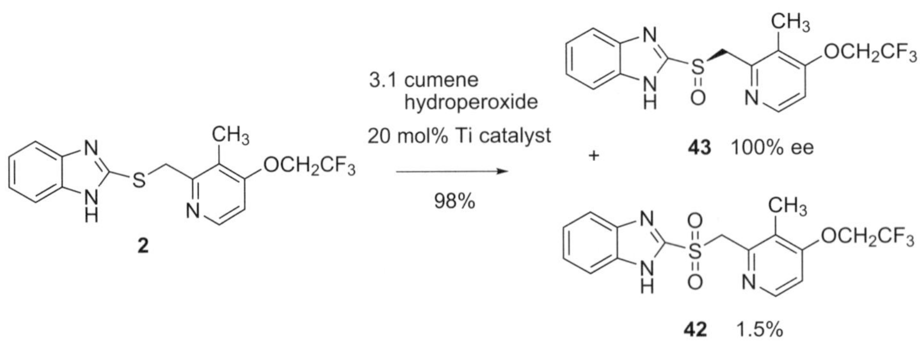


**SCHEME 7.15** (*R*)-Lansoprazole (**43**) by oxidation with cumene hydroperoxide using a water-modified titanium (+)-diethyl tartrate catalyst.

The careful attention to detail in defining the amount of water present in all these oxidations suggests that the amount of water is critical to the success of the oxidation. Controlling a small water charge in a lab procedure is relatively straightforward: we can analyze for water in the reagents and solvents and the glassware can be carefully oven dried. While we can certainly dry a large reactor for this process by charging toluene, distilling the toluene–-water azeotrope, and analyzing the solvent left in the pot for water content, a fixed reactor in a pilot plant or plant typically has valves and piping. Elimination of all traces of water from valves and piping will be both time- and labor-intensive.

### 7.3.11 Hydrogen Peroxide with a Catalyst

The oxidation of lansoprazole sulfide (**2**) with hydrogen peroxide can be catalyzed by complexes of rhenium, molybdenum, tungsten, and vanadium. Hydrogen peroxide is both corrosive and an irritant. The TLV for hydrogen peroxide is 1 ppm TWA (TLV ACGIH-USA, OSHA PEL, and NIOSH REL). The SADT for aqueous solutions of hydrogen peroxide 20–60 wt% from Solvay Chemicals is 60°C.

### 7.3.12 Methyltrioxorhenium

Methyltrioxorhenium(VII) (MTO) is commercially available but expensive.[78] Catalyst loadings for complete conversion at 5°C are as low as 0.05 mol%. It is suggested that MTO activates peroxide toward a nucleophilic attack by lansoprazole sulfide (**2**). At catalyst loadings of as low as 0.05 mol%, the oxidation goes to completion at 5°C in 4 h using 1.2 equivalents of hydrogen peroxide. When ethanol or methanol is used as the solvent, lansoprazole (**1**) precipitates and is conveniently isolated during the workup procedure. Since MTO is unstable under alkaline conditions and lansoprazole (**1**) is unstable under acidic conditions, the oxidation is best run under neutral conditions.[79]

Aqueous 30% hydrogen peroxide (1.0 equivalent) is added to lansoprazole sulfide (**2**) and methyltrioxorhenium (4.1 mol%) in 95% ethanol at −20 to −30°C. After aging 5 h, excess oxidant is quenched with aqueous sodium thiosulfate. Isopropanol is added and the suspension is cooled to 0°C and filtered. The solid is washed with isopropanol–water (1:1) and dried at an unspecified temperature to afford lansoprazole (**1**) (94%, 99.5% pure by HPLC) containing 0.05% lansoprazole *N*-oxide (**41**). The yield and purity are comparable using 0.5 mol% catalyst (90% yield, 99.8% pure by HPLC).[80]

Complete conversion and high selectivity are also possible with a lower catalyst loading at a higher temperature (5°C). Aqueous 33% hydrogen peroxide (1.2 equivalents) is added to lansoprazole sulfide (**2**) and methyltrioxorhenium (0.10 mol%) in methanol at 5°C. The mixture is aged 4 h at 5°C. Water is added and the 5°C suspension aged for 1 h. The solid is filtered and washed with water. Recrystallization of the wet solid from 9:1 ethanol–water affords lansoprazole (**1**) (75%, >99.5% pure by HPLC). Lansoprazole (**1**) with the same yield and purity is also produced using just 0.05 mol% methyltrioxorhenium. Assuming a 90% lansoprazole yield can be achieved at 0.05 mol% catalyst loading, 360 mg of methyltrioxorhenium, at a cost of $35–40, is required to produce 1 kg of lansoprazole (**1**).[25]

### 7.3.13 Vanadium(V) Oxide, Sodium Metavanadate, Vanadium(IV) Acetylacetonate

Vanadium(V) oxide ($V_2O_5$), sodium metavanadate ($NaVO_3$), and vanadium(IV) acetylacetonate catalyze the oxidation of lansoprazole sulfide (**2**) by hydrogen peroxide. These catalysts are all commercially available and relatively inexpensive. With catalyst loadings of 0.2–0.6 mol%, the oxidation goes to completion at 25°C in <8 h using 1.2 equivalents of hydrogen peroxide. When ethanol is used as the solvent, lansoprazole (**1**) precipitates and is conveniently isolated during the workup procedure. A vanadium catalyst with a tartrate ligand catalyzes the oxidation of lansoprazole sulfide (**2**) to (*S*)-lansoprazole (**44**).[81]

Vanadium(V) oxide is considered to be a carcinogen.[82] All vanadium compounds should be considered to be toxic.[83] The toxicity depends on the valence state and the solubility of the compound. For example, vanadium(V) oxide ($V_2O_5$) is considered to be five times as toxic as vanadium(II) oxide ($V_2O_3$). The first concern in handling these vanadium catalysts is exposure to dust. For vanadium(V) oxide, the OSHA permissible exposure limit for vanadium respirable dust is 0.5 mg/m$^3$ (ceiling) and for vanadium fume is 0.1 mg/m$^3$ (ceiling), and the ACGIH threshold limit value is 0.05 mg/m$^3$. Perhaps a concern about dust exposure prompted the *in situ* preparation of vanadium oxytriisopropoxide from (liquid) vanadium oxytrichloride and isopropanol in an oxidation with *tert*-butyl hydroperoxide discussed earlier.[66]

A solution of hydrogen peroxide (1.2 equivalents) and vanadium(V) oxide (0.6 mol%) in *tert*-butanol is added to lansoprazole sulfide (**2**) in dichloromethane at 20–25°C. After aging for 1 h, excess oxidant is quenched with aqueous sodium thiosulfate. The layers are separated, and the organic layer washed with water and concentrated at reduced pressure. The residue is taken up in cold ethanol–water (9:1) and the suspension is filtered. The solid is washed, presumably with cold ethanol–water (9:1), and then dissolved in ethanol–water (9:1) at 65–70°C. The solution is polish filtered and the filtrate is cooled to 5°C. The suspension is filtered and the solid is dried at an unspecified temperature to afford lansoprazole (**1**) (93%, 99.3% pure by HPLC).[45]

A solution of 35% hydrogen peroxide (1.2 equivalents) and sodium metavanadate (0.5 mol%) is added to lansoprazole sulfide (**2**) in ethanol at 20–25°C. After aging for 8 h, excess oxidant is quenched with aqueous sodium thiosulfate. The suspension is filtered and the solid is washed, presumably with cold ethanol–water (9:1), and then dissolved in ethanol–water (9:1) at 65–70°C. The solution is polish filtered and the filtrate is cooled to 5°C. The suspension is filtered and the solid is dried at an unspecified temperature to afford lansoprazole (**1**) (91%, 99.6% pure by HPLC).[45]

A solution of 35% hydrogen peroxide (1.2 equivalents) and vanadium(V) oxide (0.3 mol%) is added to lansoprazole sulfide (**2**) in ethanol at 20–25°C. After aging for 2.5 h, excess oxidant is quenched with aqueous sodium thiosulfate. The suspension is filtered and the solid is washed, presumably with cold ethanol–water (9:1), and then dissolved in ethanol–water (9:1) at 70–80°C. The solution is polish filtered and the filtrate is cooled to 5°C. The suspension is filtered and the solid is dried at an unspecified temperature to afford lansoprazole (**1**) (90%, 99.6% pure by HPLC) (Scheme 7.16).[45]

A solution of 35% hydrogen peroxide (1.2 equivalents) is added to lansoprazole sulfide (**2**) and vanadium(IV) acetylacetonate (0.2 mol%) in ethanol at 20–25°C. After aging for 5 h, excess oxidant is quenched with aqueous sodium thiosulfate. The suspension is filtered and the solid is washed, presumably with cold ethanol–water (9:1), and then dissolved in ethanol–water (9:1) at 60–70°C. The solution is polish filtered and the filtrate is cooled to 5°C. The suspension is filtered and the solid is dried at an unspecified temperature to afford lansoprazole (**1**) (91%, 99.7% pure by HPLC). In all four cases, the level of lansoprazole *N*-oxide was 0.1% or less. Assuming a 90% lansoprazole yield at 0.3 mol% catalyst loading, 1.6 g of vanadium(V) oxide, at a cost of $0.15–0.20, is required to produce 1 kg of lansoprazole (**1**).[45]

The workup procedures all generate an aqueous waste stream containing the vanadium catalyst. The cost for producing lansoprazole (**1**) by a vanadium-catalyzed oxidation should include the cost for aqueous waste treatment to remove the catalyst. Four from the many available treatment options are offered to illustrate that removing vanadium from an aqueous waste stream is feasible. The best treatment option is likely to be site specific. Zinc chloride-activated carbon from coconut coir pith is effective for the removal of vanadium(V) from water.[84] Commercially available metal oxide adsorbents GTO (Dow), E-33 (Seven Trents), and GFH (U.S. Filter) also remove vanadium, with GFH having the highest adsorption capacity.[85] Sodium carbonate softening at pH 10.3 removes vanadium from drinking water.[86] Finally, a metal sludge waste from the electroplating industry adsorbs vanadium with a capacity of 24.8 mg/g at 25°C. The metal-laden sludge is then used to produce cement.[87]

In all cases, the chemoselectivity is high and the catalyst loading is low. A low catalyst loading minimizes the potential for operator exposure during vanadium solids handling and minimizes the cost for vanadium removal from the aqueous waste stream. The age times (2.5–8 h) after all the oxidant is added suggest that the trade-off for using a low catalyst loading is use of a high oxidant inventory at the start to maintain an acceptable oxidation rate in ethanol at 20–25°C.

### 7.3.14 Tungsten(VI) Oxide

A heterogeneous catalyst produced from tungsten(VI) oxide, hydrogen peroxide, and a cinchona alkaloid-based ligand mediates an asymmetric oxidation of lansoprazole sulfide (**2**). Aqueous (30%) hydrogen peroxide (1.1 equivalents) is added to sulfide **2**, tungsten(VI) oxide (5 mol%) and $(DHQD)_2$-PYR (10 mol%) at 0°C in THF. After aging for 50 h, the catalyst is filtered off. The filtrate is diluted with ethyl acetate, washed with water and brine, dried and concentrated at reduced pressure. The residue is chromatographed to afford (*R*)-lansoprazole (**43**) (84%) with 88% ee. The long reaction time suggests a high oxidant inventory is present at the start of the reaction (Scheme 7.17).[88]

### 7.3.15 Other Catalysts

A solution of 36% hydrogen peroxide (1.1 equivalents) in *tert*-butanol is added to a solution of lansoprazole sulfide (**2**) and benzeneseleninic acid in dichloromethane at 10°C. The mixture is aged at 15–20°C for 5 h. Excess oxidant is quenched with aqueous sodium thiosulfate. The organic layer is separated, washed with water, dried, and concentrated at reduced pressure. The residue is crystallized from ethanol to afford lansoprazole (**1**) (95%).[38]

The oxidation of lansoprazole sulfide (**2**) by hydrogen peroxide and *tert*-butyl hydroperoxide can also be catalyzed by copper and iron.[89]

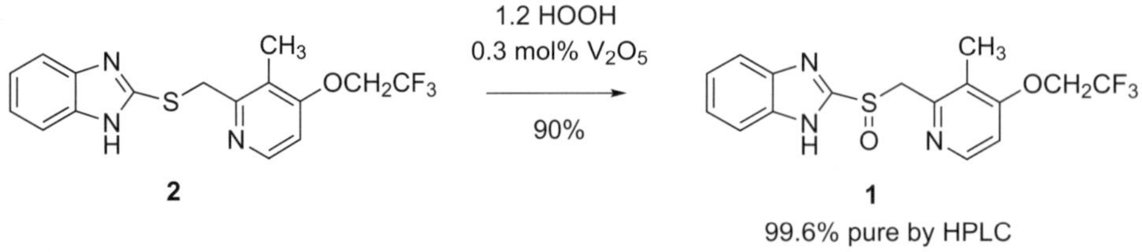


**SCHEME 7.16** Lansoprazole (**1**) by oxidation of lansoprazole sulfide (**2**) with hydrogen peroxide and a vanadium catalyst.

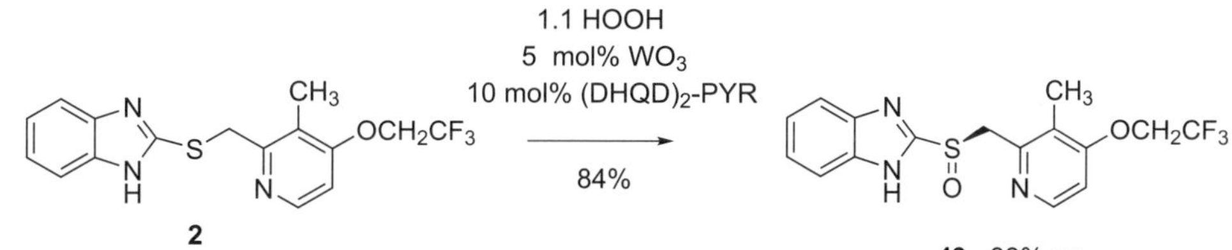


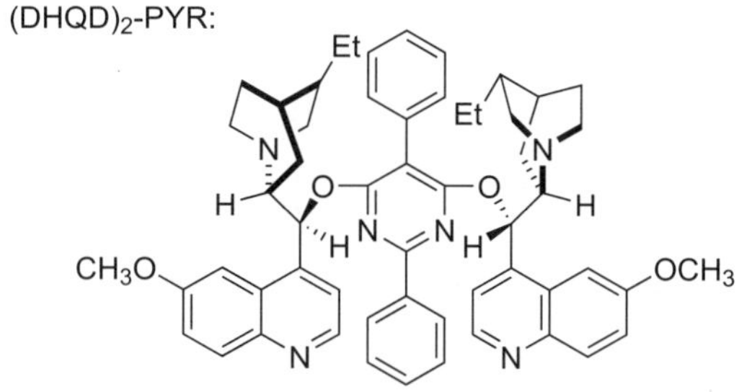


**SCHEME 7.17** (*R*)-Lansoprazole (**43**) by oxidation of lansoprazole sulfide (**2**) with hydrogen peroxide using a heterogeneous tungsten–cinchona alkaloid complex.

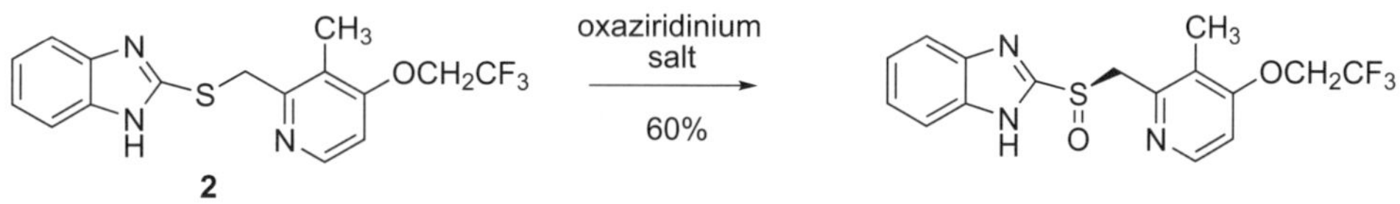


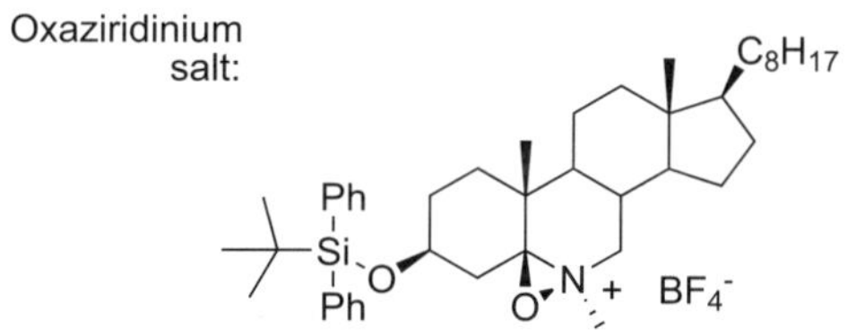


**SCHEME 7.18** (*R*)-Lansoprazole (**43**) by oxidation of lansoprazole sulfide (**2**) with a cholesterol-derived oxaziridinium salt.

### 7.3.16 Oxaziridinium Salts

Returning to the earlier discussion of asymmetric oxidation methods, an asymmetric oxidation can also be accomplished using a stoichiometric quantity of a nonmetallic chiral electrophilic oxidant. One example of a chiral oxidant is an oxaziridinium salt prepared in a multistep sequence from cholesterol. Reaction of the oxaziridinium salt (1.0 equivalent) with lansoprazole sulfide (**2**) in dichloromethane at −70 to 0°C affords (*R*)-lansoprazole (**43**) (60%) with 97% ee (Scheme 7.18).[90]

### 7.3.17 Biooxidation

Biooxidations are particularly useful for producing chiral sulfoxides. Enantioselective enzymatic oxidations of prochiral sulfides by fungi, bacteria, and isolated enzymes are well established.[91] Seven microorganisms that converted pantoprazole sulfide to pantoprazole did not convert lansoprazole sulfide (**2**) to lansoprazole (**1**).[92]

## 7.4 ALTERNATIVE SYNTHETIC STRATEGIES TO LANSOPRAZOLE SULFIDE (2) AND LANSOPRAZOLE (1)

The vast majority of methods for constructing lansoprazole sulfide (**2**) and lansoprazole (**1**) involve a nucleophilic displacement by 2-mercaptobenzimidazole to create the thioether bridge. Alternative but currently not competitive strategies for creating the bridge will be presented in this section (Scheme 7.19).

In one alternative route, the reaction of 2-mercaptobenzimidazole with cyclohexylisocyanate affords *N*-cyclohexyl-2,3-dihydro-2-thioxo-1*H*-benzimidazole-1-carboxamide (**45**). This is reacted with bromine to produce 2-cyclohexyl-1,2,4-thiadiazolo[4,5-*a*]benzimidazole-3(2*H*)-one (**46**), which is then converted to the 1-oxide **47** with MCPBA (1.0 equivalent) in chloroform at 3–8°C (85% for two steps). Lansoprazole *N*-oxide (**41**) is produced by reaction of **47** (1.7 equivalents) with 2,3-dimethyl-4-(2,2,2-trifluoroethoxy)

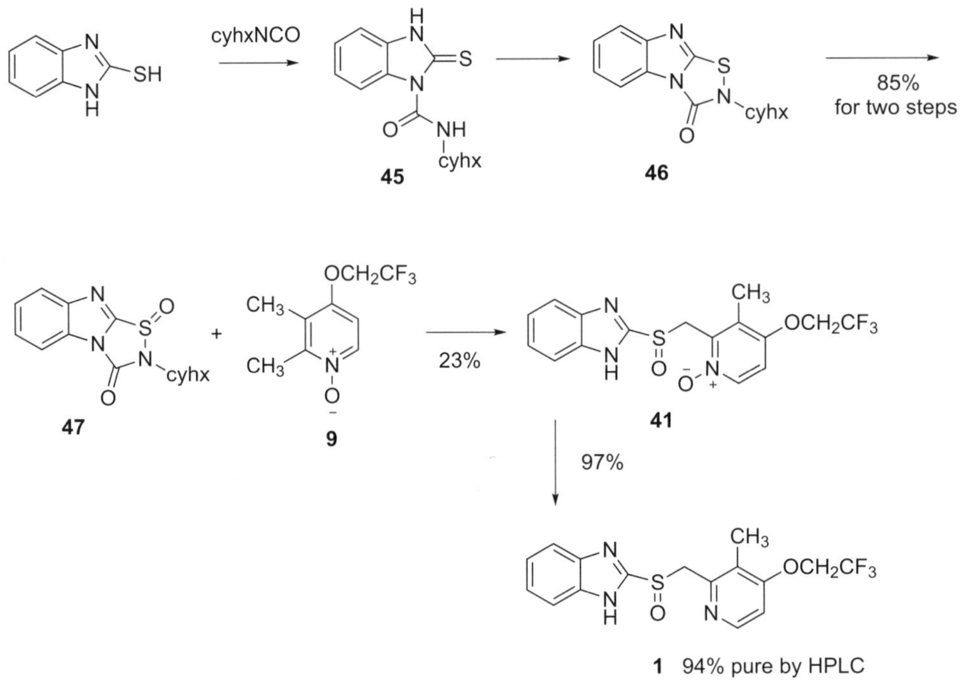

**SCHEME 7.19** Lansoprazole (**1**) from 2-cyclohexyl-1,2,4-thiadiazolo[4,5-a]benzimidazole-3(2*H*)-one (**46**).

pyridine 1-oxide (**9**) and potassium *tert*-butoxide (4.2 equivalents) in THF at 0°C (23%). Reduction with 4,4′-thiobismorpholine converts lansoprazole *N*-oxide (**41**) to lansoprazole (**1**) (97%, 94% pure by HPLC).[93]

Decarboxylation of a carboxylic acid adjacent to the sulfoxide is claimed as an alternative final step in the synthesis of lansoprazole (**1**) (Scheme 7.20). Two new strategies are enabled by introducing the carboxyl group: nucleophilic displacement of the 2-chloropyridine **48** by a 2-(1*H*-benzo[*d*]imidazole-2-ylthio)acetate (**49**) and reaction of the sulfenyl chloride **50** or disulfide derived from 2-mercaptobenzimidazole with the 2-(pyridin-2-yl)acetamide (**51**). There are experimental details for creating omeprazole (**52**) using this decarboxylation methodology but no procedure for lansoprazole (**1**).[94] Recall that 2-chlorobenzimidazole is expensive and a synthesis of 2-chloro-3-methyl-4-(2,2,2-trifluoroethoxy)pyridine (**48**) has yet to be published.

2-Chloro-3-methyl-4-(2,2,2-trifluoroethoxy)pyridine (**48**) could be a pivotal intermediate in another approach to lansoprazole (**1**). The reaction with 2-(methylsulfinyl)-1*H*-benzo[*d*]imidazole (**52**) or, better, with an N-protected derivative of 2-(methylsulfinyl)-1*H*-benzo[*d*]imidazole (**53**), would parallel reactions used to produce a 3,5-dimethylpyridyl analog **55** and RP73163 (**56**), a metabolite of a potent systematically available ACAT inhibitor (Scheme 7.21).[41,95]

## 7.5 PROCESSES FOR UPGRADING THE PURITY OF CRUDE LANSOPRAZOLE (1)

### 7.5.1 Processes for Reducing Levels of Lansoprazole *N*-Oxide (41) and Lansoprazole Sulfone (42)

Lansoprazole sulfide (**2**) oxidation has the potential to produce a number of side products, most notably lansoprazole sulfone (**42**) and lansoprazole *N*-oxide (**41**). The workup and final crystallization procedure must reduce the levels of side products below 0.1%. Two polymorphs of lansoprazole (**1**) are easily produced using different crystallization conditions. Lansoprazole (**1**) undergoes an acid-catalyzed rearrangement that generates highly colored impurities at elevated temperature. The generation of color is particularly an issue when drying water-wet lansoprazole (**1**). Even after developing a process to overcome these many challenges, the lansoprazole (**1**) pure produced is unstable when stored at high humidity. These issues are typically not addressed in the oxidation methods patents and not rigorously addressed in patents focused on the crude to pure transition presented in this section.

The available numbers suggest that lansoprazole *N*-oxide (**41**) levels of <0.1% are achievable. Levels are already at <0.1% in crude lansoprazole (**1**) produced using hydrogen peroxide and the vanadium or rhenium catalysts. Higher levels of the *N*-oxide **41** present in crude lansoprazole (**1**)

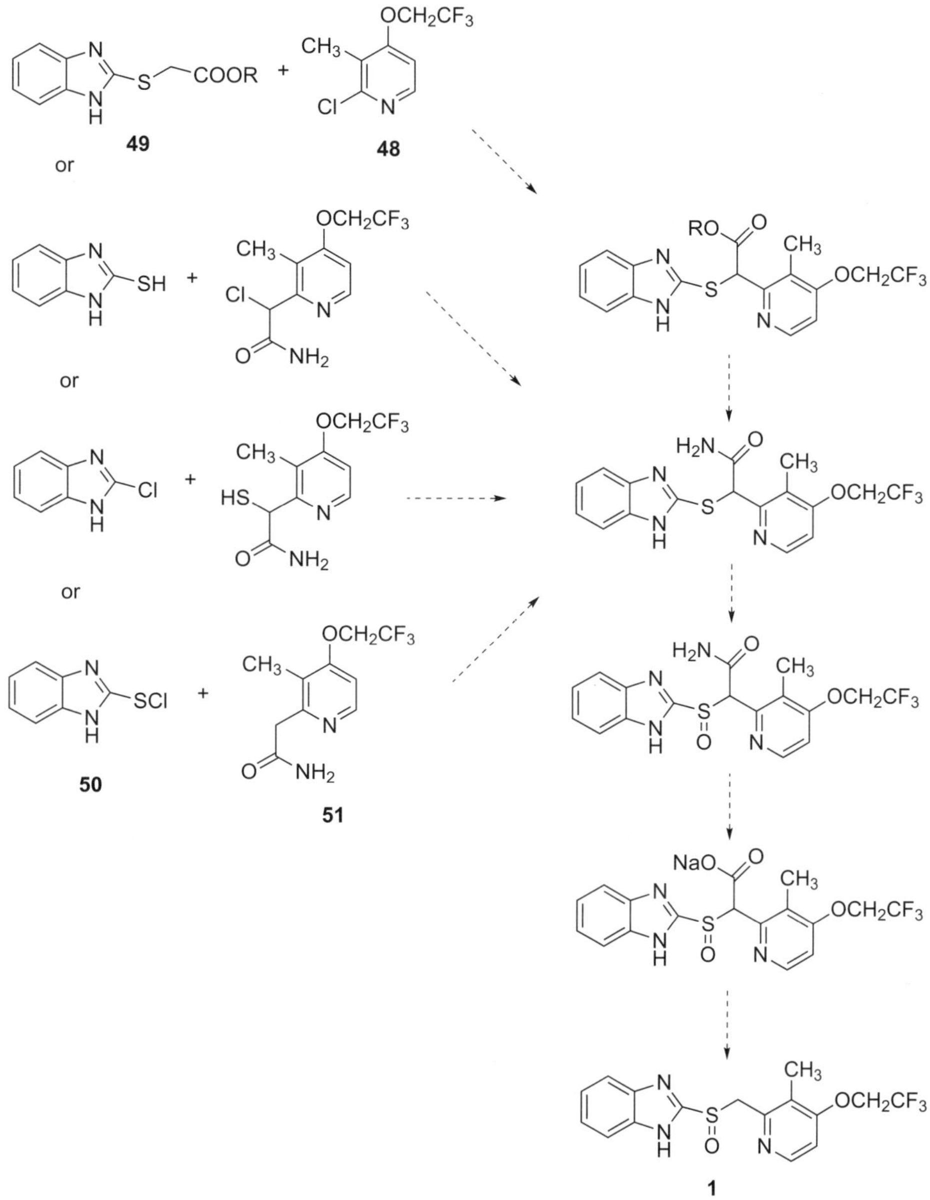


**SCHEME 7.20** Potential approaches to lansoprazole (**1**) by a decarboxylation in the final step.

produced using MCPBA are effectively reduced (0.27% reduced to 0.07%) during crystallization to remove lansoprazole sulfone (**42**) and lansoprazole sulfide (**2**).[47] A reduction to convert the *N*-oxide **41** to lansoprazole at this late stage would be unattractive.[93]

Procedures for reducing high levels (1–5%) of lansoprazole sulfone **42** capitalize on the difference in p$K_a$ between the sulfide, sulfoxide, and sulfone. After an MCPBA oxidation in dichloromethane, the sulfone is extracted with aqueous hydroxide at pH 9.5–12.0 and then the sulfoxide is extracted with aqueous base at pH 13.0. The sulfide remains in and is separated with the organic layer. The sulfoxide, lansoprazole (**1**), is precipitated from the second aqueous extract by adding acid.[96]

In another example, a mixture of lansoprazole (**1**), lansoprazole sulfide (**2**) (0.8%), and lansoprazole sulfone (**42**) (3.3%) produced using MCPBA in chloroform is extracted with aqueous potassium carbonate (pH 9.9). The layers are separated and the organic layer concentrated at reduced pressure. The residue is taken up in ethanol–water (9:1) at 60°C and the solution is saturated with potassium carbonate. The suspension is hot filtered and the filtrate is cooled to −5°C. The resulting suspension is filtered and the solid is dissolved in ethyl acetate at 55°C. The solution is treated with carbon, concentrated to a smaller volume at reduced pressure, and cooled to −5°C. The resulting suspension is filtered to afford lansoprazole (**1**) containing 0.1% lansoprazole sulfide (**2**) and 0.3% lansoprazole sulfone (**42**).[46]

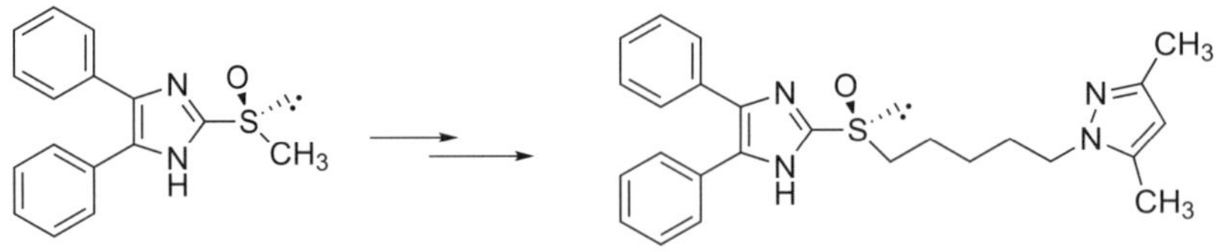

56

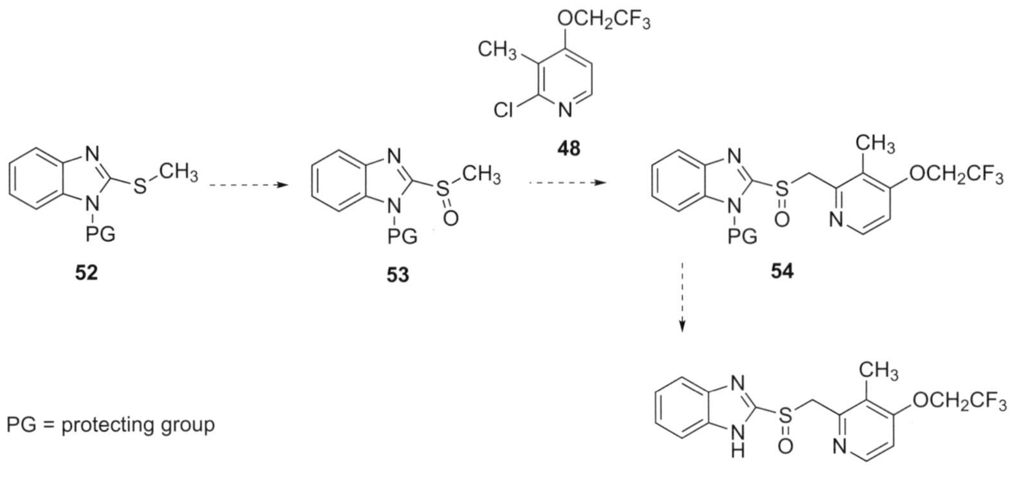


**SCHEME 7.21** A potential route to lansoprazole (**1**) from a 2-(methylsulfinyl)-1*H*-benzo[*d*]imidazole (**53**).

Of course, the best option is to keep lansoprazole sulfone (**42**) formation to a minimum and rely on an efficient crystallization to produce lansoprazole (**1**) of acceptable quality. Lansoprazole (**1**) is most often recrystallized from ethanol–water. In a preferred procedure, crude lansoprazole (**1**), produced using hydrogen peroxide and vanadyl acetylacetonate, containing 0.3 wt% lansoprazole sulfide (**2**) and 0.15 wt% lansoprazole sulfone (**42**), is dissolved in 95% ethanol–water–24% ammonia (volumes 6.8:4.5:1.5 mL/g crude **1**) at 52°C. The solution is treated with carbon, filtered, and cooled to 25°C. The carbon cake is washed with 95% ethanol and water (volumes 1.4 and 1.2 mL/g) crude **1**. Acetic acid (3.8 mL/g) is added to the combined filtrates. The suspension is filtered and the solid is washed with ethanol and water and dried at reduced pressure and 50°C to afford lansoprazole (**1**) (89%) containing 0.05 wt% lansoprazole sulfone (**42**) and no detectable lansoprazole sulfide (**2**). Similar recoveries and low lansoprazole sulfone (**42**) levels are observed in crystallizations using isopropanol or *n*-propanol in place of ethanol and triethylamine in place of ammonium hydroxide.[67]

### 7.5.2 Polymorphs of Lansoprazole (1)

Several lansoprazole polymorphs are known. Forms I and II, A, B, D, E, and F, forms solvated with water, water–ethanol, and water–acetonitrile, and an amorphous form prepared by spray-drying. Polymorph A is produced under conditions identical to polymorph I and polymorph B is produced under conditions identical to polymorph II. Each polymorph is characterized by X-ray diffraction (XRD), IR spectroscopy (KBr pellet), and differential scanning calorimetry (DSC). The XRD patterns for polymorphs I and A are identical in the $2\theta$ region between 10 and 40. Our target is lansoprazole polymorph I or A.[53,97]

Polymorph B is prepared by dissolving lansoprazole in ethanol–water (9:1) at 55°C. The solution is polish filtered and the liquors are cooled to 0°C. The crystals formed are filtered, washed with ethanol–water (1:1), and dried at an unspecified vacuum and <50°C to afford lansoprazole polymorph B (90%). Polymorph B is then dissolved in acetone at reflux. The solution is polish filtered and the liquors slowly cooled to 0°C. The suspension is filtered and the solid is dried for 2 h at an unspecified vacuum and 50°C to afford lansoprazole polymorph A (95%).[98]

We can bypass the isolation of polymorph B. Lansoprazole is dissolved in 95% ethanol at 40°C (50 g in 1 L). Carbon and sodium hydrosulfite are added and the pH is adjusted to 8 with potassium carbonate. After stirring for 30 min, the suspension is filtered and the liquors diluted by dropwise addition of water (2 L) over 30 min at 30°C. After cooling to 20°C and aging for 2 h, the suspension is filtered and the solid is dried at <40°C to afford lansoprazole polymorph A (95%, 99.9% pure).[53]

Lansoprazole crude is converted to polymorph I in an alternative two-isolation process via lansoprazole ethanol hydrate. The crude lansoprazole produced from 142 kg of lansoprazole sulfide (**2**) is dissolved by heating in 864 L of 9:1 ethanol–water with 0.9 L of 25% ammonium hydroxide added. Lansoprazole ethanol hydrate crystallizes on cooling the solution to 25°C. The suspension is filtered and the solid is suspended in water (1043 L) and 25% ammonium hydroxide (17 L). After stirring at 25°C for 60 min, the suspension is filtered and the solid is washed and air dried to produce 108 kg of lansoprazole polymorph I (76% from **2**).[24]

The throughputs for the two-isolation process via polymorph B are 115 and 67 g/L. The throughput for the process bypassing the polymorph B isolation is 15 g/L. The throughputs for the two-isolation process via lansoprazole ethanol hydrate are 125 and 102 g/L. The two-isolation process via polymorph B has two drying operations. The two-isolation process via lansoprazole ethanol hydrate has just one drying operation.

Polymorph A is more stable thermodynamically than polymorph B. Drying polymorph B at or above 50°C will result in some conversion to polymorph A. There can also be some conversion of polymorph B to polymorph A during grinding or compression. The storage stability of the polymorphs can be checked by IR spectroscopy. Polymorph B is stable for 1 year when stored at <0°C. Polymorph A is stable for 1 year when stored at ambient or even elevated temperatures.[53,98]

Polymorph B has excellent solubility but poor storage stability. Polymorph A has excellent storage stability but lower solubility. Screening for other polymorphs of lansoprazole might provide a new polymorph that has excellent storage stability and high solubility. A new polymorph might also have other superior physical properties, such as flowability of the milled solid and behavior on compaction, both critically important when developing a robust formulation protocol. Polymorph screening yielded polymorphs D–F and extended the range of conditions for producing polymorph A. Lansoprazole polymorph A can be produced by crystallization from methanol, 1-butanol, DMSO, and DMF, all of which may contain some water.[98]

Polymorph D is produced from polymorph A using isopropanol–water. For example, polymorph A is slowly converted to polymorph D by stirring in isopropanol–water (99.9:0.1) at 25°C. After 70 h, the filtered solid is a nearly 1:1 mixture of polymorph A and polymorph D. When polymorph A is dissolved in isopropanol–water (2.5–40% water) at reflux, the crystals produced on cooling the solution are polymorph D. Polymorph D is converted to polymorph E during drying under vacuum at 25°C (98% from polymorph A) or when polymorph D is ground using a mortar and pestle. Polymorph D is converted to amorphous lansoprazole when dried under vacuum at 40°C.[98]

Crystals of polymorph F are produced when polymorph A is dissolved in methanol–water (1:1) and the solution stored in a closed vessel at 25°C under a saturated methanol–water atmosphere for 14 days. Polymorph I or A remains the preferred polymorph.[98]

### 7.5.3 Processes for Producing Dry Lansoprazole (1) with Acceptable Storage Stability

Lansoprazole ethanol hydrate cannot be dried to meet specifications (<0.1% water). To illustrate, water (2700 ppm) remains when the solid is dried under an unspecified vacuum at 40°C for 17 h. Water (2300 ppm) remains when the solid is dried under an unspecified vacuum at 44°C for 24 h. Water (2200 ppm) remains when the solid is dried under an unspecified vacuum at 50°C for 17 h. Water (2900 ppm) remains when lansoprazole ethanol hydrate is dried by azeotropic distillation with toluene. When lansoprazole is crystallized from isopropanol–water, water (2500 ppm) remains in the isolated solid after drying under an unspecified vacuum at 45°C. The solid produced by vacuum drying lansoprazole ethanol hydrate is not stable during storage at 40°C. After 6 months, the lansoprazole content of three samples dropped from 99.3–99.1% to 90.6–84.1%.[24]

The mechanisms for decomposition of lansoprazole and for acid activation of lansoprazole in the enzyme active site may both involve an acid-catalyzed rearrangement. The rearrangement cascade for acid activation begins with protonation of the benzimidazole and nucleophilic attack by the pyridine ring nitrogen on the benzimidazole 2-position. An elimination regenerates the benzimidazole, now with a pyridinium salt at the 2-position. This salt can undergo a reversible dehydration or form a disulfide with a sulfhydryl group in the enzyme active site. The mechanism suggests the cascade can be blocked at the start by adding a base (Scheme 7.22).[23,99]

There are at least three processes for producing lansoprazole polymorph I that can be dried to meet specifications and that, when dry, is stable under reasonable storage conditions. What is common to all three processes is the contact of lansoprazole polymorph I with ammonium hydroxide prior to drying. In the first process, the dried lansoprazole is crystallized from acetone, acetone–water, 2-butanone, or dimethyl carbonate. The suspension is filtered and the solid is washed with cold acetone and dried under an unspecified vacuum at 50°C (95% recovery). This lansoprazole polymorph I becomes slightly brown when stored for 3 months at 40°C and 75% relative humidity. The storage stability is improved when the filtered solid is washed with cold acetone containing ammonium hydroxide (pH 8–10) and then dried under an unspecified vacuum at 45°C in the presence of a weak flow of ammonia gas (91% recovery). This solid remains white when stored for 3 months at 40°C and 75% relative humidity.[67]

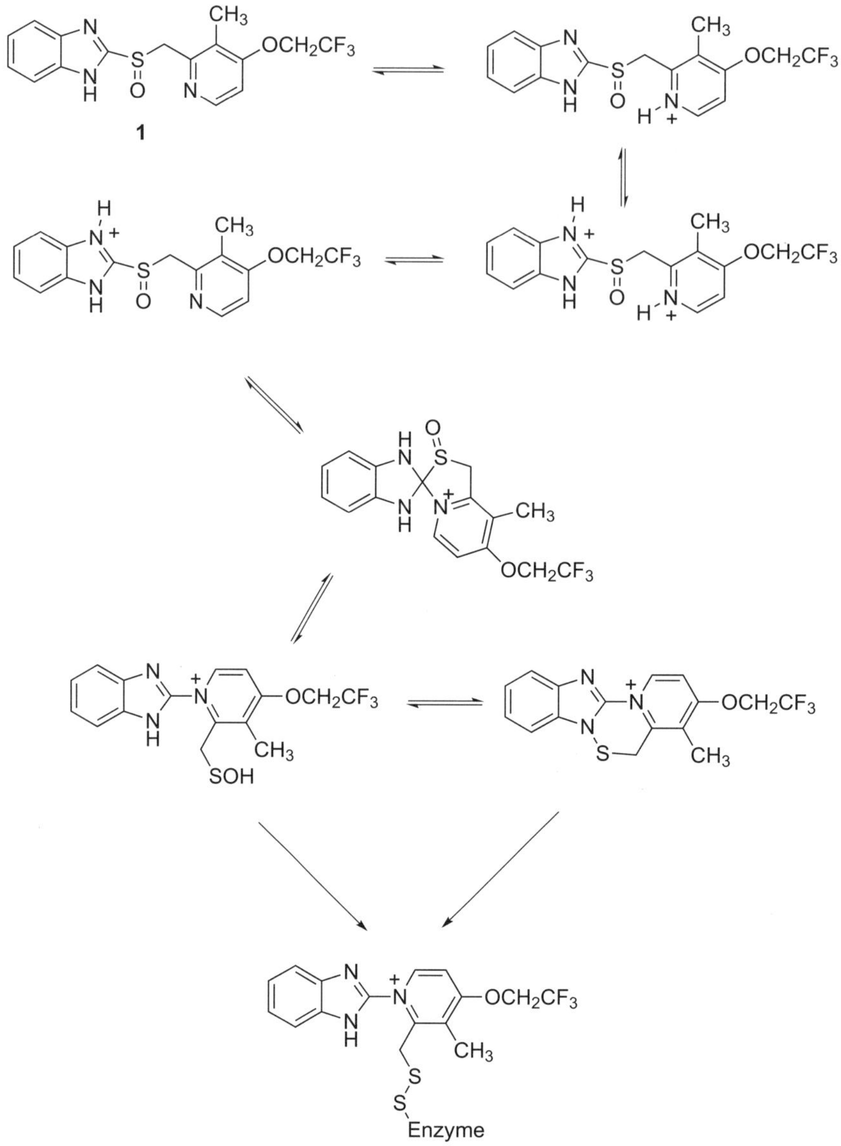


**SCHEME 7.22** The mechanism for acid activation of lansoprazole (**1**).

In the second process, lansoprazole monohydrate monoethanolate is suspended in water and stirred for 1 h at 30°C. The suspension is filtered and the solid is dried under an unspecified vacuum at 40°C to afford lansoprazole polymorph I (88%) containing 0.01% water and 63 ppm ethanol. There is some concern that lansoprazole in the water suspension or during the drying operation may turn yellow. Addition of ammonium hydroxide and some ethanol results in improved stability of the solid in water suspension, during the drying, and in storage after drying. Suspension using a mixture of water, 25% aqueous ammonium hydroxide solution, and ethanol (weight ratio 19.0:0.008:1) gives excellent results (94% recovery). The lansoprazole content (99.8%) of this material does not change during storage for 6 months at 40°C.[13,24]

The third process for dry lansoprazole polymorph I does not provide stability data. The process begins with the preparation of lansoprazole hydrate acetonitrile solvate. Lansoprazole crude (340 g) is dissolved in acetonitrile, water, and 30% aqueous sodium hydroxide (1.1 equivalents) (volumes 1530:85:76 mL) at 25°C. Carbon and dicalcite are added and the suspension aged 30 min. The solids are filtered and washed with 340 mL of 9:1 acetonitrile–water. The pH is adjusted to 9 with acetic acid and the solution seeded with lansoprazole polymorph I. After aging at 20°C for 30 min and at 5°C for 20 h, the solid is filtered, washed with acetonitrile–water (9:1), and dried at an unspecified vacuum and 40°C to afford light yellow lansoprazole hydrate acetonitrile solvate (74% recovery). The solid (250 g) is suspended in acetonitrile, water, and 30% aqueous ammonia (volumes 750:20:20 mL with total water content 3.5–4.0%). The suspension is aged at 35°C for 30 min then at 5°C for 2 h. The solid is filtered, washed with acetonitrile–water–30% ammonium hydroxide, and dried under an unspecified vacuum at 40°C to afford lansoprazole polymorph I (95% recovery) containing 0.05% water. This material still contains lansoprazole sulfide (**2**) (0.22%) and lansoprazole sulfone (**42**) (0.08%). This material has a remarkably fine particle size ($D_{50} = 9$ and $D_{90} = 20$). Data is not available to compare this particle size with the particle size of material made by the other drying processes.[100]

The stability of lansoprazole is also a concern in the environment of dosage forms. The degradation rate constant ($k$) for lansoprazole decreases with increasing pH. The water solubility ($S$) of lansoprazole remains low until pH 9 and increases rapidly with increasing pH. Thus, the degradation of lansoprazole is at a minimum at approximately pH 9. Magnesium carbonate is a suitable stabilizer since the pH of a saturated solution of magnesium carbonate is 9.1.[101]

## 7.6 TRADE SECRETS AND THE CONTINUING REFINEMENT OF ANALYTICAL METHODS

While analytical methods are often not provided in process patents, there are no shortage of methods offered for separation of lansoprazole (**1**), lansoprazole sulfide (**2**), and lansoprazole sulfone (**42**). There is one interesting analytical development not discussed previously. An ORP (oxidation reduction potential) meter can rapidly signal completion of the oxidation reaction. The oxidation–reduction potential of the oxidation solution is monitored and the end point position determined in the lab in advance. In the case of the vanadium-catalyzed lansoprazole sulfide (**2**) oxidation with hydrogen peroxide, after reaching the highest electrical potential, the end point of the oxidation corresponds to a drop in electrical potential of 70 mV. When this drop is observed, and after confirming reaction completion by HPLC analysis, aqueous sodium thiosulfate is quickly added to quench the excess hydrogen peroxide, minimizing the over-oxidation of lansoprazole (**1**) to lansoprazole sufone (**42**).[24]

## 7.7 THE BEST PROCESS AVAILABLE TODAY

The process outline is clear. Lansoprazole sulfide (**2**), produced in high yield and with excellent storage stability, is the penultimate intermediate. Lansoprazole sulfide (**2**) is best produced from available and inexpensive 2-mercaptobenzimidazole and an appropriately substituted pyridine. The only viable routes to the appropriately substituted pyridine start with 2,3-lutidine and utilize the chemistry of pyridine *N*-oxides. What remains are the process details. How do we transition from step to step and which intermediates should be isolated?

2,3-Lutidine is converted to the 1-oxide **6**, probably using peracetic acid. The workup should involve a quench of excess oxidant, analytical verification that there is no remaining oxidant, and an extraction into dichloromethane. Sulfuric acid is added and the dichloromethane distilled at reduced pressure. Reaction with nitric acid in sulfuric acid produces the 4-nitropyridine 1-oxide **7**. The reaction is quenched into water and neutralized, and the 4-nitropyridine 1-oxide **7** is extracted into dichloromethane (90% contained yield from 2,3-lutidine).

After solvent exchange to 2,2,2-trifluoroethanol, the nitro group is displaced by adding a solution of potassium *tert*-butoxide in 2,2,2-trifluoroethanol. The bulk of the trifluoroethanol is distilled at reduced pressure and the resulting solution separated between water and dichloromethane. The dichloromethane solution of 2,3-dimethyl-4-(2,2,2-trifluoroethoxy)pyridine 1-oxide (**9**) (95–97% contained yield) is carried on.

The dichloromethane solution will contain some 2,2,2-trifluoroethanol that will react with acetic anhydride and DMAP to produce 2,2,2-trifluoroethyl acetate. The dichloromethane and 2,2,2-trifluoroethyl acetate (bp 78°C) will distill during the heat up to reaction temperature (90–95°C). When the reaction is complete, the mixture is cooled and quenched with water. Water and acetic acid are distilled at reduced pressure to afford the acetate ester **14** as an oil. The ester is hydrolyzed with sodium hydroxide in methanol at 25°C. The solution is then neutralized with hydrochloric acid and extracted with dichloromethane. The dichloromethane solution containing (3-methyl-4-(2,2,2-trifluoroethoxy)pyridin-2-yl)methanol (**5**) (89–93% contained yield) is dried by azeotropic distillation (dichloromethane–water 98.5:1.5, bp 38°C) and carried on.

Thionyl chloride is added. After aging at 25°C for 1 h, the batch is quenched into water. Dichloromethane is distilled at reduced pressure to produce an aqueous solution of the 2-(chloromethyl)pyridine (**3**) hydrochloride (99% contained yield) for use in the next step.

The aqueous solution is added to a solution of 2-mercaptobenzimidazole in methanol. Aqueous 30% sodium hydroxide is added (to pH 11.0–11.5). After 30 min age time, the mixture is diluted with water. Hydrochloric acid is added (to pH to 8.5–10) and the suspension is filtered. The solid lansoprazole sulfide (**2**) is washed, presumably with water or methanol–water, and dried (97%).

There are seven transformations in the linear conversion of 2,3-lutidine to lansoprazole sulfide (**2**). Dichloromethane is uniquely suited to be used in six of the seven steps. It is a good solvent for the process intermediates. Dichloromethane is not reactive with sulfuric acid, acetic anhydride, or thionyl chloride. Dichloromethane distills at low temperature, separates from water, and can be dried by azeotropic distillation. Carrying dichloromethane solutions from step to step avoids the isolation of the carcinogenic and mutagenic 4-nitropyridine 1-oxide **7**, leaving only the isolation of lansoprazole sulfide (**2**) at the end of the seven-step sequence. On the minus side, carrying dichloromethane solutions from step to step makes campaigning the process much more of a challenge and suggests that the entire sequence, including the first two steps, must be run in one manufacturing facility. There is just one potential hold point that is not a dichloromethane solution: perhaps the acetate ester **14** can be held as an oil after the water–acetic acid distillation and prior to the ester hydrolysis.

> Are there any alternative solvents we can use in place of dichloromethane for all or part of this seven step sequence? Consider replacing dichloromethane with benzotrifluoride at each step of the process.[102] Benzotrifluoride is likely to be a good solvent for all the intermediate extractions. The problems are encountered *after* the extractions, in the transitions from one step into the next. In the first transition, the distillation of benzotrifluoride from a mixture with 98% sulfuric acid in the transition into the nitration is likely to be problematic. In the second transition, replacing the higher boiling benzotrifluoride with the lower boiling trifluoroethanol may be at best inefficient. In the third transition, distillation of benzotrifluoride from acetic anhydride during the heat up to a reaction temperature of 100°C is possible. In the fifth transition, distillation of the benzotrifluoride–water azeotrope ensures a dry benzotrifluoride solution for the thionyl chloride reaction. In the sixth and final transition, distillation of the benzotrifluoride–water azeotrope could produce the aqueous solution of the hydrochloride salt ready to be carried on. Azeotrope data on mixtures of benzotrifluoride–trifluoroethanol[103] and benzotrifluoride–water and data on reactivity of benzotrifluoride with sulfuric acid, acetic anhydride, and thionyl chloride would complete this paper exercise.[104]

With a single well-defined route to lansoprazole sulfide (**2**), recent process research and development has focused on oxidation methods. The available options can be quickly evaluated using three selection criteria. First, since it would be preferable to slowly feed the oxidant to minimize oxidant inventory in the reactor, the preferred oxidant will be a liquid. This eliminates MCPBA, magnesium monoperoxyphthalate, sodium perborate, potassium peroxymonosulfate, and sodium percarbonate, Second, the oxidation should be chemoselective to simplify the operations required to convert lansoprazole (**1**) crude to lansoprazole (**1**) that meets specifications. Processes with lower lansoprazole (**1**) yields, workup procedures designed to reduce impurity levels, and use of less than an equivalent of oxidant suggest that the chemoselectivity using peracetic acid, sodium hypochlorite, or solutions of *N*-chlorosuccinimide is lower. Cumene hydroperoxide and oxaziridinium salts are only used to produce (*R*)-lansoprazole (**43**). This leaves *tert*-butyl hydroperoxide and hydrogen peroxide as the remaining options. Both are inexpensive and have comparable transport and storage stability. These two oxidants can be differentiated by revisiting the oxidation reaction conditions and results for both.

The oxidation with *tert*-butyl hydroperoxide (1.5 equivalents) is catalyzed by vanadyl acetylacetonate (2.0 mol%) in 95% ethanol at 5°C. After aging for 6 h, excess oxidant is quenched with aqueous sodium sulfite. Lansoprazole (**1**) (85%) containing 0.15 wt% lansoprazole sulfone (**42**) and 0.3 wt% lansoprazole sulfide (**2**) is isolated by filtering the resulting suspension. The yield drops to 79% using lower catalyst charges and a higher reaction temperature.

The oxidation with hydrogen peroxide (1.2 equivalents) is catalyzed by vanadium(V) oxide (0.3 mol%) in ethanol at 20–25°C. After aging for 2.5 h, excess oxidant is quenched with aqueous sodium thiosulfate. Lansoprazole (**1**) (90%, 99.6% pure by HPLC) is isolated by filtering the resulting suspension and crystallization of the isolated solid from ethanol–water. Comparing these two processes suggests the third of the selection criteria, a low vanadium catalyst charge to minimize operator exposure during solids charging and to minimize cost associated with aqueous waste treatment to remove vanadium. Hydrogen peroxide is the preferred oxidant.

*But what about using another catalyst for the oxidation with hydrogen peroxide?* The toxicity and carcinogenicity of vanadium(V) oxide will no doubt continue to drive innovation in the search for a superior catalyst for the oxidation using hydrogen peroxide. Methyltrioxorhenium, ammonium molybdate, and sodium tungstate dihydrate might all be viable options and new catalysts appear in the patent literature every day. Since catalyst chemoselectivity and cost are not issues with vanadium(V) oxide, a superior catalyst can only offer to simplify the solids handling in the catalyst charge and the aqueous waste treatment after isolating lansoprazole (**1**).

The solids handling and waste treatment would be simplified using MTO. Lansoprazole (**1**) (75%, >99.5% pure by HPLC) can be produced using just 0.05% MTO. Even if a 90% yield at 0.05 mol% MTO loading were possible, and we were willing to pay the $35–40/kg catalyst charge, there does not appear to be a reliable bulk supply of MTO to support the production of metric ton quantities of lansoprazole (**1**). Green chemistry-inspired improvements in the process for manufacture of MTO, for example eliminating the highly toxic tin-containing methylating agents of the current process, may improve the supply situation in the near future.[105]

The oxidation using hydrogen peroxide and an ammonium molybdate tetrahydrate catalyst was described in 1993.[106] This oxidation produces 8–20% of lansoprazole *N*-oxide (**41**) and lansoprazole sulfone (**42**) side products. The exposure limits (PEL) are 5 mg/m$^3$ for soluble molybdenum compounds as Mo and 15 mg/m$^3$ for insoluble molybdenum compounds as Mo and ACGIH threshold limit values are 10 mg/m$^3$ for metal and insoluble compounds, inhalable fraction, as Mo, 3 mg/m$^3$ for metal and insoluble compounds, respirable fraction, as Mo, and 0.5 mg/m$^3$ for soluble compounds, respirable fraction, as Mo. These numbers suggest the solids handling for ammonium molybdate or other molybdenum-based catalysts will require similar protocols to minimize operator exposure as would be used when handling the vanadium catalyst.

The oxidation using hydrogen peroxide and sodium tungstate was published while this chapter was in preparation. This oxidation affords lansoprazole (**1**) in 91% yield (purity by HPLC >95%).[107] The ACGIH threshold limit values are 1 mg/m$^3$ (TWA) and 0.3 mg/m$^3$ (STEL) for soluble tungsten compounds. Again, these numbers suggest the solids handling for sodium tungstate or other tungsten-based catalysts will require similar protocols to minimize operator

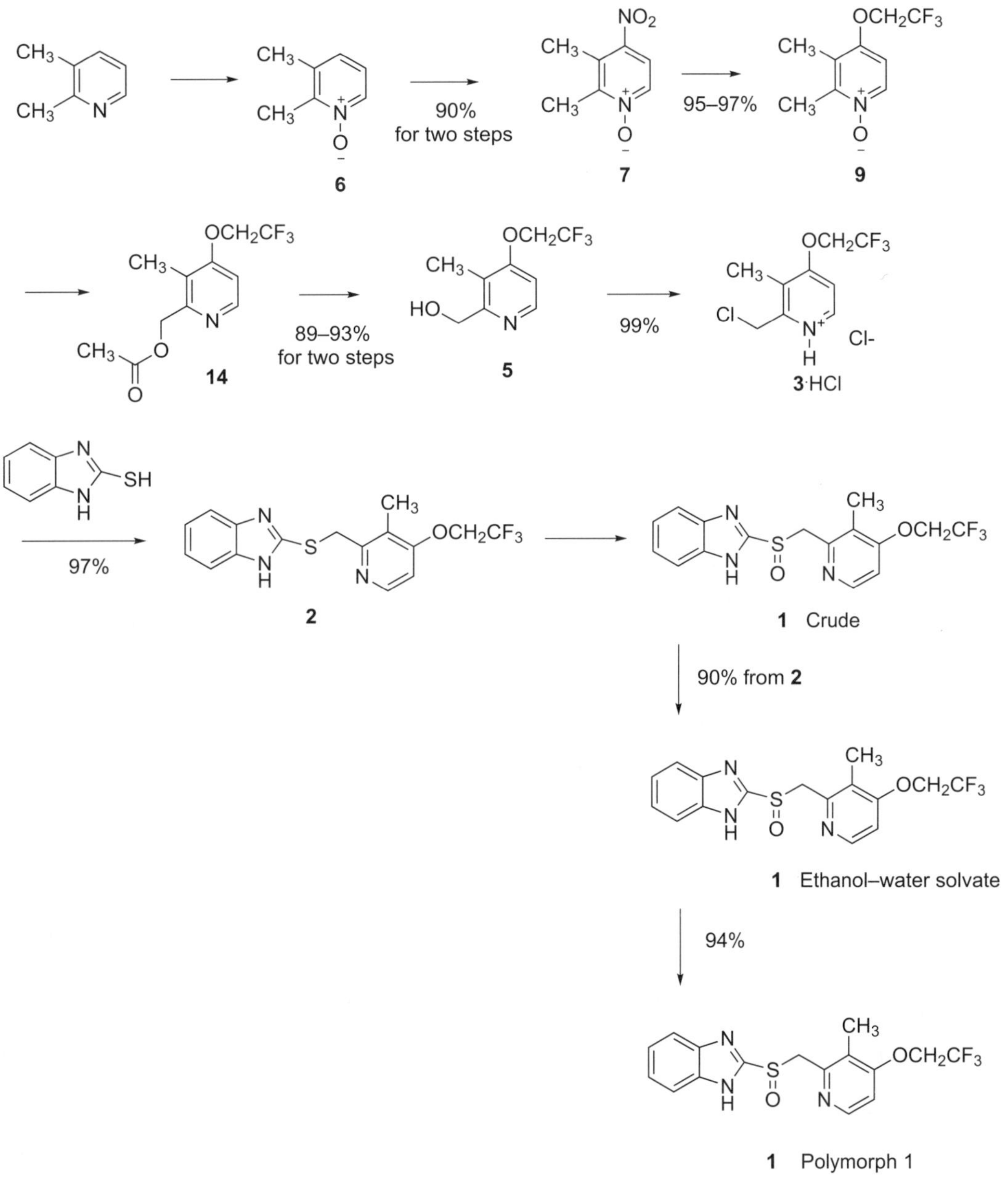


**SCHEME 7.23** The best process for manufacture of lansoprazole (**1**) in 2009.

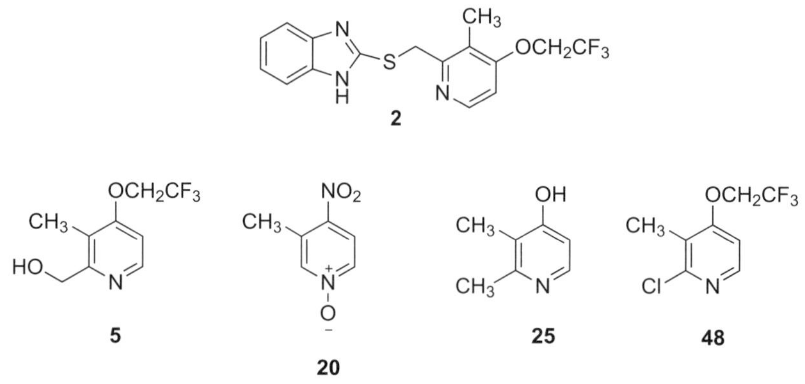


**FIGURE 7.4** Structures searched for lansoprazole (**1**) presentation.

exposure as would be used when handling the vanadium catalyst.

The process for converting lansoprazole crude to lansoprazole polymorph I is selected to minimize time-consuming drying operations and avoid the use of toxic acetonitrile as a solvent in the final isolation. The process begins with crystallization of lansoprazole from 9:1 ethanol–water. To minimize decomposition and color generation during the crystallization, some ammonium hydroxide is added and the suspension is probably heated to only 50–60°C to dissolve the solid. The lansoprazole ethanol–water solvate is then isolated and suspended in water at 25–30°C. Again, some ammonium hydroxide is added to minimize decomposition and color generation. An optimal solvent mixture for the suspension also contains some ethanol (weight ratio of water, 24% ammonium hydroxide, and ethanol of 19.0:0.008:1). The lansoprazole polymorph I isolated from this water suspension can be dried under vacuum at 40°C to meet the residual water specification (94% recovery). The dried lansoprazole polymorph I is stable when stored at 40°C for at least 6 months. Some increase in storage stability at 40°C and high (75%) relative humidity may be attained by passing ammonia gas over the wet lansoprazole during the drying operation.

The process is 10 steps from 2,3-lutidine to dry lansoprazole polymorph I with an overall yield of 62–66%. Since the first 6 steps are carried out with no solid isolations, the 10-step process has just four isolations, all at the back end. The process solvents are dichloromethane, methanol, ethanol, and water. The process weaknesses are all health and safety issues: the back-to-back oxidation and nitration, the carcinogenic and mutagenic 4-nitropyridine 1-oxide intermediate, the extensive use of dichloromethane, the handling of a toxic and carcinogenic solid vanadium catalyst, and the removal of the vanadium catalyst from an aqueous waste stream (Scheme 7.23).

## 7.8 STRUCTURES SEARCHED

Five structure searches were used to generate all the information presented in this chapter (Figure 7.4).

## REFERENCES


1. Accessed at www.drugs.com/top200.html.
2. Accessed at seekingalpha.com/article/73563-validity-of-takeda-s-prevacid-patents-upheld-teva-s-obviousness-arguments-rejected.
3. (a) Crownover, B.K. *J. Manag. Care Pharm.* **2006**, *12*, 578 [letter].(b) Wahlqvist, P. *J. Manag. Care Pharm.* **2006**, *12*, 581 [letter].
4. Accessed at www.takeda.com/press/article_32521.html.
5. Juurlink, D.N.; Gomes, T.; Ko, D.T.; Szmitko, P.E.; Austin, P. C.; Tu, J.V.; Henry, D.A.; Kopp, A.; Mamdani, M.M. *CMAJ* **2009**, *180* (7), DOI: 10.1503/cmaj.082001.
6. Van Allan, J.A.; Deacon, B.D. *Org. Syn. Coll.* Vol. *4*, **1963**, 569.
7. The Aldrich 2007–2008 catalog price for 500 g is $45.10.
8. The Aldrich 2007–2008 catalog price for 25 g (90% tech grade) is $179.50.
9. The Aldrich 2007–2008 catalog price for 500 mL is $178.50.
10. The Alfa-Aesar 2006–2007 price for 1 kg of 80% (tech) is $248.00.
11. The Aldrich 2007–2008 catalog price for 4 L of 35 wt% in $H_2O$ is $96.20.
12. (a) Kotar-Jordan, B.; Vrecer, F.; Segula Zakelj, M.; Ritlop, G. WO 2006074952 (7/20/2006). (b) Kotar-Jordan, B.; Vrecer, F.; Segula Zakelj, M.; Ritlop, G. EP 1681056 (7/19/2006).
13. (a) Kato, M.; Ishida, T.U.S. Patent 6,002,011 (12/14/1999). (b) Kato, M.; Ishida, T. WO 199821201 (5/22/1998).
14. Kubo, K.; Oda, K.; Kaneko, T.; Satoh, H.; Nohara, A. *Chem. Pharm. Bull.* **1990**, *38*, 2853.

15. The Aldrich 2007–2008 catalog price for 500 g is $101.50.
16. (a) Nohara, A.; Maki, Y.U.S. Patent 4,689,333 (8/25/1987). (b) Nohara, A.; Maki, Y.U.S. Patent 4,628,098 (12/9/1986). (c) Nohara, A.; Maki, Y. EP 174726 (3/19/1986).
17. Nohara, A.; Maki, Y. EP 233760 (8/26/1987).
18. Makino, T.; Tabata, T.; Hirai, S.-I. EP 237200 (9/16/1987).
19. Makino, T.; Tabata, T.; Hirai, S.-I. EP 446961 (9/18/1991).
20. Accessed at www.webwiser.nlm.nih.gov/getHomeData.do.
21. Ahn, K.-H.; Kim, H.; Kim, J.R.; Jeong, S.C.; Kang, T.S.; Shin, H.T.; Lim, G.J. *Bull. Korean Chem. Soc.* **2002**, *23*, 626.
22. (a) National Toxicology Program. *Report on Carcinogens*, 11th edition, Public Health Service, U.S. Department of Health and Human Services.,(b) Chemicals Known to the State to Cause Cancer or Reproductive Toxicity, State of California, March 21, 2008.
23. (a) Senn-Bilfinger, J.; Kruger, U.; Sturm, E.; Figala, V.; Klemm, K.; Kohl, B.; Rainer, G.; Schaefer, H.; Blake, T.J.; Darkin, D.W.; Ife, R.J.; Leach, C.A.; Mitchell, R.C.; Pepper, E.S.; Salter, C.J.; Viney, N.J.; Huttner, G.; Zsolnai, L. *J. Org. Chem.* **1987**, *52*, 4582. (b) Sturm, E.; Kruger, U.; Senn-Bilfinger, J.; Figala, V.; Klemm, K.; Kohl, B.; Rainer, G.; Schaefer, H.; Blake, T.J.; Darkin, D.W.; Ife, R.J.; Leach, C.A.; Mitchell, R.C.; Pepper, E.S.; Salter, C.J.; Viney, N.J.; Huttner, G.; Zsolnai, L. *J. Org. Chem.* **1987**, *52*, 4573.
24. Kaneko, M.; Yabuno, S.; Iwaoka, K.; Nishimura, H. EP 1790647 (5/30/2007).
25. (a) Turchetta, S.; Massardo, P.; Tuozzi, A.U.S. Patent 7,105,681 (9/12/2006).(b) Turchetta, S.; Massardo, P.; Tuozzi, A. WO 2004056803 (7/8/2004).
26. Lim, G.-J.; Kim, D.-S.; Yoon, N.-M. WO 2000027841 (5/18/2000).
27. Deng, J.; Zhu, J.; Wu, J.; Wang, Q.; Huang, Q.; Zhu, J. CN 1803794 (7/19/2006).
28. The Aldrich 2007–2008 catalog price for 1 L is $38.00.
29. The Aldrich 2007–2008 catalog price for 500 g is $106.50.
30. Taylor, E.C.; Crovetti, A.J. *Org. Syn. Coll.* Vol. *4*, **1963**, 654.
31. Matsumura, E.; Ariga, M.; Ohfuji, T. *Bull. Chem. Soc. Jpn.* **1970**, *43*, 3210.
32. Takahashi, K.; Huang, G.-F.; Araki, M.; Kawazoe, Y. *Gann* **1979**, *70*, 799.
33. Araki, M.; Koga, C.; Kawazoe, Y. The repeated subcutaneous administration of 3-methyl-4-nitropyridine-*N*-oxide (total dose = 42 mg/mouse) produced tumors, mostly fibrosarcomas, at the injection site, *Gann* **1971**, *62*, 325.
34. (a) Kim, J.-G.; Ahn, B.-N.; Lee, H.-W.; Yoon, S.-W.; Yoon, Y.-A.; Lee, C.-H.; Cha, M.-H.; Kang, H.-I. WO 2006038773 (4/13/2006).(b) Wittek, P.J.; Liao, T.K.; Cheng, C.C. *J. Org. Chem.* **1979**, *44*, 870.
35. Coppi, L.; Berenguer Maimo, R. EP 1411053 (5/30/2007).
36. The Aldrich 2007–2008 catalog price for 25 g is $48.10.
37. Rane, R.A.; Pathak, R.K.; Kaushik, C.P.; Prasad Rao, K.V.V.; Kumar, A. *Synth. Commun.* **2002**, *32*, 1211.
38. Kim, W.-J.; Kim, K.S.; Kim, M.H.; Baek, Y.G.; Park, J.Y.; Jang, J.M.; Choi, J.W.; Yoo, Y.S. WO 2002074766 (9/26/2002).
39. The Aldrich 2007–2008 catalog price for 100 g of a 40 wt% solution in toluene is $104.00.
40. Moon, Y.-H.; Lee, K.-I.; Lee, G.-S.U.S. Patent 6,423,846 (7/23/2002).
41. (a) Junggren, U.K.; Sjostrand, S.E. CA 1127158 (7/6/1982). (b) Junggren, U.K.; Sjostrand, S.E. EP 005129 (4/3/1979).
42. Madesclaire, M. *Tetrahedron* **1986**, *42*, 5459.
43. Accessed at www.cameochemicals.noaa.gov/chemical/2902.
44. The Aldrich 2007–2008 catalog price for 500 g (77% purity) is $206.50.
45. (a) Kato, M.; Toyoshima, Y.; Iwano, N.U.S. Patent 5,578,732 (11/26/1996).(b) Kato, M.; Toyoshima, Y.; Iwano, N. EP 0302720 (2/8/1989).(c) Kato, M.; Toyoshima, Y.; Iwano, N. CA 1263119 (8/4/1987).
46. Unsal, S. WO 2003062223 (7/31/2003).
47. Ludescher, J.; Khan, R.A.R.; Das, T.C. WO 2007017244 (2/15/2007).
48. Accessed at www.cameochemicals.noaa.gov/chemical/16016.
49. Buchi Reddy, R.; Om Prakash, G.; Madhusudhan, G. WO 2007066202 (6/14/2007).
50. Siddiqui, M.J.M.; Kulkarni, D.G.; Supekar, P.R.; Shinde, P.S.; Deshmukh, V.V. WO 2007138468 (12/6/2007).
51. The Alfa-Aesar 2006–2007 catalog price for 1 kg (tech. 80%) is $248.00.
52. Accessed at www.tana.de/anhang.php?id=2&type=1.
53. Moon, S.C.; Song, I.W.; Kang, D.S.; Oh, S.S.; Lee, S.J. WO 2007078154 (7/12/2007).
54. Kankan, R.N.; Rao, D.R.; Srinivas, P.L. WO 2004063188 (7/29/2004).
55. Kotagiri, V.K.; Vankawala, P.J.; Neelam, U.K.; Baddam, S.R.; Elati, R.R.C.; Kolla, N.K.; Dayabhai, J.L.; Pingili, R.R.; Gangula, S.; Jonnalagadda, V.R.; Thakur, P.; Neredla, A.; Kammili, V.R. WO 2008045777 (4/17/2008).
56. Bandari, M.; Purna, C.R.; Madhuresh, K.S.; Chebolu Veera, V.L.N.; Kamatam, R. WO 2008087665 (7/24/2008).
57. (a) Jaweed Mukarram, S.M.; Naithani, P.K.; Purohit, M.; Kulkarni, D.G.; Pokalwar, R.U. WO 2007026188 (3/8/2007).(b) Rajnikantvyas, J.; Mallik Yadav, K. WO 2006049486 (5/11/2006).
58. Harville, R.; Reed, S.F., Jr., *J. Org. Chem.* **1968**, *33*, 3976.
59. Tagami, K.; Niikawa, N.; Kayano, A.; Kuroda, H.U.S. Patent 6,313,303 (11/6/2001).
60. Accessed at www.hongyechemical.com/pdf/sodium_perborate/sodium_perborate_1MSDS.pdf.
61. Brennan, J.P.; Turner, A.T. WO 1999047514 (9/23/1999).
62. Accessed at www.dupont.com/oxone/techinfo.
63. Accessed at www.hansagroup.de/download=products/safetysheet/EN-E005222.pdf.
64. (a) Avrutov, I.; Mendelovici, M.U.S. Patent 7,129,358 (10/31/2006).(b) Avrutov, I.; Mendelovici, M. WO 2002062786 (8/15/2002).
65. (a) Berenguer Maimo, R.; Campon Pardo, J.; Coppi, L. EP 1270555 (8/25/2004).(b) Berenguer Maimo, R.; Campon Pardo, J.; Coppi, L.U.S. Patent 6,603,009 (8/5/2003).

(c) Berenguer Maimo, R.; Campon Pardo, J.; Coppi, L. WO 2001068594 (9/20/2001).

66. Finkelstein, N. WO 2004011455 (2/5/2004).
67. (a) Singer, C.; Liberman, A.; Veinberg, I.; Finklestein, N.; Nidam, T. EP 1501824 (10/10/2007).(b) Singer, C.; Liberman, A.; Veinberg, I. EP 1465890 (4/4/2007).(c) Singer, C.; Liberman, A.; Veinberg, I.; Finkelstein, N.; Nidam, T.U.S. Patent 7,060,837 (6/13/2006).(d) Singer, C.; Liberman, A.; Veinberg, I.; Finkelstein, N.; Nidam, T.U.S. Patent 7,022,859 (4/4/2006).(e) Singer, C.; Liberman, A.; Veinberg, I.; Finkelstein, N.; Nidam, T.U.S. Patent 6,909,004 (6/21/2005).(f) Liberman, A.; Singer, C.; Veinberg, I.U.S. Patent 20040215021 (10/28/2004).(g) Singer, C.; Liberman, A.; Veinberg, I. WO 2004072061 (6/26/2004).(h) Singer, C.; Liberman, A.; Veinberg, I. WO 2004046135 (6/3/2004). (i) Singer, C.; Liberman, A.; Veinberg, I.; Finklestein, N.; Nidam, T. WO 2004018454 (3/4/2004).
68. Kagan, H.B. Asymmetric oxidation of sulfides, in: Ojima, I. ed., *Catalytic Asymmetric Synthesis*, 2nd edition, Wiley-VCH, 2000, pp. 327–356.
69. The Aldrich 2007–2008 catalog price for 500 mL ($d = 1.038$ g/mL) (88% purity) is $53.00.
70. Accessed at www.aiha.org/1documents/Committees/WEEL-WEELSLevels2007.pdf.
71. Pitchen, P.; Dunach, E.; Deshmukh, M.N.; Kagan, H.B. *J. Am. Chem. Soc.* **1984**, *106*, 8188.
72. Zhao, S.H.; Samuel, O.; Kagan, H.B. *Tetrahedron* **1987**, *43*, 5135.
73. Larsson, M.E.; Stenhede, U.J.; Sorensen, H.; von Unge, S.P. O.; Cotton, H.K.U.S. Patent 5,948,789 (9/7/1999).
74. Hashimoto, H.; Urai, T. EP 1277752 (11/22/2006).
75. Hashimoto, H.; Urai, T.U.S. Patent 7,285,668 (10/23/2007).
76. Hashimoto, H.; Maruyama, H. EP 1293507 (11/28/2007).
77. (a) Kamiyama, K.; Hashimoto, H.U.S. Patent 7,271,182 (9/18/2007).(b) Kamiyama, K.; Hashimoto, H. EP 1306375 (8/3/2001).
78. The Aldrich 2007–2008 catalog price for 2 g is $221.00.
79. Espenson, J.H. *Chem. Commun.* **1999**, 479.
80. Choi, S.J.; Moon, S.C.; Byun, Y.S. WO 2001021617 (3/29/2001).
81. (a) Wang, Q.; Cheng, M. CN 101012141 (8/8/2007).(b) Wang, Q.; Cheng, M. CN 1995037 (7/11/2007).
82. Accessed at www.oehha.ca.gov/prop65/prop65_list/files/P65single121908.pdf.
83. Opresko, D.M. *Toxicity Summary for Vanadium and Vanadium Compounds*, Oak Ridge National Laboratory, 1991.
84. Namasivayam, C.; Sangeetha, D. *Adsorption*, **2006**, *12*, 103.
85. Naeem, A.; Westerhoff, P.; Mustafa, S. *Water Res.* **2007**, *41*, 1596.
86. Parks, J.L.; Edwards, M. *J. Environ. Eng.* **2006**, *132*, 489.
87. Bhatnagar, A.; Minocha, A.K.; Pudasainee, D.; Chung, H.-K.; Kim, S.-H.; Kim, H.-S.; Lee, G.; Min, B.; Jeon, B.-H. *Chem. Eng. J.* **2008**, *144*, 197.
88. Thakur, V.V.; Sudalai, A. *Tetrahedron Asym.* **2003**, *14*, 407.
89. Yang, G. CN 1381443 (11/27/2002).
90. del Rio, R.E.; Wang, B.; Achab, S.; Bohe, L. *Org. Lett.* **2007**, *9*, 2265.
91. (a) Fernandez, I.; Khiar, N. *Chem. Rev.* **2003**, *103*, 3651.(b) Holland, H.L. *Chem. Rev.* **1988**, *88*, 473.
92. Holt, R.; Lindberg, P.; Reeve, C.; Taylor, S.U.S. Patent 5,840,552 (11/24/1998).
93. Clausen, F.P.; McCluskey, K.K.; Preikschat, H.F.; Pedersen, S.B. WO 1998040378 (9/17/1998).
94. (a) Siemon, C.; Macel, B.U.S. Patent 5,502,195 (3/26/1996). (b) Siemon, C.; Macel, B.U.S. Patent 5,470,983 (11/28/1995). (c) Siemon, C.; Macel, B.U.S. Patent 5374730 (12/20/1994).
95. Pitchen, P.; France, C.J.; McFarlane, I.M. *Tetrahedron Lett.* **1994**, *35*, 485.
96. Broeckx, R.L.A.; De Smaele, D.; Leurs, S.M.H. WO 2003008406 (1/30/2003).
97. Piechaczek, J.; Glice, M.; Cichy, B.; Serafin, J.; Koziol, A.; Cybulski, J.; Chilmonczyk, Z. WO 2000078729 (12/28/2000).
98. Finkelstein, N.; Wizel, S. WO2003082857 (10/9/2003).
99. Shin, J.M.; Cho, Y.M.; Sachs, G. *J. Amer. Chem. Soc.* **2004**, *126*, 7800.
100. Segnalini, F.; Tuozzi, A.; Magnanti, S.; Mastrangeli, C.; Casellato, P. WO 2008077866 (7/3/2008).
101. Tabata, T.; Makino, T.; Kashihara, T.; Hirai, S.; Kitamori, N.; Toguchi, H. *Drug Dev. Ind. Pharm.* **1992**, *18*, 1437.
102. Ogawa, A.; Curran, D.P. *J. Org. Chem.* **1997**, *62*, 450.
103. Atik, Z. *J. Chem. Eng. Data* **2007**, *52*, 1128.
104. Maul, J.J.; Ostrowski, P.J.; Ublacker, G.A.; Linclau, B.; Curran, D.P. *Top. Curr. Chem.* **1999**, *206*, 80.
105. Tosh, E.; Mitterpleininger, J.K.M.; Rost, A.M.J.; Veljanovski, D.; Herrmann, W.A.; Kuhn, F.E. *Green Chem.* **2007**, *9*, 1296.
106. Palomo Coll, A. ES 2036948 (1993).
107. Fan, C.-T.U.S. Patent 20090005570 (1/1/2009).
108. The 4-nitropyridine-N-oxide **7** can be directly converted to the 4-chloropyridine-N-oxide **28**: Waser, M.; Obermuller, R.; Wiegand, J. M.; Schiek, W.; Fierz, H.; Skranc, W. *Org. Proc. Res. Dev.* **2010**, *14*, 562.

# 8

# ADVAIR DISKUS® (SALMETEROL XINAFOATE)

## 8.1 ADVAIR DISKUS® AND THE ASTHMA AND COPD MARKETS

Advair Diskus®, also known as Seretide Accuhaler®, is a multidose dry powder inhaler containing the inhaled corticosteroid fluticasone propionate (100, 250, or 500 μg) and the long-acting β-agonist (LABA) salmeterol (50 μg) as the xinafoate salt (Figure 8.1). The combination of fluticasone propionate (50, 125, 250 μg) and salmeterol (25 μg) as the xinafoate salt is also available in a metered-dose hydrofluoroalkane inhaler (Seretide Evohaler®). Serevent Diskus® is a multidose dry powder inhaler containing only the long-acting β-agonist salmeterol (50 μg) as the xinafoate salt. The Diskus® technology is a specially designed plastic inhalation system containing a double-foil blister strip of active ingredient(s) formulated with lactose powder (12.5 mg).

Advair Diskus® is prescribed for treatment of asthma and chronic obstructive pulmonary disease (COPD). The definition, causes, prevalence, and treatment goals for asthma were discussed in Chapter 6. In this chapter, the discussion will be limited to the definition, causes, prevalence, and treatment goals for COPD.

Chronic obstructive pulmonary disease (COPD is a chronic inflammatory pulmonary disorder that, in contrast to asthma, is not fully reversible. The airflow limitation is usually progressive and accompanied by significant extrapulmonary effects. Patients diagnosed with emphysema or chronic bronchitis are now considered to have COPD. The primary cause of COPD is exposure to tobacco smoke. While a detailed discussion of the challenges associated with assessing the prevalence of COPD is beyond the scope of this book, it is estimated that 210 million people currently suffer from COPD. More than 3 million people died from COPD in 2005 and projections are that the total number of deaths from COPD will increase by 30% in the next 10 years without intervention to reduce tobacco use.[1]

The treatment goals for COPD are the prevention of disease progression, relief of symptoms, improvement of exercise tolerance and health status, prevention and treatment of complications, prevention and treatment of exacerbations, and reduction of mortality.[2] The $B_2$-agonists, anticholinergics, theophylline, or combinations of these are the first-tier treatment options. When these are inadequate, the second-tier options are Advair Diskus® or Symbicort®. Advair Diskus® 250/50 was approved by the FDA in 2003 for maintenance treatment of airflow obstruction in patients with COPD/chronic bronchitis. Advair Diskus® 250/50 is also the only treatment approved by the FDA (in 2008) to reduce COPD exacerbations. Advair Diskus® 500/50 is approved in the EU for patients with severe COPD, patients who continue to have significant symptoms and exacerbations despite bronchodilator therapy.

The global respiratory market was estimated at \$30 billion in 2006 and Seretide/Advair Diskus® has dominated that market since its launch in 2001.[3] Sales of Seretide/Advair Diskus® for asthma and COPD rose 8% to £4.1 billion in 2008. Advair sales in the United States rose 6% to £2.2 billion in 2008. Advair sales in Emerging Markets rose 26% to £215 million in 2008 and, in Japan, rose >50% to £83 million since the 2007 launch.[4]

Why has Advair Diskus® been so successful? What is the advantage of taking fluticasone propionate and salmeterol

*Pharmaceutical Process Chemistry for Synthesis: Rethinking the Routes to Scale-Up*, By Peter J. Harrington

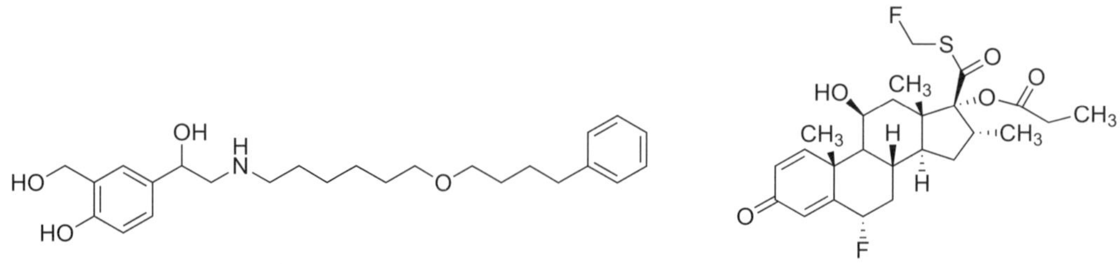


**FIGURE 8.1** Salmeterol (**1**) and fluticasone, active ingredients in Advair Diskus®.

together, rather than taking them separately? In three words, the advantages are convenience, safety, and synergy. First, it is convenient because patients need only remember to use a single inhaler twice daily. When using two inhalers, patients are more likely to forget the fluticasone propionate inhaler since inhalation does not provide immediate relief of symptoms.

Second, *it is not safe to use salmeterol without also using an ICS such as fluticasone propionate*. The long-acting β-agonists salmeterol and formoterol may reduce the number of asthma eposides but also increase their severity. The label black box warning for Advair Diskus® reads:

> Long-acting beta2-adrenergic agonists, such as salmeterol, one of the active ingredients in ADVAIR DISKUS, may increase the risk of asthma-related death. A U.S. study showed an increase in asthma-related deaths in patients receiving salmeterol (13 deaths out of 13,176 patients treated for 28 weeks on salmeterol versus 3 out of 13,179 patients on placebo).

The SMART (Salmeterol Multicenter Asthma Research Trial) study, published in 2006, revealed that the risks were even greater for African-Americans than for Whites.[5] These concerns recently (December 2008) prompted an FDA advisory committee to recommend that Serevent® and Foradil® no longer be used for asthma. The available data suggests the risk of severe asthma attack is considerably reduced when salmeterol is coadministered with an ICS, as in Advair Diskus®.

Third, at some concentrations, salmeterol and fluticasone propionate are synergistic. A detailed discussion of the observed synergistic effects for patients with asthma or COPD is beyond the scope of this introduction.[6,7]

The future for Advair Diskus® is uncertain. On the plus side, there have been many recent positive clinical results for Advaid Diskus®. Genetic differences in how people respond to short-acting β-agonists such as albuterol were not observed in a study of patients treated with Advair Diskus® or Serevent Diskus®. All genotypes showed sustained improvement. In a 3-year study of patients with COPD, patients treated with Advair Diskus® 500/50 experience the fewest adverse cardiac events compared with patients on other treatments or placebo (placebo 21%, fluticasone propionate 20%, salmeterol 19%, and Advair Diskus® 500/50 17%). Finally, in a 3-year study of patients with COPD, Advair Diskus® 250/50 showed no adverse effect on bone mineral density measured at lumbar spine and total hip compared to salmeterol alone.

On the minus side, key patents are set to expire in 2010 in the United States and 2013 in Europe, opening the door for generic competition, and many alternative combination therapies are or will soon be available. These include Symbicort® (budesonide/formoterol) from AstraZeneca, mometasone-indacaterol and mometasone/formoterol from Novartis-Schering-Plough, fluticasone/formoterol from SkyePharma-KOS, ciclesonide/formoterol from Altana-Sanofi Aventis, and beclometasone/formoterol from Chiesi. Another challenge to Advair Diskus® comes from the GSK-Theravance "Beyond Advair" collaboration to bring to market a once-daily combination of a LABA (either GSK642444 from GSK or GSK159757 from Theravance) and an inhaled corticosteroid (likely fluticasone furoate).

Also on the minus side, there is likely to be a decrease in the number of prescriptions written for Advair Diskus® for patients with asthma. The label black-box warning for Advair Diskus® also reads:

> When treating patients with asthma, only prescribe ADVAIR DISKUS for patients not adequately controlled on other asthma-controller medications or whose disease severity clearly warrants initiation of treatment with two maintenance therapies.

Physicians continue to prescribe Advair Diskus® as a frontline therapy for asthma despite this recommendation. More than 25% of U.S. revenues from Advair Diskus® sales in 2005 came from prescriptions as frontline therapy. As physicians heed the label recommendation, many new asthma patients started on a frontline therapy may not need Advair Diskus® as a second-line therapy.

The Diskus® technology was a timely advance in asthma therapy as its introduction coincided with the eventual phaseout of chlorofluorocarbon-containing inhalers. But the

Diskus® technology may pose a risk to young patients with milk allergy. Some lots of USP lactose, obtained from skim milk using a process designed to eliminate milk protein, contained variable amounts of milk proteins.[8]

## 8.2 SALMETEROL TAIL BROMIDES AND ALDEHYDES BY WILLIAMSON ETHER SYNTHESIS

At first glance, salmeterol (**1**) has a hydrophilic "head" and a hydrophobic "tail." There is one chiral center and our objective is only to produce the racemate. The (*R*)-enantiomer (**2**) is a potent bronchodilator and much recent activity is focused on developing an efficient route to this enantiomer. On the other hand, the (*S*)-enantiomer (**3**) is also a potent bronchodilator that has a better selectivity for $\beta_2$-receptors and perhaps reduced side effects. Routes to the enantiomers **2** and **3** are pertinent to the objective and included in the discussion.[9,10]

A rapid assembly of salmeterol (**1**) from three components suggests disconnections at a carbon–nitrogen bond of the amine and at a carbon–oxygen bond of the ether in the tail. The ultimate source of the amine nitrogen could be ammonia or an ammonia surrogate. The ammonia surrogate could be attached to the head or tail prior to completing the chain. From these disconnections, 1,6-dibromohexane, hexane-1,6-diol, 4-phenyl-1-butanol, and 1-bromo-4-phenylbutane emerge as promising raw material candidates for the tail. 1,6-Dibromohexane and hexane-1,6-diol are both inexpensive and available in bulk. 4-Phenyl-1-butanol is commercially available, yet surprisingly expensive. Thus, one of the challenges in selecting a process for salmeterol (**1**) will be selecting a process for the manufacture of 4-phenyl-1-butanol from the myriad of options. The presentation will be expanded beyond these four raw materials to include construction of the six-carbon chain and the four-carbon chains of the two tail fragments by reduction of alkynes or alkenes. Alkyne and alkene intermediates may take on greater significance if the hydrogenation to produce the saturated chain and a second transformation, perhaps a late-stage deprotection, can be accomplished in a single operation.

The strategies for constructing the trisubstituted aromatic head all ultimately begin with a phenol. The longest linear sequence from raw materials to salmeterol (**1**) originates in the construction of this fragment (Figure 8.2). Transformations used to convert the phenol to an appropriately functionalized fragment may include aromatic acylation, alkylation, and bromination by electrophilic substitution, aromatic acylation or alkylation by capture of an arylmetal intermediate, reduction of an aldehyde, ketone, or benzoate ester, halogenation of an aryl methyl ketone, and alcohol protection and deprotection. The discussion of options for constructing the aromatic head will focus on selecting (1) a commercially available and inexpensive phenol raw material, (2) an appropriately protected and functionalized fragment ready for the carbon–nitrogen bond formation, and (3) the choreography for the transformations converting the raw material to the protected and functionalized fragment.

### 8.2.1 (4-(6-Bromohexyloxy)butyl)benzene (4)

Williamson ether synthesis joining 4-phenyl-1-butanol and 1,6-dibromohexane can be run in THF or in toluene using sodium hydride as the base. The issues associated with handling sodium hydride can be avoided using powdered potassium hydroxide or concentrated aqueous sodium hydroxide as the base under phase transfer conditions. The statistical problem associated with displacement of one bromide from the dibromohexane is addressed by using excess (2.0–3.0 equivalents) of this less expensive reactant. The separation of the product from the excess dibromohexane (bp 243°C) is accomplished by chromatography and/or high-vacuum distillation (bp150–170°C at 0.1 mmHg for (4-(6-bromohexyloxy)butyl)benzene) (**4**).

The reaction of 4-phenyl-1-butanol[11] with sodium hydride (1.3 equivalents) and 1,6-dibromohexane[12] (2.0 equivalents) in THF is refluxed for 27 h. After a routine water–ethyl ether workup, (4-(6-bromohexyloxy)butyl)benzene (**4**) is separated from the residue by chromatography and distillation at 0.1 mmHg (48%).[13]

The reaction of 4-phenyl-1-butanol with sodium hydride (1.1 equivalents) and 1,6-dibromohexane (3.0 equivalents) in THF is stopped after 7 h at 25°C. After a routine water–ethyl ether workup, (4-(6-bromohexyloxy)butyl)benzene (**4**) is separated from the residue by chromatography (81%).[14]

The reaction of 4-phenyl-1-butanol with sodium hydride (0.9 equivalents), 1,6-dibromohexane (1.2 equivalents), tetrabutylammonium bromide and sodium iodide (catalytic quantities) in THF is aged at 45–50°C for 15–20 h. Crude (4-(6-bromohexyloxy)butyl)benzene (**4**) is isolated after routine water–toluene workup. No yield is available for this step. The yield for a two-step sequence including this step is 70%.[15]

The reaction of 4-phenyl-1-butanol with powdered potassium hydroxide (3.7 equivalents), 1,6-dibromohexane (2.0 equivalents), and tetrabutylammonium hydrogen sulfate (10 mol%) is stirred for 20 h at 25°C. After a routine water–ethyl ether workup, (4-(6-bromohexyloxy)butyl)benzene (**4**) is separated from the residue by distillation (bp 150°C at 0.1 mmHg) (74%).[16]

The reaction of 4-phenyl-1-butanol with powdered potassium hydroxide (3.7 equivalents), 1,6-dibromohexane (2.0 equivalents), and tetrabutylammonium hydrogen sulfate (10 mol%) in toluene is stirred for 2 h at 25°C. After a routine water–toluene workup, (4-(6-bromohexyloxy)butyl)benzene (**4**) is separated from the residue by fractional distillation at 0.1 mmHg (71%).[17]

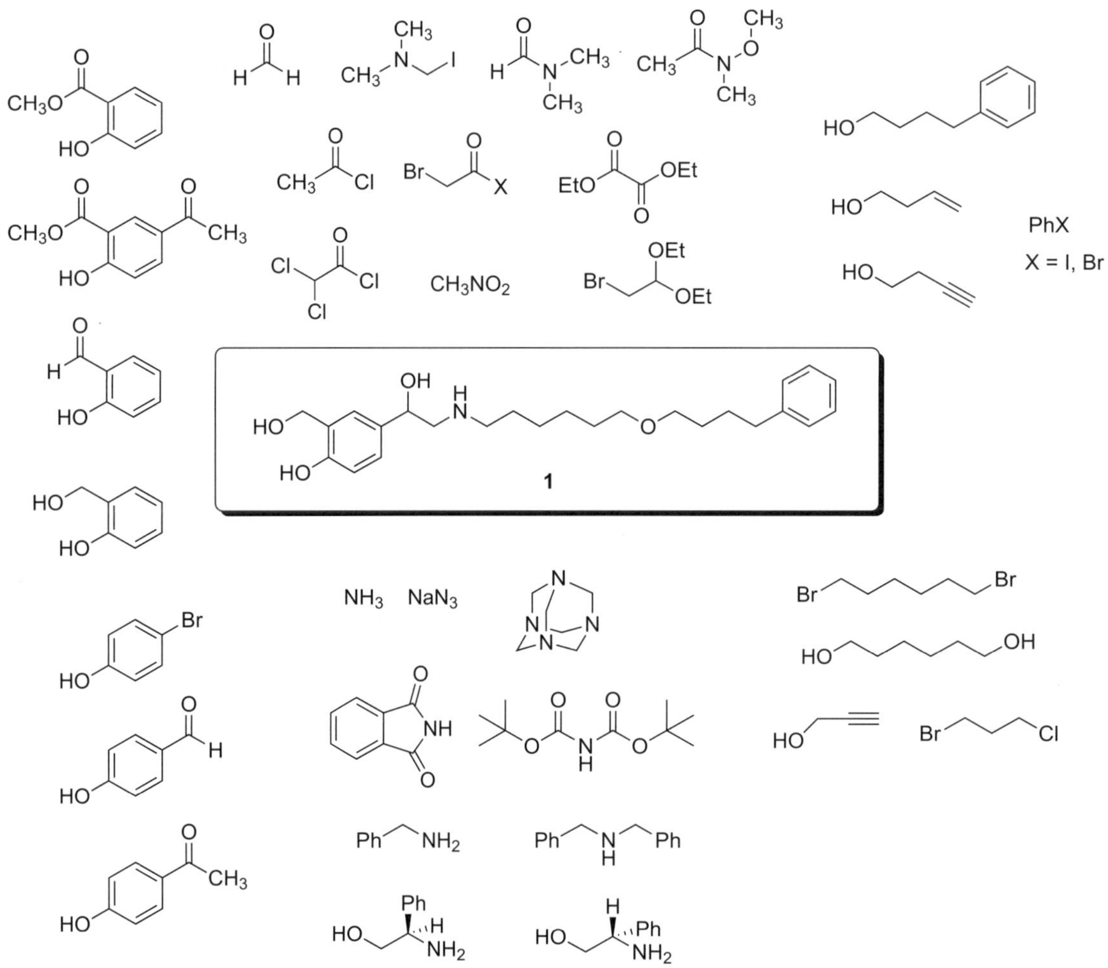


**FIGURE 8.2** Raw materials for salmeterol (**1**).

A small-scale reaction of 4-phenyl-1-butanol with 10 M sodium hydroxide (1.0 equivalent), 1,6-dibromohexane (2.0 equivalents), and benzyltriethylammonium chloride (5 mol%) in dichloromethane is stirred for 9 h at 25°C. After a routine water–dichloromethane workup, (4-(6-bromohexyloxy)butyl)benzene (**4**) is separated from the residue by chromatography (51%).[18]

Perhaps the most scaleable process and near-optimal results are demonstrated using excess 12.5 M sodium hydroxide as the base and tetrabutylammonium hydrogen sulfate as the phase transfer catalyst. A process description can be extrapolated from the process for (3-(6-bromohexyloxy)propyl)benzene (**5**). The reaction of 3-phenyl-1-propanol with 12.5 M sodium hydroxide (9.1 equivalents), 1,6-dibromohexane (3.0 equivalents), and tetrabutylammonium hydrogen sulfate (6.7 mol%) is stirred at 25°C for 30 h. After a routine water–ethyl ether workup, (3-(6-bromohexyloxy)propyl)benzene (**5**) is separated from the residue by chromatography (81%). The process using 4-phenyl-1-butanol, concentrated sodium hydroxide (12 equivalents), 1,6-dibromohexane (3.0 equivalents), and tetrabutylammonium hydrogen sulfate (8.8 mol%) at 25°C for 26 h affords (4-(6-bromohexyloxy)butyl)benzene (**4**) (bp 160–163°C at 0.1 mmHg). No yield is available (Scheme 8.1).[13,19]

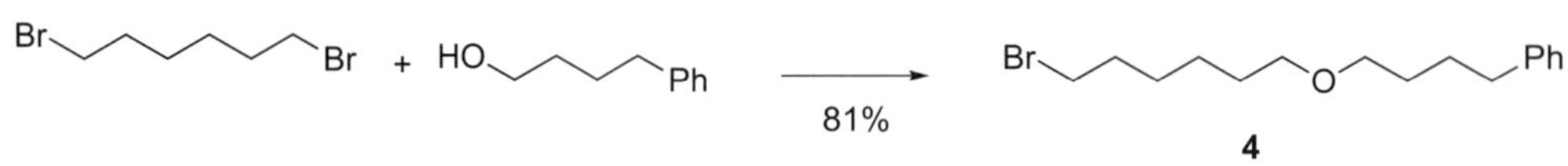


**SCHEME 8.1** (4-(6-Bromohexyloxy)butyl)benzene (**4**) from 4-phenyl-1-butanol.

Lower yields are expected using 50% aqueous sodium hydroxide (19 M) as the base. The reaction of 4-(3-bromophenyl)butan-1-ol (**6**) with 19 M sodium hydroxide (120 equivalents), 1,6-dibromohexane (4.0 equivalents), and tetrabutylammonium bromide (5.9 mol%) is stirred at 20°C for 48 h. After a routine water–ethyl acetate workup, 1-bromo-3-(4-(6-bromohexyloxy)butyl)benzene (**7**) is separated from the residue by chromatography (58%).[20]

In every available process the workup involved chromatography, high-vacuum/high-temperature distillation, or both. If (4-(6-bromohexyloxy)butyl)benzene (**4**) is a key intermediate in salmeterol (**1**) manufacturing, identifying an alternative workup procedure for this intermediate is a high priority.

A good place to begin design of a workup procedure is with pencil and paper. List the starting materials and potential side products. Separate the side products into two categories. First-generation side products are generated by a single reaction of the starting materials or product. Second-generation side products are generated by two reactions of the starting materials or product. Next highlight the starting materials and side products in each category which will react as the product reacts in the next chemical step. Focus the development effort on eliminating the highlighted starting materials and first-generation side products (Figure 8.3).

First-generation side products of this simple Williamson ether synthesis include the terminal alkenes from elimination and the alcohols from bromide displacement by hydroxide. Since the next reaction is a bromide displacement, the workup procedure should eliminate 1,6-dibromohexane and first-generation side products that contain a bromide. A distillation at 10 mmHg to separate excess 1,6-dibromohexane (bp 243°C or 112–114°C at 11 mmHg) should also remove 6-bromo-1-hexene (**8**) (bp 47–51°C at 16 mmHg) and 6-bromo-1-hexanol (**9**) (bp 106–106°C at 5 mmHg). Perhaps the crude material containing higher boiling side products can be used in the next step. We should anticipate that the quality required is likely to increase as the number of steps from the bromide (**4**) to salmeterol (**1**) decreases.

### 8.2.2 6-(4-Phenylbutoxy)hexanal (10)

6-(4-Phenylbutoxy)hexanal (**10**) is produced by oxidation of the alcohol **11** with pyridinium chlorochromate (PCC) or Kornblum oxidation of the bromide **4**. There are two routes to alcohol **11**. 6-(4-Phenylbutoxy)hexan-1-ol (**11**) is produced from (4-(6-bromohexyloxy)butyl)benzene (**4**) by bromide displacement by acetate and hydrolysis of the ester **12**. The reaction with sodium acetate trihydrate (8.0 equivalents) and trioctylpropylammonium chloride (15 mol%) in water is complete in 2.5 h at 100°C. Sodium hydroxide (3.1 equivalents) and ethanol are added. After aging for 30 min at 25°C, a routine water–ethyl ether workup affords the crude alcohol **11** (90%) (Scheme 8.2).[13]

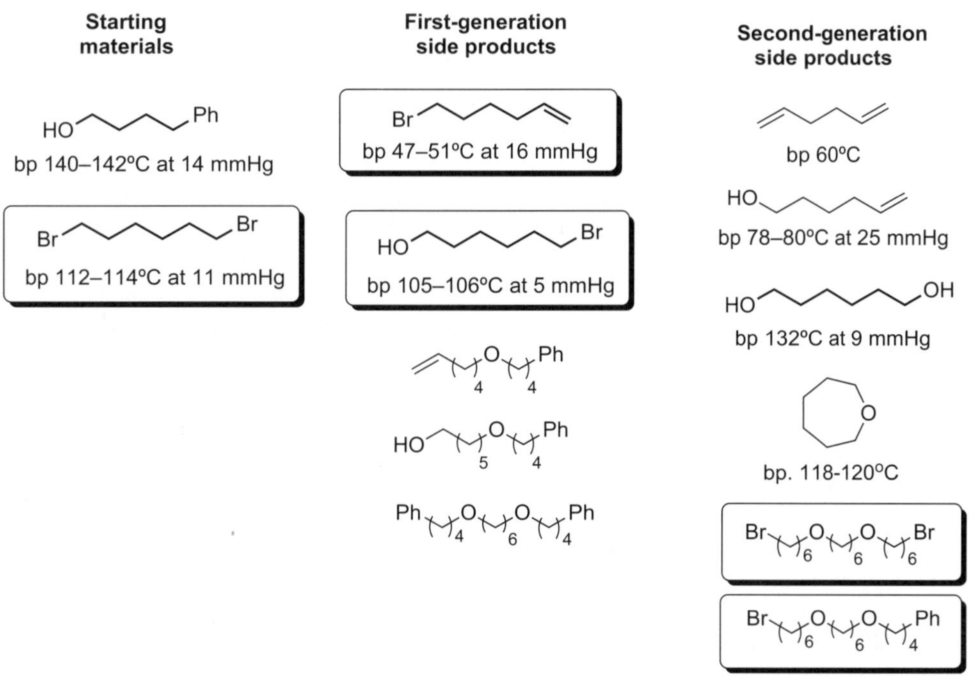


**FIGURE 8.3** Side products from the reaction of 4-phenyl-1-butanol with 1,6-dibromohexane.

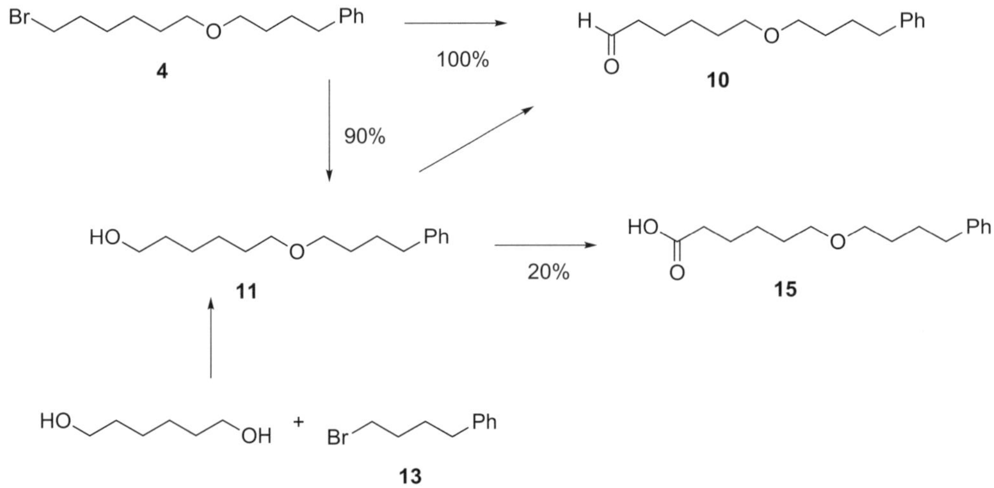

**SCHEME 8.2** 6-(4-Phenylbutoxy)hexanal (**10**) and 6-(4-phenylbutoxy)hexanoic acid (**15**) from bromide **4** and alcohol **11**.

In the second route to alcohol **11**, 4-phenyl-1-butanol is converted to the bromide **13** or methanesulfonate **14** and this is reacted with excess 1,6-hexanediol[21] and sodium hydride in toluene at 80°C. Alcohol oxidation with PCC affords 6-(4-phenylbutoxy)hexanal (**10**) in 81–92% yield for the three step sequence. Experimental details are not available.[22]

In the Kornblum oxidation, the mixture produced by addition of (4-(6-bromohexyloxy)butyl)benzene (**4**) to a 150°C degassed mixture of sodium bicarbonate (5.0 equivalents) in DMSO (7 mL/g of **4**) is heated at 150°C for 5–6 min. A routine water–ethyl ether workup affords aldehyde **10** in quantitative yield.[23,24]

### 8.2.3 6-(4-Phenylbutoxy)hexanoic Acid (15)

6-(4-Phenylbutoxy)hexan-1-ol (**11**) is oxidized by reaction with pyridinium dichromate (3.5 equivalents) in DMF at 25°C for 15 h. The mixture is diluted with water and extracted with ethyl ether. The extracts are washed with water, filtered through silica gel, and concentrated at reduced pressure to afford the acid **15** (20%).[13]

### 8.2.4 (4-(6-Iodohex-2-ynyloxy)butyl)benzene (16)

Starting with 1,6-dibromohexane leaves no handle for modification of the six-carbon chain. In one early route, an alternative sequence to a six-carbon chain via alkyne alkylation uses inexpensive materials and provides a handle for introducing functionality along the chain. 4-Phenyl-1-butanol is converted to the bromide **13**. A mixture of bromide **13**, propargyl alcohol[25] (1.0 equivalent), 50% aqueous sodium hydroxide (8.5 equivalents), and tetrabutylammonium hydrogen sulfate (1.7 mol%) is aged 72 h at 25°C. After a routine water–ethyl ether workup, (4-(prop-2-ynyloxy) butyl)benzene (**17**) is isolated by chromatography (55%). The alkyne alkylation is accomplished using lithium amide (1.1 equivalents) and 1-bromo-3-chloropropane[26] (1.1 equivalents) in liquid ammonia–ethyl ether over 3 h at −33°C. After allowing the ammonia to evaporate, water is cautiously added and the product is extracted with ethyl ether. The combined extracts are concentrated at reduced pressure and (4-(6-chlorohex-2-ynyloxy)butyl)benzene (**18**) is isolated from the residue by distillation (bp140–150°C at 0.3 mmHg) (61%). Finkelstein reaction of chloride **18** with sodium iodide (3.0 equivalents) in 2-butanone is complete in 6 h at reflux or 48 h at 25°C. A routine workup with aqueous sodium thiosulfate–ethyl ether affords the iodide **16** (73%) (Scheme 8.3).[13]

### 8.2.5 1-Bromo-6-(but-3-ynyloxy)hexane (19), 6-(But-3-ynyloxy)hexanal (20), and 1-Bromo-6-(but-3-enyloxy)-hexane (21)

GSK developed two late-stage salmeterol intermediates that can serve as hubs for rapid synthesis of a library of analogs with the possibility of functionality in the four-carbon chain and an array of aromatics replacing the phenyl at the end of the tail. The aromatics are linked to the tail by the Sonogashira coupling of an alkyne or Heck arylation of an alkene with an aryl bromide or iodide. The saturated 4-carbon chain is then produced by hydrogenation of the alkyne or alkene (Scheme 8.4).

The tail precursors for the alkyne approach are 1-bromo-6-(but-3-ynyloxy)hexane (**19**) and 6-(but-3-ynyloxy)hexanal (**20**). The reaction of 3-butyn-1-ol[27] with 1,6-dibromohexane (1.5 equivalents), 50% aqueous sodium hydroxide (6.6 equivalents), and tetrabutylammonium bromide (1.1 mol%) is aged at 50°C overnight. After a routine water–methyl *tert*-butyl ether workup, 1-bromo-6-(but-3-ynyloxy) hexane (**19**) is separated from the residue by fractional distillation at 92–98°C and 0.4 mmHg (31%).[20]

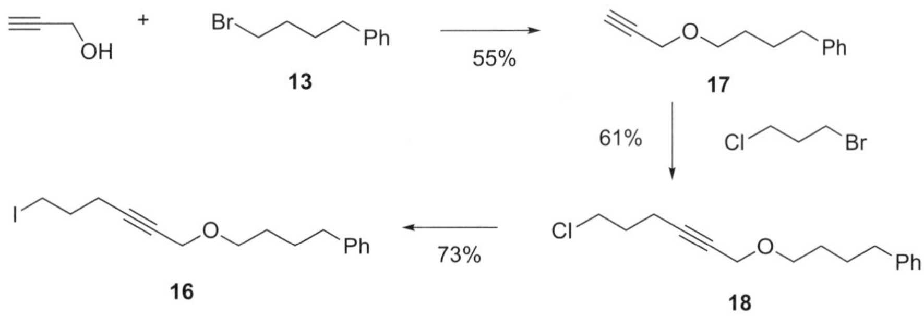


**SCHEME 8.3** (4-(6-Iodohex-2-ynyloxy)butyl)benzene (**16**) from 4-phenyl-1-butanol.

The reaction of 3-butyn-1-ol with 1,6-dibromohexane (3.0 equivalents), 50% aqueous sodium hydroxide (6.8 equivalents), and tetrabutylammonium hydrogen sulfate (1.3 mol%) is aged at 25°C for 72 h. After a routine water–dichloromethane workup, 1-bromo-6-(but-3-ynyloxy)hexane (**19**) is separated from the residue by chromatography (79%).[20, 28–33]

6-(But-3-ynyloxy)hexanal (**20**) is produced by Kornblum oxidation of the bromide **19**. The mixture produced by addition of 1-bromo-6-(but-3-ynyloxy)hexane (**19**) to a 150°C degassed mixture of sodium bicarbonate (5.3 equivalents) in DMSO (15 mL/g of **19**) is heated at 150°C for 20 min. A routine water–ethyl ether workup affords the aldehyde **20** (86%).[20,33]

The tail precursor for the alkene approach is 1-bromo-6-(but-3-enyloxy)hexane (**21**). The reaction of 3-buten-1-ol[34] with 1,6-dibromohexane (1.5 equivalents), 50% aqueous sodium hydroxide (4.5 equivalents), and tetrabutylammonium bromide (0.97 mol%) is aged at 50–55°C for 4–6 h. After a routine water–methyl *tert*-butyl ether workup, 1-bromo-6-(but-3-enyloxy)hexane (**21**) is separated from the residue by chromatography (48%).[28]

The reaction of 3-buten-1-ol with 1,6-dibromohexane (3.0 equivalents), 10 M aqueous sodium hydroxide (10 equivalents), and tetrabutylammonium bromide (10 mol%) requires 18 h at 25°C. After a routine water–ethyl acetate workup, 1-bromo-6-(but-3-enyloxy)hexane (**21**) is separated from the residue by chromatography (77%). (Scheme 8.5)[35]

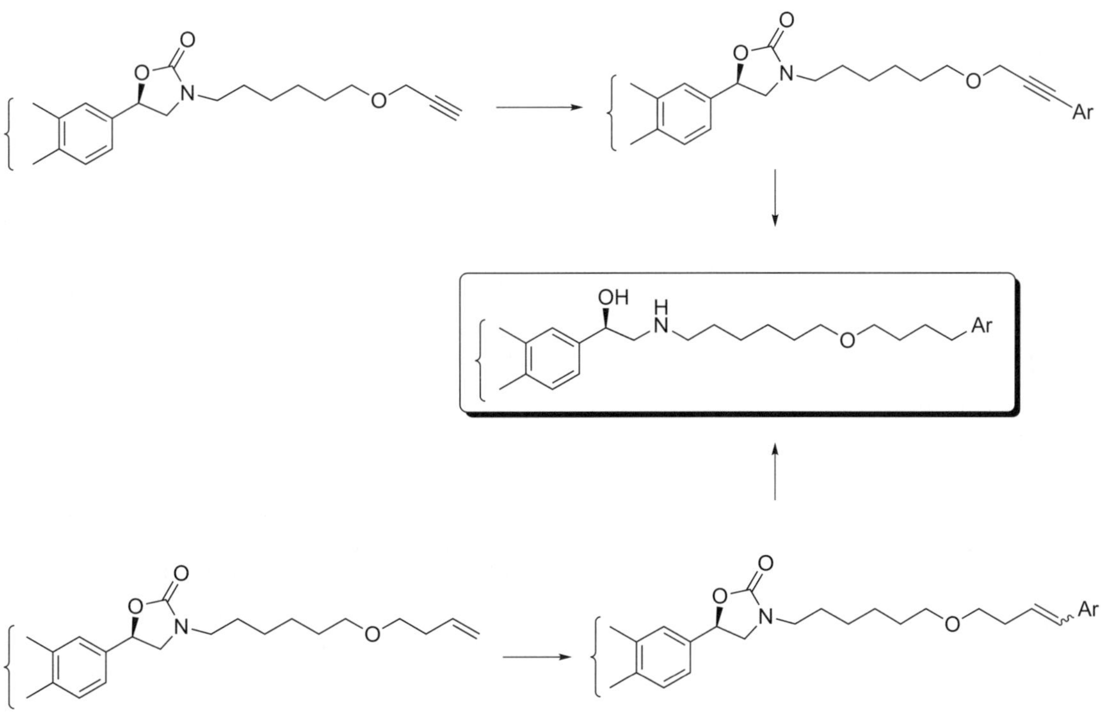


**SCHEME 8.4** Processes for generating a salmeterol (**1**) analog library by the metal-mediated coupling of an aryl halide with an alkyne or alkene.

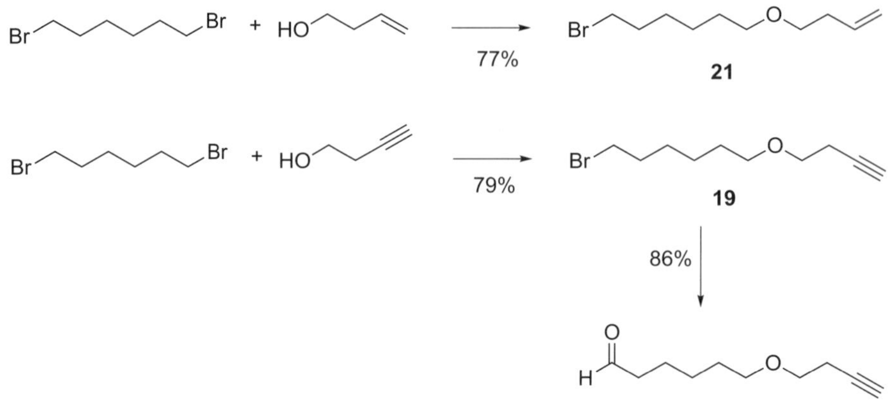

**SCHEME 8.5** Routes to 1-bromo-6-(but-3-ynyloxy)hexane (**19**), 6-(but-3-ynyloxy)hexanal (**20**), and 1-bromo-6-(but-3-enyloxy)hexane (**21**).

## 8.3 SALMETEROL TAIL AMINES BY NUCLEOPHILIC DISPLACEMENT OF A BROMIDE, IODIDE, OR METHANESULFONATE

### 8.3.1 *N*-Benzyl-6-(4-phenylbutoxy)hexan-1-amine (22)

The conversion of (4-(6-bromohexyloxy)butyl)benzene (**4**) to *N*-benzyl-6-(4-phenylbutoxy)hexan-1-amine (**22**) is well documented. Bromide **4** is purified by chromatography or distillation or, preferably, is used crude. The reaction can run neat or in a polar aprotic solvent (DMSO, DMF, acetonitrile, THF). Sodium iodide can be added in catalytic amount, presumably to increase the reaction rate. In principle, high selectivity for the product will require using a large excess of benzylamine. In practice, 2–4 equivalents of benzylamine are used to reduce the amount of benzylamine (bp184–185°C, 90°C at 12–mmHg) that might have to be distilled during a workup. A second base ($Et_3N$, $Cs_2CO_3$) is occasionally used for the same reason. While the isolation of *N*-benzyl-6-(4-phenylbutoxy)hexan-1-amine (**22**) free base requires chromatography, the chromatography can be avoided by isolation of **22** as a hydrobromide or hydrochloride salt.

The reaction of (4-(6-bromohexyloxy)butyl)benzene (**4**) (purified by chromatography) with benzylamine[36] (1.1 equivalents) and cesium carbonate (1.2 equivalents) in DMF is heated at 60–80°C for 5 h. After routine workup with 1 M aqueous sodium bicarbonate and dichloromethane, *N*-benzyl-6-(4-phenylbutoxy)hexan-1-amine (**22**) is isolated from the residue by chromatography (38%).[18]

The reaction of (4-(6-bromohexyloxy)butyl)benzene (**4**) (purified by chromatography) with benzylamine (2.0 equivalents), triethylamine (1.0 equivalent), and sodium iodide (6.2 mol%) in DMSO is aged at 25°C for 17 h. After routine workup with 10% aqueous sodium hydroxide, brine, and ethyl ether, *N*-benzyl-6-(4-phenylbutoxy)hexan-1-amine (**22**) is isolated from the residue by chromatography (70%).[14]

The reaction of (4-(6-bromohexyloxy)butyl)benzene (**4**) (distilled) with benzylamine (4.3 equivalents) in THF is aged at 25°C for 4 days. After dilution with ethyl ether, the suspension is filtered and the filtrate concentrated at reduced pressure. *N*-Benzyl-6-(4-phenylbutoxy)hexan-1-amine (**22**) is isolated from the residue by chromatography (81%).[13]

The neat reaction of (4-(6-bromohexyloxy)butyl)benzene (**4**) (distilled) with benzylamine (4.0 equivalents) is heated at 125°C for 8 h. The excess benzylamine is distilled, presumably at reduced pressure, and the residue is dissolved in 2-butanone. Aqueous hydrobromic acid (2.5 equivalents) is added at 50°C. The suspension is decanted and the decantate phases are separated. The organic phase is concentrated at reduced pressure and the residue is triturated with ethyl ether to afford *N*-benzyl-6-(4-phenylbutoxy)hexan-1-amine (**22**) hydrobromide (60%).[16]

The neat reaction of (4-(6-bromohexyloxy)butyl)benzene (**4**) (distilled) with benzylamine (6.1 equivalents) is heated at 130°C for 2 h. The mixture is cooled to 25°C and 2 M hydrochloric acid is added. After aging for 1 h, the precipitate is filtered, washed with water and ethyl ether, and dried to afford *N*-benzyl-6-(4-phenylbutoxy)hexan-1-amine (**22**) hydrochloride (mp 90–102°C). No yield is provided.[19]

A solution of benzylamine (3.0 equivalents), triethylamine (2.0 equivalents), and sodium iodide catalyst in acetonitrile is heated at 45–50°C. (4-(6-Bromohexyloxy)butyl) benzene (**4**) (crude) is added slowly. After the addition is complete, the mixture is aged at 45–50°C until a completion check (TLC) shows no remaining **4**. Acetonitrile, triethylamine, and benzylamine are distilled at an unspecified temperature and reduced pressure. The residue is separated between water and dichloromethane. The dichloromethane layer is washed with 5 M hydrochloric acid and with water and then concentrated at reduced pressure. Heptane is added

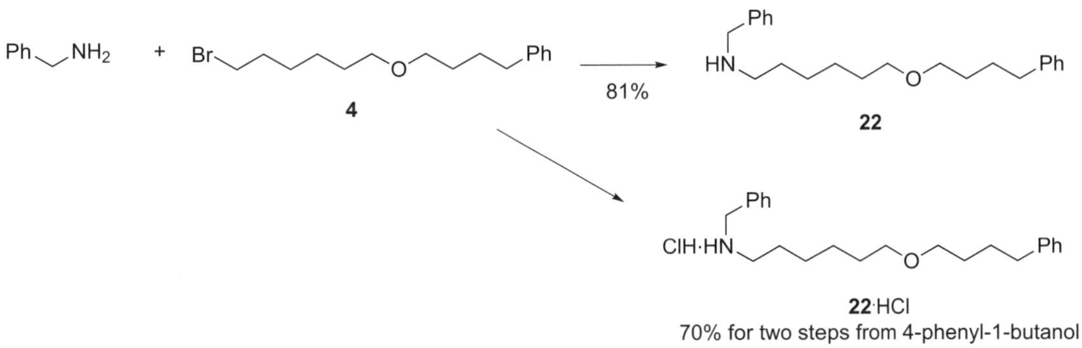


**SCHEME 8.6** *N*-Benzyl-6-(4-phenylbutoxy)hexan-1-amine (**22**) from the bromide **4**.

to the syrup residue and the resulting suspension is filtered. The solid is dissolved in isopropanol at 55°C. The solution is cooled to 40°C, diluted with heptane, and cooled to 10–15°C. The resulting suspension is filtered and the solid is dried at 50–55°C and unspecified pressure to afford *N*-benzyl-6-(4-phenylbutoxy)hexan-1-amine (**22**) hydrochloride (mp 135–140°C) (70% for two steps from 4-phenyl-1-butanol) (Scheme 8.6).[15]

There are concerns with scaling this process. First, in a distillation at reduced pressure, pot temperatures and distillation times tend to increase with scale. The stability of the amine hydrobromide salt should be evaluated and the distillation designed to minimize decomposition. Second, in theory, the first distillation could proceed until only the amine hydrobromide salt remains in the pot (distillation to dryness). A distillation end point must be identified. At the end point, sufficient contaminants should remain in the pot to prevent crystallization of the salt and loss of agitation or a broken agitator. Third, heptane should be added *during* the dichloromethane distillation.

Routine testing of pot bottoms from small-scale lab distillations is highly recommended. A distillation can be performed reproducibly 20 times in the lab. As the distillation procedure is refined by repetition, the range of conditions (particularly jacket temperature and elapsed time) the mixture is exposed to decreases. When the refined process is transferred to the pilot plant, it is almost a certainty that the first distillation at pilot plant scale will take much longer and require a hotter jacket temperature, testing the mixture as it had not been tested previously. Stability testing of the bottoms provides invaluable peace of mind during those long hours standing next to what seems like a very large reactor waiting for a distillation to reach the end point.

### 8.3.2 6-(4-Phenylbutoxy)hexan-1-amine (23)

6-(4-Phenylbutoxy)hexan-1-amine (**23**) can be prepared from (4-(6-bromohexyloxy)butyl)benzene (**4**) using sodium azide, phthalimide, or benzylamine as ammonia surrogates (Scheme 8.7). The azide process can be extrapolated from a procedure for 7-(3-phenylpropoxy)heptan-1-amine (**24**). The procedure for azide **26** from bromide **25** is not provided. The crude azide **26** is dissolved in THF and triphenylphosphine (1.0 equivalent) is added in portions at 25°C. After aging at 25°C for 2 h, water is added and aging at 25°C is continued for 18 h. The volatile components are removed, presumably at reduced pressure, and the residue is taken up in hexanes. The solid is filtered and the filtrate liquors washed with methanol–water (3:7) and concentrated at reduced pressure. The residue is dissolved in dilute hydrochloric acid and washed with ethyl ether. Sodium hydroxide is added (to pH 6) and the resulting solution is washed with ethyl ether. Sodium hydroxide is added (to pH 10) and the product is extracted with ethyl ether. Hydrogen chloride in dioxane (1.0 M) is added to the ether extract and the solvents are distilled at reduced pressure to afford the amine **24** hydrochloride (46% for two steps from bromide **25**).[37]

The reaction of (4-(6-bromohexyloxy)butyl)benzene (**4**) with potassium phthalimide (1.0 equivalents) in acetone requires 70 h at reflux. The suspension is cooled to 25°C and filtered and the liquors are concentrated at reduced pressure. Dichloromethane and 40% aqueous methylamine are added to the residue and the resulting mixture is aged at 25°C overnight. The phases are separated and the aqueous phase extracted with dichloromethane. The combined organic phases are dried and concentrated at reduced pressure. The residue is taken up in ethyl ether and polish filtered. The ether solution is concentrated at reduced pressure and the residue distilled to afford 6-(4-phenylbutoxy)hexan-1-amine (**23**) (bp 143–145°C at 0.2 mmHg). No yield is provided.[38]

The azide and phthalimide methods are both suitable for generating gram quantities of the amine. Larger quantities of

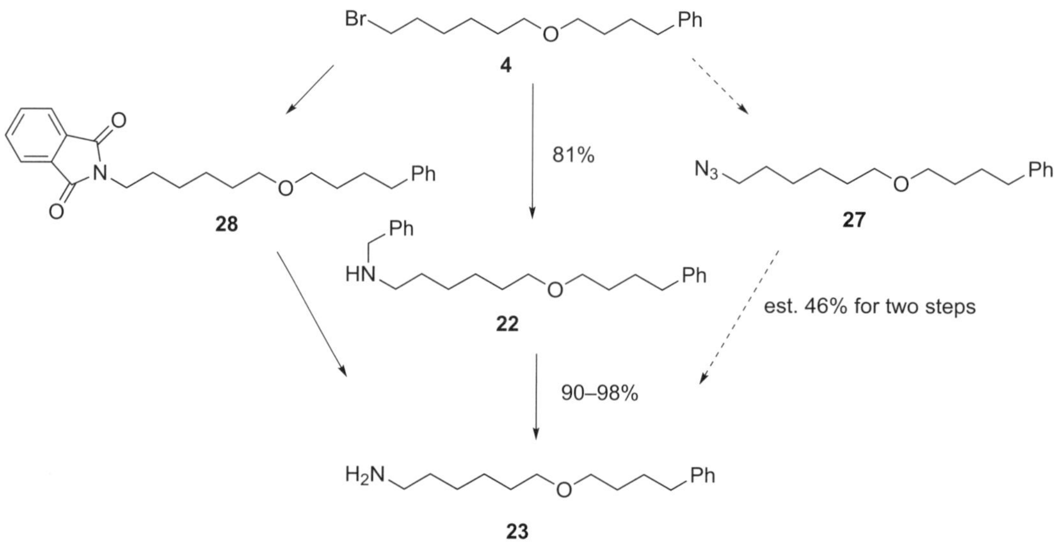


**SCHEME 8.7** 6-(4-Phenylbutoxy)hexan-1-amine (**23**) from bromide **4**.

the amine are best produced by the benzylamine route. The hydrogenolysis can be accomplished using 5% palladium on carbon or a mixed palladium and platinum on carbon catalyst in ethanol at 25°C and atmospheric pressure. After the theoretical amount of hydrogen is absorbed, the suspension is filtered and the filtrate concentrated at reduced pressure to afford 6-(4-phenylbutoxy)hexan-1-amine (**23**) (90–98%).[13,16]

### 8.3.3 (*R*)-2-Phenyl-2-(6-(4-phenylbutoxy)hexylamino)-ethanol (29)

The reaction of 6-(4-phenylbutoxy)hexyl methanesulfonate (**30**) with (*R*)-(−)-2-phenylglycinol[39] (1.1 equivalents) and diisopropylethylamine (1.2 equivalents) in acetonitrile is heated at 75°C for 20 h. (*R*)-2-Phenyl-2-(6-(4-phenylbutoxy)hexylamino)ethanol (**29**) is isolated by routine water–toluene workup and chromatography (64%).[40]

## 8.4 SALMETEROL HEAD ELECTROPHILES

Salmeterol head electrophiles used in the carbon–nitrogen bond formation with tail amines can have an adjacent ketone, alcohol, or silyl ether. The phenol hydroxyl group can be unprotected or protected as an ester, ether, or ketal. The functionality at the position adjacent to the phenol oxygen can be an ester, aldehyde, or alcohol protected as an ester, ether, or ketal.

### 8.4.1 α-Bromoketones

***8.4.1.1 Methyl 5-(2-Bromoacetyl)-2-hydroxybenzoate (31)*** Bromine (1.0–1.1 equivalents) in chloroform is added over 30 min to a solution of methyl 5-acetylsalicylate[41] in chloroform at 25°C. Solid sodium bicarbonate (7.0 equivalents) is carefully added. The salts are filtered and the solvent distilled at reduced pressure to afford crude bromide **31** (99%). Crystallization from petroleum ether or ligroin affords white crystals (mp 90–92°C) (71–72%). A similar bromination procedure starting with the ethyl ester affords ethyl 5-(2-bromoacetyl)-2-hydroxybenzoate **32** in 95% crystallized yield, suggesting that the bromination is highly selective for the monobromoketone and that the crystallization of this early intermediate is not necessary.[18,42,43]

***8.4.1.2 Methyl 2-(Benzyloxy)-5-(2-bromoacetyl)benzoate (33)*** Reaction of methyl 5-acetylsalicylate with benzyl bromide and potassium carbonate in refluxing 2-butanone affords the benzyl ether **34** that can be crystallized from benzene–petroleum ether (mp 71–72°C) (51%). Bromine (1.0 equivalent) in chloroform is added to benzyl ether **33** in chloroform at 25°C. The solvent is distilled at reduced pressure to afford the crude bromide **33** that is crystallized from benzene–cyclohexane (mp 127–128.5°C) (71%) (Scheme 8.8).[43,44]

***8.4.1.3 5-(2-Bromoacetyl)-2-hydroxybenzaldehyde (35)*** 5-(2-Bromoacetyl)-2-hydroxybenzaldehyde (**35**) is produced from salicylaldehyde using bromoacetyl chloride or bromoacetyl bromide and aluminum chloride in dichloromethane. The reaction requires 12–18 h at 35–40°C. A quench into ice water and routine workup affords a mixture of bromoacetyl and chloroacetyl products **35** and **36**, which is separated by crystallization. While the published procedures are all very similar, the reported yield range is 55–80%. This broad range likely reflects the recovery in crystallization and not the actual efficiency of the acylation reaction.

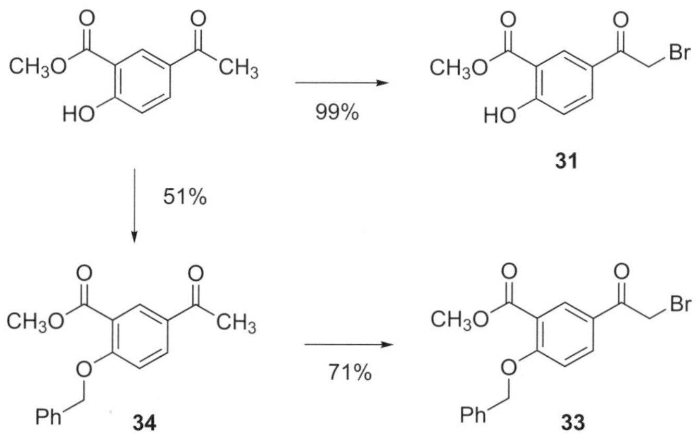


**SCHEME 8.8** Methyl 5-(2-bromoacetyl)-2-hydroxybenzoate (**31**) and methyl 2-(benzyloxy)-5-(2-bromoacetyl)benzoate (**33**) from methyl 5-acetylsalicylate.

Perhaps yield lost to crystallization liquors can be minimized by using the 95:5 mixture of products in the next step.[45–47]

Bromoacetyl bromide[48] (1.2 equivalents) in dichloromethane is added to a suspension of aluminum chloride (4.0 equivalents) in dichloromethane at 10°C. The mixture is aged at 30°C for 1 h. Salicylaldehyde[49] in dichloromethane is added at 30°C. The mixture is aged at 35–40°C for 12–15 h then carefully quenched into ice water and dichloromethane. After a routine workup, the residue is suspended in heptane and the solid filtered, washed with heptane, and dried at 50°C and unspecified vacuum to afford 5-(2-bromoacetyl)-2-hydroxybenzaldehyde (**35**) (55% yield, 97–99% pure by HPLC).[15]

Bromoacetyl chloride[50] (1.2 equivalents) in dichloromethane is added to a suspension of aluminum chloride (4.0 equivalents) in dichloromethane at 25°C. The mixture is aged at 25°C for 30 min. Salicylaldehyde in dichloromethane is added at reflux. The mixture is refluxed for 14–18 h then carefully quenched into ice water and dichloromethane. After a routine workup, the residue is crystallized from dichloromethane–hexane to afford 5-(2-bromoacetyl)-2-hydroxybenzaldehyde (**35**) (66%).[14]

Bromoacetyl chloride (1.3 equivalents) in dichloromethane is added to a suspension of aluminum chloride (5.0 equivalents) in dichloromethane at reflux. After the addition, the mixture is refluxed for 30 min. Salicylaldehyde in dichloromethane is added at reflux. The mixture is refluxed for 16 h and then carefully quenched into ice water and dichloromethane. After a routine workup, the oily solid residue is suspended in dichloromethane–ethyl ether (1:9) and the solid is filtered and dried at 50°C and reduced pressure to afford a 6:1 mixture of 5-(2-bromoacetyl)-2-hydroxybenzaldehyde (**35**) and 5-(2-chloroacetyl)-2-hydroxybenzaldehyde (**36**) (70%).[51–54]

Pure 5-(2-bromoacetyl)-2-hydroxybenzaldehyde (**35**) (mp 117–118°C) can be prepared using bromoacetyl bromide and aluminum bromide in dibromomethane (10%). Pure 5-(2-chloroacetyl)-2-hydroxybenzaldehyde (**36**) (mp 133–134°C) can be prepared from chloroacetyl chloride and aluminum chloride in dichloromethane. Chloroacetyl chloride[55] (1.3 equivalents) in dichloromethane is added to a suspension of aluminum chloride (3.8 equivalents) in dichloromethane at reflux. The mixture is refluxed for 30 min. Salicylaldehyde in dichloromethane is added at reflux. The mixture is refluxed for 18 h and then carefully quenched into ice water and dichloromethane. After a routine workup, the oily solid residue is crystallized from 2-butanone to afford 5-(2-chloroacetyl)-2-hydroxybenzaldehyde (**36**) (40%).[51–54]

Aluminum chloride (4.0 equivalents) and bromoacetyl bromide (1.2 equivalents) are mixed in dichloromethane at 10°C and the suspension is aged at 30°C for 1 h. Salicylaldehyde in dichloromethane is added at 30°C. The mixture is aged at 35°C for 15 h and then carefully quenched into ice water and dichloromethane. After a routine workup, heptane is added to the oily solid residue. The suspension is filtered and the solid is washed with heptane, dried at 50°C, and crystallized from dichloromethane–hexane to afford 5-(2-bromoacetyl)-2-hydroxybenzaldehyde (**35**) (80%).[56]

***8.4.1.4 2-Bromo-1-(4-hydroxy-3-(hydroxymethyl)phenyl)ethanone (37) and 2-Acetoxy-5-(2-bromoacetyl)benzyl Acetate (38)*** There are two multistep routes to 2-bromo-1-(4-hydroxy-3-(hydroxymethyl)phenyl)ethanone (**37**), one route beginning with salicylaldehyde and the other with 4-hydroxyacetophenone. Addition of sodium borohydride (1.0 equivalent) to a solution 5-(2-bromoacetyl)-2-hydroxybenzaldehyde (**35**) in glacial acetic acid at 8–10°C first produces sodium triacetoxyborohydride. During a 1 h age at 25°C, this mild reducing agent selectively reduces the aldehyde in the presence of the ketone.[57] Quench with water, neutralization with saturated aqueous potassium bicarbonate, and a routine workup affords **37** that is crystallized from dichloromethane–hexane (85%) (Scheme 8.9).[56]

The second route begins with chloromethylation of 4-hydroxyacetophenone using aqueous formaldehyde or paraformaldehyde and hydrochloric acid. In one preferred process, paraformaldehyde is added to a mixture of 4-hydroxyacetophenone[58] in concentrated hydrochloric acid at 45–50°C. After the addition is complete, the mixture is aged at 45–50°C for 4.5 h. The resulting precipitate is filtered, washed with water, and crystallized from ethyl acetate to afford 1-(3-chloromethyl)-4-hydroxyphenyl)ethanone (**39**) (75%). A higher yield is reported on 10 kg scale when 4-hydroxyacetophenone is added to a mixture of 40% aqueous formaldehyde and concentrated hydrochloric acid at 45–50°C. After aging at 50°C for 2 h, the mixture is cooled

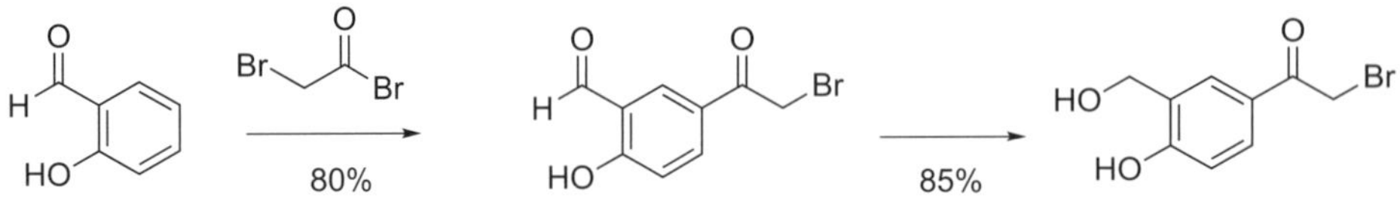


**SCHEME 8.9** 5-(2-Bromoacetyl)-2-hydroxybenzaldehyde (**35**) and 2-bromo-1-(4-hydroxy-3-(hydroxymethyl)phenyl)ethanone (**37**) from salicylaldehyde.

and diluted with water. The solid is filtered, washed with water, and dried at 60°C in air (mp 164°C) (88%).[59,60]

1-(3-Chloromethyl)-4-hydroxyphenyl)ethanone (**39**) is reacted with sodium acetate (1.0–1.1 equivalent) and acetic anhydride (2.0–2.2 equivalents) in glacial acetic acid at 100°C for 2 h. The solution is concentrated at an unspecified temperature and reduced pressure. The diacetate **40** is isolated from the residue by a routine water–chloroform or water–benzene workup and distillation (bp 145°C at 0.5 mmHg) (80–88%). Note that the condenser jacket must be kept warm to prevent crystallization of the distillate (mp 50°C).[60,61]

After selective bromination of **40** with bromine (1.0 equivalent) in chloroform at room temperature, the chloroform solution can be distilled at reduced pressure (80% yield) or washed with water, dried, then concentrated at reduced pressure to afford **38** (100%). The esters are cleaved by reaction with excess 47% hydrobromic acid in methanol at 25°C over 16 h. The solution is cooled to 0–5°C and water is added. The precipitate is suction filtered and washed with water and hexane to afford 2-bromo-1-(4-hydroxy-3-(hydroxymethyl)phenyl)ethanone (**37**) (mp 117–119°C) (69%) (Scheme 8.10).[60,61]

A parallel sequence using supported reagents and scavenging agents also converts 4-hydroxyacetophenone to 2-bromo-1-(4-hydroxy-3-(hydroxymethyl)phenyl)ethanone (**37**). An aminomethylation with expensive *N,N*-dimethylmethyleneiminium iodide[62] and a polymer-bound quaternary ammonium carbonate base in dichloromethane replaces the chloromethylation. The acetate esters are introduced using acetic anhydride and a polymer-bound sulfonic acid catalyst in toluene at 80°C. The bromination is accomplished with a polymer-bound pyridinium bromide perbromide in dichloromethane (75% for three steps from 4-hydroxyacetophenone). The esters are cleaved in 1 M HBr:THF (1:1) (>81%). Experimental details of this sequence are not available.[22]

*8.4.1.5 1-(4-(Benzyloxy)-3-(hydroxymethyl)phenyl)-2-bromoethanone (41)* It is clear that the α-bromoketone **41** is prepared by bromination of the ketone **42** but details for the procedure and yield of α-bromoketone **41** are not available. Details for the multistep synthesis of 1-(4-(benzyloxy)-3-(hydroxymethyl)phenyl)ethanone (**42**) are also not available. Perhaps it is produced from salicylaldehyde by acetylation, alkylation of the phenol with benzyl bromide, and selective reduction of the aldehyde with sodium triacetoxyborohydride (Scheme 8.11).[63]

*8.4.1.6 2-Bromo-1-(2,2-dimethyl-4H-benzo[d][1,3]dioxin-6-yl)ethanone (45)* A well-established GSK process protects the two hydroxyl groups of 2-bromo-1-(4-hydroxy-3-(hydroxymethyl)phenyl)ethanone (**37**) as a ketal (Scheme 8.12). The protection can be accomplished with 2-methoxypropene or 2,2-dimethoxypropane and is typically catalyzed by *p*-toluenesulfonic acid monohydrate. Crystallization from petroleum ether affords the ketal as a low-melting solid.

The reaction of the diol **37** with 2,2-dimethoxypropane[64] (29 equivalents) and *p*-toluenesulfonic acid monohydrate (0.90 mol%) in acetone is stopped after 30 min at reflux. Saturated aqueous bicarbonate is added and the mixture is

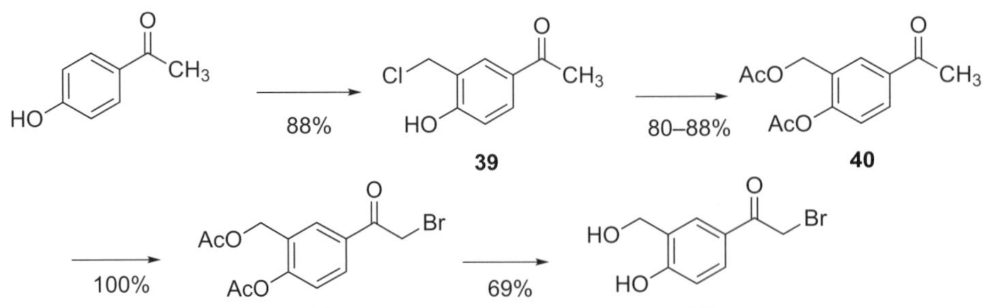


**SCHEME 8.10** 2-Acetoxy-5-(2-bromoacetyl)benzyl acetate (**38**) and 2-bromo-1-(4-hydroxy-3-(hydroxymethyl)phenyl)ethanone (**37**) from 4-hydroxyacetophenone.

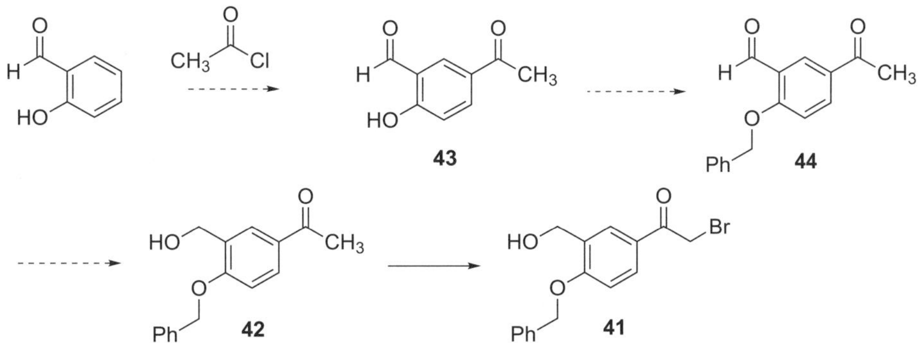

**SCHEME 8.11** Proposed route to 1-(4-(benzyloxy)-3-(hydroxymethyl)phenyl)-2-bromoethanone (**41**) from salicylaldehyde.

extracted with ethyl acetate. The combined extracts are washed with water and brine, dried, and concentrated at reduced pressure. Ketal **45** is isolated from the residue by chromatography (mp 52–54°C) (59%).[61]

The reaction of diol **37** with 2-methoxypropene[65] (6.8 equivalents) and *p*-toluenesulfonic acid (12.8 mol%) in dichloromethane is stopped after 3 h at 23°C. The mixture is filtered through triethylamine-deactivated silica and concentrated at reduced pressure. Ketal **45** is isolated from the residue by chromatography (82%). Crystallization from petroleum ether affords white crystals (mp 47–48°C).[66]

The reaction of diol **37** with 2,2-dimethoxypropane (1.1 equivalents) and *p*-toluenesulfonic acid monohydrate (0.5 mol%) in dichloromethane is aged at 25°C. When a completion check (TLC) reveals no remaining diol **37**, aqueous sodium bicarbonate is added and the organic phase is separated and concentrated at reduced pressure. Ketal **45** is isolated from the residue by chromatography (99%).[56]

The reaction of diol **37** with 2-methoxypropene (1.2 equivalents) and *p*-toluenesulfonic acid monohydrate (0.59 mol%) in dichloromethane is aged at 25°C for 30 min. More 2-methoxypropene (0.38 equivalents) is added and the mixture is aged for 10 min. The mixture is washed twice with aqueous sodium bicarbonate and concentrated at reduced pressure to afford ketal **45** as a beige solid (100%). A sulfonic acid resin can also be used as the catalyst.[67,22]

Theravance developed an alternative multistep process from 5-bromo-2-hydroxybenzyl alcohol[68] through an aryllithium intermediate to the same ketal. 5-Bromo-2-hydroxybenzyl alcohol is expensive and only available in gram quantities. This starting material can be produced from salicylaldehyde via 5-bromosalicylaldehyde[69] or 2-hydroxybenzyl alcohol[70] or from 4-bromophenol.[71–76]

The sequence begins with protection of the diol as a ketal. The ketal formation can be driven using phosphorus pentoxide, zinc chloride, aluminum chloride, montmorillonite K10, or *p*-toluenesulfonic acid. Phosphorus pentoxide (1.4 equivalents) is added to 5-bromo-2-hydroxybenzyl alcohol and acetone (2.8 equivalents) in benzene at <15°C. The suspension is aged at 25°C for 20 h. The solution is decanted from the solid and the solid is washed with benzene. The combined liquors are washed with dilute aqueous bicarbonate and with water and then dried and concentrated at reduced pressure. The residual oil is distilled (bp 100–103°C at 0.5 mmHg) to afford 6-bromo-2,2-dimethyl-4*H*-benzo[*d*][1,3]dioxine (**46**) (88%).[76]

Zinc chloride (2.7 equivalents) is added to 5-bromo-2-hydroxybenzyl alcohol and 2,2-dimethoxypropane (35 equivalents) in acetone at 25°C. After aging for 18 h at 25°C, 1.0 M sodium hydroxide is added until the aqueous layer is basic. Ethyl ether is added to the suspension and the organic phase is decanted, washed with brine, dried, and concentrated at reduced pressure to afford 6-bromo-2,2-dimethyl-4*H*-benzo[*d*][1,3]dioxine (**46**). No yield data is provided.[77]

In a similar procedure, aluminum chloride (0.35 equivalents) in ethyl ether is added to 5-bromo-2-hydroxybenzyl alcohol in acetone at 0°C. After aging 1 h at 25°C, the mixture is again cooled to 0°C and 10% aqueous sodium hydroxide is added at 0–5°C. The organic phase is decanted, washed with water, dried, and concentrated at reduced pressure to afford 6-bromo-2,2-dimethyl-4*H*-benzo[*d*][1,3]dioxine (**46**) (81%).[16]

The diol protection can be catalyzed by Montmorillonite K10 or *p*-toluenesulfonic acid monohydrate. The yield from the reaction with 2,2-dimethoxypropane (5.1 equivalents) and *p*-toluenesulfonic acid (10 mol%) in acetone is 95%.[72,75]

Slow addition of 2.14 M butyllithium (1.1 equivalents) in hexanes to a solution of 6-bromo-2,2-dimethyl-4*H*-benzo[*d*][1,3]dioxine (**46**) in THF at −78°C produces the aryllithium intermediate. *N*-Methoxy-*N*-methylacetamide[78] (1.5 equivalents) is added and the mixture aged, presumably at −78°C, for 2 h. The mixture is quenched with water, diluted with pH 7 phosphate buffer, and extracted with ethyl ether. The ether extract is washed with brine, dried, and concentrated at reduced pressure. The residue is dissolved in ethyl acetate and hexanes is added to produce a precipitate. The suspension is filtered and the solid is presumably dried to afford

1-(2,2-dimethyl)-4*H*-benzo[*d*][1,3]dioxin-6-yl)ethanone (**47**). No yield data is provided.[77]

Slow addition of sodium bis(trimethylsilyl)amide (1.2 equivalents) in THF to the ketone **47** in THF at −78°C generates an enolate. Addition of chlorotrimethylsilane (1.1 equivalents) is followed by addition of bromine (1.0 equivalent). After 10 min, the reaction is diluted with ethyl ether and quenched into premixed 5% aqueous sodium sulfite and 5% aqueous sodium bicarbonate. The organic phase is washed with brine, dried, and concentrated at reduced pressure to afford 2-bromo-1-(2,2-dimethyl-4*H*-benzo[*d*][1,3]dioxin-6-yl)ethanone (**45**). No yield data is provided.[77]

A third route to 2-bromo-1-(2,2-dimethyl-4*H*-benzo[*d*][1,3]dioxin-6-yl)ethanone (**45**) begins with the reduction of salicylaldehyde to 2-hydroxybenzyl alcohol (**48**). The diol **48** is protected with acetone (*p*-toluenesulfonic acid catalyst) in benzene. Friedel-Crafts acylation of 2,2-dimethyl-4*H*-benzo[*d*][1,3]dioxine (**49**) with acetyl chloride and aluminum chloride affords 1-(2,2-dimethyl)-4*H*-benzo[*d*][1,3]dioxin-6-yl)ethanone (**47**). Bromination is accomplished using bromine and benzoyl peroxide in chloroform. No experimental details or yields are provided.[79]

A third route to 1-(2,2-dimethyl)-4*H*-benzo[*d*][1,3]dioxin-6-yl)ethanone (**45**) begins with the chloromethylation of 4-hydroxyacetophenone. Formalin solution (40%) and 4-hydroxyacetophenone are added to concentrated hydrochloric acid. A strong stream of hydrogen chloride gas is passed through the mixture at 20°C for 3 h and the resulting mixture is aged at 20°C for 22 h. Water is added and the precipitate is filtered and washed with hot water and hot benzene. The isolated solid is dissolved in THF. Water and calcium carbonate are added and the mixture is aged at 25°C for 4 h. A routine water–ethyl ether workup affords the crude 1-(4-hydroxy-3-(hydroxymethyl)phenyl)ethanone (**50**) (59%) that is crystallized from ethanol–chloroform (1:2) (18%).The diol **50** is protected using acetone and phosphorus pentoxide (1.0 equivalent). After stirring for 3 h at 0°C, the suspension is warmed to 25°C and the acetone phase is decanted. The solid residue is washed with additional acetone. The combined acetone liquors are concentrated at reduced pressure. A routine water–dichloromethane workup of the residue affords crude 1-(2,2-dimethyl)-4*H*-benzo[*d*][1,3]dioxin-6-yl)ethanone (**47**) (54%), which is purified by distillation (bp 121°C at 0.2 mmHg) or chromatography.[76]

The Theravance route to 2-bromo-1-(2,2-dimethyl-4*H*-benzo[*d*][1,3]dioxin-6-yl)ethanone (**45**) is longer than the GSK process (five steps versus three) and uses more expensive reagents. The Friedel-Crafts acylation of 2,2-dimethyl-4*H*-benzo[*d*][1,3]dioxine (**49**) likely suffers from poor selectivity. Why allocate precious time and resources to develop an alternative route to the well-established GSK intermediate? The target and intermediates in the GSK process are all α-bromoketones. These intermediates are lacrymators and potent alkylating agents (Figure 8.4).[67]

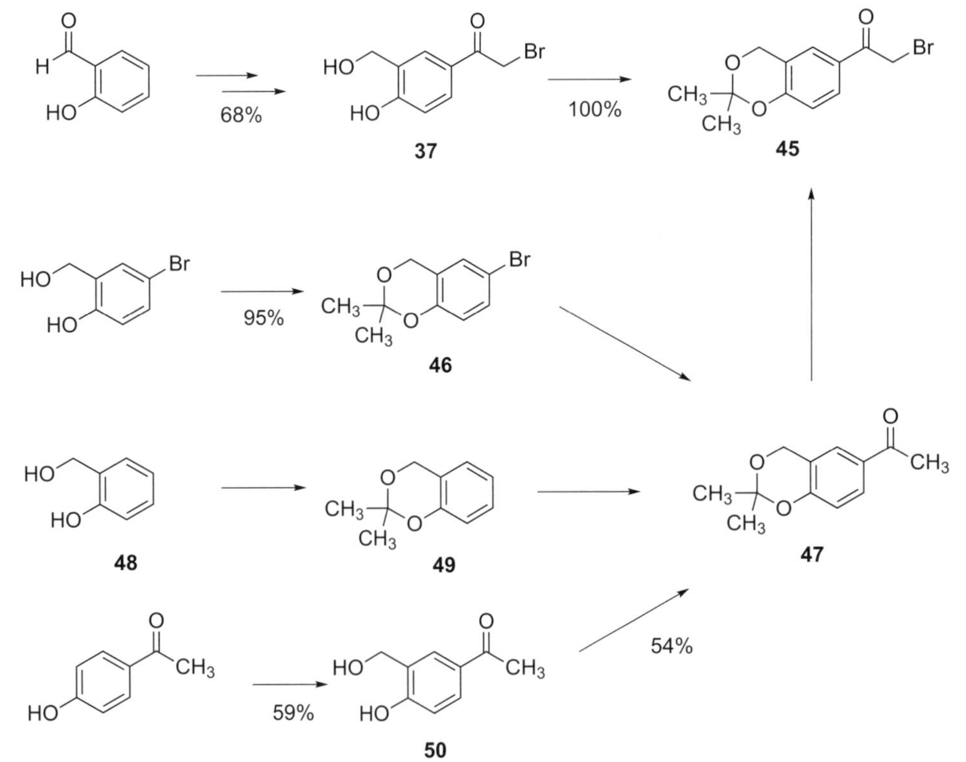


**SCHEME 8.12** 2-Bromo-1-(2,2-dimethyl-4*H*-benzo[*d*][1,3]dioxin-6-yl)ethanone (**45**) from salicylaldehyde or 4-hydroxyacetophenone.

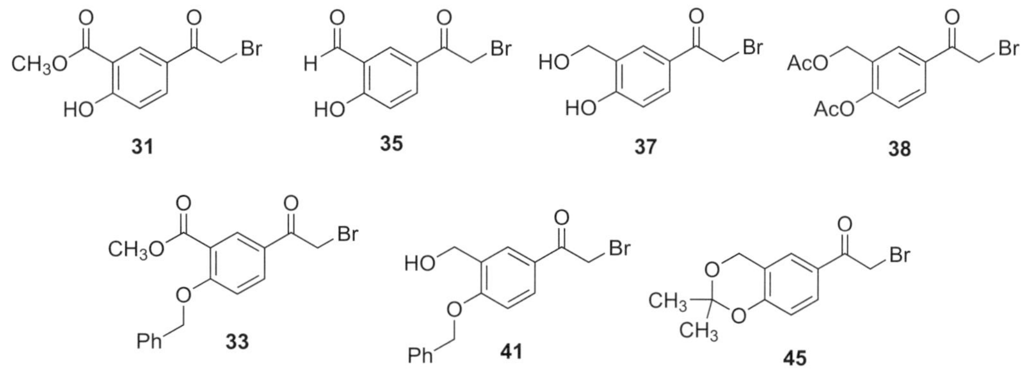

FIGURE 8.4 α-Bromoketones for construction of salmeterol (**1**).

### 8.4.2 Epoxides, Bromohydrins, and *O*-Trialkylsilyl Bromohydrins

While there are at least six α-bromoketones that are candidates for elaboration to salmeterol (**1**), the sensitivity of epoxides to phenols limits the field of epoxide candidates to those with a protected phenol. The epoxides can be produced by reduction of an α-bromoketone to a bromohydrin and cyclization of the bromohydrin with potassium carbonate or sodium hydride or by condensation of an aldehyde with dimethylsulfonium methylide. The *O*-trialkylsilylbromohydrins are prepared by reduction of an α-bromoketone to a bromohydrin and silylation of the bromohydrin.

***8.4.2.1 (R)-Methyl 2-(Benzyloxy)-5-(oxiran-2-yl)benzoate (51)*** Methyl 2-(benzyloxy)-5-(2-bromoacetyl)benzoate (**33**) is reduced to the (*R*)-bromohydrin (**52**) with borane–THF (0.6 equivalents) using Corey's (*R*)-oxazaborolidine catalyst (10 mol%). (*R*)-Bromohydrin **52** is isolated by chromatography or, on larger scale, by crystallization from ethyl acetate–hexane (95%, >94% ee). Reaction of (*R*)-bromohydrin **52** with sodium hydride in THF affords the (*R*)-epoxide **51** (85%).[44]

***8.4.2.2 (R)-(2-(Benzyloxy)-5-(oxiran-2-yl)phenyl)methanol (53)*** 1-(4-(Benzyloxy)-3-(hydroxymethyl)phenyl)-2-bromoethanone (**41**) is reduced with *Rhodotorula rubra* to the (*R*)-bromohydrin **54** (78–80%, 95% ee). Bromohydrin **54** is cyclized with potassium carbonate (1.5 equivalents) in dry THF over 2 h at reflux. The suspension is cooled and THF distilled at reduced pressure. After a routine water–ethyl acetate workup, the (*R*)-epoxide **53** is isolated by chromatography (67%). Many other microorganisms may be suitable for this general approach to (*R*)-epoxide **53**.[63,80]

***8.4.2.3 (R)-6-(Oxiran-2-yl)-4H-benzo[d][1,3]dioxine (55)*** 1-(4*H*-Benzo[*d*][1,3]dioxin-6-yl)-2-bromoethanone (**56**) is also reduced with *Rhodotorula rubra* to the (*R*)-bromohydrin **57**. Bromohydrin **57** is cyclized with potassium carbonate (1.5 equivalents) in dry THF over 2 h at reflux to afford 6-(oxiran-2-yl)-4*H*-benzo[*d*][1,3]dioxine (**55**). No yield is provided.[63]

***8.4.2.4 2,2-Dimethyl-6-(oxiran-2-yl)-4H-benzo[d][1,3]dioxine (58)*** Sodium borohydride (1.3 equivalents) is added to a solution of 2-bromo-1-(2,2-dimethyl-4*H*-benzo[*d*][1,3]dioxin-6-yl)ethanone (**45**) in ethanol at 0°C. After aging for 1 h at 0°C, excess hydride is quenched by careful addition of water and the resulting mixture is extracted with ethyl acetate. The combined extracts are washed with brine and then dried and concentrated at reduced pressure. Hexanes is added, the suspension is filtered, and the solid is dried at an unspecified temperature and reduced pressure to afford bromohydrin **59** (92%). Bromohydrin **59** is cyclized with potassium carbonate (1.2 equivalents) in methanol at 23°C over 2 h. Dilution with ethyl ether, filtration through a silica pad, and concentration of the liquors affords 2,2-dimethyl-6-(oxiran-2-yl)-4*H*-benzo[*d*][1,3]dioxine (**58**) (84%).[13]

2-Bromo-1-(2,2-dimethyl-4*H*-benzo[*d*][1,3]dioxin-6-yl)ethanone (**45**) is also reduced by rhodium-catalyzed asymmetric transfer hydrogenation to the (*R*)-bromohydrin **60** (93% yield, 98% ee). (*R*)-Bromohydrin **60** is cyclized with potassium carbonate (3.0 equivalents) in THF–methanol (1:1) over 2 h at 25°C. The suspension is filtered and the filtrate concentrated at reduced pressure. The residue is separated by chromatography to afford (*R*)-2,2-dimethyl-6-(oxiran-2-yl)-4*H*-benzo[*d*][1,3]dioxine (**61**) (93%) (Scheme 8.13).[56]

2,2-Dimethyl-6-(oxiran-2-yl)-4*H*-benzo[*d*][1,3]dioxine (**58**) is also prepared from 2,2-dimethyl-4*H*-benzo[*d*][1,3]dioxine-6-carbaldehyde (**62**) and dimethylsulfonium methylide, the Corey–Chaykovsky reagent (Scheme 8.14). There are two routes to aldehyde **62**. The first begins with chloromethylation of 4-hydroxybenzaldehyde. Formalin solution (40%) and 4-hydroxybenzaldehyde[81] are added to concentrated hydrochloric acid. A strong stream of hydrogen chloride gas is passed through the mixture with ice cooling (<40°C) for 3.5 h. The suspension is filtered and the solid

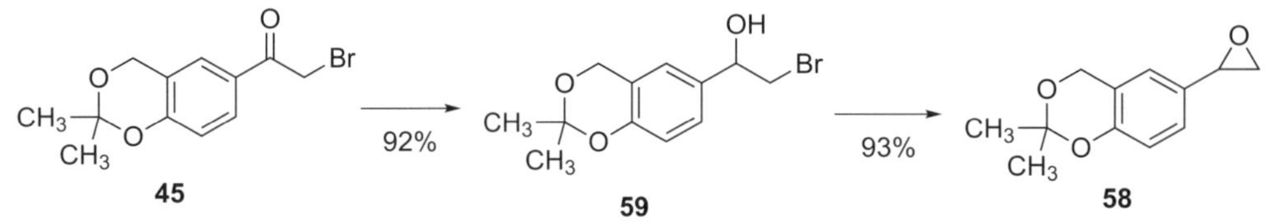

**SCHEME 8.13** 2,2-Dimethyl-6-(oxiran-2-yl)-4*H*-benzo[*d*][1,3]dioxine (**58**) from 2-bromo-1-(2,2-dimethyl-4*H*-benzo[*d*][1,3]dioxin-6-yl)ethanone (**45**).

is washed with water and dissolved in THF. Water and calcium carbonate are added to the solution and the mixture is aged at 25°C for 12 h. A routine water–ethyl ether workup affords crude 4-hydroxy-3-(hydroxymethyl)benzaldehyde (**64**) (71%) that can be crystallized from water or THF–dichloromethane (1:2).

The diol **64** is protected using acetone (23 equivalents) and phosphorus pentoxide (1.7 equivalents) in DMF–benzene. After stirring at 25°C for 15 h, more phosphorus pentoxide (1.4 equivalents) is added and the suspension aged at 25°C for 1 h. The liquid is decanted and the solid washed with additional benzene. The combined benzene liquors are washed with dilute aqueous sodium bicarbonate and with dilute brine, then dried and concentrated at reduced pressure. Distillation of the residue at 0.15 mmHg affords 2,2-dimethyl-4*H*-benzo[*d*][1,3]dioxine-6-carbaldehyde (**62**) as an oil that slowly crystallizes (mp 58°C) (51%).[76]

Aldehyde **62** can also be prepared from 4-bromophenol via 6-bromo-2,2-dimethyl-4*H*-benzo[*d*][1,3]dioxine (**46**). Butyllithium (1.6 M in hexane) (2.2 equivalents) is added to a solution of 6-bromo-2,2-dimethyl-4*H*-benzo[*d*][1,3]dioxine (**46**) in THF at −78°C. After aging for 2 h at −78°C, a solution of DMF (14.7 equivalents) in THF is added at −78°C. The resulting mixture is aged at −78°C for 45 min then warmed to 25°C. Ethyl ether and water are added and the layers separated. The organic layer is washed with water and with brine and then dried and concentrated at reduced pressure. Aldehyde **62** is isolated from the residue by chromatography (94%). A much lower yield (36%) is achieved in a similar process using 2.5 M butyllithium in hexane (1.1 equivalents) and 1.0 equivalent of DMF.[72,74]

The epoxide preparation from aldehyde **62** begins with preparation of dimethylsulfonium methylide. DMSO is added to sodium hydride (2.3 equivalents). The mixture is heated to 65–70°C, aged 1 h at 65–70°C, cooled to 25°C, diluted with THF, and cooled to 0°C. A solution of trimethylsulfonium iodide[82] (2.2 equivalents) in DMSO is added at 0°C to produce the methylide. A solution of 2,2-dimethyl-4*H*-benzo[*d*][1,3]dioxine-6-carbaldehyde (**62**) in THF is then added at 0°C. The mixture is allowed to warm

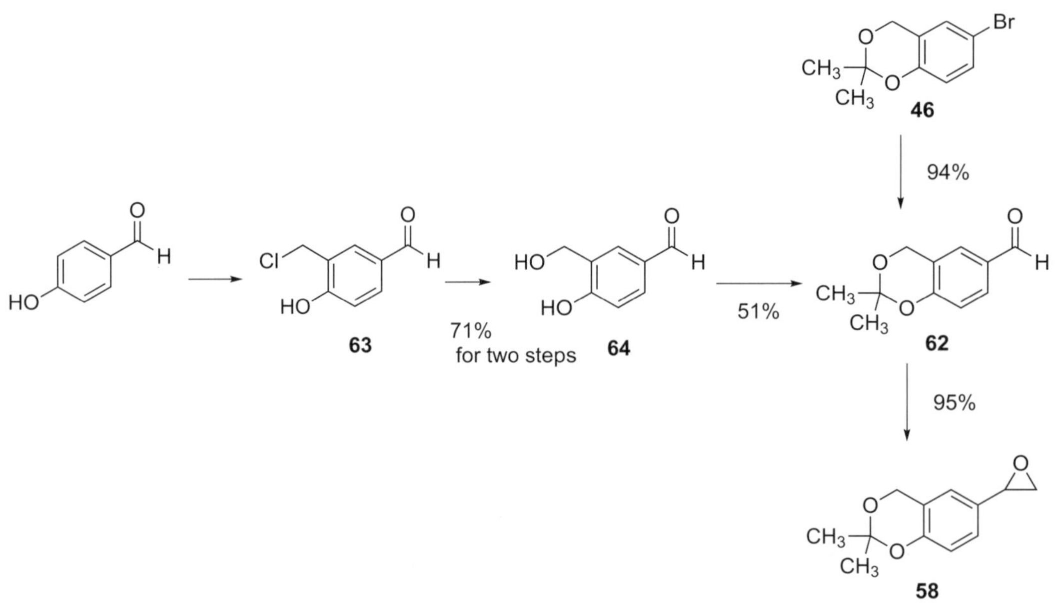

**SCHEME 8.14** 2,2-Dimethyl-6-(oxiran-2-yl)-4*H*-benzo[*d*][1,3]dioxine (**58**) from 2,2-dimethyl-4*H*-benzo[*d*][1,3]dioxine-6-carbaldehyde (**62**).

(*Note*: the procedure says to cool) to 25°C, and age for 1 h. The mixture is transferred into water and the phases are separated. The aqueous phase is extracted several times with an organic solvent, presumably ethyl ether. The combined organic layers are washed with water, dried, and concentrated at reduced pressure to afford 2,2-dimethyl-6-(oxiran-2-yl)-4*H*-benzo[*d*][1,3]dioxine (**58**) (95%) as an oil, which can be distilled at 118.5–120°C and 0.1 mmHg.[76]

A similar sequence from 4-hydroxybenzaldehyde to the epoxide uses dioxane as the chloromethylation solvent and 2,2-dimethoxypropane for the diol protection. No experimental details or yields are available.[83]

***8.4.2.5 (R)-(2-Bromo-1-(2,2-dimethyl-4H-benzo[d][1,3]dioxin-6-yl)ethoxy)triethylsilane (65)*** A chiral reducing agent is prepared by slow addition of a borane–THF solution (1.5 equivalents) to (*R*)-(+)-α,α-diphenyl-2-pyrrolidinemethanol (2.2 mol%) in THF at 20°C. After aging the reducing agent solution at 30–35°C for 1 h, the solution is cooled to 5°C and a solution of 2-bromo-1-(2,2-dimethyl-4*H*-benzo[*d*][1,3]dioxin-6-yl)ethanone (**45**) in THF is added at 0–5°C. The mixture is aged at 0°C for 30 min then methanol is added at 0°C to quench excess hydride. The solvents are removed at reduced pressure and the (*R*)-bromohydrin **60** is isolated from the residue by chromatography (84%). Chlorotriethylsilane (1.1 equivalents) is added to the (*R*)-bromohydrin **60** and imidazole (1.3 equivalents) in DMF at 5°C. The mixture is then warmed to 15°C and aged for 1 h. Hexane is added and the mixture is washed several times with water. The organic layer is dried and concentrated at reduced pressure to afford the triethylsilyl ether **65** (quantitative).[20,33]

***8.4.2.6 (2-Bromo-1-(2,2-dimethyl-4H-benzo[d][1,3]dioxin-6-yl)ethoxy)(tert-butyl)dimethylsilane (66)*** Sodium borohydride (1.0 equivalent) is added in portions to the α-bromoketone **45** in methanol at 0–5°C. The mixture is aged at 0–5°C for 30 min then quenched by addition of saturated aqueous ammonium chloride. A routine water– ethyl ether workup affords the bromohydrin **59** (79%). Bromohydrin **59** and imidazole (2.0 equivalents) are dissolved in DMF. A solution of (*tert*-butyl)chlorodimethylsilane (2.0 equivalents) in DMF is added at 25°C. After aging at 25°C for 20 h, the solvent is concentrated at an unspecified temperature and pressure. A routine water–ethyl ether workup of the residue affords the (*tert*-butyl)dimethylsilyl ether **66** (98%).[84]

A solution of borane–THF (1.0 equivalent) in THF is added to a solution of the α-bromoketone **45** and, presumably, Corey's (*R*)-oxazaborolidine catalyst (10 mol%) in THF at −20 to −10°C. The mixture is warmed to 25°C and aged at 25°C for 30 min. Excess hydride is quenched by adding methanol. The (*R*)-bromohydrin **60** is isolated by concentration at reduced pressure and chromatography of the residue. (*R*)-Bromohydrin **60** and imidazole (2.0 equivalents) are dissolved in DMF. A solution of (*tert*-butyl)chlorodimethylsilane (1.1 equivalents) in DMF is added. After aging at for 18 h, a routine brine–ethyl ether workup followed by chromatography affords the (*R*)-(*tert*-butyl)dimethylsilyl ether **67**. The temperatures during addition of (*tert*-butyl)chlorodimethylsilane and during the subsequent age time are both presumed to be 25°C. Yields for the (*R*)-bromohydrin **60** and the (*R*)-(*tert*-butyl)dimethylsilyl ether **67** are not provided.[77]

### 8.4.3 Glyoxals and Glyoxylates

A glyoxal can be prepared from a methyl, bromomethyl, or dichloromethyl ketone (Scheme 8.15). A glyoxylate is available by condensation of the aryllithium reagent derived from 6-bromo-2,2-dimethyl-4*H*-benzo[*d*][1,3]dioxine (**46**) with diethyl oxalate.

***8.4.3.1 Methyl 2-Hydroxy-5-(2-oxoacetyl)benzoate Hydrate (68)*** The reaction of methyl 5-(2-bromoacetyl)-2-hydroxybenzoate (**31**) with DMSO at 25°C is stopped after 1 week and poured into water. The precipitate is filtered and continuously extracted (Soxhlet extraction) with water. The solid that separates from the water solution on cooling is filtered and dried at an unspecified temperature to afford the ketoaldehyde hydrate **68** (64%).[43]

Methyl 5-acetylsalicylate is directly converted to the ketoaldehyde hydrate **68** by reaction with DMSO and hydrogen bromide. Isopropanol is added to a solution of methyl 5-acetylsalicylate in dichloromethane and the dichloromethane is distilled. When the internal temperature reaches 77°C, DMSO (8.6 equivalents) is added and the temperature is increased to 80°C. A solution of hydrogen bromide (1.4 equivalents) in isopropanol is added over 20 min at 85–90°C. Distillation of dimethylsulfide and isopropanol commences as half the hydrogen bromide is added. Additional isopropanol is charged to maintain a suitable volume. When complete conversion is observed by HPLC, dilute sulfuric acid is added and the residual isopropanol is distilled while maintaining the internal temperature at 70–75°C. Water is added and the suspension is cooled to 15°C and filtered. The solid is washed with water and dried at 50°C and reduced pressure to afford the ketoaldehyde hydrate **68** (85%). When 48% aqueous hydrobromic acid is used in place of hydrogen bromide in isopropanol, an identical workup with a final dilution with acetonitrile and water (3:7) also affords the ketoaldehyde hydrate **68** (85%). The reaction with 48% hydrobromic acid in DMSO with no isopropanol cosolvent resulted in a lower yield (59%).[85,86]

***8.4.3.2 Methyl 2-(Benzyloxy)-5-(2-oxoacetyl)benzoate Hydrate (69)*** Methyl 2-(benzyloxy)-5-(2,2-dichloroacetyl)benzoate (**70**) is added to sodium methoxide in methanol. The solution is refluxed for 15 min. Dilute hydrochloric acid

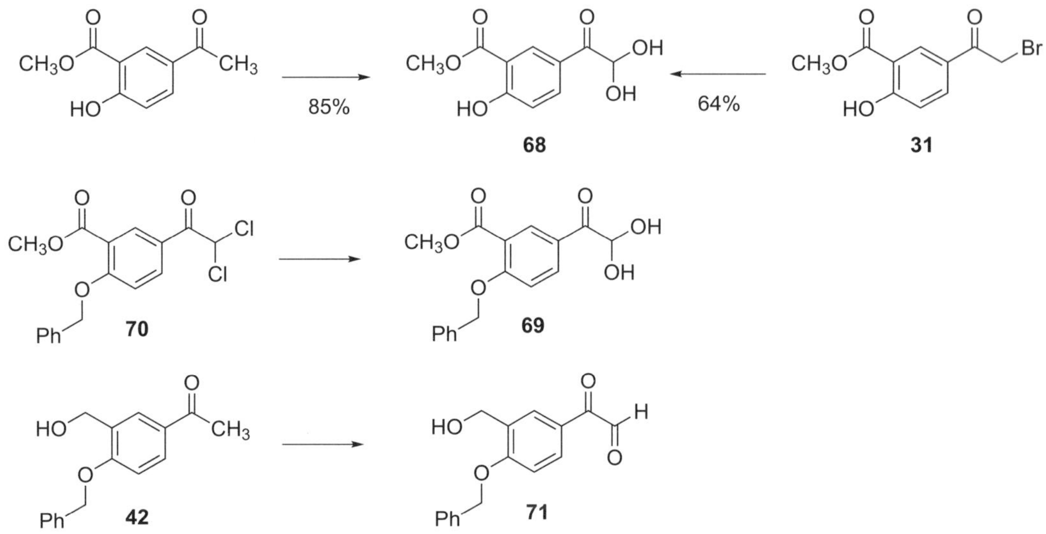

**SCHEME 8.15** Glyoxal intermediates from methyl 5-acetylsalicylate, methyl 2-(benzyloxy)-5-(2,2-dichloroacetyl)benzoate (**70**), and 1-(4-(benzyloxy)-3-(hydroxymethyl)phenyl)ethanone (**42**).

is added and the solution again refluxed for 15 min. Methanol and water are removed at an unspecified temperature and reduced pressure. A routine water–ethyl ether workup of the residue affords the ketoaldehyde hydrate **69**. No yield is provided.[87]

***8.4.3.3 2-(4-(Benzyloxy)-3-(hydroxymethyl)phenyl)-2-oxoacetaldehyde (71)*** 1-(4-(Benzyloxy)-3-(hydroxymethyl) phenyl)ethanone (**42**) is added to a solution of selenium dioxide (1.2 equivalents) in water–dioxane at 60°C. The mixture is refluxed for 4 h and filtered while hot. The filtered liquors are allowed to cool to 25°C and the precipitated aldehyde (**71**) dimer is filtered and crystallized from dioxane (mp 180–181°C). No yield is provided.[88]

***8.4.3.4 Ethyl 2-(2,2-Dimethyl)-4H-benzo[d][1,3]dioxin-6-yl)-2-hydroxyacetate (72)*** An aryllithium reagent is prepared by slow addition of 1.6 M butyllithium in hexanes (1.7 equivalents) to 6-bromo-2,2-dimethyl-4*H*-benzo[*d*][1,3]dioxine (**46**) in THF at −80°C. After aging for 15 min, a solution of diethyl oxalate[89] (1.2 equivalents) in THF is added at −80°C. After aging for 1 h at −80°C, ethanol and a solution of sodium borohydride (0.3 equivalents) in ethanol are added. The mixture is aged at −30°C for 30 min then excess borohydride is quenched by adding acetic acid. A solution of potassium bicarbonate in water is added and the mixture is concentrated at an unspecified temperature and reduced pressure. The adduct **72** is isolated from the residue by a routine water–ethyl acetate workup and chromatography (76%) (Scheme 8.16).[61,90]

## 8.5 SALMETEROL HEAD AMINES AND CARBAMATES

The longest sequence begins in the construction of the head fragment. All the head amines and carbamates are constructed from head electrophiles by adding steps to the longest sequence. For these reasons, the preferred more convergent sequences join head electrophiles and tail amines. The head amines and carbamates were primarily developed to expedite synthesis of libraries of (*R*)-salmeterol analogs with modified tail fragments in the search for a superior long-acting β-agonist (LABA).

### 8.5.1 Salmeterol Head Amines

Since a primary amine would be incompatible with an unprotected aldehyde or an ester, the range of salmeterol

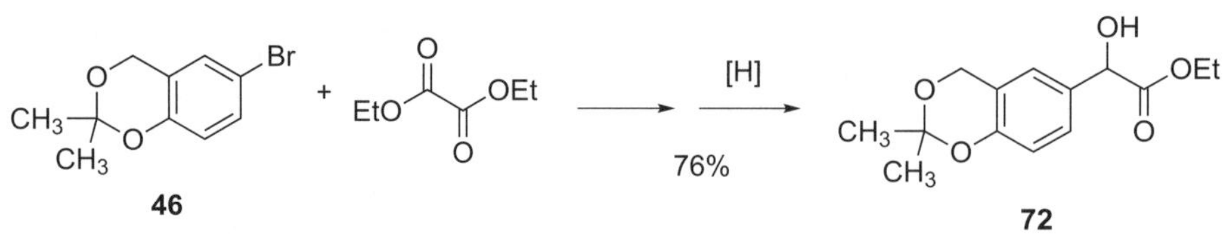

**SCHEME 8.16** Ethyl 2-(2,2-dimethyl)-4*H*-benzo[*d*][1,3]dioxin-6-yl)-2-hydroxyacetate (**72**) from 6-bromo-2,2-dimethyl-4*H*-benzo[*d*][1,3]dioxine (**46**).

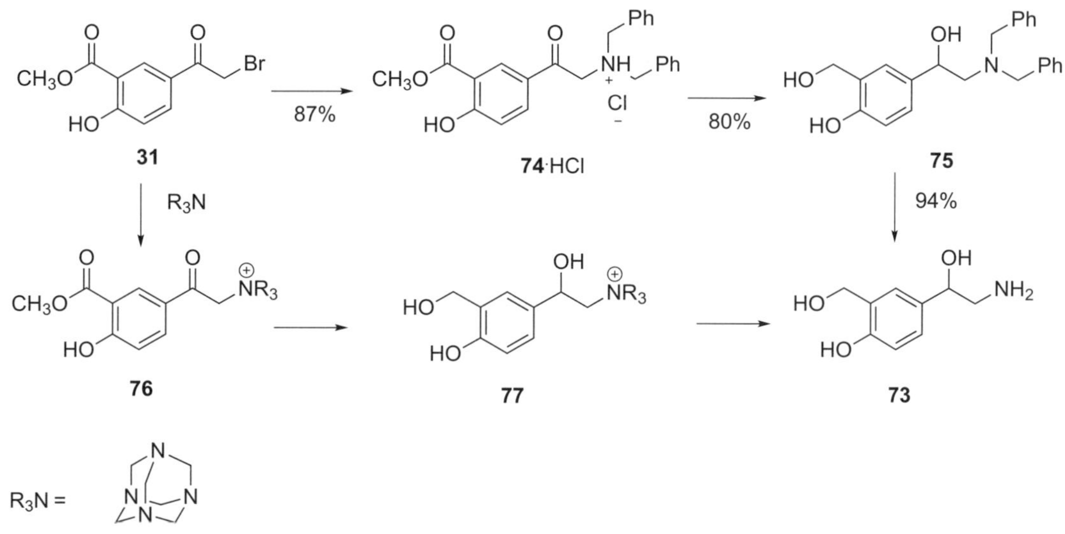


**SCHEME 8.17** 4-(2-Amino-1-hydroxyethyl)-2-(hydroxymethyl)phenol (**73**) from methyl 5-(2-bromoacetyl)-2-hydroxybenzoate (**31**).

head amines is limited to the 2-hydroxymethylphenols and protected (ether or acetal) analogs.

***8.5.1.1 4-(2-Amino-1-hydroxyethyl)-2-(hydroxymethyl)phenol (73)*** 4-(2-Amino-1-hydroxyethyl)-2-(hydroxymethyl)phenol (**73**) is a well-established intermediate first described in the 1960s by the group at Allen and Hamburys, Ltd. This intermediate is produced in three steps from methyl 5-(2-bromoacetyl)-2-hydroxybenzoate (**31**). The bromide is displaced by reaction with dibenzylamine[91] (1.8 equivalents) in 2-butanone at reflux over 2–3 h. The suspension is cooled and filtered. The liquors are concentrated at reduced pressure. The residue is taken up in ethyl ether and amine **74** is converted to the hydrochloride salt. The suspension is filtered and the solid crystallized from methanol–ethyl acetate (mp 167–169°C) (87%).

Reaction of amine **74** with lithium aluminum hydride reduces both the ketone and the ester. Amine **74** hydrochloride is first converted into the free base by an unspecified procedure. The free base in THF is added to a suspension of lithium aluminum hydride (3.4 equivalents) in THF. After refluxing the mixture for 2 h, the suspension is cooled and excess hydride is quenched by careful addition of water. The suspension is concentrated at reduced pressure. The residue is dissolved in dilute hydrochloric acid, basified with aqueous bicarbonate, and continuously extracted with ethyl ether. The extracts are concentrated at reduced pressure and the residue crystallized from ethyl acetate–cyclohexane to afford the dibenzylaminoalcohol **75** (mp 110–111°C) (80%). The yield is lower using less lithium aluminum hydride (1.8 equivalents) and a longer reflux time (73%).

The *N*-debenzylation of **75** by reaction with hydrogen and prereduced 10% palladium on carbon in ethanol at 25°C requires 6 h. The catalyst is filtered and the liquors concentrated at reduced pressure. The residue is crystallized from methanol–ethyl acetate to afford 4-(2-amino-1-hydroxyethyl)-2-(hydroxymethyl)phenol (**73**) (mp 149–151°C) (94%).[43,60,92,93] Amino alcohol **73** can also be prepared in an similar sequence using hexamethylenetetramine (HMTA) in place of dibenzylamine (Scheme 8.17).[94]

***8.5.1.2 2-Amino-1-(4-benzyloxy)-3-(hydroxymethyl)phenyl)ethanol (78)*** 2-Amino-1-(4-benzyloxy)-3-(hydroxymethyl)phenyl)ethanol (**78**) is produced in two steps from methyl 2-(benzyloxy)-5-(2-bromoacetyl)benzoate (**33**). The bromide is displaced by reaction with sodium azide (1.5 equivalents) in DMF at 25°C for 24 h (in the dark). The mixture is diluted with ethyl acetate, washed several times with brine, dried and concentrated at reduced pressure. The α-azidoketone **79** is separated from the residue by chromatography (74%).

α-Azidoketone **79** is added to a suspension of lithium aluminum hydride (5.1 equivalents) in THF at 0°C. The suspension is aged at 0°C for 1 h, at 25°C for 16 h, and at 75°C for 3 h. The suspension is cooled to 0–5°C and excess hydride is quenched by careful addition of 10% aqueous sodium hydroxide. The solids are filtered and rinsed with 5% methanol in THF. The combined liquors are concentrated at reduced pressure and the aminoalcohol **78** isolated from the residue by chromatography (66%) (Scheme 8.18).[86]

The (*R*)-aminoalcohol **80** can be produced by ring opening of (*R*)-(2-(benzyloxy)-5-(oxiran-2-yl)phenyl)methanol (**53**) by ammonia in water–methanol at 40°C. The regioselectivity of the epoxide opening is just 4.9:1 (Scheme 8.19).[80]

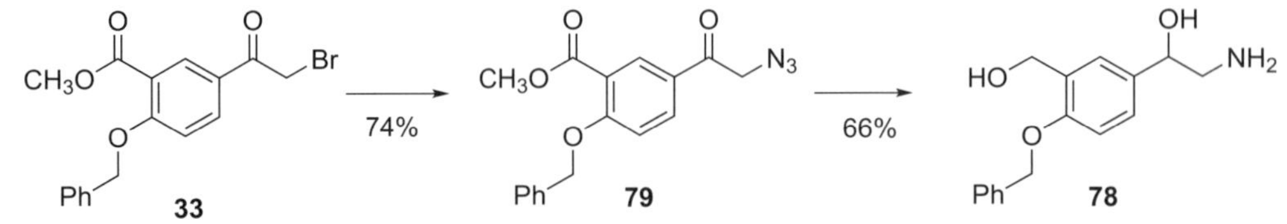

**SCHEME 8.18** 2-Amino-1-(4-benzyloxy)-3-(hydroxymethyl)phenyl)ethanol (**78**) from methyl 2-(benzyloxy)-5-(2-bromoacetyl)benzoate (**34**).

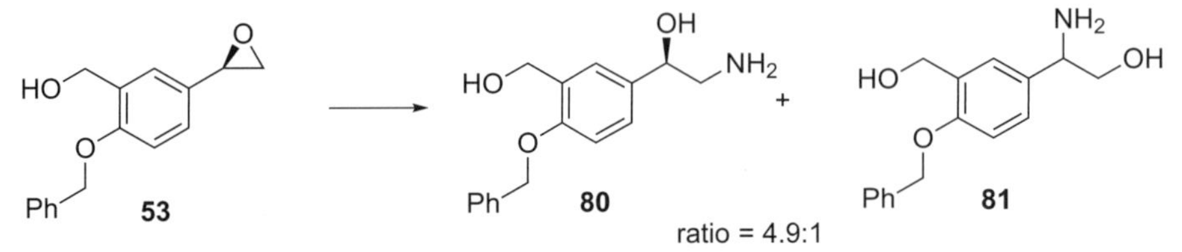

**SCHEME 8.19** (*R*)-2-Amino-1-(4-benzyloxy)-3-(hydroxymethyl)phenyl)ethanol (**80**) from (*R*)-(2-(benzyloxy)-5-(oxiran-2-yl)phenyl)methanol (**53**).

***8.5.1.3 (R)-2-Amino-1-(2,2-dimethyl-4H-benzo[d][1,3] dioxin-6-yl)ethanol (82)*** A preferred GSK route to the (*R*)-aminoalcohol **82** proceeds via the azide (Scheme 8.20). The reaction of 2-bromo-1-(2,2-dimethyl-4*H*-benzo[*d*][1,3] dioxin-6-yl)ethanone (**45**) with sodium azide (1.0 equivalent) in DMF requires 2 h at 20°C. The mixture is diluted with ethyl acetate, washed with water, dried and concentrated at reduced pressure to afford the α-azidoketone **83** (87–95%).[20,28,29,33,67]

The α-azidoketone **83** is a good substrate for asymmetric reduction using Corey's proline-derived oxazaborolidine. The reduction is accomplished by slow addition of a solution of α-azidoketone **83** to a 5°C solution of borane–THF (1.3 equivalents) and *R*-(+)-2-methyl-CBS-oxazaborolidine (7.5 mol%) in THF–toluene. After aging for 1 h at 5°C, excess hydride is quenched by adding 2 M hydrochloric acid. The mixture is extracted with ethyl ether. The combined extracts are washed with dilute hydrochloric acid, with dilute bicarbonate, and with brine and then dried and concentrated at reduced pressure. The (*R*)-azidoalcohol **84** is isolated from the residue by chromatography (68%).[20,28,29,33] The α-azidoketone **83** is reduced to the (*S*)-aminoalcohol **85** by *Pichia angusta* (94%).[67]

The (*R*)-azido alcohol **84** is reduced by catalytic hydrogenation using 10% palladium on carbon in ethanol. The temperature and hydrogen pressure for the hydrogenation are not specified but are presumed to be 25°C and 1 atm. The suspension is filtered and the liquors are concentrated at reduced pressure. The (*R*)-aminoalcohol **82** is isolated from the residue by chromatography (86%).[20,28,29,33,67]

A second approach to the (*R*)-aminoalcohol **82** is based on the asymmetric Henry reaction of 2,2-dimethyl-4*H*-benzo[*d*] [1,3]dioxine-6-carbaldehyde (**62**) with nitromethane mediated by Shibasaki's Sm-Li-6,6′-disubstituted BINOL catalysts.[95]

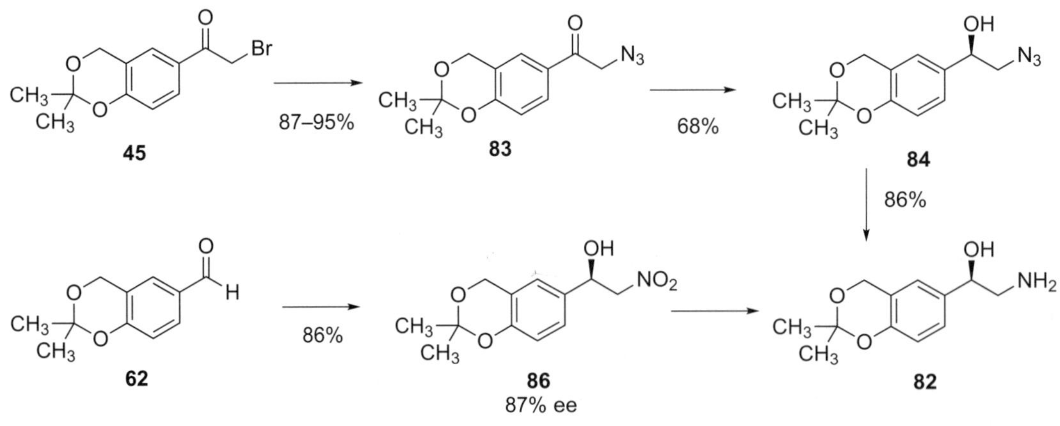

**SCHEME 8.20** Two routes to (*R*)-2-amino-1-(2,2-dimethyl-4*H*-benzo[*d*][1,3]dioxin-6-yl)ethanol (**82**).

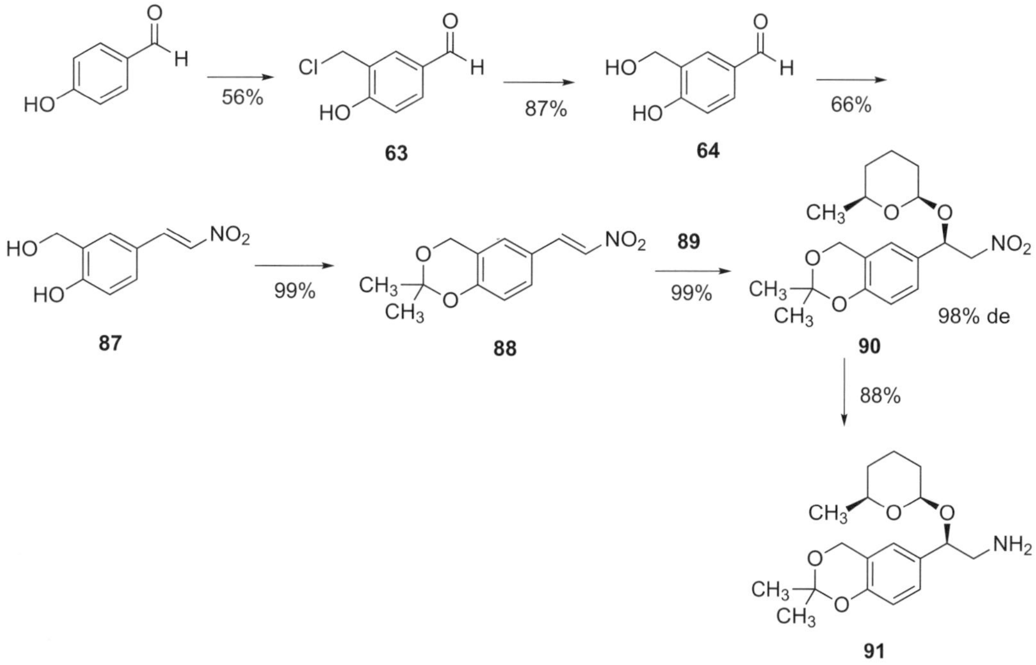


**SCHEME 8.21** A protected form of (*R*)-2-amino-1-(2,2-dimethyl-4*H*-benzo[*d*][1,3]dioxin-6-yl)ethanol produced by an asymmetric conjugate addition.

A suitable catalyst is prepared by addition of a solution of samarium(III) isopropoxide (6.6 mol%) in THF to a solution of dried (*S*)-6,6′-bis(trimethylsilylethynyl)-1,1′-dihydroxy-2,2′-binaphthalene (20 mol%) in THF at 0°C. After aging for 30 min at 0°C, butyllithium in hexanes (20 mol%) is added at 0°C. The solution is aged at 25°C overnight then water (6.7 mol%) is added. This catalyst solution is added to aldehyde **62** in THF at −40°C. After aging for 40 min at −40°C, nitromethane[96] (10 equivalents) is added and the mixture is aged at −40°C for 61 h. The workup details are not available for this 0.5 mmol scale procedure (86% yield, 87% ee). Hydrogenation of the nitroalcohol **86** using 10% palladium on carbon in methanol affords the (*R*)-aminoalcohol **82**. Experimental details and yield are not available.[97]

A protected form of the (*R*)-aminoalcohol can be produced via an asymmetric conjugate addition to (*E*)-2,2-dimethyl-6-(2-nitrovinyl)-4*H*-benzo[*d*][1,3]dioxine (**88**) (Scheme 8.21). The nitroalkene **88** is produced in four steps from 4-hydroxybenzaldehyde. (6*S*)-6-Methyltetrahydro-2*H*-pyran-2-ol (**89**) (1.5 equivalents) is added to potassium bis(trimethylsilyl)amide (1.5 equivalents) and 18-crown-6 (1.5 equivalents) in THF at −78°C. Addition of the nitroalkene **88** in THF is followed by quench with acetic acid to afford the protected (*R*)-nitroalcohol **90** (99%, 98% ee). The nitro group is reduced using nickel boride, which is generated by slow addition of sodium borohydride to the protected (*R*)-nitroalcohol **90** and nickel chloride hexahydrate in THF–methanol (1:1) at 0°C (88%).[98]

While nitromethane is inexpensive and available in bulk, it is unattractive as a raw material for a production scale process. Nitromethane is shock- and heat-sensitive and contact with organic amine bases and some metal oxides may increase the shock-sensitivity. Nitromethane is also an animal carcinogen and potential human carcinogen. The human cancer potency estimate is 0.18 $(\text{mg/kg/day})^{-1}$ and the NSRL (no significant risk level) is 39 μg/day.[99]

***8.5.1.4 (R)-2-(Benzylamino)-1-(2,2-dimethyl-4H-benzo[d][1,3]dioxin-6-yl)ethanol (92)*** Bromide displacement from (*R*)-(2-bromo-1-(2,2-dimethyl-4*H*-benzo[*d*][1,3]dioxin-6-yl)ethoxy)triethylsilane (**65**) by benzylamine (5.0 equivalents) requires refluxing in dioxane (105°C) overnight. The mixture is cooled and separated between water and ethyl ether. The organic layer is washed with saturated aqueous ammonium chloride, saturated aqueous sodium bicarbonate, and brine and then dried and concentrated at reduced pressure to afford the crude amine **93** (94%). Desilylation is accomplished by slow addition of a THF solution of tetrabutylammonium fluoride (1.3 equivalents) in THF at 5°C. After aging at 5°C for 15 min, water is added. The resulting suspension is diluted with ethyl ether and filtered. The liquors are washed with water and brine and then dried and concentrated at reduced pressure. Diisopropyl ether is added, the suspension is filtered, and the solid is dried at an unspecified temperature and pressure to afford

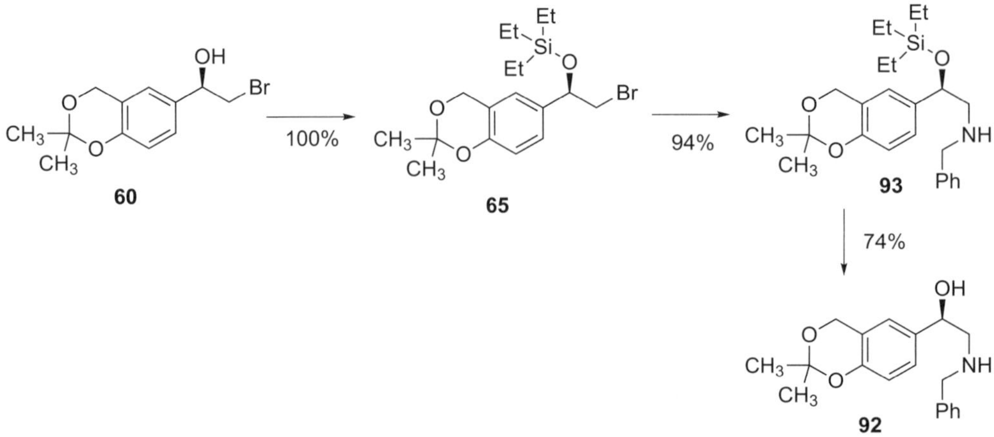

**SCHEME 8.22** (*R*)-2-(Benzylamino)-1-(2,2-dimethyl-4*H*-benzo[*d*][1,3]dioxin-6-yl)ethanol (**92**) from the (*R*)-bromohydrin **60**.

(*R*)-2-(benzylamino)-1-(2,2-dimethyl-4*H*-benzo[*d*][1,3]dioxin-6-yl)ethanol (**92**) (74%) (Scheme 8.22).[20,33]

> The reaction of a bromohydrin with an amine to produce the amino alcohol may proceed via TWO MECHANISMS TO THE SAME PRODUCT. It would appear that we need not concern ourselves with HOW the product forms. This may be an attractive approach for quickly producing gram quantities of material in the lab. However, when scale-up is the ultimate objective, it is important to strive for a single path to product to expedite process development. Consider, for example, how a change in a single variable (reaction temperature, solvent polarity, reagent stoichiometry, or concentration) affects the product impurity profile in a single-mechanism reaction. Now consider how much more challenging it would be to deconvolute the product impurity profile where the impurities result from side reactions of two different mechanisms. Rather than prepare (*R*)-2-(benzylamino)-1-(2,2-dimethyl-4*H*-benzo[*d*][1,3]dioxin-6-yl) ethanol (**92**) by the reaction of the (*R*)-bromohydrin **60** with benzylamine, the single path objective is accomplished by alcohol protection–deprotection using expensive chlorotriethylsilane.

***8.5.1.5 (R)-1-(2,2-Dimethyl-4H-benzo[d][1,3]dioxin-6-yl)-2-((S)-2-hydroxy-1-phenethylamino)ethanol (94)*** Displacing the bromide by an amine before reducing the ketone bypasses the (*R*)-bromohydrin **60** and the costly alcohol protection and deprotection. When the amine is (*S*)-(+)-2-phenylglycinol, the introduced chirality can then be called upon to direct the ketone reduction (Scheme 8.23).

The bromide displacement is accomplished by reaction of 2-bromo-1-(2,2-dimethyl-4*H*-benzo[*d*][1,3]dioxin-6-yl) ethanone (**45**) with (*S*)-(+)-2-phenylglycinol[100] (1.5 equivalents) and diisopropylethylamine (2.0 equivalents) in THF at 25°C. After 4 h, the suspension is filtered and the liquors are concentrated at reduced pressure. The residue is suspended in acetonitrile and the suspension is filtered and pulverized. The solid is suspended in cold (−15 to −20°C) acetonitrile, the suspension is filtered, and the solid is washed with cold acetonitrile and dried at an unspecified temperature and pressure (80%).

Calcium chloride dihydrate (2.0 equivalents) is added to the α-amino ketone **95** in methanol at 2°C. After the exotherm subsides, the mixture is cooled to 0°C and sodium borohydride (2.1 equivalents) is added. After aging for 2 h at 0°C, the mixture is concentrated at reduced pressure and the residue is suspended in ethyl acetate. The suspension is filtered through Hyflow Supercel and the liquors are

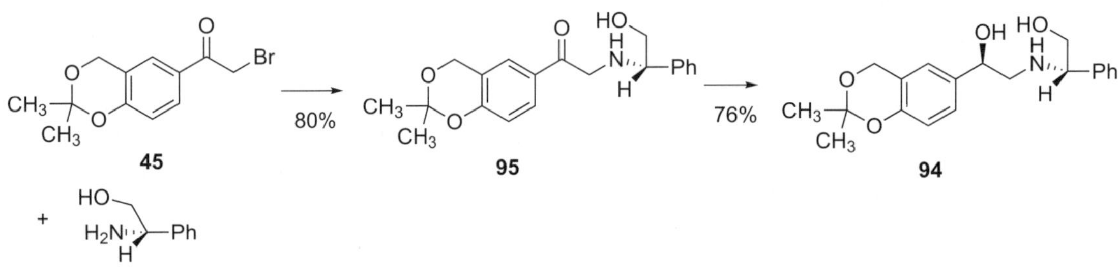

**SCHEME 8.23** (*R*)-1-(2,2-Dimethyl-4*H*-benzo[*d*][1,3]dioxin-6-yl)-2-(*S*)-2-hydroxy-1-phenethylamino)ethanol (**94**) by reduction of the α-amino ketone **95**.

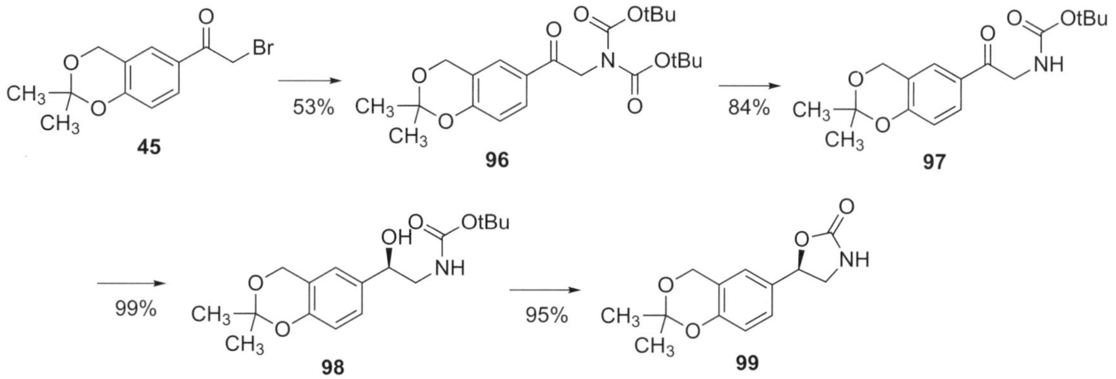

**SCHEME 8.24** (*R*)-5-(2,2-Dimethyl-4*H*-benzo[*d*][1,3]dioxin-6-yl)oxazolidin-2-one (**99**) from 2-bromo-1-(2,2-dimethyl-4*H*-benzo[*d*][1,3]dioxin-6-yl)ethanone (**45**).

concentrated at reduced pressure. The oil is dissolved in acetonitrile and the solution is cooled to 5°C and aged for 18 h. The suspension is filtered and the solid is washed with acetonitrile and dried at an unspecified temperature and pressure to afford the (*R*)-amino alcohol **94** (68%). Additional (*R*)-amino alcohol **94** (8%) can be recovered from the liquors. When the ketone is reduced using Amberlite IRA-400 ($BH_4^-$ form) in place of sodium borohydride, a simple filtration, concentration at reduced pressure, and crystallization from acetonitrile affords (*R*)-amino alcohol **94** (76%).[22–24]

### 8.5.2 Salmeterol Head Carbamates

A chiral oxazolidin-2-one suitable for elaboration to (*R*)-salmeterol and analogs can be prepared from 2-bromo-1-(2,2-dimethyl-4*H*-benzo[*d*][1,3]dioxin-6-yl)ethanone (**45**) (Scheme 8.24). The bromide is displaced by reaction with di-*tert*-butyl-iminodicarboxylate[101] (1.0 equivalent) and cesium carbonate (1.0 equivalent) in acetonitrile at 21°C over 24 h. The mixture is diluted with water and extracted with ethyl ether. The combined organic layers are washed with brine, dried, and concentrated at reduced pressure. The solid residue is crystallized from ethyl ether (28%). Concentration of the crystallization liquors gives additional product (20%). Concentration of the second crystallization liquors and chromatography of the residue yields additional product (5%).

Reaction of the alkylated iminodicarboxylate **96** with trifluoroacetic acid (1.2 equivalents) in dichloromethane at 20°C over 4 h results in cleavage of one *tert*-butyl carbamate and decarboxylation. The trifluoroacetic acid is neutralized with aqueous sodium hydroxide. The organic phase is separated, washed with water, dried and concentrated at reduced pressure. The residue is crystallized from ethyl ether (43%). Additional product is obtained by concentration of the crystallization liquors at reduced pressure and chromatography of the residue (41%).

The 2-oxoethylcarbamate **97** is another good substrate for asymmetric reduction using Corey's proline-derived oxazaborolidine. A solution of borane–dimethylsulfide (1 M) (1.0 equivalents) in THF is added to a solution of (*R*)-tetrahydro-1-methyl-3,3-diphenyl-1*H*,3*H*-pyrrolo[1,2-*c*][1,3,2]oxazaborole (10 mol%) at 5°C. After aging for 15 min, a solution of the 2-oxoethylcarbamate **97** in THF is then added at <5°C. After aging for 2 h at 25°C, excess hydride is quenched by careful addition of 2 M hydrochloric acid at 5°C. The resulting mixture is separated between water and ethyl acetate. A routine water–ethyl acetate workup followed by chromatography affords the (*R*)-2-hydroxyethylcarbamate **98** (99%).

A solution of the (*R*)-2-hydroxyethylcarbamate **98** in DMF is added to a suspension of sodium hydride (60% oil dispersion) (1.1 equivalents) in DMF at 0°C. After aging at 21°C for 2 h, the mixture is again cooled to 0°C, quenched by careful addition of 2 M hydrochloric acid, diluted with water, and extracted with ethyl acetate. The combined organic layers are washed with brine, dried, and concentrated at reduced pressure to afford (*R*)-5-(2,2-dimethyl-4*H*-benzo[*d*][1,3]dioxin-6-yl) oxazolidin-2-one (**99**) (95%).[20,28–33,35,102–105]

## 8.6 SALMETEROL (1) FROM TAIL ELECTROPHILES AND HEAD NUCLEOPHILES

We have available seven tail electrophiles, three tail nucleophiles, sixteen head electrophiles, and six head nucleophiles. Which combinations of these are known to produce salmeterol (**1**)?

### 8.6.1 Salmeterol (1) by Alkylation with (4-(6-Bromohexyloxy)butyl)benzene (4)

***8.6.1.1 4-(2-Amino-1-hydroxyethyl)-2-(hydroxymethyl) phenol (73)*** Salmeterol (**1**) is isolated in very low yield

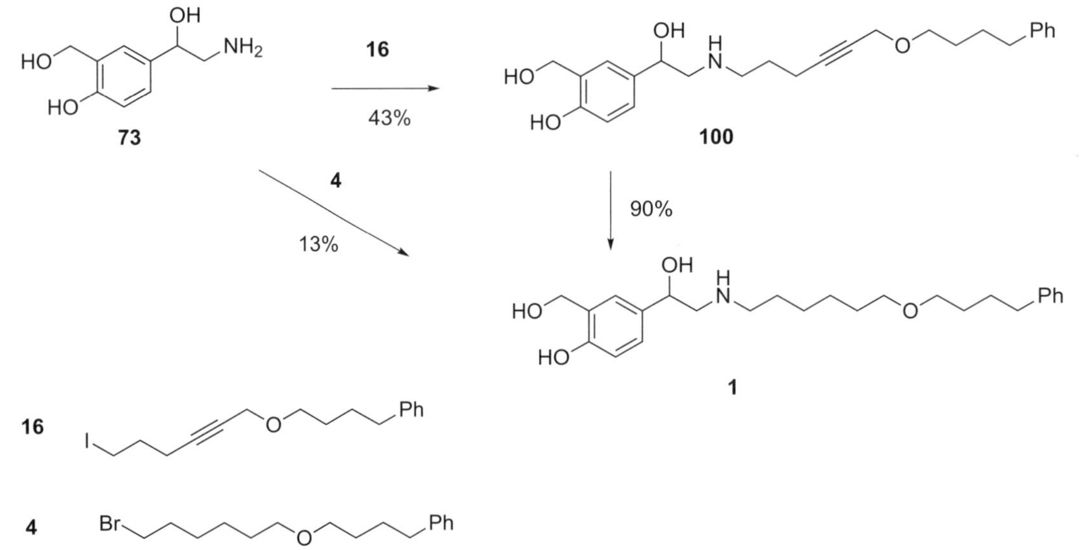

**SCHEME 8.25** Salmeterol (**1**) by alkylation of 4-(2-amino-1-hydroxyethyl)-2-(hydroxymethyl)phenol (**73**).

from the reaction of 4-(2-amino-1-hydroxyethyl)-2-(hydroxymethyl)phenol (**73**) with (4-(6-bromohexyloxy)butyl)benzene (**4**) (4.9 equivalents), triethylamine (0.74 equivalents), and potassium iodide (0.50 equivalents) in DMF at 70°C (13%) (Scheme 8.25).[13]

The alkylation of 4-(2-amino-1-hydroxyethyl)-2-(hydroxymethyl)phenol (**73**) with (3-(6-iodohex-2-ynyloxy)butyl)benzene (**16**) is also inefficient. The aminoalcohol **73** (1.5 equivalents) is reacted with iodide **16** and diisopropylethylamine (1.3 equivalents) in DMF at 70°C for 2 h. The DMF is distilled at an unspecified temperature and reduced pressure. The secondary amine **100** is isolated from the residue by routine workup with dilute aqueous bicarbonate and ethyl acetate followed by chromatography (43%). The alkyne is reduced by hydrogenation using 10% palladium on carbon (670 mg/g **100**). The temperature and hydrogen pressure are not specified but are presumably 25°C and 1 atm. The suspension is filtered and the liquors concentrated at reduced pressure to afford salmeterol (**1**) (90%).[13]

***8.6.1.2 (S)-2-Amino-1-(2,2-dimethyl-4H-benzo[d][1,3]dioxin-6-yl)ethanol (85)*** The (*S*)-aminoalcohol (**85**) (1.8 equivalents) is reacted with (4-(6-bromohexyloxy)butyl)benzene (**4**) in DMF at 25°C for 67 h. The DMF is distilled at an unspecified temperature and reduced pressure. The secondary amine **101** is isolated by a routine water–ethyl acetate workup and chromatography (70%). The ketal is cleaved by reaction with acetic acid–water at 71°C for 30 min. The mixture is concentrated at reduced pressure to afford (*S*)-salmeterol (**3**) as the acetate salt (100%) (Scheme 8.26).[67]

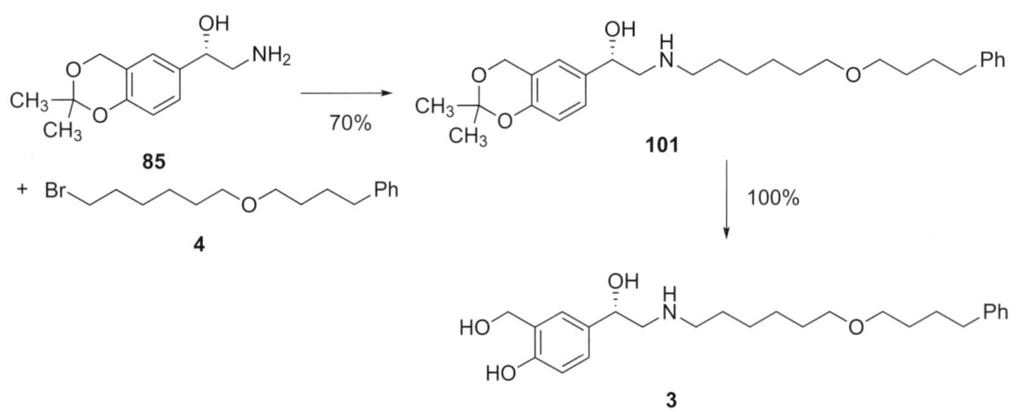

**SCHEME 8.26** (*S*)-Salmeterol (**3**) by alkylation of (*S*)-2-amino-1-(2,2-dimethyl-4*H*-benzo[*d*][1,3]dioxin-6-yl)ethanol (**85**).

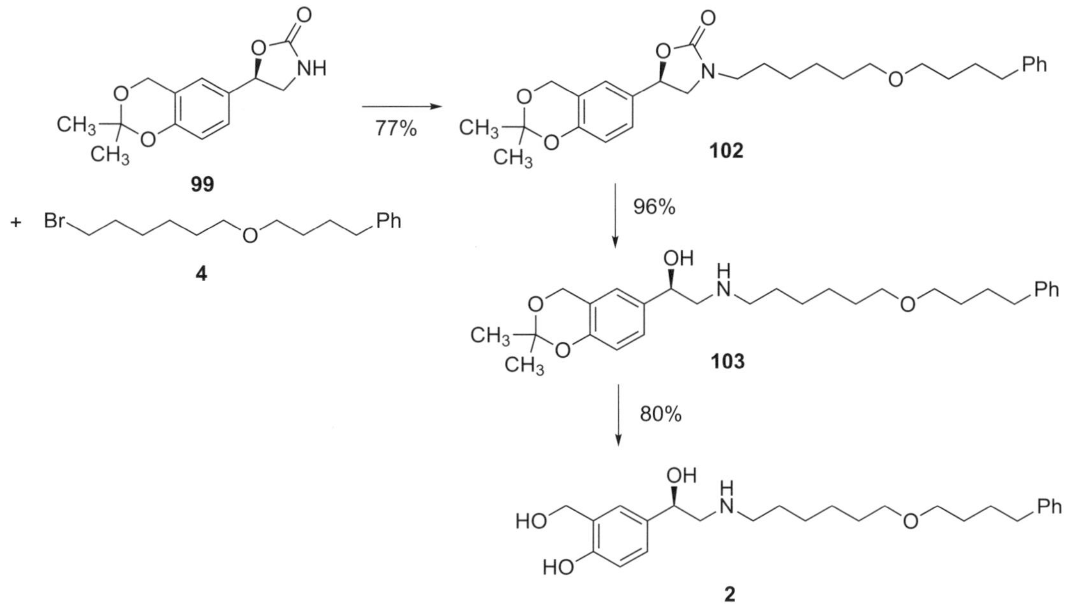

**SCHEME 8.27** (*R*)-Salmeterol (**2**) by alkylation of (*R*)-5-(2,2-dimethyl-4*H*-benzo[*d*][1,3]dioxin-6-yl)oxazolidin-2-one (**99**).

***8.6.1.3 (R)-5-(2,2-Dimethyl-4H-benzo[d][1,3]dioxin-6-yl)oxazolidin-2-one (99)*** A solution of (*R*)-5-(2,2-dimethyl-4*H*-benzo[*d*][1,3]dioxin-6-yl)oxazolidin-2-one (**99**) in DMF is added to a suspension of sodium hydride (1.1 equivalents) in DMF at 5°C. After aging for 5 min at 5°C, a solution of (4-(6-bromohexyloxy)butyl)benzene (**4**) (1.5 equivalents) in DMF is added. The mixture is aged at 5°C for 1 h and at 20°C for 3 h, then again cooled to 5°C and quenched by careful addition of 2 M hydrochloric acid. A routine water–ethyl acetate workup followed by chromatography affords the *N*-alkylated oxazolidinone **102** (77%).

The oxazolidinone is cleaved with potassium trimethylsilanolate (1.9 equivalents) in THF at reflux. A routine water–ethyl acetate workup affords the crude (*R*)-aminoalcohol **103** (96%). The ketal (20 mg) is cleaved with acetic acid in methanol at 20°C over 3 days. A routine water–ethyl acetate workup affords (*R*)-salmeterol (**2**) (80%) (Scheme 8.27).[105]

## 8.6.2 Salmeterol (1) by Reductive Amination of 6-(4-Phenylbutoxy)hexanal (10)

***8.6.2.1 4-(2-Amino-1-hydroxyethyl)-2-(hydroxymethyl) phenol (73)*** A solution of 4-(2-amino-1-hydroxyethyl)-2-(hydroxymethyl)phenol (**73**) (1.4 equivalents) and 6-(4-phenylbutoxy)hexanal (**10**) in methanol is maintained at 23°C for 30 min. Sodium borohydride (6.6 equivalents) is added and the solution aged at 23°C for 7 h. A routine water–ethyl acetate workup affords an oil residue. Salmeterol (**1**), separated from the residue by chromatography, is suspended in ethyl ether and the solid is filtered and dried at an unspecified temperature and reduced pressure (30%) (Scheme 8.28).[13]

4-(2-Amino-1-hydroxyethyl)-2-(hydroxymethyl)phenol (**73**) is likely generated *in situ* when *N*-debenzylation and aldehyde reductive amination are carried out simultaneously. A mixture of 4-(2-(dibenzylamino)-1-hydroxyethyl)-2-(hydroxymethyl)phenol (**75**) (1.5 equivalents) and 6-(4-phenylbutoxy)hexanal (**10**) is hydrogenated using 10% palladium on carbon (200 mg/g of aldehyde **10**) in ethanol at 25°C and 1 atm pressure. The suspension is filtered and the liquors are concentrated at reduced pressure. Salmeterol (**1**), separated from the residue by chromatography, is suspended in ethyl ether and the solid is filtered and dried at an unspecified temperature and reduced pressure (46%).[13]

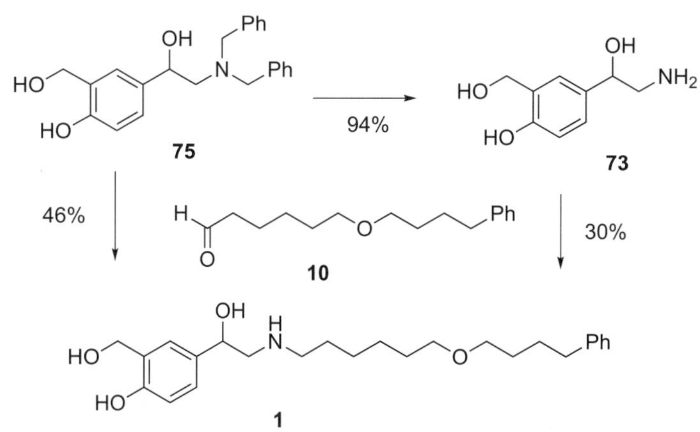

**SCHEME 8.28** Salmeterol (**1**) from 4-(2-(dibenzylamino)-1-hydroxyethyl)-2-(hydroxymethyl)phenol (**75**) and 6-(4-phenylbutoxy)hexanal (**10**).

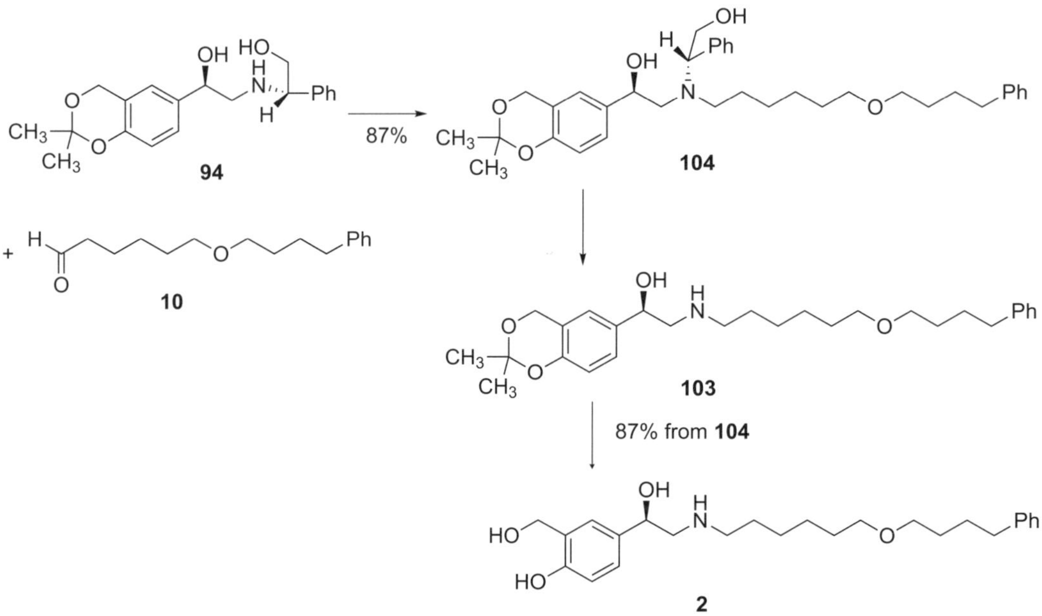


**SCHEME 8.29** (*R*)-Salmeterol (**2**) from (*R*)-1-(2,2-dimethyl-4*H*-benzo[*d*][1,3]dioxin-6-yl)-2-(*S*)-2-hydroxy-1-phenethylamino)ethanol (**94**).

*8.6.2.2 (R)-2-Amino-1-(2,2-dimethyl-4H-benzo[d][1,3] dioxin-6-yl)ethanol (82)* Experimental details for the reductive amination of (*R*)-2-amino-1-(2,2-dimethyl-4*H*-benzo[*d*][1,3]dioxin-6-yl)ethanol (**82**) with 6-(4-phenylbutoxy) hexanal (**10**) are not available.[97]

*8.6.2.3 (R)-1-(2,2-Dimethyl-4H-benzo[d][1,3]dioxin-6-yl)-2-((S)-2-hydroxy-1-phenethylamino)ethanol (94)* Sodium triacetoxyborohydride (1.4 equivalents) is added to 6-(4-phenylbutoxy)hexanal (**10**) (1.1 equivalents) and (*R*)-1-(2,2-dimethyl-4*H*-benzo[*d*][1,3]dioxin-6-yl)-2-(*S*)-2-hydroxy-1-phenethylamino)ethanol (**94**) in dichloromethane at 25°C. After aging for 18 h at 25°C, the reductive amination product **104** is isolated by a routine water–ethyl acetate workup and chromatography (87%) (Scheme 8.29).

Hydrogenolysis in ethanol using Pearlman's catalyst (200 mg $Pd(OH)_2$/g **104**) at 25°C requires 18 h. The hydrogen pressure is not specified but is presumed to be 1 atm. The suspension is filtered and the liquors are concentrated at reduced pressure. The hydrogenolysis in ethyl acetate using Pearlman's catalyst (250 mg $Pd(OH)_2$/g **104**) is complete in just 2 h at 25°C and 1 atm.

The ketal is cleaved by contact with an ion exchange resin in alcohol (methanol or ethanol) at 25°C. The residue from the hydrogenolysis workup is dissolved in alcohol and transferred to a SCX-2 ion exchange column. The column is eluted with a solution of 10% aqueous ammonia in alcohol. The eluent is concentrated at reduced pressure to afford (*R*)-salmeterol (**2**) (87%). Salt formation with 1-hydroxy-2-naphthoic acid in methanol affords (*R*)-salmeterol (**2**) xinafoate (100%).[22–24,33]

*8.6.2.4 (2R)-2-(2,2-Dimethyl-4H-benzo[d][1,3]dioxin-6-yl)-2-(tetrahydro-2H-pyran-2-yloxy)ethanamine (91)* The procedure for reductive amination of 6-(4-phenylbutoxy) hexanal (**10**) with the THP-protected aminoalcohol **91** and sodium triacetoxyborohydride is not available (72%).The ketal and THP ether in the amine **105** are removed by contact of a methanol solution with an SCX-2 ion exchange column (100%) (Scheme 8.30).[98]

### 8.6.3 Salmeterol (1) by Amide Formation and Reduction from 6-(4-Phenylbutoxy)hexanoic Acid (15)

6-(4-Phenylbutoxy)hexanoic acid (**15**) is converted to the acid chloride with thionyl chloride and catalytic DMF in dichloromethane. After aging at 25°C for 2.5 h, the solution is concentrated at reduced pressure and acid chloride **106** is dissolved in THF. Ethyl trimethylsilylacetate (5.8 equivalents) is added, presumably at 25°C, to 4-(2-amino-1-hydroxyethyl)-2-(hydroxymethyl)phenol (**73**) (1.5 equivalents) in THF. Tetrabutylammonium fluoride solution (solvent and equivalents are not specified) is added at 0°C. After aging at 25°C for 2 h, the solution is added, presumably at 25°C, to the acid chloride–THF solution. Triethylamine (7.2 equivalents) is then added and the solution is aged at 25°C overnight. The reaction mixture is added to 2 M hydrochloric acid and extracted with ethyl acetate. The extracts are washed with water, aqueous bicarbonate, and brine, and then dried and concentrated at reduced pressure. The amide **107** is isolated from the residue by chromatography (mp 96–97.5°C) (73%).

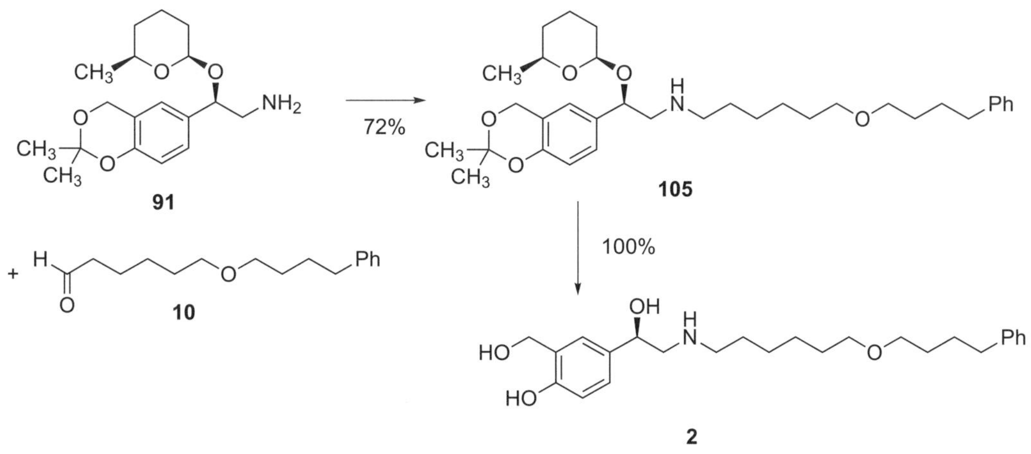


**SCHEME 8.30** (*R*)-Salmeterol (**2**) from (2*R*)-2-(2,2-dimethyl-4*H*-benzo[*d*][1,3]dioxin-6-yl)-2-(tetrahydro-2*H*-pyran-2-yloxy)ethanamine (**91**).

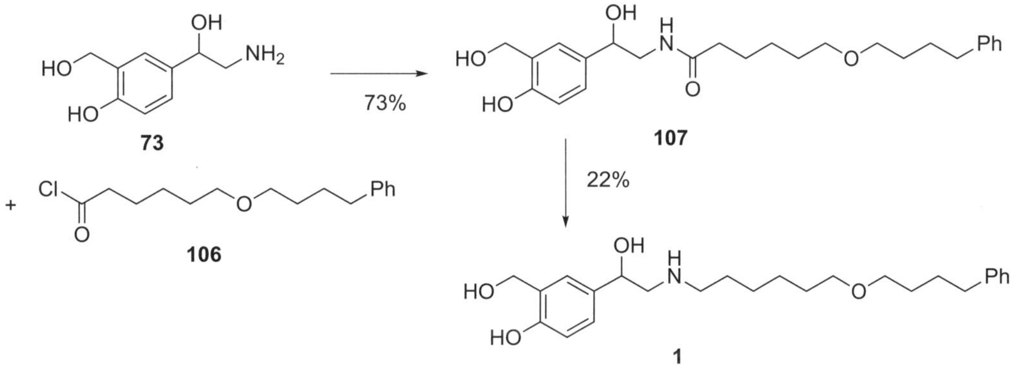


**SCHEME 8.31** Salmeterol (**1**) from 4-(2-amino-1-hydroxyethyl)-2-(hydroxymethyl)phenol (**73**) and 6-(4-phenylbutoxy)hexanoyl chloride (**106**).

A solution of the amide **107** in THF is added to a suspension of lithium aluminum hydride (9.8 equivalents) in THF at 0°C. The suspension is aged at 23°C for 18 h then excess hydride is quenched by careful addition on water. Hydrochloric acid is added (to pH 5), aqueous bicarbonate is added (to pH 8), and the mixture is extracted with ethyl acetate. The extracts are washed with brine and then dried and concentrated at reduced pressure. Salmeterol (**1**), separated from the residue by chromatography, is suspended in ethyl ether and the solid filtered and dried at an unspecified temperature and reduced pressure (mp 75–76.5°C) (22%) (Scheme 8.31).[13]

## 8.7 SALMETEROL (1) FROM TAIL NUCLEOPHILES AND HEAD ELECTROPHILES

### 8.7.1 Head Bromides

***8.7.1.1 Methyl 5-(2-Bromoacetyl)-2-hydroxybenzoate (31)*** A solution of 6-(4-phenylbutoxy)hexan-1-amine (**23**) (1.0 equivalent) in THF is added to a solution of methyl 5-(2-bromoacetyl)-2-hydroxybenzoate (**31**) and diisopropylethylamine (1.8 equivalents) in THF at 0°C. After aging at 0°C for 2 h, the mixture is diluted with ethyl ether, washed with 0.5 M hydrochloric acid, aqueous bicarbonate, and brine, then dried and concentrated at reduced pressure. The α-amino ketone **108** is isolated from the residue by chromatography (37%) (Scheme 8.32).[13]

A solution of 6-(4-phenylbutoxy)hexan-1-amine (**23**) (1.0 equivalent), methyl 5-(2-bromoacetyl)-2-hydroxybenzoate-$^{13}$C (**31**) and diisopropylethylamine (1.5 equivalents) in THF is refluxed for 5 h. The suspension is cooled and filtered, and the liquors are concentrated at reduced pressure. The residue is suspended in ethyl ether and the solid is filtered and dried at an unspecified temperature and reduced pressure to afford the α-amino ketone ($^{13}$C-**108**) (103%). α-Amino ketone **108** is unstable. It should be stored cold and used as soon as possible in the next reaction.[18,42]

A solution of the α-amino ketone **108** in THF is added to a suspension of lithium aluminum hydride (4.8 equivalents) in THF at 23°C. After aging at 23°C for 18 h, excess hydride is quenched by careful addition of water. Hydrochloric acid

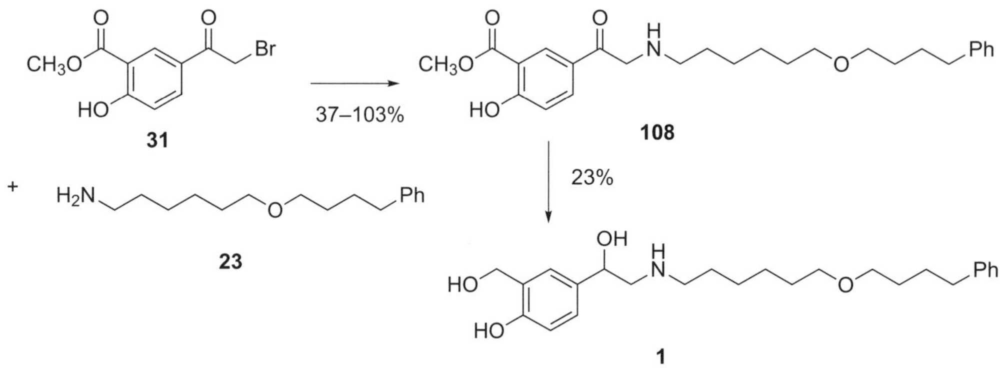

**SCHEME 8.32** Salmeterol (**1**) from methyl 5-(2-bromoacetyl)-2-hydroxybenzoate (**31**).

(2 M) is added (to pH 5), sodium carbonate is added (to pH 8), and the mixture is extracted with ethyl acetate. The extracts are dried and concentrated at reduced pressure. Salmeterol (**1**) is isolated from the residual oil by chromatography (23%).[13]

A solution of methyl 5-(2-bromoacetyl)-2-hydroxybenzoate (**31**), (*R*)-2-phenyl-2-(6-(4-phenylbutoxy)hexylamino)ethanol (**29**) (1.0 equivalent), and diisopropylethylamine (1.5 equivalents) in 1,2-dimethoxyethane is refluxed for 4 h. The suspension is cooled and filtered. Lithium borohydride (4.0 equivalents) is added to the liquors and the mixture is refluxed for 4 h. The mixture is cooled and 2 M hydrochloric acid is carefully added to quench excess hydride. Aqueous sodium carbonate (1 M) is added to neutralize the acid. A routine water–ethyl acetate workup followed by chromatography affords the (*R*)-amino alcohol **110** (40%).

The (*R*)-amino alcohol **110** is hydrogenated using water-wet 10% palladium on carbon (250 mg dry weight/g **110**) in ethanol at 25°C and 1 atm hydrogen pressure. The suspension is filtered and the liquors are concentrated at reduced pressure. (*R*)-Salmeterol (**2**) is isolated from the residue by chromatography (83%) (Scheme 8.33).[40]

#### *8.7.1.2 5-(2-Bromoacetyl)-2-hydroxybenzaldehyde (35)*

*N*-Benzyl-6-(4-phenylbutoxy)hexan-1-amine (**22**) hydrochloride salt is converted to the free base by contact of a dichloromethane solution with aqueous sodium carbonate. The organic layer is separated and concentrated at reduced pressure. A solution of the free base (1.2

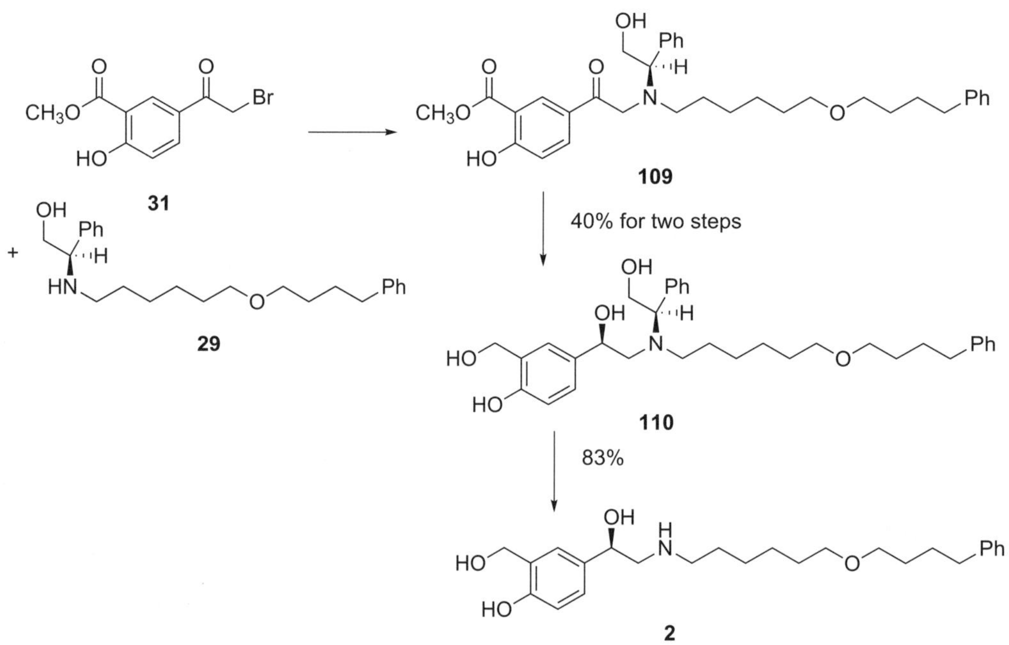

**SCHEME 8.33** (*R*)-Salmeterol (**2**) from methyl 5-(2-bromoacetyl)-2-hydroxybenzoate (**31**) and (*R*)-2-phenyl-2-(6-(4-phenylbutoxy)hexylamino)ethanol (**29**).

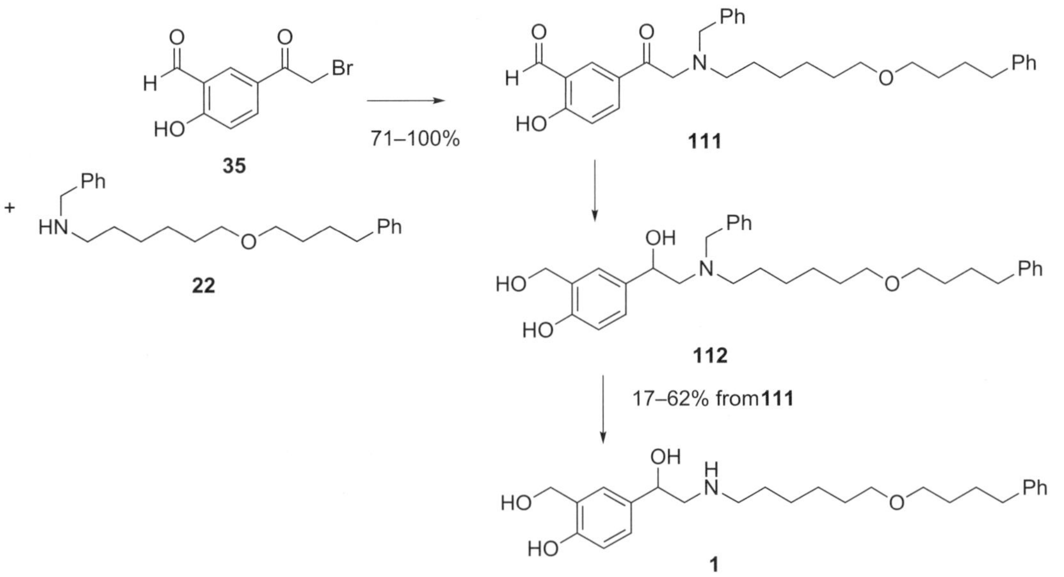


**SCHEME 8.34** Salmeterol (**1**) from 5-(2-bromoacetyl)-2-hydroxybenzaldehyde (**35**) and *N*-benzyl-6-(4-phenylbutoxy)hexan-1-amine (**22**).

equivalents) and diisopropylethylamine (1.2 equivalents) in 2-butanone is prepared. A solution of 5-(2-bromoacetyl)-2-hydroxybenzaldehyde (**35**) in 2-butanone is then added at 0–5°C. The resulting mixture is aged at 5–10°C for 10 h and then diluted with water and extracted with heptane. The combined extracts are polish filtered and the liquors containing the α-amino ketone **111** (100% yield, 98.5% pure by HPLC) is used directly in the next step (Scheme 8.34).

The hexane solution is diluted with methanol and the biphasic mixture cooled to −10°C. Sodium borohydride (7 equivalents) is added at 0–10°C and the mixture aged at 20–25°C until the reduction is complete (monitored by HPLC). Additional borohydride (2–3 equivalents) may be necessary to achieve complete conversion. The methanol layer is separated and concentrated at reduced pressure. Excess hydride is quenched by addition of water and some ethyl acetate. Dilute hydrochloric acid is added (to pH 2–3) and the mixture is aged at 25–30°C for 1 h. The aqueous layer is made basic with aqueous bicarbonate (pH 9) and extracted with ethyl acetate. The ethyl acetate extract is washed with water and concentrated at reduced pressure to afford **112** as a gum (83% yield, 96% pure by HPLC).

The gum is dissolved in methanol and hydrogenated using 20% palladium on carbon at 25–30°C and 1 atm. The reaction time is not specified. The suspension is filtered and the liquors concentrated at reduced pressure. Ethyl acetate is added and the suspension is filtered. The solid is dissolved in methanol and treated with carbon. The suspension is filtered and the liquors concentrated at reduced pressure. Ethyl acetate is added and the suspension is cooled to 5–10°C and filtered. The solid is dried at 40°C and reduced pressure to afford salmeterol (**1**) (20%, 98.0% pure by HPLC). 1-Hydroxy-2-naphthoic acid (1.0 equivalent) is added to salmeterol (**1**) in methanol at 25–30°C. The suspension is filtered and the xinafoate salt dried at reduced pressure and an unspecified temperature (116%, 99.5% pure by HPLC).[15]

In alternative processes, the bromide displacement can be carried out in isopropanol using triethylamine or excess *N*-benzyl-6-(4-phenylbutoxy)hexan-1-amine (**22**) or in acetonitrile using excess *N*-benzyl-6-(4-phenylbutoxy)hexan-1-amine (**22**).[14,106]

The bromide displacement with 2.0 equivalents of amine **22** in isopropanol or acetonitrile requires 1 h at reflux. The suspension is cooled, diluted with ethyl ether, and filtered. α-Amino ketone **111** is isolated by concentration of the liquors at reduced pressure and chromatography of the residue (71%).

Sodium borohydride (3.0 equivalents) is added to a solution of α-amino ketone **111** in methanol at 0°C. After aging at 25°C for 36 h, excess hydride is quenched by adding 3 M hydrochloric acid. The methanol is removed by concentration at reduced pressure, aqueous sodium carbonate is added (to pH 8), and the mixture is extracted with ethyl acetate. The extracts are washed with water, dried, and concentrated at reduced pressure. The α-amino alcohol **112** is hydrogenated in methanol using 10% palladium on carbon (100 mg/g **112**), most likely at 25°C and 1 atm hydrogen pressure. The suspension is filtered and the liquors concentrated at reduced pressure. The solid is dried at an unspecified temperature and reduced pressure to afford salmeterol (**1**) (62% from **111**).[14]

*8.7.1.3* ***2-Acetoxy-5-(2-bromoacetyl)benzyl Acetate*** *(38)* After a selective bromination of 2-acetoxy-5-

acetylbenzyl acetate (**40**) with bromine (1.0 equivalent) in chloroform at room temperature, the chloroform solution is washed with water and dried (100% contained yield). The bromide is displaced by reaction with *N*-benzyl-6-(4-phenylbutoxy)hexan-1-amine (**22**) (1.0 equivalent) in chloroform at reflux for 24 h. Chloroform is distilled at reduced pressure. A routine water–toluene workup of the residue affords the crude α-amino ketone **113** (Scheme 8.35).

The α-amino ketone **113** is dissolved in ethanol and 4.5 M hydrochloric acid (3.3 equivalents) is added at <20°C. After aging at 0°C for 48 h, the solution is diluted with ethanol and 50% aqueous sodium hydroxide (3.0 equivalents) is added at <15°C. An aqueous solution of sodium borohydride (2.0 equivalents) and sodium hydroxide (0.36 equivalents) is added and the mixture is aged, presumably at 25°C, for 24 h. A second charge of sodium borohydride (1.7 equivalents) is added and the mixture is aged at 25°C for 48 h. The mixture is neutralized with dilute sulfuric acid and concentrated at reduced pressure. The α-amino alcohol **112** is isolated from the suspension by a routine aqueous sodium carbonate–ethyl acetate workup and chromatography (36% from **40**).

Hydrogenolysis using water-wet 10% palladium on carbon (140 mg dry weight/g **112**) in ethanol is likely carried out at 25°C and 1 atm pressure. The suspension is filtered and the liquors concentrated at reduced pressure. The residue is suspended in ethyl ether, decanted from some insoluble gum, and aged at 25°C overnight. The precipitated salmeterol (**1**) is filtered and dried at an unspecified temperature and reduced pressure (78%) (Scheme 8.35).[13]

> An overall yield of 28% for five steps from 2-acetoxy-5-acetylbenzyl acetate (**40**) to salmeterol (**1**) is a very respectable average of 77–78% per step. But an overall yield of 28% also means that, for every kilogram of salmeterol (**1**) produced, 72% of the raw materials 2-acetoxy-5-acetylbenzyl acetate (**40**) and *N*-benzyl-6-(4-phenylbutoxy)hexan-1-amine (**22**) end up in waste streams. A multistep manufacturing sequence typically has several very efficient (>90% yield) steps and converts at least 50% of key raw materials to product.

*8.7.1.4* ***2-Bromo-1-(4-hydroxy-3-(hydroxymethyl)phenyl)ethanone (37)*** The reaction of 2-bromo-1-(4-hydroxy-3-(hydroxymethyl)phenyl)ethanone (**37**) with *N*-benzyl-6-(4-phenylbutoxy)hexan-1-amine (**22**) (1.0 equivalent) and diisopropylethylamine (1.5 equivalents) in THF is aged at

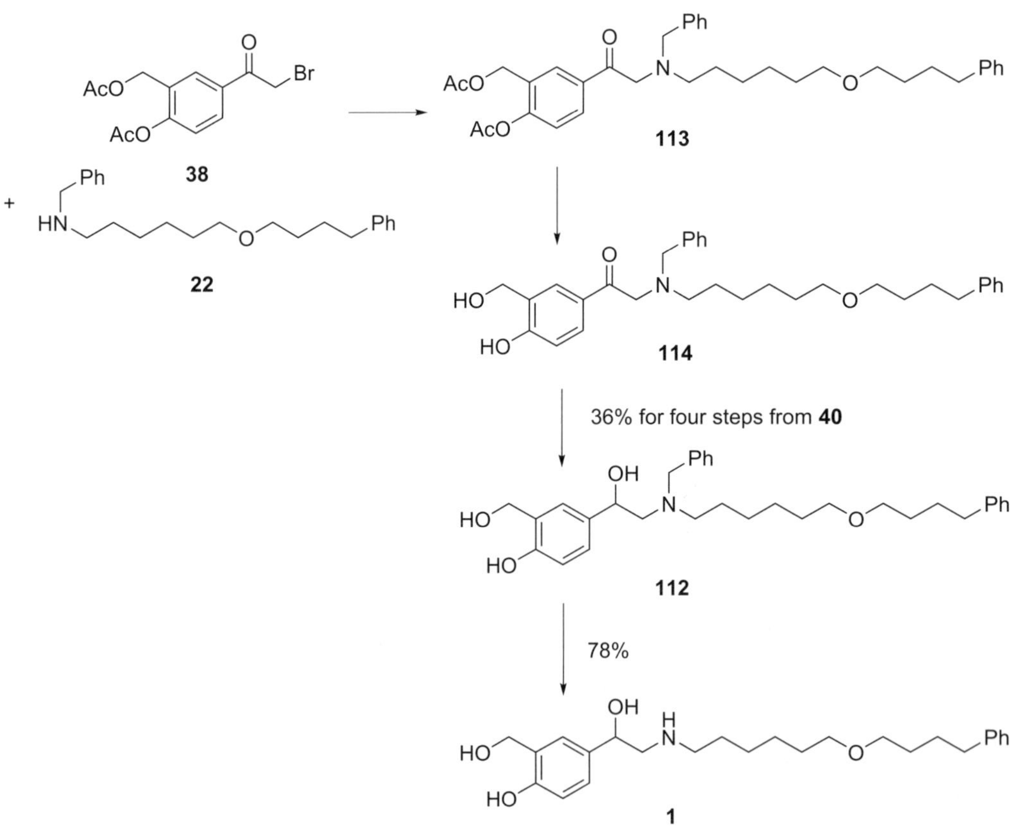


**SCHEME 8.35** Salmeterol (**1**) from 2-acetoxy-5-(2-bromoacetyl)benzyl acetate (**38**) and *N*-benzyl-6-(4-phenylbutoxy)hexan-1-amine (**22**).

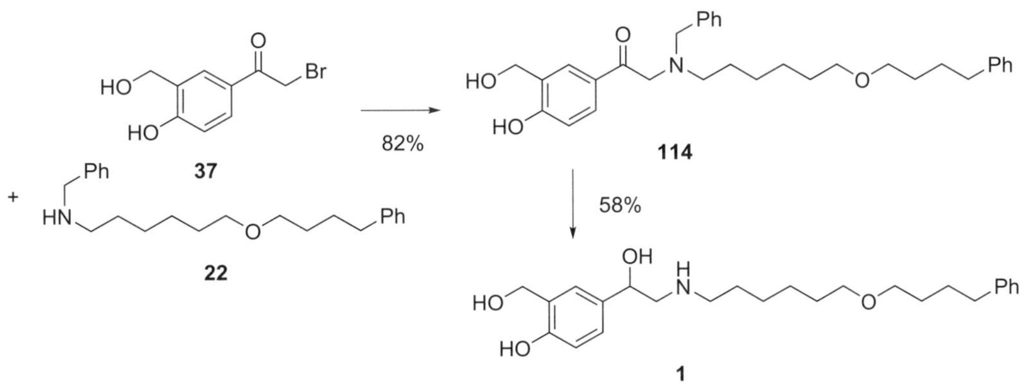

**SCHEME 8.36** Salmeterol (**1**) from 2-bromo-1-(4-hydroxy-3-(hydroxymethyl)phenyl)ethanone (**37**) and *N*-benzyl-6-(4-phenylbutoxy)hexan-1-amine (**22**).

23°C for 72 h. The mixture is diluted with ethyl acetate, washed with saturated aqueous sodium bicarbonate and with brine, and then dried and concentrated at reduced pressure. The α-amino ketone **114** is isolated from the residue by chromatography (82%).

Hydrogenation using 10% palladium on carbon and 10% platinum on carbon (each 500 mg/g **114**) in methanol at 23°C and 1 atm hydrogen pressure results in reduction of the ketone and hydrogenolysis of the *N*-benzyl group. The suspension is filtered and the liquors are concentrated at reduced pressure. Salmeterol (**1**), separated from the residue by chromatography, is suspended in ethyl ether and the solid is filtered and dried at an unspecified temperature and reduced pressure (58%) (Scheme 8.36).[13]

***8.7.1.5 2-Bromo-1-(2,2-dimethyl-4H-benzo[d][1,3]dioxin-6-yl)ethanone (45)*** The reaction of 2-bromo-1-(2,2-dimethyl-4*H*-benzo[*d*][1,3]dioxin-6-yl)ethanone (**45**) with *N*-benzyl-6-(4-phenylbutoxy)hexan-1-amine (**22**) (1.1 equivalents) and diisopropylethylamine (1.7 equivalents) in THF is aged at 23°C for 48 h. The mixture is diluted with ethyl acetate, washed with saturated aqueous sodium bicarbonate and with brine, and then dried and concentrated at reduced pressure. The α-amino ketone **115** is isolated from the residue by chromatography (72%) (Scheme 8.37).

Hydrogenation using 10% palladium on carbon (100 mg/g **115**) in methanol at 25°C and 1 atm hydrogen pressure results in reduction of the ketone and hydrogenolysis of the *N*-benzyl group. The suspension is filtered and the liquors are concentrated at reduced pressure. Hexane is added, the suspension is filtered, and the solid is dried at an unspecified temperature and pressure to afford the protected α-amino alcohol **116** (mp 68–70°C) (86%).

The ketal is deprotected with hydrochloric acid (3.0 equivalents) in methanol–water (volume ratio 1:1) at 23°C over 5 h. The mixture is diluted with ethyl acetate, washed with dilute aqueous sodium carbonate and with brine, and then dried and concentrated at reduced pressure. Ethyl ether is added, the suspension is filtered, and the solid is dried at an unspecified temperature and reduced pressure to afford salmeterol (**1**) (84%).[13]

### 8.7.2 Head Epoxides, Bromohydrins, and *O*-Trialkylsilylbromohydrins

***8.7.2.1 (R)-Methyl 2-(Benzyloxy)-5-(oxiran-2-yl)benzoate (51)*** Reaction of (*R*)-methyl 2-(benzyloxy)-5-(oxiran-2-yl)benzoate (**51**) with *N*-benzyl-6-(4-phenylbutoxy) hexan-1-amine (**22**) in THF at reflux affords the α-amino alcohol **117**, which is isolated by chromatography (60%, 94% ee). Ester reduction with lithium aluminum hydride (90%) followed by hydrogenolysis of the benzyl protecting groups affords (*R*)-salmeterol (**2**) (55%) (Scheme 8.38). The alternative reaction of the bromohydrin **52** with *N*-benzyl-6-(4-phenylbutoxy)hexan-1-amine (**22**) resulted in only low yields of α-amino alcohol **117**.[44]

***8.7.2.2 (R)-(2-(Benzyloxy)-5-(oxiran-2-yl)phenyl)methanol (53)*** An *N*-trimethylsilylamine is prepared by aging a DMSO solution of 6-(4-phenylbutoxy)hexan-1-amine (**23**) (1.0 equivalent) and *N*,*O*-bis(trimethylsilyl)acetamide (2.3 equivalents) at 25°C for 30 min. A solution of the (*R*)-epoxide **53** in DMSO is added and the mixture is aged at 85°C for 50 h. The mixture is cooled and DMSO is distilled at an unspecified temperature and 0.1 mmHg. The α-amino alcohol **118** is isolated from the residue by chromatography (66%). Hydrogenolysis using 15% palladium on carbon (125 mg/g **119**) in ethanol at 25°C and 50 psi hydrogen pressure requires 12 h. The suspension is filtered and the recovered catalyst washed. (*R*)-Salmeterol (**2**) is isolated by concentration of the combined liquors at reduced pressure and chromatography of the residue (61%). (*R*)-6-Oxiran-2-yl)-4*H*-benzo[*d*][1,3]dioxine (**55**) is converted to salmeterol (**1**) by a similar sequence (Scheme 8.39).[63]

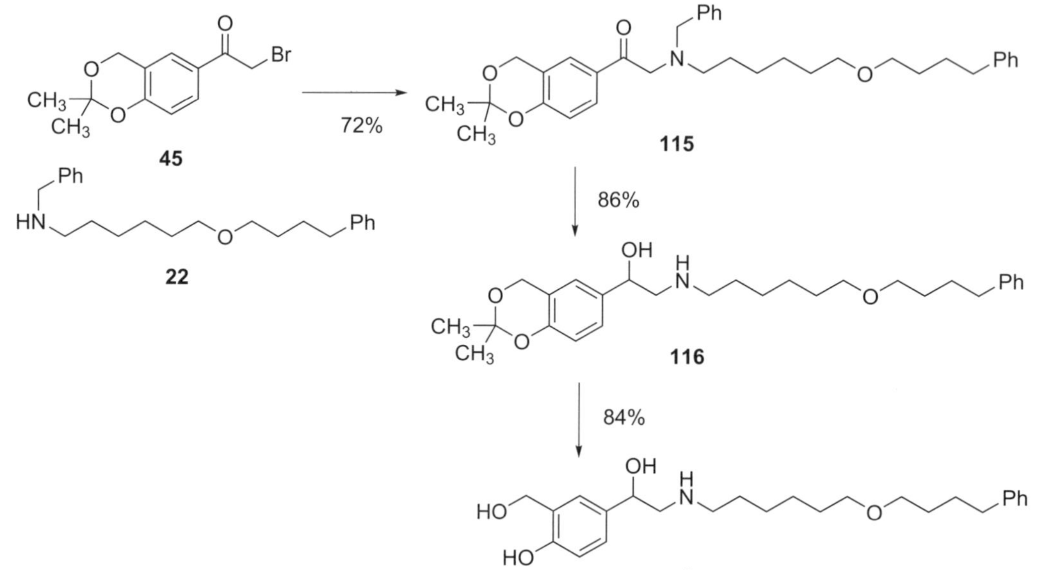


1

**SCHEME 8.37** Salmeterol (**1**) from 2-bromo-1-(2,2-dimethyl-4*H*-benzo[*d*][1,3]dioxin-6-yl)ethanone (**45**) and *N*-benzyl-6-(4-phenylbutoxy)hexan-1-amine (**22**).

***8.7.2.3 2,2-Dimethyl-6-(oxiran-2-yl)-4H-benzo[d][1,3] dioxine (58)*** A solution of 2,2-dimethyl-6-(oxiran-2-yl)-4*H*-benzo[*d*][1,3]dioxine (**58**) and *N*-benzyl-6-(4-phenylbutoxy)hexan-1-amine (**22**) (2.0 equivalents) in THF is refluxed for 24 h. The α-amino alcohol **119** is isolated by concentration at reduced pressure and chromatography of the residue (28%) (Scheme 8.40).[13]

An equimolar mixture of (*R*)-2,2-dimethyl-6-(oxiran-2-yl)-4*H*-benzo[*d*][1,3]dioxine (**61**) and *N*-benzyl-6-(4-phenylbutoxy)hexan-1-amine (**22**) is heated at 120°C for 12 h. The (*R*)-α-amino alcohol **120** is isolated by chromatography (83%). The benzyl group is removed by hydrogenolysis using 10% palladium on carbon (100 mg/g **120**) in methanol at 25°C and 150 psi hydrogen pressure over 24 h. Workup involves filtering the catalyst and concentrating the liquors at reduced pressure (95%). The residue is aged in glacial acetic acid and methanol (volume ratio 1:1) at 25°C over 48 h. Acid neutralization with aqueous potassium

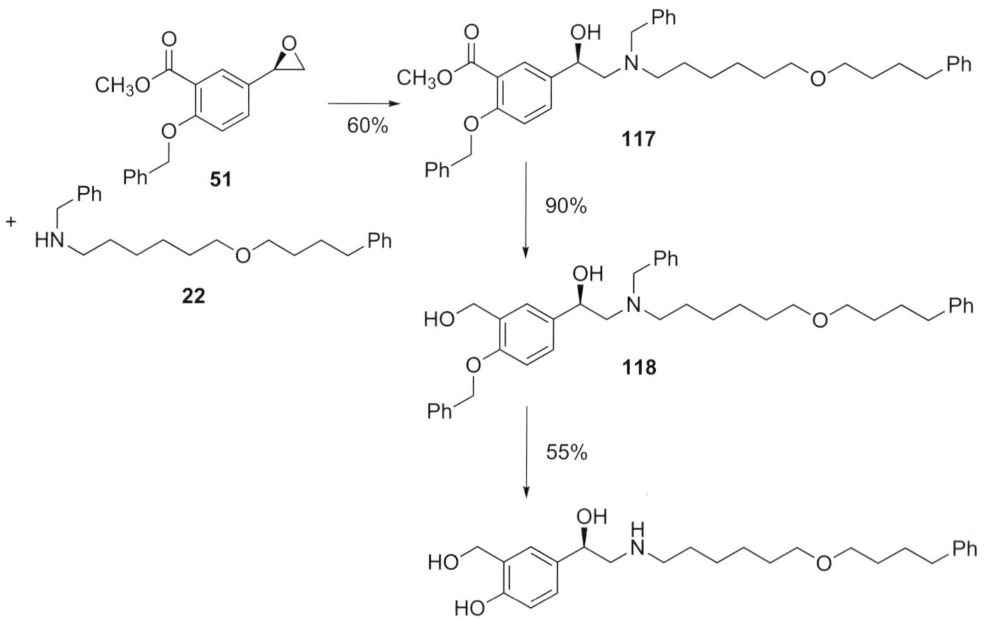


2

**SCHEME 8.38** (*R*)-Salmeterol (**2**) from (*R*)-methyl 2-(benzyloxy)-5-(oxiran-2-yl)benzoate (**51**).

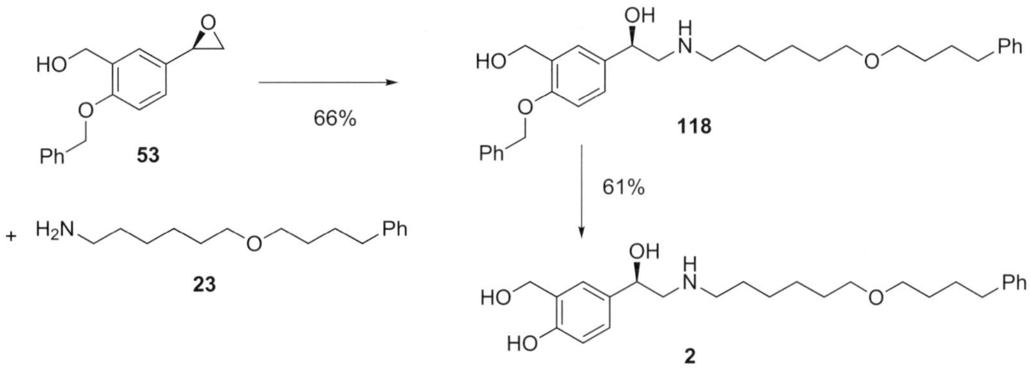

**SCHEME 8.39** (*R*)-Salmeterol (**2**) from (*R*)-(2-(benzyloxy)-5-(oxiran-2-yl)phenyl)methanol (**53**) and 6-(4-phenylbutoxy)hexan-1-amine (**23**).

carbonate, extraction with ethyl acetate, and concentration of the extract at reduced pressure affords (*R*)-salmeterol (**2**) (93%).[56]

***8.7.2.4 2-Bromo-1-(2,2-dimethyl-4H-benzo[d][1,3]dioxin-6-yl)ethanol (59)*** A solution of 2-bromo-1-(2,2-dimethyl-4*H*-benzo[*d*][1,3]dioxin-6-yl)ethanol (**59**) and *N*-benzyl-6-(4-phenylbutoxy)hexan-1-amine (**22**) (3.0 equivalents) in THF is refluxed for 18 h. The mixture is diluted with ethyl ether, washed with aqueous bicarbonate and brine, and then dried and concentrated at reduced pressure. The α-amino alcohol **119** is isolated from the residue by chromatography (24%).[13]

***8.7.2.5 (2-Bromo-1-(2,2-dimethyl-4H-benzo[d][1,3]dioxin-6-yl)ethoxy)(tert-butyl)dimethylsilane (66)*** There are apparently no reports of conversion of an *O*-silyl bromohydrin to salmeterol (**1**). In light of the low yields observed in bromide displacement from a bromohydrin *en route* to salmeterol (**1**), it would be interesting to see how much the efficiency of bromide displacement by a primary or secondary amine is increased by *O*-silylation. The reaction of (2-bromo-1-(2,2-dimethyl-4*H*-benzo[*d*][1,3]dioxin-6-yl)ethoxy)(*tert*-butyl)dimethylsilane (**66**) with 6-(2-((2,6-dichlorobenzyl)(methyl)amino)ethoxy)hexan-1-amine (**121**) (1.0 equivalent) and potassium carbonate (1.5 equivalents) in dioxane at reflux followed by an aqueous workup and

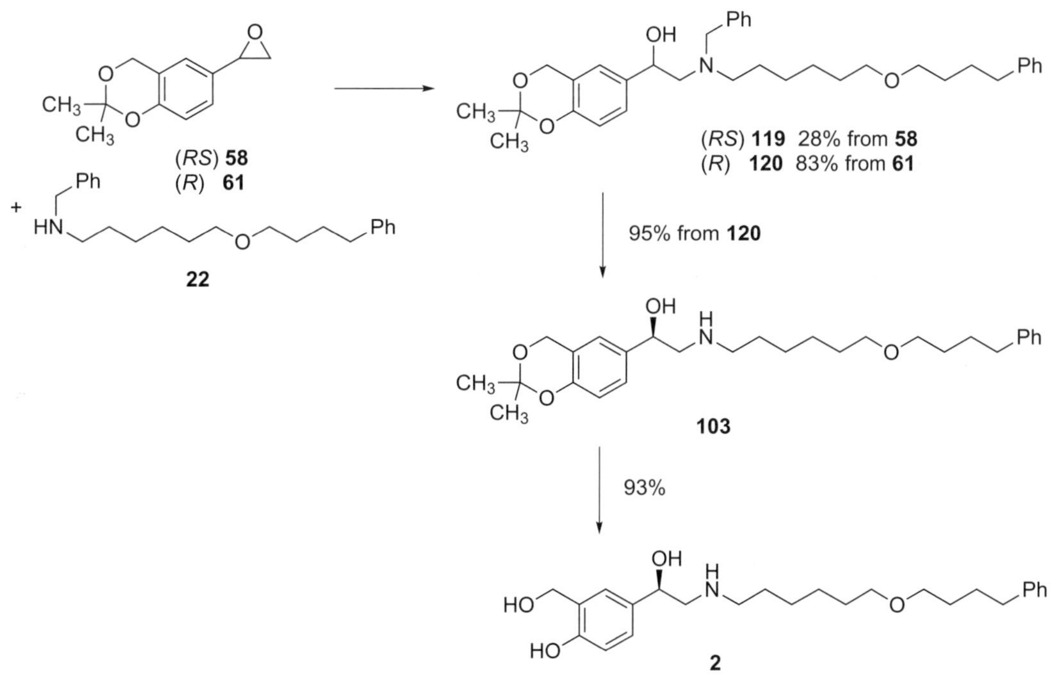

**SCHEME 8.40** (*R*)-Salmeterol (**2**) from (*R*)-2,2-dimethyl-6-(oxiran-2-yl)-4*H*-benzo[*d*][1,3]dioxine (**61**) and *N*-benzyl-6-(4-phenylbutoxy)hexan-1-amine (**22**).

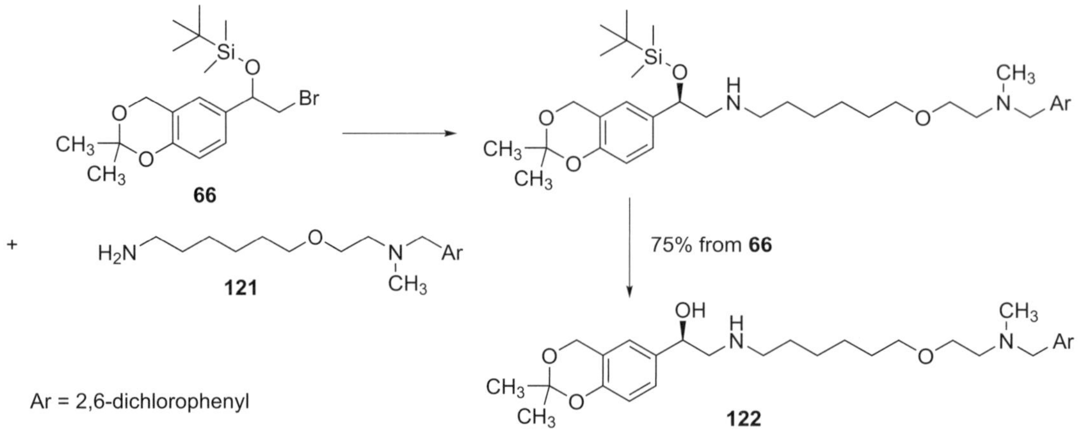


**SCHEME 8.41** Bromide displacement from (2-bromo-1-(2,2-dimethyl-4*H*-benzo[*d*][1,3]dioxin-6-yl) ethoxy)(*tert*-butyl)dimethylsilane (**66**) by amine **121**.

desilylation affords the α-amino alcohol **122** (75%) (Scheme 8.41).[84]

In a second well-established procedure from Theravance, (*R*)-(2-bromo-1-(2,2-dimethyl-4*H*-benzo[*d*][1,3]dioxin-6-yl)ethoxy)(*tert*-butyl)dimethylsilane (**67**) (1.3 equivalents) is aged with sodium iodide (1.3 equivalents) in THF at 25°C for 15 min. Biphenyl-2-ylcarbamic acid 1-(9-aminononyl)piperidin-4-yl ester (**123**) and sodium bicarbonate (3.0 equivalents) are added and the suspension is refluxed for 24 h. A routine workup followed by chromatography affords the *O*-silylated α-amino alcohol **124** (60%) (Scheme 8.42).[107]

### 8.7.3 Head Glyoxals

A solution of 6-(4-phenylbutoxy)hexan-1-amine (**23**) in methanol is added to a suspension of methyl 2-hydroxy-5-(2-oxoacetyl)benzoate (**68**) hydrate (equivalents not specified) in methanol at 23°C. After aging for just 10 min at 23°C, the *unstable* imine **125** is isolated by concentration at reduced pressure and chromatography of the residue (71%). A solution of imine **125** in THF is added to a suspension of lithium aluminum hydride (10 equivalents) in THF at 0°C. The suspension is aged at 23°C for 18 h then excess hydride is quenched by careful addition of water. Hydrochloric acid is added (to pH 5), aqueous bicarbonate is added (to pH 8), and the mixture is extracted with ethyl acetate. The extracts are washed with brine then dried and concentrated at reduced pressure. Salmeterol (**1**), separated from the residue by chromatography, is suspended in ethyl ether and the solid is filtered and dried at an unspecified temperature and reduced pressure (20%) (Scheme 8.43).[13]

## 8.8 ALTERNATIVE STRATEGIES FOR JOINING THE HEAD AND TAIL

### 8.8.1 Condensation of (2,2-Dimethyl-4H-benzo[*d*][1,3]dioxin-6-yl)magnesium Bromide (126) with a Tail Aldehyde

A functionalized aldehyde tail is prepared in three steps from 6-(4-phenylbutoxy)hexane-1-amine (**23**) (Scheme 8.44). The reaction of bromoacetaldehyde diethyl acetal[108] with amine **23** (1.0 equivalent) and potassium carbonate (1.0 equivalent) in DMF requires 20 h at 80°C. The DMF is distilled at unspecified temperature and reduced pressure. The aminoacetaldehyde acetal **127** is isolated from the residue by a routine water–dichloromethane workup and chromatography (43%). The amine is protected by reaction with (*N*-benzyloxycarbonyloxy)succinimide (1.1 equivalents) in acetone

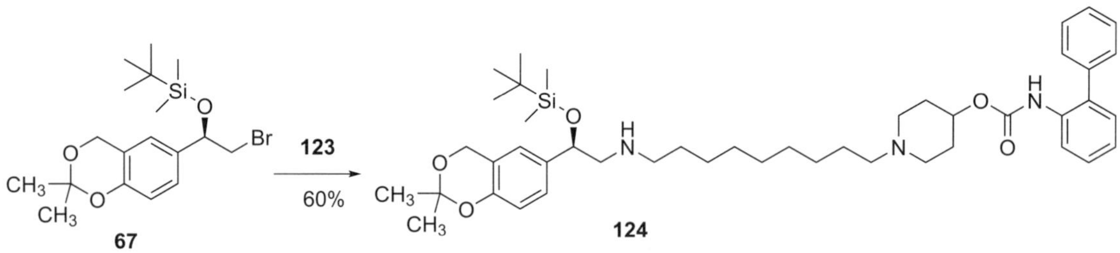


**SCHEME 8.42** Bromide displacement from (*R*)-(2-bromo-1-(2,2-dimethyl-4*H*-benzo[*d*][1,3]dioxin-6-yl)ethoxy)(*tert*-butyl)dimethylsilane (**67**) by amine **123**.

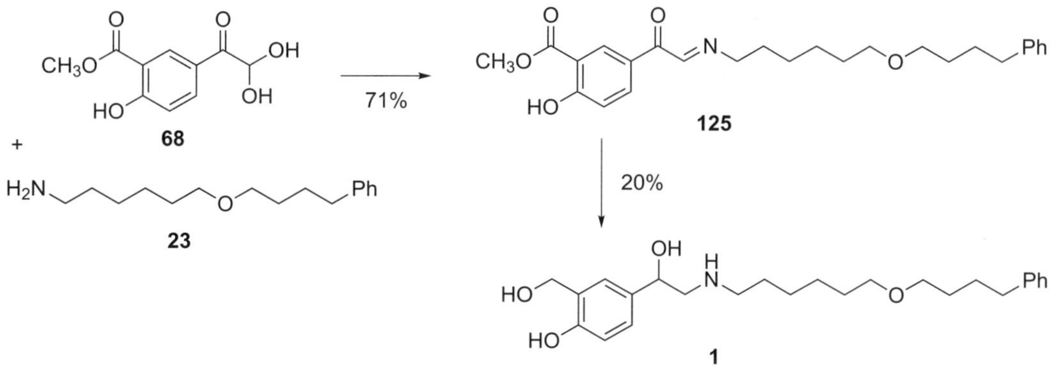

**SCHEME 8.43** Salmeterol (**1**) from methyl 2-hydroxy-5-(2-oxoacetyl)benzoate (**68**) and 6-(4-phenylbutoxy)hexan-1-amine (**23**).

at 25°C for 2 h. Concentration at reduced pressure and a routine water–ethyl acetate workup affords the carbamate **128** (99%). The aldehyde is then deprotected by reaction with *p*-toluenesulfonic acid monohydrate (0.5 equivalents) in acetone at 25°C over 4 h. Concentration at reduced pressure, a routine water–ethyl acetate workup, and chromatography affords benzyl 2-oxoethyl(6-(4-phenylbutoxy)hexyl)carbamate (**129**) (76%).

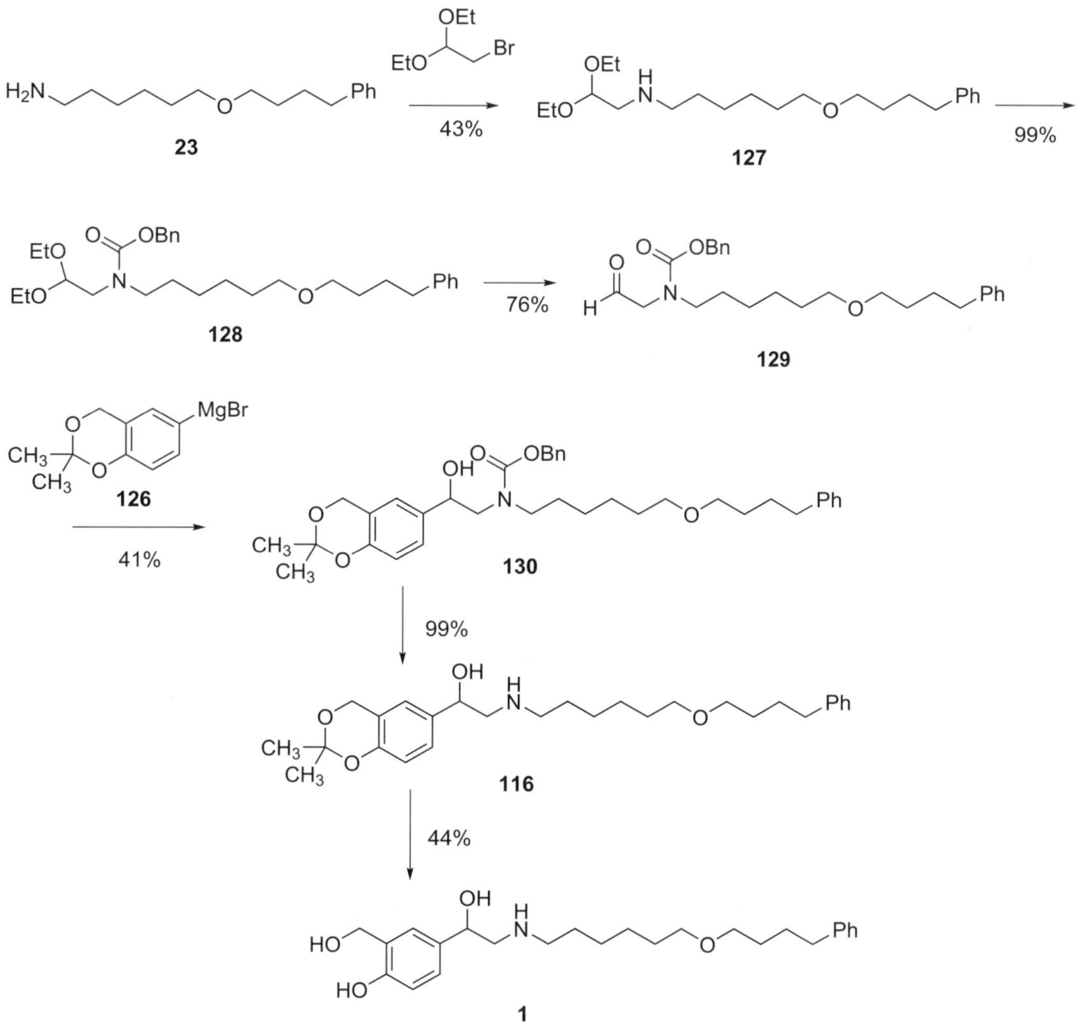

**SCHEME 8.44** Salmeterol (**1**) from (2,2-dimethyl-4*H*-benzo[*d*][1,3]dioxin-6-yl)magnesium bromide (**126**).

The Grignard reagent **126** is prepared by slow addition of a THF solution of 6-bromo-2,2-dimethyl-4*H*-benzo[*d*][1,3] dioxine (**46**) to a suspension of magnesium (1.1 equivalents) in THF at 40°C. 1,2-Dibromoethane is used as an initiator. The suspension is aged at 40°C for 30 min, then cooled to −7°C, and a solution of the aldehyde **129** (1.0 equivalent) in THF is added. After aging at 0°C for 1 h, the mixture is warmed to 25°C, quenched with saturated aqueous ammonium chloride, and extracted with ethyl acetate. The extract is washed with water, dried, and concentrated at reduced pressure. The carbamate-protected α-amino alcohol **130** is isolated from the residue by chromatography (41%).

The carbamate is cleaved by hydrogenolysis using 10% palladium on carbon (110 mg/g **130**) in methanol at 25°C and 1 atm hydrogen pressure. The catalyst is filtered and the liquors concentrated at reduced pressure to afford the α-amino alcohol **116** (99%). The ketal is deprotected with hydrochloric acid (1.3 equivalents) in methanol–water (volume ratio 1:1) at 25°C over 48 h. After removing the methanol at reduced pressure, the acid is neutralized with aqueous bicarbonate and the mixture extracted with dichloromethane. The extract is washed with aqueous bicarbonate and with water, and then dried and concentrated at reduced pressure to afford salmeterol (**1**) (44%).[16]

### 8.8.2 Hydrocyanation of 2-Methyl-4H-benzo[*d*][1,3] dioxine-6-carbaldehyde (131) and Elaboration of the Cyanohydrin

An aldehyde hydrocyanation strategy has been demonstrated for the synthesis of salbutamol (**132**) (Scheme 8.45). Hydrocyanation of 2-methyl-4*H*-benzo[*d*][1,3]dioxine-6-carbaldehyde (**131**) with (*R*)-oxynitrilase in isopropyl ether at 25°C affords the (*R*)-cyanohydrin (**133**) (96% ee at 60% conversion after 17 h). Hydrocyanation of 2,2-dimethyl-4*H*-benzo[*d*][1,3]dioxine-6-carbaldehyde (**62**) gave variable yields of the cyanohydrin **134** (<50% ee).[109]

### 8.8.3 Elaboration of the Tail at the End of the Sequence

Sodium hydride dispersion in mineral oil (1.1 equivalents) is added to a solution of (*R*)-5-(2,2-dimethyl-4*H*-benzo[*d*][1,3] dioxin-6-yl)oxazolidin-2-one (**99**) and 1,6-dibromohexane (3.0 equivalents) in DMF at 0°C. The mixture is aged at 0°C for 30 min and at 20°C for 2.5 h. Excess hydride is quenched by careful addition of phosphate buffer solution. A routine water–ethyl ether workup followed by chromatography affords the *N*-alkylated oxazolidinone **138** (86%). A Williamson ether synthesis with 4-phenyl-1-butanol and the known cleavages of the oxazolidinone and ketal would lead to (*R*)-salmeterol (**2**) (Scheme 8.46).[102,104]

A solution of (*R*)-5-(2,2-dimethyl-4*H*-benzo[*d*][1,3]dioxin-6-yl)oxazolidin-2-one (**99**) in DMF is added to a suspension of sodium hydride (1.5 equivalents) in DMF at 0°C. The mixture is aged at 0–5°C for 1 h then a solution of 1-bromo-6-(but-3-ynyloxy)hexane (**19**) (1.6 equivalents) in DMF is added at an unspecified temperature. After aging at 20–30°C for 2 h, excess hydride is quenched by careful addition of 2 M hydrochloric acid. A routine water–ethyl ether workup followed by chromatography affords the *N*-alkylated oxazolidinone **139** (86%). A Sonogashira coupling with iodobenzene, hydrogenation of the alkyne, and the known cleavages of the oxazolidinone and ketal would lead to (*R*)-salmeterol (**2**) (Scheme 8.47).[20,28,30–33]

A similar proposed sequence to (*R*)-salmeterol begins with 6-(but-3-ynyloxy)hexanal (**20**) and (*R*)-1-(2,2-dimethyl-4*H*-benzo[*d*][1,3]dioxin-6-yl)-2-(*S*)-2-hydroxy-1-

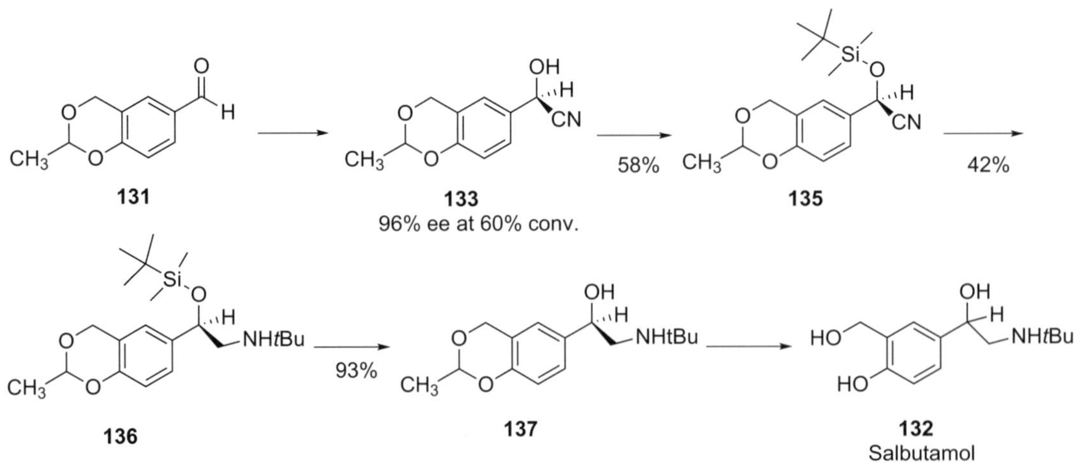


**SCHEME 8.45** (*R*)-Salbutamol (**132**) via the (*R*)-cyanohydrin **133**.

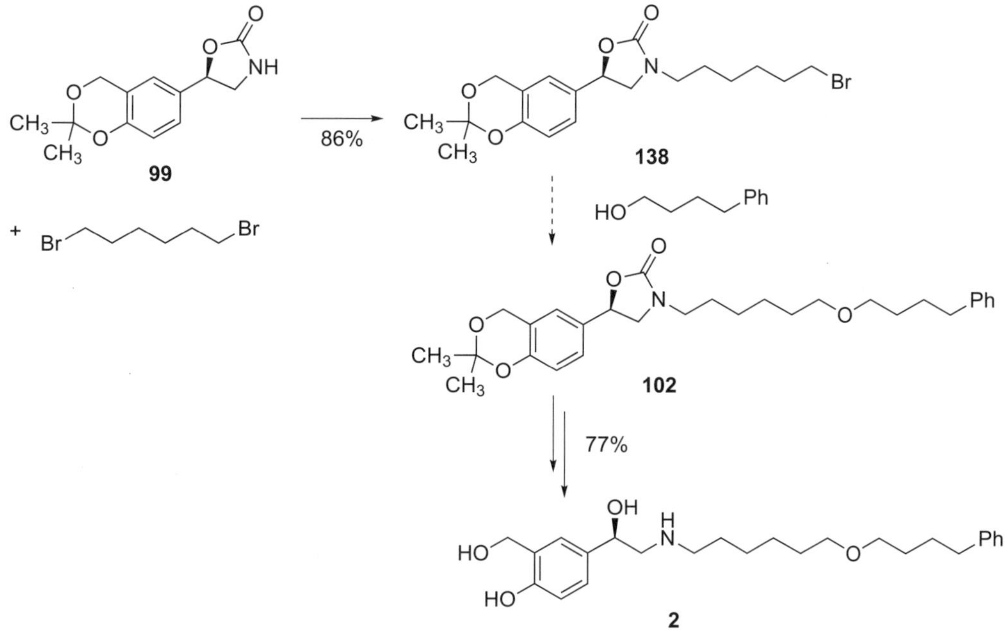

**SCHEME 8.46** Proposed alternative route to (*R*)-salmeterol (**2**) via late-stage Williamson ether synthesis.

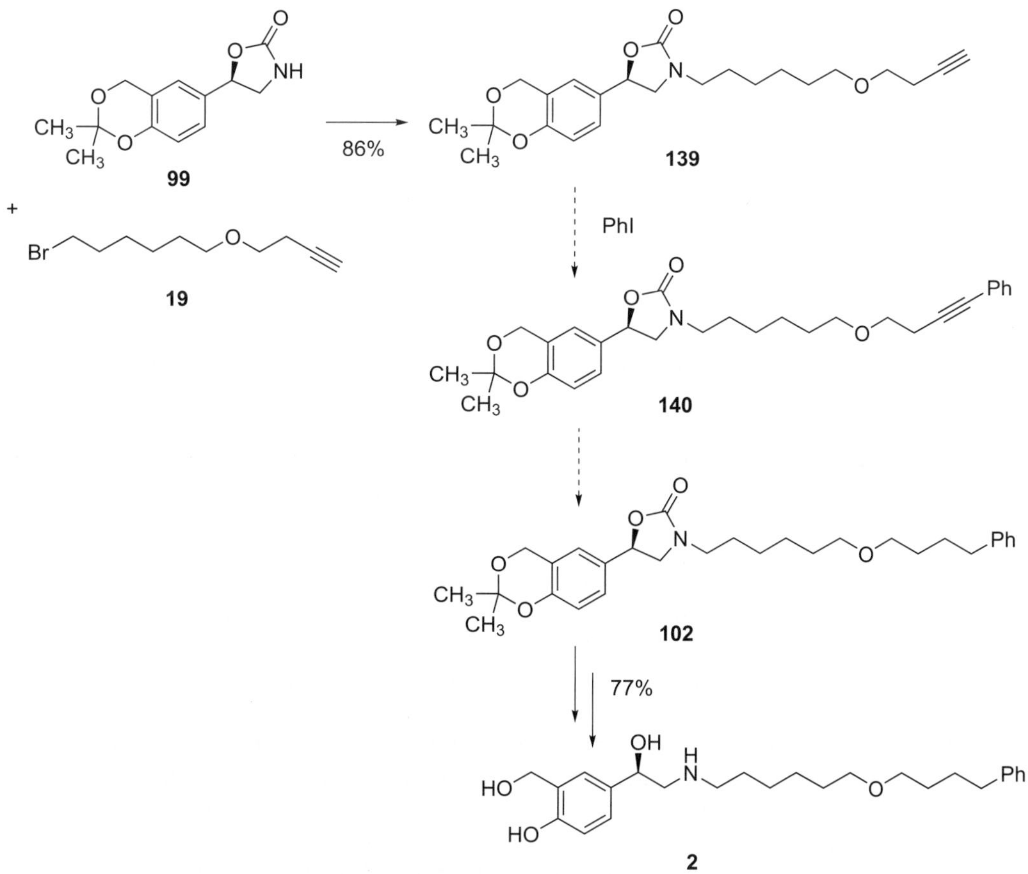

**SCHEME 8.47** Proposed alternative route from (*R*)-5-(2,2-dimethyl-4*H*-benzo[*d*][1,3]dioxin-6-yl) oxazolidin-2-one (**99**) to (*R*)-salmeterol (**2**) via a late-stage Sonogashira coupling.

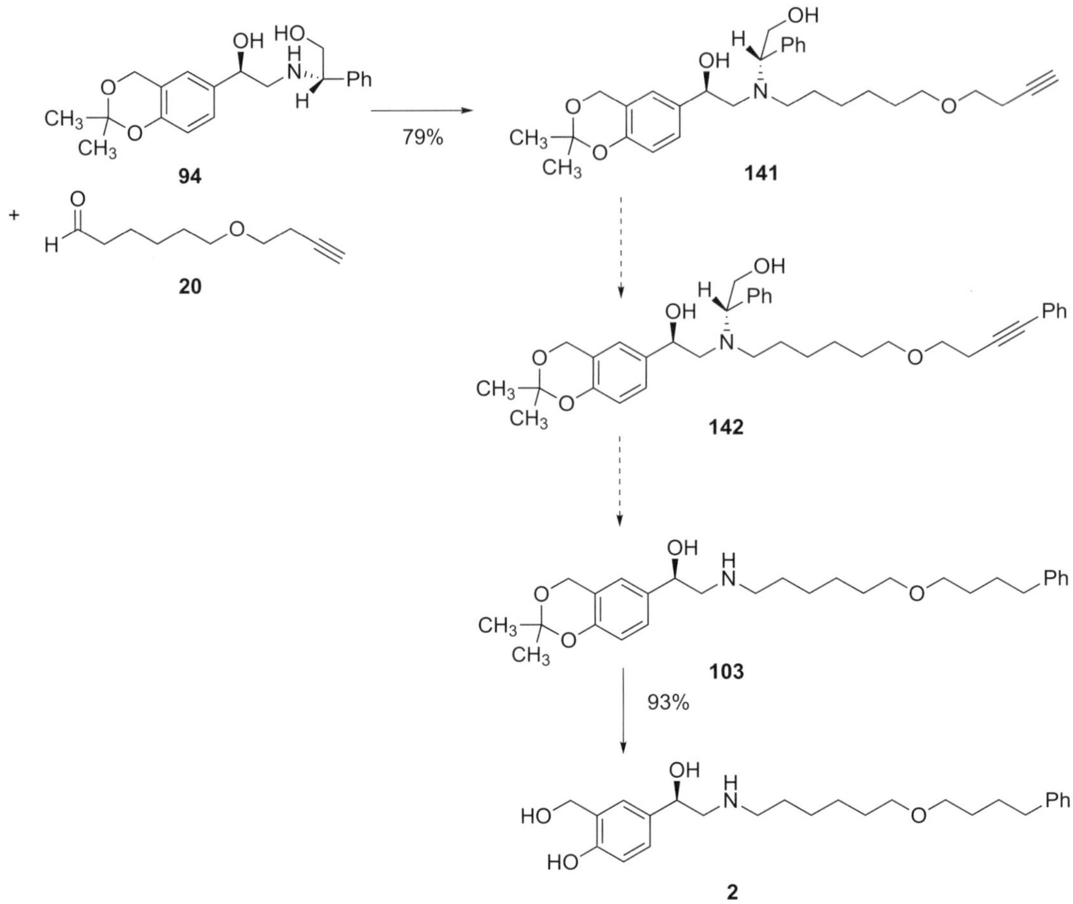


**SCHEME 8.48** Proposed alternative route from (*R*)-1-(2,2-dimethyl-4*H*-benzo[*d*][1,3]dioxin-6-yl)-2-(*S*)-2-hydroxy-1-phenethylamino)ethanol (**94**) to (*R*)-salmeterol (**2**) via a late-stage Sonogashira coupling.

phenethylamino)ethanol (**94**). The reductive alkylation of the aldehyde **20** (1.2 equivalents) with the amine **94** and sodium triacetoxyborohydride (2.0 equivalents) in chloroform at 20°C requires 48 h. A routine water–ethyl acetate workup followed by chromatography affords **141** (79%). A Sonogashira coupling with iodobenzene, simultaneous hydrogenation of the alkyne and hydrogenolysis of the *N*-benzyl group, and the known ketal cleavage would lead to (*R*)-salmeterol (**2**) (Scheme 8.48).[20,33]

Sodium hydride dispersion in mineral oil (1.2 equivalents) is added to a solution of (*R*)-5-(2,2-dimethyl-4*H*-benzo[*d*][1,3]dioxin-6-yl)oxazolidin-2-one (**99**) in DMF at 25°C. After aging at 25°C for 15 min, a solution of 1-bromo-6-(but-3-enyloxy)hexane (**21**) (1.1 equivalents) in DMF is added. After aging at 25°C for 3 h, excess hydride is quenched by careful addition of water. A routine water–ethyl acetate workup followed by chromatography affords the *N*-alkylated oxazolidinone **143** (82%). Heck arylation with bromo- or iodobenzene, hydrogenation of the alkene, and the known cleavages of the oxazolidinone and ketal would lead to (*R*)-salmeterol (**2**) (Scheme 8.49).[35]

## 8.9 POLYMORPHS AND PARTICLE SIZE

The particle size and size distribution are critical for an inhalation/insufflation powder such as salmeterol (**1**). A suitable particle size distribution (1–10 μm) is usually produced by micronization and a successful micronization is critically dependent on the flow characteristics of the bulk salmeterol. The final step in the manufacturing process should be a crystallization producing salmeterol with a high bulk density, low cohesivity, and uniform particle size distribution.

Conventional crystallization by seeding and slow cooling affords salmeterol that is not suitable for micronization. Salmeterol is dissolved in hot (>60°C) isopropanol. A solution of 1-hydroxy-2-naphthoic acid (1.0 equivalent) in hot isopropanol (70°C) is added. The resulting hot solution is seeded, cooled to 40°C and aged for 2 h, then cooled to 5°C and aged for 2 h. The suspension is filtered, and the solid is washed with cold isopropanol and dried at an unspecified temperature and reduced pressure to afford salmeterol (bulk density 0.16 g/L, cohesivity 82%, mean particle size 26 μm).

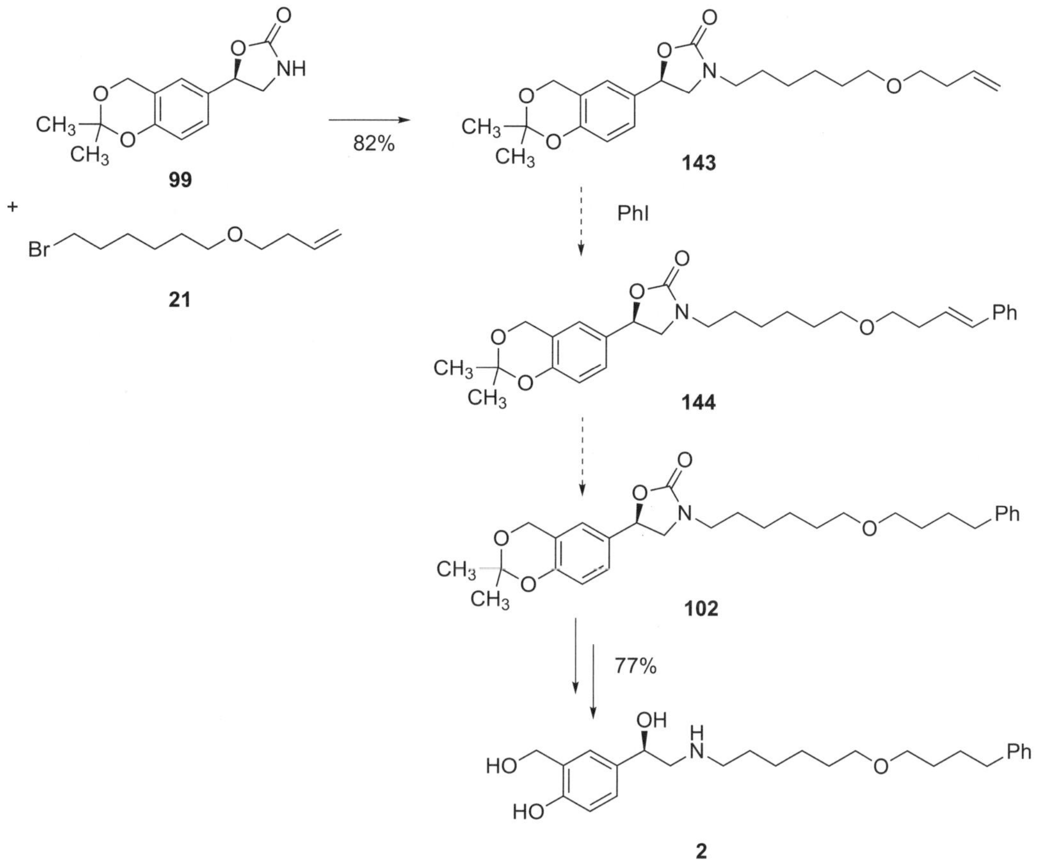


**SCHEME 8.49** Proposed alternative route to (*R*)-salmeterol (**2**) via a late-stage Heck arylation.

While fast crystallization typically produces a solid with a small particle size and many fines, fast crystallization unexpectedly produces salmeterol as spherical accretions of microcrystals with superior flow characteristics. A solution of salmeterol and 1-hydroxy-2-naphthoic acid (1.0 equivalent) in hot methanol is prepared. The hot solution is added to cold isopropanol at 12–17°C over 30 min. The suspension is aged at 15°C for 1 h and then filtered. The solid is washed with cold isopropanol and dried at 40°C and reduced pressure to afford salmeterol (bulk density 0.30 g/L, cohesivity 1.3%, mean particle size 156 μm).[110]

## 8.10 THE BEST PROCESS AVAILABLE TODAY

Twenty-four demonstrated routes to salmeterol (**1**) and six proposed routes based on demonstrated synthesis of salmeterol analogs have been discussed. The references often cited as describing the manufacturing process[13] describe many options and offer procedures, but on a small scale and with yields that are often not representative of the reaction efficiency. Where should the selection process begin?

The selection process begins with the carbon–nitrogen bond formation that joins the head and the tail. *Neither component in the carbon–nitrogen bond formation can be used in large excess* (>1.1–1.2 equivalents) because both are value-added and chromatography would likely be required to separate excess reactant from the product. Routes generating the carbon–nitrogen bond by alkylation of a primary amine head with a tail bromide or iodide use excess amine (e.g., 1.5 equivalents of amine **82** to keep overalkylation to a minimum. Routes generating the carbon–nitrogen bond by reductive alkylation of a primary amine head with a tail aldehyde also use excess amine (e.g., 1.4 equivalents of amine **73**) to keep overalkylation to a minimum.

Because the two reagents are value-added and cleanup of a low-yielding reaction would likely require chromatography, *the carbon–nitrogen bond formation must be efficient, ideally providing the target in >90% yield.* Routes with the ketone reduction before the carbon–nitrogen bond formation, via a bromohydin or epoxide, are not suitable for the manufacturing process. The bromohydrin reaction is likely to proceed via the epoxide and the regioselectivity of the epoxide opening is not high enough. A mixture of two regioisomers with identical functionality will be very difficult to separate. The solution to the epoxide regioselectivity problem is to use an *O*-trialkylsilyl bromohydrin. However, the silylating agents are too expensive and treatment and disposal of a waste stream containing the released trialkylsilyl group only adds to the overall cost for this approach.

This leaves just two head electrophiles as possibilities, a glyoxal hydrate or an α-bromoketone, and leads to the third of the selection criteria: *all process intermediates should have good storage stability at ambient temperature*. The condensation of a glyoxal hydrate with 6-(4-phenylbutoxy) hexan-1-amine (**23**) affords an unstable imine. The alkylation of an α-bromoketone with 6-(4-phenylbutoxy)hexan-1-amine (**23**) similarly affords an unstable α-aminoketone. These two approaches, and all reductive amination approaches using the aliphatic aldehyde 6-(4-phenylbutoxy) hexanal (**10**), are not suitable for the manufacturing process.

Routes using head nucleophiles and tail electrophiles generally use a head nucleophile produced from a head electrophile. Carbon–nitrogen bond formation to produce the head nucleophile is accomplished using an ammonia surrogate such as *tert*-butyliminodicarboxylate, sodium azide, benzylamine, or dibenzylamine. This leads to the fourth and fifth of the selection criteria: *the head nucleophile should be produced using an inexpensive ammonia surrogate and the route leading from the alkylated ammonia surrogate to the head amine should have no intermediates that would raise safety concerns*. All routes to salmeterol (**1**) from the oxazolidinone intermediate **99** are not suitable for the manufacturing process because the oxazolidinone **99** is produced using expensive *tert*-butyliminodicarboxylate. There are two small-molecule azide intermediates in the sequence using sodium azide. These intermediates would certainly raise some red flags, as would intermediates in the routes producing a head nucleophile using nitromethane as a raw material.

Benzylamine and dibenzylamine remain as possible ammonia surrogates for a manufacturing process. Displacement of an α-bromoketone with benzylamine produces an α-benzyaminoketone, an unstable intermediate. Displacement of an α-bromoketone with dibenzylamine affords a stable α-dibenzylaminoketone that is hydrogenated to produce an α-aminoalcohol. But an α-aminoalcohol is only converted to salmeterol (**1**) by alkylation with (4-(6-bromohexyloxy)butyl)benzene (**4**) or reductive amination of 6-(4-phenylbutoxy)hexanal (**10**). These alkylations are not suitable for the manufacturing process for reasons already discussed.

The answer is to join a head α-bromoketone and the tail amine, *N*-benzyl-6-(4-phenylbutoxy)hexan-1-amine (**22**), to produce a stable α-amino ketone. Six examples of this carbon–nitrogen bond formation, with yields ranging from 71% (isolated) to 100% (contained), have been discussed.

With the strategy for generating the carbon–nitrogen bond selected, the focus now shifts to the other side of the head aromatic ring and the sixth of the selection criteria: *the intermediate produced by joining the head and tail should be converted to salmeterol (**1**) xinafoate in the minimum number of operations and the highest possible yield*. The α-amino ketone **115**, produced from 2-bromo-1-(2,2-dimethyl-4*H*-benzo[*d*][1,3]dioxin-6-yl)ethanone (**45**), is converted to salmeterol (**1**) in the highest yield. The ketone reduction and *N*-benzyl hydrogenolysis can be accomplished in a single step (86%). Ketal deprotection with hydrochloric acid in methanol-water then affords salmeterol (**1**) (84%, overall 72% from **115**).[13] We anticipate that the yield of salmeterol (**1**) from α-amino ketone **115** can be greater than 85% when the ketone reduction (with sodium borohydride, estimate >90% yield) is separated from the *N*-benzyl hydrogenolysis (95%) and when the ketal is deprotected with aqueous acetic acid (93%) or using an ion exchange column (100%).

With the process from 2-bromo-1-(2,2-dimethyl-4*H*-benzo[*d*][1,3]dioxin-6-yl)ethanone (**45**) to salmeterol (**1**) selected, the route to α-bromoketone **45** is next on the agenda. Selecting from among the options requires the last of the selection criteria: *intermediates that are lacrymators and/or potent alkylating agents should be kept to a minimum*. The GSK routes from salicylaldehyde and from 4-hydroxyacetophenone proceed through two α-bromoketone intermediates to α-bromoketone **45** while the Theravance route converts the methyl ketone, 1-(2,2-dimethyl-4*H*-benzo[*d*][1,3]dioxin-6-yl)ethanone (**47**), directly to the α-bromoketone **45**. However, the Theravance route to methyl ketone **47** involves capture of an aryllithium by expensive *N*-methoxy-*N*-methylacetamide at −78°C and the Theravance bromination requires low temperature (−78°C) and expensive sodium bis(trimethylsilyl)amide. There are certainly less costly alternative routes to methyl ketone **47**. An inexpensive bromination would complete a much-preferred sequence to α-bromoketone **45**. Until that bromination is available, the best process is a compromise.

The best process available converts salicylaldehyde to 2-bromo-1-(2,2-dimethyl-4*H*-benzo[*d*][1,3]dioxin-6-yl)ethanone **45** in three steps and 65–70% yield and to salmeterol (**1**) in seven steps in approximately 40% yield. The overall yield from salicylaldehyde can be as high as 50% if the carbon–nitrogen bond formation is more efficient (assume a 90% yield rather than the 72% yield reported on gram scale) (Scheme 8.50).

The process from salicylaldehyde to 2-bromo-1-(2,2-dimethyl-4*H*-benzo[*d*][1,3]dioxin-6-yl)ethanone (**45**) likely has just the isolation of the α-bromoketone **45** from hexane or heptane (mp 47–49°C). Avoiding isolation of the earlier α-bromoketone intermediates **35** and **37** would minimize operator exposure to these lacrymators. The process solvents are acetic acid, dichloromethane, ethyl acetate, heptane, and hexane.

The process from 4-phenyl-1-butanol to *N*-benzyl-6-(4-phenylbutoxy)hexan-1-amine (**22**) hydrochloride has just the isolation of the amine hydrochloride salt (mp

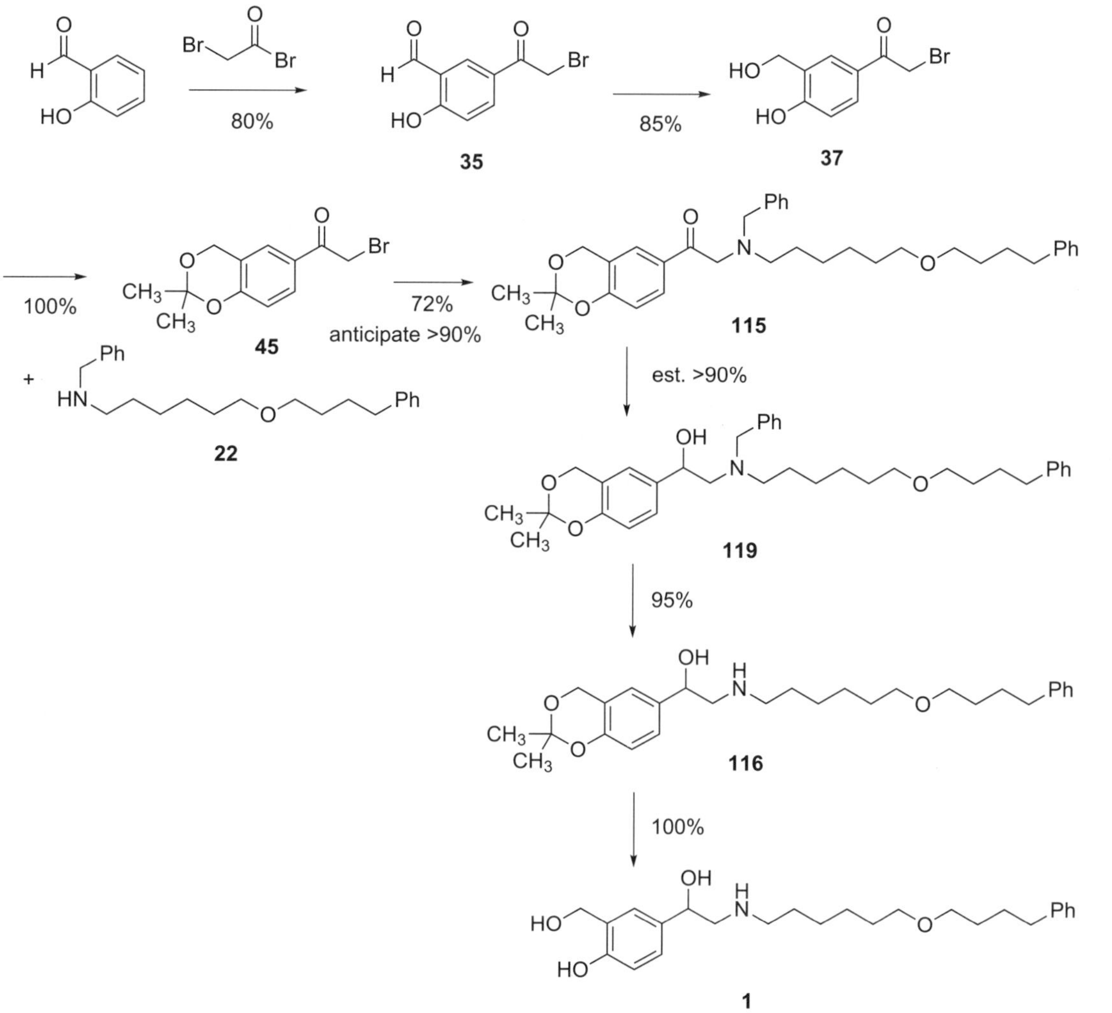


**SCHEME 8.50** The best process available today.

135–140°C). The process solvents are acetonitrile, dichloromethane, ethyl acetate or toluene, and heptane.

Four steps are required to convert 2-bromo-1-(2,2-dimethyl-4*H*-benzo[*d*][1,3]dioxin-6-yl)ethanone (**45**) and *N*-benzyl-6-(4-phenylbutoxy)hexan-1-amine (**22**) hydrochloride to salmeterol (**1**) xinafoate. The process likely has three solid isolations, the α-amino ketone **115**, perhaps the ketal-protected salmeterol **116** (mp 68–70°C), and salmeterol (**1**) xinafoate. The process solvents are acetic acid, isopropanol, methanol, THF, and ethyl acetate.

A process is only as good as its weakest link and this process has many. One of the key starting materials, 4-phenyl-1-butanol, is expensive relative to the others. Procedures for the isolation of low-melting solids **45** and **116** and for transitioning from one step to the next when a solid isolation is not necessary are not well defined. There are three α-bromoketone intermediates and one of these is isolated. We could do much better.

## 8.11 MANUFACTURING 4-PHENYL-1-BUTANOL

4-Phenyl-1-butanol is considerably more expensive (50 g for \$260) than the other raw materials. Do salmeterol manufacturers produce this key raw material in-house or have an exclusive supply contract at a more favorable price? In either case, what is the manufacturing process? Of the many routes to 4-phenyl-1-butanol from commercially available raw materials, we will consider just six (Figure 8.5).

Two routes are based on previous discussions of constructing the tail fragment. Sonogashira coupling of iodobenzene with expensive 3-butyn-1-ol followed by hydrogenation of the alkyne is not suitable. Heck arylation of expensive 3-buten-1-ol with iodo- or bromobenzene followed by reduction of the alkene and/or aldehyde is also not suitable.

Limiting the discussion to disconnection at the ring junction, we require a less expensive reagent delivering the

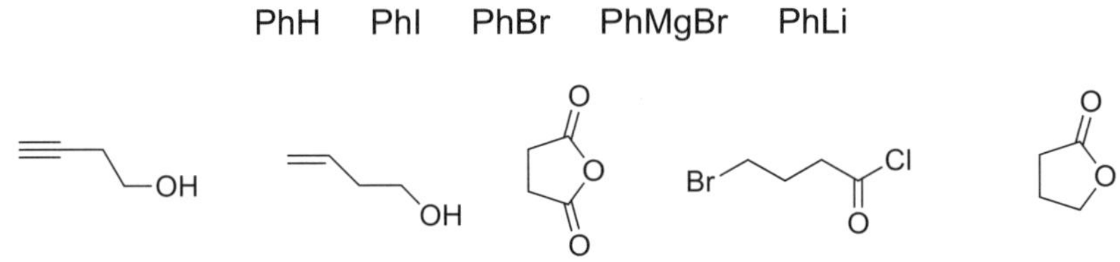


**FIGURE 8.5** Possible raw materials for manufacture of 4-phenyl-1-butanol.

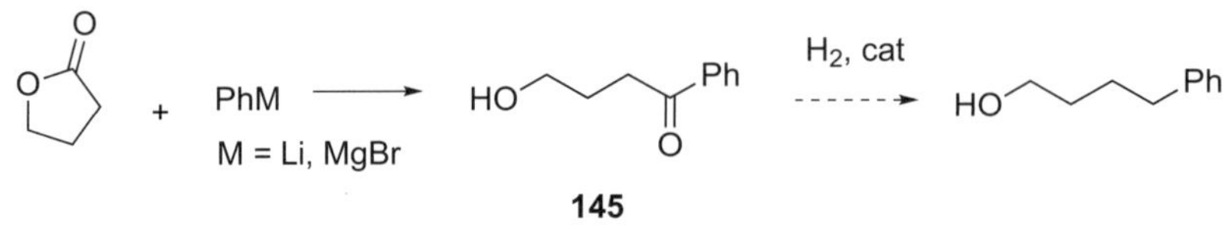


**SCHEME 8.51** Proposed low-cost route to 4-phenyl-1-butanol.

four-carbon chain. Commercially available 3-benzoylpropionic acid[111] is produced by Friedel-Crafts acylation of benzene with succinic anhydride.[112,113] This keto acid could be converted to 4-phenyl-1-butanol by catalytic reduction of the ketone and hydride reduction of the acid.

The hydride reduction is avoided in a related process. Friedel-Crafts acylation of benzene by 4-bromobutyryl chloride[114] affords 4-bromobutyrophenone.[115] 4-Bromobutyrophenone might be converted to 4-phenyl-1-butanol by catalytic reduction of the ketone and hydrolysis of the bromide. This route is unattractive because the reagent delivering the four-carbon chain, 4-bromobutyryl chloride, is too expensive.

4-Bromobutyryl chloride is produced from 4-bromobutyric acid that is produced from γ-butyrolactone. Perhaps inexpensive γ-butyrolactone is the best starting material. Phenylmagnesium bromide[116] (0.5 M in THF) (8.0 equivalents) is added to a slurry of γ-butyrolactone,[117] *N*,*O*-dimethylhydroxylamine hydrochloride[118] (1.2 equivalents), and sodium methoxide (0.25 equivalents) in THF at −15 to −20°C. The mixture is aged at −20°C for 2 h and at 25°C for 8 h. Excess Grignard reagent is quenched by adding dilute hydrochloric acid. After aging at 25°C for 2 h, the THF is distilled at reduced pressure. A routine water–dichloromethane workup and chromatography affords 4-hydroxy-1-phenylbutan-1-one (**145**) (84%). Some diol from addition to the ketone is also observed. Catalytic reduction of the ketone would then afford 4-phenyl-1-butanol. This two-step route is unattractive because it requires excess Grignard reagent, which is converted to carcinogen benzene during the quench, and because *N*,*O*-dimethylhydroxylamine hydrochloride is too expensive.[119]

Finally, phenyllithium[120] (1.67 M in cyclohexane–ethyl ether) is added at −78°C to γ-butyrolactone (1.7 equivalents) in ethyl ether. After aging at −78°C for 2 h, 10% aqueous ammonium chloride is added, the mixture is warmed to

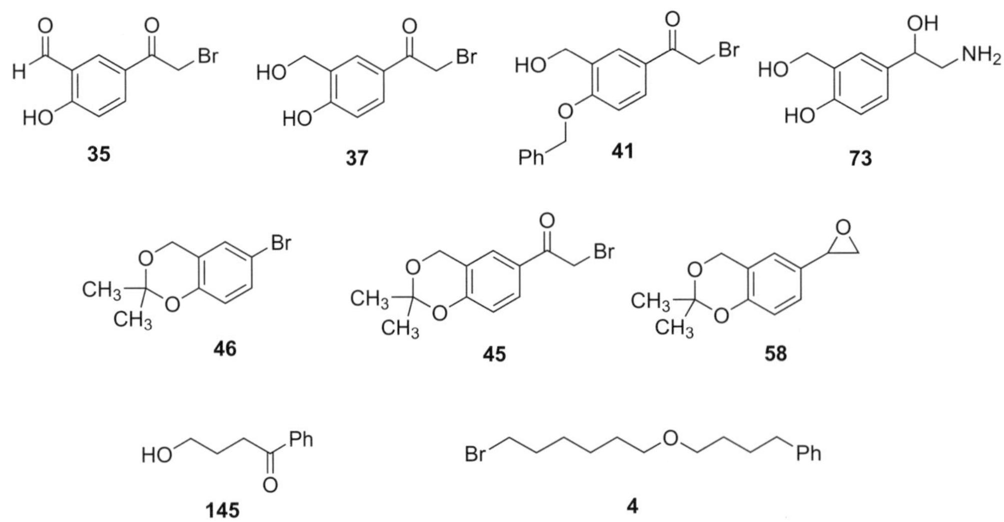


**FIGURE 8.6** Structures searched for the salmeterol (**1**) xinafoate presentation.

25°C, and the layers are separated. The aqueous layer is extracted with ethyl ether. The ether extracts are washed with water, dried, and concentrated at reduced pressure. 4-Hydroxy-1-phenylbutan-1-one is isolated from the residue by chromatography (97%).[121] While phenyllithium is expensive, ethyl ether is an unacceptable solvent, and −78°C is difficult to achieve in some manufacturing settings, producing 4-hydroxy-1-phenylbutan-1-one by adding a phenylmetal reagent to excess γ-butyrolactone followed by catalytic reduction of the ketone is one approach that could provide 4-phenyl-1-butanol at an acceptable price (Scheme 8.51).

## 8.12 STRUCTURES SEARCHED

Nine structure searches were used to generate all the information presented in this chapter (Figure 8.6).

## REFERENCES


1. Accessed at www.who.int/mediacentre/factsheets/fs315/en/print.html.
2. Global Initiative for Chronic Obstructive Pulmonary Disease (updated 2008) Accessed at www.goldcopd.com.
3. Oversteegen, L.; Rovini, H.; Belsey, M.J. *Nat. Rev. Drug Discov.* **2007**, *6*, 695.
4. GSK 2008 Annual Report, April 2009.
5. Nelson, H.S.; Weiss, S.T.; Bleecker, E.R.; Yancey, S.W.; Dorinsky, P.M. *Chest* **2006**, *129*, 15.
6. Reynolds, N.A.; Lyseng-Williamson, K.A.; Wiseman, L.R. *Drugs* **2005**, *65*, **1715**.
7. Fenton, C.; Keating, G.M. *Drugs* **2004**, *64*, 1975.
8. *Pediatric Alert* **2004**, *29*, 37.
9. Barberich, T.J.; Young, J.W. U.S. Patent 5,919,827 (7/6/1999).
10. (a) Jerussi, T.P. U.S. Patent 6,254,882 (7/3/2001). (b) Jerussi, T.P. WO 1999013867 (3/25/1999).
11. The Aldrich 2007–2008 catalog price for 50 g is $260.00.
12. The Aldrich 2007–2008 catalog price for 500 g is $66.80.
13. (a) Skidmore, I.F.; Lunts, L.H.C.; Finch, H.; Naylor, A. U.S. Patent 5,243,076 (9/7/1993). (b) Skidmore, I.F.; Lunts, L.H.C.; Finch, H.; Naylor, A. U.S. Patent 5,225,445 (7/6/1993). (c) Skidmore, I.F.; Lunts, L.H.C.; Finch, H.; Naylor, A. U.S. Patent 4,992,474 (2/12/1991).
14. Rong, Y.; Ruoho, A.E. *Synth. Commun.* **1999**, *29*, 2155.
15. Gore, V.G.; Gaikwad, A.C.; Gadkar, M.; Shukla, V.K. WO 2007045857 (4/26/2007).
16. (a) Bessa Belmunt, J.; Dalmases Barjoan, P.; Marquillas Olondriz, F. EP 1132373 (11/19/2003).(b) Bessa Belmunt, J.; Dalmases Barjoan, P.; Marquillas Olondriz, F. WO 2000018722 (4/6/2000).
17. Meyer, O.; Heitger, H. U.S. Patent 20030171620 (9/11/2003).
18. Molinski, T.F.; Stanley, S.D. *J. Label. Cmpd. Radiopharm.* **2002**, *45*, 755.
19. Bron, J.; Sterk, G.J.; Timmerman, H.; Veerman, M.E.; van der Werf, J.F.; Windhorst, A.D. WO 1995019336 (7/20/1995).
20. (a) Biggadike, K.; Coe, D.M.; Edney, D.D.; Guntrip, S.B.; Halton, A.; Looker, B.E.; Monteith, M.J.; Moore, R.J.; Patel, R.; Procopiou, P.A. U.S. Patent 7,442,837 (10/28/2008). (b) Biggadike, K.; Coe, D.M.; Edney, D.D.; Guntrip, S.B.; Halton, A.; Looker, B.E.; Monteith, M.J.; Moore, R.J.; Patel, R.; Procopiou, P.A. U.S. Patent 7,442,836 (10/28/2008). (c) Biggadike, K.; Coe, D.M.; Edney, D.D.; Guntrip, S.B.; Halton, A.; Looker, B.E.; Monteith, M.J.; Moore, R.J.; Patel, R.; Procopiou, P.A. U.S. Patent 7,442,719 (10/28/2008). (d) Biggadike, K.; Coe, D.M.; Edney, D.D.; Guntrip, S.B.; Halton, A.; Looker, B.E.; Monteith, M.J.; Moore, R.J.; Patel, R.; Procopiou, P.A. U.S. Patent 20070004807 (1/4/2007). (e) Biggadike, K.; Coe, D.M.; Edney, D.D.; Guntrip, S.B.; Halton, A.; Looker, B.E.; Monteith, M.J.; Moore, R.J.; Patel, R.; Procopiou, P.A. U.S. Patent 20060287389 (12/21/2006). (f) Biggadike, K.; Coe, D.M.; Edney, D.D.; Guntrip, S.B.; Halton, A.; Looker, B.E.; Monteith, M.J.; Moore, R.J.; Patel, R.; Procopiou, P.A. U.S. Patent 20060287286 (12/21/2006). (g) Biggadike, K.; Coe, D.M.; Guntrip, S.B.; Looker, B.E.; Procopiou, P.A. U.S. Patent 7,135,600 (11/14/2006). (h) Biggadike, K.; Coe, D.M.; Edney, D.D.; Halton, A.; Looker, B.E.; Monteith, M.J.; Patel, R.; Procopiou, P.A. U.S. Patent 20040180876 (9/16/2004). (i) Biggadike, K.; Coe, D.M.; Edney, D.D.; Guntrip, S.B.; Halton, A.; Looker, B.E.; Monteith, M.J.; Moore, R.J.; Patel, R.; Procopiou, P.A. WO 2002066422 (8/29/2002).
21. The Aldrich 2007–2008 catalog price for 500 g (97%) is $26.50.
22. Bream, R.N.; Ley, S.V.; Procopiou, P.A. *Org. Lett.* **2002**, *4*, 3793.
23. Bream, R.N.; Ley, S.V.; McDermott, B.; Procopiou, P.A. *J. Chem. Soc., Perkin Trans. 1* **2002**, 2237.
24. Propopiou, P.A. WO 2001096278 (12/20/2001).
25. The Aldrich 2007–2008 catalog price for 1 L is $40.30.
26. The Aldrich 2007–2008 catalog price for 1 kg is $59.00.
27. The Aldrich 2007–2008 catalog price for 100 g is $324.00.
28. (a) Blake, K.; Coe, D.M.; Procopiou, P.A. U.S. Patent 20070249630 (10/25/2007). (b) Blake, K.; Coe, D.M.; Procopiou, P.A. WO 2003072539 (9/4/2003).
29. Guntrip, S.B.; Procopiou, P.A. WO 2004022547 (3/18/2004).
30. (a) Biggadike, K.; Jones, P.; Payne, J.J. WO 2003035668 (5/1/2003). (b) Biggadike, K.; Jones, P.; Payne, J.J. EP 1444248 (10/17/2002).
31. Biggadike, K.; Jones, P.; Payne, J.J. WO 2002088167 (11/7/2002).
32. Biggadike, K.; Jones, P. WO 2002100879 (12/19/2002).
33. (a) Coe, D.M.; Monteith, M.J.; Procopiou, P.A. U.S. Patent 7,144,908 (12/5/2006). (b) Coe, D.M.; Monteith, M.J.; Procopiou, P.A. U.S. Patent 20040242890 (12/2/2004). (c) Coe, D.M.; Monteith, M.J.; Procopiou, P.A. WO 2002070490 (9/12/2002). (d) Coe, D.M.; Monteith, M.J.; Procopiou, P.A. EP 1366025 (3/4/2002).

34. The Aldrich 2007–2008 catalog price for 1 kg is $1210.00.
35. (a) Chapman, A.M.; Guntrip, S.B.; Looker, B.E.; Procopiou, P.A. U.S. Patent 7,442,839 (10/28/2008). (b) Chapman, A.M.; Guntrip, S.B.; Looker, B.E.; Procopiou, P.A. WO 2004037773 (5/6/2004).
36. The Aldrich 2007–2008 catalog price for 500 g is $47.70.
37. Cooper, A.B.; Saksena, A.K.; Lovey, R.; Girijavallabhan, V.; Ganguly, A. U.S. Patent 5,236,900 (8/17/1993).
38. Pieper, H.; Engelhardt, G. EP 500853 (9/3/1991).
39. The Aldrich 2007–2008 catalog price for 25 g is $246.50.
40. Evans, B. EP 422889 (4/17/1991).
41. The Aldrich 2007–2008 catalog price for 25 g is $57.10.
42. Goodwin, T.E.; Zhou, X.; Haile, P.A.; Breen, P.; Anderson, P. J.; Hiller, F.C.; Compadre, C.M. *J. Label. Cmpd. Radiopharm.* **2000**, *43*, 65.
43. Collin, D.T.; Hartley, D.; Jack, D.; Lunts, L.H.C.; Press, J.C.; Ritchie, A.C.; Toon, P. *J. Med. Chem.* **1970**, *13*, 674.
44. Hett, R.; Stare, R.; Helquist, P. *Tetrahedron Lett.* **1994**, *35*, 9375.
45. He, W.; Zhang, B.; Li, X.; Liu, P.; Sun, X.; Zhang, S. *Zhong. Yaowu Huaxue Zazhi*, **2006**, *16*, 222.
46. He, W.; Qin, H.; Luo, X. *Huaxi Yaoxue Zazhi* **1998**, *13*, 251.
47. Ariza Aranda, J.; Serra Masia, J.; Monserrat Vidal, C. ES 2065269 (2/1/1995).
48. The Aldrich 2007–2008 catalog price for 500 g is $95.70.
49. The Aldrich 2007–2008 catalog price for 1 kg is $93.20.
50. The Aldrich 2007–2008 catalog price for 500 mL ($d = 1.89$) is $499.50.
51. Babad, E.; Carruthers, N.; Steinman, M. U.S. Patent 5,011,993 (4/30/1991).
52. Babad, E.; Carruthers, N.; Steinman, M. U.S. Patent 4,952,729 (8/28/1990).
53. Babad, E.; Carruthers, N.I.; Jaret, R.S.; Steinman, M. *Synthesis*, **1988**, 966.
54. Babad, E.; Carruthers, N.; Steinman, M. EP 259159 (9/3/1987).
55. The Aldrich 2007–2008 catalog price for 500 g is $33.10.
56. Liu, J.; Zhou, D.; Jia, X.; Huang, L.; Li, X.; Chan, A.S.C. *Tetrahedron: Asym.* **2008**, *19*, 1824.
57. Gribble, G.W. *Chem. Soc. Rev.* **1998**, *27*, 395.
58. The Aldrich 2007–2008 catalog price for 500 g is $127.50.
59. Cizmarikova, R.; Polakovicova, M.; Misikova, E. *Chem. Pap.* **2002**, *56*, 256.
60. GB 1200886 (8/5/1970).
61. (a) Kitazawa, M.; Okazaki, K.; Tamai, T.; Saito, M.; Tanaka, N.; Kobayashi, H.; Kikuchi, K.; Muranaka, H. U.S. Patent 6,136,852 (10/24/2000). (b) Kitazawa, M.; Okazaki, K.; Tamai, T.; Saito, M.; Tanaka, N.; Kobayashi, H.; Kikuchi, K.; Muranaka, H. U.S. Patent 6,133,266 (10/17/2000).
62. The Aldrich 2007–2008 catalog price for 50 g is $188.00.
63. (a) Goswami, J.; Goswami, A. *J. Indian Chem. Soc.* **2002**, *79*, 469.(b) Goswami, J.; Bezbaruah, R.L.; Goswami, A.; Borthakur, N. *Tetrahedron: Asym.* **2001**, *12*, 3343.(c) Goswami, A.; Bezbaruah, R.L.; Goswami, J.; Borthakur, N.; Dey, D.; Hazarika, A.K. *Tetrahedron: Asym.* **2000**, *11*, 3701.
64. The Aldrich 2007–2008 catalog price for 18 L ($d = 0.847$) is $364.00.
65. The Aldrich 2007–2008 catalog price for 250 g is $83.30.
66. Skidmore, I.F.; Lunts, L.H.C.; Finch, H.; Naylor, A.; Campbell, I.B. U.S. Patent 4,990,505 (2/5/1991).
67. Procopiou, P.A.; Morton, G.E.; Todd, M.; Webb, G. *Tetrahedron: Asym.* **2001**, *12*, 2005.
68. The Aldrich 2007–2008 catalog price for 5 g is $63.30.
69. The Aldrich 2007–2008 catalog price for 100 g is $105.00.
70. The Aldrich 2007–2008 catalog price for 100 g is $65.40.
71. The Aldrich 2007–2008 catalog price for 500 g is $164.00.
72. (a) Gisch, N.; Balzarini, J.; Meier, C. *J. Med. Chem.* **2007**, *50*, 1658.(b) Vukadinovic-Tenter, D.; Balzarini, J.; Meier, C. *Nucleos., Nucleot., Nucleic Acids* **2007**, *26*, 1325.(c) Gisch, N.; Balzarini, J.; Meier, C. *Nucleos., Nucleot., Nucleic Acids* **2007**, *26*, 861.
73. Adams, R.; Sloan, A.W.; Taylor, B.S. *J. Am. Chem. Soc.* **1923**, *45*, 2417.
74. Bernardon, J.-M. WO 2000026167 (5/11/2000).
75. Cossement, E.; Geerts, J.-P.; Gobert, J.; Michel, P. EP 269599 (6/1/1988).
76. (a) Amann, A.; Koenig, H.; Thieme, P.C.; Giertz, H. DE 2310142 (9/12/1974). (b) Koenig, H.; Thieme, P.C.; Giertz, H. DE 2310141 (9/12/1974). (c) Amann, A.; Koenig, H.; Thieme, P.C.; Giertz, H. DE 2310140 (9/12/1974). (d) Amann, A.; Koenig, H.; Thieme, P.C.; Giertz, H. DE 2309639 (9/12/1974). (e) GB 1458251 (12/15/1976). (f) GB 1455174 (11/10/1976).
77. (a) Moran, E.J.; Jacobsen, J.R.; Aggen, J. U.S. Patent 7,452,995 (11/18/2008). (b) Mammen, M.; Hughes, A. U. S. Patent 7,358,244 (4/15/2008). (c) Mammen, M.; Mischki, T. WO 2006023457 (3/2/2006). (d) Mammen, M.; Mischki, T.; Hughes, A.; Ji, Y.-H. U.S. Patent 20060035933 (2/16/2006). (e) Moran, E.J.; Jacobsen, J.R.; Leadbetter, M.R.; Nodwell, M.B.; Trapp, S.G.; Aggen, J.; Church, T.J. U.S. Patent 20030229058 (12/11/2003). (f) Moran, E.J.; Jacobsen, J.R.; Leadbetter, M.R.; Nodwell, M.B.; Trapp, S.G.; Aggen, J.; Church, T.J. U.S. Patent 6,670,376 (12/30/2003). (g) Moran, E.J.; Jacobsen, J.R.; Aggen, J.B. U.S. Patent 6,653,323 (11/25/2003). (h) Moran, E.J.; Jacobsen, J.R.; Leadbetter, M.R.; Nodwell, M.B.; Trapp, S.G.; Aggen, J.; Church, T.J. WO 2003042164 (5/22/2003). (i) Moran, E.J.; Jacobsen, J.R.; Aggen, J. WO 2003042160 (5/22/2003).
78. The TCI 2006–2007 catalog price for 25 g is $153.35.
79. Haj-Yehia, A.-I. WO 2003088961 (10/30/2003).
80. (a) Akamatsu, H.; Yamasaki, N. U.S. Patent 6,528,686 (3/4/2003). (b) Furukawa, K.; Abe, T.; Akamatsu, H.; Matsuyama, A.; Ito, M.; Yamasaki, N.; Miki, K.; Ikura, K.; Ishiguro, T. EP 924194 (5/14/2003).
81. The Aldrich 2007–2008 catalog price for 250 g is $46.60.
82. The Aldrich 2007–2008 catalog price for 100 g is $61.80.
83. George, M.; Hill, S.J.; Kellam, B.; Middleton, R.J. WO 2004088312 (10/14/2004).

84. Puig Duran, C.; Prat Quinones, M.; Perez Crespo, D. WO 2007124898 (11/8/2007).

85. (a) Tann, C.-H.; Thiruvengadam, T.K.; Chiu, J.; Green, M.; McAllister, T.L.; Colon, C.; Lee, J. WO 1992004314 (3/19/1992).(b) Hong, Y.; Gao, Y.; Nie, X.; Zepp, C.M. *Tetrahedron Lett.* **1994**, *35*, 5551.(c) Brawner Floyd, M.; Du, M.T.; Fabio, P.F.; Jacob, L.A.; Johnson, B.D. *J. Org. Chem.* **1985**, *50*, 5022.

86. (a) Moran, E.J.; Choi, S.-K. U.S. Patent 20080269344 (10/30/2008). (b) Moran, E.J.; Choi, S.-K. U.S. Patent 7,427,639 (9/23/2008).

87. Collins, I.; Wicks, P.D. U.S. Patent 4,154,761 (5/15/1979).

88. Ainsworth, A.T.; Smith, D.G. EP 006735 (6/21/1979).

89. The Aldrich 2007–2008 catalog price for 3 kg is $76.70.

90. Creary, X. *J. Org. Chem.* **1987**, *52*, 5026.

91. The Aldrich 2007–2008 catalog price for 500 g is $55.10.

92. Lunts, L.H.C.; Toon, P. U.S. Patent 3,644,353 (2/22/1972).

93. Jiang, Z.-G.; Hua, Z.-M.; Mai, L.-G. *Huadong Shifan Daxue Xuebao, Ziran Kexueban* **2003**, 102.

94. Zhang, Y.; Li, Z.; Sun, L. *Zhongguo Yiyao Gongye Zazhi* **2001**, *32*, 182.

95. (a) Sasai, H.; Tokunaga, T.; Watanabe, S.; Suzuki, T.; Itoh, N.; Shibasaki, M. *J. Org. Chem.* **1995**, *60*, 7388.(b) Sasai, H.; Suzuki, T.; Itoh, N.; Arai, S.; Shibasaki, M. *Tetrahedron Lett.* **1993**, *34*, 2657.

96. The Aldrich 2007–2008 catalog price for 20 L (96% reagent grade) is $591.00.

97. (a) Shibasaki, M.; Sasai, I.; Urata, Y.; Fujita, M. U.S. Patent 7,205,425 (4/17/2007). (b) Shibasaki, M.; Sasai, I.; Urata, Y.; Fujita, M. EP 947498 (9/15/2004).

98. Buchanan, D.J.; Dixon, D.J.; Looker, B.E. *Synlett* **2005** 1948.

99. Accessed at www.oehha.ca.gov/prop65/law/pdf_zip/NitromethaneNSRL120707.pdf.

100. The Aldrich 2007–2008 catalog price for 5 g is $94.50.

101. The Aldrich 2007–2008 catalog price for 25 g is $386.00.

102. (a) Box, P.C.; Coe, D.M.; Looker, B.E.; Procopiou, P.A. U.S. Patent 7,361,787 (4/22/2008). (b) Box, P.C.; Coe, D.M.; Looker, B.E.; Procopiou, P.A. WO 2003024439 (3/27/2003).

103. (a) Box, P.C.; Coe, D.M.; Looker, B.E.; Procopiou, P.A. U.S. Patent 7,271,197 (9/18/2007). (b) Box, P.C.; Coe, D.M.; Looker, B.E.; Procopiou, P.A. WO 2003091204 (11/6/2003).

104. Box, P.C.; Coe, D.M.; Looker, B.E.; Mann, I.S.; Procopiou, P. A. U.S. Patent 7,439,393 (10/21/2008).

105. Coe, D.M.; Perciaccante, R.; Procopiou, P.A. *Org. Biomol. Chem.* **2003**, *1*, 1106.

106. Ariza Aranda, J.; Serra Masia, J.; Monserrat Vidal, C. ES 2065269 (2/1/1995).

107. (a) Mammen, M.; Hughes, A.; Chen, Y. U.S. Patent 7,524,959 (4/28/2009). (b) Mammen, M.; Hughes, A.; Chen, Y. U.S. Patent 7,521,561 (4/21/2009). (c) Mammen, M.; Hughes, A.; Stangeland, E.; Chen, Y. U.S. Patent 7,514,558 (4/7/2009). (d) Mammen, M.; Hughes, A.; Chen, Y. U.S. Patent 7,507,751 (3/24/2009). (e) Mammen, M.; Hughes, A.; Chen, Y. U.S. Patent 7,345,175 (3/18/2008). (f) Mammen, M.; Dunham, S.; Hughes, A.; Lee, T.W.; Husfeld, C.; Stangeland, E. U.S. Patent 20040167167 (8/26/2004).

108. The Aldrich 2007–2008 catalog price for 1 kg is $147.00.

109. (a) Effenberger, F.; Jager, J. *J. Org. Chem.* **1997**, *62*, 3867.(b) Brown, R.F.C.; Donohue, A.C.; Jackson, W.R.; McCarthy, T. D. *Tetrahedron* **1994**, *50*, 13739.

110. Beach, S.F.; Latham, D.W.S.; Roberts, T.G.; Sidgwick, C.B. U.S. Patent 5,380,922 (1/10/1995).

111. The Aldrich 2007–2008 catalog price for 100 g is $83.60.

112. The Aldrich 2007–2008 catalog price for 1 kg (97%) is $30.50.

113. Somerville, L.F.; Allen, C.F.H. *Org. Syn. Coll.* Vol. *2* **1943**, 81.

114. The Aldrich 2007–2008 catalog price for 25 g is $69.40.

115. Seko, S.; Furutani, A. U.S. Patent 6,037,509 (3/14/2000).

116. The Aldrich 2007–2008 catalog price for 800 mL (1 M in THF) is $48.10.

117. The Aldrich 2007–2008 catalog price for 20 kg is $508.00.

118. The Aldrich 2007–2008 catalog price for 100 g is $215.00.

119. Yang, S.-B.; Gan, F.-F.; Chen G.-J.; Xu, P.-F. *Synlett* **2008**, 2532.

120. The Aldrich 2007–2008 catalog price for 800 mL (2 M in butyl ether) is $204.00.

121. (a) Hans, J.; Wallace, E.M.; Zhao, Q.; Lyssikatos, J.P.; Aicher, T.D.; Laird, E.R.; Robinson, J.; Allen, S. U.S. Patent 7,449,486 (11/11/2008). (b) Hans, J.; Wallace, E.M.; Zhao, Q.; Lyssikatos, J.P.; Aicher, T.D.; Laird, E.R.; Robinson, J.; Allen, S. WO 2006044825 (4/27/2006).

# 9

# LIPITOR® (ATORVASTATIN CALCIUM)

## 9.1 LIPITOR®, HEART DISEASE, AND STROKE

Lipitor® (atorvastatin hemicalcium trihydrate) is perhaps the most well known of the class of drugs known as statins. Lipitor® reduces the risk of myocardial infarction (MI), stroke, revascularization procedures, and angina in adult patients with multiple risk factors but without clinically evident coronary heart disease (CHD). Lipitor® also reduces the risk of nonfatal MI, fatal and nonfatal stroke, revascularization procedures, hospitalization for congestive heart failure, and angina in adult patients with clinically evident CHD.

Lipitor® reduces these risks by reducing the level of so-called "bad cholesterol," LDL-C, in the blood. It does this by inhibiting the enzyme HMG-CoA reductase and thus blocking the synthesis of cholesterol in the liver. The liver cells compensate for the reduced levels of cholesterol in the blood by generating more LDL receptors. These receptors then bind to LDL or VLDL particles, drawing them out of the blood and into the liver where they are digested.

Lipitor has the distinction of being the top-selling drug in the United States for the past several years with annual sales of $5.5 billion to $6.6 billon during the period from 2003 through 2008. Worldwide sales were $12.6 billion in 2007 and $12.4 billion in 2008. Sales figures for the third quarter of 2009 are $2.853 billion (compare with $3.142 billion for 3Q 2008) with U.S. sales of $1.379 billion ($1.567 billion for 3Q 2008) and international sales of $1.474 billion ($1.575 billion for 3Q 2008).[1,2]

The statistics on heart disease and stroke puts these remarkable sales figures in perspective. Heart disease is the leading cause of death for both men and women in the United States. In 2006, 631,636 people died of heart disease in the United States. Every year 785,000 Americans have a first heart attack and 470,000 who already had a heart attack have another. One in four men and one in three women in the United States die within a year of a recognized first heart attack. The direct and indirect costs of heart disease are estimated to be $316.4 billion in the United States for 2010.

Coronary heart disease is now the leading cause of death worldwide. Each year 16.7 million people die of cardiovascular disease (29% of all deaths globally). It is estimated that of the 15 million people who suffer a stroke each year, 5 million die and 5 million are left permanently disabled. Heart disease and stroke will become the leading cause of both death and disability worldwide by 2020. A key statistic often associated with heart disease and stroke is disability-adjusted life years, or DALYs. DALYs, healthy years of life lost to poor health or a disability, from coronary heart disease are projected to increase from 47 million in 1990 to 82 million in 2020.[3–5]

That Lipitor® is perhaps the most effective drug in its class and that heart disease and stroke are and will continue to be major world health concerns, these facts are not disputed. What is newsworthy in the Lipitor® arena are the challenges to Pfizer's Lipitor® patents by Ranbaxy and other generic manufacturers, the exact date for expiration of the Lipitor® patent in the United States, what might be described as a price tug-of-war between Pfizer, health insurers, and pharmacy benefit management companies, and the continuing search for a "Holy Grail" combination pill that will both lower LDL cholesterol and raise HDL cholesterol.

*Pharmaceutical Process Chemistry for Synthesis: Rethinking the Routes to Scale-Up,* By Peter J. Harrington

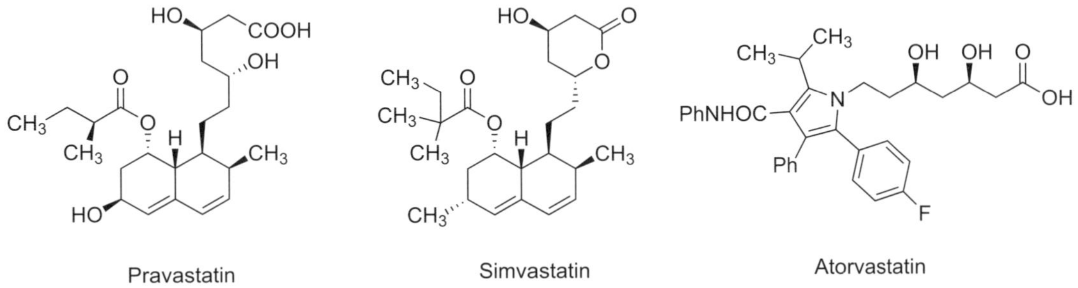


**FIGURE 9.1** Some of the statins available as generics in 2011–2012.

Ranbaxy challenged Pfizer's Lipitor® patents in 17 countries. A win-lose "litigation scorecard" would show Pfizer winning in the United States, United Kingdom, Netherlands, Denmark, Canada, Spain, and Ireland, Ranbaxy winning in Norway, and a "tie" in Australia. Looking beyond the win–loss record, many of Pfizer's patents (amorphous form, the Ca salt, single enantiomer) were invalidated.

On June 11, 2008, Ranbaxy brought in Daiichi Sankyo as a majority partner in a deal valued at $4.6 billion. One week later, Pfizer and Ranbaxy reached a settlement. Pfizer gets several months of additional exclusivity on Lipitor® sales in the United States. On *November 30, 2011*, Ranbaxy will have a license to sell generic versions of Lipitor® and Caduet® in the United States. (Caduet® combines Lipitor® and Norvasc® (amlodipine besylate) to treat high cholesterol and high blood pressure.) Ranbaxy will also have a license to use Pfizer patents necessary to make generic atorvastatin. Ranbaxy is also licensed to sell generic versions of Lipitor® on varying dates in Canada, Belgium, Netherlands, Germany, Sweden, Italy, and Australia.[6,7]

How does increasing generic competition affect Lipitor® sales right now and how will it affect Lipitor® sales in the future? Despite the fact that some less effective statins, such as pravastatin, have been off-patent and available as generic for many years in the United States, Lipitor® U.S. sales increased to $6.58 billion in 2006. In 2006, Teva launched a generic version of Merck's Zocor® (simvastatin). Health insurers such as Wellpoint and pharmacy benefit management companies such as Express Scripts offered lower copays and free initial prescriptions to promote the use of generic simvastatin as an inexpensive alternative to Lipitor®. Pfizer countered with a medical history analysis of a large U.K. primary care database, The Health Improvement Network (THIN). This observational study showed that switching patients from Lipitor® to generic simvastatin was associated with a 30% increase of major cardiovascular events. In another Pfizer study, patients who had a recent heart attack and took Lipitor (80 mg) had a 46% reduction in risk of another heart attack and a 34% reduction in risk of a major coronary event (heart attack, cardiac death, cardiac arrest) when compared with patients taking simvastatin (20–40 mg). There are also results that support the cost-driven switch to simvastatin. In 2005, a 9000 patient clinical trial showed no statistically significant difference between Lipitor® and Zocor®. In November 2009, UnitedHealth Group's pharmacy benefits unit released the results of an analysis of medical claims of 30,000 patients taking Vytorin®, Lipitor®, or simvastatin. They found no difference in the rates of heart attack or stroke among the three groups. So, should you stay on Lipitor® or switch to simvastatin to save money? We will not be asking this question in 2012 (Figure 9.1).

Looking beyond 2012, there will be atorvastatin combination drugs available to lower levels of LDL and raise levels of "good cholesterol" (HDL) in the next decade. Pfizer combined atorvastatin with the cholesterol ester transfer protein (CETP) inhibitor torcetrapib. In December 2006, a phase III trial with 15,000 patients comparing an atorvastatin–torcetrapib pill with Lipitor® was stopped when patients taking the combination pill had a 60% increase in mortality (82 patients on the combination pill died versus 51 on Lipitor®). Zetia® (ezetimibe) blocks the absorption of cholesterol in the intestine. Merck and Schering-Plough, who currently market Vytorin® (simvastatin–ezetimibe), announced their intention to pursue an atorvastatin–ezetimibe combination pill after the launch of generic atorvastatin. However, recent disappointing clinical results suggest ezetimibe is not the answer.[8] Trilipix® (fenofibrate) activates peroxisome proliferator-activated receptor (PPAR), type $\alpha$. Promising results were reported in phase III trials using an atorvastatin–fenofibrate combination pill (Figure 9.2).[9]

## 9.2 PYRROLE FORMATION BY [3 + 2]-CYCLOADDITION

At first glance, atorvastatin (**1**) offers two distinct synthetic challenges: construction of the pyrrole and generation of the two chiral centers with the correct stereochemistry in the side chain. Three of the pyrrole ring substituents (4-$FC_6H_4$, $C_6H_5$, *i*-Pr) are introduced during construction the ring. The *N*-phenylcarboxamide (CONHPh) and the side chain or side chain precursor can be introduced during the ring construction or added on later (Figure 9.3).

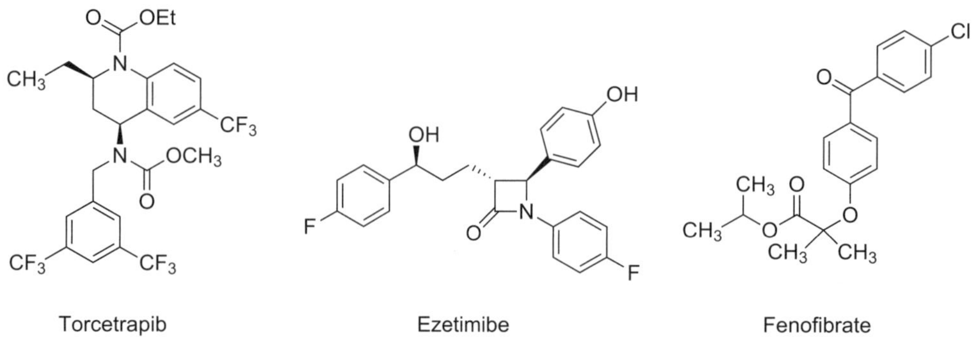


**FIGURE 9.2** Atorvastatin combination pill partners torcetrapib, ezetimibe, and fenofibrate.

Just three methods are used to construct the pyrrole ring: [3 + 2]-cycloaddition with a propiolic acid amide or ester, Paal–Knorr pyrrole synthesis from a 1,4-diketone and a primary amine, and a related pyrrole ring construction from a 1,4-diketone, an alkyl azide, and a reducing agent. These three methods all offer opportunities to maximize convergency (Scheme 9.1).

### 9.2.1 2-(*N*-(2-(1,3-Dioxolan-2-yl)ethyl)isobutyramido)-2-(4-fluorophenyl)acetic Acid (2)

A three-atom C–N–C component with a protected aldehyde suitable for elaboration to produce the side chain is prepared from three commercially available raw materials: 4-fluorophenylacetic acid, 2-(2-aminoethyl)-1,3-dioxolane, and isobutyryl chloride. 4-Fluorophenylacetic acid[10] is converted

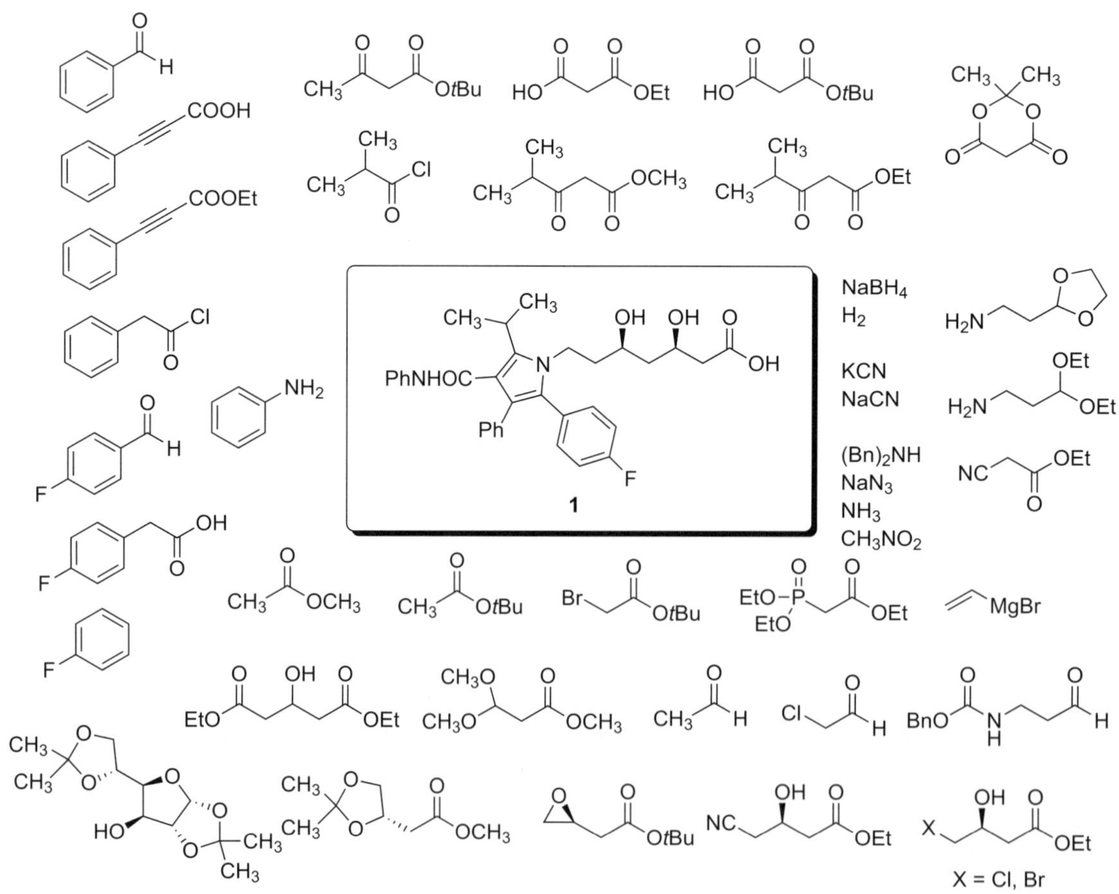


**FIGURE 9.3** Atorvastatin (**1**) raw materials.

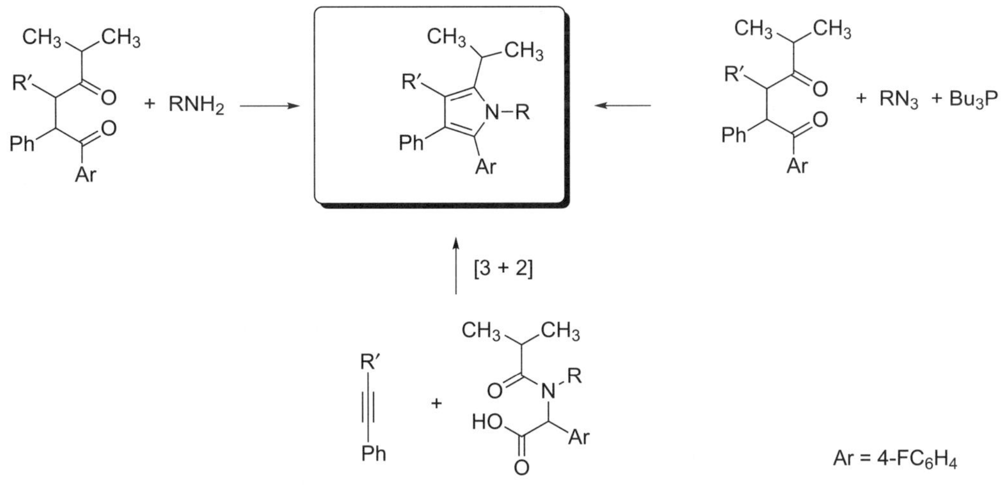

**SCHEME 9.1** Methods used to construct the pyrrole ring of atorvastatin (**1**).

to the ethyl ester **3** (94%). Hydrogen bromide (30%) in acetic acid catalyzes the α-bromination of ester **3** by *N*-bromosuccinimide (1.1 equivalents) in carbon tetrachloride at reflux (75%). The reaction of ethyl 2-bromo-2-(4-fluorophenyl)acetate (**4**) with 2-(2-aminoethyl)-1,3-dioxolane[11] (1.1 equivalents) and triethylamine (1.5 equivalents) in acetonitrile at 25°C overnight affords crude ethyl 2-(2-(1,3-dioxolan-2-yl)ethylamino)-2-(4-fluorophenyl)acetate (**5**) (83%). Isobutyryl chloride[12] (1.1 equivalents) is added to a mixture of the crude α-amino ester **5** and triethylamine (2.1 equivalents) in dichloromethane at 0°C. After aging the mixture at 0°C for 1 h, the crude amide **6** is isolated by routine water–ethyl ether workup (94%). Ester hydrolysis with sodium hydroxide (3.1 equivalents) in methanol–water (5:1) at reflux is complete in 5 h. A routine water–ethyl acetate workup affords 2-(*N*-(2-(1,3-dioxolan-2-yl)ethyl) isobutyramido)-2-(4-fluorophenyl)acetic acid (**2**) (93%) (Scheme 9.2).[13–16]

### 9.2.2 2-(*N*-(2-((4*R*,6*R*)-6-(2-*tert*-Butoxy-2-oxoethyl)-2,2-dimethyl-1,3-dioxan-4-yl)ethyl)isobutyramido)-2-(4-fluorophenyl)acetic Acid (7)

Another three atom C–N–C component with an elaborated and protected statin side chain is analogously prepared from 4-fluorophenylacetic acid, an elaborated and protected amine, and isobutyryl chloride. The reaction of ethyl 2-bromo-2-(4-fluorophenyl)acetate (**4**) with *tert*-butyl 2-((4*R*,6*R*)-6-(2-aminoethyl)-2,2-dimethyl-1,3-dioxan-4-yl)acetate (**8**) (1.1 equivalents) and triethylamine (1.5 equivalents) in acetonitrile at 25°C overnight affords the crude α-amino ester (**9**) (95%). Isobutyryl chloride (1.1 equivalents) in dichloromethane is added to a solution of the α-amino ester **9** in dichloromethane at 0–5°C. Triethylamine (2.0 equivalents) in dichloromethane is then added and the mixture is aged at 0–5°C for 1 h and at 25°C for 4 h. The amide **10** is isolated by a routine workup and chromatography (67%). Selective hydrolysis of the ethyl ester of **10** with lithium

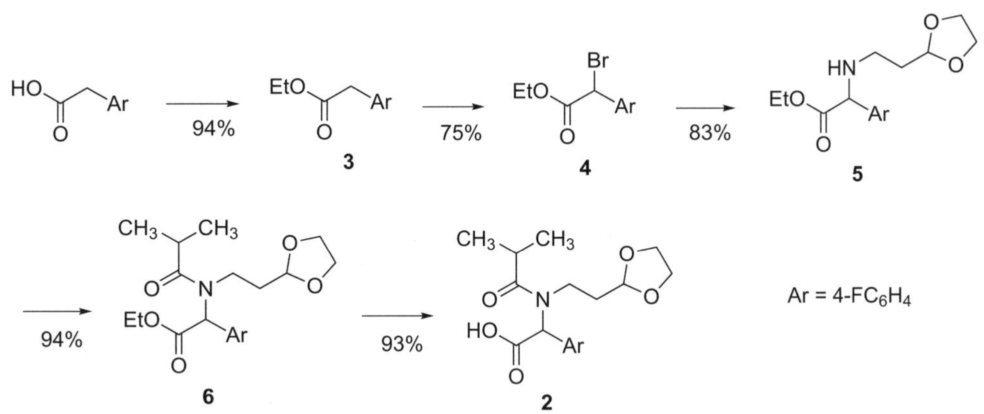

**SCHEME 9.2** 2-(*N*-(2-(1,3-Dioxolan-2-yl)ethyl)isobutyramido)-2-(4-fluorophenyl)acetic acid (**2**) from 4-fluorophenylacetic acid and isobutyryl chloride.

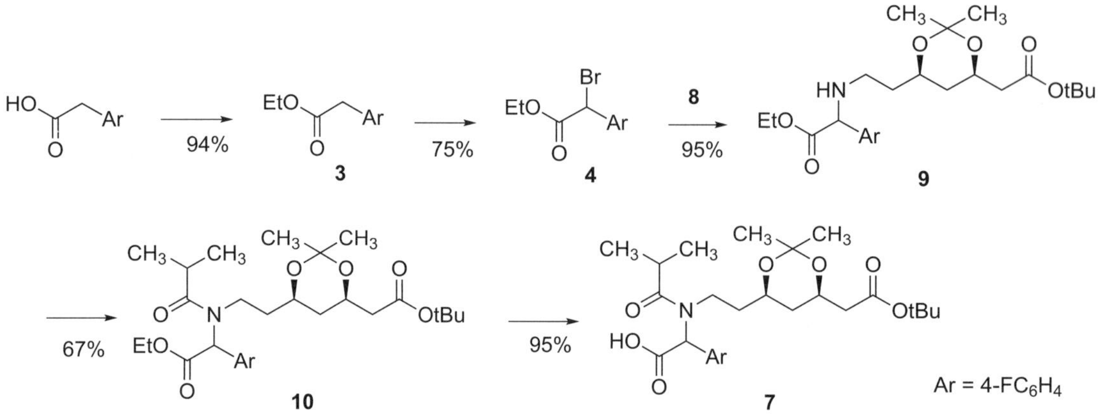


**SCHEME 9.3** 2-(*N*-(2-((4*R*,6*R*)-6-(2-*tert*-Butoxy-2-oxoethyl)-2,2-dimethyl-1,3-dioxan-4-yl)ethyl) isobutyramido)-2-(4-fluorophenyl)acetic acid (**7**) from 4-fluorophenylacetic acid and isobutyryl chloride.

hydroxide in methanol–water (3:1) requires 3 h at 25°C. A routine water–ethyl acetate workup affords 2-(*N*-(2-((4*R*,6*R*)-6-(2-*tert*-butoxy-2-oxoethyl)-2,2-dimethyl-1,3-dioxan-4-yl)ethyl)isobutyramido)-2-(4-fluorophenyl)acetic acid (**7**) (95%) (Scheme 9.3).[14]

### 9.2.3 *N*,3-Diphenylpropiolamide (11)

The alkyne for [3 + 2]-cycloaddition must be activated by an electron-withdrawing group (COOEt or CONHPh). Ethyl phenylpropiolate is commercially available. The amide **11** is prepared from phenylpropiolic acid and aniline. A solution of dicyclohexylcarbodiimide (DCC) (1.0 equivalent) in dichloromethane is added at 0°C over 2 h to a suspension of phenylpropiolic acid[17] (*Note*: The procedure reads propiolic acid.), aniline[18] (1.1 equivalents), and catalytic 4-dimethylaminopyridine (2.9 mol%) in dichloromethane. The suspension is aged at 0°C for 30 min, diluted with ethyl ether, and filtered through silica gel. The liquors are concentrated at reduced pressure and the residue is crystallized from an unspecified solvent to afford *N*,3-diphenylpropiolamide (**11**) (81%).[16]

### 9.2.4 [3 + 2]-Cycloaddition

A mixture of 2-(*N*-(2-(1,3-dioxolan-2-yl)ethyl)isobutyramido)-2-(4-fluorophenyl)acetic acid (**2**), *N*,3-diphenylpropiolamide (**11**) (1.6 equivalents), and acetic anhydride (7.6 equivalents) (*Note*: The acetic anhydride is omitted in one of two otherwise identical procedures.) is aged at 90°C for 4 h (vigorous gas evolution). The mixture is cooled to 25°C and concentrated at reduced pressure. Pyrrole **12** is isolated from the residue by chromatography (twice) followed by crystallization from diisopropyl ether (43%).[15,16] The [3 + 2]-cycloadditions using ethyl or benzyl phenylpropiolate[19] gave similar results (45–51%) (Scheme 9.4).

## 9.3 1,4-DIKETONES FOR PAAL–KNORR PYRROLE SYNTHESIS

In principle, the *N*-phenylcarboxamide substituent could be introduced after the pyrrole synthesis. In reality, introduction of the *N*-phenylcarboxamide later adds one or more steps to the linear sequence and all the methods for 1,4-diketone construction are designed with an ester or amide already in

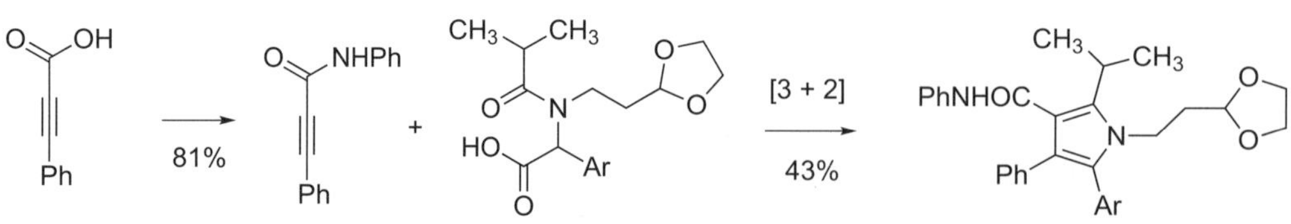


11 2 12

Ar = 4-FC6H4

**SCHEME 9.4** [3 + 2]-Cycloaddition of 2-(*N*-(2-(1,3-dioxolan-2-yl)ethyl)isobutyramido)-2-(4-fluorophenyl)acetic acid (**2**) with *N*,3-diphenylpropiolamide (**11**).

place. These methods are Stetter reaction catalyzed by a thiazolium halide or a metal cyanide, Michael addition of an acyl anion equivalent, and nucleophilic displacement of bromide from an α-bromoketone.

### 9.3.1 1-(4-Fluorophenyl)-5-methyl-2-phenylhexane-1,4-dione (13)

A mixture of methyl 4-methyl-3-oxovalerate[20], benzaldehyde[21] (1.1 equivalents), piperidine (4.1 mol%), and acetic acid (21 mol%) in toluene is refluxed for 3 h with removal of water by distillation of the toluene–water azeotrope using a Dean–Stark trap. The mixture is cooled, diluted with ethyl ether, washed with dilute hydrochloric acid, with saturated aqueous sodium bicarbonate, and with brine, and then dried and concentrated at reduced pressure. The residue is distilled (bp 127–130°C at 1 mmHg) to afford methyl 2-benzylidene-4-methyl-3-oxopentanoate (**14**) as a mixture of *E*- and *Z*-isomers (80%).[15]

The Stetter reaction with 4-fluorobenzaldehyde is catalyzed by a thiazolium salt. 3-Benzyl-5-(2-hydroxyethyl)-4-methylthiazolium chloride (15 mol%) (*Note*: The procedure reads 2-(2-hydroxyethyl)-3-methyl-4-benzylthiazolium chloride.) is added to a mixture of methyl 2-benzylidene-4-methyl-3-oxopentanoate (**14**), 4-fluorobenzaldehyde[22] (1.0 equivalent), and triethylamine (0.70 equivalents). The mixture is aged at 70°C for 24 h. After a routine water–ethyl ether workup, the residual oil **15** is dissolved in THF. Aqueous 15% sodium hydroxide (2.0 equivalents) is added and the mixture is aged at 25°C overnight. Hydrochloric acid (6 M) is added (to pH 5) and the mixture is extracted with ethyl ether. The extracts are washed with 3 M sodium hydroxide, with water, and with brine and then dried, filtered through silica gel, and concentrated at reduced pressure. The residual oil is Kugelrohr distilled (bp 145°C at 0.3 mmHg) to afford 1-(4-fluorophenyl)-5-methyl-2-phenylhexane-1,4-dione (**13**) (66%) (Scheme 9.5).[15]

### 9.3.2 Benzyl 2-(2-(4-Fluorophenyl)-2-oxo-1-phenylethyl)-4-methyl-3-oxopentanoate (16)

The benzyl ester is also prepared by Stetter reaction (Scheme 9.6). A mixture of benzyl 4-methyl-3-oxovalerate, benzaldehyde (1.1 equivalents), piperidine (4.4 mol%), and acetic acid (12 mol%) in toluene is refluxed for 4–6 h with removal of water by distillation of the toluene–water azeotrope using a Dean–Stark trap. The mixture is cooled and concentrated at reduced pressure. A routine water–dichloromethane workup followed by chromatography affords benzyl 2-benzylidene-4-methyl-3-oxopentanoate (**17**) as a mixture of *E*- and *Z*-isomers. No yield is available but the similar process for the methyl ester **14** suggests the yield is at least 80%.

3-Ethyl-5-(2-hydroxyethyl)-4-methylthiazolium bromide (20 mol%) is added to a mixture of benzyl 2-benzylidene-4-methyl-3-oxopentanoate (**17**), 4-fluorobenzaldehyde (1.1 equivalents), triethylamine (1.0 equivalent), and some ethanol. The mixture is aged at 70°C for 12–15 h. Benzyl 2-(2-(4-fluorophenyl)-2-oxo-1-phenylethyl)-4-methyl-3-oxopentanoate (**16**) is isolated after routine water–ethyl acetate workup and chromatography. No yield is available.[23]

### 9.3.3 Ethyl 2-(2-(4-Fluorophenyl)-2-oxo-1-phenylethyl)-4-methyl-3-oxopentanoate (18)

The ethyl ester **18** is prepared by another method: nucleophilic displacement of bromide from an α-bromoketone (Scheme 9.7). The α-bromoketone, 2-bromo-1-(4-fluorophenyl)-2-phenylethanone (**20**), is prepared in two steps: Friedel-Crafts acylation and ketone α-bromination. Phenylacetyl chloride[24] is added at 20°C to a mixture of aluminum chloride (1.2 equivalents) in fluorobenzene[25] (5.1 equivalents). The mixture is aged at 50°C for 5 h and at 25°C for 9 h. The deep green solution is poured into ice water. After

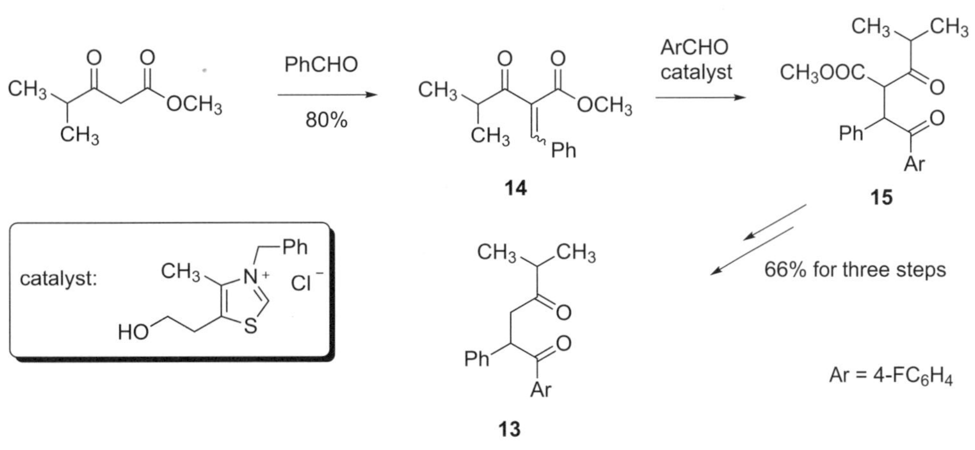


**SCHEME 9.5** 1-(4-Fluorophenyl)-5-methyl-2-phenylhexane-1,4-dione (**13**) from methyl 4-methyl-3-oxovalerate, benzaldehyde, and 4-fluorobenzaldehyde.

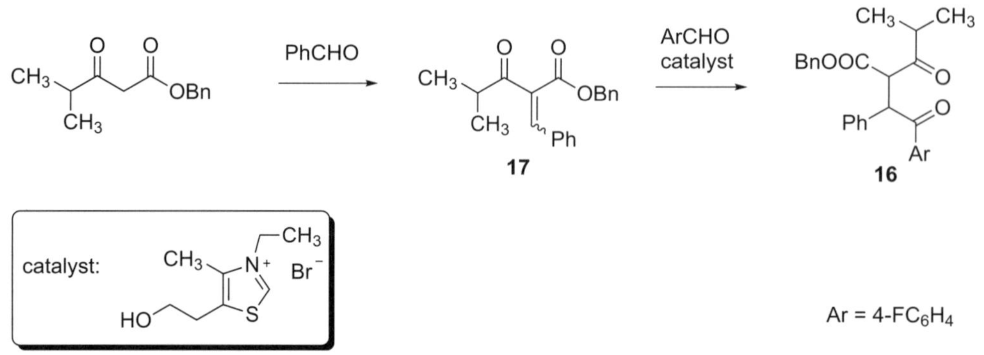


**SCHEME 9.6** Benzyl 2-(2-(4-fluorophenyl)-2-oxo-1-phenylethyl)-4-methyl-3-oxopentanoate (**16**) from benzyl 4-methyl-3-oxovalerate, benzaldehyde, and 4-fluorobenzaldehyde.

a routine water–organic workup, the residue is suspended in hexanes and the suspension is filtered. The solid is presumably washed with hexanes and dried to afford 1-(4-fluorophenyl)-2-phenylethanone (**19**) (mp 82°C) (90%).

Hydrogen bromide in acetic acid (30%) (2.7 mol%) is added to a solution of 1-(4-fluorophenyl)-2-phenylethanone (**19**) in chloroform, presumably at 25°C. A solution of bromine (1.0 equivalent) in chloroform is then added at a rate matching the rate of reaction/decolorization, again presumably at 25°C. The mixture is washed with 10% aqueous sodium sulfite, with aqueous sodium bicarbonate, with water, and with brine and then dried and concentrated at reduced pressure to afford 2-bromo-1-(4-fluorophenyl)-2-phenylethanone (**20**) (mp 46°C) (100%).[26,27]

2-Bromo-1-(4-fluorophenyl)-2-phenylethanone (**20**) and ethyl isobutyrylacetate[28] (1.4 equivalents) are dissolved in DMF at 0°C and potassium carbonate (1.4 equivalents) is added. The mixture is allowed to warm to 25°C to achieve complete conversion. The suspension is filtered and water and ethyl acetate are added to the liquors. The layers are separated and the organic layer is washed with brine and concentrated at reduced pressure. Ethyl 2-(2-(4-fluorophenyl)-2-oxo-1-phenylethyl)-4-methyl-3-oxopentanoate (**18**) is isolated from the residue by chromatography (63%).[27]

### 9.3.4 2-(2-(4-Fluorophenyl)-2-oxo-1-phenylethyl)-4-methyl-3-oxo-*N*-phenylpentanamide (21)

***9.3.4.1 4-Methyl-3-oxo-*N*-phenylpentanamide (22)*** 4-Methyl-3-oxo-*N*-phenylpentanamide (**22**) can be prepared directly from the methyl ester. A mixture of methyl 4-methyl-3-oxovalerate and ethylene diamine (9.8 mol%) in toluene is heated to 80°C and aniline (0.44 equivalents) is added. The mixture is heated to reflux and the collection of distillate is started. Three more charges of aniline (each 0.22 equivalents for 1.1 total equivalents) are charged at 40 min intervals. Distillation is continued for 1–5 h after completing the aniline charges. After cooling and aging the mixture at 25°C for 16 h, additional solvent is removed by distillation at an unspecified temperature and 85 mmHg. Water is added, the suspension is heated to 40°C, and more water is added. The remaining toluene is removed by distillation as the toluene–water azeotrope at an unspecified temperature and 20 mmHg. Water is added and the suspension is aged for 10 days. The resulting suspension is filtered and the solid is washed with hexane and dried at an unspecified temperature and reduced pressure to afford 4-methyl-3-oxo-*N*-phenylpentanamide (**22**) as a hydrate (mp 46.5–58.8°C) (63%, 98.8% pure by HPLC) (Scheme 9.8).[29,30]

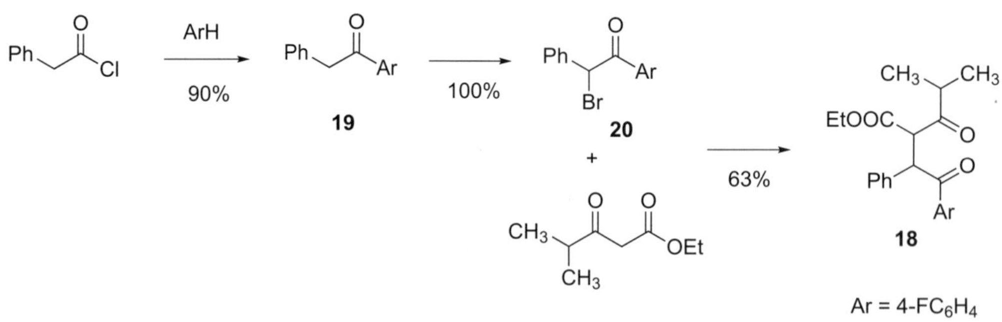


**SCHEME 9.7** Ethyl 2-(2-(4-fluorophenyl)-2-oxo-1-phenylethyl)-4-methyl-3-oxopentanoate (**18**) from phenylacetyl chloride, fluorobenzene, and ethyl isobutyrylacetate.

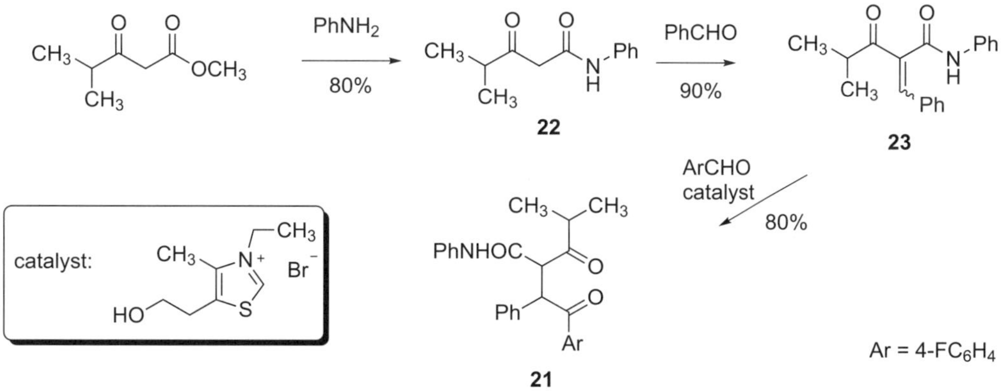


**SCHEME 9.8** 2-(2-(4-Fluorophenyl)-2-oxo-1-phenylethyl)-4-methyl-3-oxo-*N*-phenylpentanamide (**21**) from 4-methyl-3-oxo-*N*-phenylpentanamide (**22**) by a Stetter reaction.

In a preferred process, the charges of ethylene diamine and aniline are increased. Aniline (2.5 equivalents) is added to a mixture of methyl 4-methyl-3-oxovalerate and ethylene diamine (1.1 equivalents) in toluene and the mixture is refluxed for 18–20 h. The mixture is cooled to 25°C, washed with 5% hydrochloric acid and with water, and concentrated at reduced pressure to afford crude 4-methyl-3-oxo-*N*-phenylpentanamide (**22**) (80%).[31]

Pyridine is also a suitable solvent. Aniline (1.8 equivalents) is added over 2 h to a solution of methyl 4-methyl-3-oxovalerate in pyridine at 110–115°C. The mixture is aged at 110–115°C for 12 h. Pyridine and methanol are distilled at 85–90°C and reduced pressure. The residue is cooled to 35–40°C and water is added. Acid is added (to pH 1–1.5) and the suspension is cooled to 10–20°C and filtered. The solid is presumably washed with water and dried to afford 4-methyl-3-oxo-*N*-phenylpentanamide (**22**) as a hydrate (75%).[32]

***9.3.4.2 2-Benzylidene-4-methyl-3-oxo-*****N*****-phenylpentanamide (23)*** A mixture of 4-methyl-3-oxo-*N*-phenylpentanamide (**22**), presumably as the hydrate, benzaldehyde (1.0 equivalent), $\beta$-alanine (20 mol%), and acetic acid (48 mol%) in hexanes is refluxed for 20 h with removal of water by distillation of the hexane–water azeotrope using a Dean– Stark trap. More acetic acid (11 mol%) and hexanes are added and the reflux is continued for 1 h. The suspension is cooled to 25°C and filtered. The solid is suspended in hexanes at 25°C and the suspension is filtered. The solid is suspended in water at 25°C and the suspension is filtered. The water suspension wash at 25°C is repeated. The solid is dried at an unspecified temperature and pressure to afford 2-benzylidene-4-methyl-3-oxo-*N*-phenylpentanamide (**23**), presumably as a mixture of *E*- and *Z*-isomers (85%). A higher yield (90%) is accomplished in a shorter reflux time (8–12 h) using benzaldehyde (1.8 equivalents), $\beta$-alanine (50 mol%), and acetic acid (10 mol%).[29–31]

***9.3.4.3 Stetter Reaction of 2-Benzylidene-4-methyl-3-oxo-*****N*****-phenylpentanamide (23) with 4-Fluorobenzaldehyde*** In the Stetter reaction, 2-benzylidene-4-methyl-3-oxo-*N*-phenylpentanamide (**23**), 4-fluorobenzaldehyde (1.1 equivalents) and triethylamine (0.7–1.0 equivalents) are added to a concentrated solution of 3-ethyl-5-(2-hydroxyethyl)-4-methylthiazolium bromide (15–20 mol%) in ethanol. The mixture is aged at 75–80°C for 23 h. Isopropanol is added and the suspension is heated to 80°C. The hot solution is cooled, presumably to 25°C, and the suspension is filtered. The solid is washed with isopropanol and dried at an unspecified temperature and reduced pressure to afford 2-(2-(4-fluorophenyl)-2-oxo-1-phenylethyl)-4-methyl-3-oxo-*N*-phenylpentanamide (**21**) as a mixture of diastereomers (70%).[29,30]

The thiazolium halide catalyst plays a critical role: benzoin condensation of 4-fluorobenzaldehyde predominated using 3-benzyl-5-(2-hydroxyethyl)-4-methylthiazolium chloride. The Stetter reaction under rigorously anhydrous conditions using 3-ethyl-5-(2-hydroxyethyl)-4-methylthiazolium bromide or 5-(2-hydroxyethyl)-3,4-dimethylthiazolium iodide as catalyst is efficient (80%). Anhydrous conditions are best achieved on scale by rinsing the reactor with THF using of a spray ball nozzle to ensure that all surfaces of the vessel are rinsed, especially the top inner surface and the agitator.[33,34]

A higher yield (84%) is observed in just 10–14 h using 3-ethyl-5-(2-hydroxyethyl)-4-methylthiazolium bromide (97 mol%), 4-fluorobenzaldehyde (1.3 equivalents), and triethylamine (2.2 equivalents).[31]

The *N*-phenyl amide (**22**) can also be prepared from isobutyryl chloride, Meldrum's acid, and aniline. The Stetter reaction can be catalyzed by sodium cyanide. Experimental details and yields using these alternative reagents and catalysts are not available.[35–37]

***9.3.4.4 Michael Addition of an Acyl Anion Equivalent*** 4-Fluorobenzaldehyde is reacted with ethanethiol

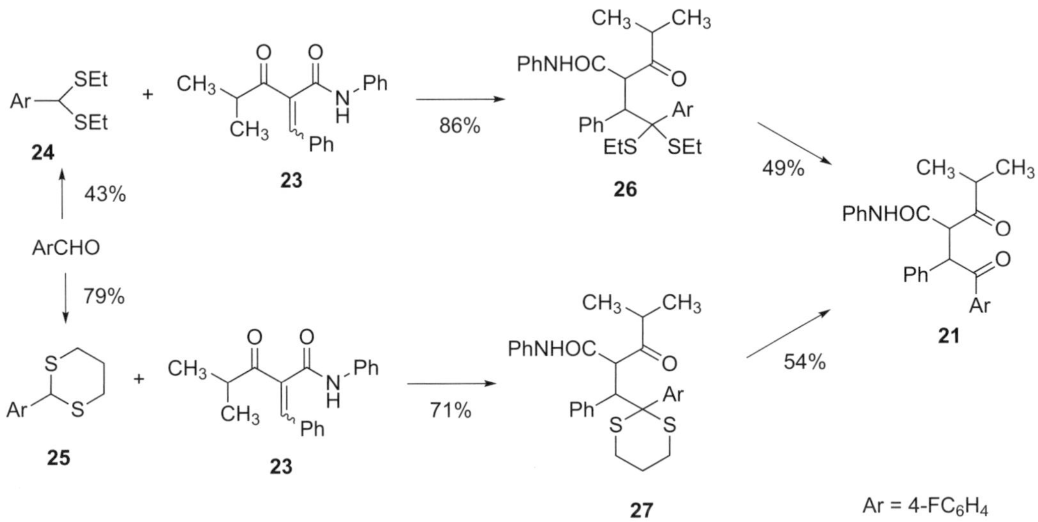


**SCHEME 9.9** 2-(2-(4-Fluorophenyl)-2-oxo-1-phenylethyl)-4-methyl-3-oxo-*N*-phenylpentanamide (**21**) from 2-benzylidene-4-methyl-3-oxo-*N*-phenylpentanamide (**23**) by Michael addition of an acyl anion equivalent.

(3.0 equivalents) and iodine (5 mol%) in THF at 25°C. A routine water–ethyl ether workup affords dithioacetal **24** (43%). Alternatively, 4-fluorobenzaldehyde is converted to 1,3-dithiane **25** using 1,3-propanedithiol (1.5 equivalents) and iodine (5 mol%) (79%) (Scheme 9.9).

An acyl anion equivalent is produced by deprotonation of dithioacetal **24** (1.0 equivalent) with *n*-butyllithium (1.0 equivalent) in THF at −20 to −25°C. A solution of 2-benzylidene-4-methyl-3-oxo-*N*-phenylpentanamide (**23**) in THF is added at −20 to −25°C and the mixture is aged at −20 to −25°C for 30 min and at 25°C for 1 h. A routine water–ethyl acetate workup affords the crude Michael adduct **26** (77%). (*Note*: Deprotonation of the amide should consume an equivalent of the carbanion.) The yield is higher (86%) when copper(II) chloride (0.5 mol%) is added.

The dithioketal **26** is cleaved by reaction with red mercury(II) oxide (2.0 equivalents) and boron trifluoride etherate (2.0 equivalents) in 15% aqueous THF, presumably at 25°C. Ethyl ether is added, the suspension is filtered, and the liquors are washed with 10% aqueous sodium carbonate and with brine and concentrated at reduced pressure to afford 2-(2-(4-fluorophenyl)-2-oxo-1-phenylethyl)-4-methyl-3-oxo-*N*-phenylpentanamide (**21**) (49%).

The Michael addition is less efficient (54%) when the acyl anion equivalent is produced from the 1,3-dithiane **25** (1.0 equivalent) and *n*-butyllithium (4.2 equivalents). A higher yield (71%) is observed when this Michael addition is catalyzed by magnesium chloride (5.0 mol%).

The 1,3-dithiane **27** is then cleaved by reaction with copper(II) chloride (2.0 equivalents) and copper(II) oxide (4.1 equivalents) in aqueous acetone at reflux for 1 h. The suspension is filtered and the liquors are concentrated at reduced pressure. The residue is suspended in ethyl ether and the suspension is filtered. The liquors are concentrated and the resulting suspension is filtered to afford 2-(2-(4-fluorophenyl)-2-oxo-1-phenylethyl)-4-methyl-3-oxo-*N*-phenylpentanamide (**21**) (54%).[38]

***9.3.4.5 Nucleophilic Displacement of Bromide from 2-Bromo-1-(4-fluorophenyl)-2-phenylethanone (20)*** 2-Bromo-1-(4-fluorophenyl)-2-phenylethanone (**20**) and 4-methyl-3-oxo-*N*-phenylpentanamide (**22**) (1.0 equivalent) are dissolved in DMF and potassium carbonate (1.0 equivalent) is added. After a completion check shows >90% conversion (temperature and time not specified), water and ethyl acetate are added and the layers are separated. The organic layer is washed with water, presumably several times, to remove DMF and then dried and concentrated at reduced pressure. The residue is suspended or dissolved in a small amount of dichloromethane and *n*-hexane is added. The suspension is filtered and the solid is presumably washed with *n*-hexane and dried. Additional product is isolated from the combined liquors by chromatography. The combined isolated and chromatographed yields of 2-(2-(4-fluorophenyl)-2-oxo-1-phenylethyl)-4-methyl-3-oxo-*N*-phenylpentanamide (**21**) is 80% (Scheme 9.10).[26,27]

Potassium carbonate (1.6 equivalents) is added to a solution of 4-methyl-3-oxo-*N*-phenylpentanamide (**22**) (1.1 equivalents) in isopropanol at 10–15°C. A solution of 2-bromo-1-(4-fluorophenyl)-2-phenylethanone (**20**) in isopropanol is then added at 10–15°C over 2–3 h. The mixture is aged at 40–45°C for 8–10 h. The mixture is concentrated at <55°C and reduced pressure. The residue is separated between ethyl acetate and water at 40–45°C. The organic

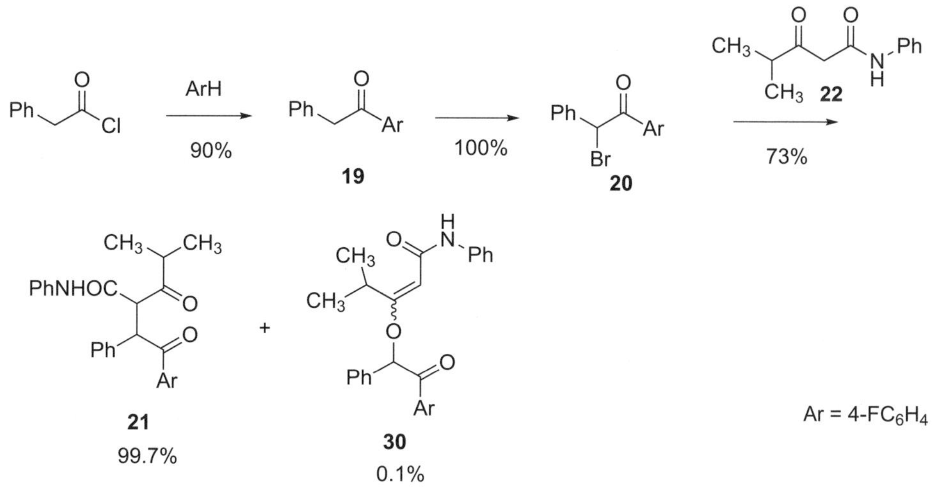


**SCHEME 9.10** 2-(2-(4-Fluorophenyl)-2-oxo-1-phenylethyl)-4-methyl-3-oxo-*N*-phenylpentanamide (**21**) from 4-methyl-3-oxo-*N*-phenylpentanamide (**22**) and 2-bromo-1-(4-fluorophenyl)-2-phenylethanone (**20**).

layer is concentrated at reduced pressure and the residue is suspended in isopropanol–methanol (ratio not specified). The suspension is filtered and the solid is presumably washed with isopropanol and dried to afford 2-(2-(4-fluorophenyl)-2-oxo-1-phenylethyl)-4-methyl-3-oxo-*N*-phenylpentanamide (**21**) (73%, 99.69% pure by HPLC). Detected impurities include the desfluoro **28** (0.047%) and difluoro **29** (nil) and the vinyl ether **30** from an alternative *O*-alkylation (0.1%). Similar results (68%, 99.5% pure by HPLC), are observed using *tert*-butanol in place of isopropanol. Incomplete conversion and an increase in the level of the vinyl ether **30** is observed with methanol or acetone as the solvent.[32]

A chloro-, iodo-, or sulfonyloxyketone can be used in place of the α-bromoketone. Sodium ethoxide can also be used as the base.[39–42]

An alternative process likely proceeds via the same bromide displacement (Scheme 9.11). The reaction of 4-methyl-3-oxo-*N*-phenylpentanamide with *N*-bromosuccinimide (1.0 equivalent) in acetone, presumably at 25°C, is complete in 3 h. The mixture is concentrated at reduced pressure and the residue is crystallized from ethyl acetate–petroleum ether to afford 2-bromo-4-methyl-3-oxo-*N*-phenylpentanamide (**31**) (92%).

Sodium carbonate (3.4 equivalents) is added to a solution of 1-(4-fluorophenyl)-2-phenylethanone (**19**) (1.0 equivalent) in DMF and the mixture is aged, presumably at 25°C, for 15 min. 2-Bromo-4-methyl-3-oxo-*N*-phenylpentanamide (**31**) is added and the mixture is aged, presumably at 25°C, for 18 h and at 90°C for 5 h. After a routine water–ethyl acetate workup, the residue is dissolved in hot isopropanol and the solution is cooled to 25°C. The suspension is filtered and the solid is presumably washed with isopropanol and dried to afford 2-(2-(4-fluorophenyl)-2-oxo-1-phenylethyl)-4-methyl-3-oxo-*N*-phenylpentanamide (**21**) (51%) containing just 0.01% of desfluoro impurity **28**.[43]

Sodium carbonate (1.0 equivalent) and diisopropylethylamine (2.0 equivalents) are added to a solution of 1-(4-fluorophenyl)-2-phenylethanone (**19**) in DMF and the mixture is aged, presumably at 25°C, for 30 min. 2-Bromo-4-methyl-3-oxo-*N*-phenylpentanamide (**31**) (1.0 equivalent) is

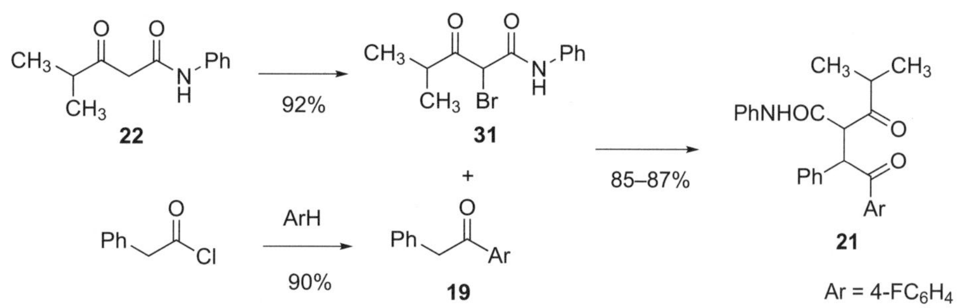


**SCHEME 9.11** 2-(2-(4-Fluorophenyl)-2-oxo-1-phenylethyl)-4-methyl-3-oxo-*N*-phenylpentanamide (**21**) from 2-bromo-4-methyl-3-oxo-*N*-phenylpentanamide (**31**) and 1-(4-fluorophenyl)-2-phenylethanone (**19**).

added and the mixture is aged at 25°C for 18 h. More 2-bromo-4-methyl-3-oxo-*N*-phenylpentanamide (**31**) (0.3 equivalents) is added and the mixture is aged at 25°C for 5 h and at 90–95°C for 6 h. After a routine water–ethyl acetate workup, the residue is dissolved in hot isopropanol and the solution is cooled to 25°C. The suspension is filtered and the solid is presumably washed with isopropanol and dried to afford 2-(2-(4-fluorophenyl)-2-oxo-1-phenylethyl)-4-methyl-3-oxo-*N*-phenylpentanamide (**21**) (62%) containing just 0.05% of desfluoro impurity **28**.[43]

Lithium diisopropylamide (LDA) is prepared by addition of *n*-butyllithium (1.2 equivalents) in hexanes to diisopropylamine (1.2 equivalents) in THF at −10 to −25°C. The LDA solution is aged at −10 to −25°C for 30 min. A solution of 1-(4-fluorophenyl)-2-phenylethanone (**19**) (1.0 equivalent) in THF is added at −60 to −78°C and the solution is aged at −60 to −78°C for 1 h. A solution of 2-bromo-4-methyl-3-oxo-*N*-phenylpentanamide (**31**) in THF is added at −60 to −78°C. The mixture is aged at −60 to −78°C for 30 min and then warmed to 10–15°C over 1 h. Quench with water and a routine water–ethyl acetate workup affords 2-(2-(4-fluorophenyl)-2-oxo-1-phenylethyl)-4-methyl-3-oxo-*N*-phenylpentanamide (**21**) (85–87%).[44]

## 9.4 PRIMARY AMINES FOR PAAL–KNORR PYRROLE SYNTHESIS

There are 14 primary amines designed to deliver all or part of the atorvastatin side chain. Two of these were already called upon for synthesis of the three-atom C–N–C fragments used for [3 + 2]-cycloaddition. The preparation of 2-(1,3-dioxolan-2-yl)ethanamine, which delivers an aldehyde for side chain elaboration after the pyrrole is constructed, is straightforward. In sharp contrast, there appears to be no limit to the possibilities for preparation of primary amines with fully elaborated and appropriately protected side chains. The elaborated and protected amines discussed in this section deliver the hydroxyl groups protected as a ketal, acetal, silyl ether, or boronate ester and the carboxylic acid protected as a *tert*-butyl ester, amide, aldehyde acetal, or alcohol. The routes to elaborated-protected amines can be evaluated based on the number of steps and overall yield of their linear synthetic sequences.

### 9.4.1 *tert*-Butyl 2-((4*R*,6*R*)-6-(2-Aminoethyl)-2,2-dimethyl-1,3-dioxan-4-yl)acetate (8)

This fully elaborated and diol-protected amine is most often prepared by reduction of the commercially available and expensive nitrile **32**. There are many approaches to nitrile **32** but the route most often taken begins with commercially available and expensive ethyl (*R*)-(−)-4-cyano-3-hydroxybutyrate. How many steps are involved in the preparation of this value-added starting material from commercially available and inexpensive raw materials? The answer to that question would divert the discussion from pharmaceutical manufacturing to specialty chemical manufacturing (Scheme 9.12).

Ethyl (*R*)-(−)-4-cyano-3-hydroxybutyrate is converted to *tert*-butyl (4*R*,6*R*)-6-(cyanomethyl)-2,2-dimethyl-1,3-dioxan-4-yl)acetate (**32**) in three steps: condensation with *tert*-butyl acetate or Meldrum's acid to extend the chain, stereoselective reduction of a *β*-hydroxy ketone, and protection of the resulting (*R*,*R*)-diol.

#### 9.4.1.1 (R)-*tert*-Butyl 6-Cyano-5-hydroxy-3-oxohexanoate (33)

*Claisen Condensation with* tert-*Butyl Acetate* tert-Butyl acetate[45] (5.0 equivalents) is added to a solution of lithium diisopropylamide (4.4 equivalents) in THF–hexanes at −50°C. A solution of ethyl (*R*)-(−)-4-cyano-3-hydroxybutyrate in THF is added, presumably at −50°C, and the mixture is allowed to warm to −20°C and age for 20 min. The mixture is quenched into 2.8 M hydrochloric acid at 0°C and the quenched mixture is extracted with ethyl acetate. The extracts are concentrated at reduced pressure to afford crude (*R*)-*tert*-butyl 6-cyano-5-hydroxy-3-oxohexanoate (**33**). This process is demonstrated on multikilogram scale (94%) (Scheme 9.13).[46–49]

The Claisen condensation must be run at low temperature to maximize the yield. The yield is low (26%) when the condensation is run at 0–5°C. Better results (46%) can be achieved in the condensation at 0–5°C when magnesium chloride (3.0 equivalents) is added. Still better results (57%) are achieved in the condensation at 0–5°C when *tert*-butylmagnesium chloride (1.0 equivalent) in THF–toluene is added to pregenerate the magnesium alkoxide. *tert*-Butylmagnesium chloride (1.0 equivalent) in THF–toluene is added to a mixture of ethyl (*R*)-(−)-4-cyano-3-

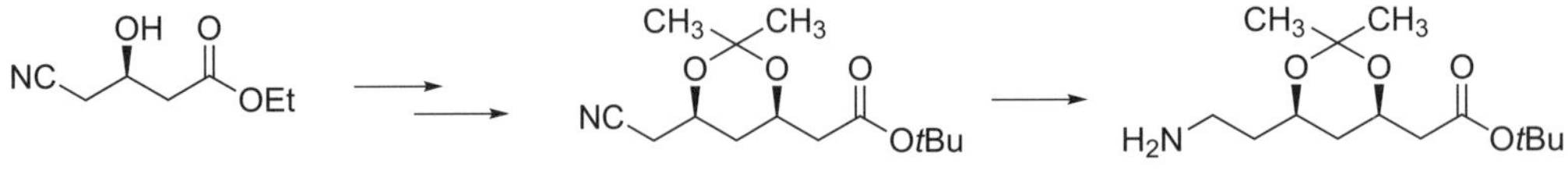


32 8

**SCHEME 9.12** Commercially available intermediates in one sequence leading to *tert*-butyl 2-((4*R*,6*R*)-6-(2-aminoethyl)-2,2-dimethyl-1,3-dioxan-4-yl)acetate (**8**).

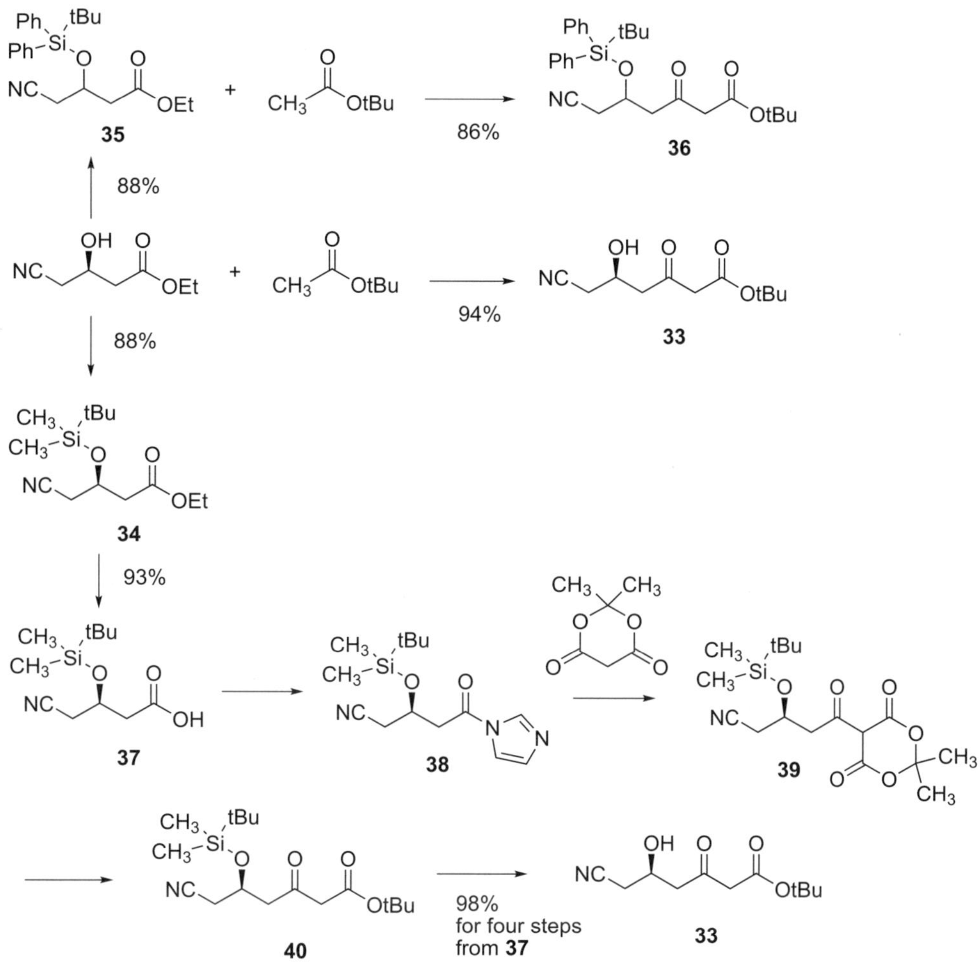

**SCHEME 9.13** (*R*)-*tert*-Butyl 6-cyano-5-hydroxy-3-oxohexanoate (**33**) from ethyl (*R*)-(−)-4-cyano-3-hydroxybutyrate.

hydroxybutyrate and *tert*-butyl acetate (2.0 equivalents) in THF at 0–5°C. The mixture is aged at 5°C for 50 min. Lithium diisopropylamide (3.5 equivalents) in THF–hexane is added, presumably at 0–5°C, and the mixture is aged at 5–20°C for 16 h. The mixture is quenched into 3.0 M hydrochloric acid and ethyl acetate. The organic layer is separated, washed with brine, dried, and concentrated at reduced pressure to afford crude (*R*)-*tert*-butyl 6-cyano-5-hydroxy-3-oxohexanoate (**33**) (57%). Perhaps changing the order of addition, charging ethyl (*R*)-(−)-4-cyano-3-hydroxybutyrate first, then adding in succession *tert*-butylmagnesium chloride, lithium diisopropylamide, and *tert*-butyl acetate, would result in less self-condensation of *tert*-butyl acetate and a higher yield of product.[50]

Lithium diisopropylamide is expensive and used in excess. Less base is required when the hydroxyl group of ethyl (*R*)-(−)-4-cyano-3-hydroxybutyrate is protected as a silyl ether. Does protection of the hydroxyl group also increase the efficiency of the Claisen condensation with *tert*-butyl acetate? While a yield for the Claisen condensation of the *tert*-butyldimethylsilyl ether **34** with *tert*-butyl acetate is not available, the overall yield of *tert*-butyl (4*R*,6*R*)-6-(cyanomethyl)-2,2-dimethyl-1,3-dioxan-4-yl)acetate (**32**) for the four-step sequence from the *tert*-butyldimethylsilyl ether **34** (60–65%) is lower than for the three-step route with no hydroxyl group protection and deprotection.[48]

A yield for the Claisen condensation with the *tert*-butyldiphenylsilyl (TBDPS) ether **35** can be derived from a sequence starting with the racemic alcohol. *tert*-Butyldiphenylsilylchloride is added to a mixture of ethyl 4-cyano-3-hydroxybutyrate (1.1 equivalents) and imidazole (1.7 equivalents) in dichloromethane at 25°C. After aging at 25°C for 4 h, water is added and the layers are separated. The aqueous layer is extracted with dichloromethane. The combined organic layers are concentrated at reduced pressure to afford ethyl 3-(*tert*-butyldiphenylsiloxy)-4-cyanobutanoate (**35**) (88%).

*tert*-Butyl acetate (4.2 equivalents) is added to lithium diisopropylamide (4.2 equivalents) in THF–hexane at −45°C. The mixture is aged at −20 to −30°C for 1 h. A solution of ethyl 3-(*tert*-butyldiphenylsiloxy)-4-cyanobutanoate (**35**) in

THF is added at −75°C and the mixture is aged at −75°C for 2 h. Methanol is added, followed by water. The organic layer is separated and the aqueous layer extracted with ethyl acetate. The combined organic layers are concentrated at reduced pressure to afford *tert*-butyl 5-(*tert*-butyldiphenylsilyloxy)-6-cyano-3-oxohexanoate (**36**) (86%).[51]

*Acylation of Meldrum's Acid* Lithium diisopropylamide can be replaced by a less expensive base using a Meldrum's acid-based method for construction of the $\beta$-ketoester. Ethyl (*R*)-(−)-4-cyano-3-hydroxybutyrate is protected as a silyl ether. The ester is hydrolyzed and the acid is converted to the acylimidazole. Acylation of Meldrum's acid (2,2-dimethyl-1,3-dioxane-4,6-dione) followed by cleavage with *tert*-butanol and desilylation with tetrabutylammonium fluoride affords (*R*)-*tert*-butyl 6-cyano-5-hydroxy-3-oxohexanoate (**33**). The alternative acylation of Meldrum's acid with the acid chloride was not successful.

*tert*-Butyldimethylsilylchloride (1.1 equivalents) is added to a mixture of ethyl 4-cyano-3-hydroxybutyrate, imidazole (1.2 equivalents), and 4-dimethylaminopyridine (1.0 mol%) in dichloromethane at 25°C. After aging at 25°C for 2 h, water is added and the layers are separated. The organic layer is concentrated at reduced pressure to afford (*R*)-ethyl 3-(*tert*-butyldimethylsilyloxy)-4-cyanobutanoate (**34**) (86%).

Lithium hydroxide monohydrate (1.7 equivalents) is added to (*R*)-ethyl 3-(*tert*-butyldimethylsilyloxy)-4-cyanobutanoate (**34**) in 1:2 THF–water at 25°C. After aging at 25°C for 1 h, the mixture is concentrated at reduced pressure. Aqueous citric acid (10%) is added to the residue (to pH 4) and the mixture is extracted with dichloromethane. The extracts are concentrated at reduced pressure to afford (*R*)-3-(*tert*-butyldimethylsilyloxy)-4-cyanobutanoic acid (**37**) (93%).

Carbonyldiimidazole (1.4 equivalents) is added, presumably at 25°C, to a solution of (*R*)-3-(*tert*-butyldimethylsilyloxy)-4-cyanobutanoic acid (**37**) in dichloromethane. The mixture is aged, presumably at 25°C, for 1 h. Meldrum's acid[52] (1.1 equivalents) and pyridine (1.4 equivalents) are added and the mixture is aged at 40°C for 5 h. The mixture is presumably cooled to 25°C and 1 M hydrochloric acid is added. The layers are separated and the organic layer is dried and concentrated at reduced pressure. *tert*-Butanol is added to the residue and the mixture is refluxed for 3 h. The mixture is presumably cooled and concentrated at reduced pressure to afford (*R*)-*tert*-butyl 5-(*tert*-butyldimethylsilyloxy)-6-cyano-3-oxohexanoate (**40**). Tetrabutylammonium fluoride (1 M in THF) is added to the crude silyl ether in THF at 0°C over 1 h. The mixture is aged at 25°C for 1 h and then concentrated at reduced pressure. The residue is separated between ethyl ether and water. The organic layer is washed with 10% aqueous sodium sulfite, dried, and concentrated at reduced pressure to afford (*R*)-*tert*-butyl 6-cyano-5-hydroxy-3-oxohexanoate (**33**) (98% for four steps from **37**, 99.6% *R*).[53]

***9.4.1.2 (3R,5R)-tert-Butyl 6-Cyano-3,5-dihydroxyhexanoate (41)*** The key step in the sequence from ethyl (*R*)-(−)-4-cyano-3-hydroxybutyrate to *tert*-butyl (4*R*,6*R*)-6-(cyanomethyl)-2,2-dimethyl-1,3-dioxan-4-yl)acetate (**32**) has to be the reduction to produce the second chiral center. A chemical reduction relies on metal chelation of the $\beta$-hydroxyketone and the enhanced selectivity possible at low temperature (<−50°C) to create a *cis*-1,3-diol. The chelating agents are diethylmethoxyborane, cerium(III) chloride, or titanium(IV) isopropoxide (1.0–1.3 equivalents). Diethylmethoxyborane is conveniently generated *in situ* from triethylborane and methanol and can be recovered by distillation and recycled with no loss of optical purity of the diol. The reducing agent in all three cases is sodium borohydride (1.0–2.4 equivalents) and this is charged in portions to the reaction mixture. Sodium borohydride can be charged as a solid or as a solution in triglyme or methanol–aqueous sodium hydroxide to avoid the exposure and transfer issues associated with charging the solid. Reduction yields of >95% with an (*R*,*R*) to (*R*,*S*) ratio of >40:1 are possible (Scheme 9.14).

*Sodium Borohydride with Diethylmethoxyborane* (*R*)-*tert*-Butyl 6-cyano-5-hydroxy-3-oxohexanoate (**33**) is dissolved in 4.7:1 THF–methanol. The solution is cooled to −85°C and a 50% solution of diethylmethoxyborane in THF is added. The mixture is cooled to −97°C and sodium borohydride[54] (2.4 equivalents) is added in portions over 3 h. The mixture is aged at −93 to −85°C for 5 h and then at 25°C for 10 h. Excess hydride is quenched by adding acetic acid (2.3 equivalents) and the mixture is concentrated at reduced pressure. The residual oil is dissolved in methanol and the solution is concentrated at reduced pressure to remove the boron-containing by-products. The methanol dissolution and distillation procedure is repeated. The residual oil is separated between ethyl acetate and water. The organic layer

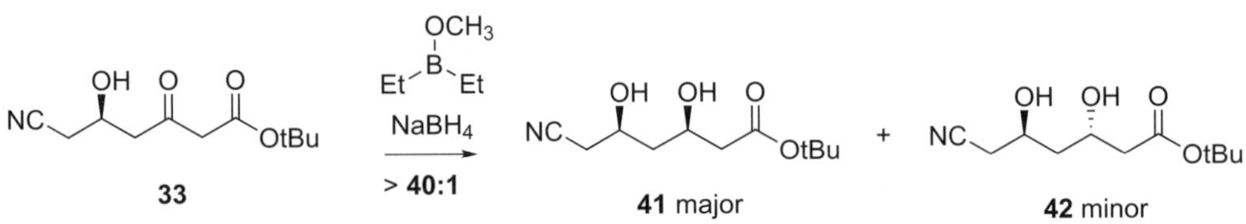


**SCHEME 9.14** Chelation of the $\beta$-hydroxyketone directs the hydride delivery.[55]

is concentrated at reduced pressure to afford crude (3*R*,5*R*)-*tert*-butyl 6-cyano-3,5-dihydroxyhexanoate (**41**). This process is demonstrated on a multikilogram scale.[46–48]

Sodium borohydride can be dissolved in triglyme and the solution added to the batch to avoid the solid charge. A solution of diethylmethoxyborane in THF is added to (*R*)-*tert*-butyl 6-cyano-5-hydroxy-3-oxohexanoate (**33**) in THF. The solution is aged at 25°C for 2 h, cooled to −75 to −70°C, and diluted with methanol. The sodium borohydride (1.1 equivalents) solution in triglyme is added at −75 to −65°C. The mixture is warmed to 15–25°C, quenched with acetic acid, and concentrated at reduced pressure. The distillate containing recovered diethylmethoxyborane is saved for recycle. The residue is dissolved in methanol and the solution is concentrated at reduced pressure. The residue is separated between water and ethyl acetate and the organic layer is concentrated at reduced pressure. The residue is dissolved in methanol and acetic acid and the solution is concentrated at reduced pressure. The residue is dissolved in ethyl acetate and the solution is concentrated at reduced pressure to afford crude (3*R*,5*R*)-*tert*-butyl 6-cyano-3,5-dihydroxyhexanoate (**41**) with an (*R*,*R*) to (*R*,*S*) ratio of 35:1. No yield is provided. Repeating the reduction process using the recovered borane in the first distillate affords crude (3*R*,5*R*)-*tert*-butyl 6-cyano-3,5-dihydroxyhexanoate with an (*R*,*R*) to (*R*,*S*) ratio of 25:1.[49]

> Using a minimal amount of acetic acid for the quench, the first distillate stream is a mixture of diethylmethoxyborane, THF, and methanol. The direct recycle of the distillate stream into the next batch is demonstrated. This suggests that *there is no need for an off-line cleanup by distillation*. This is important for two reasons. First, time in a pharmaceutical manufacturing plant is too valuable to be spent upgrading the purity of a reagent. Second, the pharmaceutical manufacturing facility may require additional certification as a *recycling facility* to do any off-line cleanup of a reagent in a process waste stream. The recovery of diethylmethoxyborane is certainly not quantitative. To make up for borane reagent lost, fresh borane reagent solution is added with the recovered borane stream to each batch.

Diethylmethoxyborane can be generated *in situ* from triethylborane and methanol. Triethylborane (1.2 equivalents) is added to 4:1 THF–methanol at 25°C and the mixture is aged at 25°C for 1 h. The solution is cooled to −78°C, (*R*)-*tert*-butyl 6-cyano-5-hydroxy-3-oxohexanoate (**33**) in THF is added, and the mixture is aged at −78°C for 1 h. Sodium borohydride (0.25 equivalents) (*Note*: More sodium borohydride is probably required for complete reduction.) is added in portions and the mixture is aged, presumably at −78°C, for 4 h. The mixture is presumably warmed to 25°C and saturated aqueous ammonium chloride is added. The layers are presumably separated and the organic layer is concentrated at reduced pressure. The residue is dissolved in ethyl acetate and the solution is concentrated at reduced pressure. The residue is dissolved in methanol and the solution is concentrated at reduced pressure to afford (3*R*,5*R*)-*tert*-butyl 6-cyano-3,5-dihydroxyhexanoate (**41**) (95%, 99.0% (*R*,*R*)). (*Note*: The (*R*,*R*)-selectivity here is high when compared with many similar reductions.)[53,55]

Sodium borohydride can also be dissolved in methanol–aqueous sodium hydroxide and the solution added to the batch to avoid the solid charge. A solution of triethylborane (1.3 equivalents, assuming 95% triethylborane is used) in THF is added to (*R*)-*tert*-butyl 6-cyano-5-hydroxy-3-oxohexanoate (**33**) in THF. The solution is aged at 25°C for 2 h, cooled to −95 to −55°C, and diluted with methanol and acetic acid. The sodium borohydride (1.4 equivalents) solution in methanol–aqueous sodium hydroxide is added, presumably at −95 to −55°C. The mixture is warmed to 25°C, quenched with methanol and acetic acid, and concentrated at reduced pressure. The distillate, identified as recovered triethylborane but probably containing recovered diethylmethoxyborane, is saved for recycle. A similar multiple-distillation workup as described for the reduction using diethylmethoxyborane affords crude (3*R*,5*R*)-*tert*-butyl 6-cyano-3,5-dihydroxyhexanoate with an (*R*,*R*) to (*R*,*S*) ratio of 30:1. No yield is provided. Repeating the reduction process using the recovered borane in the first distillate affords crude (3*R*,5*R*)-*tert*-butyl 6-cyano-3,5-dihydroxyhexanoate (**41**) with an (*R*,*R*) to (*R*,*S*) ratio of 40:1.[49]

*Sodium Borohydride with Cerium(III) Chloride or Titanium (IV) Isopropoxide* The borane reagents are expensive. Are there other reagents that will coordinate oxygen and deliver the hydride with chelation control? The borohydride reduction is accomplished using anhydrous cerium(III) chloride and titanium(IV) isopropoxide but no optical purity data for the product is provided. Titanium(IV) isopropoxide (1.0 equivalents) is added to (*R*)-*tert*-butyl 6-cyano-5-hydroxy-3-oxohexanoate (**33**) in 3:1 THF–methanol at 25°C. The mixture is aged at 25°C for 30 min and then cooled to −50°C. Sodium borohydride (1.0 equivalent) is added in four portions at −50°C and the mixture is stirred at −50°C for 1 h. The mixture is warmed to 25°C and concentrated at reduced pressure. A routine water–ethyl acetate workup affords crude (3*R*,5*R*)-*tert*-butyl 6-cyano-3,5-dihydroxyhexanoate (**41**) (74%). A higher yield (89%) is possible using cerium(III) chloride at −70 to −90°C.[56]

*Enzymatic Reduction* (*R*)-*tert*-Butyl 6-cyano-5-hydroxy-3-oxohexanoate (**33**) can also be also be converted to (3*R*,5*R*)-*tert*-butyl 6-cyano-3,5-dihydroxyhexanoate (**41**)

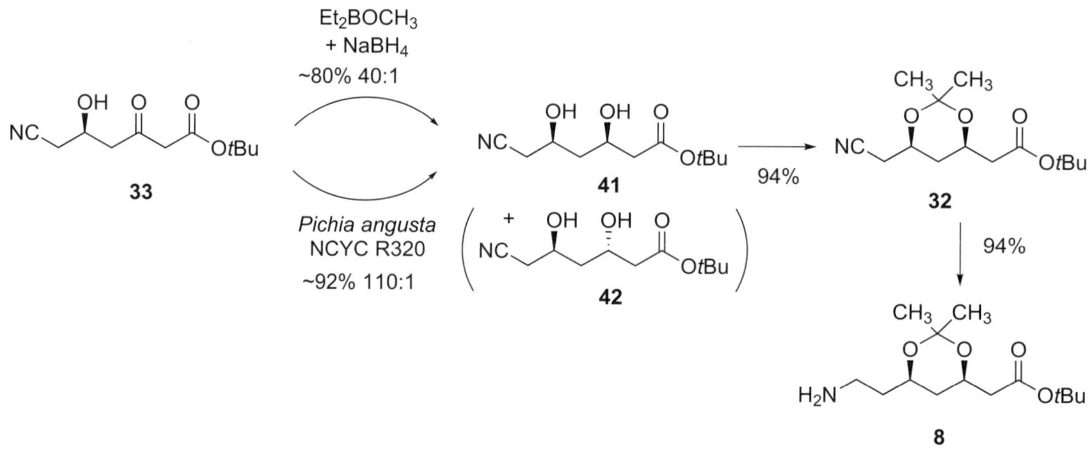


**SCHEME 9.15** *tert*-Butyl 2-((4*R*,6*R*)-6-(2-aminoethyl)-2,2-dimethyl-1,3-dioxan-4-yl)acetate (**8**) from (*R*)-*tert*-butyl 6-cyano-5-hydroxy-3-oxohexanoate (**33**).

using a reductase produced by microorganisms (*Beauveria, Candida, Kluyveromyces, Torulaspora, Pichia*). For example, a mixture of *Pichia angusta* NCYC 495 cell suspension, (*R*)-*tert*-butyl 6-cyano-5-hydroxy-3-oxohexanoate (**33**), and glucose (5 g/g of **33**) for cofactor generation is incubated on a rotary shaker at 150 rpm at 28°C for 18–24 h. Analysis by HPLC reveals 100% conversion to a 110:1 mixture of the (*R,R*) **41** and (*R,S*) **42** diols. On a preparative scale using *Pichia angusta* NCYC R320, *tert*-butyl (4*R*,6*R*)-6-(cyanomethyl)-2,2-dimethyl-1,3-dioxan-4-yl)acetate (**32**) is isolated after reduction, diol protection, and crystallizations from hexanes and from heptane in 81% yield (98.7% pure by HPLC) (Scheme 9.15).[57]

***9.4.1.3 tert-Butyl (4R,6R)-6-(Cyanomethyl)-2,2-dimethyl-1,3-dioxan-4-yl)acetate (32)*** Methanesulfonic acid (5 mol%) is added to a solution of (3*R*,5*R*)-*tert*-butyl 6-cyano-3,5-dihydroxyhexanoate (**41**), 2,2-dimethoxypropane (1.1 equivalents), and acetone and the solution is aged at 25°C for 2 h. After a routine water–ethyl acetate–hexanes workup, the residue is dissolved in warm hexane and the solution is cooled. The resulting suspension is filtered and the solid is presumably washed with hexane and dried. Crystallization from heptane (~9 L/kg) affords *tert*-butyl (4*R*,6*R*)-6-(cyanomethyl)-2,2-dimethyl-1,3-dioxan-4-yl) acetate (**32**) (mp 64.7–68°C) (53% for three steps from ethyl (*R*)-(−)-4-cyano-3-hydroxybutyrate).[46,47] Higher yields for the three-step sequence (65–70%) are achieved after process development.[48]

*p*-Toluenesulfonic acid monohydrate (5 mol%) is added to a mixture of (3*R*,5*R*)-*tert*-butyl 6-cyano-3,5-dihydroxyhexanoate (**41**), 2,2-dimethoxypropane (1.7 equivalents), and acetone and the solution is aged at 25°C for 5 h. The mixture is concentrated at reduced pressure. The residue is dissolved in ethyl acetate. The solution is washed with aqueous sodium bicarbonate and then dried and concentrated at reduced pressure to afford *tert*-butyl (4*R*,6*R*)-6-(cyanomethyl)-2,2-dimethyl-1,3-dioxan-4-yl)acetate (**32**) (94%, 99.0% (*R,R*)).[53]

Pyridinium *p*-toluenesulfonate (5 mol%) is added to a mixture of (3*R*,5*R*)-*tert*-butyl 6-cyano-3,5-dihydroxyhexanoate (**41**), 2,2-dimethoxypropane (4.0 equivalents), and acetone and the solution is aged at 25°C for 5 h. The mixture is concentrated at reduced pressure. The residue is suspended in water and then extracted with ethyl acetate. The extracts are washed with brine, dried, and concentrated at reduced pressure. *tert*-Butyl (4*R*,6*R*)-6-(cyanomethyl)-2,2-dimethyl-1,3-dioxan-4-yl)acetate (**32**) is isolated from the residue by chromatography (85%).[58]

***9.4.1.4 tert-Butyl 2-((4R,6R)-6-(2-Aminoethyl)-2,2-dimethyl-1,3-dioxan-4-yl)acetate (8)*** The nitrile of **32** is most often reduced by catalytic hydrogenation using Raney nickel. Hydrogenations using a broad range of Raney nickel charges (80–980 g/kg nitrile **32**) at 25–40°C and 50 psi hydrogen are complete in 2–20 h. The preferred solvents are methanol, isopropanol, or a toluene–methanol mixture. Ammonia is added as ammonium hydroxide or as a solution in methanol and ammonia charges also cover a broad range (2–40 equivalents). The amine **8**, an oil, is purified by chromatography or distillation or is carried on crude to the next step. The distillation requires a high temperature and high vacuum (bp 125–135°C at 0.5 mmHg). At 125–135°C, some dimerization by amide formation is likely. Along this same line, the workup procedures generally call for solvent distillation at a specified temperature (50–60°C) and reduced pressure, no doubt to keep dimer formation to a minimum. The potential for dimerization led to the selection of the *tert*-butyl ester over the methyl, ethyl, and isopropyl esters.

While recycle of the catalyst, especially at high loadings, would be a reasonable expectation, the recycle of the catalyst is not described. Crude amine **8** carried into the next step may contain some residual water or organic solvent (toluene or perhaps cyclohexane[59]).

Raney nickel #30 (103 g/kg nitrile **32**) is presumably washed with methanol. A solution of nitrile **32** in methanol (11.9 L/kg nitrile **32**) saturated with ammonia (estimate 29 equivalents) is added and the suspension is hydrogenated at 40°C and 50 psi for 10 h. The suspension is cooled and filtered and the catalyst is presumably washed with methanol. The liquors are concentrated at reduced pressure. *tert*-Butyl 2-((4*R*,6*R*)-6-(2-aminoethyl)-2,2-dimethyl-1,3-dioxan-4-yl)acetate (**8**) is isolated from the residue by chromatography (80%).[29]

Raney nickel #30 (90 g/kg nitrile **32**) is presumably washed with methanol. A solution of nitrile **32** in methanol (17.7 L/kg nitrile **32**) saturated with ammonia (estimate 43 equivalents) is added and the suspension is hydrogenated at 40°C and 50 psi for 16 h. The suspension is cooled and filtered and the catalyst is presumably washed with methanol. The liquors are concentrated at reduced pressure. *tert*-Butyl 2-((4*R*,6*R*)-6-(2-aminoethyl)-2,2-dimethyl-1,3-dioxan-4-yl)acetate (**8**) is isolated from the residue by chromatography (86%).[30]

Raney nickel (81 g/kg nitrile **32**) is presumably washed with methanol. A solution of nitrile **32** in methanol (16 L/kg nitrile **32**) saturated with ammonia (estimate 39 equivalents) is added and the suspension is hydrogenated at 40°C and 50 psi for 20 h. The suspension is cooled to 25°C and filtered and the catalyst is presumably washed with methanol. The liquors are concentrated at reduced pressure. *tert*-Butyl 2-((4*R*,6*R*)-6-(2-aminoethyl)-2,2-dimethyl-1,3-dioxan-4-yl) acetate (**8**) is isolated from the residue by chromatography (92%).[60]

Raney nickel doped with 1% molybdenum (980 g/kg nitrile **32**) is presumably washed with methanol. A solution of nitrile **32** in methanol (12 L/kg nitrile **32**) saturated with ammonia (29 equivalents) is added and the suspension is hydrogenated at 30°C and 50 psi for 6 h. The suspension is cooled to 20°C and filtered and the catalyst is presumably washed with methanol. The liquors are distilled and then further concentrated at reduced pressure. The residual oil is distilled (bp 125–135°C at 0.5 mmHg) to afford *tert*-butyl 2-((4*R*,6*R*)-6-(2-aminoethyl)-2,2-dimethyl-1,3-dioxan-4-yl) acetate (**8**) (94%, 96% pure by HPLC).[46]

Raney nickel (750 g/kg nitrile **32**) is washed with water. Methanol (3.8 L/kg nitrile **32**) is added followed by 15–18% aqueous ammonia (0.72 L/kg nitrile **32**, 1.5–1.9 equivalents). A solution of nitrile **32** in methanol (2.3 L/kg nitrile **32**) is added and the suspension is hydrogenated at 30–40°C and 43–50 psi for 6–8 h. The suspension is presumably cooled and filtered and the catalyst is presumably washed with methanol. The liquors are concentrated at <60°C and reduced pressure. Methanol is added and the solution is concentrated at <60°C and reduced pressure. The addition of methanol and concentration are repeated twice more. The final concentration affords *tert*-butyl 2-((4*R*,6*R*)-6-(2-aminoethyl)-2,2-dimethyl-1,3-dioxan-4-yl)acetate (**8**) (84–94%, >95% pure by HPLC) containing about 5 wt% water.[61]

Raney nickel (250 g/kg nitrile **32**) is presumably washed with methanol. Methanol (0.68 L/kg nitrile **32**) and 6.5 M ammonia in methanol (0.80 L/kg nitrile **32**, 1.4 equivalents) are added. Nitrile **32** in toluene (6 L/kg nitrile **32**) is added and the suspension is hydrogenated at 30–40°C and 50 psi for 2–6 h. The suspension is cooled to 25°C and filtered and the catalyst is washed with toluene. The liquors are concentrated at 55°C and reduced pressure to afford a toluene solution of the amine. The solution is washed with brine and then concentrated at reduced pressure to afford *tert*-butyl 2-((4*R*,6*R*)-6-(2-aminoethyl)-2,2-dimethyl-1,3-dioxan-4-yl)acetate (**8**) (104%) containing about 7 wt% residual toluene.[62]

Wet sponge nickel catalyst (265 g/kg nitrile **32**) is presumably washed with water. Isopropanol (7.5 L/kg nitrile **32**), 28% ammonium hydroxide (2.75 L/kg nitrile **32**, 10.9 equivalents) and nitrile **32** are added and the suspension is hydrogenated at an unspecified temperature and 50 psi. The suspension is filtered and the catalyst is presumably washed with isopropanol. The liquors are concentrated at reduced pressure. The residual oil is dissolved in warm toluene. The solution is washed with water and concentrated at reduced pressure to afford *tert*-butyl 2-((4*R*,6*R*)-6-(2-aminoethyl)-2,2-dimethyl-1,3-dioxan-4-yl)acetate (**8**). The amine is carried directly into the next step.[63]

The hydrogenation at 25°C and 65–70 psi is complete in 4–12 h. The catalyst charge is not specified.[47] The nitrile **32** can also be reduced using palladium on carbon and acetic acid.[64]

### 9.4.2 *tert*-Butyl 2-((7*R*,9*R*)-9-(2-Aminoethyl)-6,10-dioxaspiro[4.5]decan-7-yl)acetate (43)

Raney nickel (125 g/kg nitrile **44**) is presumably washed with methanol. Methanol (0.66 L/kg nitrile **44**) and 7 M ammonia in methanol (0.81 L/kg nitrile **44**, 1.7 equivalents) are added. Nitrile **44** in toluene (6 L/kg nitrile **44**) is added and the suspension is hydrogenated at 35°C and 44–58 psi for 5 h. The suspension is filtered and the catalyst is presumably washed with toluene. The liquors are concentrated at reduced pressure to afford a toluene solution of the amine. The solution is cooled to 25°C, washed with brine, and distilled to afford *tert*-butyl 2-((7*R*,9*R*)-9-(2-aminoethyl)-6,10-dioxaspiro[4.5]decan-7-yl)acetate (**43**) presumably containing some residual toluene. No yield is available.[65]

### 9.4.3 *tert*-Butyl 2-((2*R*,4*R*)-4-(2-Aminoethyl)-1,5-dioxaspiro[5.5]undecan-2-yl)acetate (45)

Raney nickel (128 g/kg nitrile **46**) is presumably washed with methanol. Methanol (0.68 L/kg nitrile) and 7 M ammonia in methanol (0.82 L/kg nitrile **46**, 1.8 equivalents) are added. Nitrile **46** in toluene (6.1 L/kg nitrile **46**) is added and the suspension is hydrogenated at 50–60°C and unspecified pressure for 23 h. The suspension is filtered and the catalyst is presumably washed with toluene. The liquors are concentrated at reduced pressure to afford a toluene solution of the amine. The solution is cooled to 25°C, washed with brine, and distilled to afford *tert*-butyl 2-((2*R*,4*R*)-4-(2-aminoethyl)-1,5-dioxaspiro[5.5]undecan-2-yl)acetate (**45**) presumably containing some residual toluene. No yield is available.[65]

### 9.4.4 (3*R*,5*R*)-*tert*-Butyl 7-Amino-3,5-bis(tert-butyldimethylsilyloxy)heptanoate (47)

Hydrochloric acid (37%) (0.96 equivalents) is added to a solution of *tert*-butyl 2-((4*R*,6*R*)-6-(2-aminoethyl)-2,2-dimethyl-1,3-dioxan-4-yl)acetate (**8**) in THF at 25°C. The mixture is aged at 25°C for 3 h and then concentrated at reduced pressure. The residue is dissolved in THF and the solution is dried over molecular sieves (to 0.8% $H_2O$). The THF solution of (3*R*,5*R*)-*tert*-butyl 7-amino-3,5-dihydroxyheptanoate (**48**) hydrochloride is diluted with additional THF. A solution of *tert*-butyldimethylsilyl chloride (3.6 equivalents) and triethylamine (5.3 equivalents) in THF is added at 25°C. The resulting suspension is aged at 25°C for 19 h and then filtered. The solid is washed with THF. The combined liquors are concentrated at reduced pressure to afford crude (3*R*,5*R*)-*tert*-butyl 7-amino-3,5-bis(*tert*-butyldimethylsilyloxy)heptanoate (**47**) that is carried directly into the next step.[66]

### 9.4.5 *tert*-Butyl 2-((4*R*,6*R*)-6-(2-Aminoethyl)-2,2-diisopropyl-1,3,2-dioxasilinan-4-yl)acetate (49)

Dichlorodiisopropylsilane (1.1 equivalents) is added to a mixture of (3*R*,5*R*)-*tert*-butyl 6-cyano-3,5-dihydroxyhexanoate (**41**), 1-hydroxybenzotriazole (HOBT) (11 mol%) and triethylamine (3.1 equivalents) in dichloromethane at 25–30°C. The mixture is refluxed for 4 h and then cooled to 25°C and filtered. The solid is washed with dichloromethane. The liquors are washed with water and with dilute hydrochloric acid and then dried and concentrated at reduced pressure to afford *tert*-butyl 2-((4*R*,6*R*)-6-(cyanomethyl)-2,2-diisopropyl-1,3,2-dioxasilinan-4-yl)acetate (**50**) (85%).

Raney nickel (amount not specified) is washed with ethanol. A solution of nitrile **50** in 7 M ammonia in methanol (12.8 L/kg nitrile **50**, 24 equivalents) is added and the suspension is hydrogenated at 25°C and 65–73 psi hydrogen for 8 h. The suspension is filtered and the catalyst is washed with methanol. The liquors are concentrated at reduced pressure to afford *tert*-butyl 2-((4*R*,6*R*)-6-(2-aminoethyl)-2,2-diisopropyl-1,3,2-dioxasilinan-4-yl)acetate (**49**) (90%).[66]

### 9.4.6 *tert*-Butyl 2-((4*R*,6*R*)-6-(2-Aminoethyl)-2-phenyl-1,3,2-dioxaborinan-4-yl)acetate (51)

A solution of phenylboronic acid (1.0 equivalent) in toluene is added to (3*R*,5*R*)-*tert*-butyl 6-cyano-3,5-dihydroxyhexanoate (**41**). The mixture is refluxed for 20 h with removal of water as the azeotrope using a Dean–Stark trap. The solution is cooled and concentrated at reduced pressure. The residue is suspended in petroleum ether. The suspension is cooled to 0°C and filtered and the solid is presumably washed with cold petroleum ether and dried to afford *tert*-butyl 2-((4*R*,6*R*)-6-(cyanomethyl)-2-phenyl-1,3,2-dioxaborinan-4-yl)acetate (**52**). No yield is provided.

Raney nickel (1 kg/kg nitrile **52**) is presumably washed with methanol. A solution of nitrile **52** in methanol saturated with ammonia (equivalents not specified) is added and the mixture is aged at (presumably) 25°C and 70 psi hydrogen pressure. The suspension is presumably filtered and the solid is washed with methanol. The liquors are concentrated at reduced pressure to afford *tert*-butyl 2-((4*R*,6*R*)-6-(2-aminoethyl)-2-phenyl-1,3,2-dioxaborinan-4-yl)acetate (**51**). Details of the procedure and the yield are not provided.[67,68]

### 9.4.7 *tert*-Butyl 2-(6-(2-Aminoethyl)-2-phenethyl-1,3,2-dioxaborinan-4-yl)acetate (53)

Phenethylboronic acid is added to a solution of *tert*-butyl 6-cyano-3,5-dihydroxyhexanoate (**41**) in toluene. The mixture is refluxed for 5 h with removal of water as the azeotrope using a Dean–Stark trap. The solution is cooled and concentrated at reduced pressure. The residue is suspended in ethyl ether at 25°C. The suspension is filtered and the solid is presumably washed with ethyl ether and dried to afford *tert*-butyl 2-(6-(cyanomethyl)-2-phenethyl-1,3,2-dioxaborinan-4-yl)acetate (**54**) (86% based on phenethylboronic acid).

Raney nickel (1 kg/kg nitrile **54**) is presumably washed with methanol. A solution of nitrile **54** in methanol saturated with ammonia (24 equivalents assuming 7 M ammonia in methanol) is added and the mixture is aged at (presumably) 25°C and 70 psi hydrogen pressure. The suspension is filtered and the solid is presumably washed with methanol. The liquors are concentrated at reduced pressure to afford *tert*-butyl 2-(6-(2-aminoethyl)-2-phenethyl-1,3,2-dioxaborinan-4-yl)acetate (**53**) (99%).[69]

### 9.4.8 *tert*-Butyl 2-(6-(2-Aminoethyl)-2-propyl-1,3,2-dioxaborinan-4-yl)acetate (55)

Propylboronic acid is added to a solution of *tert*-butyl 6-cyano-3,5-dihydroxyhexanoate (**41**) in toluene. The mixture is refluxed for 5 h with removal of water as the azeotrope using a Dean–Stark trap. The solution is cooled and concentrated at reduced pressure. The residue is suspended in ethyl ether at 25°C. The suspension is filtered and the solid is presumably washed with ethyl ether and dried to afford *tert*-butyl 2-(6-(cyanomethyl)-2-propyl-1,3,2-dioxaborinan-4-yl)acetate (**56**) (72% based on propylboronic acid).

Raney nickel (1 kg/kg nitrile **56**) is presumably washed with water. A solution of nitrile **56** in methanol saturated with ammonia (20 equivalents assuming 7 M ammonia in methanol) is added and the mixture is aged at (presumably) 25°C and 70 psi hydrogen pressure. The suspension is filtered and the solid is presumably washed with methanol. The liquors are concentrated at reduced pressure to afford *tert*-butyl 2-(6-(2-aminoethyl)-2-propyl-1,3,2-dioxaborinan-4-yl)acetate (**55**) (69%).[69]

### 9.4.9 2-((4*R*,6*R*)-6-(2-Aminoethyl)-2,2-dimethyl-1,3-dioxan-4-yl)-*N*,*N*-diphenylacetamide (57)

*N*,*N*-Diphenylacetamide is prepared from diphenylamine.[70] Lithium diisopropylamide (4.0 equivalents) in THF–heptane is added to a solution of *N*,*N*-diphenylacetamide (4.0 equivalents) in THF at −10 to −5°C. The mixture is aged at 0–20°C for 30 min. A solution of ethyl (*R*)-(−)-4-cyano-3-hydroxybutyrate in THF is added at an unspecified temperature, presumably <−20°C, and the mixture is aged at −20 to −5°C for 30 min. The mixture is then transferred into 2.2 M hydrochloric acid, presumably at 25°C, and extracted with ethyl acetate. The extracts are concentrated at reduced pressure to afford (*R*)-6-cyano-5-hydroxy-3-oxo-*N*,*N*-diphenylhexanamide (**58**) (estimated 80%).

A 50% solution of diethylmethoxyborane in THF is added to the crude (*R*)-6-cyano-5-hydroxy-3-oxo-*N*,*N*-diphenylhexanamide (**58**) in THF–methanol at −20°C. The mixture is cooled to −78°C and sodium borohydride (3.2 equivalents) is added over 30 min. The mixture is aged at −78°C for 5 h and at 25°C for 10 h. Excess hydride is quenched by adding acetic acid and the mixture is concentrated at reduced pressure. The residue is dissolved in methanol and the solution is concentrated at reduced pressure. The methanol dissolution and distillation procedure is repeated. The residual oil is separated between ethyl acetate and water. The organic layer is concentrated at reduced pressure to afford crude (3*R*,5*R*)-6-cyano-3,5-dihydroxy-*N*,*N*-diphenylhexanamide (**59**) (estimated 90%).

The crude (3*R*,5*R*)-6-cyano-3,5-dihydroxy-*N*,*N*-diphenylhexanamide (**59**) is dissolved in 2,2-dimethoxypropane (8.3 equivalents) and acetone. Methanesulfonic acid (4 mol %) is added and the mixture is aged at 25°C for 2 h. Aqueous sodium bicarbonate and ethyl acetate are added and the layers are separated. The organic layer is concentrated at reduced pressure to afford 2-((4*R*,6*R*)-6-(cyanomethyl)-2,2-dimethyl-1,3-dioxan-4-yl)-*N*,*N*-diphenylacetamide (**60**) (96%).

Raney nickel A-7000 (380 g/kg nitrile **60**) is presumably washed with methanol. A solution of nitrile **60** in methanol (15 L/kg nitrile **60**) containing anhydrous ammonia (4.9 equivalents) is added and the suspension is hydrogenated at 30°C and presumably 50 psi for 3 h. The suspension is cooled to 20°C and filtered and the catalyst is presumably washed with methanol. The liquors are concentrated at reduced pressure to afford 2-((4*R*,6*R*)-6-(2-aminoethyl)-2,2-dimethyl-1,3-dioxan-4-yl)-*N*,*N*-diphenylacetamide (**57**) (96%, overall 66% for four steps from ethyl (*R*)-(−)-4-cyano-3-hydroxybutyrate) (Scheme 9.16).[71]

Similar results (overall 60–70% for four steps from ethyl (*R*)-(−)-4-cyano-3-hydroxybutyrate) are observed for parallel sequences starting with dibenzylamine, diethylamine, *n*-butylmethylamine, benzyl-*tert*-butylamine, and piperidine. Looking ahead to a late-stage amide hydrolysis, the morpholine amide would also be a good candidate.[71]

### 9.4.10 2-((4*R*,6*R*)-6-(2,2-Dimethoxyethyl)-2,2-dimethyl-1,3-dioxan-4-yl)ethanamine (61)

Hydrolysis of methyl 3,3-dimethoxypropionate[72] affords the acid **62**. The chain is extended by converting acid **62** to the *N*-acylimidazole and reaction with monoethyl malonate[73] and magnesium. Crude ethyl 5,5-dimethoxy-3-oxopentanoate (**63**) is distilled (bp 86–92°C at 0.09 mmHg) (91%) and then reduced using an (*R*)-tolBINAP ruthenium catalyst in methanol or ethanol at 50°C and 725 psi hydrogen pressure (100%). While the target is (*S*)-ethyl 3-hydroxy-

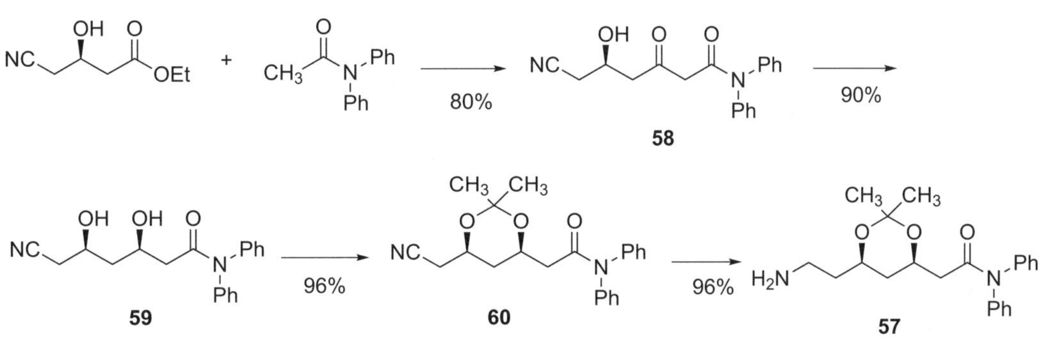


**SCHEME 9.16** 2-((4*R*,6*R*)-6-(2-Aminoethyl)-2,2-dimethyl-1,3-dioxan-4-yl)-*N*,*N*-diphenylacetamide (**57**) from ethyl (*R*)-(−)-4-cyano-3-hydroxybutyrate.

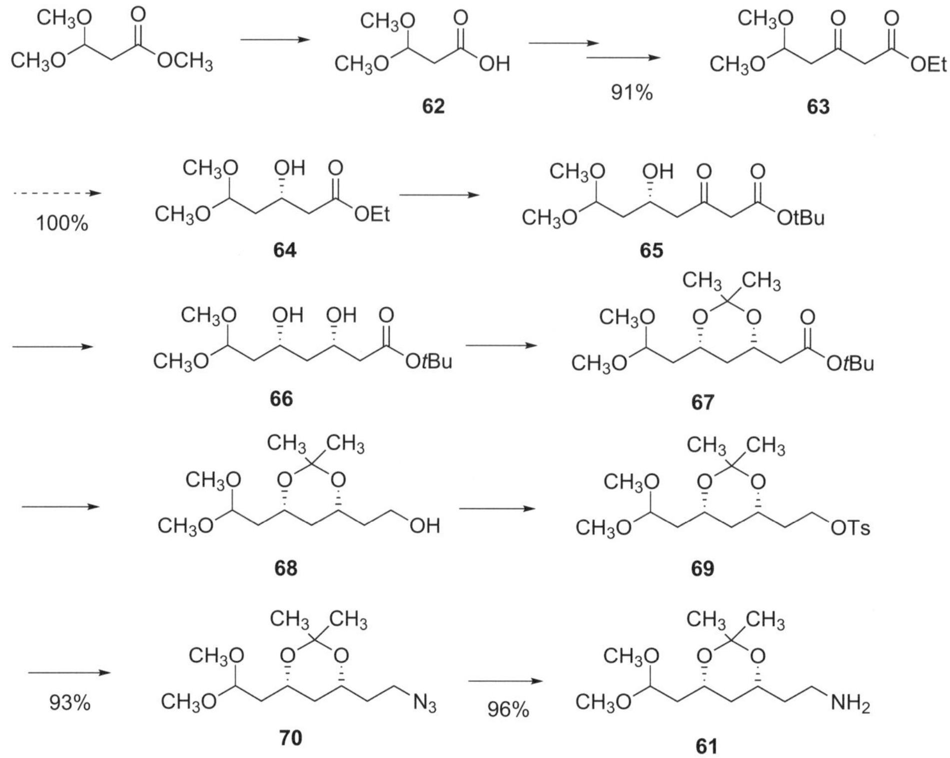

**SCHEME 9.17** 2-((4*R*,6*R*)-6-(2,2-Dimethoxyethyl)-2,2-dimethyl-1,3-dioxan-4-yl)ethanamine (**61**) from methyl 3,3-dimethoxypropionate.

5,5-dimethoxypentanoate (**64**), high enantioselectivity is only offered for the (*R*)-enantiomer (99.1% *R*). The remaining sequence is described without reference to the configuration or optical purity of the intermediates (Scheme 9.17).

The chain is extended by Claisen condensation using lithium diisopropylamide (3.9 equivalents) and *tert*-butyl acetate (4.2 equivalents). The $\beta$-hydroxyketone is reduced using diethylmethoxyborane (1.1 equivalents) and sodium borohydride (1.9 equivalents) in THF–methanol at −78°C. The diol is protected with 2,2-dimethoxypropane (5.7 equivalents) and *p*-toluenesulfonic acid monohydrate. Note that the aldehyde acetal would not be expected to remain as a spectator during the diol protection step but no side products isolated from the protection step are discussed. The ester of **67** is reduced to the alcohol with lithium aluminum hydride and the alcohol is converted to a leaving group. To this point no yield is provided since each crude intermediate is carried to the next step. The *p*-toluenesulfonate is displaced by sodium azide in DMSO at 80°C (93%) and the azide is reduced using 5% palladium on carbon, presumably at 25°C and atmospheric pressure (96%).[74]

### 9.4.11 2-((4*S*,6*R*)-6-(2-Aminoethyl)-2,2-dimethyl-1,3-dioxan-4-yl)ethanol (71)

A 1 M solution of lithium aluminum hydride (1.7 equivalents) in ethyl ether is added to a solution of *tert*-butyl (4*R*,6*R*)-6-(cyanomethyl)-2,2-dimethyl-1,3-dioxan-4-yl)acetate (**32**) in ethyl ether at 0°C. The suspension is aged at 0°C for 15 min and at 25°C for 16 h. The mixture is cooled to 0–5°C and 1:1 methanol–water is added. The suspension is filtered and the solid is washed with ethyl acetate. The combined liquors are concentrated at reduced pressure to afford 2-((4*S*,6*R*)-6-(2-aminoethyl)-2,2-dimethyl-1,3-dioxan-4-yl)ethanol (**71**) (92%) (Scheme 9.18).[66]

### 9.4.12 2-((2*R*,4*R*,6*R*)-4-(*tert*-Butyldiphenylsilyloxy)-6-methoxytetrahydro-2H-pyran-2-yl)ethanamine (72)

Another route to an elaborated side chain with the carboxylic acid masked as an aldehyde acetal begins with commercially available and expensive methyl (4*S*)-( + )-2,2-dimethyl-1,3-dioxolane-4-acetate (Scheme 9.19). A discussion of the many possible approaches to this value-added starting

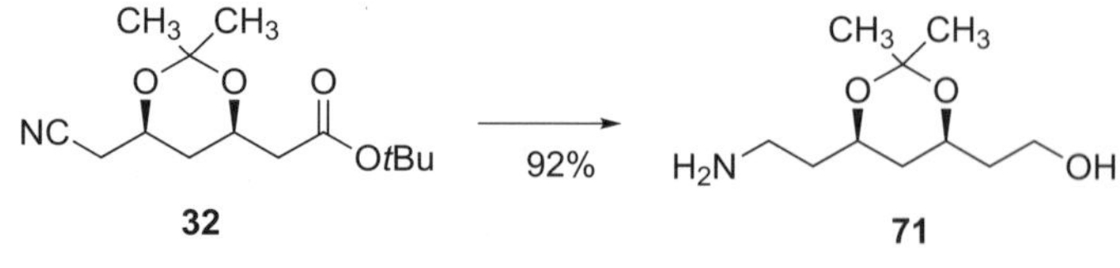

**SCHEME 9.18** 2-((4*S*,6*R*)-6-(2-Aminoethyl)-2,2-dimethyl-1,3-dioxan-4-yl)ethanol (**71**) from *tert*-butyl (4*R*,6*R*)-6-(cyanomethyl)-2,2-dimethyl-1,3-dioxan-4-yl)acetate (**32**).

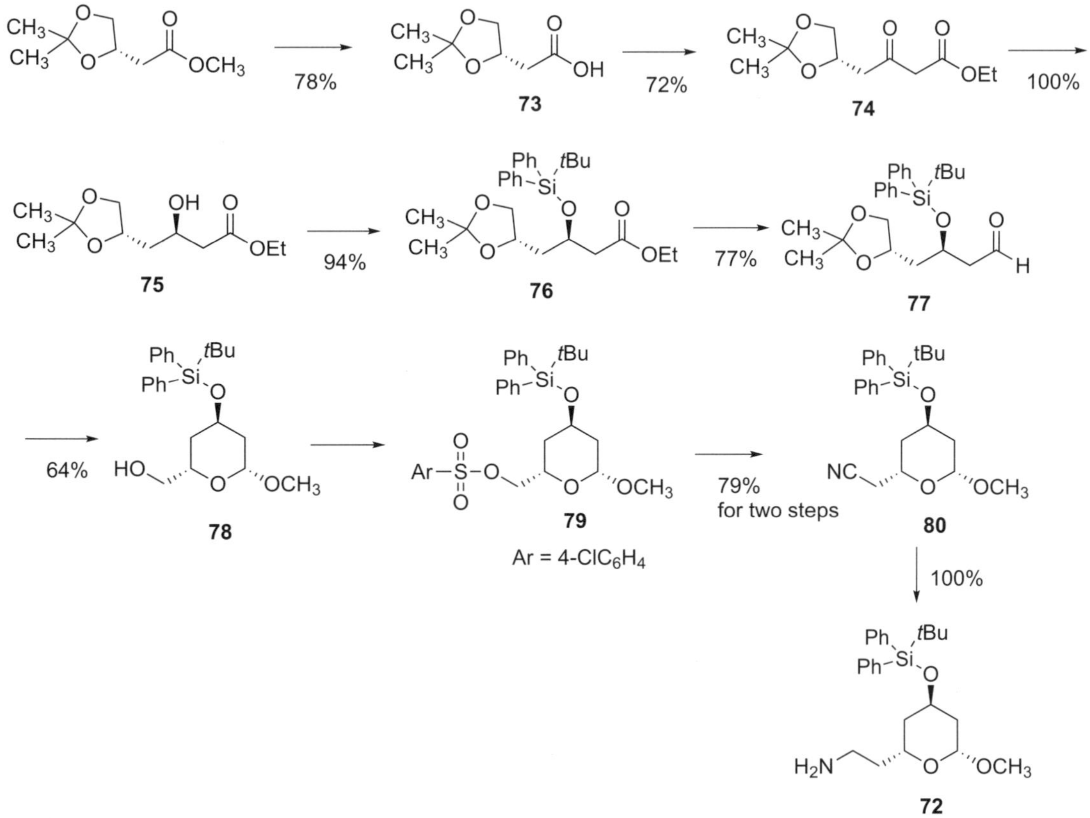

**SCHEME 9.19** 2-((2*R*,4*R*,6*R*)-4-(*tert*-Butyldiphenylsilyloxy)-6-methoxytetrahydro-2*H*-pyran-2-yl) ethanamine (**72**) from methyl (4*S*)-(+)-2,2-dimethyl-1,3-dioxolane-4-acetate.

material would divert the discussion away from pharmaceutical process chemistry to specialty chemical manufacturing.

Methyl (4*S*)-(+)-2,2-dimethyl-1,3-dioxolane-4-acetate[75] is hydrolyzed to the acid **73** by reaction with 2 M sodium hydroxide (78%). The chain is extended by converting acid **73** to the *N*-acylimidazole and reaction with monoethyl malonate and magnesium. Crude (*S*)-ethyl 4-(2,2-dimethyl-1,3-dioxolan-4-yl)-3-oxobutanoate (**74**) is distilled (bp 90–93°C at 0.06 mmHg) (72%) and then reduced using an (*R*)-tolBINAP ruthenium catalyst in methanol or ethanol at 25–50°C and 725–1400 psi hydrogen pressure (100%). The alcohol of **75** is protected as the *tert*-butyldiphenylsilyl ether (94%) and the ester of **76** is reduced with DIBAL (77%). Reaction of the aldehyde **77** with trimethyl orthoformate, methanol, and a *p*-toluenesulfonic acid catalyst affords the (6-methoxytetrahydro-2*H*-pyran-2-yl)methanol **78** after crystallization from hexane (64%). The alcohol is converted to a leaving group and the leaving group is displaced by reaction with sodium cyanide in DMSO at 25°C over 4 days. The nitrile **80** is isolated by filtration through silica using toluene–ethyl acetate and crystallization from hexane–ethanol (79% for two steps). The overall yield for the eight-step sequence from methyl (4*S*)-(+)-2,2-dimethyl-1,3-dioxolane-4-acetate is 21%.

Raney nickel catalyst (480 g/kg nitrile **80** on 500 mg scale) is washed with methanol. A solution of nitrile **80** in methanol containing ammonia (11 equivalents) is added and the suspension is hydrogenated at 25°C and 725 psi for 5 h. The suspension is filtered and the liquors are concentrated at reduced pressure to afford 2-((2*R*,4*R*,6*R*)-4-(*tert*-butyldiphenylsilyloxy)-6-methoxytetrahydro-2*H*-pyran-2-yl)ethanamine (**72**) (100%).[76]

### 9.4.13 2-((2*R*,4*R*)-4-(Benzyloxy)-6-methoxytetrahydro-2*H*-pyran-2-yl)ethanamine (81)

A closely related amine has the 4-position alcohol protected as a benzyl ether (Scheme 9.20). (4*R*,6*S*)-6-(Chloromethyl) tetrahydro-2*H*-pyran-2,4-diol (**82**) is produced by aldolase-mediated condensation of chloroacetaldehyde[77] and acetaldehyde.[78,79] The lactol is protected as the methyl acetal by sulfuric acid-catalyzed exchange with methanol at 40°C over 2 h. The mixture is presumably cooled and concentrated at reduced pressure. After a routine water–dichloromethane workup, (2*S*,4*R*)-2-(chloromethyl)-6-methoxytetrahydro-2*H*-pyran-4-ol (**83**) is isolated from the residue by chromatography (99%).

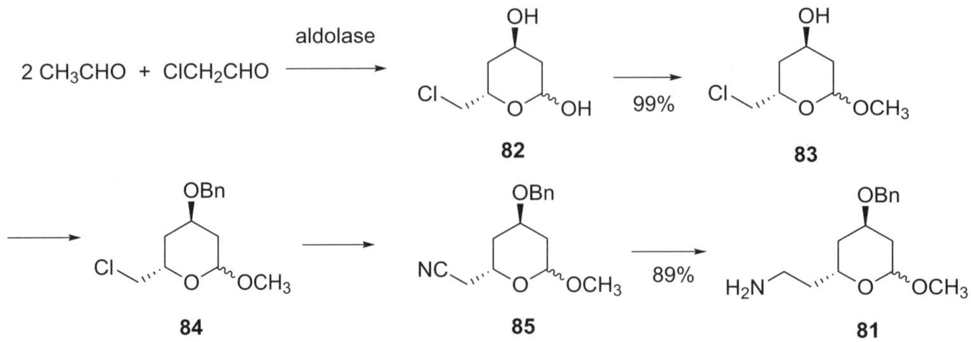


**SCHEME 9.20** 2-((2*R*,4*R*)-4-(Benzyloxy)-6-methoxytetrahydro-2*H*-pyran-2-yl)ethanamine (**81**) from chloroacetaldehyde and acetaldehyde.

Sodium hydride 60% dispersion in mineral oil (1.5 equivalents) is added to a solution of the pyran-4-ol (**83**) in THF at 25°C. Benzyl bromide (2.0 equivalents) is added, presumably at 25°C, and the mixture is refluxed for 2 h. Quench of excess hydride with methanol and a routine water–dichloromethane workup affords the benzyl ether **84**. No yield is available.

The chloride is next displaced by reaction with potassium cyanide (4.1 equivalents) in DMSO at 80°C for 4 days. The mixture is presumably cooled to 25°C and separated between water and ethyl ether. The organic layer is dried and concentrated at reduced pressure. The nitrile **85** is isolated from the residue by chromatography. No yield is available. It is likely that the configuration at the 2-position dramatically affects the mechanism and rate of substitution of the neopentyl-type chloride (Scheme 9.21).

Borane–THF complex (6.2 equivalents) is added to a solution of nitrile **85** in THF at 10°C and the mixture is refluxed for 9 h. The mixture is cooled, quenched with methanol, and concentrated at reduced pressure. The residue is dissolved in methanol and the solution is concentrated at reduced pressure. The methanol dissolution and concentration procedure is repeated once more to afford crude 2-((2*R*,4*R*)-4-(benzyloxy)-6-methoxytetrahydro-2*H*-pyran-2-yl)ethanamine (**81**) (89%).[80]

### 9.4.14 2-((3a*R*,5*R*,6a*R*)-2,2-Dimethyltetrahydrofuro[2,3-*d*][1,3]-dioxol-5-yl)ethanamine (86)

Aldehyde **87** with both chiral centers correctly set is prepared in five steps (74%) from 1,2:5,6-di-*O*-isopropylidene-α-D-glucofuranose (Scheme 9.22).[81] Horner–Emmons–Wadsworth reaction of the aldehyde with triethyl phosphonoacetate (1.3 equivalents) and sodium hydride (2.0 equivalents) affords the alkene **88** (94%). The alkene is reduced with Raney nickel (99%) and the ester is converted to the amide by reaction with aqueous ammonia (86%). Hofmann reaction with sodium methoxide (3.0 equivalents) and bromine (2.0 equivalents) affords the methyl carbamate **91** (61%) that is hydrolyzed in aqueous sodium hydroxide at 80–85°C to afford 2-((3a*R*,5-*R*,6a*R*)-2,2-dimethyltetrahydrofuro[2,3-*d*][1,3]-dioxol-5-yl)ethanamine (**86**) (74%). The overall yield of the amine for the five-step sequence from aldehyde **87** is 36% and for the 10-step sequence from 1,2:5,6-di-*O*-isopropylidene-α-D-glucofuranose is 27%.

The amine can also be produced from the aldehyde in four steps via Henry reaction with nitromethane, dehydration to an alkene, reduction of the alkene with sodium borohydride, and reduction of the nitro group using Raney nickel in ammonium hydroxide. No yield is available for this alternative sequence.[82]

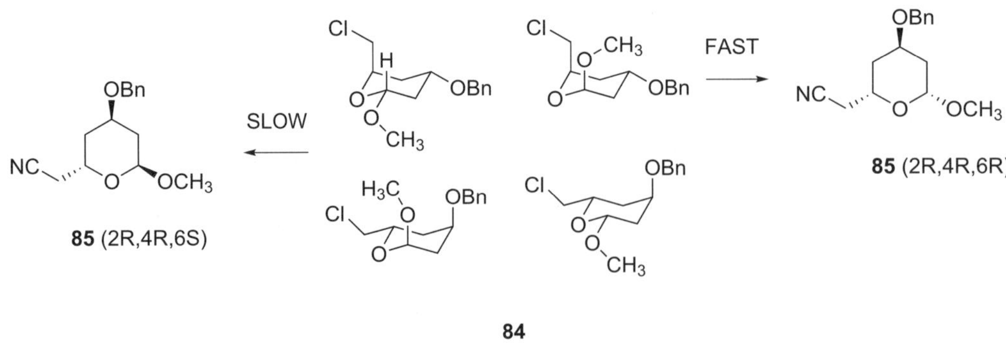


**SCHEME 9.21** Displacement of the neopentyl-type chloride of (2*S*,4*R*)-4-benzyloxy-2-(chloromethyl)-6-methoxytetrahydro-2*H*-pyran (**84**).

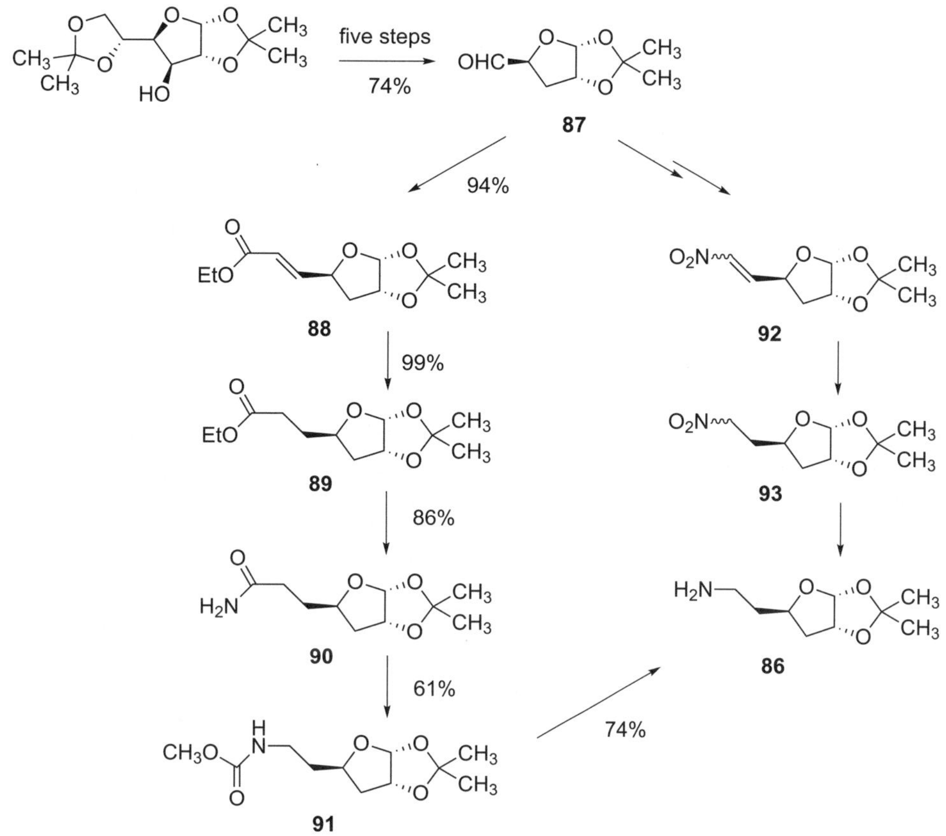

**SCHEME 9.22** 2-((3a*R*,5*R*,6a*R*)-2,2-Dimethyl(tetrahydrofuro[2,3-*d*][1,3]-dioxol-5-yl)ethanamine (**86**) from 1,2:5,6-di-*O*-isopropylidene-α-D-glucofuranose.

## 9.5 PAAL–KNORR PYRROLE SYNTHESIS

The Paal–Knorr condensation of a 1,4-diketone with an amine generates a pyrrole. Just as the pyrrole is the core of atorvastatin, this method for construction of the pyrrole is the focal point for many routes to atorvastatin. Water is produced in the condensation. There are three options for dealing with the water by-product: retain it in the reaction mixture, remove it by distillation of a water–organic solvent azeotrope using a Dean–Stark trap, and remove it by passing the distillate through molecular sieves in a Soxhlet extractor. Removing the water may reduce the time required for complete conversion but have little effect on the yield.

The Paal–Knorr condensation is typically run using a nearly 1:1 ratio of the value-added 1,4-diketone and the value-added amine. Reactions between the amine and the *tert*-butyl ester and the amine and the *N*-phenylamide are likely to produce amide side products. The generation of amide side products suggests that a small excess of amine (1.03–1.2 equivalents) will be required to consume all of the 1,4-diketone. There is no evidence that using more amine increases the pyrrole yield or decreases the time required for complete conversion of the 1,4-diketone.

While the condensation proceeds at elevated temperatures without a catalyst, higher yields and less amide side product formation are observed in condensations catalyzed by organic acid (pivalic acid, acetic acid, hexanoic acid). The catalyst charge is typically 0.6–0.7 equivalents. The condensation proceeds to completion faster and yields are slightly higher (3%) using an ammonium carboxylate catalyst (pivalic acid with $Et_3N$, $(i\text{-}Pr)_2NH$, or *N*-ethylmorpholine).

The condensation is conveniently run in a single solvent (heptane, cyclohexane, toluene, xylene[36], toluene, 2-methylTHF, THF) or in binary or ternary mixtures of solvents. The mixtures are designed to provide a specific reflux temperature and perhaps a water azeotrope. Examples of solvent mixtures are 9:1 heptane-–toluene, 3:1 toluene–THF, 1.9:1 methyl *tert*-butyl ether–THF, 20:1 cyclohexane–THF, 20:1 cyclohexane–methanol, and 2:3:2 heptane–THF–toluene. The pyrrole product is likely to crystallize from the mixture using heptane or cyclohexane as the solvent.

On small scale, the pyrrole is separated from starting materials and side products by chromatography. On large scale, the pyrrole is crystallized from isopropanol, isopropanol–water, isopropanol–acetone–water, or ethanol–water. Yields of 70–78% are to be expected. The most pressing development issue for the pyrrole ring formation step is the long reaction time, typically >24 h.

### 9.5.1 1,4-Diketone 13 with 3-Aminopropionaldehyde Diethyl Acetal

3-Aminopropionaldehyde diethyl acetal[83] (1.6 equivalents) is added to a solution of 1-(4-fluorophenyl)-5-methyl-2-phenylhexane-1,4-dione (**13**) in toluene at 25°C. An unspecified amount of *p*-toluenesulfonic acid monohydrate is added and the mixture is refluxed for 24 h with removal of water as the azeotrope using a Dean–Stark trap. The solution is cooled, diluted with ethanol, and filtered through silica gel. The liquors are concentrated at reduced pressure. The residue is suspended in diisopropyl ether and the suspension is filtered. The solid is presumably washed with diisopropyl ether and dried to afford the pyrrole **94**. Additional pyrrole **94** is isolated from the liquors (total yield 74%) (Scheme 9.23).[15]

### 9.5.2 1,4-Diketone 21 with 3-Aminopropionaldehyde Diethyl Acetal

Pivalic acid (0.70 equivalents) is added to a mixture of 2-(2-(4-fluorophenyl)-2-oxo-1-phenylethyl)-4-methyl-3-oxo-*N*-phenylpentanamide (**21**) and 3-aminopropionaldehyde diethyl acetal (1.3 equivalents) in heptane–toluene (9:1) at 25°C. The mixture is refluxed for 32 h with removal of water as the azeotrope using a Dean–Stark trap. The mixture is cooled to 60–65°C, diluted with 3:2 isopropanol–water, seeded, and cooled to 25°C. The resulting suspension is filtered and the solid is washed with isopropanol and dried to afford pyrrole **95** (81%). Pyrrole **95** can be crystallized from ethanol.[29]

### 9.5.3 1,4-Diketone 21 with 3-Amino-1-morpholinopropan-1-one (96)

The deceptively simple amines, 3-aminopropionaldehyde diethyl acetal and 2-(2-aminoethyl)-1,3-dioxolane, are commercially available but expensive. An alternative pyrrole construction uses a simple and potentially less expensive amine to deliver an amide for side chain elaboration. The amine is prepared in three steps from methyl cyanoacetate.

A mixture of morpholine (1.2 equivalents) and methyl cyanoacetate[84] in methyl *tert*-butyl ether is aged at 55°C for 12–18 h. The mixture is diluted with methyl *tert*-butyl ether and cooled to < 50°C until nucleation is apparent. Additional methyl *tert*-butyl ether is added during cooling and crystallization. The suspension is cooled to 0°C and filtered. The solid is washed with methyl *tert*-butyl ether and dried at 40°C and reduced pressure to afford 3-morpholino-3-oxopropanenitrile (**97**) (90%).

The nitrile **97** is reduced using 5% platinum on carbon, 58% water wet (2.1 g Pt/kg nitrile **97**) in methanol-12 M hydrochloric acid (1.1 equivalents) at 25°C and 50 psi hydrogen over 24 h. The catalyst is filtered and the liquors are concentrated to approximately 40% of the original volume. Isopropanol is added, presumably at 25°C, and the suspension is cooled to 0°C and filtered. The solid is washed with

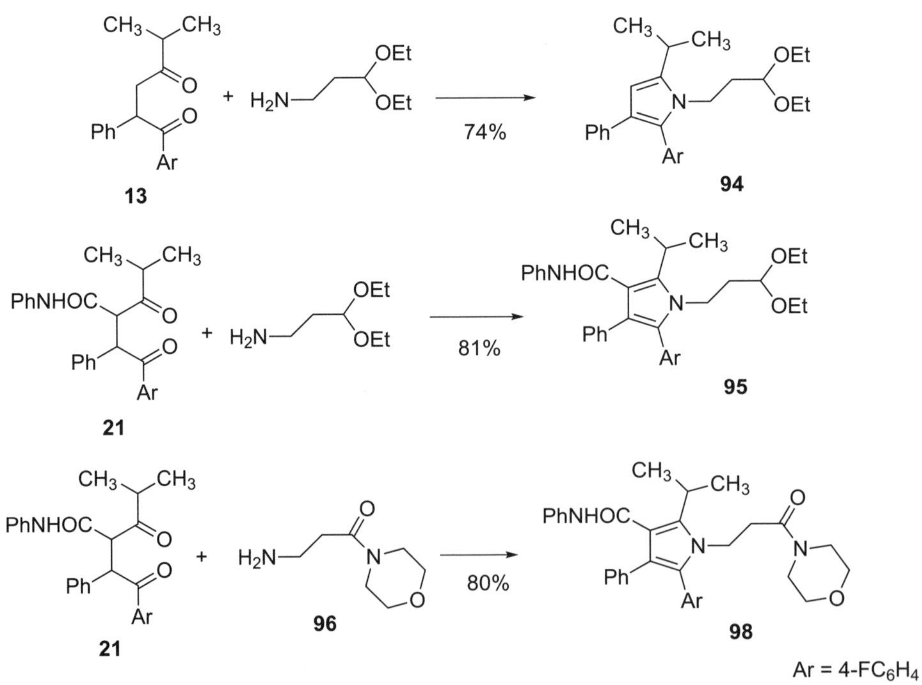


**SCHEME 9.23** Paal–Knorr syntheses of pyrroles suitable for conversion to atorvastatin (**1**) by side chain elaboration.

methyl *tert*-butyl ether and dried at 30°C and reduced pressure to afford 3-amino-1-morpholinopropan-1-one (**96**) as the hydrochloride (81%). The hydrochloride salt is converted to the phenylacetic acid salt (85%) for use in the Paal–Knorr pyrrole synthesis.

A mixture of 2-(2-(4-fluorophenyl)-2-oxo-1-phenylethyl)-4-methyl-3-oxo-*N*-phenylpentanamide (**21**) and 3-amino-1-morpholinopropan-1-one (**96**) phenylacetic acid salt (1.7 equivalents) in THF is refluxed for 24 h with removal of water by passing the condensate through freshly activated molecular sieves. The resulting suspension is presumably cooled to 25°C and aqueous sodium bicarbonate is added. The suspension is then cooled to 0°C, diluted with methyl *tert*-butyl ether, and filtered. The solid is washed with water and with methyl *tert*-butyl ether and dried at <50°C and reduced pressure to afford pyrrole **98** (80%).[85]

### 9.5.4 1,4-Diketone 21 with *tert*-Butyl 2-((4*R*,6*R*)-6-(2-Aminoethyl)-2,2-dimethyl-1,3-dioxan-4-yl)acetate (8)

A mixture of 2-(2-(4-fluorophenyl)-2-oxo-1-phenylethyl)-4-methyl-3-oxo-*N*-phenylpentanamide (**21**) and *tert*-butyl 2-((4*R*,6*R*)-6-(2-aminoethyl)-2,2-dimethyl-1,3-dioxan-4-yl) acetate (**8**) (1.3 equivalents) in heptane–toluene (9:1) is refluxed for 24 h. The solution is cooled, presumably to 75–80°C, isopropanol is added, and the solution is cooled to 25°C. The resulting suspension is filtered and the solid presumably washed and dried to afford pyrrole **99** (74%) (Scheme 9.24).[29,30,46]

Heptanoic acid (1.1 equivalents) is added to a mixture of 2-(2-(4-fluorophenyl)-2-oxo-1-phenylethyl)-4-methyl-3-oxo-*N*-phenylpentanamide (**21**) and *tert*-butyl 2-((4*R*,6*R*)-6-(2-aminoethyl)-2,2-dimethyl-1,3-dioxan-4-yl)acetate (**8**) (1.05 equivalents) in heptane–THF–toluene (~10:5:2) at 25°C. The mixture is refluxed for 8 h. The mixture is cooled to 25°C and concentrated at reduced pressure. The residue is presumably dissolved in isopropanol at 60°C and the solution is cooled to 25°C and aged at 25°C for 10 h. The resulting suspension is cooled to 15°C, diluted with hexanes, and filtered. The solid is presumably washed with isopropanol and dried to afford pyrrole **99** (57%).[86]

Pivalic acid (0.67 equivalents) is added to a mixture of 2-(2-(4-fluorophenyl)-2-oxo-1-phenylethyl)-4-methyl-3-oxo-*N*-phenylpentanamide (**21**) and *tert*-butyl 2-((4*R*,6*R*)-6-(2-aminoethyl)-2,2-dimethyl-1,3-dioxan-4-yl)acetate (**8**) (1.3 equivalents) in THF at 25°C. The mixture is refluxed for 40 h. The mixture is cooled to 25°C and concentrated at reduced pressure. The residue is dissolved in hot ethanol. Water is added and the solution is cooled to 25°C. The resulting suspension is aged at 25°C for 3 h and filtered. The solid is washed with ethanol–water and dried at 55°C and reduced pressure to afford pyrrole **99** (75–79%). The yield is lower (72%) after 48 h at reflux using a ternary solvent mixture (2:2:3 heptane–THF–toluene).[66]

Pivalic acid (0.70 equivalents) is added to a mixture of 2-(2-(4-fluorophenyl)-2-oxo-1-phenylethyl)-4-methyl-3-oxo-*N*-phenylpentanamide (**21**) and *tert*-butyl 2-((4*R*,6*R*)-6-(2-aminoethyl)-2,2-dimethyl-1,3-dioxan-4-yl)acetate (**8**)

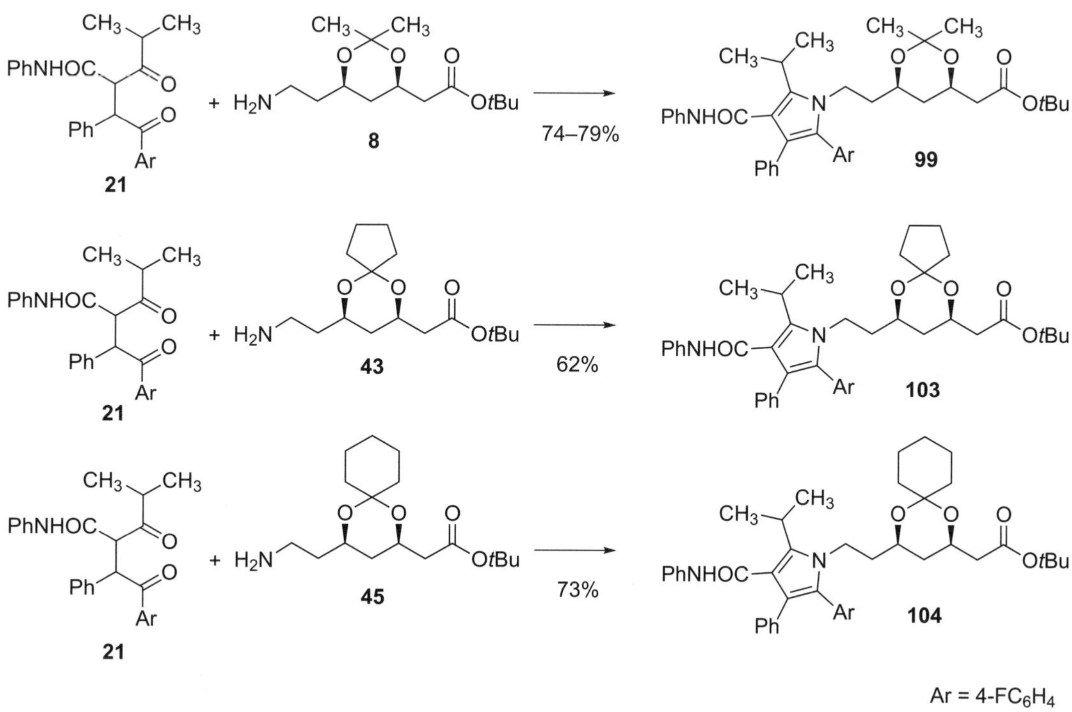


**SCHEME 9.24** Paal–Knorr syntheses of pyrroles suitable for conversion to atorvastatin (**1**) by ketal deprotection and ester hydrolysis.

(1.1 equivalents) in 2-methylTHF at 25°C. The mixture is refluxed for 30–35 h with removal of water as the azeotrope using a Dean–Stark trap. The mixture is cooled to 25°C and concentrated at reduced pressure. The residue is dissolved in isopropanol at an unspecified elevated temperature. The solution is cooled to 25°C and aged at 25°C for 2 h. The resulting suspension is cooled to 15°C, aged at 15°C for 1 h, and filtered. The solid is washed with isopropanol and dried at 60°C and reduced pressure to afford pyrrole **99** (58%, >99% pure by HPLC).[87]

Pivalic acid (0.58 equivalents) is added to a mixture of 2-(2-(4-fluorophenyl)-2-oxo-1-phenylethyl)-4-methyl-3-oxo-*N*-phenylpentanamide (**21**) and *tert*-butyl 2-((4*R*,6*R*)-6-(2-aminoethyl)-2,2-dimethyl-1,3-dioxan-4-yl)acetate (**8**) (1.05 equivalents) in cyclohexane at 25°C. The mixture is refluxed for 62 h. The mixture is cooled to 25°C, washed with aqueous sodium bicarbonate, and concentrated at reduced pressure. The residue is dissolved in isopropanol at 30–35°C and water is added. The resulting suspension is filtered and the solid is presumably washed with water or water–isopropanol and dried to afford pyrrole **99**. No yield is available.[32]

Pivalic acid (0.30 equivalents) is added to a mixture of 2-(2-(4-fluorophenyl)-2-oxo-1-phenylethyl)-4-methyl-3-oxo-*N*-phenylpentanamide (**21**) and *tert*-butyl 2-((4*R*,6*R*)-6-(2-aminoethyl)-2,2-dimethyl-1,3-dioxan-4-yl)acetate (**8**) (1.0 equivalent) in cyclohexane at 25°C. The mixture is refluxed for an unspecified time. The mixture is cooled to an unspecified temperature and the resulting suspension is filtered. The solid is crystallized from isopropanol to afford pyrrole **99** (59%).[88]

Pivalic acid (0.63 equivalents) is added to a mixture of 2-(2-(4-fluorophenyl)-2-oxo-1-phenylethyl)-4-methyl-3-oxo-*N*-phenylpentanamide (**21**) and *tert*-butyl 2-((4*R*,6*R*)-6-(2-aminoethyl)-2,2-dimethyl-1,3-dioxan-4-yl)acetate (**8**) (1.2 equivalents) in cyclohexane at 25°C. The mixture is refluxed with removal of water as the azeotrope using a Dean–Stark trap until analysis reveals complete conversion. The mixture is concentrated at <70°C and reduced pressure. The residue is presumably suspended in isopropanol and the mixture is concentrated at reduced pressure. The residue is dissolved in isopropanol at <50°C and the solution is cooled to 25°C. The resulting suspension is aged at 25°C for 6–8 h, aged at 0–10°C for 2–3 h, and filtered. The solid is washed with cold isopropanol and dried at 60–70°C and reduced pressure to afford pyrrole **99** (73%).[61]

Pivalic acid (0.3 equivalents) is added to a mixture of 2-(2-(4-fluorophenyl)-2-oxo-1-phenylethyl)-4-methyl-3-oxo-*N*-phenylpentanamide (**21**) and *tert*-butyl 2-((4*R*,6*R*)-6-(2-aminoethyl)-2,2-dimethyl-1,3-dioxan-4-yl)acetate (**8**) (1.01 equivalents) in toluene at 90°C. The mixture is refluxed for 10 h with removal of water as the azeotrope using a Dean–Stark trap. The mixture is cooled to 25°C and concentrated at reduced pressure. The residue is dissolved in ethanol at 25°C and water is added. The resulting suspension is aged at 25°C for 3 h and then filtered. The solid is washed with ethanol–water (3:1) and dried at 40°C and reduced pressure to afford pyrrole **99** (52%).[89]

Pivalic acid (0.67 equivalents) is added to a mixture of 2-(2-(4-fluorophenyl)-2-oxo-1-phenylethyl)-4-methyl-3-oxo-*N*-phenylpentanamide (**21**) and *tert*-butyl 2-((4*R*,6*R*)-6-(2-aminoethyl)-2,2-dimethyl-1,3-dioxan-4-yl)acetate (**8**) (1.03 equivalents) in heptane–THF–toluene (4:1:1) at 25°C. The mixture is refluxed for 40 h. After a routine water–-toluene workup and a carbon treatment, the concentrated toluene solution is dissolved in isopropanol and the solution is concentrated at reduced pressure. The isopropanol dissolution and concentration is repeated. The residue is then dissolved in isopropanol and water is added. The temperatures for the dissolution and water addition are not specified. The resulting suspension is filtered and the solid is washed with isopropanol–water (2:1) and dried to afford crude pyrrole **99** (78%, 94% pure by HPLC). The purity can be upgraded by crystallization from isopropanol (82% recovery, 99% pure by HPLC). This process is demonstrated on a 10–20 kg scale.[47]

Pivalic acid (0.66 equivalents) is added to a mixture of 2-(2-(4-fluorophenyl)-2-oxo-1-phenylethyl)-4-methyl-3-oxo-*N*-phenylpentanamide (**21**) and *tert*-butyl 2-((4*R*,6*R*)-6-(2-aminoethyl)-2,2-dimethyl-1,3-dioxan-4-yl)acetate (**8**) (1.05 equivalents) in heptane–THF–toluene (4.7:1:1) at 25°C. The mixture is refluxed for 50–55 h. After a routine water–toluene workup and a carbon treatment, the residue is dissolved in isopropanol at 80–85°C and the solution is cooled to 25°C. The resulting suspension is aged at 25°C for 10 h, aged at 0–5°C for 3 h, and filtered. The solid is washed with cold isopropanol and dried at 40–45°C and reduced pressure to afford pyrrole **99** (63%, 99.3% pure by HPLC).[90]

Pivalic acid (0.60 equivalents) is added to a mixture of 2-(2-(4-fluorophenyl)-2-oxo-1-phenylethyl)-4-methyl-3-oxo-*N*-phenylpentanamide (**21**) and *tert*-butyl 2-((4*R*,6*R*)-6-(2-aminoethyl)-2,2-dimethyl-1,3-dioxan-4-yl)acetate (**8**) (1.14 equivalents) in heptane–THF–toluene (4.3:1:1.3) at 25°C. The mixture is refluxed for 40 h with removal of water as the azeotrope using a Dean–Stark trap. After a routine water–toluene workup and a carbon treatment, the residue is dissolved in hot isopropanol and the solution is cooled to 25°C. Water is added and the resulting suspension is filtered. The solid is washed with isopropanol–water and dried at 40–45°C and reduced pressure to afford pyrrole **99** (70%).[91]

Pivalic acid (0.60 equivalents) is added to a mixture of 2-(2-(4-fluorophenyl)-2-oxo-1-phenylethyl)-4-methyl-3-oxo-*N*-phenylpentanamide (**21**) (1.1 equivalents) and *tert*-butyl 2-((4*R*,6*R*)-6-(2-aminoethyl)-2,2-dimethyl-1,3-dioxan-4-yl)acetate (**8**) in hexanes–THF (1:1) at 25°C. The reactor is sealed and the mixture is aged at 75°C for

96 h. The solution is cooled, diluted with methyl *tert*-butyl ether, washed with dilute sodium hydroxide and with dilute hydrochloric acid, and concentrated at reduced pressure. No yield for pyrrole **99** is provided at this point but a downstream yield suggests the yield is >70%.[63]

Perhaps the reaction efficiency with respect to 1,4-diketone **21** could be increased and the reaction time decreased by using an excess of amine **8**. This modification would only be attractive if value-added amine **8** could also be efficiently recovered. Pivalic acid (0.46 equivalents) is added to a mixture of 2-(2-(4-fluorophenyl)-2-oxo-1-phenylethyl)-4-methyl-3-oxo-*N*-phenylpentanamide (**21**) and *tert*-butyl 2-((4*R*,6*R*)-6-(2-aminoethyl)-2,2-dimethyl-1,3-dioxan-4-yl) acetate (**8**) (4.0 equivalents) in cyclohexane–THF (20:1) at 25–35°C. The mixture is refluxed for 18 h. The mixture is cooled to 25°C, diluted with water, and aged at 25°C for 20 min. Ammonium hydroxide is added (to pH 8.5–9.5) and the mixture is aged at 25°C for 30 min. The layers are separated and the aqueous layer is extracted with dichloromethane. The combined organic layers are concentrated, presumably at reduced pressure. The residue is presumably dissolved in isopropanol at 50–60°C and water is added. The mixture is cooled to 25–35°C and aged at 25°C for 5 h. The resulting suspension is filtered and the solid is washed with isopropanol–water and dried at 50–55°C and reduced pressure to afford pyrrole **99** (75%). Excess amine **8** is recovered from the liquors by adding acetic acid, extracting the acetate salt into water, adding ammonium hydroxide (to pH 9–9.5), and extracting the free base into dichloromethane. The extract is then concentrated at reduced pressure (79% recovery).[92]

It is likely that amides are produced from the amine and the pivalic acid catalyst, by amine dimerization, and by reaction of the amine with pyrrole **99** during the extended reflux period. The amount of amide impurity produced is influenced by the reaction solvent(s) and the reflux temperature (Figure 9.4).

Pivalic acid (0.40 equivalents) is added to a mixture of 2-(2-(4-fluorophenyl)-2-oxo-1-phenylethyl)-4-methyl-3-oxo-*N*-phenylpentanamide (**21**) and *tert*-butyl 2-((4*R*,6*R*)-6-(2-aminoethyl)-2,2-dimethyl-1,3-dioxan-4-yl)acetate (**8**) (1.3 equivalents) in heptane at 25°C. The mixture is refluxed for 30 h with removal of water as the azeotrope using a Dean–Stark trap. The solution is cooled to 60–70°C, washed with brine, with aqueous sodium bicarbonate, and again with brine, and concentrated at reduced pressure. The residue is crystallized from ethanol–water to afford pyrrole **99** (59%, 98.8% pure by HPLC). The crystallized pyrrole **99** contains amide impurity **102** (0.47%).[93]

Pivalic acid (0.40 equivalents) is added to a mixture of 2-(2-(4-fluorophenyl)-2-oxo-1-phenylethyl)-4-methyl-3-oxo-*N*-phenylpentanamide (**21**) and *tert*-butyl 2-((4*R*,6*R*)-6-(2-aminoethyl)-2,2-dimethyl-1,3-dioxan-4-yl)acetate (**8**) (1.3 equivalents) in cyclohexane–methanol (20:1) at 25°C. The mixture is refluxed for 35 h with removal of water as the azeotrope using a Dean–Stark trap. The solution is cooled to 50–60°C, washed with brine, with aqueous sodium bicarbonate, and again with brine, and concentrated at reduced pressure. The residue is crystallized from ethanol–water to afford pyrrole **99** (74%, 99.4% pure by HPLC). The crystallized pyrrole **99** contains less amide impurity **102** (0.16%).[93]

In addition to generating amide impurities, another significant development issue is the very long reaction time, on the order of days, typically required for complete conversion. Perhaps the reaction proceeds to completion more quickly when an tertiary amine is added with pivalic acid. Pivalic acid (0.70 equivalents) is added to a mixture of 2-(2-(4-fluorophenyl)-2-oxo-1-phenylethyl)-4-methyl-3-oxo-*N*-phenylpentanamide (**21**) (1.1 equivalents) and *tert*-butyl 2-((4*R*,6*R*)-6-(2-aminoethyl)-2,2-dimethyl-1,3-dioxan-4-yl) acetate (**8**) in THF–hexane (2:1) at 50°C. Triethylamine (0.70 equivalents) is then added at 50°C and the mixture is refluxed for an unspecified time with removal of water as the azeotrope using a Dean–Stark trap. The mixture is cooled to

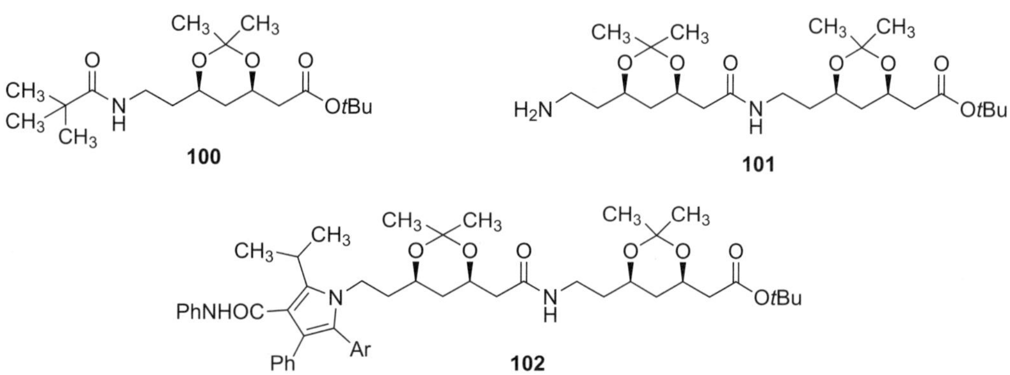


**FIGURE 9.4** Amide impurities produced during the Paal–Knorr pyrrole synthesis from 1,4-diketone **21** and *tert*-butyl 2-((4*R*,6*R*)-6-(2-aminoethyl)-2,2-dimethyl-1,3-dioxan-4-yl)acetate (**8**).

25°C and methyl *tert*-butyl ether is added. The solution is washed with 5% potassium hydroxide and with 3.5% hydrochloric acid and then concentrated at reduced pressure. The residue is dissolved in 1:1 isopropanol–acetone at 50°C. Water is added and the mixture is cooled to 0°C and aged for 1 h. The resulting suspension is filtered and the solid is washed with cold isopropanol and dried at <40°C to afford pyrrole **99**. No yield is provided. Toluene or methyl *tert*-butyl ether–THF can also be used as reaction solvent(s) and *N*-ethylmorpholine or *N,N*-diisopropylethylamine can be used as the amine base.[94]

While not specified in the procedures above, isolation from isopropanol, isopropanol–water, ethanol, acetonitrile, or diisopropyl ether likely affords pyrrole **99** polymorph I-US7186848. Isolation from cyclohexane affords pyrrole **99** polymorph II-US7186848. These are characterized by melting point (mp 140–142°C polymorph I-US7186848 and 128–130°C for polymorph II-US7186848) and XRPD.[95]

### 9.5.5 1,4-Diketone 21 with *tert*-Butyl 2-((7*R*,9*R*)-9-(2-Aminoethyl)-6,10-dioxaspiro[4.5]decan-7-yl)acetate (43)

Pivalic acid (1.4 equivalents) is added to a mixture of 2-(2-(4-fluorophenyl)-2-oxo-1-phenylethyl)-4-methyl-3-oxo-*N*-phenylpentanamide (**21**) (1.03 equivalents), *tert*-butyl 2-((7*R*,9*R*)-9-(2-aminoethyl)-6,10-dioxaspiro[4.5]decan-7-yl) acetate (**43**), and triethylamine (1.0 equivalent) in THF–methyl *tert*-butyl ether (1.7:1) at 50°C. The mixture is refluxed for 96 h with removal of water as the azeotrope using a Dean–Stark trap. The solution is cooled to 25°C and concentrated by distillation at atmospheric pressure. The residual paste is suspended in isopropanol at 60°C. The suspension is aged at 60°C for 30 min, aged at −5°C for 1 h, and filtered. The solid is presumably washed with isopropanol and dried to afford pyrrole **103** (62%).[65]

### 9.5.6 1,4-Diketone 21 with *tert*-Butyl 2-((2*R*,4*R*)-4-(2-Aminoethyl)-1,5-dioxaspiro[5.5]undecan-2-yl)acetate (45)

Pivalic acid (1.0 equivalent) is added to a mixture of 2-(2-(4-fluorophenyl)-2-oxo-1-phenylethyl)-4-methyl-3-oxo-*N*-phenylpentanamide (**21**) (1.04 equivalents) and *tert*-butyl 2-((2*R*,4*R*)-4-(2-aminoethyl)-1,5-dioxaspiro[5.5]undecan-2-yl)acetate (**45**) in methyl *tert*-butyl ether–THF (1.9:1) at 50°C. The mixture is refluxed for 96 h with removal of water as the azeotrope using a Dean–Stark trap. The solution is cooled to 25°C and concentrated by distillation at atmospheric pressure. The residual paste is suspended in isopropanol at 60°C. The suspension is aged at 60°C for 30 min, cooled to −5°C (*Note*: The procedure reads −50°C.) and aged for 1 h, and then filtered. The solid is presumably washed with isopropanol and dried to afford pyrrole **104** (73%).[65]

### 9.5.7 1,4-Diketone 21 with (3*R*,5*R*)-*tert*-Butyl 7-Amino-3,5-bis(tert-butyldimethylsilyloxy)heptanoate (47)

Pivalic acid (0.71 equivalents) is added to a mixture of 2-(2-(4-fluorophenyl)-2-oxo-1-phenylethyl)-4-methyl-3-oxo-N-phenylpentanamide (**21**) and (3R,5R)-tert-butyl 7-amino-3,5-bis(tert-butyldimethylsilyloxy)heptanoate (**47**) (2.9 equivalents) in toluene at 25°C. The mixture is refluxed for 17 h with removal of water as the azeotrope using a Dean–Stark trap. The mixture is presumably cooled to 25°C, washed with water, and carried directly into the next step. No contained yield for pyrrole **105** is provided (Scheme 9.25).[66]

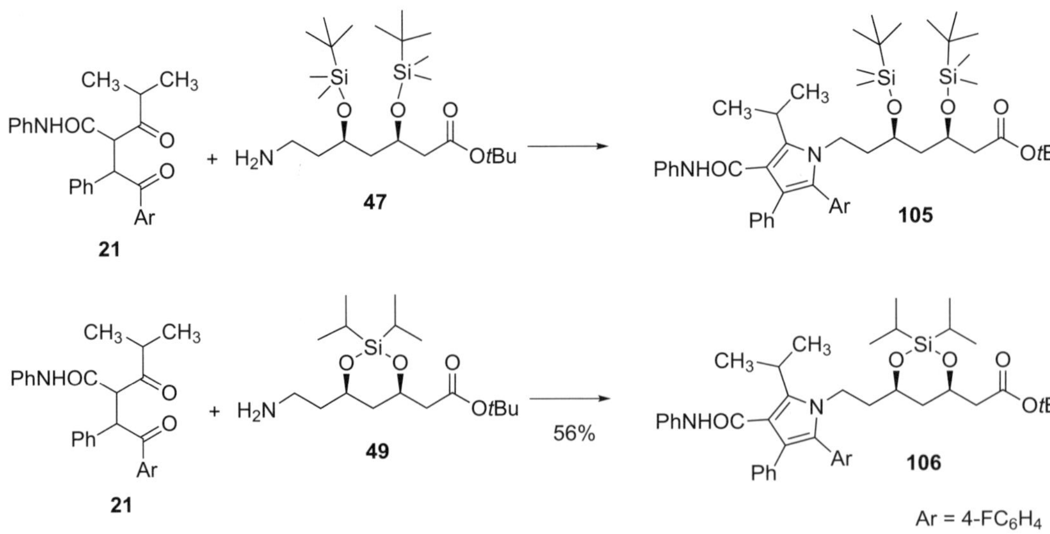


**SCHEME 9.25** Paal–Knorr syntheses of pyrroles suitable for conversion to atorvastatin (**1**) by silyl ether deprotection and ester hydrolysis.

### 9.5.8 1,4-Diketone 21 with *tert*-Butyl 2-((4*R*,6*R*)-6-(2-Aminoethyl)-2,2-diisopropyl-1,3,2-dioxasilinan-4-yl) acetate (49)

*tert*-Butyl 2-((4*R*,6*R*)-6-(2-aminoethyl)-2,2-diisopropyl-1,3,2-dioxasilinan-4-yl)acetate (**49**) is converted to the pivalate salt in heptane–toluene–THF (4:1:1) (23%). A mixture of 2-(2-(4-fluorophenyl)-2-oxo-1-phenylethyl)-4-methyl-3-oxo-*N*-phenylpentanamide (**21**) and the pivalate salt (1.4 equivalents) in heptane–THF–toluene (2:2:1) is refluxed for 25 h. The mixture is presumably cooled and then concentrated at reduced pressure. The residue is dissolved in chloroform and the solution is washed with aqueous sodium bicarbonate and concentrated at reduced pressure. Pyrrole **106** is isolated from the residue by chromatography (56%).[66]

### 9.5.9 1,4-Diketone 21 with *tert*-Butyl 2-((4*R*,6*R*)-6-(2-Aminoethyl)-2-phenyl-1,3,2-dioxaborinan-4-yl)acetate (51)

Acetic acid (equivalents not specified) is added to a mixture of 2-(2-(4-fluorophenyl)-2-oxo-1-phenylethyl)-4-methyl-3-oxo-*N*-phenylpentanamide (**21**) and *tert*-butyl 2-((4*R*,6*R*)-6-(2-aminoethyl)-2-phenyl-1,3,2-dioxaborinan-4-yl)acetate (**51**) (equivalents not specified) in xylenes at 25°C. The mixture is refluxed for 44 h. The mixture is cooled, diluted with diisopropyl ether and methanol, washed with dilute aqueous sodium hydroxide (methanol added) and with dilute hydrochloric acid, and concentrated at reduced pressure. No yield is available. The residue is carried directly into the next step (Scheme 9.26).[67,68]

### 9.5.10 1,4-Diketone 21 with *tert*-Butyl 2-(6-(2-Aminoethyl)-2-phenethyl-1,3,2-dioxaborinan-4-yl) acetate (53)

*p*-Toluenesulfonic acid monohydrate (catalytic amount) is added to a mixture of 2-(2-(4-fluorophenyl)-2-oxo-1-phenylethyl)-4-methyl-3-oxo-*N*-phenylpentanamide (**21**) and *tert*-butyl 2-(6-(2-aminoethyl)-2-phenethyl-1,3,2-dioxaborinan-4-yl)acetate (**53**) (1.03 equivalents) in toluene at 25°C. The mixture is refluxed for 30 h with removal of water as the azeotrope. The toluene volume is also gradually reduced during the 30 h reflux period. The resulting oil is separated between ethyl acetate and water. The organic layer is concentrated at reduced pressure. The residue is suspended in *n*-hexane and the suspension is filtered. The solid is presumably washed with *n*-hexane and dried to afford pyrrole **108** (85%).[69]

### 9.5.11 1,4-Diketone 21 with *tert*-Butyl 2-(6-(2-Aminoethyl)-2-propyl-1,3,2-dioxaborinan-4-yl)acetate (55)

*p*-Toluenesulfonic acid monohydrate (catalytic amount) is added to a mixture of 2-(2-(4-fluorophenyl)-2-oxo-1-phenylethyl)-4-methyl-3-oxo-*N*-phenylpentanamide (**21**) (1.1

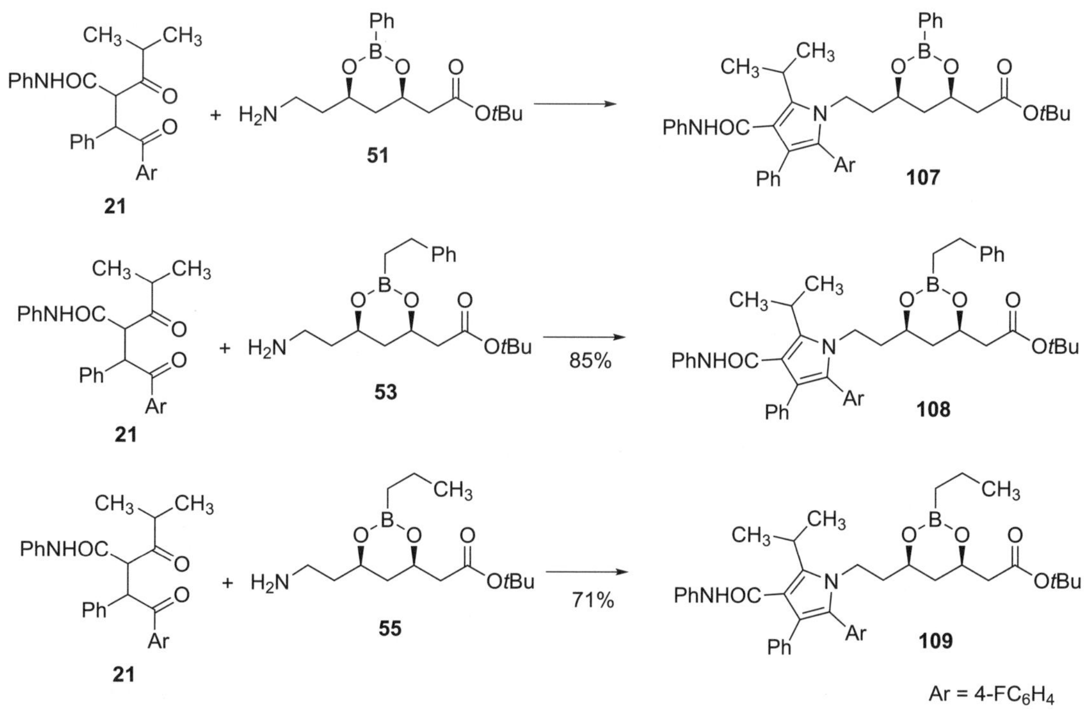


**SCHEME 9.26** Paal–Knorr syntheses of pyrroles suitable for conversion to atorvastatin (**1**) by hydrolysis of boronate and *tert*-butyl esters.

equivalents) and *tert*-butyl 2-(6-(2-aminoethyl)-2-propyl-1,3,2-dioxaborinan-4-yl)acetate (**55**) in toluene at 25°C. The mixture is refluxed for 30 h with removal of water as the azeotrope. The toluene volume is also gradually reduced during the 30 h reflux period. The resulting oil is separated between ethyl acetate and water. The organic layer is concentrated at reduced pressure. The residue is suspended in *n*-hexane and the suspension is filtered. The solid is presumably washed with *n*-hexane and dried to afford pyrrole **109** (71%).[69]

### 9.5.12 1,4-Diketone 21 with 2-((4*R*,6*R*)-6-(2-Aminoethyl)-2,2-dimethyl-1,3-dioxan-4-yl)-*N*,*N*-diphenylacetamide (57)

Pivalic acid (1.1 equivalents) is added to a mixture of 2-(2-(4-fluorophenyl)-2-oxo-1-phenylethyl)-4-methyl-3-oxo-*N*-phenylpentanamide (**21**) (1.2 equivalents) and 2-((4*R*,6*R*)-6-(2-aminoethyl)-2,2-dimethyl-1,3-dioxan-4-yl)-*N*,*N*-diphenylacetamide (**57**) in heptane–THF–toluene (2:1:1.2) at 25°C. The mixture is refluxed for 48 h. The mixture is cooled to 25°C, diluted with toluene, and washed with 0.5 M sodium hydroxide and with 0.5 M hydrochloric acid. The organic layer is concentrated at reduced pressure to afford pyrrole **110**. No yield is provided. The dibenzylamide, diethylamide, *n*-butylmethylamide, benzyl-*tert*-butylamide, and piperidine amides are also prepared (Scheme 9.27).[71]

### 9.5.13 1,4-Diketone 21 with 2-((4*R*,6*R*)-6-(2,2-Dimethoxyethyl)-2,2-dimethyl-1,3-dioxan-4-yl) ethanamine (61)

Pivalic acid (1.0 equivalent) is added to a equimolar mixture of 2-(2-(4-fluorophenyl)-2-oxo-1-phenylethyl)-4-methyl-3-oxo-*N*-phenylpentanamide (**21**) and 2-((4*R*,6*R*)-6-(2,2-dimethoxyethyl)-2,2-dimethyl-1,3-dioxan-4-yl)ethanamine (**61**) in heptane–THF–toluene (2:1:1.2), presumably at 25°C. The mixture is refluxed for 40 h. The mixture is then cooled and concentrated at reduced pressure. Pyrrole **111** is isolated from the residue by chromatography (42%).[74]

### 9.5.14 1,4-Diketone 21 with 2-((4*S*,6*R*)-6-(2-Aminoethyl)-2,2-dimethyl-1,3-dioxan-4-yl)ethanol (71)

2-((4*S*,6*R*)-6-(2-Aminoethyl)-2,2-dimethyl-1,3-dioxan-4-yl)ethanol (**71**) is converted to the pivalate salt in methyl *tert*-butyl ether (9%). A mixture of the pivalate salt (1.2 equivalents) and 2-(2-(4-fluorophenyl)-2-oxo-1-phenylethyl)-4-methyl-3-oxo-*N*-phenylpentanamide (**21**) in toluene is refluxed for 27 h. The mixture is presumably cooled to 25°C and then washed with water. Ethyl acetate and brine are added to facilitate layer separation. The layers are separated and the organic layer is dried and concentrated at reduced pressure to afford pyrrole **112** (80%).[66]

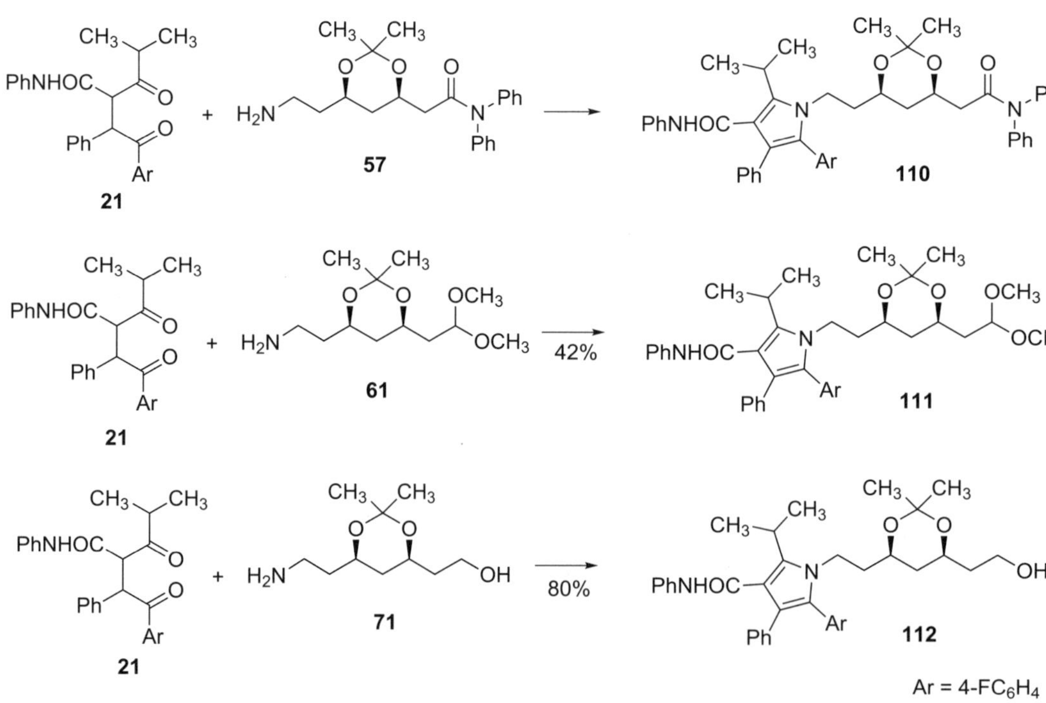


**SCHEME 9.27** Paal–Knorr syntheses of other pyrroles suitable for conversion to atorvastatin (**1**) by ketal deprotection.

### 9.5.15 1,4-Diketone 21 with 2-((2*R*,4*R*,6*R*)-4-(tert-Butyldiphenylsilyloxy)-6-methoxytetrahydro-2*H*-pyran-2-yl)ethanamine (72)

Pivalic acid (0.82 equivalents) is added to a mixture of 2-(2-(4-fluorophenyl)-2-oxo-1-phenylethyl)-4-methyl-3-oxo-*N*-phenylpentanamide (**21**) and 2-((2*R*,4*R*,6*R*)-4-(*tert*-Butyldiphenylsilyloxy)-6-methoxytetrahydro-2*H*-pyran-2-yl)ethanamine (**72**) (1.3 equivalents) in heptane–THF–toluene (2:1:1.2) at 25°C. The mixture is refluxed for 30 h. The mixture is cooled and diluted with ethyl acetate. The solution is washed with aqueous sodium bicarbonate and with brine, and then dried and concentrated at reduced pressure. Pyrrole **113** is isolated from the residue by chromatography (72%) (Scheme 9.28).[76]

### 9.5.16 1,4-Diketone 18 with 2-((2*R*,4*R*)-4-(Benzyloxy)-6-methoxytetrahydro-2*H*-pyran-2-yl)ethanamine (81)

Acetic acid (13 equivalents) is added to a mixture of ethyl 2-(2-(4-fluorophenyl)-2-oxo-1-phenylethyl)-4-methyl-3-oxopentanoate (**18**) and 2-((2*R*,4*R*)-4-(benzyloxy)-6-methoxytetrahydro-2*H*-pyran-2-yl)ethanamine (**81**) (1.3 equivalents) in THF at 25°C. The mixture is refluxed for 48 h. The mixture is concentrated at reduced pressure and the residue is separated between water and ethyl ether. The organic layer is dried and concentrated at reduced pressure. Pyrrole **114** is isolated from the residue by chromatography (21%).[80]

### 9.5.17 1,4-Diketone 21 with 2-((3a*R*,5*R*,6a*R*)-2,2-Dimethyl(tetrahydrofuro[2,3-*d*][1,3]-dioxol-5-yl) ethanamine (86)

Pivalic acid (1.3 equivalents) is added to a mixture of 2-(2-(4-fluorophenyl)-2-oxo-1-phenylethyl)-4-methyl-3-oxo-*N*-phenylpentanamide (**21**) and 2-((3a*R*,5*R*,6a*R*)-2,2-dimethyl (tetrahydrofuro[2,3-*d*][1,3]-dioxol-5-yl)ethanamine (**86**) (1.3 equivalents) in heptane–toluene (4:1) at 25°C. The mixture is refluxed for 24–30 h with removal of water as the azeotrope using a Dean–Stark trap. A routine water–ethyl acetate workup affords crude pyrrole **115**. The weight yield quoted exceeds the theoretical yield.[82]

## 9.6 PYRROLE 117 FROM 1,4-DIKETONE 21 AND ETHYL 2-((4*R*,6*R*)-6-(2-AZIDOETHYL)-2,2-DIMETHYL-1,3-DIOXAN-4-YL)ACETATE (116)

An alternative methodology is used to prepare ethyl 2-((4*R*,6*R*)-6-(2-azidoethyl)-2,2-dimethyl-1,3-dioxan-4-yl)acetate (**116**) in eight steps from diethyl 3-hydroxyglutarate. The route to azide **116** will be discussed later.

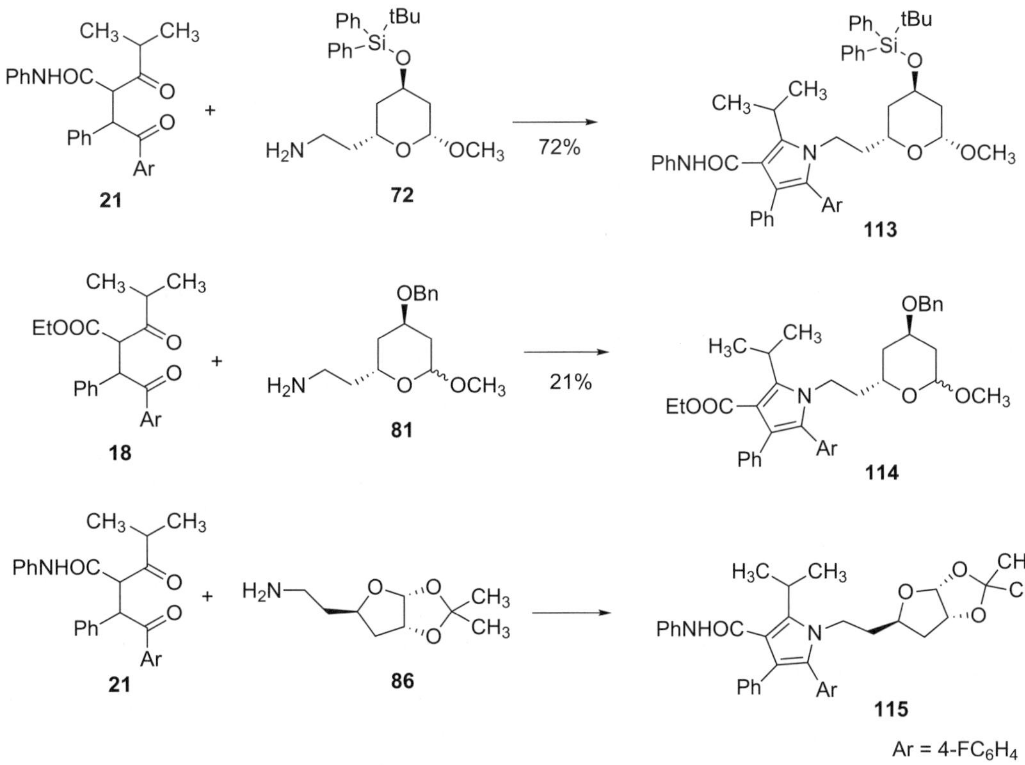


**SCHEME 9.28** Paal–Knorr synthesis of pyrroles suitable for conversion to atorvastatin (**1**) by diol and aldehyde deprotection and oxidation.

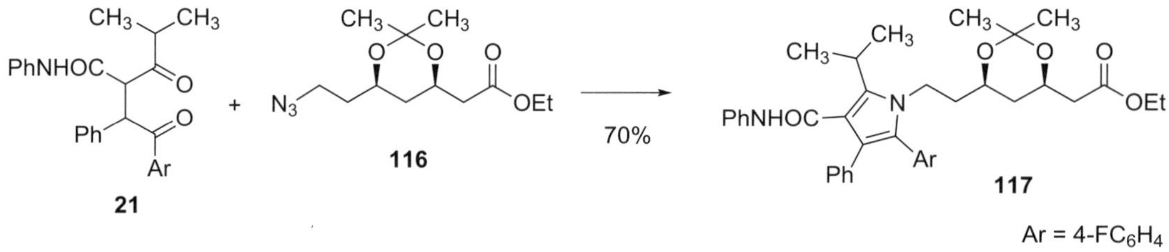


**SCHEME 9.29** Pyrrole **117** from 1,4-diketone **21** and ethyl 2-((4*R*,6*R*)-6-(2-azidoethyl)-2,2-dimethyl-1,3-dioxan-4-yl)acetate (**116**).

Tributylphosphine (1.3 equivalents) is added to a solution of ethyl 2-((4*R*,6*R*)-6-(2-azidoethyl)-2,2-dimethyl-1,3-dioxan-4-yl)acetate (**116**) (1.3 equivalents) in toluene at 25°C. After aging until nitrogen gas evolution stops, the solution is added to a mixture of 2-(2-(4-fluorophenyl)-2-oxo-1-phenylethyl)-4-methyl-3-oxo-*N*-phenylpentanamide (**21**), 2,4,6-trimethylbenzoic acid (1.3 equivalents), and molecular sieves in toluene at 60°C. At complete conversion, the mixture is decated washed with 1 M sodium hydroxide, with 1 M hydrochloric acid, and with brine and presumably concentrated at reduced pressure. Pyrrole **117** is isolated from the residue by chromatography (70%) (Scheme 9.29).[27]

## 9.7 SIDE CHAIN ELABORATION AFTER PYRROLE RING CONSTRUCTION

### 9.7.1 Elaboration of the Side Chain Aldehyde

The side chain is extended from the aldehyde by adding two acetate esters sequentially. This sequential addition strategy has the obvious disadvantage of a longer linear sequence but also offers an opportunity to use a chiral auxiliary to direct the initial acetate addition to the aldehyde. Neither step is efficient with respect to acetate and base consumption. The pyrrole amide substituent protonates/consumes some of the chiral acetate anion, increasing the probability that the chiral acetate will self-condense. The $\beta$-hydroxyester product of the first acetate addition also protonates/consumes some of the *tert*-butyl acetate anion of the second acetate addition and increases the probability that *tert*-butyl acetate will self-condense. A now-familiar asymmetric reduction (−78°C, $Et_2$-$BOCH_3$, $NaBH_4$) is used to create the second chiral center.

***9.7.1.1 5-(4-Fluorophenyl)-2-isopropyl-1-(3-oxopropyl)-N,4-diphenyl-1H-pyrrole-3-carboxamide (118)*** Dilute hydrochloric acid is added to the diethylacetal **95** in acetone and the mixture is aged at 40°C for 24 h. The resulting suspension is filtered and aldehyde **118** is washed, presumably with acetone–water, and dried (93%) (Scheme 9.30).[96]

A catalytic amount of hydrochloric acid is added to the dioxolane **12** in ethanol and the solution is refluxed for 24 h. The solution is cooled and concentrated at reduced pressure. The residue is taken up in acetone–water, a catalytic amount of *p*-toluenesulfonic acid is added, and the mixture is refluxed for 48 h. A routine water–ethyl ether workup followed by crystallization from diisopropyl ether affords aldehyde **118** (87%).[15]

***9.7.1.2 The (R)-3-Hydroxypentanoate 120*** A solution of lithium diisopropylamide (2.1 equivalents) in THF–hexane is added to a suspension of (*S*)-(−)-1,1,2-triphenyl-1,2-ethanediol 2-acetate[97] in THF at −70°C. The mixture is warmed to −10°C over 2 h. In another flask, a suspension of magnesium bromide (2.1 equivalents) in THF is prepared. The ester enolate solution is added at −78°C to the magnesium bromide suspension and the mixture is aged at −78°C for 1 h. A solution of 5-(4-fluorophenyl)-2-isopropyl-1-(3-oxopropyl)-*N*,4-diphenyl-1*H*-pyrrole-3-carboxamide (**118**) (1.1 equivalents) in THF is added at −78°C. Acetic acid is then added at −78°C. The mixture is warmed, water is added, and the volatiles are removed by concentration at 40–50°C and reduced pressure. The residual slurry is suspended in ethyl acetate–heptane and the suspension is filtered. The solid is washed with dilute hydrochloric acid, with water, and with cold ethyl acetate–heptane and then dried at 40°C and reduced pressure. The solid is crystallized from ethyl acetate to afford the (*R*)-3-hydroxypentanoate **120** (43%, 97.4% *R*).[98]

The ratio of the *R*- and *S*-adducts produced in the reaction is available from a similar process. A solution of lithium diisopropylamide (2.2 equivalents) in THF–hexane is added to a suspension of (*S*)-(−)-1,1,2-triphenyl-1,2-ethanediol 2-acetate (1.1 equivalents) in THF at 0 to −10°C. The mixture is aged at 0°C for 30 min and then cooled to −78°C. A solution of 5-(4-fluorophenyl)-2-isopropyl-1-(3-oxopropyl)-*N*,4-diphenyl-1*H*-pyrrole-3-carboxamide (**118**) in THF is then added at −78°C and the mixture is aged at −78°C for 30 min. Acetic acid is then added at an unspecified temperature, presumably −78°C. The mixture is presumably warmed, water is added, and the layers are separated. The organic layer is washed, presumably with water, and then dried and concentrated at reduced pressure. The weight yield of the residue exceeds the theoretical yield. Analysis of the residue by HPLC reveals the *R*–*S* ratio at the new chiral center is 86:14. An adduct mixture with a better *R*–*S* ratio (95:5) at that center is produced when the enolate addition is run in 1,2-dimethoxyethane–THF–hexane (~5:3:1).[96]

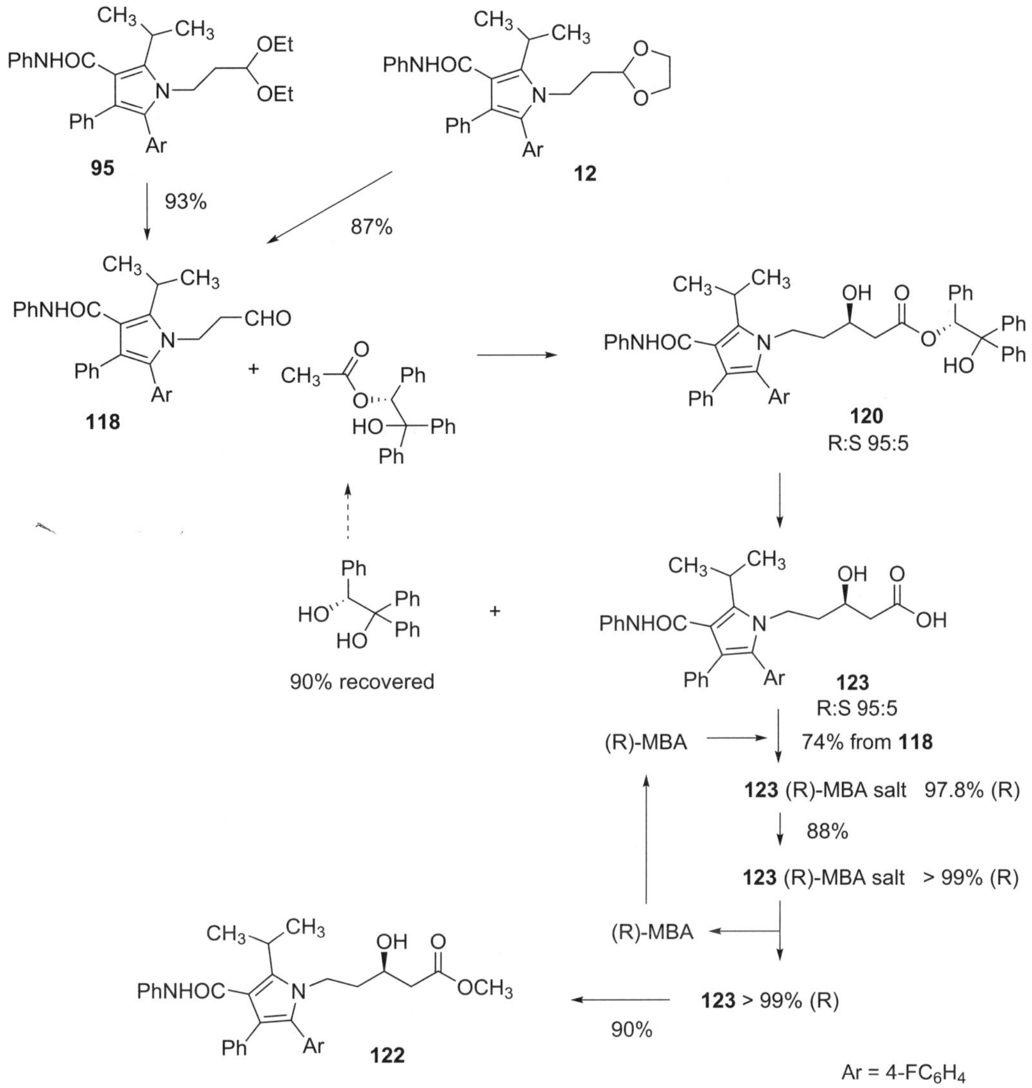


**SCHEME 9.30** The (*R*)-methyl 3-hydroxypentanoate **122** from the diethylacetal **95**.

Transesterification with methanol or ester hydrolysis releases the chiral auxiliary. The auxiliary is recovered by chromatography from the transesterification. In the hydrolysis, the chiral auxiliary precipitates and is recovered by a simple filtration. The transesterification affords the methyl ester that is then used directly in the condensation step. The hydrolysis affords the acid that must be converted to the *N*-acylimidazole or methyl ester before proceeding to the condensation step.

The chiral auxiliary is cleaved by reaction of the ester of **120** with sodium methoxide (1.1 equivalents) in THF–methanol at 0°C. After aging at −20°C overnight, the mixture is warmed, quenched with acetic acid, and concentrated at 40°C and reduced pressure. After a routine water–ethyl acetate workup, the methyl (*R*)-3-hydroxypentanoate **122** is separated from the chiral auxiliary by chromatography and then crystallized from ethyl ether–heptane (75%).[15,98]

The expensive chiral auxiliary can be recovered without chromatography when the ester of **120** is hydrolyzed to the acid with potassium hydroxide in methanol–water at 25°C. When the hydrolysis is complete, water is added and the suspension is filtered. The solid is washed with methanol–water and dried at 45–50°C and reduced pressure to recover (*S*)-(−)-1,1,2-triphenyl-1,2-ethanediol (92%). The liquors are concentrated at reduced pressure and hydrochloric acid is added to the residual aqueous solution (to pH 2–2.5). The resulting suspension is filtered. The solid is washed with water and dried at 45–50°C and reduced pressure to afford the (*R*)-3-hydroxypentanoic acid **123** (91%).[99]

The *R–S* 95:5 reaction mixture in 1,2-dimethoxyethane–THF–hexane can be hydrolyzed directly by adding methanol and water and refluxing the mixture for 3 h. The organic solvents are then removed by concentration at reduced pressure. After cooling to 25°C, the suspension is filtered and the solid is washed with methanol–water and dried to recover (*S*)-(−)-1,1,2-triphenyl-1,2-ethanediol (90%).

The (*R*)-3-hydroxypentanoic acid **123** in the liquors is isolated as the (*R*)-(+)-methylbenzylamine (MBA) salt. Methanol is recovered from the liquors by concentration at reduced pressure. Ethyl acetate is added to the remaining aqueous solution, hydrochloric acid is added (presumably to pH < 2–2.5), and the layers are separated. The aqueous layer is extracted with ethyl acetate. The organic layers are washed with brine and concentrated at reduced pressure. The residue is dissolved in methyl *tert*-butyl ether and methanol and (*R*)-(+)-methylbenzylamine (1.2 equivalents) is added. The mixture is refluxed for 6 h and then cooled to 25°C. The suspension is filtered and the solid is presumably washed and dried to afford the (*R*)-3-hydroxypentanoic acid **123** as the MBA salt (74% from aldehyde **118**, 97.8% *R*). The MBA salt can be further upgraded (>99% *R*) using methyl *tert*-butyl ether–methanol (88% recovery).[96]

***9.7.1.3 The (R)-tert-Butyl 5-Hydroxy-3-oxoheptanoate 121*** The (*R*)-3-hydroxypentanoic acid **123** is converted to the *N*-acylimidazole in THF and the second acetate is delivered as magnesium mono *tert*-butyl malonate[100] (5.5 equivalents). A routine water–ethyl acetate workup affords the (*R*)-*tert*-butyl 5-hydroxy-3-oxoheptanoate **121** (42%) (Scheme 9.31).[99]

The *N*-acyl imidazole would certainly not be the first choice for an acylating agent in a low-cost manufacturing process. The acid or, better, the acid MBA salt can be converted to the (*R*)-methyl 3-hydroxypentanoate **122** with hydrogen chloride in methanol (90% from acid **123**). (*Note*: From the acid MBA salt, the weight yield offered exceeds the theoretical yield.) The alcohol is protected as the trimethylsilyl ether with chlorotrimethylsilane (0.97 equivalents) and imidazole (1.1 equivalents). A routine workup affords a solution of the silyl ether **124** in toluene. An enolate solution is prepared by adding *tert*-butyl acetate (3.1 equivalents) to a solution of lithium diisopropylamide (3.1 equivalents) in THF–hexane–ethylbenzene at −35°C. The ester solution is then added at −35 to −40°C and the mixture is aged at −35°C for 45 min. Acetic acid is added, presumably at −35°C, the mixture is presumably warmed to 0°C, and water is added. After aging the mixture at 25°C for 8 h, aqueous sodium hydroxide is added (to pH 8) and the layers are separated. The aqueous layer is extracted with toluene. The organic layers are washed with brine, concentrated at reduced pressure to a small volume, and then diluted with heptane. The resulting suspension is aged at 25°C for 3 h and filtered. The solid is washed with toluene–heptane and dried to afford the (*R*)-*tert*-butyl 5-hydroxy-3-oxoheptanoate **121** (61% from the (*R*)-3-hydroxypentanoic acid **123**).[96,99]

A solution of *tert*-butyl acetate (3.0 equivalents) in THF is added to lithium diisopropylamide (2.9 equivalents) in THF–hexane at −40°C. The solution is aged at −40°C for 30 min and then more butyllithium (2.0 equivalents) is added at −40°C. A solution of the methyl (*R*)-3-hydroxypentanoate **122** in THF is added at −40°C and the mixture is aged at −70°C for 4 h. The mixture is quenched with acetic acid, warmed to 25°C, and concentrated at reduced pressure. The residue is suspended in ethyl acetate and washed with water,

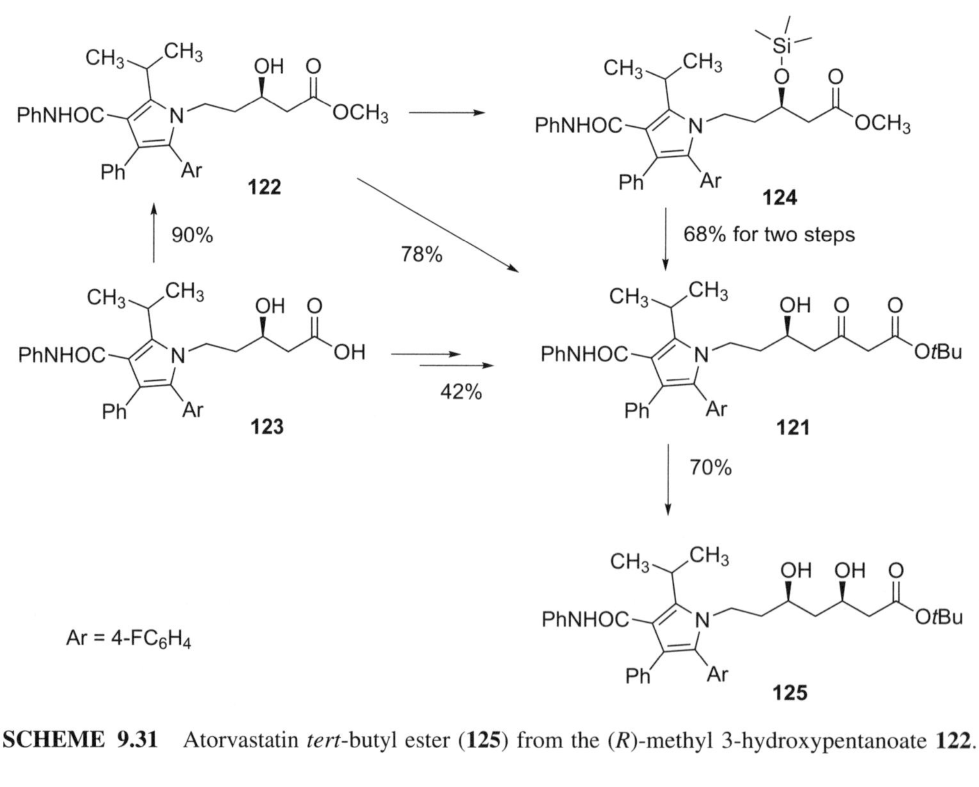


**SCHEME 9.31** Atorvastatin *tert*-butyl ester (**125**) from the (*R*)-methyl 3-hydroxypentanoate **122**.

saturated aqueous ammonium chloride, aqueous sodium bicarbonate, and brine. The organic layer is dried and concentrated at reduced pressure to afford the (*R*)-*tert*-butyl 5-hydroxy-3-oxoheptanoate (**121**) (78%). (*Note*: The two descriptions have different accounts of the workup procedure past this point.)[15,98]

***9.7.1.4 Atorvastatin tert-Butyl Ester (125)*** Triethylborane (1.0 equivalent) is added to a solution of the (*R*)-*tert*-butyl 5-hydroxy-3-oxoheptanoate **121** in THF, presumably at 25°C. Pivalic acid (4.9 mol%) is then added and the mixture is aged at 25°C for 10 min, cooled to −78°C, and diluted with methanol. Sodium borohydride (1.0 equivalent) is added in portions at −78°C and the mixture is aged at −78°C for 6 h. The cold mixture is transferred into cold aqueous hydrogen peroxide and then allowed to warm to 25°C over 16 h. The mixture is extracted with chloroform and the extracts are washed with water, dried, and concentrated at reduced pressure. Atorvastatin *tert*-butyl ester (**125**) is isolated from the residue by chromatography (70%).[15,98]

***9.7.1.5 Alternative Approaches to the (R)-3-Hydroxypentanoic Acid*** There are two other approaches to the (*R*)-3-hydroxypentanoic acid **123**: by resolution of the racemic acid with (*R*)-(+)-MBA or by stereoselective reduction of a chiral acetoacetate with sodium borohydride.

*Resolution of the Racemic 3-Hydroxypentanoic Acid* Methyl acetate (1.9 equivalents) is added to lithium diisopropylamide (2.0 equivalents) in heptane–THF–ethylbenzene at −78°C. After aging the solution at −78°C for 1 h, a solution of aldehyde **118** in THF–dichloromethane is added at −78°C and the mixture is aged at −78°C for 1 h. The mixture is then warmed to 25°C and quenched with methanol. Water and sodium hydroxide are added and the mixture is refluxed for 2 h. The mixture is cooled to 25°C, 2 M hydrochloric acid is added (to pH 2–2.5), and the layers are separated. The aqueous layer is extracted with ethyl acetate. The combined organic layers are washed with brine, dried, and concentrated at reduced pressure. The residue is suspended in toluene, the suspension is filtered, and the solid is washed with heptane and dried to afford the racemic 3-hydroxypentanoic acid **123** (84%).

In the resolution step, (*R*)-(+)-MBA (1.0 equivalent) is added to a solution of racemic 3-hydroxypentanoic acid **123** in 2:1 methyl *tert*-butyl ether–methanol. After 10 h at reflux, the mixture is cooled to 25°C and the suspension is filtered. The solid is washed with methyl *tert*-butyl ether–methanol and dried to afford (*R*)-3-hydroxypentanoic acid **123** as the (*R*)-(+)-MBA salt (28%, 99% *R*) (Scheme 9.32).[96]

*Reduction of a Chiral Acetoacetate* Chiral 3-oxopentanoate **126** might be prepared by condensation of a pyrrole with an ester or amide on the side chain with (*R*)-(+)-1,1,2-triphenyl-1,2-ethanediol 2-acetate (see elaboration of the morpholino amide below). Sodium borohydride (4.2 equivalents) is added to a solution of

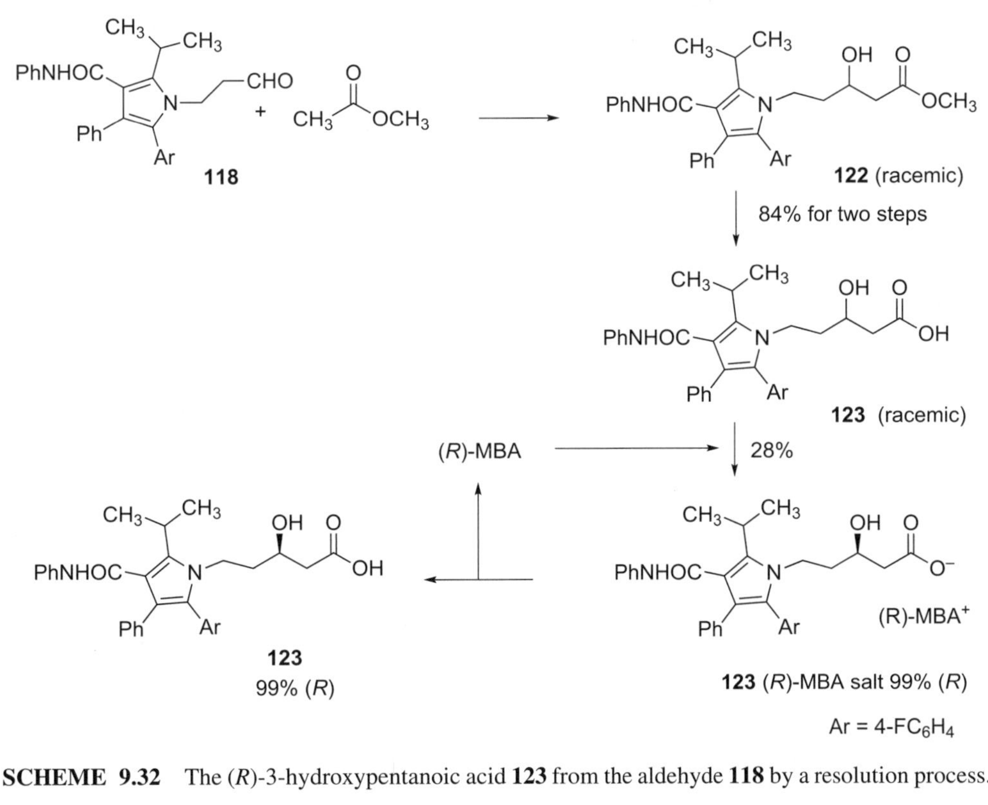


**SCHEME 9.32** The (*R*)-3-hydroxypentanoic acid **123** from the aldehyde **118** by a resolution process.

chiral acetoacetate **126** in THF–methanol at −78°C. The mixture is aged at −78°C for 1 h, quenched with saturated aqueous ammonium chloride, presumably warmed to 25°C, and extracted with dichloromethane. The organic extracts are washed with brine, dried, and concentrated at reduced pressure to afford a 3:1 mixture of the 3(*R*)- and 3(*S*)-3-hydroxypentanoates **120** (94%) (Scheme 9.33).[99]

### 9.7.2 Elaboration of Morpholino Amide 98

The side chain is extended from the morpholino amide using *tert*-butyl acetoacetate (Scheme 9.34). The condensation is inefficient with respect to acetoacetate and base consumption since the pyrrole amide substituent protonates/consumes some of the acetoacetate dianion. A ruthenium-catalyzed hydrogenation of the product *tert*-butyl 3,5-dioxoheptanoate delivers the 3,5-dihydroxyheptanoate with the correct stereochemistry at the 5-position. The stereochemistry at the 3-position is then set using an elimination–addition strategy.

*tert*-Butyl acetoacetate[101] (4.0 equivalents) is converted to the sodium salt. The salt is dissolved in toluene–THF and the solution is cooled to −10°C. Butyllithium (4.2 equivalents) in hexanes is added at <1°C and the dienolate solution is aged at 0 to −6°C for 20–30 min. The dienolate solution is then added to a suspension of amide **98** in THF at −5°C and the suspension is aged at −5°C for 2 h. Water is carefully added at 0°C followed by 12 M hydrochloric acid, also at 0°C. The biphasic mixture is concentrated at reduced pressure to reduce the volume of the organic layer by >50%. The layers are separated, water is added to the organic layer, and the concentration at reduced pressure is continued until most of the organic solvent is removed. Isopropanol and water are then added and the resulting suspension is aged at 25°C for 6 h and then filtered. The solid is washed with isopropanol–water and dried at 35°C and reduced pressure to afford the *tert*-butyl 3,5-dioxoheptanoate **127** (92%). The ethyl, methyl, and isopropyl esters and the dimethyl and morpholino amides are prepared by analogous processes.

The *tert*-butyl 3,5-dioxoheptanoate **127** is reduced using a ruthenium-MeO-BIPHEP catalyst in toluene and acetic acid at 65°C and 50 psi hydrogen pressure. The *tert*-butyl 3,5-dihydroxyheptanoate **128** in solution is likely to be >98.5% (*R*) at the 5-position. The preparative details, yield, and optical purity data are not provided.[85]

The *tert*-butyl 3,5-dioxoheptanoate **127** is reduced with formic acid (1.9 equivalents) using a ruthenium catalyst with chiral diamine ligand in toluene–triethylamine at 25°C. After 24 h, HPLC analysis of the mixture reveals a 6:1 ratio of the *syn*- and *anti*-diols.[102]

The lack of stereocontrol in the ketone reduction at the 3-position necessitates one or more additional steps to set this center. In the single-step conversion, a mixture of the *tert*-butyl 3,5-dihydroxyheptanoate **128**, benzaldehyde dimethyl acetal (4.4 equivalents), and *p*-toluenesulfonic acid monohydrate (10 mol%) in toluene is aged at 25°C at reduced pressure for 20 h. The solution is cooled to −5°C and a 1.0 M solution of potassium *tert*-butoxide in THF is added in portions. The solution is aged at 0°C for 12–14 h. Hydrochloric acid (1 M) is added, the mixture is allowed to warm to 15°C, and the layers are separated. The organic layer is washed with brine, dried, and concentrated at reduced pressure to afford atorvastatin *tert*-butyl ester benzaldehyde acetal (**129**). The yield and optical purity data are not provided.[85]

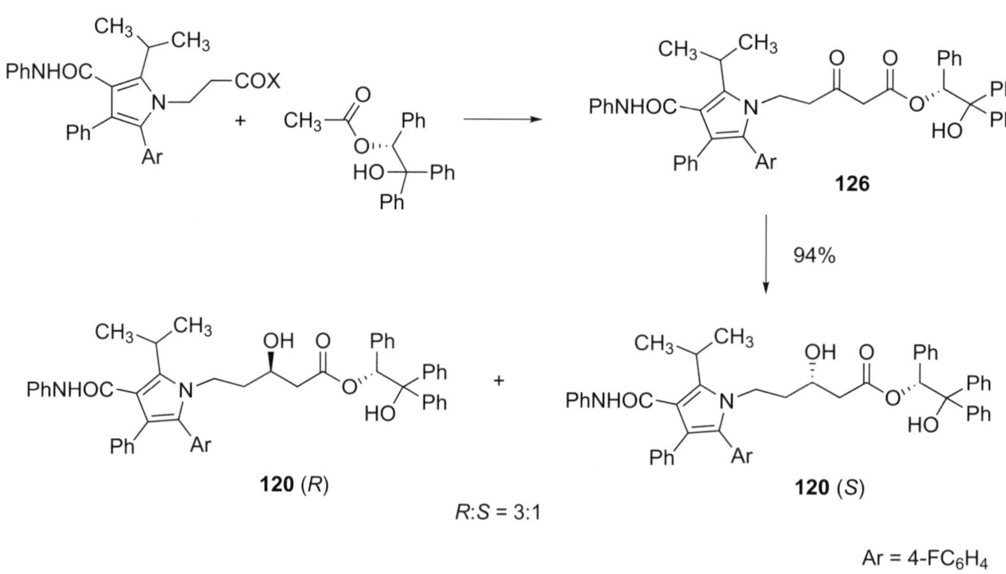


**SCHEME 9.33** Borohydride reduction of chiral acetoacetate **126**.

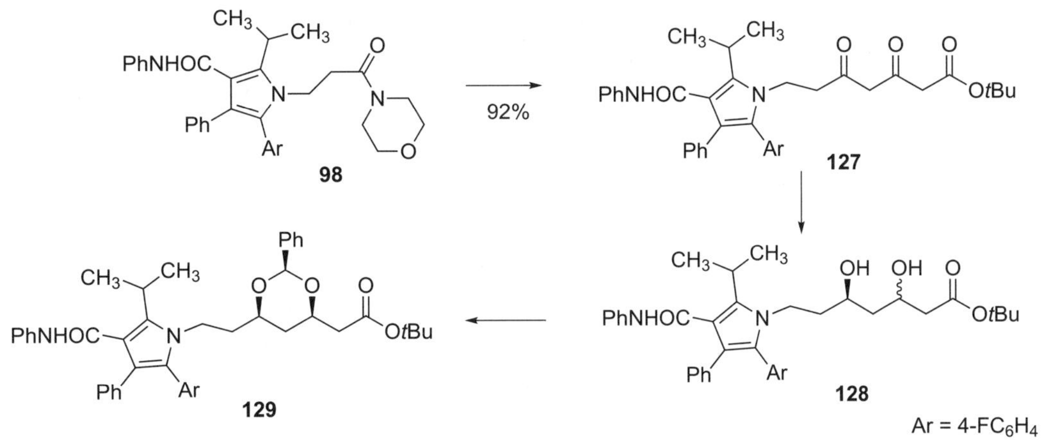

**SCHEME 9.34** Atorvastatin *tert*-butyl ester benzaldehyde acetal (**129**) from the morpholino amide **98**.

### 9.7.3 Elaboration of Ethyl 5-(4-Fluorophenyl)-2-isopropyl-4-phenyl-1*H*-pyrrole-3-carboxylate 130

Butyllithium (1.1 equivalents) is added to a solution of ethyl 5-(4-fluorophenyl)-2-isopropyl-4-phenyl-1*H*-pyrrole-3-carboxylate (**130**) in THF at −78°C. The solution is aged at −78°C for 1 h and then warmed to 25°C. The *p*-toluenesulfonate **69** (1.2 mmol) is added and the mixture is refluxed for 24 h. The mixture is cooled to 25°C and concentrated at reduced pressure. The pyrrole **131** is isolated from the residue by chromatography (58%). Details of the preparation of ethyl 5-(4-fluorophenyl)-2-isopropyl-4-phenyl-1*H*-pyrrole-3-carboxylate (**130**) are not provided (Scheme 9.35).[74]

## 9.8 ATORVASTATIN (1) CALCIUM

### 9.8.1 From Atorvastatin *tert*-Butyl Ester Acetone Ketal (99)

The syntheses of atorvastatin presented deliver the diol protected as a ketal, acetal, silyl ether, or boronate ester and the acid protected as an ester, amide, aldehyde acetal, or alcohol. The ketals are cleaved with aqueous acid. The *tert*-butyl ester or amide is then hydrolyzed in a separate step with aqueous hydroxide. The silyl ethers and the boronate esters and their *tert*-butyl esters can be cleaved in a single step with aqueous hydroxide.

Acid catalyzed cleavage of atorvastatin *tert*-butyl ester acetone ketal (**99**) with 1–2 M hydrochloric acid (0.2–2.0 equivalents) in methanol at 25°C produces a mixture containing atorvastatin *tert*-butyl ester (**125**) (72%), atorvastatin methyl ester (**132**) (21%), atorvastatin lactone (**133**) (3%), and atorvastatin (**1**) (1%).[103] Other water-miscible organic solvents (*i*-PrOH, EtOH, $CH_3CN$, THF, 1,4-dioxane) can be used in place of methanol. Aqueous acetic acid can be used in place of hydrochloric acid–methanol.

Since release of the diol facilitates the transesterification and hydrolysis of the ester, atorvastatin *tert*-butyl ester (**125**) is typically carried directly into the ester hydrolysis. The hydrolysis using sodium hydroxide is run in an organic–aqueous solvent mixture. When hydrolysis is nearly complete, the remaining atorvastatin ester(s) are removed by washing with an organic solvent. Acidification of the aqueous layer then affords atorvastatin. Atorvastatin can be converted to atorvastatin lactone (**133**) by extraction into toluene followed by distillation of the toluene–water azeotrope.

Hydrolysis of the lactone (**133**) with an exact charge of sodium hydroxide affords an aqueous solution of atorvastatin sodium. Addition of a solution of a calcium salt ($Ca(OH)_2$, $CaCl_2$, $Ca(OAc)_2$, Ca(2-ethyl hexanoate)$_2$, Ca (gluconate)$_2$) in water produces a suspension of atorvastatin calcium. The exact hydroxide charge is important to minimize formation and precipitation of calcium hydroxide that may result in a suspension that is difficult to filter and dry.

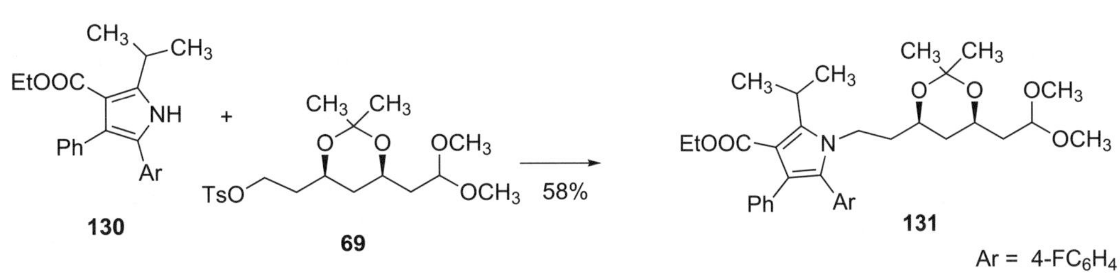

**SCHEME 9.35** Pyrrole **131** from ethyl 5-(4-fluorophenyl)-2-isopropyl-4-phenyl-1*H*-pyrrole-3-carboxylate (**130**).

Lactone hydrolysis with an exact hydroxide charge is likely to leave some atorvastatin lactone (**133**) unreacted. The aqueous solution is saturated with an organic solvent to keep the lactone and other trace organic–soluble impurities in solution during the calcium salt precipitation.

An alternative strategy avoids atorvastatin lactone (**133**) and the exact charge of hydroxide. Atorvastatin produced by acidification of the aqueous layer is extracted into an organic solvent and then precipitated as a salt with ammonia or a volatile amine. The salt is then dissolved in water and converted to atorvastatin calcium by adding a solution of a calcium salt in water.

Atorvastatin calcium is directly isolated from the deprotection, hydrolysis, and salt exchange sequence as a crude solid (morphology not specified), as an amorphous solid, or as a crystalline solid (polymorph I-US5969156, trihydrate with 4.5 wt% $H_2O$). When an organic solution of atorvastatin calcium is dried (with $MgSO_4$ or $CaSO_4$ or by water–organic azeotrope) and then added to a nonpolar solvent, the amorphous atorvastatin calcium isolated is assumed to be anhydrous. When atorvastatin calcium is precipitated from aqueous–organic solvent mixture, it is assumed to be the trihydrate. This is not often confirmed in the procedure descriptions. Volume throughputs are provided for each atorvastatin calcium isolation step. The conversion of isolated atorvastatin calcium to amorphous atorvastatin calcium is discussed in Section 9.9. Since much of the global process research and development effort is focused on producing amorphous atorvastatin, the conversion of isolated atorvastatin calcium to (many) other specific polymorphs will not be included in this chapter.

Atorvastatin *tert*-butyl ester acetone ketal (**99**) is a pivotal intermediate. If atorvastatin *tert*-butyl ester acetone ketal meets exacting specifications, the diol deprotection, ester hydrolysis, and sodium–calcium salt exchange can proceed with no intermediate isolation directly to atorvastatin calcium. Since a more complex mixture is carried into the salt exchange step, avoiding calcium hydroxide coprecipitation now becomes more difficult to orchestrate. Calcium hydroxide can be filtered from a hot aqueous or aqueous–alcohol solution prior to cooling and precipitation of atorvastatin calcium. Alternatively, atorvastatin calcium can be extracted into ethyl acetate. The extract is filtered to remove any suspended calcium hydroxide and added to a nonsolvent (hexane).

Finally, while reviewing all the deprotection, hydrolysis, and salt exchange options, keep in mind that the core of atorvastatin calcium is a pyrrole and that pyrroles are prone to air oxidation.

***9.8.1.1 Atorvastatin tert-Butyl Ester (125)*** Hydrochloric acid (1 M) (1.0 equivalent) is added to a solution of atorvastatin *tert*-butyl ester acetone ketal (**99**) in methanol at 25°C. The mixture is heated at 35°C for 3 h and then cooled to 25°C. The resulting suspension is filtered and the solid is washed with water and dried to afford atorvastatin *tert*-butyl ester **125** (74%) (Scheme 9.36).[104]

Hydrochloric acid (1 M) (1.4 equivalents) is added to atorvastatin *tert*-butyl ester acetone ketal (**99**) and isopropanol at 25°C. The mixture is heated at 40–45°C for 4–5 h and then cooled to 25°C. Aqueous sodium bicarbonate is added (to pH 7.5) and the mixture is concentrated at reduced pressure. The residue is separated between water and dichloromethane. The aqueous layer is extracted with dichloromethane. The combined organic layers are washed with water and with brine and then concentrated at reduced pressure to afford amorphous atorvastatin *tert*-butyl ester (**125**) (85%).[90]

Hydrochloric acid (1.2 M) (3.1 equivalents) is added to a solution of atorvastratin *tert*-butyl ester acetone ketal (**99**) in acetonitrile at 25–30°C and the mixture is aged at 25°C for an unspecified time. Water is added followed by aqueous sodium carbonate at 10–15°C (to pH 7–8). The resulting suspension is filtered and the solid is washed with water. The solid is dissolved in acetonitrile–water at 75°C and the solution is cooled to 25°C. The suspension is filtered and the solid is washed with acetonitrile and with water and then dried at 50–60°C and reduced pressure to afford atorvastatin *tert*-butyl ester (**125**) (82%).[61]

Hydrochloric acid (0.11 M) (0.16 equivalents) is added to a mixture of atorvastratin *tert*-butyl ester acetone ketal (**99**) in acetonitrile, presumably at 25°C, and the mixture is aged at 45°C for 4 h. The suspension is cooled to 25°C, diluted with water, and filtered. The solid is washed with 1:1 acetonitrile–water and dried at 25°C and reduced pressure to afford atorvastatin *tert*-butyl ester (90%, >99% pure by HPLC). Atorvastatin *tert*-butyl ester (**125**) can be crystallized from isopropanol or, better, acetonitrile (92% recovery) (mp 146–148°C).[105]

***9.8.1.2 Atorvastatin Lactone (133)*** Aqueous sodium hydroxide (1 M) (1.2 equivalents) is added to a solution of atorvastatin *tert*-butyl ester (**125**) in THF–methanol at 25°C. The solution is aged at 25°C for 4 h and then concentrated at reduced pressure. The residue is taken up in water and washed twice with ethyl ether. Hydrochloric acid (1 M) is added (to pH < 7) and the mixture is extracted with ethyl acetate. The extracts are washed with water, dried, and concentrated at reduced pressure. The residue is dissolved in toluene and the solution is refluxed for 20 min with water removed as the azeotrope using a Dean–Stark trap. The mixture is cooled to 25°C and concentrated at reduced pressure. The dehydration-lactonization procedure, is then repeated. The mixture in toluene is concentrated at reduced pressure. Atorvastatin lactone (**133**) is isolated from the residue by chromatography. No yield is available.[15]

Hydrochloric acid (2.7 M) (6.0 equivalents) is added to a solution of atorvastatin *tert*-butyl ester acetone ketal (**99**) in

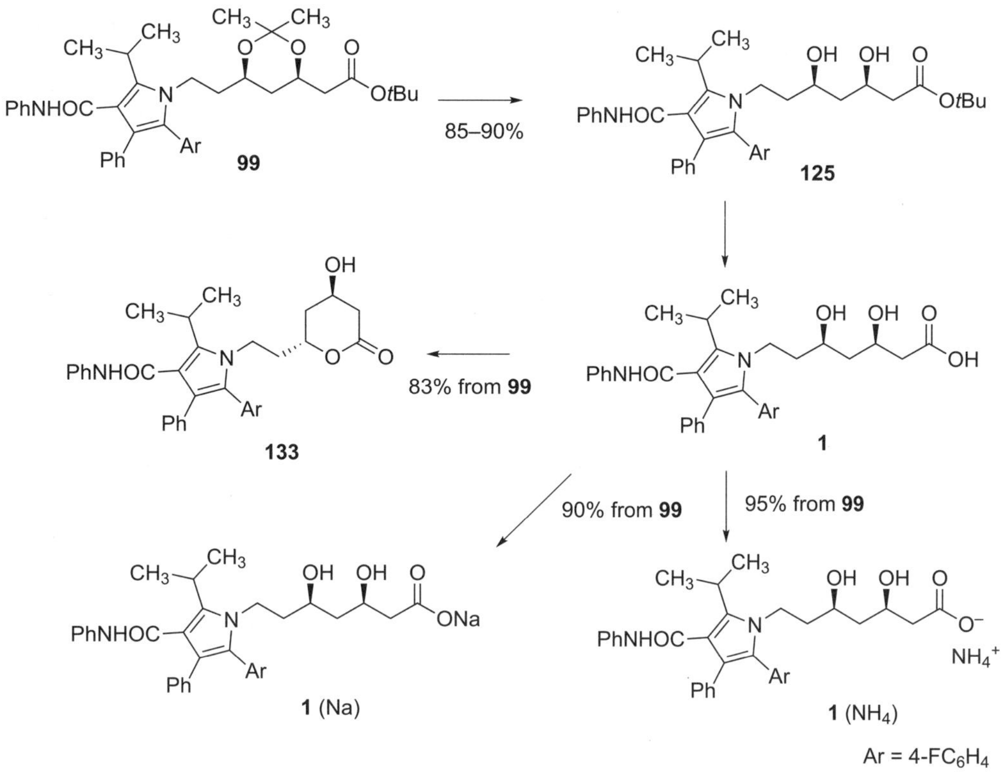


**SCHEME 9.36** Atorvastatin *tert*-butyl ester (**125**), atorvastatin lactone (**133**), atorvastatin (**1**) ammonium salt, and atorvastatin (**1**) sodium salt from atorvastatin *tert*-butyl ester acetone ketal (**99**).

THF at 25°C and the solution is aged at 25°C for 15 h. Solid sodium hydroxide (13.5 equivalents) is then added and the mixture is aged at 25°C for 30 h. Water and hexane are added and the layers are separated. Dilute hydrochloric acid is added to the aqueous layer (to pH < 7). The mixture is aged for 3 h at 25°C and then extracted with ethyl acetate. Hydrochloric acid (catalytic amount) is added to the organic extracts and the solution is aged at 25°C for 18 h and then concentrated at reduced pressure. The residue is dissolved in ethyl acetate and hydrochloric acid (catalytic amount) is added. The solution is aged at 25°C for 2 h and then concentrated at reduced pressure. The residue is crystallized from toluene to afford atorvastatin lactone (**133**) (83%).[29,30,46]

Hydrochloric acid (0.6 M) (0.27 equivalents) is added to a solution of atorvastatin *tert*-butyl ester acetone ketal (**99**) in methanol at 25°C and the solution is aged at 30°C for an unspecified time. Methyl *tert*-butyl ether and aqueous sodium hydroxide (1.7 equivalents) are added and the mixture is aged at 30°C for an unspecified time. The mixture is diluted with water and washed with methyl *tert*-butyl ether. Methyl *tert*-butyl ether and methanol are distilled from the aqueous layer at atmospheric pressure and pot temperature up to 99°C (methanol content is <0.4% (w/v) at that point). The aqueous solution is then aged at 75–85°C for 18 h. The solution is cooled to 25°C, acid is added (to pH < 7), and the mixture is extracted with toluene. The toluene extract is refluxed for 4 h with removal of water as the azeotrope using a Dean–Stark trap. The mixture is cooled to 25°C, the resulting suspension is filtered, and the solid is washed with toluene. The solid is then crystallized from toluene to afford atorvastatin lactone (**133**) (63% from nitrile **32**). HPLC analysis reveals that the atorvastatin lactone (**133**) contains 0.1% atorvastatin methyl ester (**132**). Higher levels of atorvastatin methyl ester (**132**) are associated with higher levels of residual methanol left in the aqueous solution of the sodium salt.[63]

A mixture from Paal–Knorr reaction of the 1,4-diketone **21** with amine **8** (1.3 equivalents) in 9:1 heptane–toluene can be carried into the diol deprotection. The mixture is cooled and poured into THF and saturated aqueous ammonium chloride. The layers are separated. Hydrochloric acid (2.7 M) (5.5 equivalents) is added to the organic layer and the mixture is aged at 25°C for 15 h. Solid sodium hydroxide (12.5 equivalents) is added and the mixture is aged at 25°C for 30 h. Water and hexane are added and the layers are separated. Dilute hydrochloric acid is added to the aqueous layer (to pH < 7). The mixture is aged for 3 h at 25°C and then extracted with ethyl acetate. Hydrochloric acid (catalytic amount) is added to the combined ethylacetate extracts. The solution is aged at 25°C for 18 h and concentrated at reduced pressure. The residue is dissolved in ethyl acetate and hydrochloric acid (catalytic amount) is added. The solution is aged at 25°C for 2 h and then concentrated at reduced pressure. The residue is crystallized from toluene to afford atorvastatin lactone (**133**) (75% from 1,4-diketone **21**).[29,30,46]

Atorvastatin lactone (**133**) can also be crystallized from ethyl acetate–toluene or ethyl acetate–hexane (mp 147–148°C).[15,16,98]

***9.8.1.3 Atorvastatin (1) Ammonium Salt*** Hydrochloric acid (1 M) (1.0 equivalent) is added to a solution of atorvastatin *tert*-butyl ester acetone ketal (**99**) in methanol at 25°C and the solution is aged at 50°C for 10 h. The solution is cooled to 25°C, 1 M aqueous sodium hydroxide (2.0 equivalents) is added, and the solution is aged at 60°C for 10 h. The solution is cooled to 25°C, dilute hydrochloric acid is added (to pH < 7), and the mixture is extracted with methyl *tert*-butyl ether. After washing the extracts with brine, 28–30% ammonium hydroxide (1.7 equivalents) is added, presumably at 25°C. The resulting suspension is aged, presumably at 25°C, for 3 h and then filtered. The solid is washed with methyl *tert*-butyl ether and dried (95%, 99.25% pure). The same yield (94–95%) is achieved using amines (1.2 equivalents) ($CH_3NH_2$, $(CH_3)_2NH$, $(CH_3)_3N$, $NH_2CH_2CH_2NH_2$, *i*-$Pr_2NH$, cyhx$_2$NH) in place of ammonia.[104]

***9.8.1.4 Atorvastatin (1) Triethylammonium Salt*** Hydrochloric acid (2.7 M) (6.2 equivalents) is added to a solution of atorvastatin *tert*-butyl ester acetone ketal (**99**) in THF at 25°C and the solution is aged at 25°C for 24 h. Aqueous sodium hydroxide (30%) (13.5 equivalents) is added, and the solution is aged at 25°C for 15 h. The mixture is diluted with water and washed with hexane. Hydrochloric acid (2.7 M) is added (to pH 4) and the mixture is extracted with dichloromethane. The organic extracts are washed with brine and dried. Triethylamine (1.6 equivalents) is added and the mixture is concentrated at reduced pressure. Dichloromethane is added and the solution is concentrated at reduced pressure. Dissolution in dichloromethane and concentration is repeated once to afford atorvastatin (**1**) triethylammonium salt (95%).[106]

***9.8.1.5 Atorvastatin (1) Sodium Salt*** Hydrochloric acid (2.7 M) (6.0 equivalents) is added to a solution of atorvastatin *tert*-butyl ester acetone ketal (**99**) in THF at 25°C and the solution is aged at 25°C for 24 h. Aqueous sodium hydroxide (40%) (19 equivalents) is added, and the solution is aged at 25°C for 17 h. The mixture is diluted with water and washed with hexane, washed with hexane–THF, and then extracted with ethyl acetate. The ethyl acetate extracts are concentrated at reduced pressure to afford atorvastatin (**1**) sodium (90%). Atorvastatin sodium can be crystallized from ethanol.[106]

#### *9.8.1.6 Atorvastatin (1) Calcium*

*From Atorvastatin **tert**-Butyl Ester (125)*

Atorvastatin (1) Calcium (Crystal Form Not Specified) Aqueous sodium hydroxide (0.65 M) (1.2 equivalents) is added to atorvastatin *tert*-butyl ester (**125**) in methanol and the mixture is aged at 50°C for 1 h. The mixture is presumably cooled to 25°C and concentrated at reduced pressure. The residue is dissolved in ethyl acetate and water and a solution of calcium acetate (0.61 equivalents, assuming monohydrate) in water is added. After aging at 50°C for 1 h, the mixture is cooled to 25°C and the layers are separated. The organic layer is washed with water and concentrated at reduced pressure to afford atorvastatin calcium. No yield is provided. Assuming a quantitative yield, the volume throughput is 21 g/L.[107]

Sodium hydroxide (1.1 equivalents), perhaps as a 0.72 M solution in water, is added to atorvastatin *tert*-butyl ester (**125**) in acetonitrile. The mixture is heated to 40–45°C for 2 h. A solution of calcium acetate (0.56 equivalents, assuming monohydrate) in water is added at 40–45°C. Aqueous sodium hydroxide (0.08 equivalents) is added and the mixture is aged at 70–80°C for 1 h. The mixture is polish filtered at 60–65°C and the solid is washed with acetonitrile. The liquors are cooled to 25°C and diluted with water. Aqueous sodium hydroxide (0.03 equivalents) is added and the mixture is aged at 70–80°C for an additional 8 h. The mixture is cooled to 25°C over 2 h and the suspension is filtered. The solid is washed with water and dried to afford atorvastatin calcium. No yield is provided at this point in the process. Assuming a quantitative yield, the volume throughput is 39 g/L.[61]

Sodium hydroxide (2.5 equivalents) and water are added to a solution of atorvastatin *tert*-butyl ester (**125**) in methanol and the mixture is aged, presumably at 25°C, for an unspecified time. Water is added and the mixture is washed with 1:1 cyclohexane–ethyl acetate. A solution of calcium chloride[108] (0.86 equivalents) in water is added at 55–60°C. The suspension is cooled, presumably to 25°C, and filtered. The solid is presumably washed with water and dried at 50°C to afford atorvastatin calcium. No yield is available. Assuming a quantitative yield, the volume throughput is 31 g/L.[109]

Sodium hydroxide (1.1 equivalents), perhaps as a 0.28 M solution in water, is added to a mixture of atorvastatin *tert*-butyl ester (**125**) in methanol and methyl *tert*-butyl ether. The mixture is refluxed for 4 h. Aqueous sodium hydroxide (2 M) is added (to pH 11) and the reflux is continued for 3 h. After cooling to 25°C, hydrochloric acid is added (to pH 8.0). The mixture is washed with methyl *tert*-butyl ether three times and then polish filtered. A solution of calcium chloride hexahydrate (0.49 equivalents) in water is added in portions at 25°C. The resulting thick suspension is aged at 25°C for 30 min, diluted with water, and filtered. The solid is washed with water–methanol and with water and dried to afford atorvastatin calcium (82%). The volume throughput is 26–30 g/L.[105]

Amorphous Atorvastatin (1) Calcium Sodium hydroxide solid (1.1 equivalents) is added to atorvastatin *tert*-butyl ester (**125**) in water. The mixture is heated to 75–80°C and then cooled, presumably to 25°C. Calcium acetate[110] (0.70 equivalents, assuming monohydrate) is added, the mixture is aged, presumably at 25°C, for 1 h. The pH is maintained at 8. The mixture is then extracted with ethyl acetate. The extract is washed with water, dried, and concentrated at reduced pressure. The residue is suspended in isopropanol, diisopropyl ether, cyclohexane, or methyl *tert*-butyl ether and the suspension is filtered. The solid is presumably washed with the suspension solvent and dried to afford amorphous atorvastatin calcium. No yield is provided. Assuming a quantitative yield, the volume throughput is 52 g/L. Calcium acetate can be added as an aqueous solution. Calcium 2-ethylhexanoate[111] can be used in place of the acetate. Ester **125** can be hydrolyzed in 2-butanone–water at 60°C for 2 h. 2-Butanone or xylenes can replace ethyl acetate as the extraction solvent.[88]

Aqueous sodium hydroxide (0.26 M) (1.1 equivalents) is added to atorvastatin *tert*-butyl ester (**125**) in acetonitrile and the mixture is heated to 45–50°C for 6 h. A solution of calcium acetate (0.54 equivalents, assuming monohydrate) in water is added at 45–50°C and the mixture is aged at 45–50°C for 1 h. The mixture is presumably cooled to 25°C, 2-methyl THF is added, and the layers are separated. The aqueous layer is extracted with 2-methyl THF. The organic layers are combined and polish filtered. The liquors are concentrated at reduced pressure and the resulting concentrated solution is added to *n*-hexane–methyl *tert*-butyl ether, presumably at 25°C. The suspension is filtered and the solid is washed, presumably with *n*-hexane–methyl *tert*-butyl ether, and dried to afford amorphous atorvastatin calcium (85–87%, assuming anhydrous). The volume throughput is 31 g/L. The same yield (87–89%) is observed using calcium D-gluconate[112] (0.64 equivalents) in place of calcium acetate or n-heptane in place of *n*-hexane–methyl *tert*-butyl ether.[87]

Calcium hydroxide can provide the hydroxide for ester hydrolysis. Water and calcium hydroxide[113] (2.3 equivalents) are added to a solution of atorvastatin *tert*-butyl ester (**125**) in methanol and the mixture is aged, presumably at 25°C, for an unspecified time. The mixture is filtered and the liquors concentrated at reduced pressure. The residue is dissolved in 1,4-dioxane at 40°C, the solution is polish filtered, and the liquors are added to cyclohexane–methyl *tert*-butyl ether. The resulting suspension is aged at 25°C for 2 h and filtered. The solid is presumably washed with cyclohexane–methyl *tert*-butyl ether and dried to afford amorphous atorvastatin calcium. No yield is available. Assuming a quantitative yield, the throughput is 17 g/L.[109]

Atorvastatin (1) Calcium Polymorphs Sodium hydroxide (1.2 equivalents) perhaps as a 0.76 M solution in water, is added to atorvastatin *tert*-butyl ester (**125**) in acetonitrile. The mixture is heated to 25–55°C for 4.5 h. A solution of calcium acetate hemihydrate (0.59 equivalents) in water is added, presumably at 25°C, and the mixture is aged at 30–50°C for 1 h. The suspension is filtered, the solid is washed with acetonitrile, and the liquors are concentrated at reduced pressure. Acetonitile–water (1:1) is added to the residue. The suspension is refluxed for 13 h and then filtered while hot. The solid is washed with 1:1 acetonitrile–water and dried at 60–70°C and reduced pressure to afford atorvastatin calcium polymorph VI-US7074818 (99.71% pure by HPLC). No yield is available. Assuming a quantitative yield, the volume throughput is 49 g/L.[114]

Sodium hydroxide flakes (1.2 equivalents) are added to atorvastatin *tert*-butyl ester in acetonitrile. The mixture is heated to 30–45°C for 6 h. A solution of calcium acetate hemihydrate (0.59 equivalents) in water is added, presumably at 30°C, and the mixture is aged at 30–50°C for 1 h. The suspension is filtered, the solid is washed with acetonitrile, and the liquors are concentrated at reduced pressure. Acetonitile–water (1:1) is added to the residue and the suspension is refluxed for 15 min. Acetonitile–water (1:1) is added to the suspension and reflux is continued for 30 min. Acetonitile–-water (1:1) is added to the suspension and reflux is continued for 1 h. The suspension is then cooled to 0°C and filtered. The solid is presumably washed with 1:1 acetonitrile–water and dried at 50–60°C and reduced pressure to afford atorvastatin calcium polymorph VII-US7074818 (99.81% pure by HPLC). No yield is available. Assuming a quantitative yield, the volume throughput is 55 g/L.[114]

*From Atorvastatin Lactone (133)*

Atorvastatin Calcium (Crystal Form Not Specified) Sodium hydroxide (1.0 equivalent), as pellets or an aqueous solution, is added to a solution of atorvastatin lactone (**133**) in methanol–water at 25°C. The solution is aged at 25°C until analysis (HPLC) indicates <2% of lactone **133** and methyl ester **132** remain. The solution is diluted with water and washed several times with 1:1 ethyl acetate–hexane. If analysis indicates the presence of impurities or if the solution is yellow, carbon is added and the suspension is aged for 2 h and filtered. The liquors are assayed for atorvastatin sodium. Calcium chloride (0.5 equivalents) is dissolved in water and the solution is heated to 60°C. The atorvastatin sodium solution is also heated to 60°C and the calcium chloride solution is then slowly added with efficient agitation. The resulting suspension is cooled to 15°C and filtered. The solid is washed with water and dried at 50°C and reduced pressure to afford atorvastatin calcium. No yield is provided. Assuming a quantitative yield, the volume throughput is 12 g/L.[98]

Aqueous sodium hydroxide (0.15 M) (1.03 equivalents) is added to a solution of atorvastatin lactone (**133**) in methyl *tert*-butyl ether and methanol. The mixture is aged at 52°C for 1 h. The mixture is cooled to 34°C and the layers are separated. The aqueous layer is washed with methyl *tert*-butyl ether at 33°C. The aqueous layer is diluted with some methyl *tert*-butyl ether and heated to 52°C. A solution of calcium acetate (0.51 equivalents) in water is added. After starting the calcium acetate solution addition, the mixture is seeded with atorvastatin calcium (1.1%). The resulting suspension is aged at 52°C for 15 min and then cooled to 20°C and filtered. The solid is washed with 2:1 methanol–water and dried at 70°C and reduced pressure to afford atorvastatin calcium (97%). The volume throughput is 52 g/L.[85]

Amorphous Atorvastatin Calcium Aqueous sodium hydroxide (2.5 M) (1.1 equivalents) is added to a solution of atorvastatin lactone (**133**) in methanol at 25°C. The mixture is aged at 50°C for 2 h. The mixture is presumably cooled to 25°C and diluted with water. A solution of calcium acetate monohydrate (0.49 equivalents) in water is added at 30°C. Dichloromethane is added and the mixture is aged at 30°C for 30 min. The layers are separated and the aqueous layer is extracted with dichloromethane. The organic layers are distilled at atmospheric pressure to produce a concentrated solution. The solution is polish filtered and then added to diisopropyl ether at 0°C. The resulting suspension is aged for 30 min at 0°C and then filtered. The solid is presumably washed with diisopropyl ether–dichloromethane and dried at 50°C and reduced pressure to afford amorphous atorvastatin calcium (89%, 99.3% pure by HPLC). The yield offered suggests this material is anhydrous. The volume throughput is just 17 g/L.[115]

Calcium hydroxide can be used as the hydroxide source for the lactone hydrolysis. A suspension of calcium hydroxide (1.0 equivalent) in water is added to a solution of atorvastatin lactone (**133**) in THF at 25°C. The mixture is aged at 50°C for 2 h. Water and dichloromethane are added and the mixture is aged at 30°C for 30 min. The layers are separated and the aqueous layer is extracted with dichloromethane. The organic layers are distilled at atmospheric pressure to produce a concentrated solution. The solution is polish filtered and then added to diisopropyl ether at 0°C. The resulting suspension is aged for 30 min at 0°C and then filtered. The solid is presumably washed with diisopropyl ether–dichloromethane and dried at 50°C and reduced pressure to afford amorphous atorvastatin calcium (94%, 99.3% pure by HPLC). The yield suggests this material is anhydrous. The volume throughput is 22 g/L.[115]

When lactone **133** is hydrolyzed in methanol–water, the hydrolysis mixture is usually washed at least twice with 1:1 ethyl acetate–hexane or methyl *tert*-butyl ether prior to calcium salt formation. The washes are not necessary when the lactone is hydrolyzed in THF–water. Aqueous sodium hydroxide (1 M) (1.03 equivalents) is added to a solution of atorvastatin lactone (**133**) in THF–water, presumably at 25°C. The solution is aged, presumably at 25°C, for an unspecified time. The solution is diluted with water and ethyl acetate and then added to a solution of calcium chloride (0.55 equivalents) in water, presumably at 25°C. The resulting suspension may be aged at 25°C and then filtered. The solid is washed with water and dried to afford amorphous atorvastatin calcium (94%). The water content of the wet cake is high (55 wt%). Assuming the ethyl acetate charge (unspecified) is 0.10 L/kg atorvastatin lactone, the throughput is 96 g/L.[116]

Atorvastatin (1) Calcium Polymorphs Aqueous sodium hydroxide (0.15 M) (1.03 equivalents) is added to a solution of atorvastatin lactone (**133**) in methanol–methyl *tert*-butyl ether, presumably at 25–30°C. The mixture is aged at 48–58°C for 40–60 min. The mixture is cooled to 25–35°C and the layers are separated. The aqueous layer is washed with methyl *tert*-butyl ether. The reactor is then sealed, and the aqueous solution (saturated with methyl *tert*-butyl ether) is heated to 47–57°C under pressure. A solution of calcium acetate hemihydrate (0.51 equivalents) in water is then added. After starting the calcium acetate solution addition, the mixture is seeded with a suspension of atorvastatin calcium, polymorph I-US5969156 (1–2%) in methanol–water. The calcium acetate solution addition is then completed. The resulting suspension is heated to 51–57°C, aged for >10 min, cooled to 15–40°C, and filtered. The solid is washed with water, with methanol, and again with water and then dried in a stainless steel agitation pan dryer at 60–70°C and reduced pressure to afford atorvastatin calcium polymorph I-US5969156 (86%). The volume throughput is 46 g/L. This process is described on 70 kg and 250 kg scales.[117–121]

Aqueous sodium hydroxide (0.31 M) (2.1 equivalents) is added to a solution of atorvastatin lactone (**133**) in methanol–methyl *tert*-butyl ether, presumably at 25–30°C. (*Note:* The methyl *tert*-butyl ether is missing from the process description.) The mixture is aged at 50–55°C for 45 min. The mixture is cooled to 25°C and the layers are separated. The aqueous layer is washed with methyl *tert*-butyl ether. The methyl *tert*-butyl ether dissolved in the aqueous layer is distilled at 70–80°C. The solution is cooled to 60–70°C and a solution of calcium acetate hemihydrate (0.51 equivalents) in water is added. After starting the calcium acetate solution addition, the mixture is seeded with a suspension of atorvastatin calcium, polymorph IV-US5969156 (1–2%) in methanol–water. (*Note:* The procedure reads polymorph I.) The resulting suspension is aged at 65–75°C for 5 min and then cooled to 50–55°C and filtered. The solid is suspended in methanol at 55–60°C. The suspension is cooled to 25°C and filtered. The solid is presumably washed with methanol and dried at 66–70°C and reduced pressure to afford atorvastatin calcium, polymorph IV-US5969156 (22%).[119]

A suspension of calcium hydroxide (0.73 equivalents) in water is added to a solution of atorvastatin lactone (**133**) in acetone and the mixture is aged at 45°C for 2 h. Water is added at 45°C and the solution is cooled to 25°C. The resulting suspension is filtered and the solid is presumably washed with water and dried in a vacuum tray drier at 50°C to afford atorvastatin calcium, polymorph VI-WO2004022053 (90–92%, 99.2–99.3% pure by HPLC). Processes with volume throughputs of 5, 67, and 500 g/L (questionable) are described.[122]

*From Atorvastatin (1) Ammonium Salt* A solution of the atorvastatin (**1**) ammonium salt in THF–water is added to a solution of calcium acetate monohydrate (0.54 equivalents) in water, presumably at 25°C. The resulting suspension is aged, presumably at 25°C, for 12 h and then filtered. The solid is washed with water and dried to afford atorvastatin calcium (86%, 99.18% pure by HPLC). The volume throughput is 79 g/L.[104]

*From Atorvastatin (1) Triethylammonium Salt* A solution of calcium acetate monohydrate (0.54 equivalents) in water is added to a solution of atorvastatin (**1**) triethylammonium salt in ethyl acetate at 25°C. The organic layer is separated, washed with water, dried, and diluted with toluene. Hexane is added and the mixture is heated to 50°C to produce a clear solution. The solution is then cooled to 30°C and more hexane is added. The resulting suspension is aged at 25°C for 1 h and filtered. The solid is washed with hexane and dried to afford amorphous atorvastatin calcium (77%, assuming anhydrous). The volume throughput is 17 g/L. Pentane, heptane, cyclohexane, diethyl ether, diisopropyl ether, and methyl *tert*-butyl ether can be used in place of hexane.[106]

*From Atorvastatin (1) L-Lysine Salt* A solution of atorvastatin (**1**) L-lysine salt in acetone–water is filtered and a solution of calcium acetate hydrate (0.51 equivalents, assuming monohydrate) in water is added, presumably at 25°C. The resulting suspension is aged at 15–20°C for 1 h and filtered. The solid is washed with 1:1 acetone–water and dried at 50°C and reduced pressure to afford amorphous atorvastatin calcium (88%, 99.87% pure by HPLC, anhydrous). The volume throughput is 73 g/L. This process is demonstrated on a 10 kg scale. The salt exchange can also be run using ethanol or isopropanol in place of acetone. The same yield (88%) is observed using the L-arginine salt in water–isopropanol. Many other atorvastatin amine salts are known and could presumably be converted to amorphous atorvastatin calcium by a similar process.[123,124]

*From Atorvastatin (1) Sodium* A solution of atorvastatin (**1**) sodium in ethyl acetate is washed three times with a solution of calcium acetate (0.82 equivalents, assuming monohydrate) in water. The ethyl acetate solution is washed with water, dried, and concentrated to a smaller volume at reduced pressure. The solution is then added to hexane. The resulting suspension is aged for 20 min at 25°C and filtered. The solid is washed with hexane and dried to afford amorphous atorvastatin calcium (78%, assuming anhydrous). The volume throughput is 9 g/L.[106]

A solution of atorvastatin (**1**) sodium salt in methanol–water is washed twice with 1:1 ethyl acetate–hexane. Carbon is added and the suspension is aged at 50°C for 2 h. The suspension is filtered and the solid is washed with methanol. A solution of calcium acetate (0.5 equivalents) in water at 60°C is added to the liquors at 63°C. The solution is then cooled to 13°C. The resulting suspension is filtered and the solid is dried over anhydrous silica for 5 days to afford atorvastatin calcium, polymorph V-US7411075.[125]

*Directly from Atorvastatin tert-Butyl Ester Acetone Ketal (99)*

ATORVASTATIN (1) CALCIUM (CRYSTAL FORM NOT SPECIFIED) A suspension of atorvastatin *tert*-butyl ester acetone ketal (**99**) and moist silica in dichloromethane is aged at 25°C for 18 h. The suspension is then presumably filtered. Aqueous sodium hydroxide is added to the liquors and the biphasic mixture is aged at 25°C for 4 h. Water and diisopropyl ether are added and the layers are separated. Hydrochloric acid is added to the aqueous layer (to pH $< 7$) and the mixture is extracted with diisopropyl ether. The extract is presumably concentrated at reduced pressure. The residue is dissolved in ethyl acetate, ammonia gas is bubbled in, and the solution is concentrated at reduced pressure. The residual atorvastatin (**1**) ammonium salt is dissolved in diisopropyl ether–isopropanol and a solution of calcium acetate, presumably in water, is added at 25°C. The resulting suspension is filtered and the solid is presumably washed and dried to afford atorvastatin calcium. A detailed procedure and the yield are not provided.[36]

A suspension of atorvastatin *tert*-butyl ester acetone ketal (**99**) and Indion 525 ($H^+$ form) (1 kg/kg ketal **99**) in acetonitrile is aged at 25°C for an unspecified time. The suspension is filtered and aqueous sodium hydroxide (1.6 equivalents) is added to the liquors. The mixture is aged at 25°C for an unspecified time and then concentrated at reduced pressure. The residue is dissolved in methanol at 50°C, then water is added and the solution is polish filtered. A solution of calcium acetate (0.65 equivalents) in water is added to the liquors at 55°C. The mixture is aged at 55°C for 30 min and then cooled to 15°C. The resulting suspension is filtered and the solid is washed with water and dried at 50°C and reduced pressure to afford atorvastatin calcium (76%). The volume throughput is 27 g/L.[86]

A mixture of atorvastatin *tert*-butyl ester acetone ketal (**99**) in 80% aqueous acetic acid is aged at 25°C for 16 h. The resulting clear solution is concentrated at reduced pressure. The residue is dissolved in toluene and the solution concentrated, presumably at reduced pressure, to eliminate residual acetic acid. The dissolution–concentration procedure is repeated twice more. The residue is dissolved in ethanol–water, calcium hydroxide (5.5 equivalents) and tetrabutylammonium bromide (9 mol%) are added, and the mixture is aged at 45°C for 3 h. The mixture is filtered while hot to remove the excess calcium hydroxide. The liquors are cooled to 25°C and water is added. The resulting suspension is filtered and the solid is presumably washed with water and dried at 65°C and reduced pressure to afford atorvastatin calcium (84%, anhydrous).[126,127]

Hydrochloric acid (2.7 M) (0.68 equivalents) is added to a solution of atorvastatin *tert*-butyl ester acetone ketal (**99**) in methanol, presumably at 25°C. The mixture is aged at 35°C and reduced pressure for 3 h with continuous removal of acetone by distillation of an acetone, methanol, and water mixture. Makeup methanol is added every 30 min. Calcium hydroxide (1.5 equivalents), water, and methanol are added and the mixture is heated at 70°C for 2 h. The mixture is filtered while hot to remove the excess calcium hydroxide. Water is added to the liquors at 65°C. The mixture is heated to 78°C and then cooled and aged at 20°C for 20 h. The resulting suspension is filtered and the solid is presumably washed with water and dried at 65°C and reduced pressure to afford atorvastatin calcium (96%) (polymorph II-US7501450) containing 3.2 wt% $H_2O$. The volume throughput is 33 g/L.[127–129]

Hydrochloric acid (0.4 M) (0.67 equivalents) is added to a solution of atorvastatin *tert*-butyl ester acetone ketal (**99**) in ethanol, presumably at 25°C. The mixture is aged at 40°C and reduced pressure for 11 h with continuous removal of acetone by distillation of an acetone, ethanol, and water mixture. Makeup ethanol is added every hour. Calcium hydroxide (1.5 equivalents) is added and the mixture is heated at 70°C for 4–5 h. The mixture is filtered while hot to remove the excess calcium hydroxide. Water is added to the liquors at 65°C. The mixture is heated to 84°C to produce a clear solution and then cooled and aged at 20°C for 20 h. The resulting suspension is filtered and the solid is presumably washed with water to afford a wet cake of atorvastatin calcium (polymorph XII-US7501450). The wet cake is dried at 65°C and reduced pressure to afford atorvastatin calcium (95%) (polymorph V-US7501450) containing 2.8–6.6 wt% $H_2O$. The volume throughput is 26 g/L.[127–129]

Hydrochloric acid (1 M) (amount not specified) is added to a solution of atorvastatin *tert*-butyl ester acetone ketal (**99**) in methanol at 50–55°C and the mixture is aged at 50–55°C for 2 h. Aqueous sodium hydroxide (2.8 M) (0.36 equivalents) is added at 50–55°C and the mixture is aged at 50–55°C for 30 min. The mixture is cooled to 25°C and water and 1 M hydrochloric acid are added (to pH 8–8.5). The mixture is washed with diisopropyl ether. The aqueous layer is then added to a solution of calcium chloride in water at 50–55°C. The resulting suspension is aged at 50°C for 1 h, at 25°C for 1 h, and at 10°C for 2 h, and then filtered. The solid is washed with water and dried at 50–55°C to afford crude atorvastatin calcium (98%). The volume throughput is 36 g/L. The purity can be upgraded by dissolving the solid in methanol at 30°C, treating the solution with carbon, filtering the suspension, and then precipitating atorvastatin calcium by transferring the methanol solution into water (98% recovery).[92]

A 1 M solution of hydrogen chloride (2.0 equivalents) in 96% ethanol is added to atorvastatin *tert*-butyl ester acetone ketal (**99**) and the solution is aged at 50°C for 30 min. Aqueous 2 M sodium hydroxide (3.8 equivalents) is then added and the solution is aged at 50°C for 30 min. The solution is presumably cooled to 25°C and water and toluene are added. The layers are separated and 2 M hydrochloric acid is added to the aqueous layer (to pH 7.8). The solution is heated to 60°C and a solution of calcium acetate (0.56 equivalents) in water is added. The resulting suspension is cooled to 20°C, aged at 20°C for 3 h, aged at 0°C for 1 h, and filtered. The solid is washed with water and dried at 40°C to afford amorphous atorvastatin calcium (83%, 98% pure by HPLC). The volume throughput is 19 g/L. The atorvastatin calcium purity is lower (89% pure by HPLC) when 37% hydrochloric acid (3.5 equivalents) in 1,4-dioxane is used for the diol deprotection.[89]

Hydrochloric acid (1 M) (1.1 equivalents) is added to a suspension of atorvastatin *tert*-butyl ester acetone ketal (**99**) in methanol at 25°C and the mixture is aged at 25°C for 6 h. Aqueous sodium hydroxide (10%) is added at 25°C and the mixture is aged at 25°C for 6 h. The solution is polish filtered and distilled to reduce the volume. Water, methanol, and methyl *tert*-butyl ether are added and the layers are separated. The aqueous layer is washed with methyl *tert*-butyl ether and polish filtered. Hydrochloric acid (6 M) is added (to pH 7.9–8.1) and the solution is heated to 48°C. A solution of calcium hydroxide (0.55 equivalents) in water is added at 51°C. The solution is seeded with atorvastatin calcium polymorph I-US20090216029. Methyl *tert*-butyl ether is distilled at reduced pressure. The mixture is heated to 58°C and water and additional seeds are added. The resulting suspension is cooled to 30°C and filtered. The solid is washed with methanol–water and dried at 45–50°C and reduced pressure to afford crude atorvastatin calcium (87%, 97.5% pure by HPLC). The volume throughput is 27 g/L. The purity can be upgraded by crystallization from methanol–THF–water (92% recovery). This process is demonstrated on a 10 kg scale. The mixture should contain some methyl *tert*-butyl ether at the start of atorvastatin calcium salt precipitation.[47]

Hydrochloric acid (1.3 M) (2.4 equivalents) is added to a suspension of atorvastatin *tert*-butyl ester acetone ketal

(**99**) in methanol at 30–35°C and the mixture is aged at 30–35°C for 15 h. Aqueous sodium hydroxide (3.3 equivalents) is added, presumably at 25°C, and the mixture is aged, presumably at 25°C, for 5 h. Hydrochloric acid (1.3 M) is added (to pH 8.5) and the volume is reduced by half by methanol distillation at <50°C and reduced pressure. Some methanol and water are added and the solution is washed with diisopropyl ether (methanol–water–ether volume ratio is 1.4:1:1.4). The aqueous layer is heated to 50–55°C, the pH is adjusted to 8.5 if necessary, and a solution of calcium chloride (0.83 equivalents) in water is added at 50–55°C. The resulting suspension is aged at 50°C for 1 h, aged at 15°C for 1 h, and filtered. The solid is washed with water and dried at 50–60°C and reduced pressure to afford crude atorvastatin calcium (95%). The volume throughput is 26 g/L. This process is demonstrated on a 20 kg scale. Calcium hydroxide in the crude atorvastatin calcium is removed by crystallization from ethyl acetate (containing 4% $H_2O$) and *n*-hexane (96% recovery). The crude atorvastatin calcium assay will be influenced by the calcium ion source. In one series of experiments, the w/w assays are with calcium acetate >98%, with calcium chloride 95%, with calcium hydroxide 90%.[103,130]

AMORPHOUS ATORVASTATIN (1) CALCIUM Hydrochloric acid (2.2 M) (1.5 equivalents) is added to a suspension of atorvastatin *tert*-butyl ester acetone ketal (**99**) in methanol at 25°C and the mixture is aged at 25°C for 6–9 h. Aqueous sodium hydroxide (12%) is added (to pH 7). Carbon is added and the suspension is aged for 30 min at 25°C and filtered. Aqueous sodium hydroxide (12%) is added to the liquors (to pH 12–12.5) and the mixture is aged, presumably at 25°C, for an unspecified time. The mixture is polish filtered, concentrated at reduced pressure to reduce the volume, and diluted with 1:3:6 methanol, water, and methyl *tert*-butyl ether. The layers are separated and the organic layer is extracted with water. The combined aqueous layers are washed with methyl *tert*-butyl ether. Dilute hydrochloric acid is added (to pH 7.5–8) followed by ethyl acetate and a solution of calcium acetate (0.56 equivalents, assuming monohydrate) in water. The mixture is aged, presumably at 25°C, for 2 h and the layers are separated. The aqueous layer is extracted twice with ethyl acetate. The combined organic layers are washed with water, again treated with carbon, and filtered. The liquors are dried (molecular sieves) and concentrated at 40–45°C and reduced pressure. The resulting ethyl acetate solution is diluted with THF. The solution is then added to petroleum ether at 25°C. The resulting suspension is aged at 25°C for 1 h and filtered. The solid is washed with petroleum ether and dried at 40–45°C and reduced pressure to afford amorphous atorvastatin calcium (99%, 99.51% pure by HPLC). The volume throughput is 16 g/L. The yield is lower (80–83%) using cyclohexane or methylcyclohexane in place of petroleum ether.[91]

Hydrochloric acid (2.7 M) (5.9 equivalents) is added to a suspension of atorvastatin *tert*-butyl ester acetone ketal (**99**) in THF at 25°C and the mixture is aged at 25°C for 15 h. Sodium hydroxide (13 equivalents) is added and the mixture is aged, presumably at 25°C, for 30 h. The mixture is concentrated at reduced pressure to remove some of the THF and then diluted with water and washed with hexane. Hydrochloric acid (5 M) is added (to pH 7–7.5). A 35°C solution of calcium acetate (0.64 equivalents) in water is added at 35°C. The solution is aged at 25°C for 1 h and at 0°C for 2 h. The resulting suspension is filtered and the solid is washed with water and dried at 40°C and reduced pressure to afford amorphous atorvastatin calcium (93%). The volume throughput is 17 g/L. The yield is lower (81%) in a similar procedure using more calcium acetate (0.94 equivalents).[131]

Hydrochloric acid (2.7 M) (2.0 equivalents) is added to a suspension of atorvastatin *tert*-butyl ester acetone ketal (**99**) in THF at 25°C and the mixture is aged at 25°C for 15 h. Sodium hydroxide (4.5 equivalents) is added and the mixture is aged, presumably at 25°C, for 3 h. Hydrochloric acid (5 M) is added (to pH 7.8). Cyclohexane and brine are added and the layers are separated. A solution of calcium acetate (0.65 equivalents) in water is added to the organic layer at 30°C, the biphasic mixture is aged at 30°C for 1 h, and the layers are separated. The organic layer is dried and concentrated at reduced pressure. The solution is then transferred into diisopropyl ether, presumably at 25°C. The resulting suspension is filtered and the solid is washed with ethyl ether and dried at 45°C and reduced pressure to afford amorphous atorvastatin calcium (88%). The volume throughput is 27 g/L. Higher yields (90–93%) are observed using chloroform in place of cyclohexane and omitting the first layer separation.[132]

Hydrochloric acid (2.7 M) (6.0 equivalents) is added to a solution of atorvastatin *tert*-butyl ester acetone ketal (**99**) in THF at 25°C and the solution is aged at 25°C for 6 h. Aqueous sodium hydroxide (40%) (19 equivalents) is added, and the solution is aged at 25°C for 15 h. The mixture is diluted with water and washed with hexane, washed with hexane–THF, and then extracted with ethyl acetate. The combined ethyl acetate extracts are then aged with a solution of calcium acetate (0.75 equivalents, assuming monohydrate) in water at 25°C. The layers are separated and the aging with aqueous calcium acetate and layer separation are repeated twice. The organic layer is washed with water, dried, and concentrated at reduced pressure to a smaller volume. The solution is added to hexanes at 25°C. The resulting suspension is aged at 25°C for 20 min and filtered. The solid is washed with hexanes and dried to afford amorphous atorvastatin calcium (93%, assuming anhydrous). The volume throughput is 14 g/L.[106]

Hydrochloric acid (2.7 M) (6.0 equivalents) is added to a solution of atorvastatin *tert*-butyl ester acetone ketal (**99**) in THF at 25°C and the solution is aged at 25°C for 24 h.

Aqueous sodium hydroxide (40%) (19 equivalents) is added, and the solution is aged at 25°C for 17 h. The mixture is diluted with water and washed with hexane. Hydrochloric acid (2.7 M) is added (to pH 3) and the mixture is extracted with ethyl acetate. The extracts are then aged with a solution of calcium hydroxide (1.5 equivalents) in water at 25°C for 20 min. The organic layer is washed with water and dried. Hexane is then added and the resulting suspension is aged for 1 h at 25°C and filtered. The solid is washed with hexane and dried to afford amorphous atorvastatin calcium (75%, assuming anhydrous). The volume throughput is 11 g/L.[106]

### 9.8.2 From Atorvastatin *tert*-Butyl Ester Cyclopentanone Ketal (103)

Hydrochloric acid (1 M) (0.22 equivalents) is added to a mixture of atorvastatin *tert*-butyl ester cyclopentanone ketal (**103**) in methyl *tert*-butyl ether and methanol at 50°C and the mixture is aged at 50°C for 5 h. Aqueous sodium hydroxide (10%) (0.14 equivalents) is added at 50°C (to pH > 13) (*Note*: The amount of hydroxide added is not sufficient to neutralize the remaining hydrochloric acid catalyst and hydrolyze the ester.) and the mixture is aged at 50°C for 1 h. The mixture is cooled to 25°C and more aqueous hydroxide is added if the pH is <10. The layers are separated and the aqueous layer is washed three times with methyl *tert*-butyl ether. The aqueous layer is then heated to 47–57°C and a solution of calcium acetate hemihydrate (0.51 equivalents) in water is added. After starting this addition, the mixture is seeded with atorvastatin calcium. After completing the addition, the suspension is aged at 47–57°C for 30 min, cooled to 15–25°C, and filtered. The solid is washed with water–methanol and with water and dried at 60–70°C and reduced pressure to afford atorvastatin calcium (94%). The volume throughput is 122 g/L (Scheme 9.37).[65]

### 9.8.3 From Atorvastatin *tert*-Butyl Ester Cyclohexanone Ketal (104)

Hydrochloric acid (1.1 M) (0.83 equivalents) is added to a mixture of atorvastatin *tert*-butyl ester cyclohexanone ketal (**104**) in methanol, presumably at 25°C, and the mixture is refluxed for 18 h. The solution is cooled to 25°C, 33% aqueous sodium hydroxide (0.83 equivalents) is added, and the solution is concentrated to a paste. The paste is taken up in methanol and methyl *tert*-butyl ether, more aqueous sodium hydroxide (0.97 equivalents) is added, and the mixture is aged at 47–52°C for 1 h. The mixture is cooled to 25°C and more aqueous hydroxide is added if the pH is <10. The layers are separated and the aqueous layer is washed three times with methyl *tert*-butyl ether. The aqueous layer is then heated to 47–57°C and a solution of calcium acetate hemihydrate (0.76 equivalents) in water is added. After starting this addition, the mixture is seeded with atorvastatin calcium. After completing the addition, the suspension is aged at 47–57°C for 30 min, cooled to 15–25°C, and filtered. The solid is washed with water–methanol and with water and dried at 60–70°C and reduced pressure to afford atorvastatin calcium (90%). The volume throughput is 48 g/L.[65]

### 9.8.4 From Atorvastatin *tert*-Butyl Ester Benzaldehyde Acetal (129)

A mixture of atorvastatin *tert*-butyl ester benzaldehyde acetal (**129**), 5% palladium on carbon (50% water wet) (2.3 g Pd/kg benzaldehyde acetal **129**), hydrogen chloride

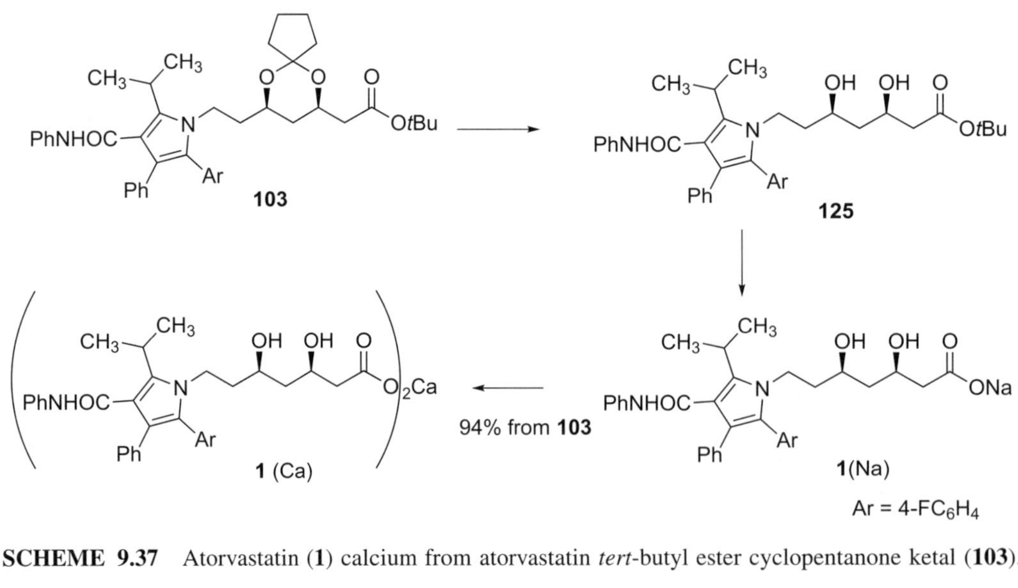


**SCHEME 9.37** Atorvastatin (**1**) calcium from atorvastatin *tert*-butyl ester cyclopentanone ketal (**103**).

(0.53 equivalents) in methanol, and toluene is hydrogenated at 40°C and 50 psi hydrogen pressure for 2.5 h. The suspension is cooled and filtered and the solid is washed with methanol. A solution of potassium hydroxide (1.5 equivalents) in water is added to the liquors and methanol is distilled from the mixture at atmospheric pressure to a pot temperature of 90°C. The resulting biphasic mixture is cooled to 70°C and the layers are separated. The aqueous layer is washed with toluene at 70°C. Toluene and 2 M hydrochloric acid (1.4 equivalents) are added to the aqueous layer, presumably at 25°C, and the layers are separated. The aqueous layer is extracted with toluene. The two toluene extracts are refluxed for 2.5 h with water removed as the azeotrope using a Dean–Stark trap. The mixture is cooled to 65°C, seeded with atorvastatin lactone (**133**), aged for 2 h at 65°C, and then cooled to 25°C. The resulting suspension is cooled to 0°C and filtered. The solid is washed with cold toluene, crystallized again from toluene, and dried at 70°C and reduced pressure to afford atorvastatin lactone (**133**). No yield is provided.[85]

### 9.8.5 From Atorvastatin *tert*-Butyl Ester 2,2-Diisopropylsilyl-1,3,2-Dioxasilinane (106)

Solid sodium hydroxide (86 equivalents) is added to a solution of atorvastatin *tert*-butyl ester 2,2-diisopropylsilyl-1,3,2-dioxasilinane (**106**) in ethanol at 30°C. The mixture is aged at 55°C for 25 h. The mixture is cooled to 25°C and 8% hydrochloric acid (6.7 equivalents) is added (to pH 2–2.5). (*Note*: The amount of acid suggests the sodium hydroxide charge is incorrect.) Calcium hydroxide (2.1 equivalents) is added. The suspension is aged at 70°C for 4 h and then filtered while hot. The liquors are diluted with water at 70°C and allowed to cool to 25°C. The resulting suspension is filtered and the solid is presumably washed with water and dried at 60°C and reduced pressure to afford atorvastatin calcium (56%). The volume throughput of this 200 mg-scale procedure is 10 g/L.[66]

### 9.8.6 From Atorvastatin *tert*-Butyl Ester Phenylboronate (107)

A mixture of crude atorvastatin *tert*-butyl ester phenylboronate (**107**) and moist silica in dichloromethane is aged, presumably at 25°C, for 18 h. The suspension is presumably filtered. Aqueous sodium hydroxide is added to the liquors and the biphasic mixture is aged, presumably at 25°C, for 4 h. The mixture is diluted with water and the layers are presumably separated. The aqueous layer is washed with diisopropyl ether. Hydrochloric acid is then added to the aqueous layer (to pH $< 7$) and the mixture is extracted with diisopropyl ether. The extract is presumably concentrated at reduced pressure and the residue is dissolved in ethyl acetate. Ammonia gas is bubbled in and the mixture is concentrated at reduced pressure to afford atorvastatin ammonium salt. The salt is dissolved in diisopropyl ether–isopropanol and a solution of calcium acetate, presumably in water, is added, presumably at 25°C. The resulting suspension is filtered and the solid is washed and dried to afford atorvastatin calcium. Many of the details of the procedure and the yield are not provided.[67,68]

Aqueous sodium hydroxide (1.0 M) (4.8 equivalents) is added to a solution of atorvastatin *tert*-butyl ester phenylboronate (**107**) in THF, presumably at 25°C, and the mixture is refluxed for 4 h. The mixture is cooled to 25°C and concentrated at reduced pressure. The aqueous mixture is diluted with water and washed with methyl *tert*-butyl ether. The aqueous layer is maintained at 25–28°C and reduced pressure to remove residual organic solvent. The resulting suspension is filtered and the solid atorvastatin sodium is dissolved in ethyl acetate. A solution of calcium acetate (0.77 equivalents, assuming monohydrate) in water is added at 40–45°C. The layers are separated and the organic layer is washed with water and concentrated at reduced pressure to afford amorphous atorvastatin calcium (80%). The volume throughput is $\leq$36 g/L. This process is demonstrated on 10 kg scale.[133]

Atorvastatin calcium is produced directly by ester hydrolysis with calcium hydroxide. A mixture of atorvastatin *tert*-butyl ester phenylboronate (**107**) and calcium oxide (12.9 equivalents) in 1:1 THF–water is aged at 50–60°C for 8 h. The suspension is filtered (temperature not specified) and the liquors are concentrated to a smaller volume, presumably at reduced pressure. The mixture is then washed with methyl *tert*-butyl ether and concentrated at reduced pressure. The residual solid is dissolved in THF and the solution is polish filtered and concentrated at reduced pressure to afford atorvastatin calcium. No yield is provided. Assuming a quantitative yield, the volume throughput is 11 g/L. Methanol or acetonitrile can be used in place of THF.[134]

### 9.8.7 From Atorvastatin *tert*-Butyl Ester Phenethylboronate (108)

Aqueous sodium hydroxide (1 M) (2.4 equivalents) is added to a solution of atorvastatin *tert*-butyl ester phenethylboronate (**108**) (*Note*: The (*R,R*)-stereochemistry is not specified.) in ethyl acetate. The mixture is aged, presumably at 25°C, for an unspecified time. Hydrochloric acid (1 M) is then added (to pH 1–2) and the layers are separated. Calcium acetate (*Note*: The procedure reads potassium acetate.) is added to the organic layer and the mixture is aged, presumably at 25°C, for 2 h. (*Note*: The salt exchange is more likely preformed using an ethyl acetate solution of an atorvastatin (**1**) salt. Perhaps several steps of the procedure are missing.) The resulting suspension is filtered and the solid is washed, presumably with ethyl acetate, and dried to afford atorvastatin calcium (56%). The volume throughput is 35 g/L.[69]

### 9.8.8 Atorvastatin Calcium from Atorvastatin Diphenylamide Acetone Ketal (110)

Hydrochloric acid (1 M) (equivalents not specified) is added to a solution of atorvastatin diphenylamide acetone ketal (**110**) in methanol at 25°C and the solution is aged at 25°C for 12 h. The resulting suspension is filtered and the solid is presumably washed with methanol–water and dried to afford atorvastatin diphenylamide (**134**). No yield is provided.

A mixture of atorvastatin diphenylamide (**134**) in 2.0 M aqueous sodium hydroxide (21 equivalents) and methanol is heated at 70°C for 4 h. The suspension is cooled to 25°C and filtered. Methyl *tert*-butyl ether is added to the liquors and the layers are separated. Hydrochloric acid (2 M) is added (to pH 2) and the mixture is extracted with methyl *tert*-butyl ether. Water–methanol (10:1) and 2.0 M aqueous sodium hydroxide are added (to pH 12) to the organic layer and the layers are separated. The aqueous layer containing atorvastatin sodium is diluted with water and washed with methyl *tert*-butyl ether.

A solution of calcium acetate (1.2 equivalents) in water is added to the aqueous layer and the mixture is aged at 25°C for 2 h and at 10°C for 3 h. The suspension is filtered and the solid is washed with cold water and dried to afford atorvastatin calcium. No yield is provided. Assuming a quantitative yield, the volume throughput is <10 g/L. The diol deprotection, amide hydrolysis, and salt exchange process is also demonstrated for other atorvastatin amide acetone ketals.[71]

### 9.8.9 From Atorvastatin *O*-Benzyl Lactol Methyl Acetal (135)

The ethyl ester of **114** is converted to the *N*-phenylamide by reaction with aniline (22 equivalents) in DMF at 80°C. The mixture is cooled and concentrated at reduced pressure. A routine water–ethyl ether workup affords atorvastatin *O*-benzyl lactol methyl acetal (**135**) (81%) (Scheme 9.38).

A solution of the atorvastatin *O*-benzyl lactol methyl acetal (**135**) in acetic acid–water is aged at 40°C for 24 h. The solution is cooled to 25°C, 1 M sodium hydroxide is added (to pH 7–8), and the mixture is extracted with dichloromethane. The extracts are washed with saturated aqueous sodium bicarbonate and with brine and then dried and concentrated at reduced pressure to afford the lactol benzyl ether **136** (90%).

A suspension of the lactol benzyl ether **136** and manganese dioxide (64 equivalents) in dichloromethane is aged at 25°C for 2 h. The suspension is filtered and the solid is washed with dichloromethane. The liquors are concentrated at reduced pressure to afford atorvastatin lactone benzyl ether **137** (79%).[74]

Atorvastatin lactone (**133**) can also be produced by debenzylation, acid-catalyzed cleavage of the lactol acetal, and lactol oxidation with Dess–Martin periodinane. This sequence is demonstrated on milligram scale[80].

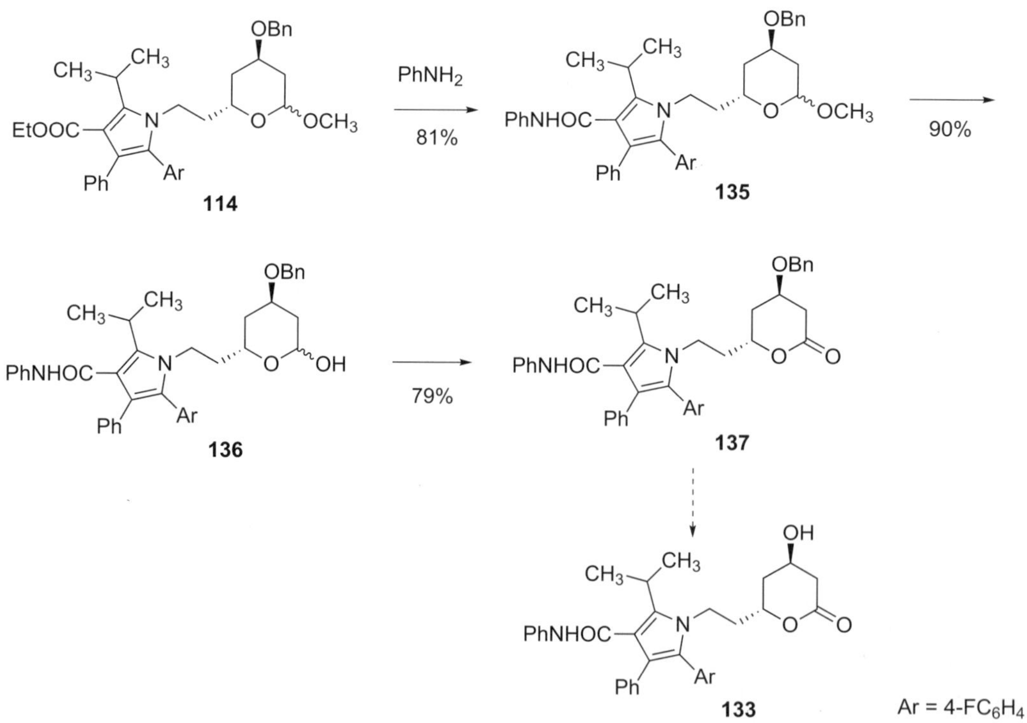


**SCHEME 9.38** Atorvastatin lactone (**133**) from atorvastatin *O*-benzyl lactol methyl acetal (**135**).

### 9.8.10 From Atorvastatin Alcohol Acetone Ketal (112)

Dess–Martin periodinane (1.5 equivalents) is added to a solution of atorvastatin alcohol acetone ketal (**112**) in dichloromethane at 0–5°C and the mixture is aged, presumably at 5°C, for 5 h. Water is added and the layers are separated. The aqueous layer is extracted with chloroform. The organic layers are washed with aqueous sodium metabisulfite, with 5% sodium bicarbonate, and with brine and then dried and concentrated at reduced pressure to afford atorvastatin aldehyde acetone ketal (**138**) (94%).

A solution of sodium chlorite (3.0 equivalents) and sodium monophosphate (2.0 equivalents) in water is added to a solution of aldehyde **138** and 2-methyl-2-butene (2.0 equivalents) in *tert*-butanol at 25°C. The mixture is aged at 25°C for 4 h. Saturated aqueous sodium sulfite is added and the mixture is then added to dilute hydrochloric acid. The resulting mixture is extracted with dichloromethane. The extracts are washed with brine, dried, and concentrated at reduced pressure to afford atorvastatin acetone ketal (**139**) (34%) (Scheme 9.39).[66]

## 9.9 AMORPHOUS ATORVASTATIN CALCIUM FROM ATORVASTATIN CALCIUM

As synthetic organic chemists, our first inclination is to generate crystalline intermediates and products whenever possible. An optimal crystallization from solution affords a solid that will consistently meet purity and particle size range specifications. With these specifications met, the filtration and washing times will be consistent from batch to batch. The filtration and washing will deliver a *wet cake* with a consistent *volume*. The same drying operation (time, temperature, and agitation program) will deliver the dry solid with consistent physical properties including *bulk density* and *storage stability*. Why is this important? If one batch just fills the filter the next batch with a lower bulk density may require a second (and unavailable) filter. If the bulk density of a batch of the dry solid is lower than the bulk density of material used to develop the formulation procedure, the solid may not fit in the hopper used to deliver it to the formulation mixture or, worse, the formulation may be unsuccessful.

Why is there so much interest in producing amorphous atorvastatin calcium? First, atorvastatin calcium in Lipitor® is polymorph I-US5969156. The U.S. patent on this polymorph does not expire for several more years. Second, perhaps there is a form of amorphous atorvastatin calcium has a faster dissolution rate and/or better bioavailability than polymorph I-US5969156.

The processes delivering amorphous atorvastatin calcium in the preceding section generally fall into two categories. There are many processes with lower volume throughput involving addition of a solution in a polar organic solvent to a nonpolar organic antisolvent. These are likely to produce consistently amorphous material in the lab and on scale-up. There are also some higher volume throughput processes involving salt exchange in an organic–aqueous solvent mixture. These salt exchange processes are more likely to produce mixtures of amorphous and crystalline forms.

There are many solvent–antisolvent processes used to convert crystalline atorvastatin calcium to the amorphous form. Polar organic solvents (DMF, DMSO, methanol, ethanol, acetone, THF, 1,4-dioxane, or dichloromethane) are mixed and matched with water or nonpolar organic antisolvents (ethyl ether, diisopropyl ether, cyclohexane, *n*-hexane or *n*-heptane). In most cases the amorphous form isolated is

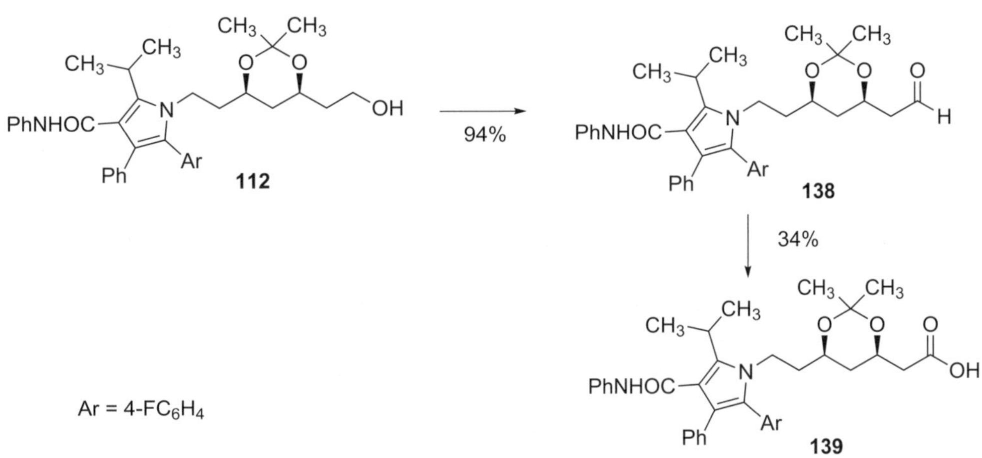


**SCHEME 9.39** Atorvastatin acetone ketal (**139**) from atorvastatin alcohol acetone ketal (**112**).

anhydrous. In cases where this is not specifically stated, the amorphous form is presumed to be anhydrous. Crystalline atorvastatin calcium is also converted to the amorphous form by spray drying, freeze-drying, or grinding.

Crude amorphous atorvastatin calcium is dissolved in DMF or DMSO (667 g/L), presumably at 25°C, and the solution is added to water at 25°C. The resulting suspension is aged at 25°C for 1 h, aged at 10°C for 1 h, and filtered. The solid is washed with cold water and dried at 40°C and reduced pressure to afford amorphous atorvastatin calcium (88–91%). The volume throughput is 76–79 g/L.[135]

Atorvastatin calcium polymorph I-US5969156 is dissolved in isopropanol (23 g/L). The solution is hot filtered, cooled to 4°C, and aged at 4°C for 4 h. The resulting suspension is filtered and the solid is washed with cold isopropanol and dried at 25°C and reduced pressure to afford amorphous atorvastatin calcium (91%). Amorphous atorvastatin calcium is also produced by crystallization from a mixture of *tert*-butanol and methanol.[109,136]

Atorvastatin calcium polymorph I-US5969156 is dissolved in refluxing ethanol (100 g/L). The solution is hot filtered into 100 mL of hot isopropanol. The resulting suspension is then cooled to 4°C and aged at 4°C for 4 h, and filtered. The solid is washed with cold isopropanol and dried at 25°C and reduced pressure to afford amorphous atorvastatin calcium (87%).[136]

Atorvastatin calcium polymorph I-US5969156 is dissolved in methanol at 25°C. The solution is concentrated at reduced pressure. The concentrated solution is diluted with ethyl ether and then added to ethyl ether. The resulting suspension is filtered and the solid is dried on a rotary evaporator at 50°C and reduced pressure to afford amorphous atorvastatin calcium (97%). Similar results are achieved using ethanol or acetone in place of methanol. The addition can be reversed but this requires rapid addition of the nonsolvent (ethyl ether) to the solution to minimize the potential for atorvastatin calcium to crystallize.[137]

Atorvastatin calcium (polymorph I-US5969156 or a mixture of crystalline and amorphous forms) is dissolved in acetone at 45°C. The solution is concentrated at 45°C and reduced pressure. The resulting concentrated solution is dried in a vacuum oven at 50°C and reduced pressure to afford amorphous atorvastatin calcium.[138]

Atorvastatin calcium is dissolved in THF (50 g/L) at 55°C. The solution is polish filtered, presumably at 25°C, and concentrated at reduced pressure. The resulting concentrated solution is presumably added to diisopropyl ether at 25°C. The resulting suspension is aged for 30 min and then filtered. The solid is washed with diisopropyl ether and dried at 55°C and reduced pressure to afford amorphous atorvastatin calcium. The volume throughput is 22 g/L.[139]

Atorvastatin calcium polymorph I-US5969156 is dissolved in THF (333 g/L) at 25°C and the solution is added to cyclohexane at 25°C. The resulting suspension is centrifuged and the solid is presumably washed with cyclohexane and dried at 60°C and reduced pressure to afford amorphous atorvastatin calcium (95%). The volume throughput is 25 g/L. Residual solvent levels are low (0.01 wt% THF, 0.6 wt% cyclohexane). The process is demonstrated on a 10 kg scale. The same results are achieved when the addition is nonsolvent to solution and when *n*-hexane or *n*-heptane are used in place of cyclohexane.[140]

Isopropanol (1.5% by volume) can be added to keep the amorphous atorvastatin calcium from sticking to the reactor walls. Atorvastatin calcium is dissolved in THF (100 g/L) at 25°C and the solution is added to heptane–isopropanol at 15–25°C. The resulting suspension is aged at 25°C for 1 h, aged at 0–5°C for 1 h, and filtered. The solid is dried at 50–60°C and reduced pressure to afford amorphous atorvastatin calcium (89%). The volume throughput is 22 g/L. The yield drops to 85% when the isopropanol charge is increased to 3% by volume.[141,142]

Crude amorphous atorvastatin calcium is dissolved in THF (333 g/L), presumably at 25°C, and the solution is added to water at 25°C. The resulting suspension is aged at 10°C for 1 h and filtered. The solid is washed with water to afford a wet cake weighing 2.7 times the atorvastatin calcium charged. The cake is dried at 40°C and reduced pressure to afford amorphous atorvastatin calcium (95%). The volume throughput is 42 g/L.[135]

Atorvastatin calcium is dissolved in 1,4-dioxane (333 g/L) at 45–50°C and the solution is added to methyl *tert*-butyl ether, presumably at 25°C. The resulting suspension is filtered and the solid is presumably washed with methyl *tert*-butyl ether and dried at reduced pressure to afford amorphous atorvastatin calcium. No yield is available. Cyclohexane, *n*-heptane, and diisopropyl ether can be used in place of methyl *tert*-butyl ether. Assuming a quantitative yield, the volume throughputs are 9–19 g/L.[109]

Atorvastatin calcium polymorph I-US5969156 is dissolved in dichloromethane (33 g/L) at 36–38°C and the solution is distilled. The resulting concentrated solution (125 g/L) is added to diisopropyl ether at 0–2°C. The suspension is aged at 0–2°C for 30 min and filtered. The solid is dried at 45–50°C and reduced pressure to afford amorphous atorvastatin calcium (90%, 98.94% pure by HPLC). The volume throughput is 27 g/L. The residual solvent levels are all <0.1%.[115]

Atorvastatin calcium polymorph I-US5969156 is dissolved in THF (400 g/L) and toluene is added to produce a solution in 1.5:1 THF–toluene. The solution is polish filtered during transfer to a vertical pan dryer. The solution is concentrated at 35°C and reduced pressure. Near the end of the concentration, the agitator is raised out of the mixture. At the end of the concentration, the agitator is lowered again to convert the brittle glass foam to a free-flowing powder. The powder is dried at 85°C and reduced pressure to afford amorphous atorvastatin (95%). The volume throughput is

211 g/L. The residual solvent levels are low (0.01% THF, 0.29% toluene).[121]

Amorphous atorvastatin calcium is produced by spray-drying a solution in methanol (50–150 g/L), ethanol (50 g/L), acetone (40–55°C, 100–800 g/L), 2-butanone (50–55°C, 340–430 g/L), ethyl acetate (60–75°C, 180–225 g/L), or dichloromethane (40–45°C, 460 g/L).[90,141,143–146] Amorphous atorvastatin calcium is produced from atorvastatin calcium polymorph I-US5969156 by freeze-drying a solution in methanol or DMSO.[144,147] Amorphous atorvastatin calcium is produced from atorvastatin calcium polymorph I-US5969156 by pulverizing using a mortar and pestle or a mechanical grinder.[145]

Many mind-set shifts are required to anticipate how a process must change when moving from the laboratory to the pilot plant. There is perhaps no greater shift than the "clean and dry vessel" mind-set shift. For each batch of material prepared in the laboratory, we start with carefully inspected clean and perhaps oven-dried equipment. Since clean and dry equipment is a core principle of laboratory work, it is rarely specified in a laboratory procedure. In contrast, in the pilot plant, the batch procedure will clearly state that the first reagent is charged to a clean and dry vessel. Cleaning and drying a pilot plant vessel may require many hours. The status as clean and dry must be verified by analyzing samples of the rinse solvent(s). *The cleaning and verification of a vessel (and the entire equipment train) is only performed before the first batch.* After moving the first batch through each piece of equipment in the process train, the equipment is usually held at ambient temperature to receive the next batch. What problems can arise when the equipment is not cleaned, dried, and verified for each batch? Problems can arise when the material moving through the equipment is not stable over time at ambient temperature. This may be the case when converting atorvastatin calcium to the amorphous form. Amorphous material left in contact with solvent at ambient temperature will eventually crystallize. Once crystals form in any piece of equipment in the process train, cleaning the train to again produce consistently amorphous material may be a monumental challenge.

## 9.10 THE CONTINUING REFINEMENT OF ANALYTICAL METHODS: IDENTIFYING PROCESS IMPURITIES AND EVALUATING DRUG PRODUCT STABILITY DURING STORAGE AND FORMULATION

Many atorvastatin calcium impurities can be resolved by a single HPLC method.[47] The potential impurities can be categorized based on their point of origin. In a Paal–Knorr synthesis of the pyrrole, impurities can be traced back to (1) the 1,4-diketone synthesis, (2) the synthesis of the amine, (3) the pyrrole ring formation, (4) the deprotection–hydrolysis sequence, or (5) the inherent instability of atorvastatin calcium (Figure 9.5). For example, desfluoro **140** and *ortho*-fluoro **141** impurities originate in the 1,4-diketone synthesis. The (3*S*,5*S*) **142** and (3*S*,5*R*) **143** epimers and the 3-oxo-5-hydroxy impurity **144** originate in the side chain elaboration sequence. Amide impurities **102** and **145** originate in the Paal–Knorr pyrrole ring synthesis. Atorvastatin *tert*-butyl ester **125** and methyl ester **132** are carried over from the deprotection–hydrolysis sequence. Finally, pyrrole oxidation products result from exposure to air and light.[148]

The amide impurities 100–102 and 145 are especially difficult to quantify since they are produced from amine **8** that has no chromophore. A qualitative detection of the amide impurities is possible by $^{13}C$ NMR spectroscopy. A $^{13}C$-labeled version of nitrile **32** is produced using labeled potassium cyanide. The nitrile is reduced, the $^{13}C$-labeled amine is used in Paal–Knorr pyrrole synthesis, and the $^{13}C$-labeled atorvastatin *tert*-butyl ester acetone ketal **99** is converted to crude $^{13}C$-labeled atorvastatin lactone **133**. The amide impurities can be detected and identified in the $^{13}C$ NMR spectrum of the lactone.[149]

The atorvastatin calcium process research and development effort continues beyond confirming the yield and purity of final drug product. Atorvastatin calcium is unstable to heat, moisture, low pH, and UV and fluorescent light. When exposed to intense simulated sunlight, particularly the "low visible" wavelengths (365 nm), atorvastatin is rapidly converted to several decomposition products. It is suggested that the decomposition begins with photostimulated formation of an endoperoxide. Ultraviolet irradiation of the eventual oxidation product **146** generates phenanthrene **147** that then serves as a sensitizer to accelerate the endoperoxide formation (Scheme 9.40).[150]

Operations involving the final drug substance should be designed to minimize exposure to light. In the pilot plant and in manufacturing, this may become an issue while transferring solid to and from a dryer.

When air-oxidation of atorvastatin calcium is evaluated under stress conditions ($O_2$ atmosphere, 80°C for 6 days), the level of oxidation products increases with an increase in the oxygen content of the atmosphere. Crystalline atorvastatin calcium (presumably polymorph I-US5969156) fares much better than a form of amorphous atorvastatin calcium in this study: the increase in degradation products is 0.05% for crystalline and 0.89% for amorphous. Four of the decomposition products are isolated by preparative chromatography and characterized by mass spectrometry and $^{1}H$ and $^{13}C$ NMR spectroscopy. One of these decomposition products is diketo epoxide **148** also produced by photodecomposition (Figure 9.6).[151,152]

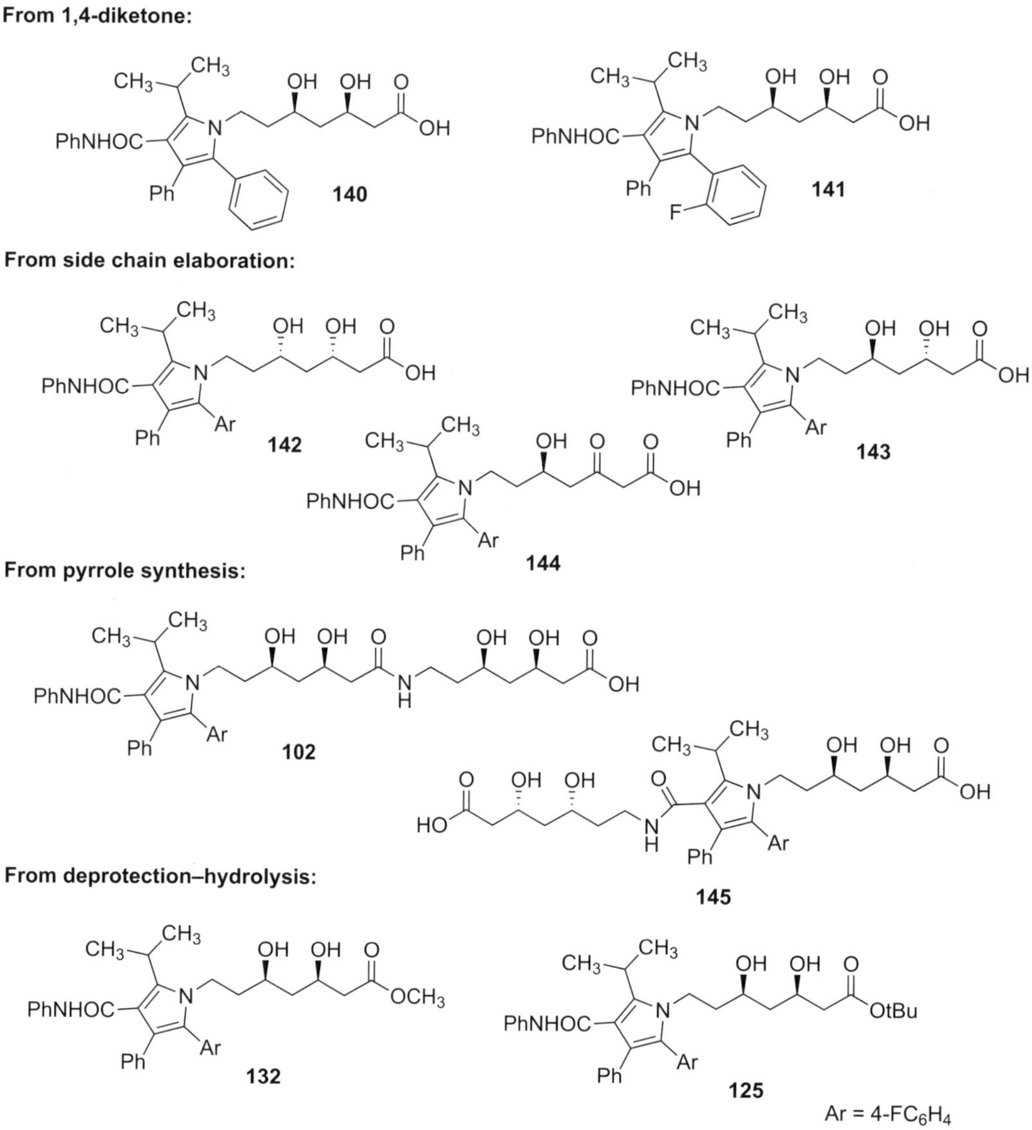


**FIGURE 9.5** Potential atorvastatin impurities.

The potential for air oxidation suggests the last item on the process chemist's checklist is to ensure that the bulk drug is packaged in an inert gas atmosphere to maximize storage stability. The bulk drug could be packaged in an airtight bag that is filled with nitrogen and sealed. The sealed bag is then placed in a second airtight bag. Oxygen-absorbing and moisture-absorbing packets might be added to the second bag before it is filled with nitrogen and sealed. The second bag is then packaged in a rigid container suitable for storage and shipping.[153]

Contact with acid converts atorvastatin (**1**) calcium to atorvastatin (**1**) and catalyzes cyclization to atorvastatin lactone (**133**) and elimination of the $\beta$-hydroxy group. The instability to low pH is addressed during formulation by adding a base stabilizer. Lipitor® is formulated with calcium carbonate.[154] Amorphous atorvastatin calcium might be stabilized with *N*-methylglucamine,[155] sodium carbonate,[156] dibasic sodium phosphate,[156,157] tribasic sodium phosphate,[158] crospovidone (a cross-linked polyvinylpyrrolidone),[159] zinc carbonate, and other water-insoluble alkaline excipients.[160] The list of effective stabilizers and robust formulation recipes appears to be endless.

## 9.11 THE BEST PROCESS AVAILABLE

The best process discussion begins with the method for pyrrole ring construction. There are just three methods to consider: [3 + 2]-cycloaddition, Paal–Knorr synthesis, and the related 1,4-diketone-azide condensation. In the [3 + 2]-cycloaddition, two pyrrole ring carbons are derived from expensive phenylpropiolic acid. *N*,3-Diphenylpropiolamide

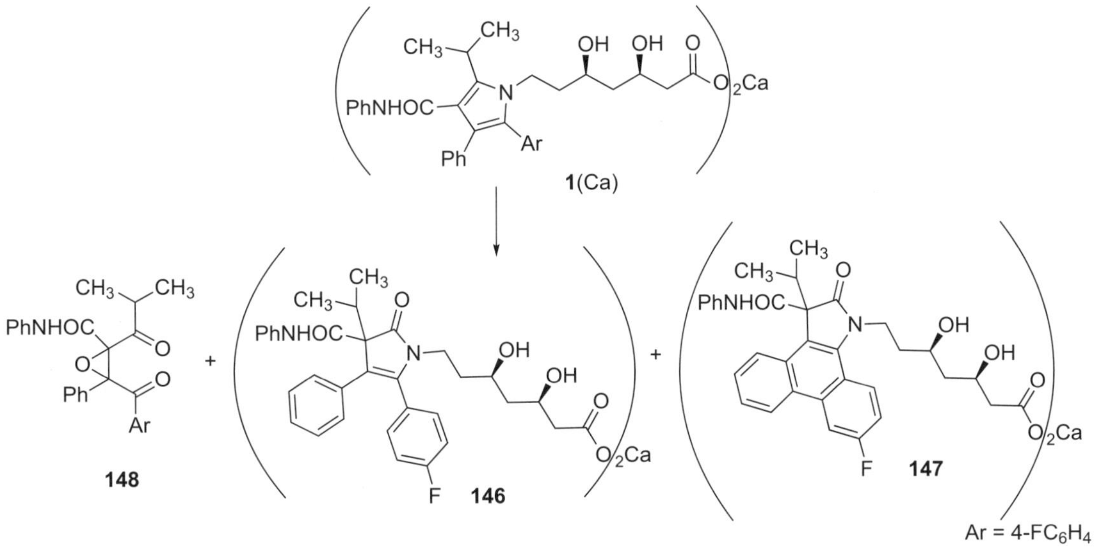


**SCHEME 9.40** Photodecomposition of atorvastatin calcium.

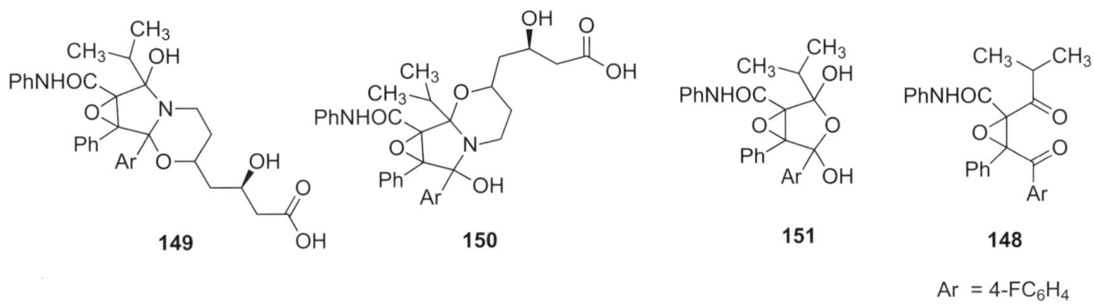


**FIGURE 9.6** Products from exposure of atorvastatin (**1**) calcium to oxidative stress ($O_2$ atmosphere, 80°C for 30 days).

(**11**) is used in excess (1.6 equivalents) in the cycloaddition and excess **11** must be separated from the pyrrole product by chromatography. The yields in the cycloaddition are only 45–55%. Consider the maximum convergency route using the elaborated-protected statin side chain. The acetal and *tert*-butyl ester protecting groups on the side chain are not sufficiently robust: with a 60% yield to the dipolarophile **7** and assuming a yield of 50% in the cycloaddition, the yield of atorvastatin *tert*-butyl ester acetone ketal (**99**) from the expensive amine **8** would be just 30%.

Pyrrole formation from 1,4-diketone **21** and azide **116** requires the use of excess tributylphosphine. Tributylphosphine is a potent lacrymator, has a garlic-like stench, and is pyrophoric. Pyrrole **117** is separated from the phosphine and phosphine oxide by chromatography. The use of tributylphosphine and a chromatography are too high a price to pay.

Which of the 14 amines is the best choice for the Paal–Knorr synthesis? To answer this question we must first decide if the side chain should be elaborated before or after the pyrrole construction. This is a complex question with advantages and disadvantages to both options (Schemes 9.41–9.43).

When assigned a new target in process research, the first objective is to identify all the available pertinent data and assemble that data in some logical pattern. The second objective is to bridge the knowledge gaps with well-designed experiments to create a complete and efficient sequence. When there is a lot of available data, it is easy to "get lost." We may become entrenched in reviewing the data and defining and redefining the well-defined experiments and lose sight of the big picture. One way to rediscover the big picture is to strip away the data. This approach may come at the request of manufacturing management, management uncomfortable with mountains of data and the time spent reviewing it, perhaps management with an engineering or business background. Management may look at a process research presentation and see only aromatic ring "chicken wire." To rediscover the big picture, redraw the scheme(s) representing process starting materials, intermediates that are likely stopping points, and the target by boxes. Sound process decisions may come from examining the "no chicken wire" representation. Pertinent information of interest to management, for example, the cost for a key material, can be added inside the boxes where appropriate. For the specific analysis at hand, the materials associated with each box are identified by number.

The route utilizing 2-(2-aminoethyl)-1,3-dioxolane to produce aldehyde **118** has a longer linear sequence (13 boxes in linear sequence versus 7 or 8 for the other options). The sequence represented is for the asymmetric addition to the aldehyde. A sequence involving addition of methyl acetate to the aldehyde and a downstream resolution/racemization to produce acid **123** would have the same number of boxes. The route utilizing potentially less expensive amine **96** affords the 3,5-dioxoheptanoate **127**. Without a doubt, this is a key strategic precursor to atorvastatin (**1**). However, the current less-than-perfect reduction methodology affords a mixture of diastereomers. Both side chain elaboration approaches require a late-stage upgrade to meet optical purity specifications. For these reasons, the best available option in 2009 is to elaborate the side chain before pyrrole construction.

All three routes require 1,4-diketone **21**. What is the best route to this key intermediate? Michael addition of an acyl anion equivalent is unattractive. The conversion of 4-fluorobenzaldehyde to the dithioacetal **24** (43%) or 1,3-dithiane **25** (79%) is inefficient, as are the cleavage of dithioacetal **26** (49%) and 1,3-dithiane **27** (54%). The dithioacetal and 1,3-dithiane cleavage generates a heavy metal waste stream. The ethanethiol or 1,3-propanedithiol will no doubt lead to odor complaints.

Focusing only on the cost of the fluorine-containing reagent, fluorobenzene versus 4-fluorobenzaldehyde, the α-bromoketone route appears to be the better option. The Friedel-Crafts acylation of fluorobenzene is run in fluorobenzene as the solvent. Can the fluorobenzene distillate be recycled without additional distillation? But there is another important consideration: α-bromoketone **20** is a lacrymator. The lacrymator can be avoided by generating the α-bromoketone *in situ* from the bromine exchange between ketone **19** and 2-bromo-4-methyl-3-oxo-*N*-phenylpentanamide (**31**).

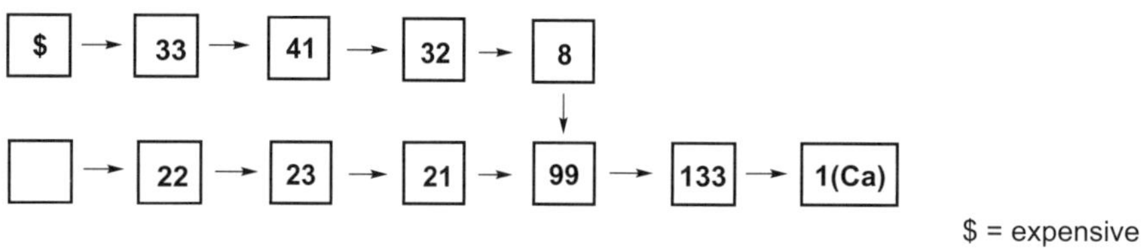


**SCHEME 9.41** "No chicken wire" process for atorvastatin (**1**) calcium by Paal–Knorr synthesis using amine **8**.

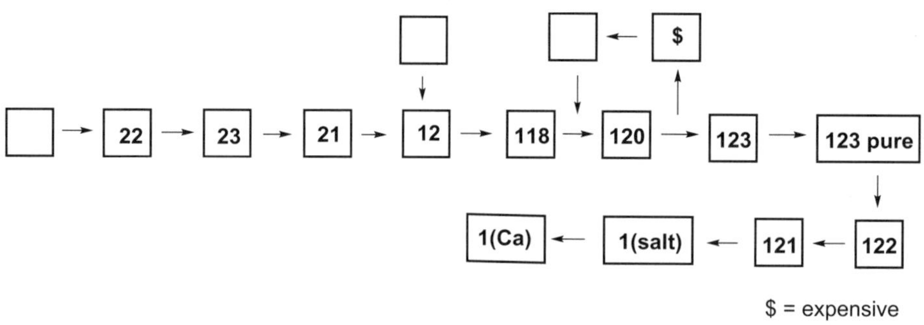


**SCHEME 9.42** "No chicken wire" process for atorvastatin (**1**) calcium by Paal–Knorr synthesis using 2-(2-aminoethyl)-1,3-dioxolane.

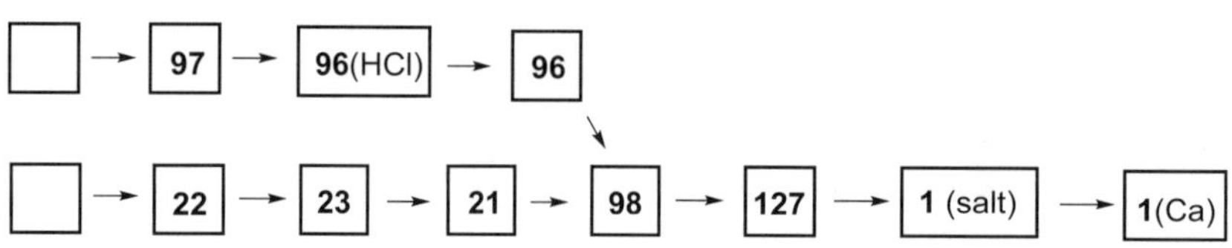


**SCHEME 9.43** "No chicken wire" process for atorvastatin (**1**) calcium by Paal–Knorr synthesis using amine **96**.

But the yields of 1,4-diketone **21** using this approach are only 51–62% using sodium carbonate as the base. Higher yields (85–87%) are only achieved using lithium diisopropylamide at low temperature (−60 to −78°C).

The Stetter route to 1,4-diketone **21** is preferred. This 1,4-diketone construction includes an uncommon back-to-back incorporation of two reagents, each with the same (aldehyde) functional group. Benzaldehyde is added first, the product 2-benzylidene-4-methyl-3-oxo-*N*-phenylpentanamide (**23**) is filtered and washed, and then 4-fluorobenzaldehyde is added second. It is critically important that the wash of **23** remove all traces of benzaldehyde before proceeding to the Stetter reaction with 4-fluorobenzaldehyde. This is a tall order, considering that the suggested wash solvent is water and the solubility of benzaldehyde in water is just 6 g/L at 20°C. In addition, the highest yields in the Stetter reaction with 4-fluorobenzaldehyde are achieved under anhydrous conditions, so the water-washed 2-benzylidene-4-methyl-3-oxo-*N*-phenylpentanamide must be rigorously dried. Use of a wash solvent other than water could address both these issues. The thiazolium bromide catalyst (15–20 mol%) can be produced at the start of the Stetter procedure from 4-methyl-5-thiazoleethanol[161] and ethyl bromide. 1,4-Diketone **21**, crystallized from isopropanol, is assayed to ensure that the specification for the desfluoro impurity **28** will be met before proceeding.

Diketone **21** is assembled from four commercially available materials: methyl 4-methyl-3-oxovalerate, aniline, benzaldehyde, and 4-fluorobenzaldehyde. The overall yield for three steps from methyl 4-methyl-3-oxovalerate is 58%. There are two isolations. The process solvents are hexanes, toluene, ethanol, and isopropanol.

Which elaborated-protected amine is the preferred partner? The amines with silyl protecting groups are too expensive. The amines with ketals and boronate esters offer the highest yields in the pyrrole construction and do not require a late-stage oxidation. *tert*-Butyl 2-((4*R*,6*R*)-6-(2-aminoethyl)-2,2-dimethyl-1,3-dioxan-4-yl)acetate (**8**) is selected to continue the discussion. We will revisit this decision after completing the process description to atorvastatin calcium.

Amine **8** is prepared in four steps from ethyl (*R*)-(−)-4-cyano-3-hydroxybutyrate and *tert*-butyl acetate. The highest yield in the initial condensation (94%) is achieved by using 4.4 equivalents of lithium diisopropylamide and low temperature (−50°C). There is a great deal of published information on the reduction of ketone **33** with sodium borohydride and diethylmethoxyborane, even a demonstrated recycle of the expensive diethylmethoxyborane. The yield and selectivity expected for the borohydride reduction (80%, 40:1) do no match with the results achieved using *Pichia angusta* NCYC R320 (92%, 110:1). The diol **41** is protected as the ketal (94%). Ketal **32** is crystallized from heptane and assayed to ensure that the specification for side chain epimer impurities will be met. The overall yield from ethyl (*R*)-(−)-4-cyano-3-hydroxybutyrate is 81%. There is one solid isolation at the end of this sequence. The process solvents are hexanes, THF, ethyl acetate, and acetone.

Nitrile **32** is reduced using Raney nickel (250 g/kg nitrile **32**) in methanol saturated with ammonia at 30°C and 50 psi hydrogen pressure. It is important to maintain a low maximum temperature during the reduction and workup to minimize formation of amine dimer **101**. After filtering and washing the catalyst, the liquors are concentrated at reduced pressure. The concentrated methanol solution is diluted with toluene and residual methanol is then chased by distillation at reduced pressure. The concentrated solution of amine **8** (99%) in toluene is carried into the next step. Rather than filtering the catalyst, the suspension could be decanted. The catalyst in the hydrogenation vessel is then ready for the next batch of nitrile.

The 1,4-diketone **21**, the toluene solution of the amine **8** (1.05 equivalents), THF, perhaps hexane, and pivalic acid (0.67 equivalents) are charged and the mixture is heated at 60–70°C for 40–48 h. From here there are many workup options leading to a solution of the pyrrole in isopropanol or ethanol. Adding water and cooling affords pyrrole **99** (74–79%).

Ketal deprotection with aqueous hydrochloric acid (0.27 equivalents) in methanol is followed by ester hydrolysis with sodium hydroxide. The aqueous solution of atorvastatin sodium is washed with methyl *tert*-butyl ether to remove atorvastatin *tert*-butyl ester (**125**) and methyl ester (**132**).

The conversion of atorvastatin (**1**) sodium to atorvastatin lactone (**133**) and the conversion of the lactone to atorvastatin (**1**) calcium are well established. With the manufacturing cost soon to be an important consideration, it appears that many groups consider this two-step approach a weakness and an opportunity for process improvement. Three key factors to consider in selecting salt exchange conditions are the *filterability of the suspension*, the *efficiency of the drying operation*, and the *purity of the atorvastatin (1) calcium* isolated.

> Few of the salt exchange procedures address filterability of the suspension and these offer only qualitative evaluations such as "it filters quickly." As a process chemist, the data you can offer to describe a filtration includes batch size, type of filter media (sintered glass funnel porosity A/B/C or porcelain Buchner funnel with paper), circumference of the filter plate, height of the washed wet cake, vacuum used (if any), and time to complete the filtration and washing. Pictures of the crystals from several lab batches demonstrate reproducibility of the crystallization, isolation, and washing processes and suggest the crystal morphology and size distribution to expect.

Very few of the "end-game" procedures address the fate of the amide impurities produced during the pyrrole synthesis. The atorvastatin *tert*-butyl ester acetone ketal (**99**) from the pyrrole synthesis can be crystallized from isopropanol. There is no data available on reducing the levels of the amide impurities by this crystallization. Atorvastatin *tert*-butyl ester acetone ketal (**99**) can be converted to atorvastatin lactone (**133**) and the lactone can be crystallized from toluene. This crystallization does reduce the levels of the amide impurities. Atorvastatin *tert*-butyl ester acetone ketal (**99**) can be converted to the ammonium salt. There is no data available on the levels of amide impurities in the isolated ammonium salt. Atorvastatin *tert*-butyl ester acetone ketal (**99**) can also be converted directly to atorvastatin (**1**) calcium. There is no data available on the levels of amide impurities in the atorvastatin calcium. With no purity data available for the alternative processes, the conservative atorvastatin lactone route must be selected.

Perhaps the interest in the other atorvastatin *tert*-butyl ester ketals and the boronate esters comes from data suggesting that their crystallization can reduce the amide impurities to acceptable levels? If so, a minor protecting group change could create an opportunity to bypass the lactone isolation.

The atorvastatin (**1**) sodium salt solution is acidified and atorvastatin (**1**) is extracted into toluene. The extracts are refluxed with removal of water as the azeotrope. The mixture is cooled and the resulting suspension is filtered. The solid is crystallized from toluene to afford atotvastatin lactone (**133**) (83%).

An exact charge of 0.15 M sodium hydroxide (1.03 equivalents) is added to atorvastatin lactone (**133**) in methanol–methyl *tert*-butyl ether. After the hydrolysis is complete, the layers are separated and an exact charge of calcium acetate is added to the aqueous layer. Recall the aqueous layer should be saturated with methyl *tert*-butyl ether (reactor sealed). The mixture is seeded with polymorph I-US5969156 and then cooled and aged to produce atorvastatin calcium polymorph I-US5969156 (97%).

1,4-Diketone **21** is converted to atorvastatin (**1**) calcium in three steps and overall 64% yield. There are three isolations. The solvents are toluene, methyl *tert*-butyl ether, THF, methanol, and isopropanol.

The overall process from methyl 4-methyl-3-oxovalerate to atorvastatin (**1**) calcium has eight steps, six isolations, and an overall yield of 37%. The overall process from ethyl (*R*)-(−)-4-cyano-3-hydroxybutyrate has nine steps and an overall yield of 46%. The process solvents, start-to-finish are hexanes, toluene, methyl *tert*-butyl ether, THF, ethyl acetate, acetone, methanol, ethanol, and isopropanol (Scheme 9.44).

The process weaknesses are the long reaction time, low yield, and amide impurities associated with the pyrrole construction step, the high cost for ethyl (*R*)-(−)-4-cyano-3-hydroxybutyrate, and the low temperature and excess of lithium diisopropylamide required for the *tert*-butyl acetate condensation.

## 9.12 ALTERNATIVE APPROACHES TO *TERT*-BUTYL 2-((4*R*,6*R*)-6-(2-AMINOETHYL)-2,2-DIMETHYL-1,3-DIOXAN-4-YL)ACETATE (8)

With the best route to atorvastatin (**1**) calcium identified, many groups see an opportunity to significantly reduce the manufacturing costs by reducing the price of amine **8**. Amine **8** is prepared in four steps from expensive ethyl (*R*)-(−)-4-cyano-3-hydroxybutyrate. The Claisen condensation with *tert*-butyl acetate (−50°C) requires low-temperature capability and an excess of expensive lithium diisopropylamide. The wealth of available information suggests each step is optimized, leaving little hope of decreasing the cost by an incremental process improvement. Thus, it is time to start with a "clean sheet," to consider alternative strategies that target elimination of the process weaknesses. The following discussion of alternative strategies is intended to be both forward-looking and thought-provoking. No attempt is made to consider all of the available options.

### 9.12.1 Cyanide Displacement to Produce *tert*-Butyl 2-((4*R*,6*R*)-6-(Cyanomethyl)-2,2-dimethyl-1,3-dioxan-4-yl)acetate (32)

Introducing the nitrile downstream would eliminate expensive ethyl (*R*)-(−)-4-cyano-3-hydroxybutyrate as the starting material. Consider taking any one of the many routes leading to the alcohol, *tert*-butyl 2-((4*R*,6*S*)-6-(hydroxymethyl)-2,2-dimethyl-1,3-dioxan-4-yl)acetate (**152**). This alcohol is a branch point in routes to other statins. Conversion of the alcohol to a leaving group and displacement with cyanide will produce *tert*-butyl (4*R*,6*R*)-6-(cyanomethyl)-2,2-dimethyl-1,3-dioxan-4-yl)acetate (**32**) (Scheme 9.45). But how efficient is this leaving group displacement by cyanide? A review of countless examples drives home the point that this displacement is difficult and the yield of the nitrile is low. For example, methanesulfonyl chloride (0.65 equivalents) is added to a mixture of the alcohol and triethylamine (0.89 equivalents) in toluene at 25°C and the mixture is aged at 25°C for 1 h. A routine water–toluene workup affords the crude methanesulfonate **153** (54–56% based on alcohol **152**). Potassium cyanide (1.1 equivalents) and 18-crown-6 (15 mol%) are added to a suspension of methanesulfonate **153** in water at 35°C. The mixture is aged at 80°C for 24 h. The mixture is then extracted with toluene. The extract is passed through a Fuller's earth column twice to remove the crown ether and some color, and the eluent is concentrated, presumably at reduced pressure. The residue is crystallized from hexane to afford *tert*-butyl 2-((4*R*,6*R*)-6-(cyanomethyl)-2,2-dimethyl-1,3-dioxan-4-yl)acetate (**32**) (60%). The displacement can be accomplished in DMSO without the crown ether. After 192 h at 45°C, nitrile **32** is isolated in 51% yield. A similar methanesulfonate displacement on the phenylboronate ester by potassium cyanide (1.5 equivalents)

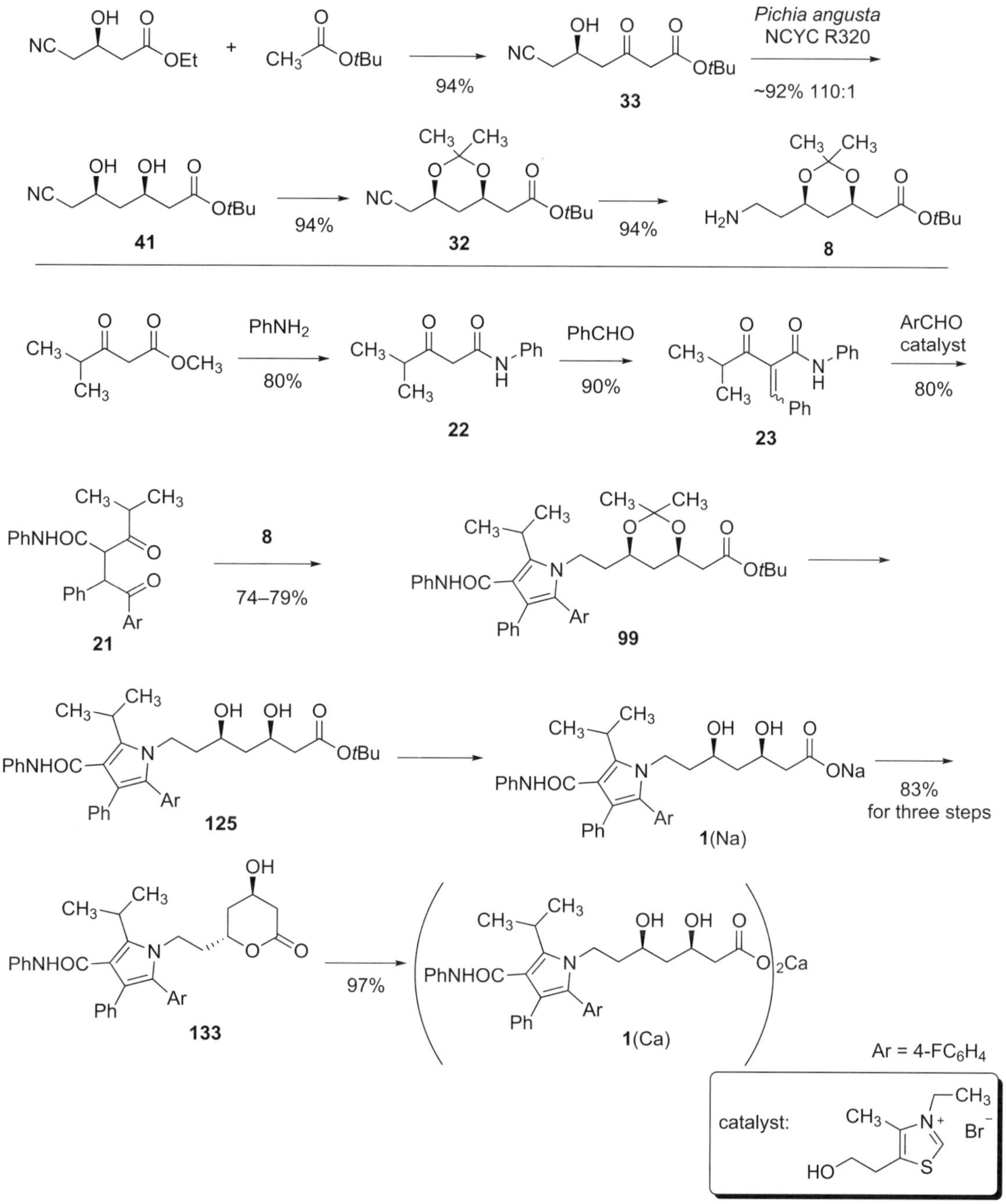


**SCHEME 9.44** The best process for atorvastatin (**1**) calcium in 2009.

required 18–22 h in refluxing DMSO. No yield is available.[162,163]

Iodolactonization is used to establish the *syn*-relationship of the side chain alcohols in a route starting with expensive 1,6-heptadien-4-ol. Reaction of the lithium alkoxide with carbon dioxide and then iodine affords the carbonate **154** (91%). The protecting group is exchanged, carbonate for acetone ketal, by reaction with catalytic *p*-toluenesulfonic acid in acetone at 25°C (90%). Here, again, the cyanide displacement of the neopentyl-type iodide of **155** is slow in DMSO (125 h at 40°C) (75–80%). The alkene of **156** is converted to the aldehyde **157** by ozonolysis with a reductive workup or by dihydroxylation and cleavage of the diol (65–70%). The aldehyde is oxidized with chromium trioxide to the acid **158** (70%). Resolution of the acid was unsuccessful. The acid is converted to the *tert*-butyl ester with *tert*-butanol, dicyclohexylcarbodiimide, and *N*,*N*-dimethylaminopyridine (Scheme 9.46).[164]

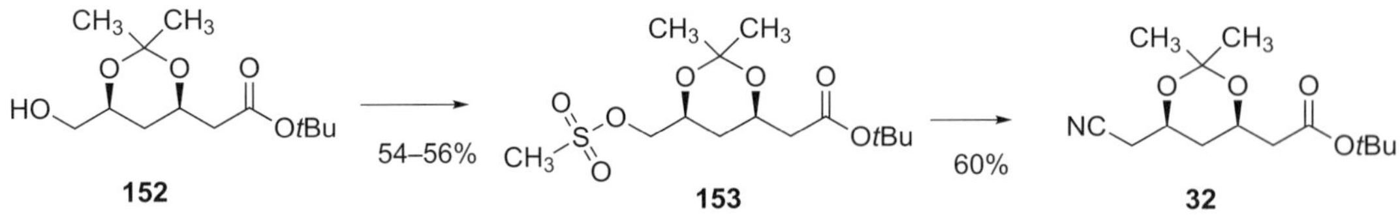


**SCHEME 9.45** *tert*-Butyl 2-((4*R*,6*R*)-6-(cyanomethyl)-2,2-dimethyl-1,3-dioxan-4-yl)acetate (**32**) from *tert*-butyl 2-(((4*R*,6*S*)-6-(hydroxymethyl)-2,2-dimethyl-1,3-dioxan-4-yl)acetate (**152**).

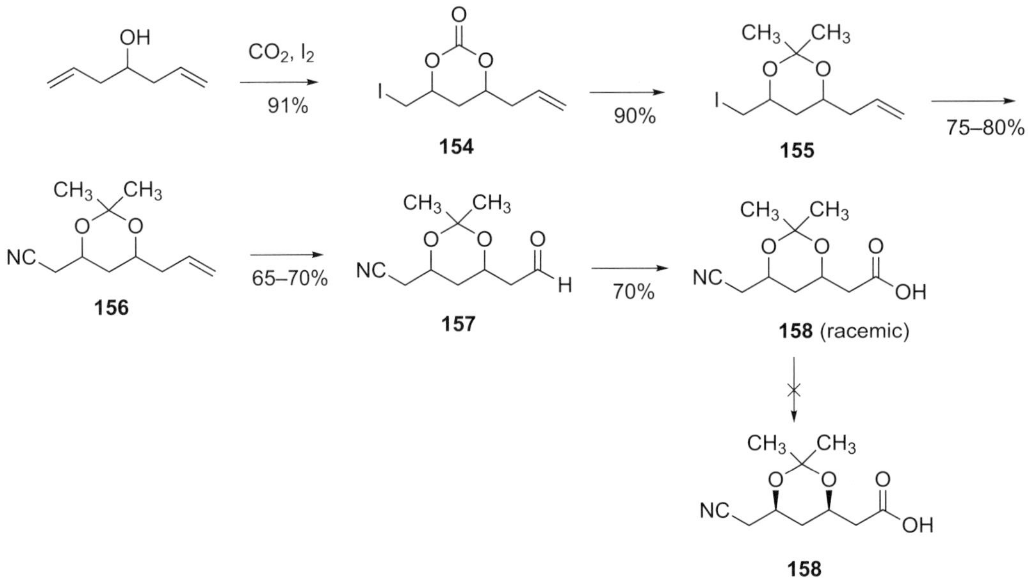


**SCHEME 9.46** A potential route to *tert*-butyl 2-((4*R*,6*R*)-6-(cyanomethyl)-2,2-dimethyl-1,3-dioxan-4-yl)acetate (**32**) from 1,6-heptadien-4-ol.

### 9.12.2 (3*R*,5*R*)-*tert*-Butyl 6-Cyano-3,5-dihydroxyhexanoate (41) by Epoxide Ring Opening

Taking one step back in the sequence, consider an alternative route via (3*R*,5*S*)-*tert*-butyl-6-chloro-3,5-dihydroxyhexanoate (**159**) (Scheme 9.47). The condensation of ethyl (*S*)-4-chloro-3-hydroxybutyrate and *tert*-butyl acetate can be accomplished at 20°C instead of −50°C when *n*-butyllithium is replaced by *n*-butylmagnesium chloride. Diisopropylamine (5.5 equivalents) is added to *n*-butylmagnesium chloride (5.0 equivalents) in toluene–THF at 40°C. The solution is then added to a mixture of ethyl (*S*)-4-chloro-3-hydroxybutyrate and *tert*-butyl acetate (2.5 equivalents) in 1,2-dimethoxyethane at 0–5°C. The mixture is aged at 20°C for 16 h. The mixture is quenched into dilute hydrochloric acid and ethyl acetate. The layers are separated and the organic layer is washed with brine, dried, and concentrated at reduced pressure. *tert*-Butyl (*S*)-6-chloro-5-hydroxy-3-oxohexanoate (**160**) is isolated from the residue by chromatography (80%). *tert*-Butyl (*S*)-6-chloro-5-hydroxy-3-oxohexanoate (**160**) can also be produced by Blaise reaction of (*S*)-4-chloro-3-(trimethylsilyloxy)butanenitrile (**161**) with *tert*-butyl bromoacetate and zinc dust (87%).[165,166]

Reduction using *Candida magnoliae* converts the $\beta$-ketoester to (3*R*,5*S*)-*tert*-butyl 6-chloro-3,5-dihydroxyhexanoate (**159**) (79%, 100% (3*R*,5*S*)).[165] Along similar lines, the chirality can be introduced by reduction of *tert*-butyl 6-chloro-3,5-dioxohexanoate (**162**) using *Lactobacillus brevis* or *Lactobacillus kefir*. The propensity of *tert*-butyl 6-chloro-3,5-dioxohexanoate (**162**) to cyclize to *tert*-butyl 2-(4-oxo-4,5-dihydrofuran-2-yl)acetate (**163**) is a significant issue for this alternative approach.[167]

Aqueous basic conditions used to convert the chlorohydrin of **159** to the epoxide are likely to cause hydrolysis of the *tert*-butyl ester. The reaction of (4*R*,6*S*)-6-(chloromethyl)-4-hydroxytetrahydro-2*H*-pyran-2-one (**164**) with potassium hydroxide (0.98 equivalents) probably generates the same carboxylate salt **165**. Reaction of carboxylate salt **165** with potassium cyanide (1.6 equivalents) at 50°C affords 2-((2*R*,4*R*)-4-hydroxy-6-oxotetrahydro-2*H*-pyran-2-yl)acetonitrile (**166**) (76%). Reaction with 2,2-dimethoxypropane

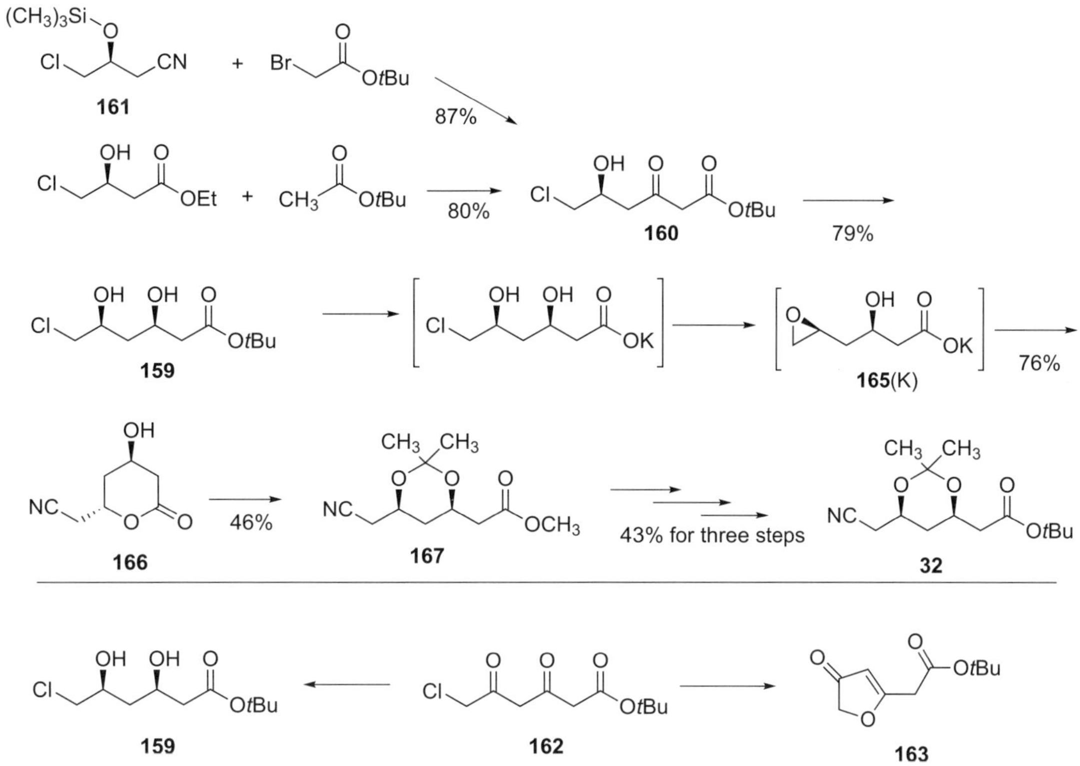


**SCHEME 9.47** *tert*-Butyl 2-((4*R*,6*R*)-6-(cyanomethyl)-2,2-dimethyl-1,3-dioxan-4-yl)acetate (**32**) from (3*R*,5*S*)-*tert*-butyl-6-chloro-3,5-dihydroxyhexanoate (**159**).

and catalytic *p*-toluenesulfonic acid monohydrate opens the lactone and converts the diol to the acetone ketal (46%). The methyl ester of **167** can be converted to the *tert*-butyl ester via the sodium salt and acid chloride (43% for three steps).[168]

Generating and opening the epoxide under anhydrous conditions would avoid hydrolysis of the *tert*-butyl ester. The epoxide generation and the ring opening by cyanide are demonstrated under anhydrous conditions in an alternative process starting with specialty chemical (*S*)-*tert*-butyl 2-(oxiran-2-yl)acetate. Reaction with vinylmagnesium bromide (1.3 equivalents) and copper(I) iodide (0.99 equivalents) produces the secondary alcohol **168**, which is then converted to the *tert*-butyl carbonate **169** (56% chromatographed for two steps). Reaction with iodine monobromide (1.5 equivalents) results in iodolactonization to a 22:1 mixture of the (4*R*,6*S*) and (4*R*,6*R*) products **170**. The (4*R*,6*S*) product **170** is isolated by chromatography (68%). Conversion to the epoxide **171** is accomplished under anhydrous conditions by reaction with potassium carbonate in methanol at −30 to 0°C (80%). Finally, the epoxide opening under anhydrous conditions (in benzene) utilizes titanium (IV) isopropoxide (2.0 equivalents), 18-crown-6 (1.0 equivalent) and potassium cyanide (1.0 equivalent) (77%) (Scheme 9.48).[60]

### 9.12.3 *tert*-Butyl 2-((4*R*,6*R*)-6-(2-Aminoethyl)-2,2-dimethyl-1,3-dioxan-4-yl)acetate (8) from Benzyl 3-Oxopropylcarbamate (172) and Acetaldehyde

(4*R*,6*S*)-6-(Chloromethyl)-4-hydroxytetrahydro-2*H*-pyran-2-one (**164**) introduced above is produced in two steps: an aldolase-mediated condensation of chloroacetaldehyde with 2 mol of acetaldehyde and oxidation of the resulting lactol with bromine.[169,170] The powerful aldolase-mediated condensation, which simultaneously creates two carbon–carbon bonds and the two chiral centers with the correct (*R*,*R*) configuration, can be extended to incorporate other 2-substituted acetaldehydes (Scheme 9.49).

For example, deoxyribose aldolase DERA 04 mediates the condensation of benzyl 3-oxopropylcarbamate (**172**) with acetaldehyde (2.0 equivalents) in DMSO–phosphate buffer at 25°C. At 20–24 h and 95% conversion, the mixture is extracted with ethyl acetate. The layers are separated by centrifugation and the organic layer is concentrated at reduced pressure to afford the lactol **173** (76%). Lactol **173** is oxidized by air using a platinum on carbon catalyst in 1:1 *tert*-butanol–water to produce (3*R*,5*R*)-7-(benzyloxycarbonylamino)-3,5-dihydroxyheptanoic acid (**174**). The diol is protected by reaction with 2,2-dimethoxypropane

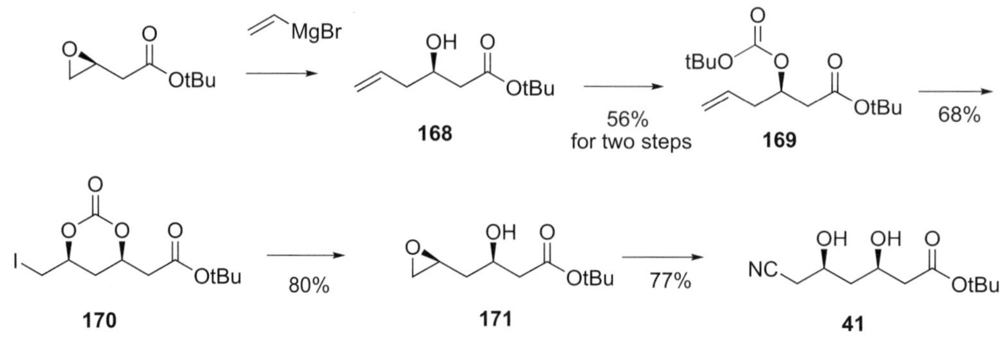

**SCHEME 9.48** (3*R*,5*R*)-*tert*-Butyl 6-cyano-3,5-dihydroxyhexanoate (**41**) from (*S*)-*tert*-butyl 2-(oxiran-2-yl)acetate.

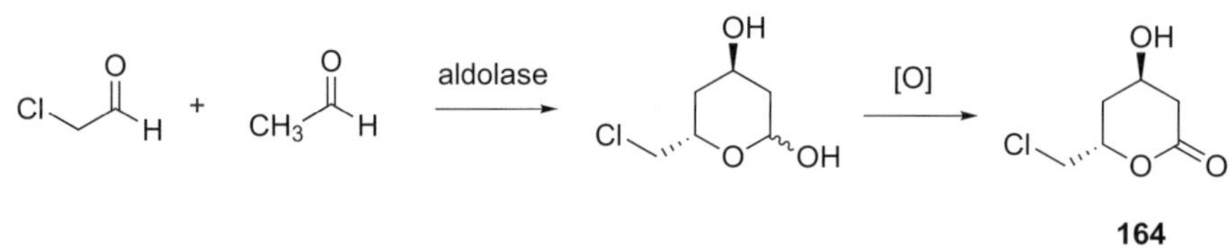

**SCHEME 9.49** (4*R*,6*S*)-6-(Chloromethyl)-4-hydroxytetrahydro-2*H*-pyran-2-one (**164**) from chloroacetaldehyde and acetaldehyde.

and catalytic methanesulfonic acid in acetone. The acid of **175** is converted to the *tert*-butyl ester with *tert*-butanol and dicyclohexylcarbodiimide in acetonitrile. Finally, hydrogenolysis of the benzyl carbamate of **176**, catalyzed by palladium on carbon, in methanol affords *tert*-butyl 2-((4*R*,6*R*)-6-(2-aminoethyl)-2,2-dimethyl-1,3-dioxan-4-yl)acetate (**8**) (Scheme 9.50). A suspension of *E. coli* cells containing DERA 102 mediates the condensation of 3-(1,3-dioxoisoindolin-2-yl)propanal (**177**) with acetaldehyde. The eventual deprotection of the phthalimide in the presence of the *tert*-butyl ester presents a challenge.[171,172]

### 9.12.4 Routes to the Amine Based on Desymmetrization of Diethyl 3-Hydroxyglutarate

Two dramatically different strategies are based on creating a chiral center by desymmetrization of diethyl 3-hydroxyglutarate using α-chymotrypsin (Scheme 9.51). In the first route, the hydroxyl group is protected by reaction with acetic anhydride in pyridine (99%). Hydrolysis of one ethyl ester of **178** is then mediated by α-chymotrypsin (92%). The acid **179** is converted to the acid chloride **180** (99%). Friedel-Crafts reaction with aluminum chloride and dry ethylene gas affords the chain-extended chloride **181** (99%). The chloride

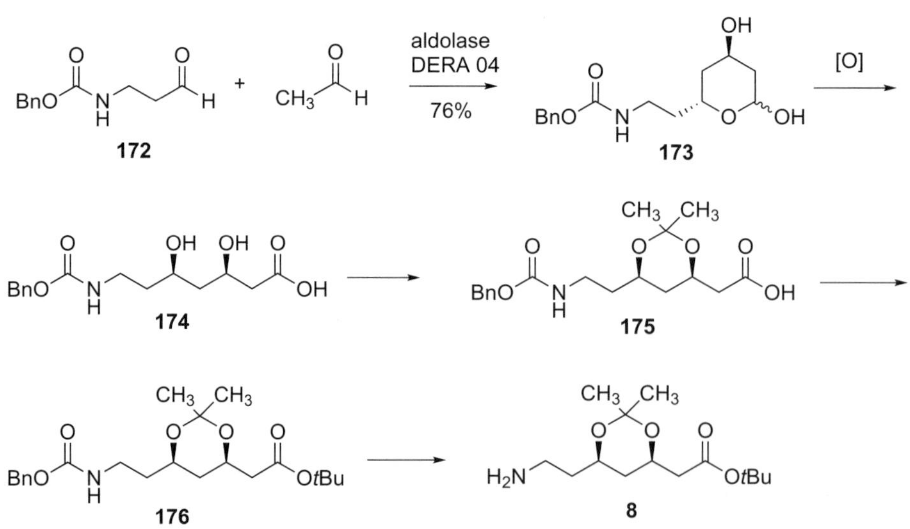

**SCHEME 9.50** *tert*-Butyl 2-((4*R*,6*R*)-6-(2-aminoethyl)-2,2-dimethyl-1,3-dioxan-4-yl)acetate (**8**) from benzyl 3-oxopropylcarbamate (**172**) and acetaldehyde.

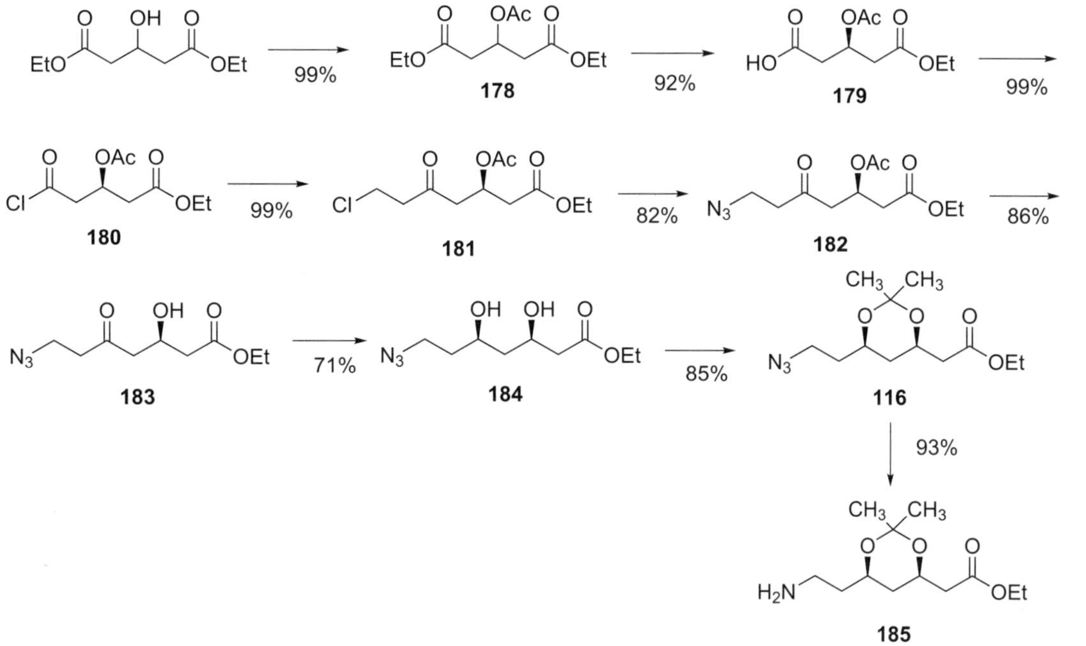

**SCHEME 9.51** Ethyl 2-((4*R*,6*R*)-6-(2-aminoethyl)-2,2-dimethyl-1,3-dioxan-4-yl)acetate (**185**) from diethyl 3-hydroxyglutarate.

is displaced by reaction with sodium azide and 18-crown-6 (82%). The acetate of **182** is removed using Chirazyme E1 (PLE, Roche) (86%) and the ketone of the $\beta$-hydroxy-$\delta$-ketoester **183** is reduced with sodium borohydride and triethylborane in THF–methanol at −65°C (71%). Diol **184** is protected by reaction with 2,2-dimethoxypropane and catalytic *p*-toluenesulfonic acid in THF (85%). Finally, azide reduction is catalyzed by palladium on carbon in ethanol to afford ethyl 2-((4*R*,6*R*)-6-(2-aminoethyl)-2,2-dimethyl-1,3-dioxan-4-yl)acetate (**185**) (93%). The overall yield for the nine-step sequence is 35%.[27,173]

Using the same construction methods but introducing the azide last eliminates several potentially problematic azide intermediates. The overall yield for this alternative nine-step sequence, demonstrated using the methoxyacetoxy group for protection of the alcohol, is 21%. Use of the azide **116** to deliver the side chain in an alternative pyrrole ring construction is discussed in Section 9.6.[27,173]

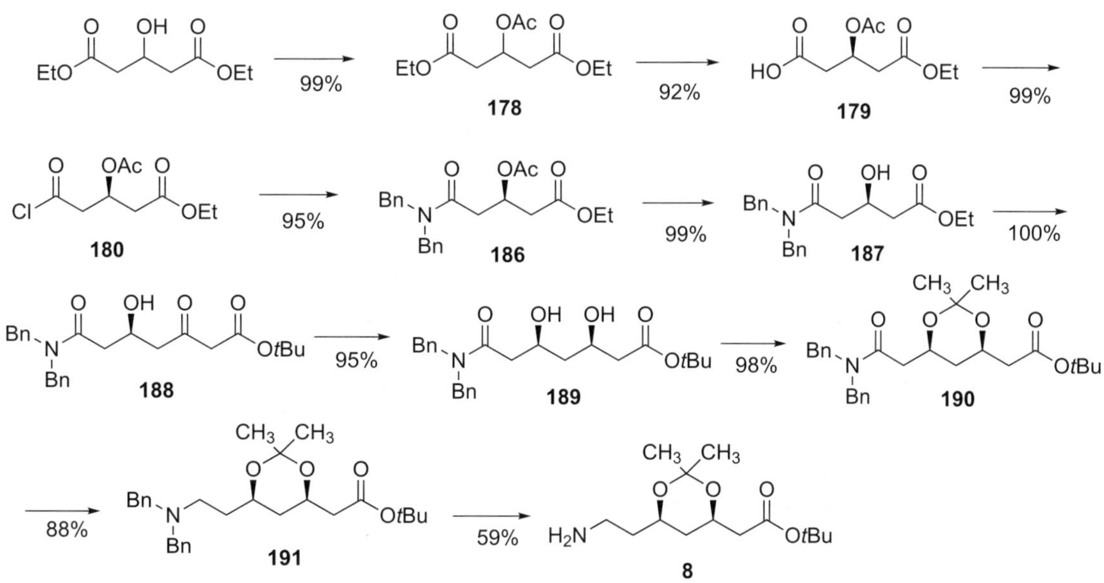

**SCHEME 9.52** *tert*-Butyl 2-((4*R*,6*R*)-6-(2-aminoethyl)-2,2-dimethyl-1,3-dioxan-4-yl)acetate (**8**) from diethyl 3-hydroxyglutarate.

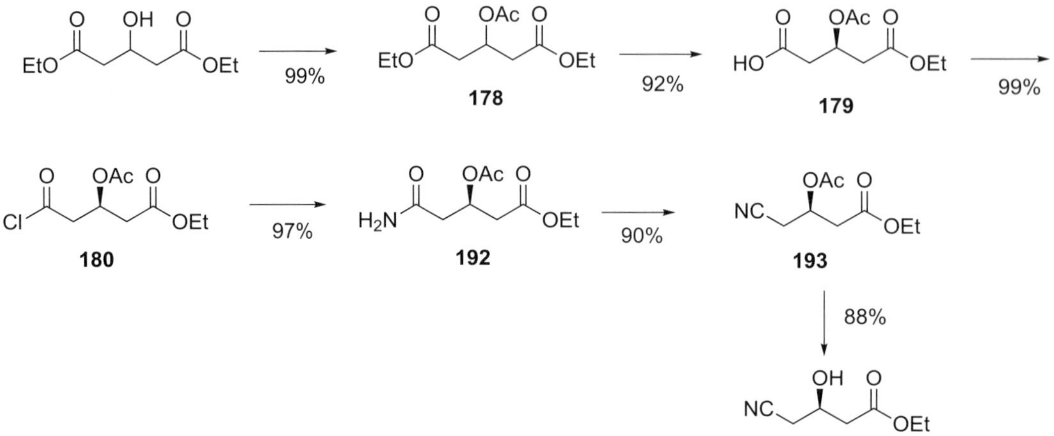


**SCHEME 9.53** Ethyl (*R*)-(−)-4-cyano-3-hydroxybutyrate by desymmetrization of diethyl 3-hydroxyglutarate.

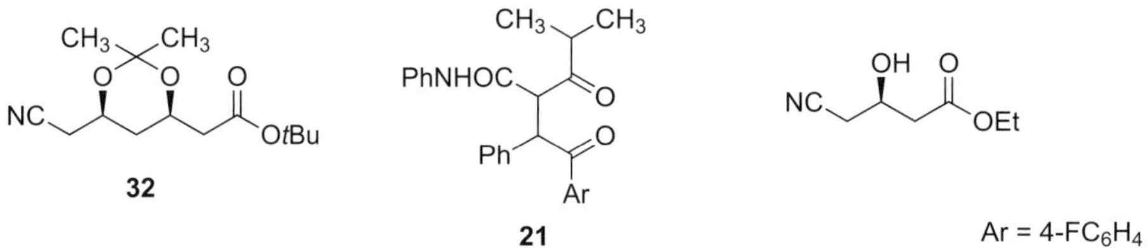


**FIGURE 9.7** Structures searched for atorvastatin (**1**) presentation (December 2009).

The second strategy also utilizes the acid chloride **180**. Reaction of the acid chloride **180** with dibenzylamine affords the amide **186** (95%) (Scheme 9.52). Reaction with 2 M hydrogen chloride in ethanol cleaves the acetate ester (99%). Condensation with *tert*-butyl acetate affords the *tert*-butyl δ-hydroxy-*β*-ketoheptanoate **188** (100%) that is reduced with sodium borohydride and triethylborane at −70°C to afford the *β*,δ-dihydroxyheptanoate **189** (95%). The diol is protected by reaction with 2,2-dimethoxypropane and catalytic hydrogen chloride (98%). The amide is reduced with sodium borohydride–chlorotrimethylsilane in diglyme (88%). Debenzylation of **191** using palladium on carbon in methanol– water at 70°C and 140 psi hydrogen affords *tert*-butyl 2-((4*R*,6*R*)-6-(2-aminoethyl)-2,2-dimethyl-1,3-dioxan-4-yl)acetate (**8**) (59%). The amide reduction and debenzylation are demonstrated on 1–2 kg scale. The yield for the 10-step sequence from diethyl 3-hydroxyglutarate is a remarkable 41%.[174,175]

The desymmetrization of diethyl 3-hydroxyglutarate can also deliver ethyl (*R*)-(−)-4-cyano-3-hydroxybutyrate. Reaction of acid chloride **180** with ammonia in methyl *tert*-butyl ether at 0°C affords amide **192** (97%). Amide **192** is converted to nitrile **193** with cyanuric chloride (90%) and the acetate ester of **193** is cleaved with hydrogen chloride in ethanol (88%). The overall yield for this six-step sequence is 69%.[174,175] While every step of this process is high yielding, diethyl 3-hydroxyglutarate[176] is an expensive starting material and this is just one of many approaches to specialty chemical ethyl (*R*)-(−)-4-cyano-3-hydroxybutyrate (Scheme 9.53).

## 9.13 STRUCTURES SEARCHED

Three structure searches were used to generate all the information presented in this chapter (Figure 9.7).

## REFERENCES

1. www.drugs.com/top200.html.
2. www.pfizer.com/files/annualreport/2008/financial/financial2008.pdf.
3. www.cdc.gov/heartDisease/facts.com.
4. www.americanheart.org/downloadable/heart/1140811583642InternationalCVD.pdf.
5. (a) www.who.int/cardiovascular_diseases/en/cvd_atlas_13_coronaryHD.pdf. (b) www.who.int/cardiovascular_diseases/en/cvd_atlas_14_deathHD.pdf. (c) www.who.int/cardiovascular_diseases/en/cvd_atlas_15_burden_stroke.pdf. (d) www.who.int/cardiovascular_diseases/en/cvd_atlas_16_ death_from_stroke.pdf. (e) www.who.int/cardiovascular_diseases/en/cvd_atlas_17_economics.pdf.
6. www.reuters.com/article/idUSN18418654.
7. www.dsi.com/news/pdfs/Ranbaxy_press_release_final.pdf.

8. Mitka, M. *JAMA* **2008**, *299*, 2266.
9. Yang, L.P.H.; Keating, G.M. *Am. J. Cardiovasc. Drugs* **2009**, *9*, 401.
10. The Aldrich 2007–2008 catalog price is 100 g for $161.50.
11. The TCI 2006–2007 catalog price for 5 g is $148.85.
12. The Aldrich 2007–2008 catalog price for 500 g is $64.80.
13. Epstein, J.W.; Brabander, H.J.; Fanshawe, W.J.; Hofmann, C. M.; McKenzie, T.C.; Safir, S.R.; Osterberg, A.C.; Cosulich, D. B.; Lovell, F.M. *J. Med. Chem.* **1981**, *24*, 481.
14. Kennedy, R.M.; Park, W.K.-C.; Roth, B.D.; Song, Y.; Trivedi, B.K.U.S. Patent 7,250,444 (7/31/2007).
15. Roth, B.D.; Blankley, C.J.; Chucholowski, A.W.; Ferguson, E.; Hoefle, M.L.; Ortwine, D.F.; Newton, R.S.; Sekerke, C.S.; Sliskovic, D.R.; Stratton, C.D.; Wilson, M.W. *J. Med. Chem.* **1991**, *34*, 357.
16. Roth, B.D.U.S. Patent 4,681,893 (7/21/1987).
17. The Aldrich 2007–2008 catalog price for 25 g is $120.50.
18. The Aldrich 2007–2008 catalog price for 18 kg is $304.50.
19. The Aldrich 2007–2008 catalog price for 26 g (ethyl) is $93.90.
20. The Aldrich 2007–2008 catalog price for 25 kg is $1921.10.
21. The Aldrich 2007–2008 catalog price for 18 kg is $220.00.
22. The Aldrich 2007–2008 catalog price for 250 g is $161.50.
23. Sattigeri, J.A.; Salman, M.; Rawat, S.; Sethi, S. WO 2005118536 (12/15/2005).
24. The Aldrich 2007–2008 catalog price for 500 g is $104.50.
25. The Aldrich 2007–2008 catalog price for 500 g is $54.00.
26. Schoning, K.-U.; Hartwig, J. WO 2003004457 (1/16/2003).
27. Ohrlein, R.; Baisch, G. WO 2003044011 (5/30/2003).
28. The Aldrich 2007–2008 catalog price for 10 g is $65.90.
29. (a) Butler, D.E.; Deering, C.F.; Millar, A.; Nanninga, T.N.; Roth, B.D.U.S. Patent 5,280,126 (1/18/1994). (b) Butler, D. E.; Deering, C.F.; Millar, A.; Nanninga, T.N.; Roth, B.D.U.S. Patent 5,245,047 (9/14/1993). (c) Butler, D.E.; Deering, C.F.; Millar, A.; Nanninga, T.N.; Roth, B.D.U.S. Patent 5,216,174 (6/1/1993). (d) Butler, D.E.; Deering, vC.F.; Millar, A.; Nanninga, T.N.; Roth, B.D.U.S. Patent 5,149,837 (9/22/1992). (e) Butler, D.E.; Deering, C.F.; Millar, A.; Nanninga, T.N.; Roth, B.D.U.S. Patent 5,124,482 (6/23/1992). (f) Butler, D.E.; Deering, C.F.; Millar, A.; Nanninga, T.N.; Roth, B.D.U. S. Patent 5,097,045 (3/17/1992). (g) Butler, D.E.; Deering, C. F.; Millar, A.; Nanninga, T.N.; Roth, B.D.U.S. Patent 5,003,080 (3/26/1991). (h) Butler, D.E.; Deering, C.F.; Millar, A.; Nanninga, T.N.; Roth, B.D. EP 330172 (8/30/1989). (i) Butler, D.E.; Deering, C.F.; Millar, A.; Nanninga, T.N.; Roth, B.D. WO 1989007598 (8/24/1989).
30. (a) Millar, A.; Butler, D.E.U.S. Patent 5,248,793 (9/28/1993). (b) Millar, A.; Butler, D.E.U.S. Patent 5,103,024 (4/7/1992).
31. Sagyam, R.R.; Vurimidi, H.; Padi, P.R.; Ghanta, M.R. *J. Het. Chem.* **2007**, *44*, 923.
32. Pai, G.G.; Kishore, K.N.; Chaudhari, N.P.; Anjaneyulu, N.; Ghogare, B.N.U.S. Patent 20090298907 (12/3/2009).
33. Baumann, K.L.; Butler, D.E.; Deering, C.F.; Mennen, K.E.; Millar, A.; Nanninga, T.N.; Palmer, C.W.; Roth, B.D. *Tetrahedron Lett.* **1992**, *33*, 2283.
34. (a) O'Sullivan, S.; Fox, E. WO 2007029217 (3/15/2007). (b) O'Sullivan, S.; Fox, E. EP 1922301 (3/15/2007).
35. Fan, P.; Jiang, N. *Yaowu Huaxue Zazhi* **2005**, *15*, 48.
36. Sambasivan, G.; Sridharan, M.; Padudevastana, S.; Poornaprajna, A.; Mathew, J.; Srinath, S.; Nair, S. WO 2001072706 (10/4/2001).
37. Anon *IP. com J.* **2008**, *8*, 2.
38. Mathew, J.; Puthiaparampil, T.T.; Sridharan, M.; Shanker, P. S.; Ganesh, S. WO 2004103957 (12/2/2004)
39. Li, R.; Xie, Z.; Luan, G. CN 101307009 (11/19/2008)
40. Joung, S.S.; Kim, S.J.; Lee, T.A. KR 2006078351 (7/5/2006)
41. Ahn, S.H.; Kim, M.H.; Lee, G.J.; Lee, S.H.; Ryu, U.S.; Sim, Y.G. KR 2004001435 (1/7/2004)
42. Yin, D.; Li, C. CN 1325844 (12/12/2001)
43. Mathew, J.; Puthiaparampil, T.T.; Chandrappa, R.; Ganesh, S. WO 2006021968 (3/2/2006)
44. (a) Mathew, J.; Puthiaparampil, T.T.; Sridharan, M.; Shanker, P.S.; Ganesh, S.U. S. Patent 7,179,942 (2/20/2007). (b) Mathew, J.; Puthiaparampil, T.T.; Sridharan, M.; Shanker, P.S.; Ganesh, S. WO 2004108660 (12/16/2004).
45. The Aldrich 2007–2008 catalog price for 500 mL ($d = 0.866$ g/mL) is $138.50.
46. (a) Butler, D.E.; Le, T.V.; Millar, A.; Nanninga, T.N.U.S. Patent 5,155,251 (10/13/1992). (b) Butler, D.E.; Le, T.V.; Millar, A.; Nanninga, T.N. WO 199307115 (4/15/1993). (c) Butler, D.E.; Le, T.V.; Millar, A.; Nanninga, T.N. EP 643689 (12/30/1998).
47. (a) Kumar, Y.; Kumar, S.M.D.; Sathyanarayana, S.U.S. Patent 20090216029 (8/27/2009). (b) Kumar, Y.; Kumar, S.M.D.; Sathyanarayana, S.H. EP 1659110 (5/24/2006). (c) Kumar, Y.; Kumar, S.M.D.; Sathyanarayana, S. WO 2005092852 (10/6/2005). (d) Kumar, Y.; Kumar, S.M.D.; Sathyanarayana, S. WO 2005090301 (9/29/2005). (e) Kumar, Y.; Kumar, S.M.D.; Sathyanarayana, S. EP 1577297 (9/21/2005).
48. Brower, P.L.; Butler, D.E.; Deering, C.F.; Le, T.V.; Millar, A.; Nanninga, T.N.; Roth, B.D. *Tetrahedral Lett.* **1992**, *33*, 2279.
49. (a) Bosch, R.L.; McCabe, R.J.; Nanninga, T.N.; Stahl, R.J.U. S. Patent 7,232,915 (6/19/2007). (b) McCabe, R.J.; Nanninga, T.N.; Bosch, R.L.; Stahl, R.J. EP 1054860 (4/25/2007). (c) Bosch, R.L.; McCabe, R.J.; Nanninga, T.N.; Stahl, R.J.U. S. Patent 6,962,994 (11/8/2005). (d) Bosch, R.L.; McCabe, R.J.; Nanninga, T.N.; Stahl, R.J.U.S. Patent 6,596,879 (7/22/2003). (e) Bosch, R.L.; McCabe, R.J.; Nanninga, T.N.; Stahl, R.J.U.S. Patent 6,433,213 (8/13/2002). (f) McCabe, R.J.; Nanninga, T.N.; Bosch, R.L.; Stahl, R.J. WO 1999032434 (7/1/1999).
50. (a) Nishiyama, A.; Inoue, K.U.S. Patent RE39333 (10/10/2006). (b) Nishiyama, A.; Inoue, K. EP 1104750 (4/14/2004).
51. (a) Srinath, S.; Puthiaparampil, T.T.; Ganesh, S.U.S. Patent 7,557,238 (7/7/2009). (b) Srinath, S.; Puthiaparampil, T.T.; Ganesh, S. WO 2005026107 (3/24/2005).
52. The Aldrich 2007–2008 catalog price for 100 g is $86.10.

53. Cho, Y.-H.; Roh, K.R.; Shin, J.H.; Chun, J.P.; Yu, H.S.; Cho, C.-W. WO 2002096915 (12/5/2002).
54. The Aldrich 2007–2008 catalog price for 2 kg (VenPure AF granular) is $481.00.
55. Chen, K.-M.; Gunderson, K.G.; Hardtmann, G.E.; Prasad, K.; Repic, O.; Shapiro, M.J. *Chem. Lett.* **1987**, 1923.
56. Puthiaparampil, T.T.; Poornaprajna, A.; Aswathanarayanappa, C.; Sridharan, M.; Ganesh, S.U.S. Patexnt 7,232,920 (6/19/2007).
57. (a) Reeve, C.D. EP 833938 (3/19/2003). (b) Reeve, C.D.U.S. Patent 6,001,615 (12/14/1999). (c) Reeve, C.D. WO 1997000968 (1/9/1997).
58. Mitsuda, M.; Miyazaki, M.; Inoue, K.U.S. Patent 6,344,569 (2/5/2002).
59. Gokhale, U.B.; Gundap, R.T. IN 2006MU00871 (7/11/2008).
60. (a) Lee, I.; Lee, S.U.S. Patent 7,205,418 (4/17/2007). (b) Lee, I.; Lee, S. WO 2003053950 (7/3/2003).
61. Gudipati, S.; Katkam, S.; Komati, S.; Kudavalli, S.J. WO 2006039441 (4/13/2006).
62. Miller, J.M. WO 2007049121 (5/3/2007).
63. O'Sullivan, S.; O'Neill, J. WO 2007029216 (3/15/2007).
64. Pathi, S.L. IN 2006MU01343 (7/11/2008).
65. Chopra, T.J.; O'Neill, P.M.; Wilkes, P.B. WO 2008075165 (6/26/2008).
66. Oren, J.; Dolitzky, B.-Z.; Harel, Z.; Perlman, N.; Lidor-Hadas, R. WO 2004046105 (6/3/2004).
67. Srinath, S.; Mathew, J.; Ujire, S.; Sridharan, M.; Sambasivam, G.U.S. Patent 6,867,306 (3/15/2005).
68. Srinath, S.; Mathew, J.; Ujire, S.; Sridharan, M.; Sambasivam, G. WO 2002057274 (7/25/2002).
69. Chung, I.H. WO 2008103016 (8/28/2008).
70. The Aldrich 2007–2008 catalog price for 1 kg is $86.90.
71. (a) Butler, D.E.; Le, T.V.; Nanninga, T.N.U.S. Patent 5,510,488 (4/23/1996). (b) Butler, D.E.; Le, T.V.; Nanninga, T.N.U.S. Patent 5,489,691 (2/6/1996). (c) Butler, D.E.; Le, T. V.; Nanninga, T.N.U.S. Patent 5,489,690 (2/6/1996). (d) Butler, D.E.; Le, T.V.; Nanninga, T.N.U.S. Patent 5,470,981 (11/28/1995). (e) Butler, D.E.; Le, T.V.; Nanninga, T.N.U.S. Patent 5,446,054 (8/29/1995). (f) Butler, D.E.; Le, T. V.; Nanninga, T.N.U.S. Patent 5,397,792 (3/14/1995). (g) Butler, D.E.; Le, T.V.; Nanninga, T.N. WO 1994020492 (9/15/1994). (h) Butler, D.E.; Le, T.V.; Nanninga, T.N.U.S. Patent 5,342,952 (8/30/1994). (i) Butler, D.E.; Le, T.V.; Nanninga, T.N.U.S. Patent 5,298,627 (3/29/1994).
72. The Aldrich 2007–2008 catalog price for 25 g is $54.50.
73. The Aldrich 2007–2008 catalog price for 5 g is $21.20.
74. Korostylev, A.; Tararov, V.; Borner, A.; Konig, G.; Bobal, P.; Frantisek, J.; Stohandl, J.; Denike, K.; Jeker, N. EP 1834944 (9/19/2007).
75. The Aldrich 2007–2008 catalog price for 5 g is $177.50.
76. (a) Tararov, V.; Borner, A.; Konig, G.; Korostylev, A. WO 2006122644 (11/23/2006). (b) Tararov, V.I.; Andrushko, N.; Andrushko, V.; Konig, G.; Spannenberg, A.; Borner, A. *Eur. J. Org. Chem.* **2006**, 5543.
77. The Aldrich 2007–2008 catalog price for 1 L (50 wt% in $H_2O$) ($d = 1.326$ g/mL) is $76.50.
78. The Aldrich 2007–2008 catalog price for 1 L ($d = 0.785$ g/mL) is $148.00.
79. Greenberg, W.; Burk, M.J.; Varvak, A.; Wong, K.U.S. Patent 7,414,119 (8/19/2008).
80. Moody, D.J.; Wiffen, J.W.U.S. Patent 7,414,141 (8/19/2008).
81. The Aldrich 2007–2008 catalog price for 25 g is $34.30.
82. Pradhan, B.S. WO 2008152836 (12/18/2008).
83. The TCI 2006–2007 catalog price is 25 g for $80.60.
84. The Aldrich 2007–2008 catalog price for 2 kg is $131.00.
85. (a) Nelson, J.D.; Pamment, M.G.U.S. Patent 7,399,773 (7/15/2008). (b) Nelson, J.D.; Pamment, M.G.; Stuk, T.L. U.S. Patent 7,371,855 (5/13/2008). (c) Butler, D.E.; Dejong, R.L.; Nelson, J.D.; Pamment, M.G.; Stuk, T.L.U.S. Patent 7,361,771 (4/22/2008). (d) Butler, D.E.; Dejong, R.L.; Nelson, J.D.; Pamment, M.G.; Stuk, T.L.U.S. Patent 7,183,408 (2/27/2007). (e) Butler, D.E.U.S. Patent 6,933,393 (8/23/2005). (f) Butler, D.E.; Dejong, R.L.; Nelson, J.D.; Pamment, M.G.; Stuk, T.L.U.S. Patent 6,545,153 (4/8/2003). (g) Butler, D.E.; Dejong, R.L.; Nelson, J.D.; Pamment, M.G.; Stuk, T.L. U.S. Patent 6,476,235 (11/5/2002).
86. Joshi, N.S.; Bhirud, S.B.; Damle, S.V. WO 2006032959 (3/30/2006).
87. Ambalal, M.I.; Pratima, J.; Rajput, A.L.; Tekade, P.M.; Joshi, A.C.; Ravi, P.; Mafatlal, K.B. WO 2008053312 (5/8/2008).
88. Gupta, R.; Ram, K.; Bhadwal, P.; Thaper, R.K.; Dubey, S.K. WO 2007088553 (8/9/2007).
89. Faja Genoves, M.; Vilarrasa Llorens, J.; Asensio Dominguez, R.; Garcia Chapinal, F.; Cruzado Rodriguez, M.C. WO 2007034012 (3/29/2007).
90. Dixit, G.; Khile, A.S.; Pradhan, N.S.; Valgeirsson, J. WO 2009007856 (1/15/2009).
91. Vasantray, V.A.; Pranlal, D.V. EP 2075246 (7/1/2009).
92. Patel, D.J.; Kumar, R.; Dwivedi, S.P.D. WO 2007096751 (8/30/2007).
93. Kansal, V.K.; Chaurasia, B.P.; Patel, H.K.; Rao, V.G.; Daund, V.S. WO 2009023260 (2/19/2009).
94. Davies, S.; Denholm, A.A.; Kelleher, P.; Tully, W.E. WO 2006097909 (9/21/2006).
95. (a) Barkoczy, J.; Simig, G.; Greff, Z.; Nagy, P.K.; Kirallyi, Z. S.; Bartha, F.; Donath, G.V.; Nagy, K.U.S. Patent 7,186,848 (3/6/2007). (b) Barkoczy, J.; Simig, G.; Greff, Z.; Nagy, P.K.; Kirallyi, Z.S.; Bartha, F.; Donath, G.V.; Nagy, K. WO 2003024959 (3/27/2003).
96. (a) Wang, F.; Che, D.; Guntoori, B.R.; Zhao, Y.; Kinsman, A. C.; Faught, J.; Chow, A.U.S. Patent 7,429,613 (9/30/2008). (b) Wang, F.; Che, D.; Guntoori, B.R.; Zhao, Y.; Kinsman, A. C.; Faught, J.; Chow, A.U.S. Patent 20060252816 (11/9/2006). (c) Wang, F.; Che, D.; Guntoori, B.R.; Zhao, Y.; Kinsman, A.C.; Faught, J.; Chow, A.U.S. Patent 7,112,604 (9/26/2006).
97. The Aldrich 2007–2008 catalog price for 5 g (*R*)-isomer) is $173.00.

98. (a) Roth, B.D.U.S. Patent RE40667 (3/17/2009). (b) Roth, B. D. EP 409281 (10/31/2001). (c) Roth, B.D.U.S. Patent 5,273,995 (12/28/1993).
99. (a) Guntoori, B.R.; Che, D.; Wang, F.; Zhao, Y.; Murthy, K.S. K.; Horne, S.E.U.S. Patent 7,193,090 (3/20/2007). (b) Guntoori, B.R.; Che, D.; Wang, F.; Zhao, Y.; Murthy, K.S.K.; Horne, S.E. WO 2005087723 (9/22/2005). (c) Guntoori, B.R.; Che, D.; Wang, F.; Zhao, Y.; Murthy, K.S.K.; Horne, S.E. EP 1725527 (9/22/2005).
100. The Aldrich 2007–2008 catalog price for 10 g is $94.30.
101. The Aldrich 2007–2008 catalog price for 4 L is $434.00.
102. (a) Nelson, J.D.; Pamment, M.G.U.S. Patent 7,084,282 (8/1/2006). (b) Nelson, J.D.; Pamment, M.G.U.S. Patent 6,777,560 (8/17/2004).
103. (a) Agarwal, V.K.; Vakil, M.H.; Pandita, K.; Ramakrishna, N.V.S.; Patel, P.R.; Manakiwala, S.C. WO 2002083638 (10/24/2002). (b) Agarwal, V.K.; Vakil, M.H.; Pandita, K.; Ramakrishna, N.V.S.; Patel, P.R. WO 2002083637 (10/24/2002).
104. Murthy, K.S.K.; Zhao, Y.; Che, D.; Guntoori, B.R.; Duncan, S.C.; Horne, S.E.U.S. Patent 7,615,647 (11/10/2009).
105. Stimac, A.; Zupet, R.; Grcman, M.; Smrkolj, M.; Jakse, R. WO 2005097742 (10/20/2005).
106. (a) Radl, S.; Stach, J.U.S. Patent 7,208,608 (4/24/2007). (b) Radl, S.; Stach, J.U.S. Patent 20050131055 (6/16/2005). (c) Radl, S.; Stach, J. WO 2003068739 (8/21/2003).
107. Dabak, K.; Keskin, H. WO 2005005384 (1/20/2005).
108. The Aldrich 2007–2008 catalog price for 5 kg (dihydrate) is $207.50.
109. Bhatraju, S.R.; Jain, P.; Thaper, R.K.; Dubey, S.K. WO 2006048888 (5/11/2006).
110. The Aldrich 2007–2008 catalog price for 2.5 kg (hydrate) is $255.00.
111. The Aldrich 2007–2008 catalog price for 25 g is $87.70.
112. The Aldrich 2007–2008 catalog price for 1 kg is $55.60.
113. The Aldrich 2007–2008 catalog price for 2.5 kg ($\geq$95%) is $96.30.
114. (a) Manne, S.R.; Chakilam, N.; Gudipati, S.; Katkam, S.; Sagyam, R.R.U.S. Patent 7,074,818 (7/11/2006). (b) Manne, S.R.; Chakilam, N.; Gudipati, S.; Katkam, S.; Sagyam, R.R. WO 2003011826 (2/13/2003).
115. (a) Suri, S.; Singh, J.; Sarin, S.G.; Bansal, R.B.U.S. Patent 7,563,911 (7/21/2009). (b) Suri, S.; Singh, J.; Sarin, S.G.; Bansal, R.B.U.S. Patent 20050119493 (6/2/2005). (c) Suri, S.; Singh, J.; Sarin, S.G.; Bansal, R.B. WO 2003018547 (3/6/2003).
116. (a) Che, D.; Kinsman, A.C.; Guntoori, B.R.; Murthy, K.S.K.; Zhao, Y.; Horne, S.E. WO 2005073187 (8/11/2005). (b) Che, D.; Kinsman, A.C.; Guntoori, B.R.; Murthy, K.S.K.U.S. Patent 20050165242 (7/28/2005).
117. (a) O'Connell, J.; Tully, W.; Madigan, E.U.S. Patent 6,730,797 (5/4/2004). (b) O'Connell, J.; Tully, W.; Madigan, E.U.S. Patent 6,605,728 (8/12/2003).
118. Tully, W.U.S. Patent 6,600,051 (7/29/2003).
119. (a) Briggs, C.A.; Jennings, R.A.; Wade, R.; Harasawa, K.; Ichikawa, S.; Minohara, K.; Nakagawa, S.U.S. Patent 5,969,156 (10/19/1999). (b) Briggs, C.A.; Jennings, R.A.; Wade, R.; Harasawa, K.; Ichikawa, S.; Minohara, K.; Nakagawa, S. WO 1997003959 (2/6/1997).
120. (a) McKenzie, A.T.U.S. Patent 6,121,461 (9/19/2000). (b) McKenzie, A.T. WO 1997003958 (2/6/1997).
121. (a) Lin, M.; Schweiss, D.U.S. Patent 6,274,740 (8/14/2001). (b) Lin, M.; Schweiss, D. EP 839132 (3/7/2001). (c) Lin, M.; Schweiss, D.U.S. Patent 6,087,511 (7/11/2000). (d) Lin, M.; Schweiss, D. WO 1997003960 (2/6/1997).
122. Suri, S.; Singh, J.; Grewal, M.S.; Raj, B. WO 2004022053 (3/18/2004).
123. Czibula, L.; Dobay, L.; Werkne Papp, E.; Deutschne Juhasz, I.; Balint, S. WO 2004085391 (10/7/2004).
124. Campeta, A.M.; Krzyzaniak, J.F.; Leonard, J.A.U.S. Patent 7,534,810 (5/19/2009).
125. (a) Ayalon, A.; Levinger, M.; Roytblat, S.; Niddam, V.; Lifshitz, R.; Aronhime, J.U.S. Patent 7,411,075 (8/12/2008). (b) Ayalon, A.; Levinger, M.; Roytblat, S.; Niddam, V.; Lifshitz, R. WO 2001036384 (5/25/2001).
126. (a) Niddam-Hildesheim, V.; Lifshitz-Liron, R.; Lidor-Hadas, R.U.S. Patent 6,777,552 (8/17/2004). (b) Niddam-Hildesheim, V.; Lifshitz-Liron, R.; Lidor-Hadas, R. WO 2003016317 (2/27/2003).
127. (a) Niddam, V.; Lidor-Hadas, R.; Lifshitz, R.; Ishai, E.U.S. Patent 6,528,661 (3/4/2003). (b) Lidor-Haddas, R.; Lifshitz, R.; Ishai, E.; Niddam, V. WO 2002043667 (6/6/2002).
128. (a) Tessler, L.; Aronhime, J.; Lifshitz-Liron, R.; Maidan-Hanoch, D.; Hasson, N.U.S. Patent 7,501,450 (3/10/2009). (b) Tessler, L.; Aronhime, J.; Lifshitz-Liron, R.; Maidan-Hanoch, D.; Hasson, N. WO 2004043918 (5/27/2004). (c) Tessler, L.; Aronhime, J.; Lifshitz-Liron, R.; Maidan-Hanoch, D.; Hasson, N.U.S. Patent 20030212279 (11/13/2003).
129. (a) Aronhime, J.; Lidor-Hadas, R.; Niddam, V.; Lifshitz, R.; Wizel, S.U.S. Patent 7,488,750 (2/10/2009). (b) Aronhime, J.; Lidor-Hadas, R.; Niddam, V.; Lifshitz, R.; Wizel, S.U.S. Patent 7,468,444 (12/23/2008). (c) Tessler, L.; Aronhime, J.; Lifshitz-Liron, R.; Maidan-Hanoch, D.; Hasson, N.U.S. Patent 7,456,297 (11/25/2008). (d) Aronhime, J.; Lidor-Hadas, R.; Niddam, V.; Lifshitz, R.; Wizel, S.U.S. Patent 7,342,120 (3/11/2008). (e) Tessler, L.; Aronhime, J.; Lifshitz-Liron, R.; Maidan-Hanoch, D.; Hasson, N.U.S. Patent 7,256,212 (8/14/2007). (f) Tessler, L.; Aronhime, J.; Lifshitz-Liron, R.; Maidan-Hanoch, D.; Hasson, N.U.S. Patent 7,189,861 (3/13/2007). (g) Tessler, L.; Aronhime, J.; Lifshitz-Liron, R.; Maidan-Hanoch, D.; Hasson, N.U.S. Patent 7,161,012 (1/9/2007). (h) Tessler, L.; Aronhime, J.; Lifshitz-Liron, R.; Maidan-Hanoch, D.; Hasson, N.U.S. Patent 7,151,183 (12/19/2006). (i) Aronhime, J.; Lidor-Hadas, R.; Niddam, V.; Lifshitz, R.; Wizel, S.U.S. Patent 7,144,916 (12/5/2006). (j) Lidor-Hadas, R.; Niddam, V.; Lifshitz, R.; Kovalevski-Ishai, E.U.S. Patent 6,992,194 (1/31/2006). (k) Tessler, L.; Aronhime, J.; Lifshitz-Liron, R.; Maidan-Hanoch, D.; Hasson, N.U.S. Patent 20050267198 (12/1/2005). (l)

Tessler, L.; Aronhime, J.; Lifshitz-Liron, R.; Maidan-Hanoch, D.; Hasson, N.U.S. Patent 20050261360 (11/24/2005). (m) Tessler, L.; Aronhime, J.; Lifshitz-Liron, R.; Maidan-Hanoch, D.; Hasson, N.U.S. Patent 20050261359 (11/24/2005). (n) Aronhime, J.; Lidor-Hadas, R.; Niddam, V.; Lifshitz, R.; Ishai, E.; Sambursky, G.U.S. Patent 20020183378 (12/5/2002).

130. Venkatachalam, R.; Dixit, G.; Babu Prasad, B.R.R.; Singh, J.; Chahal, A.S. WO 2006011155 (2/2/2006).
131. (a) Sorsak, G.U.S. Patent 7,074,940 (7/11/2006). (b) Sorsak, G.U.S. Patent 6,750,353 (6/15/2004). (c) Sorsak, G.U.S. Patent 20,030,109,569 (6/12/2003). (d) Sorsak, G. WO 2002059087 (8/1/2002).
132. Antoncic, L.; Sorsak, G.; Copar, A. WO 2004089895 (10/21/2004).
133. Srinath, S.; Mathew, J.; Puthiaprampil, T.T.; Ganesh, S. WO 2006021969 (3/2/2006).
134. Mathew, J.; Sridharan, M.; Ganesh, S.; Puthiaparampil, T.T.U.S. Patent 7,361,772 (4/22/2008).
135. (a) Turchetta, S.; Massardo, P.; Tuozzi, A.U.S. Patent 20,060,128,971 (6/15/2006). (b) Turchetta, S.; Massardo, P.; Tuozzi, A. WO 2003093233 (11/13/2003).
136. (a) Greff, Z.; Kotay Nagy, P.; Barkoczy, J.; Simig, G.; Balazs, L.; Doman, I.; Ratkai, Z.; Seres, P.; Szent Kirallyi, Z.; Bartha, F.; Vereczkeyne Donath, G.; Nagy, K. EP 1235800 (11/15/2006). (b) Greff, Z.; Kotay Nagy, P.; Barkoczy, J.; Simig, G.; Balazs, L.; Doman, I.; Ratkai, Z.; Seres, P.; Szent Kirallyi, Z.; Bartha, F.; Vereczkeyne Donath, G.; Nagy, K.U.S. Patent 6,646,133 (11/11/2003). (c) Greff, Z.; Kotay Nagy, P.; Barkoczy, J.; Simig, G.; Balazs, L.; Doman, I.; Ratkai, Z.; Seres, P.; Szent Kirallyi, Z.; Bartha, F.; Vereczkeyne Donath, G.; Nagy, K. WO 2001028999 (4/26/2001).
137. (a) Pflaum, Z.U.S. Patent 6,891,047 (5/10/2005). (b) Pflaum, Z.U.S. Patent 20040024046 (2/5/2004). (c) Pflaum, Z.U.S. Patent 6,613,916 (9/2/2003). (d) Pflaum, Z.U.S. Patent 20020183527 (12/5/2002). (e) Pflaum, Z. EP 1237864 (6/14/2001). (f) Pflaum, Z. WO 2001042209 (6/14/2001).
138. Agarwal, V.K.; Vakil, M.H.; Pandita, K.; Manakiwala, S.C.; Patel, P.R. WO 2002099785 (12/4/2003).
139. Joy, M.; Ganesh, S. WO 200257228 (7/25/2002).
140. (a) Kumar, Y.; Thaper, R.K.; Kumar, S.M.D. EP 1185264 (5/23/2007). (b) Kumar, Y.; Thaper, R.K.; Kumar, S.M.D.U.S. Patent 20030149279 (8/7/2003). (c) Kumar, Y.; Thaper, R.K.; Kumar, S.M.D.U.S. Patent 6,528,660 (3/4/2003).
141. Berchielli, A.; Waterman, K.C.; Daugherity, P.D. EP 1635791 (7/2/2008).
142. Rose, P.R.; Leonard, J.A. WO 2005100313 (10/27/2005).
143. (a) Lorenz, D.; Waterman, K.C.U.S. Patent 20050032880 (2/10/2005). (b) Lorenz, D.; Waterman, K.C. WO 2004110407 (12/23/2004).
144. (a) Poornaprajna, A.; Mathew, J.; Chandrappa, R.; Sridharan, M.; Ganesh, S.EP 1485343 (12/23/2009). (b) Poornaprajna, A.; Mathew, J.; Chandrappa, R.; Sridharan, M.; Ganesh, S.U.S. Patent 7,230,120 (6/12/2007). (c) Poornaprajna, A.; Mathew, J.; Chandrappa, R.; Sridharan, M.; Ganesh, S.U.S. Patent 20040186162 (9/23/2004). (d) Poornaprajna, A.; Mathew, J.; Chandrappa, R.; Sridharan, M.; Ganesh, S. WO 2003078379 (9/25/2003).
145. Sebastian, S.; Devarasetty, S.; Dandala, R.; Sivakumaran, M. U.S. Patent 20040242670 (12/2/2004).
146. (a) Pinchasov, M.; Maidan-Hanoch, D.U.S. Patent 20060106230 (5/18/2006). (b) Pinchasov, M.; Maidan-Hanoch, D. WO 2006045018 (4/27/2006).
147. Debettencourt, A.J., III; Rose, P.R.; Shalaev, E.Y.; Quallich, G.J.; Ziegler, C.B. WO 2006046109 (5/4/2006).
148. Syntheses of the desfluoro impurity, the (3*S*,5*S*)- and (3*S*,5*R*)-epimers, and one of the amide impurities have been published: Stach, J.; Havlicek, J.; Placek, L.; Radl, S. *Coll. Czech. Chem. Commun.* **2008**, *73*, 229.
149. Wade, R.A.; Zennie, T.M.; Briggs, C.A.; Jennings, R.A.; Nanninga, T.N.; Palmer, C.W.; Clay, R.J. *Org. Proc. Res. Dev.* **1997**, *1*, 320.
150. Hurley, T.R.; Colson, C.E.; Clipper, S.A.; Uhlendorf, S.E.; Reily, M.D. *Tetrahedron* **1993**, *49*, 1979.
151. Bastarda, A.; Salobir, M. WO 2004032920 (4/22/2004).
152. Grahek, R.; Kocjan, D.; Bastarda, A.; Kocijan, A.; Kracun, M. WO 2006008091 (1/26/2006).
153. Sebek, P.; Prokopova, A.; Svoboda, E.; Radl, S.; Stach, J.; Svoboda, M.; Kovacik, A.; Danko, A.; Petro, R.; Strizinec, M. WO 2005011638 (2/10/2005).
154. (a) Mills, N.; Muhammad, N.A.; Weiss, J.; Nesbitt, R.U.U.S. Patent 6,126,971 (10/3/2000). (b) Mills, N.; Muhammad, N. A.; Weiss, J.; Nesbitt, R.U.U.S. Patent 5,686,104 (11/11/1997).
155. Hedge, D.; Kulkarni, S. WO 2002089788 (11/14/2002).
156. Singh, R.B.; Kumar, P.M.; Malik, R. EP 1336405 (8/20/2003).
157. (a) Kerc, J. EP 1148872 (12/5/2007). (b) Kerc, J.U.S. Patent 6,680,341 (1/20/2004). (c) Kerc, J. WO 2000035425 (6/22/2000).
158. Sherman, B.C. WO 2003097039 (11/27/2003).
159. (a) Fox, M.; Dorossiev, I.U.S. Patent RE39502 (3/6/2007). (b) Fox, M.; Dorossiev, I. WO 2001076566 (10/18/2001).
160. Huang, C.G.; Huang, H.M.H.U.S. Patent 20080038332 (2/14/2008).
161. The Aldrich 2007–2008 catalog price for 100 g is $68.00.
162. Blacker, A.J.; Houson, I.N.; Wiffen, J.W. WO 2003004459 (1/16/2003).
163. (a) Puthiaparampil, T.T.; Srinath, S.; Sridharan, M.; Ganesh, S.U.S. Patent 7,301,046 (11/27/2007). (b) Puthiaparampil, T. T.; Srinath, S.; Sridharan, M.; Ganesh, S.U.S. Patent 7,238,826 (7/3/2007).
164. Radl, S.; Stach, J.; Hajicek, J. *Tetrahedron Lett.* **2002**, *43*, 2087.
165. (a) Kizaki, N.; Yamada, Y.; Yasohara, Y.; Nishiyama, A.; Miyazaki, M.; Mitsuda, M.; Kondo, T.; Ueyama, N.; Inoue, K. U.S. Patent 6,903,225 (6/7/2005). (b) Kizaki, N.; Yamada, Y.; Yasohara, Y.; Nishiyama, A.; Miyazaki, M.; Mitsuda, M.; Kondo, T.; Ueyama, N.; Inoue, K.U.S. Patent 20030040634 (2/27/2003). (c) Kizaki, N.; Yamada, Y.; Yasohara, Y.; Nishiyama, A.; Miyazaki, M.; Mitsuda, M.; Kondo, T.;

Ueyama, N.; Inoue, K.U.S. Patent 6,472,544 (10/29/2002). (d) Mitsuda, M.; Miyazaki, M.; Inoue, K.; Yasutomi, Y. WO 1999057109 (11/11/1999).

166. Shin, H.-I.; Choi, B.-S.U.S. Patent 7,081,546 (7/25/2006).
167. (a) Muller, M.; Wolberg, M.; Hummel, W.; Wandrey, C.U.S. Patent 6,689,591 (2/10/2004). (b) Wolberg, M.; Muller, M.; Hummel, W.U.S. Patent 6,399,339 (6/4/2002).
168. (a) Mink, D.; Wolberg, M.; Boesten, W.H.J.; Sereinig, N. EP 1620423 (2/20/2008). (b) Mink, D.; Wolberg, M.; Boesten, W. H.J.; Sereinig, N. WO 2004096788 (11/11/2004).
169. Wong, C.-H.; Gijsen, H.J.M. WO 1996031615 (10/10/1996).
170. (a) Kooistra, J.H.M.H.; Zeegers, H.J.M.; Mink, D.; Mulders, J.M.C.A. EP 1317440 (9/27/2006). (b) Kooistra, J.H.M.H.; Zeegers, H.J.M.; Mink, D.; Mulders, J.M.C.A. WO 2002006266 (1/24/2002).
171. Hu, S.; Tao, J.; Xie, Z. WO 2006134482 (12/21/2006).
172. Bauer, D.W.; Hu, S.; O'Neill, P.M.; Watson, T.J.N. WO 2009019561 (2/12/2009).
173. (a) Ohrlein, R.; Baisch, G.; End, N.; Burkhardt, S.; Studer, M. U.S. Patent 7,420,078 (9/2/2008). (b) Ohrlein, R.; Baisch, G.; End, N.; Burkhardt, S.; Studer, M. WO 2003004450 (1/16/2003).
174. (a) Ohrlein, R.; Baisch, G.; Kirner, H.J.; Burkhardt, S.; Studer, M.; Bienewald, F.U.S. Patent 7,504,532 (3/17/2009). (b) Ohrlein, R.; Baisch, G.; Kirner, H.J.; Burkhardt, S.; Studer, M.; Bienewald, F.U.S. Patent 20040242916 (12/2/2004).
175. Ohrlein, R.; Baisch, G.; Kirner, H.J.; Bienewald, F.; Burkhardt, S.; Studer, M. WO 2003004456 (1/16/2003).
176. The Aldrich 2007–2008 catalog price is 25 g for $92.70.

# INDEX

**Index.** This book has unique core content: detailed descriptions for processes intended to be used for large scale manufacturing. Precedent for the use of name reactions, reagents, solvents, and azeotropic distillations in manufacturing processes can be accessed using the indices below.

## NAME REACTION INDEX



## REAGENT INDEX

Some reagents are so common they are not included in this index. These are acids (hydrochloric acid, sulfuric acid, nitric acid) and bases (sodium hydroxide, sodium carbonate, sodium bicarbonate, sodium chloride (brine), potassium hydroxide, potassium carbonate, and potassium bicarbonate). An analysis of the reagent index suggests there are many "workhorse" reagents. These are: *p*-toluenesulfonic acid, butyllithium, lithium diisopropylamide, sodium hydride, sodium methoxide, diisopropylamine, triethylamine, N,N-diisopropylethylamine, thionyl chloride, acetic anhydride, methanesulfonyl chloride, aluminum chloride, tetrabutylammonium bromide, palladium on carbon, and sodium borohydride.




*Pharmaceutical Process Chemistry for Synthesis: Rethinking the Routes to Scale-Up,* By Peter J. Harrington

## SOLVENT INDEX

Solvents used in processes described in this book can be separated into four groups: 1) common solvents routinely used in pharmaceutical manufacturing, 2) common solvents which are undesirable for scale up from an E H & S perspective, 3) common solvents often used as a catalyst or reagent, and 4) solvents used less frequently to achieve a specific solubility, stability, or reaction temperature objective. Solvents in groups 2, 3, and 4 are indexed in this section.

One surprising result of the analysis of the "hits" is that group 2 solvents chloroform, diisopropyl ether, and ether ether are still being used in some process R & D groups.

**Group 1**. Common solvents routinely used in pharmaceutical manufacturing



**Group 2**. Common solvents which are undesirable from an environmental health and safety perspective



**Group 3**. Common solvents often used as catalyst or reagent



**Group 4**. Common solvents used less often to achieve a specific solubility, hydrolytic stability, or temperature objective.

## AZEOTROPE INDEX

This index keys on the term azeotrope used in process descriptions in the book.

The azeotropic distillations described can be separated into two groups: Group 1 distillations used to dry (separate water from) organic solutions and Group 2 distillations used to exchange one organic solvent for another. Group 1 azeotropic distillations are listed by organic solvent component. If there is more than one organic solvent, the mixture is indexed by the major component. Group 2 azeotropic distillations are indexed by initial solvent.

**Group 1**. Separation of water from organic solvents and organic solvent mixtures by azeotropic distillation



**Group 2**. Separation of organic solvent mixtures by azeotropic distillation



1. The challenging separation of mixtures of ethyl acetate and ethanol can be achieved by adding a third solvent. Suitable third solvents include methyl *tert*-butyl ether and methyl formate. See: Berg, L. US 5,993,610 (11/30/1999).